Current Topics in Pathology
83

G. Seifert (Ed.)

Cell Receptors

Morphological Characterization
and Pathological Aspects

Contributors

U. Beisiegel · F. Buck · G. V. Childs · I. Damjanov
M. Dietel · H. Griesser · R. D. Hesch · H. Höfler
E. V. Jensen · W. Jonat · H. Jüppner · P. Karlson
H. Kreipe · J. Lloyd · G. Mengod · W. Meyerhof
S. D. Morley · A. Niendorf · J. M. Palacios
M. R. Parwaresch · A. Probst · H. J. Radzun
D. Richter · G. Seitz · H.-E. Stegner · R. Tibolt
M. Vierbuchen · N. Wernert

Springer-Verlag
Berlin Heidelberg New York
London Paris Tokyo
Hong Kong Barcelona

G. SEIFERT, Professor Dr.,
Institut für Pathologie der Universität Hamburg,
Martinistraße 52,
W-2000 Hamburg 20, Federal Republic of Germany

C.L. BERRY, Professor, M.D., Ph.D., F.R.C. Path.,
Department of Morbid Anatomy, The London Hospital,
London E1 1BB, United Kingdom

E. GRUNDMANN, Professor Dr.,
Gerhard-Domagk-Institut für Pathologie
der Universität Münster, Domagkstr. 17
W-4400 Münster, Federal Republic of Germany

With 142 Figures and 47 Tables

ISBN-13: 978-3-642-75517-0   e-ISBN-13: 978-3-642-75515-6
DOI: 10.1007/978-3-642-75515-6

Library of Congress Cataloging-in-Publication Data. Cell receptors : morphological characterization and pathological aspects / Gerhard Seifert (ed.) ; contributors, U. Beisiegel . . . [et al.].
p. cm. – (Current topics in pathology ; 83) Includes bibliographical references.
                                   Cell receptors. 2. Pathology, Molecular. 3.
Cancer-Pathophysiology. I. Seifert, Gerhard, 1921–   . II. Beisiegel, U. (Ulrike) III. Series.
IV. Series: Current topics in pathology ; v. 83. [DNLM: 1. Cell Membrane–anatomy &
histology. 2. Cell Membrane–pathology. 3. Receptors, Endogenous Substances. W1 CU821H
v. 83 / QH 603.C43 C393] RB1.E6 vol. 83 [QH603.C43] 616.07 s–dc20 [611'.0181]
DNLM/DLC for Library of Congress 90-9896 CIP

10/3130-543210 – Printed on acid-free paper

List of Contributors

BEISIEGEL, U.,
Priv.-Doz. Dr.

1. Medizinische Klinik
der Universität Hamburg, UKE,
Martinistraße 52, W-2000 Hamburg 20,
Federal Republic of Germany

BUCK, F.,
Dr.

Institut für Zellbiochemie
und klinische Neurobiologie,
Universität Hamburg, UKE,
Martinistraße 52, W-2000 Hamburg 20,
Federal Republic of Germany

CHILDS, G. V.,
Prof. Dr.

Department of Anatomy
and Neurosciences,
The University of Texas Medical Branch,
Galveston, TX 77551, USA

DAMJANOV, I.,
Prof. Dr.

Department of Pathology and Cell Biology,
Jefferson Medical College,
Thomas Jefferson University,
Philadephia, PA 19107, USA

DIETEL, M.,
Prof. Dr.

Institut für Pathologie der Universität Kiel,
Michaelisstraße 11, W-2300 Kiel,
Federal Republic of Germany

GRIESSER, H.,
Dr.

Direktor des Instituts für Pathologie
der Universität Kiel, Michaelisstraße 11,
W-2300 Kiel, Federal Republic of Germany

HESCH, R. D.,
Prof. Dr.

Abteilung für Klinische Endokrinologie,
Zentrum Innere Medizin und Dermatologie,
Medizinische Hochschule Hannover,
Konstanty-Gutschow-Straße 8,
W-3000 Hannover 61,
Federal Republic of Germany

HÖFLER, H.,
Prof. Dr.

Direktor des Instituts für Allgemeine
Pathologie und Pathologische Anatomie
der Technischen Universität München,
Klinikum rechts der Isar,
Ismaninger Straße 27, W-8000 München 80,
Federal Republic of Germany

JENSEN, E. V.,
Dr.

The Ben May Institute,
The University of Chicago,
5841 South Maryland Avenue, Box 424,
Chicago, IL 60637, USA
Present address: New York Hospital
Cornell Medical Center, Room F-2311,
525 East 68th Street, New York, NY 10021,
USA

JONAT, W.,
Prof. Dr.

Frauenklinik der Universität Hamburg,
UKE, Martinistraße 52,
W-2000 Hamburg 20,
Federal Republic of Germany

JÜPPNER, H.
Dr.

Kinderklinik und Abt. für Klinische
Endokrinologie, Dept. Innere Medizin,
Medizinische Hochschule Hannover,
Konstanty-Gutschow-Straße 8,
W-3000 Hannover 61,
Federal Republic of Germany
Present address: Endocrine Unit,
Massachusetts General Hospital,
Boston, MA 02114, USA

KARLSON, P.,
Prof. Dr. Dr.
h.c. mult.

Direktor des Instituts für
Molekularbiologie und Tumorforschung
der Philipps-Universität,
Emil-Mannkopff-Straße 2,
W-3550 Marburg,
Federal Republic of Germany

KREIPE, H.,
Dr.

Institut für Pathologie der Unversität Kiel,
Michaelisstraße 11, W-2300 Kiel,
Federal Republic of Germany

LLOYD, J.,
Dr.

The University of Maryland,
School of Medicine,
Department of Physiology,
655 West Baltimore St., Baltimore,
MD 21201, USA

MENGOD, G., Dr.	Preclinical Research, Sandoz Ltd., CH-4002 Basel, Switzerland
MEYERHOF, W., Dr.	Institut für Zellbiochemie und klinische Neurobiologie, Universität Hamburg, UKE, Martinistraße 52, W-2000 Hamburg 20, Federal Republic of Germany
MORLEY, S. D., Dr.	Institut für Zellbiochemie und klinische Neurobiologie, Universität Hamburg, UKE, Martinistraße 52, W-2000 Hamburg 20, Federal Republic of Germany
NIENDORF, A., Dr.	Institut für Pathologie der Universität Hamburg, UKE, Martinistraße 52, W-2000 Hamburg 20, Federal Republic of Germany
PALACIOS, J. M., Dr.	Preclinical Research, Sandoz Ltd., CH-4002 Basel, Switzerland
PARWARESCH, M. R., Prof. Dr.	Institut für Pathologie der Universität Kiel, Michaelisstraße 11, W-2300 Kiel, Federal Republic of Germany
PROBST, A., Priv.-Doz. Dr.	Institut für Pathologie der Universität Basel, Schönbeinstraße 40, CH-4003 Basel, Switzerland
RADZUN, H. J., Prof. Dr.	Institut für Pathologie der Universität Kiel, Michaelisstraße 11, W-2300 Kiel, Federal Republic of Germany
RICHTER, D., Prof. Dr.	Direktor des Instituts für Zellbiochemie und klinische Neurobiologie, Universität Hamburg, UKE, Martinistraße 52, W-2000 Hamburg 20, Federal Republic of Germany
SEITZ, G., Dr.	Pathologisches Institut der Universität des Saarlandes, W-6650 Homburg/Saar, Federal Republic of Germany

STEGNER, H.-E.,   Direktor der Abteilung für
Prof. Dr.   Gynäkologische Histopathologie
                        und Elektronenmikroskopie,
                        Frauenklinik der Universität Hamburg,
                        UKE, Martinistraße 52,
                        W-2000 Hamburg 20,
                        Federal Republic of Germany

TIBOLT, R.,   Department of Anatomy
Dr.   and Neurosciences,
                        The University of Texas Medical Branch,
                        Galveston, TX 77551, USA

VIERBUCHEN, M.,   Pathologisches Institut
Prof. Dr.   der Universität Köln,
                        Joseph-Stelzmann-Straße 9,
                        W-5000 Köln 41,
                        Federal Republic of Germany

WERNERT, N.,   Pathologisches Institut der Universität
Priv.-Doz. Dr.   des Saarlandes, W-6650 Homburg/Saar,
                        Federal Republic of Germany

Preface

The methods of molecular biology, biochemistry, immunocytochemistry, and in-situ hybridization introduce new opportunities for the classification and functional characterization of cell receptors under normal conditions and for a better understanding of pathogenetic mechanisms in human diseases. The cellular localization and translocation of receptor proteins can be identified using morphological methods, and it is apparent that receptors and receptor defects play an important role in pathology, notably in genetic diseases, endocrine disorders, atherosclerosis, infections, and cancer. In this volume international experts give a current review of the morphology and pathological aspects of cell receptors.

The complex communication of multicellular organisms is coordinated by two regulatory systems: neural and humoral. Both systems function via signaling substances (ligands) and signal-recognizing and -transmitting molecules, called receptors. The historical development of the receptor concept is based upon Paul Ehrlich's theory of "receptors in the immune system," Langley's "receptive substances in postssynaptic membranes," and Earl Sutherland's discovery of "second messengers" (cAMP and Ca^{2+}).

The main topics of the volume are the morphological characterization and pathological aspects of cell receptors. Starting with the fundamentals and covering special descriptions of the different receptor systems, the first chapter provides a current review of the general classification and characterization of cell receptors with reference to the historical development of the receptor concept, cell biology, biochemistry, morphology, and molecular biology. The subsequent chapters contain specialized contributions relating to the cell surface receptors (peptide hormone receptors, growth factor receptors, LDL receptors, neurotransmitter receptors, and lectin receptors) and to the cytosol/nuclear receptors (steroid hormone receptors).

A topical aspect of cancer research is oncogene expression and interactions of growth receptors and related proteins. Morphological evidence of the presence of receptors in human tumors has led to a better understanding of tumor growth, prognosis, and treatment. This will be demonstrated in three important and frequent human tumor entities, breast cancer, prostatic cancer, and malignant lymphomas or myelomonocytic disorders.

Precise understanding of receptor functions in cellular physiology and pathology is of increasing importance, since new therapeutic measures interfering directly with receptor and postreceptor processes are now relevant.

The emphasis of this volume is on the morphological characterization of cell receptors under normal and pathological conditions. It has an interdisciplinary character because understanding the morphological observations is only possible when the results of molecular biology and biochemistry are considered simultaneously.

The volume should be seen in context with the earlier volume "Morphological Tumor Markers: General Aspects and Diagnostic Relevance," which was published as volume 77 in Current Topics in Pathology.

I would like to express my cordial appreciation to all contributors to this volume for their excellent cooperation. All the authors have given generously of their precious time. I am also most grateful to my personal secretary, Mrs. M. Teichmann, and to the staff of Springer-Verlag, especially Mrs. Ursula N. Davis for her constant and invaluable assistance.

Hamburg, Federal Republic of Germany   G. SEIFERT

Contents

Indexed in ISR

General Biochemical
and Morphological Aspects

Historical Development of the Receptor Concept

P. KARLSON

1 Definitions

The term receptor is used in different ways in physiology and in biochemistry. In physiology, it is often used in reference to sensory cells which respond to a special stimulus. In pharmacology and biochemistry, it designates certain proteins – often located at the cell surface – that bind certain substances and elicit a physiological reaction. The present review deals only with the second meaning of the term "receptor."

It should be mentioned in passing that the double usage of the term "receptor" is especially confusing in the field of olfaction. It is a long-standing practice to call the sensory cells chemoreceptors. In the field of molecular biology of olfaction, scientists are now investigating the proteins on the receptor cell surface that bind signal substances, e.g., pheromones. In biochemical terms, these proteins are receptor molecules. But they cannot simply be called receptors since use of this term has been preempted. Sometimes the term "acceptor molecule" is used – also in conflict with other uses of the word acceptor.

It should be stressed that in pharmacological and endocrinological usage the term receptor is tied to a physiological response, the recognition and transduction of a chemical signal. Proteins of the cell membrane that specifically bind and internalize, e.g., the cholesterol-LDL complex, are better termed specific transport systems. They are sometimes referred to as "class II receptors." For a detailed discussion see HOLLENBERG and CUATRECASAS (1979).

2 Ehrlich's Theory of Receptors in the Immune System

The term "receptor" was coined by Paul EHRLICH about 100 years ago to designate "special chemical groups of living protoplasma that bind foreign substances like bacterial toxins, antigenic fragments of bacterial membranes or foreign erythrocytes." This results, in competent cells, in the production of antibodies which were believed, according to EHRLICH's theory, to be receptors produced in surplus amounts and secreted into the blood plasma.

Bacterial toxins (diphtheria toxin, tetanus toxin) were the antigenic substances which were first used by EHRLICH in the experiments which led him to suggest the concept of receptors of antigens (for details, see the contemporary review by ASCHOFF 1902). These receptors were also called "*Seitenketten*" (English: side chains). The term is misleading because it did not mean a special chemical group on a protein, as we would understand it nowadays, but a special structure of the cell surface. The basic idea developed by EHRLICH was that the antibodies which arise in the blood after immunization with bacterial toxins are nothing other than the receptors which are produced in surplus amounts and secreted into the blood plasma. This process is outlined in Fig. 1. Since the receptors were able to bind a toxin, they could neutralize the substance and protect the experimental animal from the action of the toxin.

We know nowadays that essential parts of this theory have proved to be correct. According to our present knowledge, antigens like toxins or membrane fragments are recognized by antibody molecules (IgE) anchored in the membrane. As a result of this interaction, the cells divide, form a clone, develop into plasma cells, and produce the immunoglobulins (IgG) of the same specificity.

Later, EHRLICH extended his theory to explain the cytotoxic action of certain antibodies in conjunction with the complement system of the blood. He postulated more complex receptors with one specific site for interaction with the antigen and a second specific site for interaction with the complement (which was then believed to be one single substance). This is outlined in Fig. 2.

EHRLICH used his receptor theory to account for the tissue specificity of toxins. Thus, tetanus toxin attacked mainly nervous tissue because the nervous tissue expressed, at the cell surface, the receptors which bind the tetanus toxin

-->

Fig. 1. EHRLICH's "side-chain theory". On the left, the cell used for immunisation is shown. The first column outlines the action of the immunizing substances (J. S.) on the receptors of first (r I, *upper row*), second (r II, *middle*) and third (r III, *lower row*) order. Receptors of first order react with soluble toxins and enzymes ("Fermente"), receptors of second order react with insoluble agents (on the cell surface) and produce agglutinins, receptors of third order react with cell surface structures and produce, together with complement, cytotoxicity. – In the second column, the free receptors produced in surplus and released, are shown. Note the different specific sites. Antitoxins are antibodies (immunoglobulins). – The third column shows the interaction of the antibodies with the antigens or cell surface. – The last column shows the reaction of antibodies against agglutinins and complement, inhibiting the specific interaction of agglutinins and complement with the cells used for immunisation. (ASCHOFF 1902)

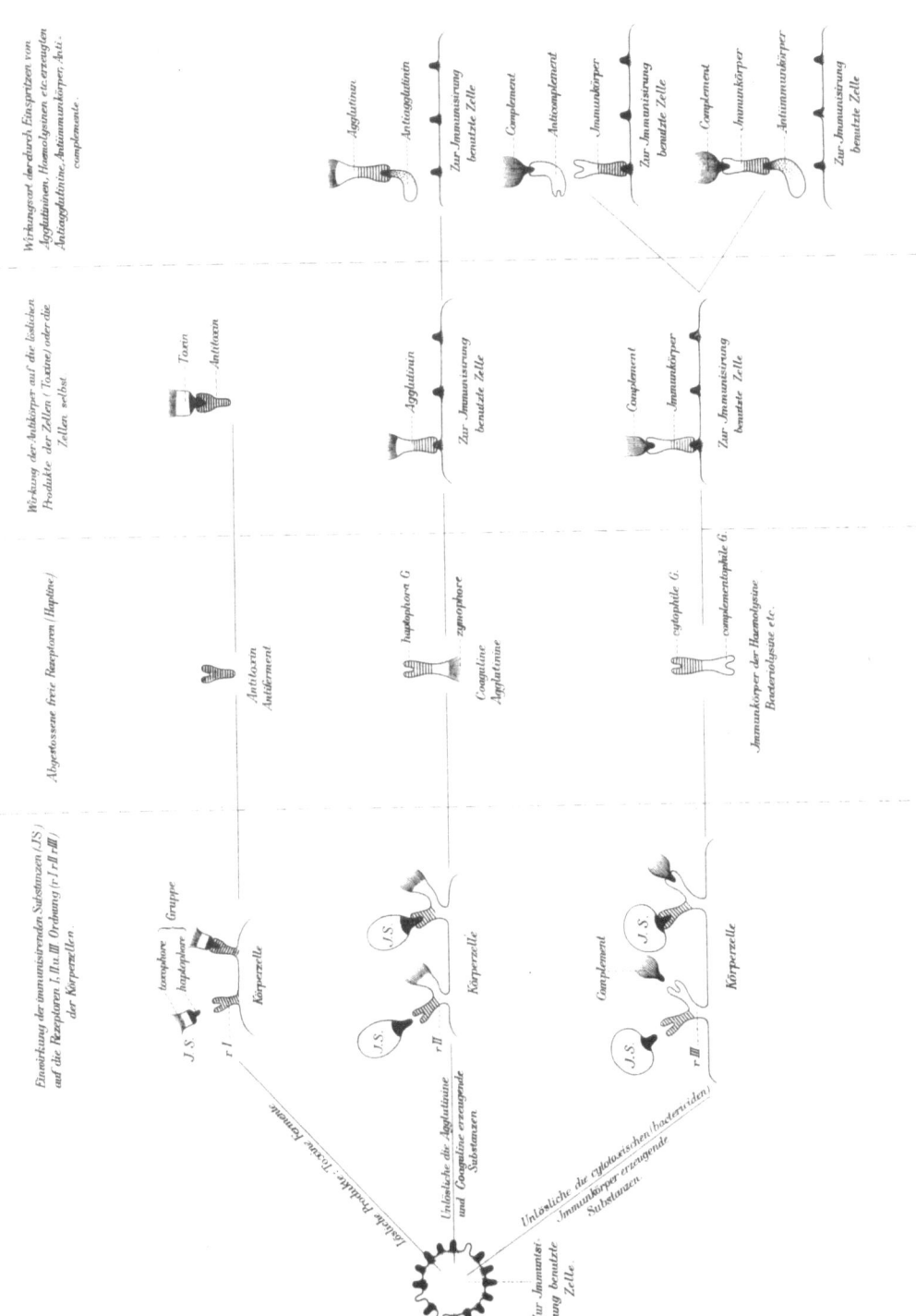

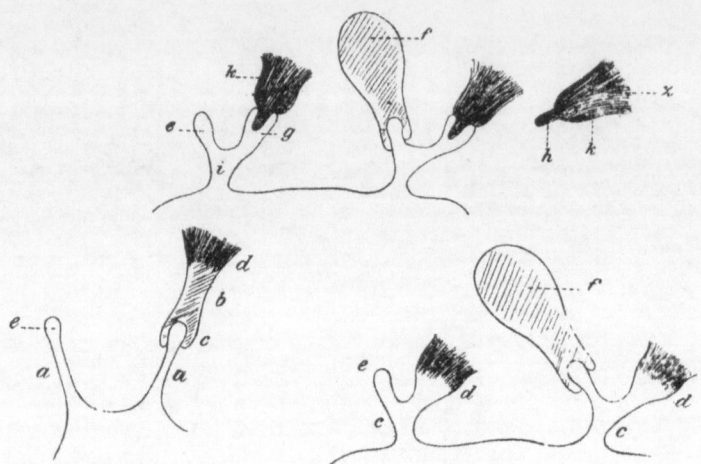

Fig. 2. Scheme of receptor structures after EHRLICH. In the *lower row to the left,* receptors of the first order (*a*) are shown. They expose a structure (*e*) that binds specifically to the haptophoric group (*c*) of the toxin (*b*) carrying also a toxophoric group (*d*). In the *lower row to the right,* receptors of the second order (*c*) are shown. They are believed to enable internalization of large molecules (proteins, *f*) and expose a haptophoric group (*e*), binding the protein (*f*), and a zymophoric group (*d*) that can attack the protein to facilitate resorption. Above, receptors of the third order (*i*) are shown. They contain, in addition to the haptophoric group (*e*), a complementophilic group (*g*) that binds the complement (*k*). The latter exposes a haptophoric group (*h*) and a zymotoxic group (*x*) responsible for the lysis of cells, e.g., erythrocytes. (ASCHOFF 1902)

circulating in the blood. The fact that turtles are insensitive to tetanus toxin is explained by the fact that tissue of the turtle does not have receptors for this toxin.

EHRLICH drew an analogy with enzymes: The receptors or antibodies (antitoxins) on the one hand and the toxin molecule on the other interacted in a way similar to that in which an enzyme interacts with a substrate. He used the lock-and-key theory of EMIL FISCHER to describe this very specific interaction.

EHRLICH started out as a histologist. He was fascinated by the fact that many artificial dyes stained some tissues very specifically. For example, methylene blue stained nerve fibers and other nervous tissues in a very specific way. The dye adhered to such tissues much stronger than to other tissues. EHRLICH concluded that it should be possible to build chemical molecules that can interact very specifically with either bacteria or protozoa that are important in infectious diseases. On this basis, he developed in 1909 the famous Salvarsan, an arsenobenzene derivative which proved to be a very effective cure for syphilis (BÄUMLER 1980).

EHRLICH sometimes also used the term "receptor" in reference to the interaction of drugs with the cell surface of microorganisms; but he was aware that this interaction was quite different from the interaction of high molecular antigens with the receptors of the cells of the immune system.

3 Langley and Dale: Receptive Substances in Postsynaptic Membranes

In pharmacology, the term receptor is used in a somewhat different way. It was introduced by LANGLEY. Based on a very detailed study of the action of nicotine and curare upon somatic nerve endings and skeleton muscle, LANGLEY (1906) came to the following conclusion:

The mutual antagonism of nicotine and curare on muscle can only satisfactorily be explained by supposing that both combine with the same radicle of the muscle, so that nicotine–muscle compounds and curare–muscle compounds are formed. Which compound is formed depends upon the mass of each poison present and the relative chemical affinities for the muscle radicle. ... Since neither curare nor nicotine, even in large doses, prevents direct stimulation of muscle from causing contraction, it is obvious that the muscle substance which combines with nicotine or curare is not identical with a substance which contracts. It is convenient to have a term for the specially excitable constituent and I have called it the receptive substance. It receives the stimulus and, by transmitting it, causes contraction.

Important is the last sentence, which clearly states that not the specific binding to the receptor is important, but the biological response of the binding of the receptor to events in the cell. In this respect, the statement of LANGLEY is closer to our present definition of a receptor for hormones, neurotransmitters, or other active agents like drugs.

Sir HENRY DALE (1906) extended the concept of Langley to the action of epinephrine (adrenaline) in relation to the sympatholytic action of ergot alkaloids, and already recognized "motor and inhibitory elements."

Quantitative aspects of the interaction of effectors (and inhibitors) with receptors were first studied by CLARK (1926a, b; 1927). Using a frog heart muscle preparation, he showed that 10^{-14} mol acetylcholine (equivalent to 20000 molecules per cell) was sufficient to elicit a response. He concluded that only small amounts of receptors were present on the surface of the responsive cells.

The classification of epinephrine receptors into α-receptors and β-receptors stems from the work of AHLQUIST (1948). He studied in detail the action of epinephrine and five related substances in a number of bioassays in which these drugs acted in either an excitatory or an inhibitory manner and found the results could only be explained by postulating two different receptors, i.e., the α- and the β-receptors (Table 1). Today, they must be subclassified into α_1- and α_2- and β_1- and β_2-receptors.

Studies with antagonists led SCHILD (1949) and ARUNLAKSHANA and SCHILD (1959) to develop

the powerful *null* hypothesis, whereby it is assumed that when the response to one concentration of the agonist in the absence of inhibitor is the same as the response to a higher concentration of agonist in the presence of competitive inhibitor, then the amount of agonist reaching the receptor in the two situations is identical. It is thus possible by bioassay to measure the concentration dependence of an antagonist's ability to prevent agonist access to the receptor site, so as to determine the antagonist equilibrium dissociation constant. (quoted from HOLLENBERG and CUATRECASAS 1979).

The concept of drug receptors as related to the neurotransmitter system has proved very successful. The postulate of a morphine receptor finally led to the

Table 1. Structures or functions containing or associated with each of the two types of adrenotropic receptors[a]

Alpha receptor	Beta-receptor
Vasoconstriction	Vasodilation
Viscera	Skeletal muscle
Skin	Coronary
Nictitating membrane	Viscera (few)
Uterus (excitatory)	Myocardium
Rabbit	Uterus (inhibitory)
Dog	Rat
Human	Cat
Intestine	Dog
Ureter	Human
Dilator pupillae	Bronchi

[a] The following structures or functions have not as yet been completely tested: spleen, *erectores pilorum*, urinary bladder, glands and glycogenolysis.

discovery of the endorphins and enkephalins (HUGHES et al. 1975; SIMANTOV and SNYDER 1975; SNYDER and INNIS 1979).

4 Earl Sutherland and the Discovery of Second Messengers

Epinephrine not only has the physiological effects mentioned above and studied by AHLQUIST; it also has a number of biochemical effects. The most striking is probably the rise in blood glucose level due to breakdown of glycogen in the liver. SUTHERLAND and CORY showed that the biochemical mechanism was an activation of the enzyme phosphorylase in liver and in muscle. Glucagon, a peptide hormone that also raises blood glucose, acts only on liver, not on muscle phosphorylase.

In vitro studies with homogenates showed that the effect of epinephrine as well as glucagon can be traced to a particulate fraction, and that the first effect of the hormones in this particulate fraction is the synthesis of a nucleotide, the well-known cyclic adenosine $3',5'$-monophosphate (cAMP). If this nucleotide is added to the supernatant – essentially a cytosol preparation –, phosphorylase activity is increased (SUTHERLAND et al. 1965). This led to the scheme of the mechanism of action given in Fig. 3.

We know today that the biochemical mechanisms involved in the hormonal control of glycogen breakdown and biosynthesis are rather complex. They involve a number of protein phosphorylations and dephosphorylations, which are controlled in an allosteric manner by cAMP.

Returning to the receptors, it was later shown that the effect of epinephrine on the production of cAMP is mediated by the β-receptors, and more strongly

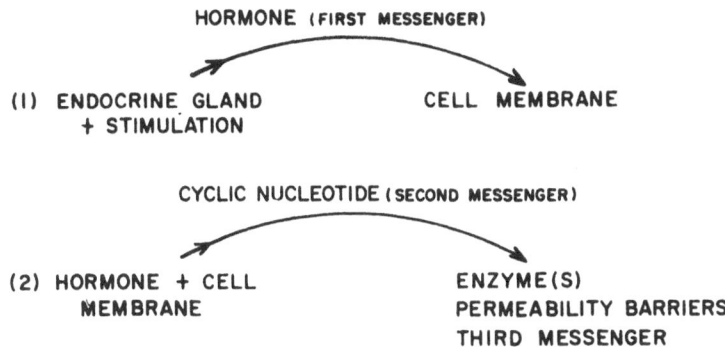

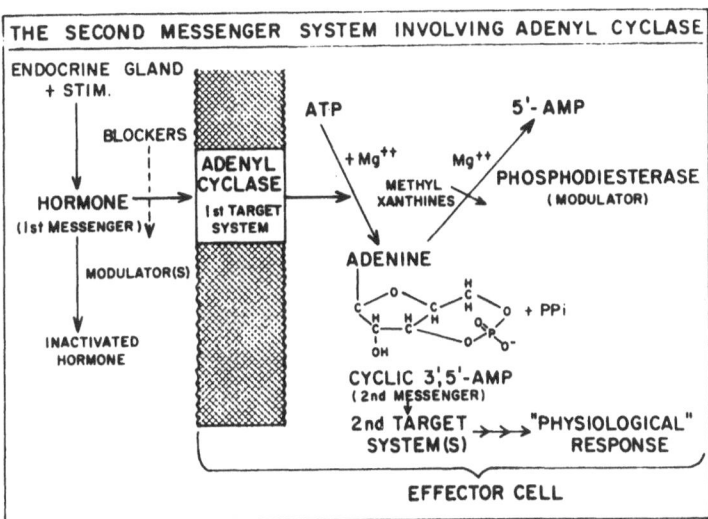

Fig. 3. The second messenger theory. *Above,* the general concept; *below,* cAMP as second messenger. (SUTHERLAND et al. 1965)

by β_2- than by β_1-receptors. According to the present view, three membrane-bound proteins are involved: the receptor protein, adenylate cyclase, and a guanosine triphosphate binding protein, a so-called G-protein. That G proteins are involved in the transduction of signals was first suggested by RODBELL et al. (1971) and later substantiated by CASSEL and SELINGER (1976, 1977); for a recent review, see GILMAN (1987).

G proteins are involved in a large number of membrane-bound receptor systems. The second messenger is not always cAMP; it may also be Ca^{2+} or the cleavage products of phospholipids, namely inositol 1,4,5-trisphosphate and diacylglycerol. This is described in detail in other chapters of this book.

In summary, the second messenger concept explains the transduction of the hormonal signal from the cell surface to the cytosol where the biochemical events take place. This is the most important lesson learned from the experiments by SUTHERLAND and co-workers, and later by other groups.

5 E. V. Jensen and the Intracellular Steroid Hormone Receptors

Around 1960, JENSEN and JACOBSON synthesized a tritium-labeled estradiol 17β of high specific activity to study the fate of the hormone in the target tissues. The labeled hormone is accumulated in the uterus and in the pituitary as well as in adenocarcinomas of the breast. This is due to the fact that the cells of these tissues contain a macromolecular, soluble protein that binds the hormone and mediates the action of estradiol (JENSEN and JACOBSON 1962).

The mechanism of action of steroid hormones has been elucidated by KARLSON. Following the observation that the steroid hormone ecdysone can induce puffs in the giant chromosomes of the salivary glands of the midge *Chironomus*, he postulated that this hormone activates certain genes. The effect is very sensitive and very specific for ascertaining visible bands on the giant chromosomes that undergo puffing (CLEVER and KARLSON 1960). The interpretation of these results are outlined in two reviews by KARLSON (1961, 1963), and presented schematically in Fig. 4.

In these early reviews, the existence of receptor proteins was not taken into account. However, it became clear that it is not the hormone itself that interacts with the chromatin of the insect chromosomes or with the chromatin in other target cells. It is the receptor–hormone complex that binds to the chromatin and recognizes certain sequences in the DNA upstream of the transcriptional unit, thereby activating the enzyme RNA polymerase to produce the first transcript (pre-mRNA) (for a recent review, see BEATO 1989).

The mechanism of action deduced from the puffing experiments holds not only for ecdysteroids in insect target tissues but for all known steroid hormones. Likewise, the role of intracellular receptors mediating this action as deduced by JENSEN et al. from their experiments with estradiol is a general mechanism for all steroid hormones. It may also apply to some other hormones and effectors (thyroid hormones, retinoic acid). This field is now under active study. The reader is referred to a symposium held in Mosbach in 1989 which has recently appeared in print (GEHRING et al. 1989).

The present state of the art is dealt with in the other chapters of this volume. It has been my task to outline the historical development in the last 100 years, and I have restricted my paper to the highlights which really represent breakthroughs in our understanding of the action on receptors in mediating the action of hormones and other effector substances.

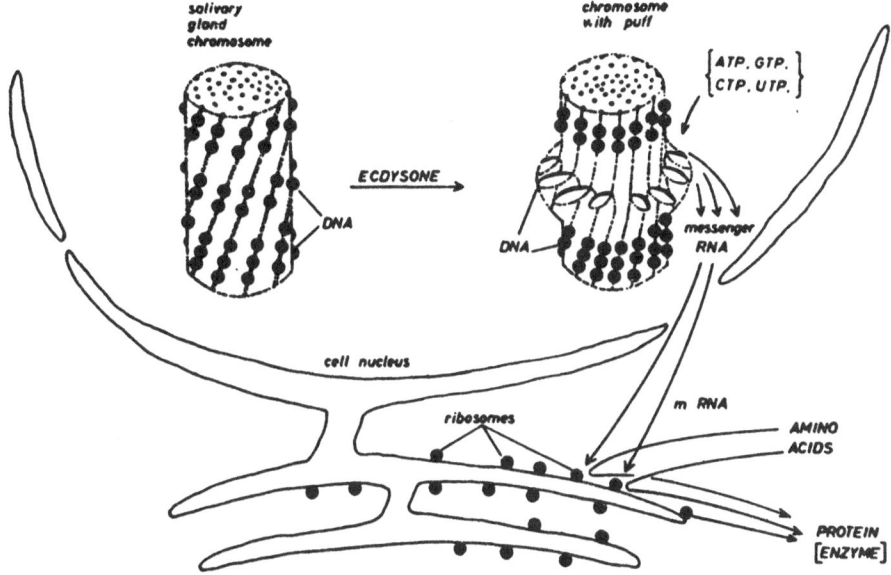

Fig. 4. Hormone action by gene activation. The hormone (ecdysone) acts first on the chromosome, producing a puff, shown to the *right* as and unwound region. In the puff, messenger RNA is synthesized from precursors and then transferred to the cytoplasm, where it directs synthesis of specific proteins, e.g., enzymes. (KARLSON 1963)

References

Ahlquist RP (1948) A study of the adrenotropic receptors. Am J Physiol 153:586–600

Arunlakshana O, Schild HO (1959) Some quantitative uses of drug antagonists. Br J Pharmacol 14:48–58

Aschoff L (1902) Ehrlich's Seitenkettentheorie und ihre Anwendung auf die künstlichen Immunisierungsprozesse. Z Allg Physiol 1:69–248

Bäumler E (1980) Paul Ehrlich, Forscher für das Leben, 2nd edn. Societäts-Verlag, Frankfurt

Beato M, Brüggemeier U, Chalepakis G, Gross B, Pina B, Schauer M, Slater EP, Truss M (1989) Interaction of steroid hormone receptors with DNA, p. 12–20 In: Gehring U, Helmreich E, Schultz G (eds) Molecular mechanisms of hormone action 40. Mosbacher Coll. Ges. Biolog. Chemie, Springer Berlin Heidelberg New York

Cassel D, Selinger Z (1976) Catecholamine-stimulated GTPase activity in turkey erythrocyte membranes. Biochim Biophys Acta 452:538–551

Cassel D, Selinger Z (1977) Mechanisms of adenylate cyclase activation by cholera toxin: inhibition of GTP hydrolysis at the regulatory site. Proc Natl Acad Sci USA 74:3307–3311

Clark AJ (1926a) The reaction between acetylcholine and muscle cells. J Physiol 61:530–546

Clark AJ (1926b) The antagonism of acetylcholine by atropine. J Physiol 61:547–556

Clark AJ (1927) The reaction between acetylcholine and muscle cells, part II. J Physiol 64:123–143

Clever U, Karlson P (1960) Induktion von Puff-Veränderungen in den Speicheldrüsenchromosomen von *Chironomus tentans* durch Ecdyson. Exp Cell Res 20:623–626

Dale HH (1906) On some physiological actions of ergot. J Physiol 34:163–176

Gehring U, Helmreich E, Schultz G (eds) (1989) Molecular mechanism of hormone action. 40. Coll. Ges. Biolog. Chemie. Springer, Berlin Heidelberg New York

Gilman AG (1987) G proteins: transducers of receptor-generated signals. Annu Rev Biochem 56:615–649

Hollenberg MD, Cuatrecasas P (1979) Distinction of receptor from nonreceptor interactions in binding studies. O'Brien R (ed) The receptors, vol 1. New York

Hughes JT, Smith TW, Kosterlitz HW, Fothergill LA, Morgan BA, Morris HR (1975) Identification of two related pentapeptides from the brain with potent opiate agonist activity. Nature 258:577–579

Jensen EV, Jacobson HI (1962) Basic guides to the mechanism of estrogen action. Recent Prog Horm Res 18:387–414

Karlson P (1961) Biochemische Wirkungsweise der Hormone. Dtsch Med Wochenschr 86:668–674

Karlson P (1963) New concepts on the mode of action of hormones. Perspect Biol Med 6:203–214

Langley JN (1905) On the reaction of cells and of nerve-endings to certain poisons, chiefly as regards the reaction of striated muscle to nicotine and to curare. J Physiol 33:374–413

Langley JN (1906) On nerve endings and on special excitable substances. Proc R Soc B 78:170–194

Rodbell M, Birnbaumer L, Pohl SL, Krans HMJ (1971) The glucagon-sensitive adenyl cyclase system in plasma membranes of rat liver. J Biol Chem 246:1877–1882

Schild HO (1949) pAx and competitive drug antagonism. Br J Pharmacol 4:277–280

Simantov R, Snyder SH (1976) Morphin-like peptides in the brain: isolation, structure elucidation, and interactions with the opiate receptor. Proc Natl Acad Sci USA 73:2515–2519

Snyder SH, Innis RB (1979) Peptide neurotransmitters. Annu Rev Biochem 48:755–782

Sutherland EW, Oye I, Butcher RW (1965) The action of epinephrine and the role of the adenylcyclase system in hormone action. Recent Prog Horm Res 21:623–646

Classification of Cell Receptors *

R. D. HESCH

1 Introduction

In classifying receptors it is necessary to consider their biological function. A receptor is an integral membrane protein which transduces biological information, conferred in signals. Two mechanisms are used for signal transformation: (a) the energy-loaded conformation of a ligand, interacting with the receptor by exchanging energy, and (b) the dynamic with which the ligand transduces its information to the receptor. Receptors exist where the transduction of information abuts on an evolutionary border. They are used to reprogram specific information by signal transduction at biological borders which exist around cellular compartments and make it possible to separate different structural information into cellular domains. Structures can be formed through compartmentalization, and information exchange at structures is equivalent to function. The function and structure of a cell hence depend on the information exchange system of signals through ligands at receptors.

---

* This article represents an expanded and translated version of a chapter taken from the textbook *Endokrinologie; Innere Medizin der Gegenwart* (editor: R. D. Hesch), Urban und Schwarzenberg, Munich, 1989 (with permission of the publisher).

These considerations show that we can observe certain consistencies in the development of biological information transduction which follow the rules of the evolution of biological systems (HESCH 1989). One of the most important discoveries in respect of such consistencies is that ligands and signal transduction on the one hand and receptors on the other have developed together genetically in a few structurally varied families. This genetically defined coevolution of ligand and receptor implies that the genetic factors controlling the synthesis of ligands and receptors always developed together, though on different chromosomes ("genetic code"). In the following, we propose a subdivision of the numerous hormone receptors into several evolutionary biological concepts in a few ligand-receptor families, usually called original families [1].

The specificity of the biological function concerning a family can vary substantially and must always be regarded in context with the surrounding systems. The dynamic control of the ligand–receptor family concerns the regulation of the kinetics of signal transduction by the ligand. Through discontinuous, time-dependent gradients in the blood, in paracrine compartments, and at synapses, ligands construct dynamic information patterns (pulse amplitude and frequency modulation) which are received by the receptor and transduced into the inside of the compartment. The conformation and the information patterns of the ligand determine the amount of thermodynamic energy and its specific use in the compartment. Receptors are energy transformers of biological information. They operate against equilibrium and help build a concentration of energy used for biological function and structure by developing special gradients. This dynamic transformation of energy and the reprogramming of specific biological information through information transduction at the ligand–receptor unit is a universal fundamental character of life; we therefore call this process a "dynamic code".

When we classify receptors we must thus also take into consideration an assessment of the dynamics of the ligand–receptor unit as well as the substructure determined by the molecular architecture.

Receptor classification and postreceptor events: Receptors employing the same ligand have different ways of directing signal transduction paths as second messenger pathways. This applies especially to peptide hormones but also to the family of hormones containing cyclic hydrophobic compounds, such as the thyroid and steroid hormones, which are known to have pleiotropic effects on the receptor. Peptide hormones like TSH, parathyroid hormone (PTH), and vasopressin are able to activate adenylate cyclase [cyclic adenosine $3',5'$-monophosphate (cAMP) production], phospholipase C (activation of the phosphatidylinositol cascade), and ion channels, via a particular peptide conformation, through their receptor. These second messenger cascades show completely different cellular pathways (cAMP activation = function and dif-

---

[1] The concept of "original families" is a translation of the German "Urfamilien." In fact in this instance, as in some others (e.g., "Urdynamik": "original dynamic"), an entirely accurate translation of the prefix "Ur-" is not possible. It should be borne in mind that apart from "original," "Ur-" also connotes "ancestral."

ferentiation; phospholipase C activation = mitosis and growth). For some ligands like epinephrine, norepinephrine, serotonin, acetylcholine, secretin, and vasopressin, different receptor types have been described by cloning the genes and giving accounts of the corresponding receptor protein sequence. These receptor types differ in structure only in some structural regions involving only a few amino acids. These differences relate to the observed biological effect and receptor specificity which is not encoded by the ligand.

Although some receptors have different molecular structures, there are also hints that hormone–receptor interactions, which use G proteins between enzymes for the synthesis of the second messenger to transduce information, can be specified by analyzing their kinetic behavior (pulse amplitude and frequency modulation). However, this principle seems not to be limited to receptors which transform their information primarily on G proteins, because growth hormone as well as insulin transports its information in the blood in episodic secretory patterns. It is known from in vitro analysis that the different dynamic application of hormones to cells can cause a change in the intracellular patterns of calcium pulsation (WOODS et al. 1986). The pulsatile information pattern of platelet-derived growth factor determines growth and transparency of the eye lens (BREWITT and CLARK 1988). Pulsatile growth hormone more efficiently stimulates the hormone receptor for growth to synthesize somatomedin C than does continuous growth hormone infusion (ISGAARD et al. 1988). Thus pulsatile PTH infusion leads to bone growth, whereas continuous PTH infusion reduces bone volume (TAM et al. 1982). The gonadotropin-releasing hormone frequency determines the preferential expression of either the α- or the β-subunit mRNA of luteinizing hormone (HAISEN-LEDER et al. 1987). Finally, the pattern of neuropeptides determines the construction of a neuronal reticulum (MARDER 1988).

Nevertheless it must be pointed out that at the moment it is useful primarily to classify receptors according to their molecular architecture in the membrane. After this we can further divide ligands and ligand receptors as well as dynamic modulation types. The question of how the different signal transduction paths of the cell (DUMONT et al. 1989) can be found by an identical hormone molecule in selective and cell-specific ways cannot be answered in terms of a classification of the hormone receptors at present.

2 General Structure of Receptors

A hormone receptor is an integral membrane protein, complex in structure, with a protein domain that protrudes from the cell membrane into the extracellular space. This protein domain produces binding interaction with hormones and ligands. It causes a conformation change which is afterwards communicated to the membrane-buried domain and runs through it. The transmembrane part of a receptor can span the membrane only with a short hydrophobic chain, which is usually about 20 amino acids long. It can, however, fold back and forth several times through the membrane to the inner and outer surface with hydrophobic and hydrophilic sequences. In this way the receptor protein is anchored in the membrane through hydrophobic interaction. Finally, the receptor protein leaves the membrane with another hydrophilic tail; this dives into the cytosol and thus is able to transfer information by contact with cytosolic structures.

The following receptor domains can thus be defined:

- Hormone-binding protein domains
- Hydrophobic sequences, serving as anchor proteins in the membrane
- Cytosol protein sequences which carry enzymatic activity, being able to take up contact with the cytoskeleton or join other enzyme protein complexes

The topology of integral receptor proteins can be shown schematically for different receptors (Fig. 1). If the internal signal sequence for anchoring the membrane protein in the membrane is not split off, the C-terminus protrudes into the extracellular space (transferrin). If it is split off, which is more common, this happens to the N-terminal end of such a protein. Some proteins pass through the membrane several times with the hydrophobic anchoring peptide sequence. These receptor proteins often form ion channels (acetylcholine).

Each receptor protein has well-defined protein domains which have a special function in protein communication. The protein structure of the insulin receptor can be shown as an example (Fig. 2). This receptor exists as a heterodimer, which means that two receptor proteins are linked in the membrane through disulfide bridges. Each receptor protein consists of three subunits, the α-, β-, and γ-subunits. The extracellular α-subunit carries a sequence which is rich in cysteine. As Fig. 3 shows, it is also found in receptors that are similar in structure, namely the low density lipoprotein (LDL) and epidermal growth factor (EGF) receptors, suggesting a genetic homology in evolution of this signal domain in several receptors with a similar function. This region permits the intermolecular binding of this kind of receptor structure or the binding of other proteins. The receptor association (capping, patching), as it develops after hormone binding at the receptor complex (considered as a preliminary phase for the internalization of the hormone−receptor complex), can, for instance, be formed by such intermolecular bindings.

Today the hormone-binding domain can be described for several receptors. The insulin receptor has a region for a glycolysation reaction, which means that corresponding amino acids can have sugar residues added enzymatically, manipulating their conformation to a substantial degree. The following N-terminal part of the β-subunit of the insulin is also glycosylated in vivo, causing a high level of glycolysation of the whole receptor. This means it can change its conformation considerably by addition or removal of different amounts of sugar. The glycolysation domain is situated near the extracellular membrane surface. Finally a hydrophobic transmembrane peptide has been described which anchors the receptor protein in the membrane.

In the cytosolic domain of these receptors, there is a consensus sequence for the binding of adenosine triphosphate which is found in protein kinases. This

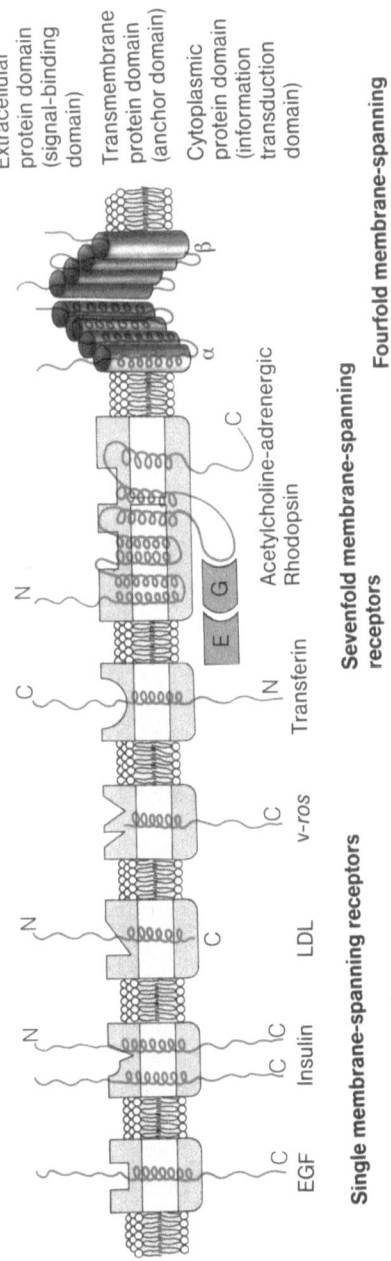

Extracellular protein domain (signal-binding domain)

Transmembrane protein domain (anchor domain)

Cytoplasmic protein domain (information transduction domain)

Single membrane-spanning receptors

EGF   Insulin   LDL   v-ros   Transferin

Sevenfold membrane-spanning receptors

Acetylcholine-adrenergic Rhodopsin

Fourfold membrane-spanning receptor ion channel

Fig. 1. Topology of receptor structures in the membrane. *E*, effector protein; *G*, G protein; *C*, C-terminus; *EGF*, epidermal growth factor; *v-ros*, viral *ros* receptor

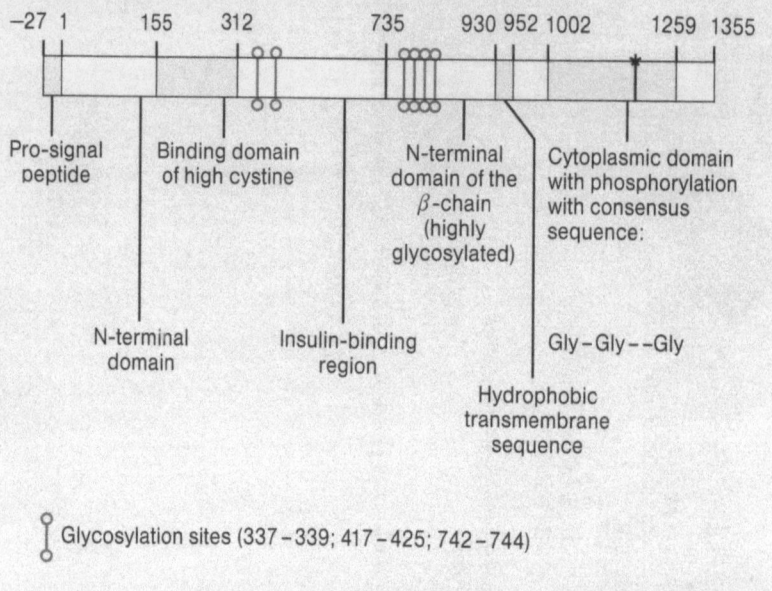

Fig. 2. Protein structure of the insulin receptor

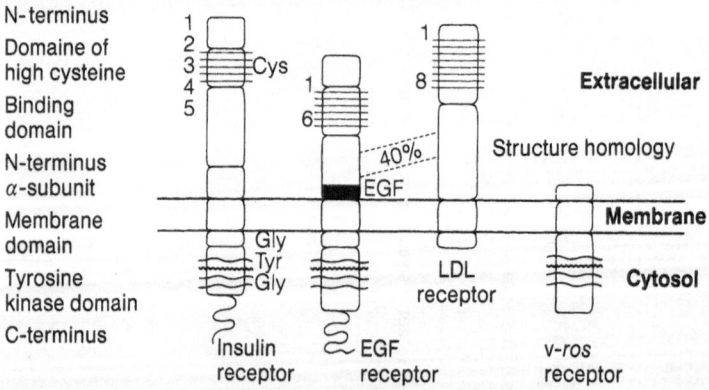

Fig. 3. Similarity of receptor structures. Abbreviations as in Fig. 1

sequence is written as (Gly-x-x-Gly) where x can be Tyr. This special amino acid region is present on the intracellular portion of these membrane proteins and is the location of an enzymatic phosphorylation which often occurs on tyrosine residues. Thus the signal is carried through the receptor via a conformational change in the protein structure which ends in the phosphorylation of a tyrosine residue. It is remarkable that the intracytosolic domain of the insulin receptor can act as a phosphotransferring enzyme which, after receiving the signal through the receptor, will autophosphorylate itself. In addition it can transfer phosphates to other protein structures. In this region of the insulin receptor there is a strong structural homology to the cytoplasmic portion of the EGF receptor, as well as to a special oncogene receptor known as the viral *ros* receptor. Although these receptors have developed distinct functions during evolution, they contain a highly homologous and functionally similar cytoplasmic domain, encoding a membrane receptor tyrosine kinase that was probably assembled as a module and then transferred to other receptor proteins via exon shuffling. Similarly, other homologies exist between receptors that function as ion channels. In order to execute certain functions, nature has evidently developed peptide sequences for communicating information. These are designed as modules and then assemble these domains into larger proteins.

Recent research in molecular biology and theoretical views of evolution allow us to classify receptors in a new way, by reference to what may be termed general principles. The latter are the few basic principles which repeatedly apply to receptors which transduce information through the binding of the hormone/ligand to the receptor, and to the development of structure and function. These principles include the:
– Genetic coevolution of the ligand–receptor unit
– Genetic construction principles of the receptor proteins, receptor families with coevolution of ligand families
– Evolution of postreceptor transduction and effector systems (original transduction and effector systems)
– Dynamic modulation types (original dynamic)
– Functional coevolution of information (original information)

When looking at the cell function, it is possible to define phylogenetic and oncogenetic pathways as subfunctions of:
– "Gestalt" (morphogenesis and differentiation)
– Growth
– Adaption and function
– Metabolic processes

In organizing receptor families, one can assume that most proteins are derived from a small number of archetypal proteins, since it is easier to duplicate, modify, or recombine genes than to construct new ones.

The original receptor families can be divided into the following protein families:
– Nuclear receptors which bind cyclic hydrophobic compounds
– Membrane receptors containing single, fourfold, sevenfold, and 24-fold membrane-spanning domains

3 Families of Cytosol Receptors which Bind Cyclic Hydrophobic Compounds

Receptors which bind cyclic hydrophobic compounds are shown in Fig. 4. It is generally known that the development of multicellular organisms depends

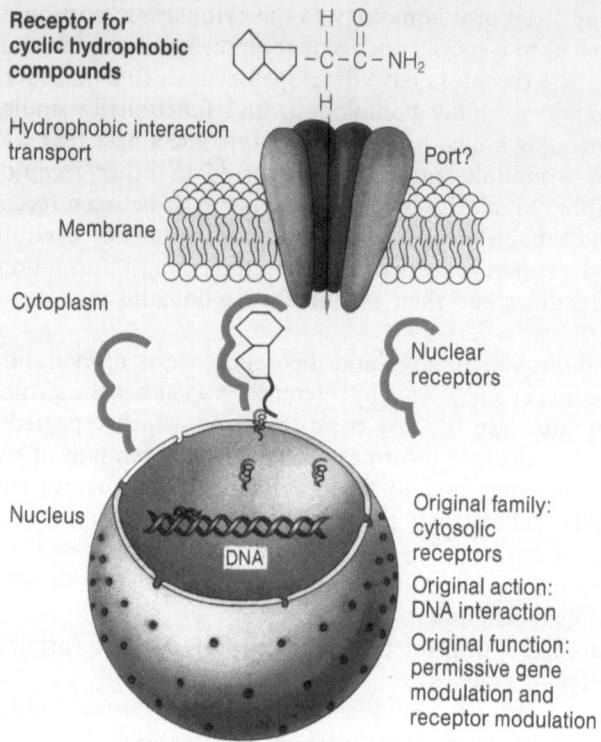

Receptor for cyclic hydrophobic compounds

Hydrophobic interaction transport

Port?

Membrane

Cytoplasm

Nuclear receptors

Nucleus

DNA

Original family: cytosolic receptors

Original action: DNA interaction

Original function: permissive gene modulation and receptor modulation

Fig. 4. Receptor binding cyclic hydrophobic compounds (e.g., adrenocortical steroids, estrogens, androgens, aldosterone, thyroid hormone, vitamin A derivate, vitamin D derivate, ligand X)

on the communication of intercellular information through morphogenic signals. These are responsible for a differential expression of genes, which finally determine form. It was assumed that there are local concentration gradients of activators and inhibitors carrying cellular positional information (MEINHARDT 1987) leading to pattern formation. Recently, the structure of a morphogen has been described which induces the development of an ameba to the fully differentiated organism (*Dyctosteleum discoidum*). The structure of this morphogen DIF1 closely resembles the structure of vitamin A, which is a morphogen for the wings of chickens. Vitamin A also induces the differentiation of numerous tumor cell lines to partly benign phenotypes. Vitamin A was recently described as an essential ligand belonging to the basic receptor family of hydrophobic ligands, so that we are now able to define a ligand–receptor system which determines "Gestalt".

In considering the high degree of structural similarity of these original receptor families, it is remarkable that the thyroid hormone T_3 functions as a morphogen in the tadpole. The original receptor family containing cyclic hydrophobic compounds consists of proteins of similar structure which have

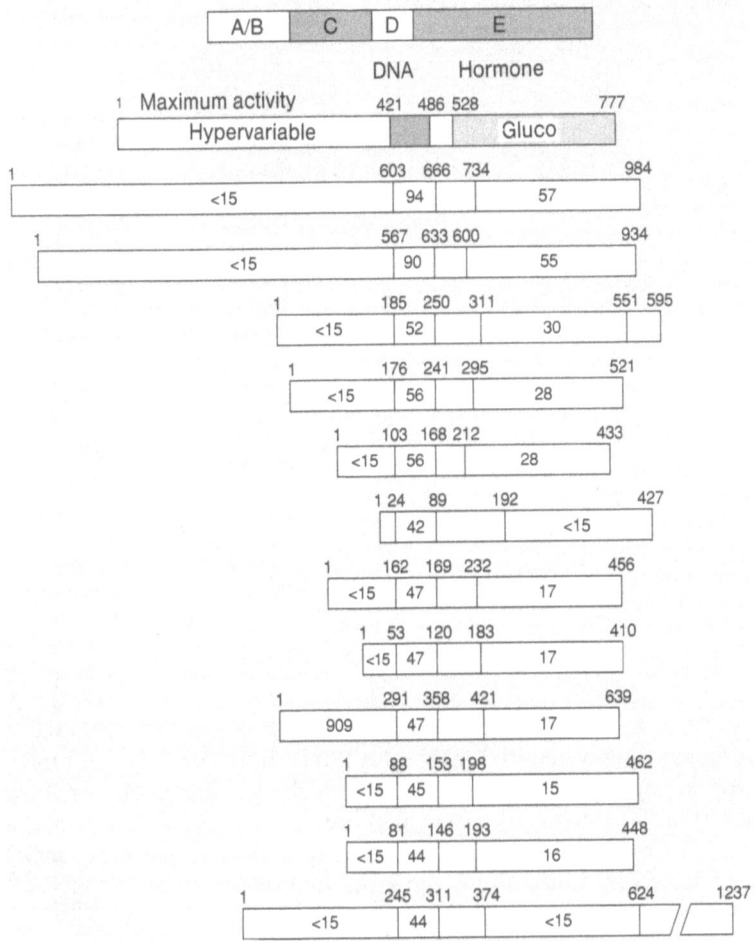

Fig. 5. Structure of the receptor family for cyclic hydrophobic compounds

binding sites in the nucleus, where they modulate DNA transcription. Their structure is shown in Fig. 5.

The function of the nonconserved region A/B and D is not known. The highly conserved region C contains the DNA-binding region, and the domain E contains the hormone-binding region. Earlier it was thought that these kinds of hydrophobic ligands served as signals from the environment that passed unhindered from the cell membrane to the nucleus. In this respect, the ligands were thought to be effective for only short periods of time. Later these ligands with endocrine function were found to be carried through the cytoplasm via transport proteins, and then into the nucleus, where they would act via nuclear receptors on selective genes, causing cellular differentiation.

Table 1. Original and specific partial functions of receptors which bind cyclic hydrophobic compounds

Receptor	Original function	Specific partial function	
Estrogens [6]	Female	Character, puberty	Fertility
Androgens [?]	Male	Character, puberty	Fertility
Progesterone [11]	?	Reproduction	Gravidity
Aldosterone [4]	?	?	Mineral and water balance
Thyroid hormone			
− α-receptor [17]	Metamorphosis	Pleiotropic receptor and gene modulation	Enzyme synthesis
− β-receptor [3]	CNS		
Vitamin D [?]	Bone composition	Pleiotropic receptor and gene modulation	Enzyme synthesis
Vitamin A [17]	Organogenesis	Rhodopsin gene modulation	Rhodopsin synthesis
Ligand X [3] hap gene }	Organ development?	?	?
Glucocorticoids [5]	Stress	Pleiotropic receptor and gene modulation	Enzyme synthesis

* Chromosome.

At the same time, one can see that this family of genes is distributed on several chromosomes and that in this receptor family hormone–receptor interaction is programmed to function in ontogenetic development. The phylogenetic specificity derives from the diversification of the ligand. The original functions and the evolution of specific partial functions of receptors which bind cyclic hydrophobic compounds are listed in Table 1.

4 Primary Structure and Function of Membrane Receptors

According to our concept of the coevolution of ligand and receptor, we can distinguish three more original receptor families which can be assigned to the original ligand families. The elucidation of the structure of some receptors through molecular cloning allows us to define two main classes: single and multiple membrane-spanning receptors. From these data, we can assume that neurotransmitters and peptide hormones transfer information via multiple membrane-spanning receptors. Since the structure of most peptide hormone receptors is not known, there are large gaps in our understanding of receptor function. For the multiple membrane-spanning receptors, the structural diversification takes place in the families of fourfold, sevenfold, and 24-fold membrane-spanning receptors. The fourfold membrane-spanning receptors have diversified into selective ion channels, and as yet no information transduction

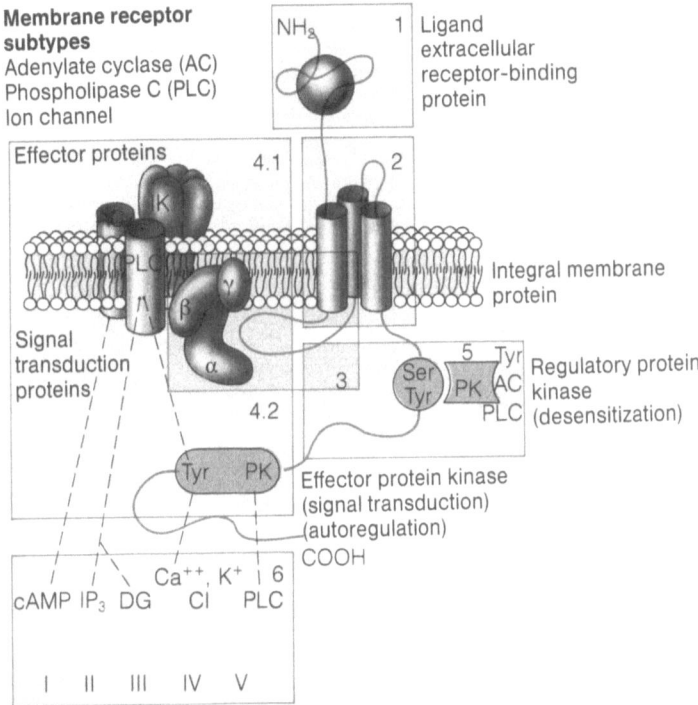

Fig. 6. Structure of membrane receptor subtypes. *1*, binding domain; *2*, membrane receptor domain; *3*, signal transduction domain; *4.1*, membrane effector domain; *4.2*, cytosol effector domain and autoregulation domain; *5*, regulator domain (heteroregulation); *6*, action types

or effector system has been found in this series. However, the specific diversification for the sevenfold membrane spanning receptors has been achieved through involvement of a signal transduction system and several effector systems. Through this diversification, at least five more action types of intracellular information transduction have been constructed (Fig. 6).

The diversification of information transduction from the receptor to the system has evolved from a protein that initially had only a channel function (Fig. 7, above). Later the information not only penetrated the cell, but also developed the first branching in the functional original dynamic of information transduction and structural branching of the receptor protein (original branching). The 24-fold membrane-spanning receptor is a typical ion channel protein, however; no information transduction is known for this receptor.

The information, already coded dynamically, is transduced to the ligand-binding site of the receptor; the membrane portion takes up the information and passes it on to compartments for further expression (Fig. 7, middle). In the next step, the gene encoding the original receptor can be duplicated, mutated, or scattered onto other chromosomes. For further information diversification, these receptor proteins must expand the ability of their membrane-spanning domains, so that a transduction system (G-proteins) can be attached. From

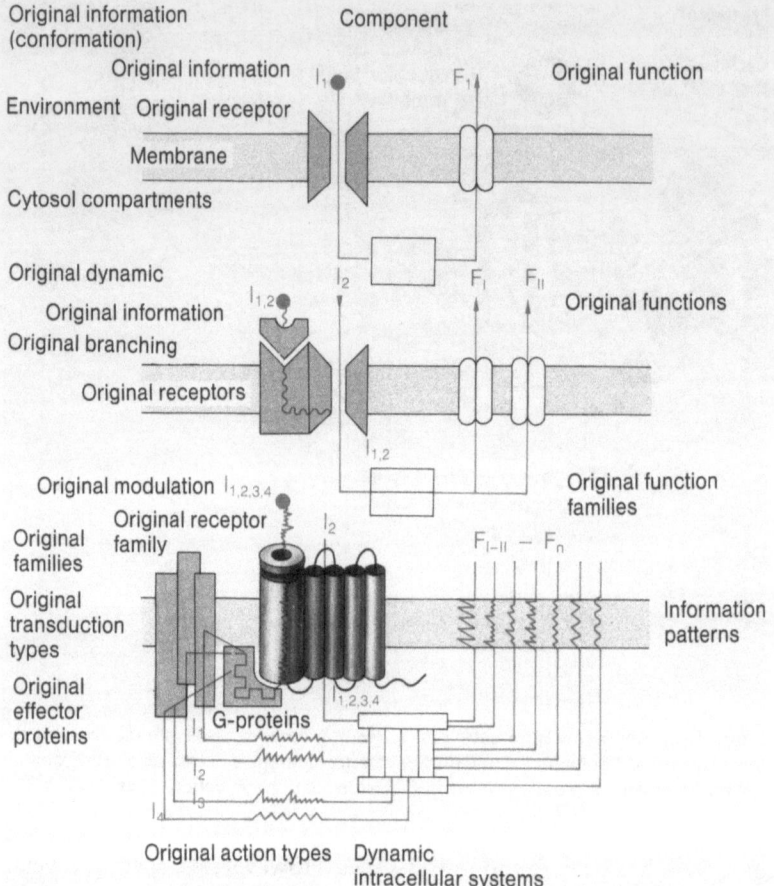

Fig. 7. Evolution of the diversification of information branching from receptor to the intracellular system

there the diversified information is reprogrammed to several effector proteins (adenylate cyclase, phospholipase C, and ion channels). The original receptor family is thus joined to a family of transduction types and to effector proteins. The information divided in this way runs in original action types and builds up into a dynamic intracellular system. The components of the system can be divided into domains (Fig. 6). In accordance with this development, we are now able to distinguish the receptor families as follows.

4.1 Single Membrane-Spanning Receptors

The family of single membrane-spanning receptors can be further divided into a monomer and a dimer class. Presently we only know of the insulin receptor and the receptor for insulin-like growth factor I (Fig. 8) (EBINA et al. 1985) in the dimer class. There are several receptors which belong to the monomer class of the single membrane-spanning receptor family: receptors for growth hormone (WALLIS 1987), for transferrin, for the transformation growth factors (RIEDEL et al. 1987), and for the lymphokines. Recently the receptor for nerve growth factor (NGF) was also cloned. In contrast to the fourfold and sevenfold membrane-spanning receptors, the single membrane-spanning receptors do not share a common genetic origin. Although these receptors do have certain characteristics in common, they do not show interprotein homologies: their shared characteristics are ascribable not to protein structure but to the level of information transduction (Fig. 9) (GOLDFINE 1987). Single membrane-spanning domains display a structural similarity in that they mostly show repetitious sequence domains, which indicates that every receptor has its origin in a relatively simple gene structure. Moreover, numerous receptors have repetitious sequences which are rich in cysteine. All of them contain a phosphorylation domain at the cytosol end of the protein. These phosphorylation sites are occupied by the amino acids tyrosine, serine, or threonine. Although an identical origin cannot be ascribed to the ligand families or to the receptor families, an original function which unites all receptors can be assigned. This is the modulation of transformation and growth. All receptors have in common a relatively similar information transduction mechanism that can be best explained by reference to the insulin receptor.

Characteristic for all partners of the single membrane-spanning receptors, which are stimulated by transformation growth factors, is the fact that endocytosis usually takes place after the ligand binds to the receptor. The hormone–receptor complex is taken into an endocytotic vesicle that is then absorbed within the intracellular compartments (Fig. 9). The signal for endocytosis is the phosphorylation of specific amino acids in the cytosol receptor domain. For growth hormones, however, an exact signal transduction chain has not yet been identified. The endocytotic vesicle, which contians the hormone–receptor complex, fuses, for example, with the Golgi apparatus, with lysosomes (as described for EGF receptor), or with nuclear structures (as described for insulin) (Fig. 9).

It is assumed that the insulin released can cause intranuclear dephosphorylation directly or maybe through secondary proteins. This in turn causes specific changes in the metabolic processing (Fig. 9). Thus, an essential transduction step for single membrane-spanning receptors is the endocytic introduction of the hormone-activated receptor and its fusion with intracellular membranes; then, taking the insulin receptor as an example, it can either dephosphorylate proteins, or the ligands themselves can interact directly with unknown DNA-regulatory proteins. The interaction of insulin with the Golgi apparatus, for example, leads to the activation and translocation of the protein glycose transporter from the Golgi apparatus into the cell membrane, thus intensifying the transport of glucose. It now seems that a specific G protein family represents another transduction system coupled to the single membrane-spanning receptors. This G protein family may couple to the phospho-

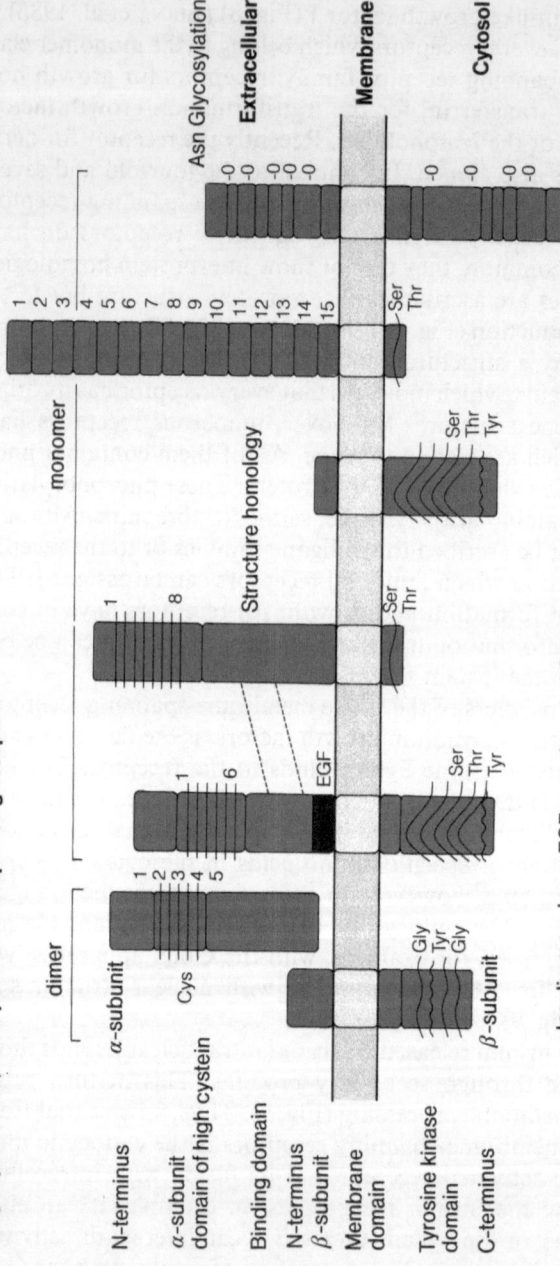

Fig. 8. Single (monomer, dimer) membrane-spanning receptor. *IGF*, insulin-like growth factor; *GH*, growth hormone. For further abbreviations, see Fig. 1

Single membrane-spanning receptor

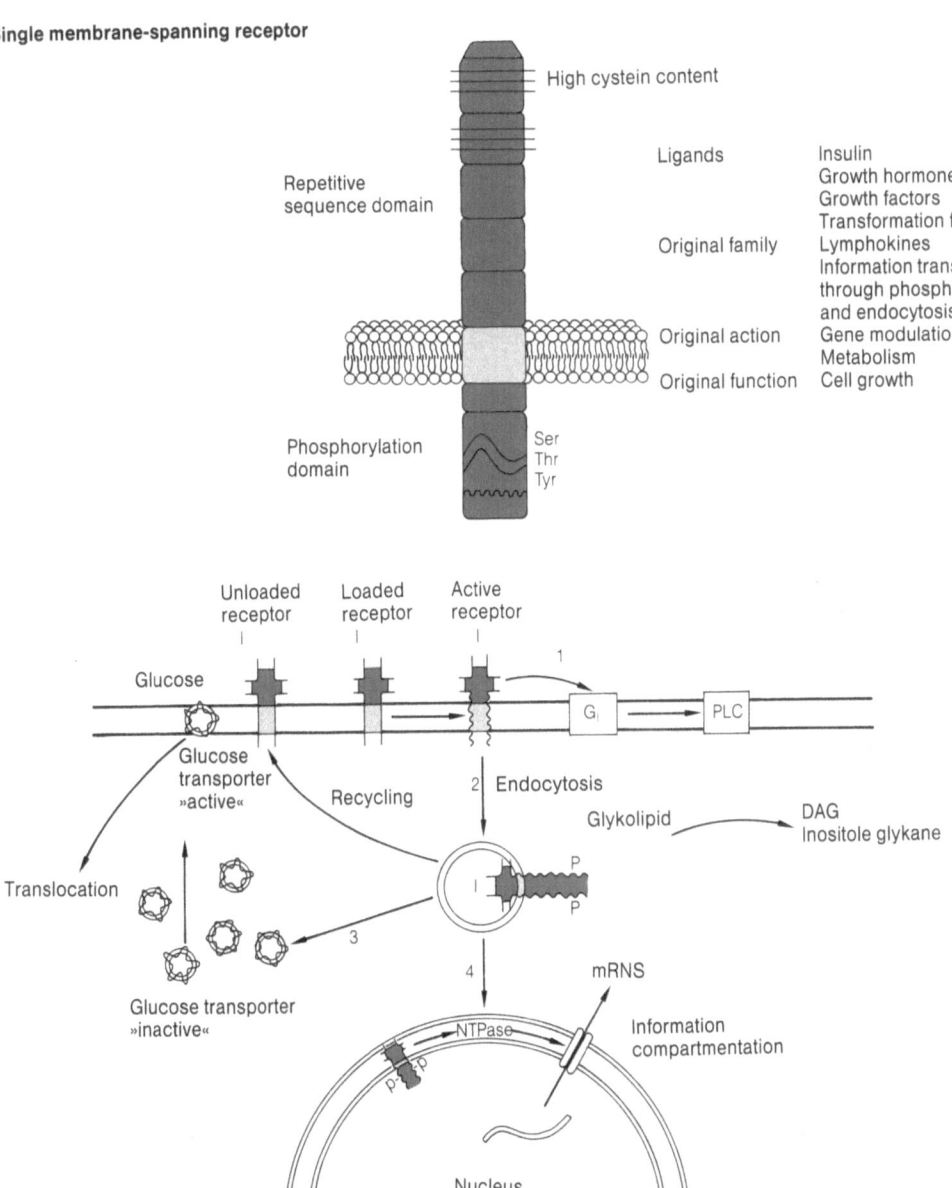

Fig. 9. Single membrane-spanning receptor

lipase C and so lead to a transduction and effector system that resembles the sevenfold membrane-spanning receptor.

The growth hormone receptor is especially interesting because a high degree of homology exists between the growth hormone receptor membrane protein and the growth hormone serum-binding protein.

The biological significance of such hormone-binding proteins is completely unknown at present. However, considering our hypothesis of the coevolution of ligand and receptor, it is not surprising that hormone-binding proteins show protein homology towards the receptor. For the receptor of NGF, which was recently cloned, it has been demonstrated that the hormone–receptor complex is internalized through receptor-induced endocytosis.

4.2 Fourfold Membrane-Spanning Receptors

The first type of membrane receptor, whose ligands are class 1 neurotransmitters, such as n-acetylcholine (cations), GABA, and glycine (anions), are remarkable because the protein spans the membrane with four conserved domains (Fig. 10). The amino acid sequences of the α-, β-, γ-, and δ-subunits of the n-acetylcholine receptor are shown in Fig. 11 (CHANGEUX et al. 1985).

Thus the receptor family in question belongs to the class 1 neurotransmitter family. The original action is the transport of anions and cations as a result of peripheral stimulation and these exhibit a centrally inhibitory action whose original function is the modulation of quick synapses. This receptor has not yet been coupled to a transduction system; rather it is directly employed as an ion channel. However, it has already been shown to possess a regulatory domain that can be phosphorylated by different protein kinases. In the case of the GABA and glycine receptors, it is assumed that four such fourfold membrane-spanning protein subunits are assembled in the membrane (two α- and two β-units), whereas it is still discussed whether the ion channel of the nicotinic acetylcholine receptor might be constructed of five fourfold protein subunits. The additional subunit is supposed to be amphipathic (STEVENS 1985).

The cloning of these receptors revealed a high structural homology (GRENNINGLOH et al. 1987; SCHOFIELD et al. 1987). In the extracellular domain, all of the receptors have a ligand-binding region residing in a protein fold which is held together by two cysteine residues forming a disulfide bridge. However, it can be seen that the GABA and the glycine receptors show a stronger degree of homology between each other (34%–38%) than either has to the acetylcholine receptor (15%–20%). GABA and glycine receptors form chloride ion channels, whereas the nicotinic acetylcholine receptor contains a potassium channel. This is one reason why the GABA and glycine receptors contain mainly a positively charged area composed of positively charged amino acids located on the extracellular and cytosolic surfaces of the receptor, whereas the nicotinic acetylcholine receptor is thought to contain mainly negatively charged areas. The M_1 protein domains present in both the four- and the five-subunit receptors are arranged in the middle to form the channel entrance

Fourfold membrane-spanning receptor

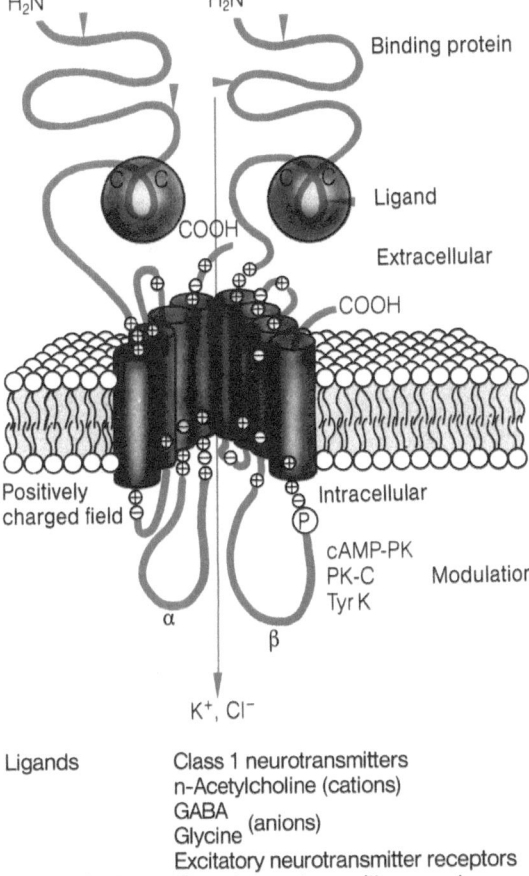

Ligands   Class 1 neurotransmitters
n-Acetylcholine (cations)
GABA
Glycine (anions)
Excitatory neurotransmitter receptors
Original family   Class 1 neurotransmitter receptors
Original actions   Anion transport and cation transport
with peripheral stimulating and
central inhibiting action
Original function   Rapid modulation of synapses

Fig. 10. Fourfold membrane-spanning receptor

for all receptor types. This domain is rich in positively charged arginine and lysine residues and can therefore concentrate anions and accelerate their flow through assembled subunits in the membrane.

The structure of the receptor in the membrane and how it functions are shown in Figs. 12 and 13. The main function of the fourfold membrane-spanning receptors is at synapses, where they modulate informational intensity. The branching of information takes place within the anatomical structures of the nervous system in the form of a neuronal network.

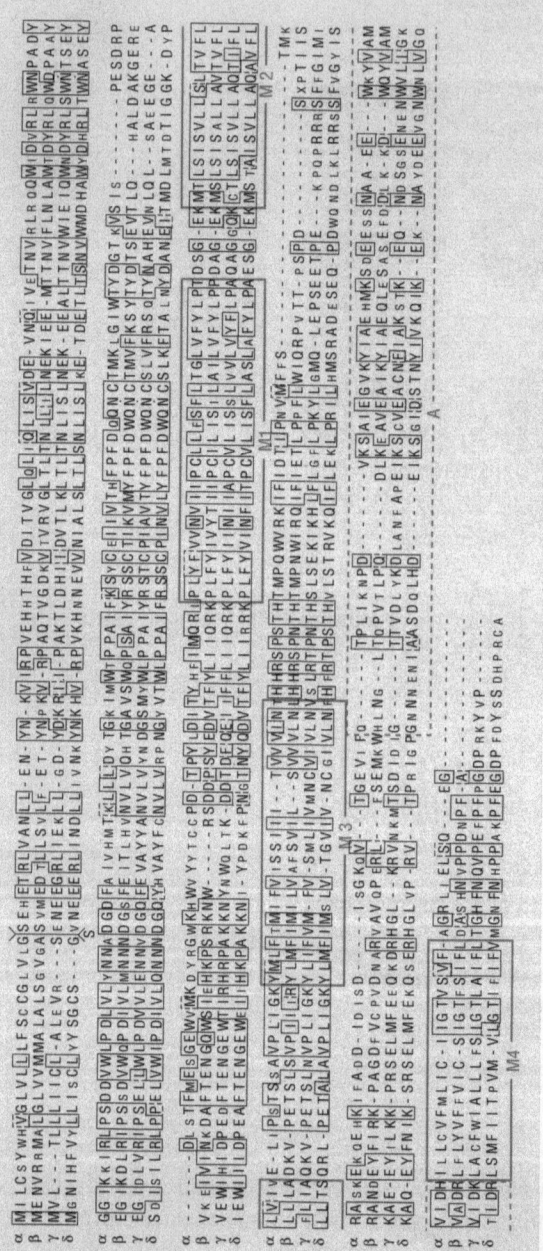

Fig. 11. Gene structure and construction of the nicotinic acetylcholine receptor

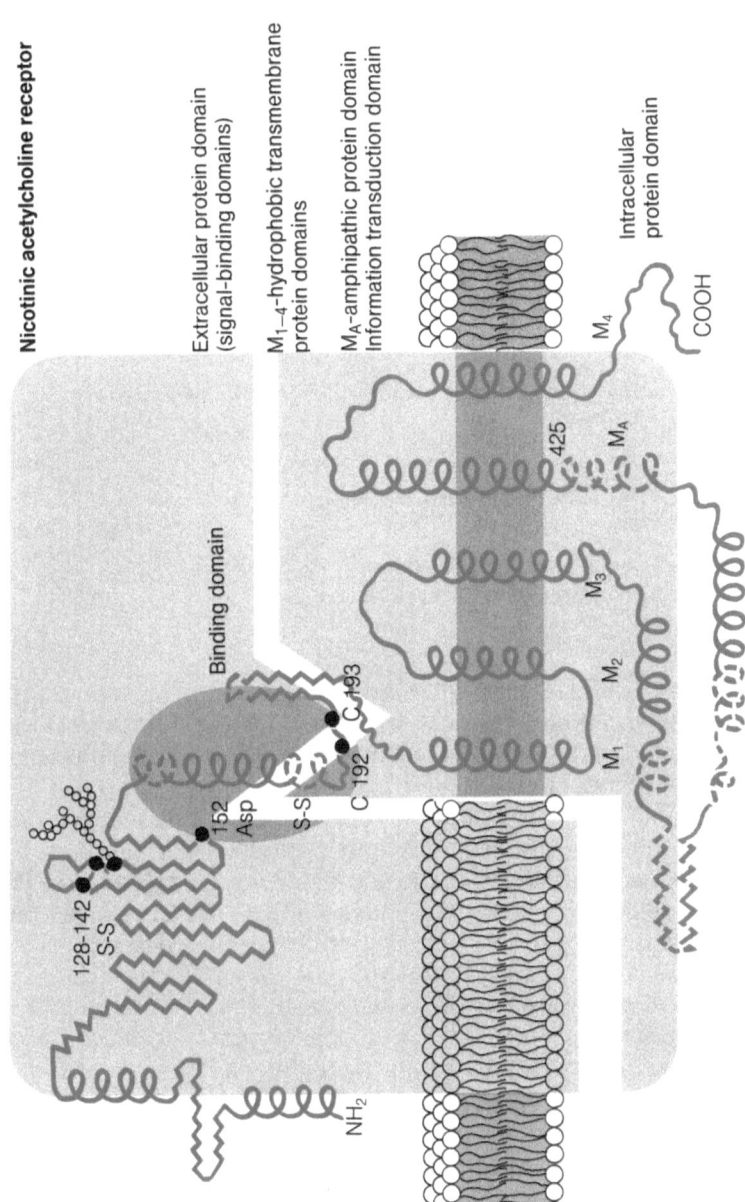

Fig. 12. Membrane structure of the nicotinic acetylcholine receptor

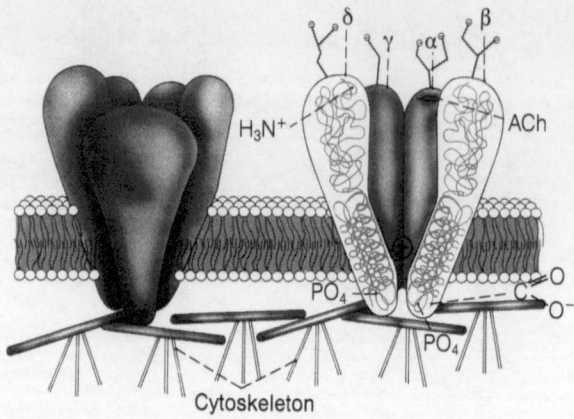

Fig. 13. Functional structure of the nicotinic acetylcholine receptor (according to HUCHO 1986)

4.3 Sevenfold Membrane-Spanning Receptors

4.3.1 Gene Structure and Receptor Assembly

The second important class of membrane-spanning receptors are proteins with seven hydrophobic membrane-spanning structures where the N-terminal end extends extracellularly into the environment and the C-terminal end extends into the cytosol (Fig. 14).

Figure 15 (KUBO et al. 1986) shows a comparison of the amino acid sequences from the homologous muscarinic acetylcholine receptor, the β-adrenergic receptor, and opsin as well as the gene structure and composition of the acetylcholine receptor. These receptors interact with a very heterogeneous family of ligands which, however, can be classified as either class II neurotransmitters or as neuropeptides. It should be pointed out that the neuropeptide substance K receptor was the first neuropeptide receptor cloned. It is remarkable that neurotransmitters and neuropeptides use an identically constructed membrane receptor structure. Furthermore, an unknown ligand X uses a receptor that is encoded by the *mas*-oncogene and which also shows the same structure (HANLEY and JACKSON 1987). In the present context, however, it is of greater significance that light also uses this receptor (Fig. 14). In accordance with the very different ligands, information is transmitted through the sevenfold membrane-spanning receptor to an original transduction system, the G protein, and subsequently to a family of original effector proteins to which adenylate cyclase, phospholipase C, and ion channels belong (Fig. 16).

The G protein family is divided into several subunits (BIRNBAUMER et al. 1987). Principally G proteins are composed of α-, β- and γ-subunits. The γ-subunit is hydrophobic and anchors the G protein complex at the membrane. The α- and β-subunits are hydrophilic. The α-subunit can be coupled to

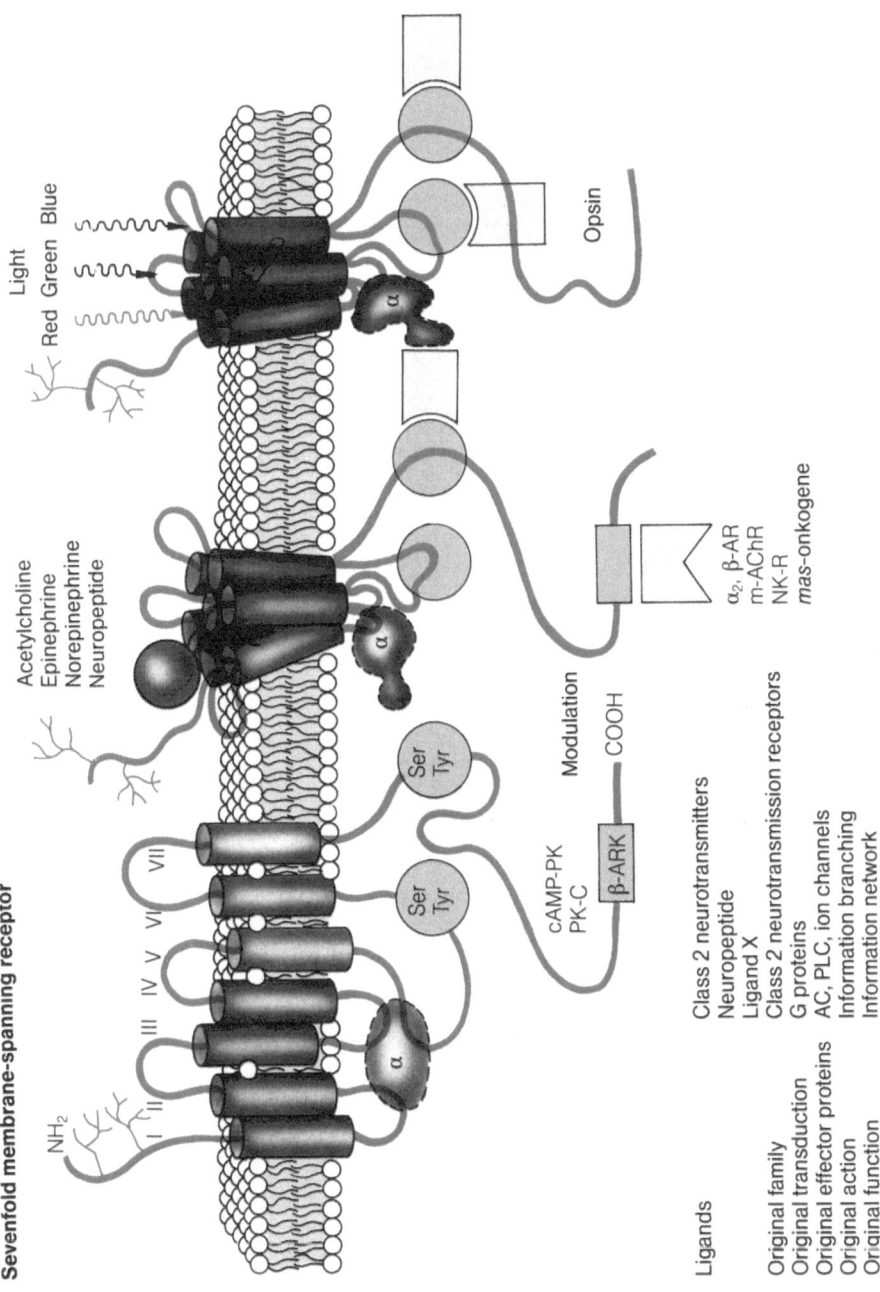

Sevenfold membrane-spanning receptor

Light
Red Green Blue

Acetylcholine
Epinephrine
Norepinephrine
Neuropeptide

NH₂

I   II   III   IV   V   VI   VII

Ser Tyr   Ser Tyr

cAMP-PK
PK-C   β-ARK   COOH

Modulation

Opsin

α₂, β-AR
m-AChR
NK-R
mas-onkogene

Ligands   Class 2 neurotransmitters
                                 Neuropeptide
                                 Ligand X
Original family   Class 2 neurotransmission receptors
Original transduction   G proteins
Original effector proteins   AC, PLC, ion channels
Original action   Information branching
Original function   Information network

Fig. 14. Sevenfold membrane-spanning domain

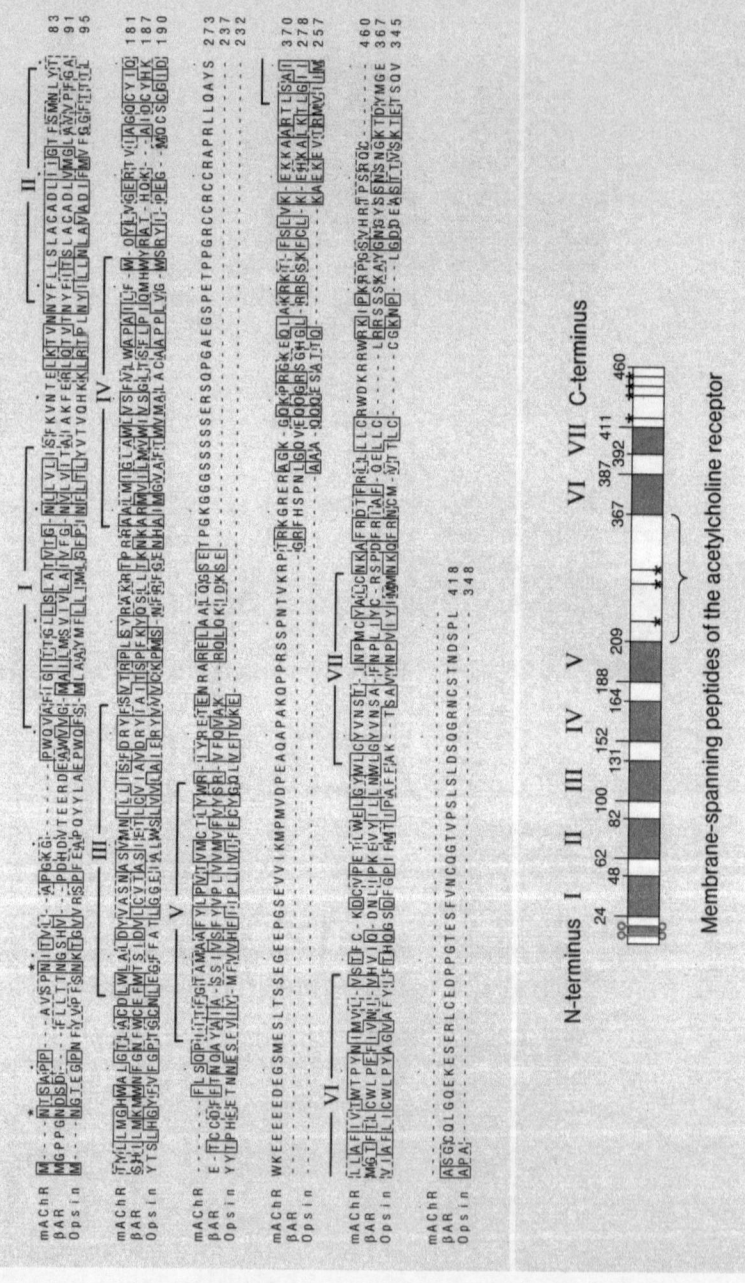

Membrane-spanning peptides of the acetylcholine receptor

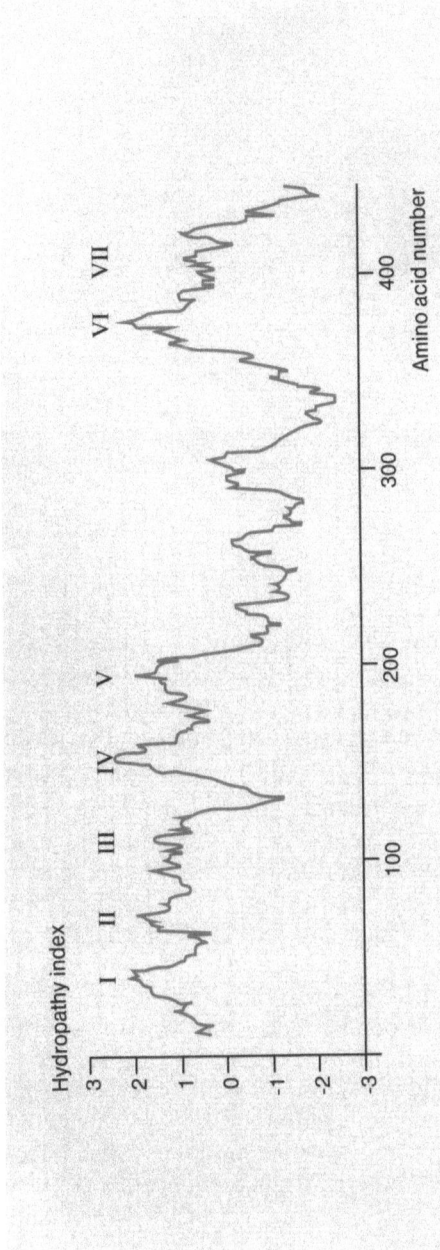

Fig. 15. Gene structure and construction of the muscarinic acetylcholine receptor (m-AChR), the β-adrenergic receptor (β-AR), and opsin

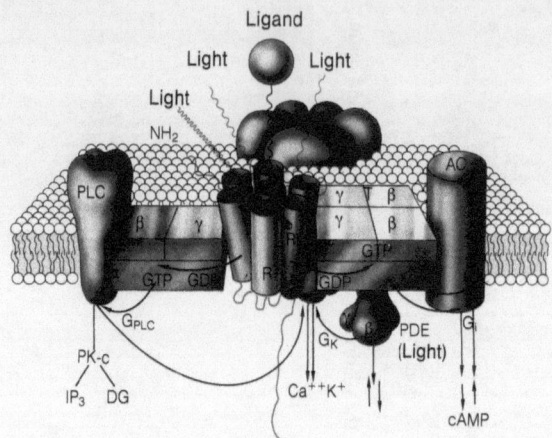

Fig. 16. Transduction and effector systems of the sevenfold membrane-spanning receptors. *R*, receptor; *AC*, adenylate cyclase; *IK*, ion channel; *PLC*, phospholipase C; *RT*, retinal; *T*, transducin (light)

the effector proteins in a cyclical manner (Fig. 17). While guanosine triphosphate (GTP) is bound in the active center of the α-subunit, the G protein complex is bound to the effector system. The conformational change of the receptor after binding ligand leads to the activation of G-α (Fig. 17) (MASTERS et al. 1986). This activates the effector protein while the G-α protein is uncoupled from the G-β and G-γ proteins. Immediately dephosphorylation takes place, the active center in the α-subunit is replaced with GTP, the G proteins associate with each other, the hormone–receptor complex dissociates again, and the hormone is released from the receptor (Fig. 18). To date we know of five different G protein trimer complexes and transducin. The information branching in the G proteins leads to a specific modulation of the effector proteins adenylate cyclase and phospholipase C as well as of ion channels and, via transducin, of phosphodiesterase.

The α-subunit of G proteins of the opsin receptors is called transducin. Transducin regulates the enzyme phosphodiesterase, the activity of which probably also directly coupled to an ion channel (Fig. 16). Thus, we can distinguish five different subunits of the regulatory G proteins. G_s is the stimulating protein for adenylate cyclase; G_i is the inhibiting regulatory protein for adenylate cyclase. The regulatory protein GPLC modulates the effector protein for phospholipase C; the functional significance of a protein G_o, which is mainly found in the central nervous system, has not been entirely elucidated, but it can presumably be classified as a modulator of a calcium channel. G-K is a G protein that modulates an ion channel transporting potassium.

It is not known which protein domains of the sevenfold membrane-spanning receptor protein are used to couple it to the G proteins. All cytoplasmic loops have been discussed in this respect. The cytoplasmic domains of opsins are highly homologous, whereas the homology within the acetylcholine, α- and

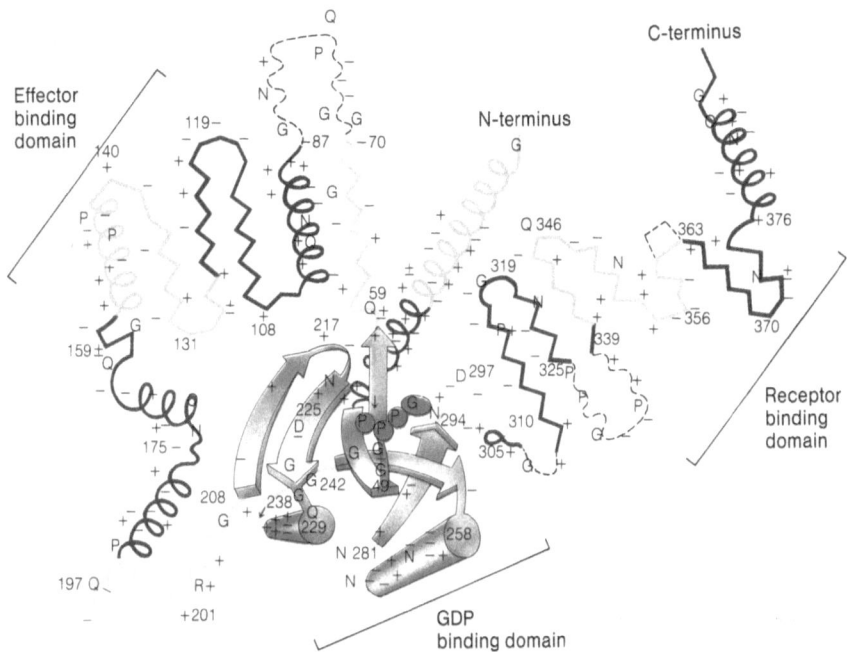

Fig. 17. Structure of the α-unit of the G protein (slightly modified from a figure supplied personally by R. BOURNE)

β-adrenergic, and substance K receptors is low. Additionally the third cytoplasmic loop has a different length. Perhaps this protein diversification permits the biophysical organization of their specific coupling to different G proteins.

To modulate the activity of the complex system of sevenfold membrane-spanning receptors, an information stop is coupled to these proteins, functioning as a regulating system (Fig. 14). The protein phosphorylation at the third loop of the receptor and on the C-terminus represents the effective regulatory protein domain, through which the sevenfold membrane-spanning receptor is activated or desensitized by phosphorylation. The autophosphorylation of the β-adrenergic receptor at a specific region at the C-terminal end of the receptor is unique to this receptor, since only the ligand occupation of the receptor ends in phosphorylation and is followed by desensitization.

Besides this, there is also heterologous phosphorylation of the β-adrenergic receptor mediated through both cAMP-activated protein kinases and protein kinase C, either of which can give rise to receptor desensitization. However, this also means that by activating their own effector proteins other hormone systems can modulate the sevenfold membrane-spanning receptors through phosphorylation. Thus, the pleiotropic processing and coupling of these systems results in a more sophisticated informational network. Since it was always assumed that all hormones which couple at G proteins can have a sevenfold membrane-spanning receptor, it is interesting to list the hormones

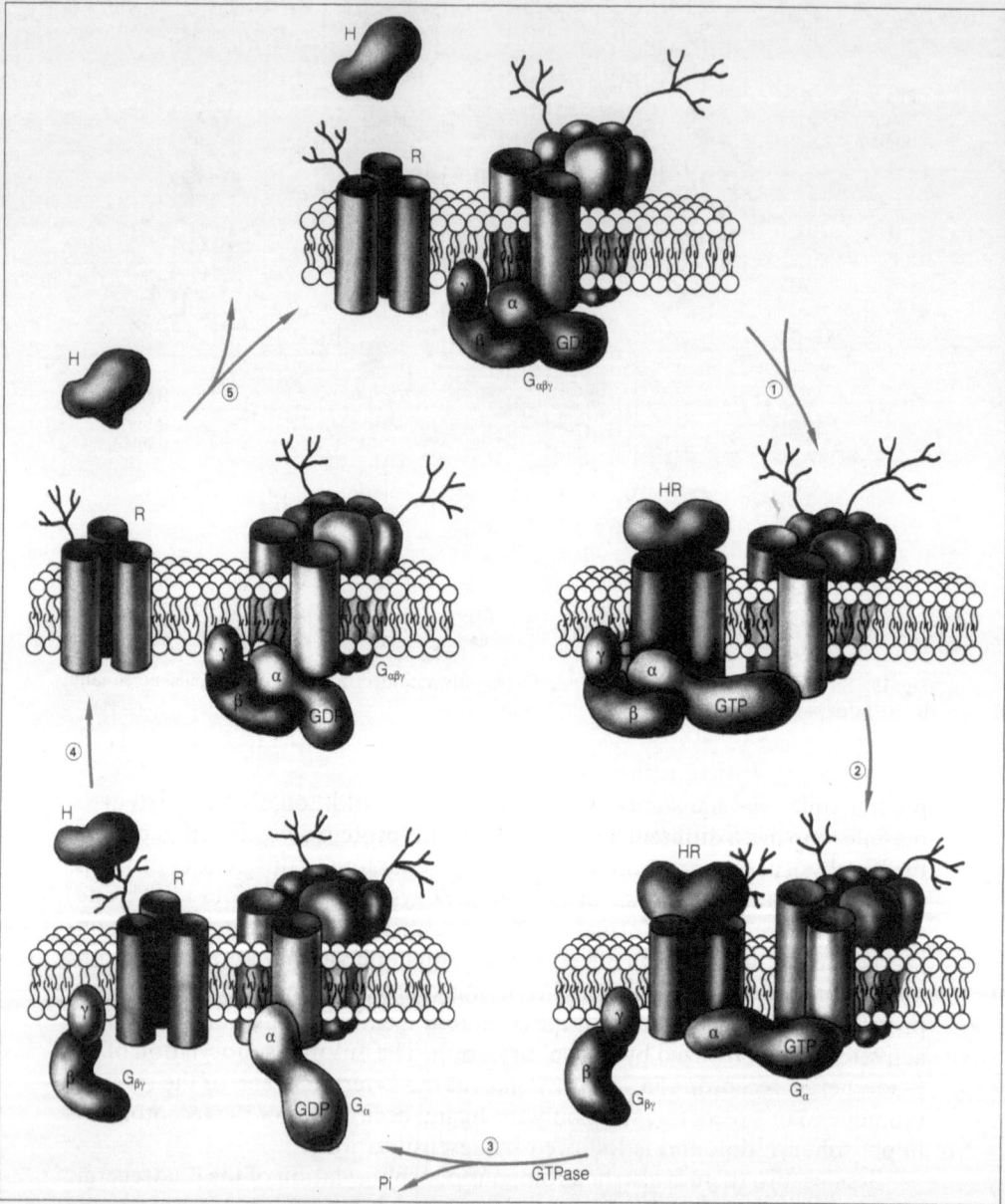

Fig. 18. The role of G protein in the receptor-mediated regulation of the effector function (see text)

that are able to couple at various G transduction systems through their receptor. Until now it has not been clarified in what form the hormones wrap their specific information such that in combination with their receptor they activate different G proteins.

4.3.2 Information Transduction

In general, a key to the understanding of information transduction is provided by the similarity of the receptor structure for the class II neurotransmitters and neuropeptides at the sevenfold membrane-spanning receptor to the information received from rhodopsin and opsins.

In the first case, the information carried by the binding of organic ring compounds and peptides is transduced to the receptor through a highly energetic conformational change of the receptor. In the second case the information is transduced by light, a ligand with two different information is transduced by light, a ligand with two different information qualities, photons and waves. Hence a biophysical understanding of this mechanism can help in the understanding of receptor modulation, given that both ligands cause changes in the receptor protein that are mechanistically identical: the coupling to the G protein. For example, comparing the structure of the turkey β_2-adrenergic receptor to that of rhodopsin and m-acetylcholine receptors, one can see the highest degrees of homology in the membrane-spanning protein sequence, especially between M_1 and M_2 and between M_3 and M_4, but also in both of the first cytoplastic loops. The lowest degree of homology is found in the N- and C-terminal domains of the protein chains. The homology between the β-adrenergic receptor of the hamster and cerebral or cardiac muscarinic acetylcholine receptor in the transmembrane protein palisades is 32% and 28% respectively, and in the first cytoplasmatic loops, 39%–42% (BIRNBAUMER et al. 1987). The homology to the proteins of the substance K receptors is around 39%, so that for the structural ligands of the class II neurotransmitters and the neuropeptides the conformation-specific ligand–palisade interaction is more relevant (BUCK et al. 1988). The homology of the opsins, however, is divided differently. Here we distinguish between opsin molecules for the colors blue, red, and green and rhodopsin molecules for dim light. The highest degree of homology exists between red and green over the whole molecule.

However, it is remarkable that the degree of homology of light-sensitive proteins is lower in the membrane-spanning protein domains, and is in fact equally pronounced in the cytoplasmic loop and even the C-terminal end of the protein. This is related to the physical structure of the information.

The class II neurotransmitter receptor and the neuropeptide receptor probably receive their information through the hydrophobic interaction of the ligand in intercalating and altering the conformation of the protein columns which are anchored in the membrane like palisades. This causes a conformational change of the receptor in the membrane.

It must be pointed out that we do not talk of a specific receptor-binding site, but of an interaction with several membrane-spanning receptor protein domains. The ligand presses between these palisades and sends conformational energy to the conformation of the membrane palisades, which is followed by a conformational change in the energy distribution of the palisade proteins in the membrane. An exact description of the ligand-binding domain, e.g., like that of receptor-binding proteins, is not yet possible, but in view of the specific

interaction of synthetic agonists and antagonists of retinal, in the opsin molecule, it should become possible (Fig. 16).

The ligand, light, passes its energy directly to the antenna molecule retinal,
which reconverts it after absorption into an electrochemically controlled conformational change of the hydrophobic interactions in the palisade proteins.
Thus, we are dealing with a complex pleiotropic information transduction
system. The conformational change of the palisade proteins leads to a coupling
of these proteins with the α-subunit of the G-protein family. The homologies
for discriminating light sources can be much more discrete inside the opsin
proteins.

Again, the principle of coevolution of ligand and receptor allows a bio-
physically convincing description of the ligand–receptor interaction. It is significant for our later observations that a ligand which is primarily defined as
a structured hormone molecule transduces identical information as a ligand
which is primarily defined as a wave-shaped ligand (in the present context,
light), and in a biologically identical way. But we have learned through the
discovery of the quantum theory that the biological quality of light appears
either wave shaped or as a quantum, depending on the method of observation.
Both forms cannot be described simultaneously; there is only one form in one
observation time period. We would like to employ this fact later for the description of hormones either as molecules with a high energy conformation or
as dynamically moving informational carriers.

4.4 Twenty-four-fold Membrane-Spanning Receptors

The protein structure of the sodium channel was clarified for the first time in
1986 through cDNA cloning and sequencing (NODA et al. 1986). This protein
shows a high molecular weight of 260000 and has four homologous protein
domains with six membrane-spanning protein segments each, constituting the
24-fold membrane-spanning protein. The ligand of this ion channel is still
unknown for humans. Recently the same group, however, succeeded in
cloning and sequencing the receptor for the calcium blocker dihydropyridine
(TANABE et al. 1987). This receptor is also a big protein with a molecular weight
of 170000, which shows a high degree of homology with the sodium channel.
Its structure is also characterized by four homologous protein segments which
were created through gene duplication; each segment contains six membrane-
spanning peptide sequences (Fig. 19). Five peptide sequences are hydrophobic
$(S_1, S_2, S_3, S_5,$ and $S_6)$; S_4, however, is positively charged. It is speculated that
the six peptide sequences of each segment assemble like a paliside cylinder in
the membrane and form an ion channel. Thus this ion channel is formed by a
big, single protein, in contrast to the fourfold membrane-spanning receptor,
where several protein subunits assemble to build the channel. The positively
charged columns can perform as a charge sensor, whereas the second columns
of each segment are supposed to line the ion channel. While natural ligands for
the sodium channel are not yet known, the group of calcium channel proteins

24-fold membrane-spanning receptor

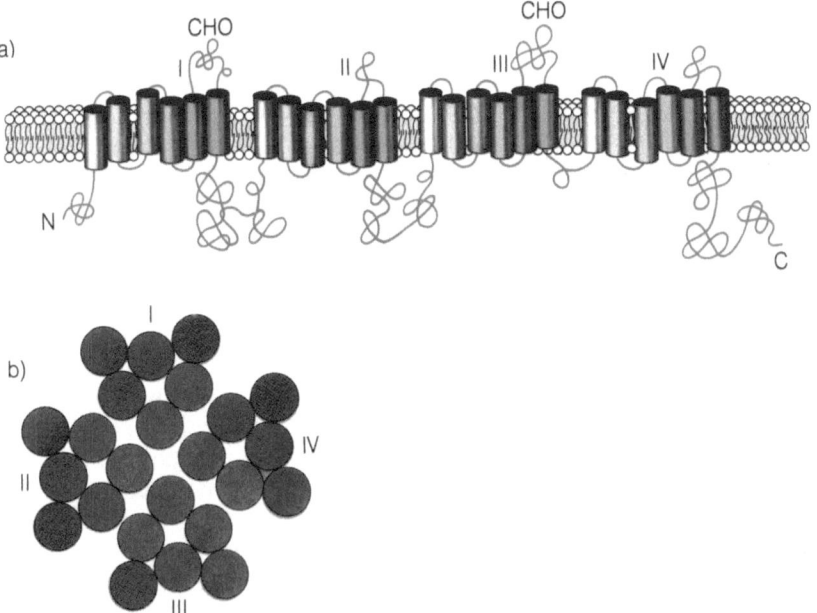

Fig. 19 a, b. Twenty-four-fold membrane-spanning receptor. **a** Linear model of the four homologous protein segments. **b** Model of the ion channel as a central pore

is significant by virtue of the fact that it is today proposed that the peptide hormone endothelin, which was recently discovered, can function as an endogenous ligand and receptor for calcium channel blockers (GORDON 1988). Endothelin belongs to a group of still unknown peptide hormones with four cysteine residues. Each forms a disulfide bridge and gives the molecule a complex group formation that is usually only found in toxins of nonmammals, in which case it interacts with the ion channel.

5 Hormone–Receptor Interactions

5.1 Primary Dynamics

Previously we mentioned that the expression "information" only makes sense for ligands and receptors when the dynamics of their interactions are considered. In this context we described – in addition to the coevolution of ligands and receptors – the coevolution of a dynamic for this information transduction. Thus, we postulate the coevolution of a biological system which is made up of three components: ligand, receptor, and dynamic. A typology of receptors remains a typology in hardware without programming if one does not consider the dynamics of information transduction. Hence we are talking about the machine and its program, and in the following the aim is to prove that a typology of dynamic programming for receptors, for the realization of life through information transduction, is of the same importance as the understanding of their structural programming. Here we encounter the long-known fundamental communication problem: How does a function develop and what is a function in contrast to a structure? We encounter the dualism of matter and "Geist", a problem that, through the indivisibility of the unliving and the living, can be solved even in our case. We must try to solve it with a kind of physicochemical explanation of biological phenomena.

The "biological system" was defined at the start of this chapter. The expression "relation" means the functional program in the biological system, the "Geist" of a program which creates, maintains, and destroys life through information generation, transduction, and exchange at the biological structure of the hormone–receptor unit. To understand structure and program in the current context, the ligand, light, can again be used as an example. Light is constructed of photons, i.e., of tiny particles, but just as much of waves. Neither can be observed separately from the other. The photon particles describe the molecular structure of light; waves, however, do not consist of substance; rather they are a dynamic energy of molecular information, the program that causes the reciprocal action of the material light in space and time. In this dualistic function, light is used by the opsin receptor. As a particle, it transduces energy to the opsin protein which changes its conformation. This is communicated to the G protein transducin and the effector proteins. From there it is transformed into cellular responses. Hence the particle supplies the energy for spatial interactions. In the nature of the wave, however, the specific information is coded; the wavelength selects the opsin protein for the colors blue, red, and green and thereby the pattern for information energy that is passed on to transducin for each color. The pattern of this information transduction is encoded by light of different wavelengths. Only this pattern will determine the pattern of the effector cascade. Figure 20 shows this ligand (light)-controlled information modulation of the sevenfold membrane-spanning receptor in a synoptic way. The colors blue, red, and green pass on their dynamic information to the transducin chain and into the cytoplasm (spectral analysis); analogously the dynamic pattern of the peptide ligands is translated to the transduction chain through spectral analysis of the pulse amplitude and frequency modulation at the receptor. By reference to the action of light at the sevenfold membrane-spanning receptor, we can understand the modulation of the identical receptor protein through hormone ligands. Hormones deliver their energy, wrapped in their conformation, to the receptor. The hormone molecule corresponds to undivided light (photon) and determines the specificity of the topographic interaction between ligand and receptor. The information delivered by hormones is then coded in a pulse amplitude and frequency modulation, analogously to the light waves. The pulse amplitude and frequency modulation regulates the information transduction into the cytoplasm on

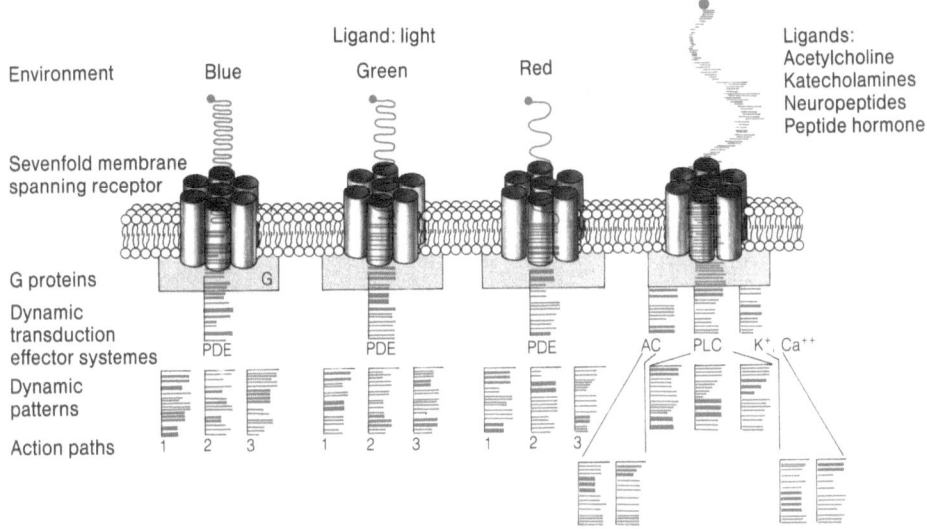

Fig. 20. Ligand (light) controlled information modulation of the sevenfold membrane-spanning receptor

the one hand, but also the metabolic destiny of the receptor itself through up- and down-regulation.

Today we must assume that the information diversification of a single hormone, which can serve several signal transduction paths at G proteins (Table 2) to modulate intracellular oscillating second-messenger systems (cAMP, Ca^{2+}, phospholipase C, see also Fig. 21), is not found in the structure of the hormone or receptor molecule but in the pattern of the hormone–receptor interaction. Hence the photon is to the light source as the hormone molecule is to pulse amplitude and frequency modulation. As in the quantum mechanics of light, one can examine only one form at a time, although both forms are parts of a unity.

The principle of pulse amplitude and frequency modulation of receptors is a general principle in biological information transduction, just like the typology of receptor structure. Figure 22 represents the dynamic primary function principles – the primary dynamic – of the receptors mentioned up until now. On the system level one can see that different dynamics of the hormone–receptor interaction are of different biological significance.

The principle of pulse amplitude and frequency modulation is one of the most important primary principles in living nature. One can assume, like Benno HESS, that the dynamic modulation of hormone–receptor interaction as an organizational principle of life is kept, like the library of the DNA code, in dynamic libraries (HESS and MEHNS 1986). Although the genetic code can be

Table 2. Examples of hormones which use different transduction systems

Hormone	Receptor	G-AC	G-PL	G ion channel	Tyrosine kinase	Target organ
Epinephrine	$\beta_{1/2}$	G_3				Heart, fat cell, liver
	β_2	G_1				Thrombocytes, fat cells
	β_1		G_{plc}			Smooth muscle, liver
	β_1		G_{pla2}	$G_{0 (p?)}$ = calcium channel closed		FRTL-1 cells
	β_2					Sympathetic presynapse
Acetylcholine	M_1		G_p (PhL C)			Pancreas
	M_1			? = potassium channel closed		CNS, sympathetic
	M_2	G_3 (inhibiting)		G_k (G_1?) = potassium channel open		Heart
	M_2					Heart, CNS
Parathormone	P_1	G_3				Kidney, bones
	P_2			? = calcium channel open		Heart, kidney bones
	P_3		$G_?$ (PhL C)			Lymphocytes
PDGF		$G_?$ (PhL C)			Fibroblasts?	
				Tyrosine		Endothelium, fibroblasts

Abbreviations: G-AC, protein for adenylate cyclase; G-PL, protein for phospholipase; G_3, G protein stimulating; G_1, G protein inhibiting.

preserved in mummies for thousands of years, a disturbance of this "dynamic code" leads to disease and finally to death if the dynamic code breaks down.

5.2 Modulation of the Receptor Proteins in the Membrane

Ligand binding to a hormone receptor not only transduces information into the cytoplasm, but also determines the metabolism of the receptor itself. In most cases receptor occupation prevents binding of further hormone molecules (reduction of the number of ligand-binding sites and perhaps of the affinity as well). Furthermore, ligand binding to receptor constitutes a signal for the aggregation of the hormone–receptor complex with other integral receptor proteins ("capping," "patching"). In a second step, the complex is internalized and interacts with lysosomes and the Golgi apparatus. The polypeptide is either degraded to amino acids, which occurs in the case of most

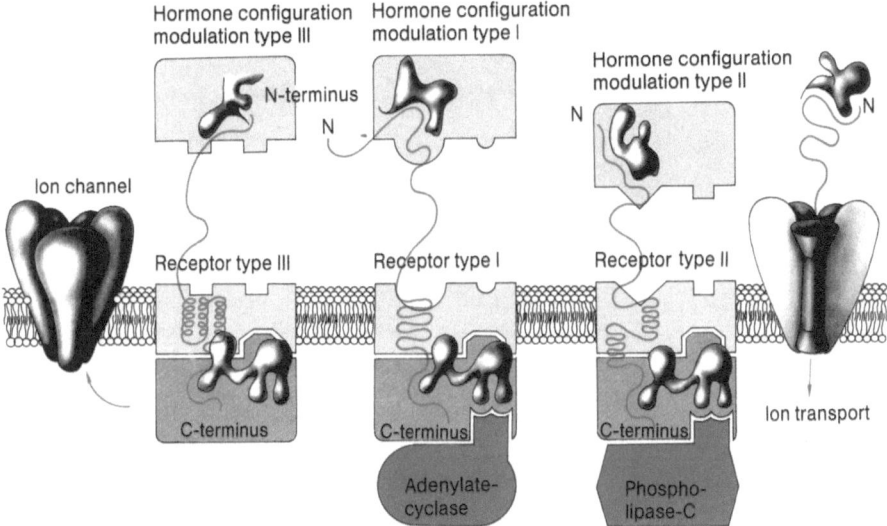

Fig. 21. Information transduction paths to integral membrane receptor proteins

hormone molecules, or the hormone molecules and hormone fragments are transported to subcellular structures such as the nucleus (e.g., insulin and parathyroid hormone), where they fulfil an unknown function. Receptor protein domains can be reassembled in the Golgi apparatus into complete receptor molecules and can be inserted in the membrane. The internalization of the hormone–receptor complex and the resynthesis of receptors determine the amount of receptors in the membrane. It is not yet known whether the different receptor domains are metabolized differently. The regulation of the presence of hormone-binding proteins, the signal transferred into the membrane, and information transduction into cytosol and subcellular structures are, however, determined by the presence of circulating hormone molecules. For the regulation of the biological ability to regulate the receptor information structure, it is necessary to control the hormone concentration (episodic and pulsatile hormone secretion) continuously. This is because only oscillating information transduction systems guarantee the necessary elasticity and stability of such complex systems.

An increased stable hormone concentration (circulating hormone pool) over long periods leads to desensitization of receptor structures. This means that the hormone-binding protein is permanently occupied by hormones. Thus, not enough receptor will be exposed to the extracellular space, and information transfer through the membrane is blocked, the cytosolic domain no longer takes part in information transduction, and the hormone–receptor complex is removed from the membrane without sufficient resynthesis of new receptor structures. Exactly the opposite can be achieved by an appropriate hormone secretion modulation: the activation of the receptor function in all its steps, including activation of the synthesis of new receptor proteins (Fig. 23).

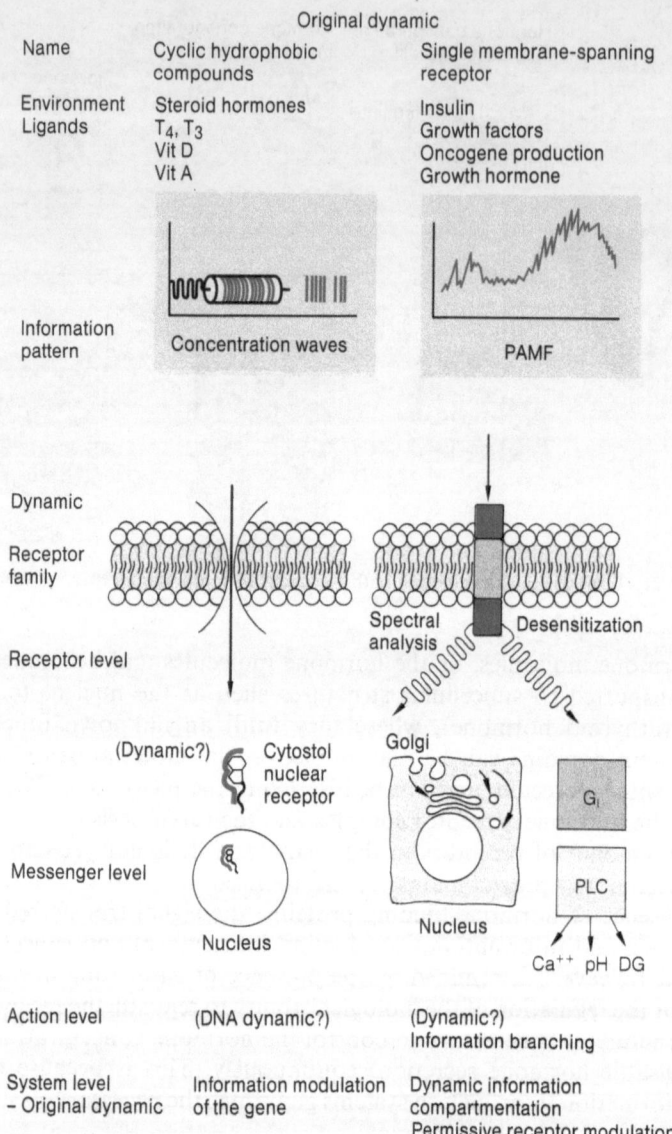

Fig. 22. Dynamic function of the receptor families. *DG*, diacylglycerol; *E*, effector pathways; *G*, G proteins; *PAMF*, pulse amplitude and frequency modulation; *PL*, phospholipase

The upper part of Fig. 23 shows to what extent different pulse amplitudes of hormonal stimulation are able to stimulate the hormone receptor unit differentially. Ultimately substantial hormone pulses will lead to a maximum stimulation of the hormone–receptor unit. It cannot be increased further. From a certain concentration on, stimulation can only lead to prolongation of

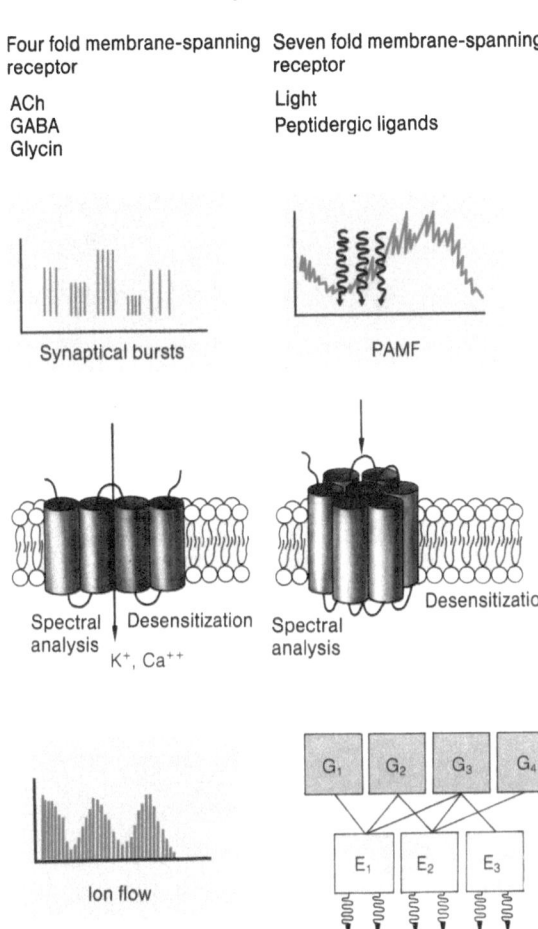

Four fold membrane-spanning receptor

ACh
GABA
Glycin

Seven fold membrane-spanning receptor

Light
Peptidergic ligands

Synaptical bursts

PAMF

Spectral analysis ↓ Desensitization
K^+, Ca^{++}

Spectral analysis

Desensitization

Ion flow

cAMP, Ca, pH, DG, Cl, K

Information intensity

Neuronal net systems

Dynamic information branching

Complex dynamic systems

the refractory period (desensitization) of the hormone–receptor unit. The lower part of the figure demonstrates the contribution of pulse amplitude and frequency modulation to the stimulation of the hormone–receptor unit. According to its pulse frequency, a single pulse is followed by a single stimulated hormone–receptor unit, with a fast regeneration of the biological response

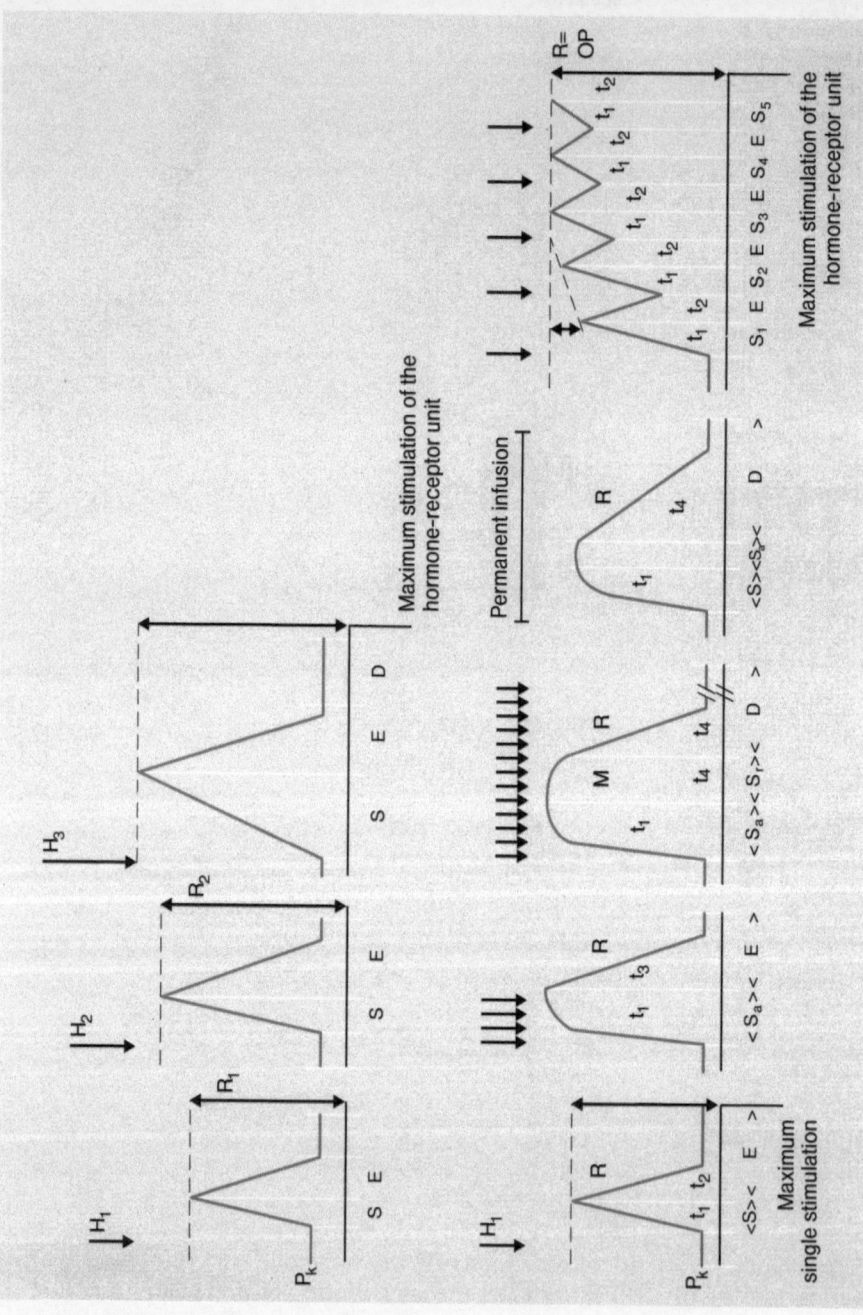

Fig. 23. Modulation of the hormone-receptor interaction. H, hormone pulse; R, response; t_1, seconds; t_2, minutes; $t_3 - t_4$, minutes to hours; M, macropulse; OP, oscillating pulses; S, stimulation; E, relaxation; D, desensitization

system. Repeated stimulating pulses are added to one single macropulse. In this case the hormone–receptor unit needs longer to regain the ability to respond.

With a pulse amplitude and frequency modulation adjusted to the system, it is possible to obtain a maximum stimulation of the receptor unit according to the biological needs of the organism. Recent research has shown that through pulse amplitude and frequency modulation, it is possible to modulate not only the kind of hormone–receptor interaction but also its specificity and the biological actions.

6 Summary

This manuscript constitutes a first attempt to categorize the cell receptors. Based on evolutionary and biological characteristics, it is possible to classify ligand–receptor units into original families. Certain dynamic patterns that could not be classified previously lead us towards the ligand–receptor unit. Biological information is transformed at the receptor by cellular transduction and effector pathways. The dynamic code is compared with the genetic code. The dynamic code controls the biological information patterns necessary for the integral function and structure of cells, tissue, organs, and organisms. The genetic code controls the assembly and molecular structure of proteins. The original families can be assigned certain original functions. A coevolution of ligand–receptor and dynamic code is revealed.

For the time being the most useful approach to receptor classification would seem to be based on the organization of the molecular and protein structure of receptors. Two classes of receptor can be described: receptors for cyclic hydrophobic compounds and membrane receptors. Until we know of the families of single membrane-spanning receptors, fourfold membrane-spanning receptors, sevenfold membrane-spanning receptors, and 24-fold membrane-spanning receptors. For the sevenfold membrane-spanning receptor the identity of conformation and dynamics of biological function has been established. Light and peptide ligands are used as examples. A systems theory for information transduction at receptors is introduced which can also describe the processes of sensitivity modulation.

References

Birnbaumer L, Codina J, Mattera R, Yatani A, Scherer W, Toro M-J, Brown AM (1987) Signal transduction by G proteins. Kidney Int 32:114 ff.

Blecher M (1984) Receptors, antibodies, and disease. Clin Chem 30:1137–1156

Brewitt B, Clark JI (1988) Growth and transparency in the lens, an epithelial tissue, stimulated by pulses of PDGF. Science 242:777–779

Buck S, Pruss RM, Krstenansky JL, Robinson P, Stauderman KA (1988) A tachykinin peptide receptor joins an elite club. Trends Biochem Sci 9:3 ff.

Changeux JP, Devillers-Thiéry A, Chemouilli P (1985) Acetylcholine receptor: an allosteric protein. Science 225:1335–1345

Cohen P, Honslay MD (1985) Molecular mechanisms of transmembrane signalling. Elsevier, Amsterdam

Doniach D, Bottazzo GF (1983) Autoimmune endocrine disorders. Hospital Update (October) 1145–1159

Dumont JE, Jauniaux JC, Roger PP (1989) The cyclic AMP-mediated stimulation of cell proliferation. Trends Biochem Sci 14:67–71

Ebina Y, Ellis L, Harmagen K et al. (1985) The human insulin receptor cDNA: the structural basis for hormone-activated transmembrane signalling. Cell 40:747–758

Goldfine I (1987) The insulin receptor: molecular biology and transmembrane signalling. Endocr Rev 8:235 ff.

Gordon J (1988) Put out to contract. Nature 332:395

Grenningloh G, Gundelfinger E, Schmitt B, Betz H (1987) Glycine versus GABA receptors. Nature 330:25 ff.

Haisenleder DJ, Khoury S, Zmeili SM et al. (1987) The frequency of gonadotropin-releasing hormone secretion regulates expression of alpha and luteinizing hormone beta-subunit messenger ribonucleic acids in male rats. Mol Endocrinol 1:834–838

Hanley MO, Jackson T (1987) Return of the mganificant seven. Nature 329:766 ff.

Hesch RD (1989) Hormonlehre. In: Hesch RD (ed) Endokrinologie. Urban und Schwarzenberg. Munich (Innere Medizin der Gegenwart)

Hess B, Mehns M (1986) Chemische Uhren. In: Dress A, Hendriks H, Küppers G (eds) Selbstorganisation. Piper, Munich

Isgaard J, Carlsson L, Isaksson OGP, Jansson JO (1988) Pulsatile intravenous growth hormone (GH) infusion to hypophysectomized rats increases insulin-like growth factor I messenger ribonucleic acid in skeletal tissues more effectively than continuous GH infusion. Endocrinology 123:2605–2610

Jüppner H, Atkinson MJ, Baethke R, Hesch RD (1984) Autoantibodies against parathyroid hormone in a patient with terminal renal insufficiency. Lancet I:1379–1381

Kubo T, Fukuda K, Mikami A et al. (1986) Cloning, sequencing and expression of complementary DNA encoding the muscarinic acetylcholine receptor. Nature 323:411–416

Lambert PH (ed) Immunopathology of idiotypic interactions. Springer Seminars in Immunopathology, vol 6. Springer, Berlin Heidelberg New York

Marder E (1988) Modulating a neuronal network. Nature 335

Masters SB, Stroud RM, Bourne HR (1986) Family of G protein alpha chains: amphipathic analysis and predicted structure of functional domains. Prol Engineering 1:47 ff.

Meinhardt H (1987) Bildung geordneter Strukturen bei der Entwicklung höherer Organismen. In: Küppers BO (eds) Ordnung aus dem Chaos. Piper, Munich

Moncada VY, Hedo JA, Serrano-Rios M, Taylor SI (1986) Insulin-receptor biosynthesis in cultured lymphocytes from an insulin-resistant patient (Rabson-Menderhall syndrome). Diabetes 35:802–807

Nathan J, Thomas D, Hogness DS (1986) Molecular genetics of human colour vision: the genes encoding blue, green, and red pigments. Science 232:193–202

Noda M, Ikeda T, Suzuki H, Takeshima H, Takahashi T, Kuno M, Numa S (1986) Expression of functional sodium channels from cloned cDNA. Nature 322:826–828

Riedel H, Dull T, Schlessinger J, Ullrich A (1986) A chimaeric receptor allows insulin to stimulate tyrosine kinase activity of epidermal growth factor receptor. Nature 324:68–70

Riedel H, Dull T, Schlessinger J, Ullrich A (1987) Insulin-like growth factor II receptor as a multifunctional binding protein. Nature 329:301 ff.

Schofield PR, Darlison MC, Fujita N et al. (1987) Sequence and functional expression of the GABA receptor shows a ligand-gated receptor super-family. Nature 328:221–227

Spratt DI, Finkelstein JS, Butler JP, Badger TM, Crowley WF (1987) Effects of increasing the frequency of low dosis of gonadotropin-releasing hormone (GnRH) on gonadotropin secretion in GnRH-deficient men. J Clin Endocrinol 64:1179

Stevens CF (1985) Molecular tinkerings that tailor the acetylcholine receptor. Nature 313:350–351

Tam CS, Heersche JNM, Murray TM, Parsons JA (1982) Parathyroid hormone stimulates the bone apposition rate independently of its resorptive action: differential effects of intermittent and continuous administration. Endocrinology 110:506–512

Tanabe T, Taheshima H, Mikami A et al. (1987) Primary structure of the receptor for calcium channel blockers from skeletal muscle. Nature 328:313–318

Walker EJ, Jeffray PD (1986) Polymyositis and molecular mimicry, a mechanism of autoimmunity. Lancet II:605

Wallis M (1987) Growth hormone receptor cloned. Nature 330:521–522

Woods NM, Cuthbertson KSR, Cobbold PH (1986) Repetitive transient rises in cytoplasmic free calcium in hormone-stimulated hepatocytes. Nature 319:600–602

Zor U (1983) Role of the cytoskeletal organization in the regulation of adenylate cyclase-cyclic adenosine monophosphate by hormones. Endocr Rev 4:1 ff.

Biochemical Characterization
of Cellular Hormone Receptors

H. JÜPPNER and R. D. HESCH

1 Introduction

A large variety of cellular functions and physiological events are modulated or controlled by hormones and neurotransmitters. The cellular receptors for these molecules can be divided into two general categories: (a) cell surface receptors for peptides, growth factors, and neurotransmitters that are linked to secondary events (some of which are still unknown) (KAHN 1989), and (b) receptors for steroids, dihydrocholecalciferol, and thyroid hormones ultimately allowing interaction with specific DNA sequences and regulation of gene expression (BAULIEU and MESTER 1989). Modifications in hormone receptors and/or their secondary events have been implicated in a steadily growing number of pathophysiological disorders, thus making detailed knowledge of their primary structure and their functional properties increasingly important. Recent advances in receptor research have led to the isolation, cloning, and expression of multiple receptor proteins, and the subsequent dissection of functionally important domains through site-directed mutagenesis and site-specific monoclonal antibodies (KRIS et al. 1985; STRADER et al. 1988; OKAMURA et al. 1989; CUNNINGHAM et al. 1989). The characterization of hormone receptors and ligand-induced secondary events in suitable cell lines or tissue membranes may be achieved through multiple techniques, e.g., specific receptor binding of labeled ligands, hormonal analogues, or antibodies, ligand-mediated regulation of receptor expression on the surface of intact cells, and alterations in secondary events such as cyclic nucleotide production, phosphatidyl inositol-4,5-diphosphate (PIP_2) metabolism, protein phosphorylation, or ligand-gated ion channels. Furthermore, recent advances in the application of homo- or heterobifunctional cross-linking reagents have allowed the physicochemical characterization of receptor proteins and evaluation of their metabolism within the cell. This chapter will focus on methods required for the biochemical characterization of hormone receptors and their secondary events.

2 Characteristic Features of Hormone Receptor Families

A rapidly increasing number of receptor proteins have recently been purified to homogeneity. Partial amino acid sequence analysis and the screening cDNA libraries with corresponding oligonucleotides have permitted their molecular cloning and deduction of the entire receptor amino acid sequences (Dixon et al. 1986; Bonner et al. 1987; Kobilka et al. 1987a; Masu et al. 1987; Julius et al. 1988; McFarland et al. 1989). The overall receptor structure and the intracellular effector systems activated through these proteins are subdivided into at least three general categories: G protein-coupled receptors with seven membrane-spanning domains that are linked to adenylate cyclase, PIP_2, or receptor-operated ion channels; tyrosine kinase receptors for growth factors; and peptide hormone receptors with a single membrane-spanning region linked to the particulate guanylate cyclase and/or unknown effector systems (Fig. 1).

A large variety of receptors for peptides and neurotransmitters are coupled to adenylate cyclase and/or PIP_2 through guanine nucleotide-binding regulatory (G) proteins (Gilman 1984, 1987; Berridge 1987). Both G protein-linked pathways are thought to be activated independently through distinct receptors. Recent data, however, suggest that the same receptor may be coupled to alternating G protein-coupled intracellular events (Ashkenazi et al. 1989). Ion channels operated by acetylcholine-occupied receptors represent another signaling mechanism with various functions in the central and peripheral nervous system (Changeux et al. 1987; Schimerlik 1989). Common features of most G protein-linked receptors include seven hydrophobic segments of membrane-spanning domains which reveal significant sequence homology among themselves. This similarity recently allowed the molecular cloning of presumably other receptors whose ligands have not yet been identified (Kobilka et al. 1987b; Libert et al. 1989). Additional features are potential phosphorylation sites (serine, threonine, and tyrosine residues) (Sibley et al. 1988) within the intracellular carboxyl-terminal amino acid sequence and a hydrophilic, probably extracellular domain with multiple N-glycosylation sites concordant with the glycoprotein nature of these receptors.

Another distinct class of receptors has also been established for growth factors such as insulin, EGF, IGF-I, and PDGF (Ullrich et al. 1984, 1985,

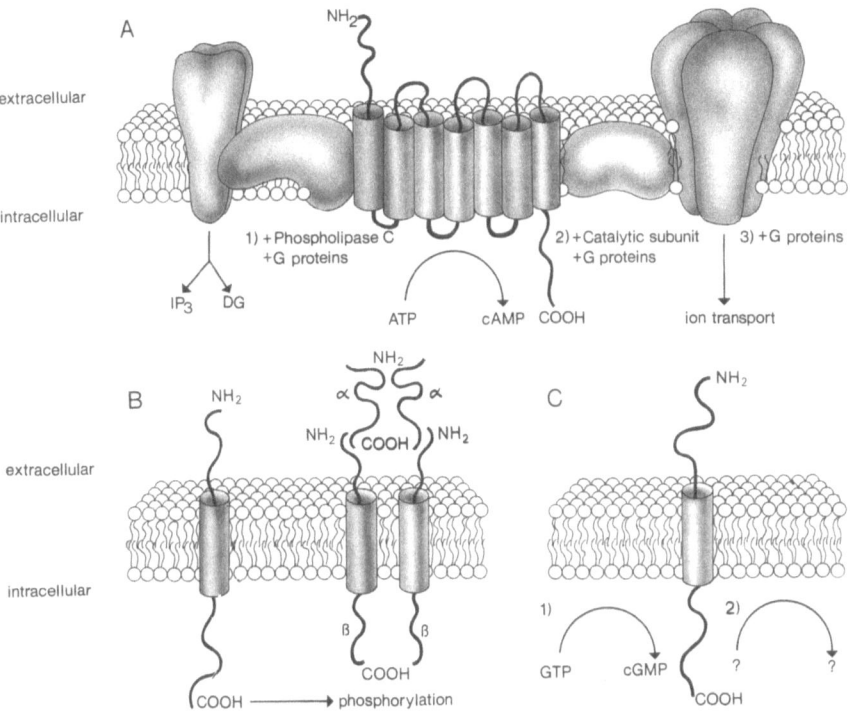

Fig. 1A–C. Schematic representation of hormone receptor families and their second messenger systems. **A** G protein-coupled receptors and their effector systems; **B** tyrosine kinase receptors; and **C** guanylate cyclase-coupled receptors and receptors with unknown second messenger systems

1986; YARDEN et al. 1986). Similar to a number of proteins encoded by retro-viral oncogenes, tyrosine-specific protein kinases are stimulated by these ligand-occupied receptors (RAMACHANDRAN and ULLRICH 1987). Unlike the family of G-linked receptors, growth factor receptors are either mono- or dimeric glycoprotein structures with α- and β-subunits that are linked through disulfide bonds. Each monomer contains a single stretch of hydrophobic amino acids which is thought to anchor the receptor within the cell membrane, an amino-terminal cysteine-rich, extracellular domain with multiple asparagine-linked glycosylation sites and a carboxyl-terminal cytoplasmic region possessing tyrosine kinase activity.

A third class of peptide hormone receptors with considerable biological importance, e.g., growth hormone and prolactin receptors, functions through unidentified cellular signaling mechanisms (LEUNG et al. 1987; BOUTIN et al. 1988). The receptor for atrial natriuretic peptide (ANP), with a molecular weight of 120000 Daltons which is linked to a particulate guanylate cyclase, appears to be an additional member of this family (CHINKERS et al. 1989). Hydropathy plots of all three receptors, the prolactin, the growth hormone, and the ANP receptor, predict a single hydrophobic, transmembranous region, an amino-terminal, extracellular ligand-binding domain, and an intracellular, carboxyl-terminal region which is thought to catalyze the biological response. Interestingly, the presumed intracellular regions of the prolactin and the growth hormone receptor are quite variable in size and amino acid composition, whereas the transmembrane regions and the extracellular domains reveal considerable homology (BOUTIN et al. 1988).

3 Characterization of Receptors by Labeled Ligands, Antagonists, and Specific Antibodies

Ligand binding to a receptor is the first step towards the initiation of all biological events. Studies to examine the interaction between ligands and their receptors were, however, not possible until the introduction of methods that permitted the radiolabeling of the hormone yet retained its biological activity (LEFKOWITZ et al. 1970; KAHN 1976). Since this was not readily achieved for every hormone, alternative approaches used either radiolabeled, affinity-purified antibodies against that part of the ligand which was not required for interaction with the receptor (MCINTOSH and HESCH 1975), or synthetic peptides with a modified primary structure to permit iodination without damaging the bioactivity (ROSENBLATT et al. 1976). Currently, radioreceptor assays using either intact cells, partially purified membrane preparations, or extracts from membranes are widely available for a large variety of hormones, growth factors, and neurotransmitters (KAHN 1989). Radioactively labeled and unlabeled ligands compete for their specific receptors in a manner similar to that of ligand-specific antibody binding in radioimmunoassays. After phase separation of receptor-bound and free ligand, the radioactivity associated with the cell membrane can provide valuable data on specificity, receptor number, and affinity.

Although antibodies to receptor structures have long been implicated in various endocrine diseases (KAHN 1989), and were shown in several cases to displace the physiological ligands from receptor sites, it was not until recently that antibodies became available in sufficient quantities.as ligands for receptor studies. Monoclonal antibodies against receptor proteins not only allowed their purification to homogeneity for microsequencing, but also served for the characterization of ligand-binding sites in three-dimensional models of the receptor (KATOH et al. 1987; OKAMURA et al. 1989; CUNNINGHAM et al. 1989).

In general, the hormone–receptor interaction is rapid and reversible, and time, temperature, and pH-dependent, the latter two aspects having a greater effect on dissociation rate than association rate. The finite number of binding sites provides saturable, specific binding of ligands, nonspecific binding being defined as the amount of labeled hormone that is bound to the receptor solution (suspension) in the presence of a large excess of unlabeled ligand. Ligand–receptor interaction is usually associated with the induction of secondary events (see Sect. 4) which reflect the ligands' biological activity. Lig-

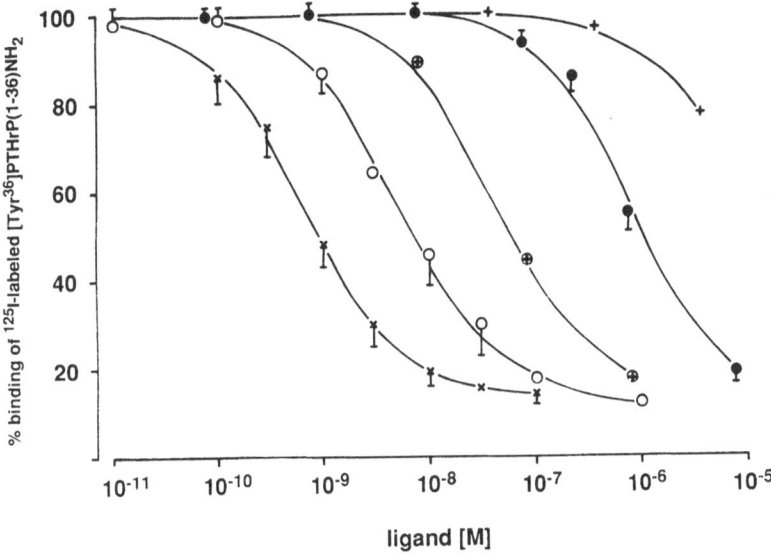

Fig. 2. Inhibition of ^{125}I-[Tyr^{36}]PTHrP(1-36)amide binding to ROS 17/2.8 cells by amino-terminally truncated PTH analogues. Increasing doses of [Nle^{8,18}, Tyr^{34}]bPTH(1-34)amide (×——×), [Nle^{8,18}, Tyr^{34}]bPTH(3-34)amide (o——o), [Tyr^{34}]bPTH(5-34)amide (⊕——⊕), [Tyr^{34}]bPTH(7-34)amide (•——•), or hPTH(13-34) (+——+), and radiolabeled [Tyr^{36}]PTHrP(1-36)amide (100 000 cpm/well) were incubated with confluent ROS 17/2.8 cells for 4 h at 15 °C as described (Jüppner et al. 1988). After rinsing and lysing the cells, receptor-bound radioactivity was counted. Data are given as means ± SD

ands may, however, differ in their intrinsic activity, that is, their capacity to initiate secondary messenger events once occupying the receptor. For example, analogues of PTH (1-34) and PTH (3-34) reveal in vitro relatively high affinity for the PTH receptor on canine renal membranes (SEGRE et al. 1979; NUSSBAUM et al. 1980) and bone-derived osteoblast-like ROS 17/2 cells (YAMAMOTO et al. 1988). Independent of whether [Nle^{8,18,}Tyr^{34}]bPTH(1-34)amide or [Tyr^{36}]PTHrP(1-36)amide is used as radioactive ligand, truncation of the amino terminus leads only to a gradual loss of affinity for the receptor (Fig. 2). Adenylate cyclase-stimulating activity is lost, however, instantly with removal or modification of the first amino acid (JÜPPNER 1989). Thus although receptor binding of truncated ligands is of sufficiently high affinity, there is no measurable intrinsic activity in vitro, at least with respect to cyclic adenosine 3',5'-monophosphate (cAMP) as the second messenger. At the extreme, a ligand would have no intrinsic activity in any second messenger system, high affinity for the receptor, and thus could serve as a competitive antagonist. For PTH, an antagonist with reasonably high affinity has previously been established (HORIUCHI et al. 1983). Recently synthesized analogues of either PTH or PTHrP have displayed improved affinity (GOLDMAN et al. 1988; CAULFIELD et al. 1989), and thus may ultimately lead to the development of clinically significant peptides.

4 Second Messenger Events for the Characterization of Hormone Receptors

After ligand binding to the receptor protein, the hormonal information requires transformation into specific biological responses within specialized tissues and the subsequent translation of the unique hormonal signal into a limited, and thus less discriminant, repertoire of intracellular second messengers which include cAMP, cyclic guanosine 3',5'-monophosphate (cGMP), inositol 1,4,5-trisphosphate (IP$_3$), diacylglycerol (DG), calcium, and other ions.

4.1 Guanine Nucleotide Regulatory (G) Proteins Coupled to Adenylate Cyclase

The discovery of the second messenger, cAMP, in 1957 (RALL et al. 1957), led to extensive efforts to characterize the mechanism(s) involved in its intracellular regulation in a large variety of cells and tissues. Due to the simplicity of cAMP measurements and the frequency of hormonal systems linked to this second messenger, adenylate cyclase-coupled receptors have been extensively studied (GILMAN 1984, 1987). Receptors and their effector systems are coupled to each other through a closely related family of heterotrimeric, guanine nucleotide regulatory (G) proteins (Fig. 3). They serve as intermediaries in transmembrane signaling pathways and consist of a guanosine triphosphate

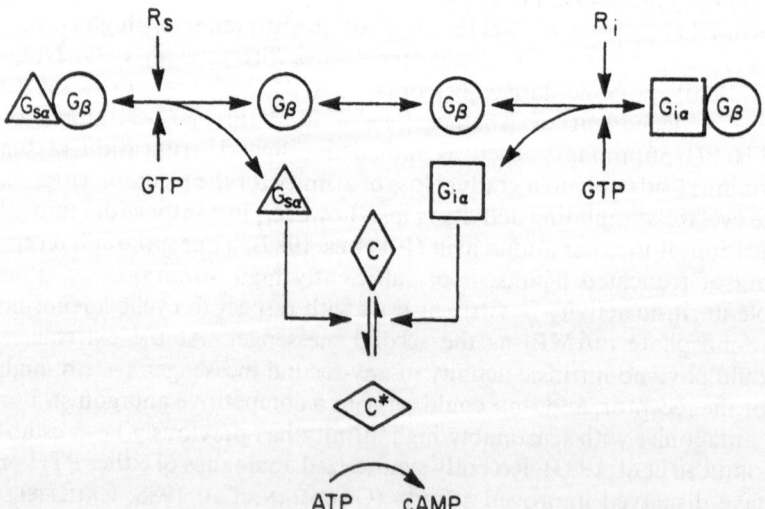

Fig. 3. Adenylate cyclase stimulation by G protein-coupled receptors through stimulatory and inhibitory pathways (from Gilman 1984, with permission)

(GTP)-binding subunit and the hydrolyzing subunits α, β, and γ. The ligand-occupied receptors alter their conformation which then allows GTP to bind to the G protein in exchange with guanosine diphosphate (GDP). Adenylate cyclase is subsequently activated, i.e., stimulated or inhibited, either by the stimulatory guanine nucleotide subunit ($G_{s\alpha}$) or by the inhibitory guanine nucleotide subunit ($G_{i\alpha}$). G_s is a mixture of two oligomers with differing α-subunits, and indistinguishable β- and γ-subunits. Cholera toxin, in the presence of a membrane-bound protein cofactor, inactivates the adenylate cyclase through adenosine diphosphate (ADP) ribosylation of G_s, which is thought to inhibit the receptor-stimulated GTPase activity of the G protein. The α-subunit of G_i is distinctly different from that of G_s. However, functional analysis, amino acid composition, and proteolytic peptide maps show both proteins to have identical β-subunits (which can deactivate $G_{s\alpha}$). Pertussis toxin abolishes the hormone-induced inhibition of adenylate cyclase through ADP ribosylation of the 41 000-dalton G_i protein, which then appears to reverse the inhibition of the adenylate cyclase. There is, however, evidence for other G_i-like proteins with a considerable degree of homology to the pertussis toxin-sensitive G proteins. These seem to lack the acceptor cysteine residue; they are thus not ADP ribosylated and may therefore serve different functions. Incubation of either G_s or G_i with nonhydrolyzable guanine nucleotides or with fluoride (in the presence of Mg^{2+} and Al^{3+}) results in the "activation" of either α-subunit by stabilizing the G proteins in their active configuration. The resulting changes in cAMP production alter the activity of cAMP-dependent protein kinases, which in turn modify the specific cellular effect(s). The amount of intracellular second messenger is, however, dependent not only on the ligand-induced changes in cAMP generation, but also on the activity of a specific phosphodiesterase (PDE), an enzyme which converts cAMP into the biologically inactive 5'-adenosine monophosphate.

Most of the proteins involved in the signal transduction cascade have recently been purified and cloned. This was largely achieved through a variety of agents that activate individual proteins without the necessity of ligand-occupied receptors. These served either for affinity purification of the respective protein, or its identification during purification. For example, the diterpene forskolin, isolated from the roots of the aromatic herb *Collus forskohlii,* and calmodulin directly stimulate cAMP production through direct interaction with at least some adenylate cyclase species (GILMAN 1987). Immobilized forskolin was, therefore, subsequently used to affinity purify a 120 000-dalton adenylate cyclase species from solubilized bovine brain membranes. After partial amino acid analysis, cDNAs from a bovine brain library were isolated and allowed the deduction of a 1134 residue protein with two alternating sets of hydrophilic and hydrophobic domains, each of which appears to span the membrane six times (KRUPINSKI et al. 1989). Interestingly, adenylate cyclase shares considerable homology with various ion channels and transporter proteins, thus possibly representing not only the catalytic unit for the generation of cAMP but having some unknown transport functions, which might include the release of cAMP from the cell.

4.2 Regulation of Cell Surface Expression of Membrane Receptors

In addition to diverting the hormonal information through the various second messenger systems, possibly through distinct receptor proteins, the responsiveness of adenylate cyclase may be regulated in some tissues through desensitizing mechanisms. Prolonged exposure of the target cell to agonists, i.e., hormones, drugs, or neurotransmitters, results in a decreased capacity of the cell to respond further to agonist stimulation.

Two forms of desensitization have to be distinguished (Sibley et al. 1988): (a) agonist-induced or homologous desensitization, characterized by the exclusive loss of responsiveness to one particular ligand [or a related ligand that binds to the same receptor (Jüppner et al. 1988)] while the responsiveness to other agonists and fluoride ions remains intact, and (b) heterologous or agonist-independent desensitization which impairs the cellular responsiveness to a variety of second messenger activators, such as guanine nucleotides and fluoride ions, that "bypass" the hormone receptor. Homologous desensitization has been attributed to the physical disappearance of the receptor protein from the cell surface and/or to an uncoupling of the catalytic subunit from the receptor. Heterologous desensitization is primarily caused by the latter mechanism and is associated not only with a loss of ligand-induced responsiveness, but also with a loss towards bypass reagents that directly activate regulatory proteins and/or the catalytic moiety of adenylate cyclase.

Ultimately, desensitization appears to be regulated for many, if not all, receptors through protein phosphorylation. The ligand-occupied β-adrenergic receptor not only stimulates a cAMP-dependent kinase and protein kinase C, but also a novel β-adrenergic receptor kinase (β-ARK) (Benovic et al. 1986) (Fig. 4). This latter kinase appears to be specific for the β-adrenergic receptor

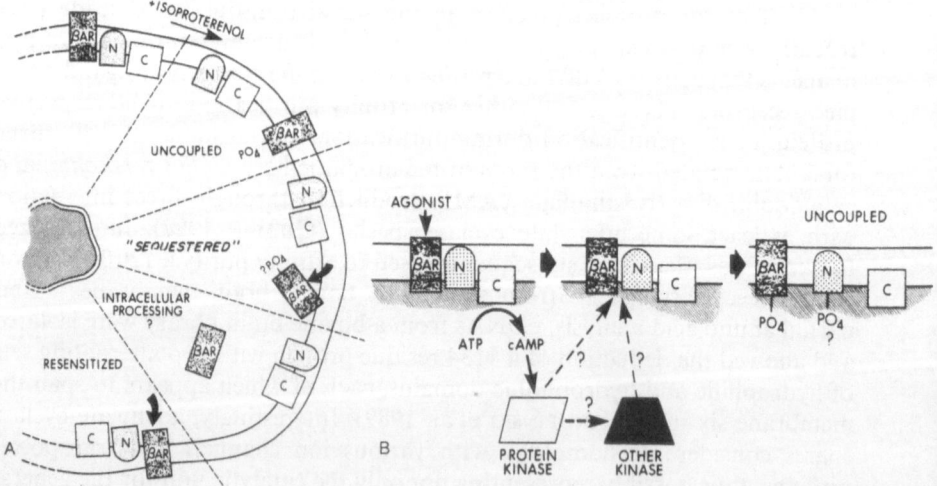

Fig. 4. Homologous (**A**) and heterologous (**B**) desensitization of the adenylate cyclase coupled to the β-adrenergic receptor (from Sibley et al. 1988, with permission)

and probably phosphorylates serine/threonine-rich receptor sites that are close to the cytosolic carboxyl terminus of the receptor protein. After ligand binding, the cytosolic β-ARK is translocated to the receptor protein, which after phosphorylation and uncoupling from adenylate cyclase becomes sequestered or internalized. Cytosolic phosphatases functionally regenerate the receptor, which is then reinserted into the plasma membrane (SIBLEY et al. 1988). Similar mechanisms have been described for light adaptation in the retina and the growth factor receptor family, and thus might represent a common regulatory mechanism for receptor availability.

4.3 Guanylate Cyclase-Coupled Receptors

The synthesis of cGMP from GTP is catalyzed by guanylate cyclase, an enzyme that functions quite differently from adenylate cyclase. Two forms of guanylate cyclase with different kinetic and physicochemical properties, soluble and membrane associated (particulate), have been identified in various tissues (MITTEL and MURAD 1982). The soluble isoenzyme, which has been purified and extensively characterized, is activated through nitrovasodilators, some porphyrins, and endothelium-dependent vasodilators (MURAD 1986).

Earlier data suggested that the particulate enzyme is a membrane-bound glycoprotein whose extracellular domain serves as a cell surface receptor for chemotactic peptides (SINGH et al. 1988) and ANP (KUNO et al. 1986; TAKAYANAGI et al. 1987; PAUL et al. 1987). The intracellular portions of these proteins share significant homology with the protein kinase receptor family and appear to be required to signaling the hormonal information (CHINKERS and GARBERS 1989). Reminiscent of the growth hormone and the prolactin receptor (LEUNG et al. 1987; BOUTIN et al. 1988), the particulate, larger ANP receptor contains a single transmembrane domain, separating the extracellular ligand binding domain from the cytoplasmic, catalytic domain (CHINKERS et al. 1989). The smaller ANP receptor with an M_r of 62 000 is cyclase free, and therefore appears to be linked to different signal transduction mechanism(s).

In addition to its apparent function as a mediator of vasodilation and the hormonal information of ANP, cGMP plays a major role in the processing of visual information which involves another G protein, G_t or transducin (LOCHREI et al. 1985). The heterotrimeric transducin, existing in two isoforms, reveals homology with the protein synthesis elongation factor and the protein encoded by the *ras* oncogene. One isoform of the α-subunit ($G_{t\alpha1}$), consisting of 350 amino acids, is found in the rod outer segments of photoreceptors, but not in the cones (GRUNWALD et al. 1986). The other 354-residue $G_{t\alpha2}$ subunit isoform was reported to be part of the cone photoreceptor outer segments (LEREA et al. 1986). Both isoforms couple the photolysis of rhodopsin to PDE-induced changes in intracellular levels of cGMP and are, therefore, crucial determinants of visual excitation (GILMAN 1987). The α-subunit binds GTP in exchange for GDP and may be ADP ribosylated by either cholera or pertussis toxins. Upon binding of GTP, the α-subunit dissociates from β and γ, activating the cGMP-specific PDE. The β-subunit of transducin is indistinguishable from those of G_s and G_i, and can deactivate $G_{s\alpha}$.

4.4 Phosphoinositol Hydrolysis

A third second messenger system that is attracting increasing attention centers around IP_3 and DG as signal transduction mechanisms (BERRIDGE 1987) (Fig. 5). The precursor for this signal pathway is an inositol lipid which is located within the plasma membrane. After agonist binding to the receptor and activation of the guanine nucleotide binding protein, PIP_2 is converted by a PDE (phospholipase C) into IP_3. The primary function of this metabolite is the mobilization of calcium from the intracellular stores. DG, which remains within the membrane compartment, activates an additional pathway by stimulating protein kinase C (PK-C). Both signaling mechanisms control short-term cellular responses, distinguishing their mode of action from that of cAMP- and steroid-dependent cellular responses, which have been implicated in a large variety of endocrine functions. It appears that the chronic treatment of intact cells with phorbol esters, which are direct activators of PK-C, may result in depletion of total PK-C activity (ABOU-SAMRA et al. 1989). Conversely, chronic agonist treatment inhibits the responsiveness of PK-C to an acute agonist challenge, but not to acute phorbol ester treatment, thus representing homologous ligand-specific desensitization.

Since G protein activity is blocked by pertussis toxin in some, but not all, cell types, it appears likely that more than one type of G protein is involved in the activation of phospholipase C (COCKROFT 1987). Recent evidence suggests that multiple G proteins that are distinguishable on the basis of their pertussis toxin sensitivity exist within one cell type and may couple selectively to different receptors (ASHKENAZI et al. 1989). The biological role for these functionally specific G proteins remains to be clarified. This mechanism may, however, serve to segregate particular cellular responses, similar to that observed for G_s and G_i in adenylate cyclase-regulated receptors.

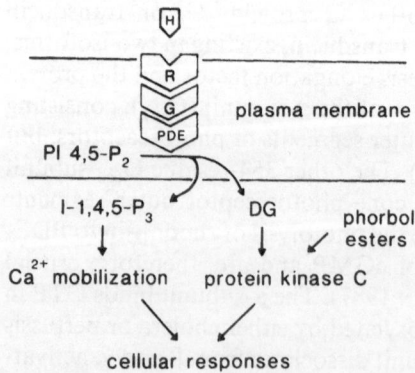

Fig. 5. Schematic representation of the ligand-activated phosphatidylinositol pathway

4.5 Ligand-Dependent Ion Transport Mechanisms

The nicotinic and the muscarinic acetylcholine receptors mediate multiple functions within the central and peripheral nervous system (CHANGEUX et al. 1987; SCHIMERLIK 1989) and are neurotransmitter-dependent ion-specific channels. The nicotinic acetylcholine receptor is a pentameric glycoprotein with a molecular weight of approximately 290000 and a complex subunit structure containing the ligand-binding domain, the agonist-operated ion channel, binding sites for various allosteric effectors, and structural elements permitting the specific interaction between these different receptor portions. Though some receptor subunits have recently been cloned, complete amino acid sequence and reconstitution of the entire recombinant receptor have not yet been achieved. Complementary DNAs of multiple members of the muscarinic acetylcholine receptor have recently been identified (BONNER et al. 1987) and reveal structural similarities with other G-linked receptor proteins. Highly conserved regions have been observed within the membrane-spanning domains, subsequently allowing the isolation of a further receptor protein, initially named G21 and later identified as being the serotonin receptor (KOBILKA et al. 1989b; JULIUS et al. 1988). In addition to the G protein-linked stimulation of adenylate cyclase and the phosphatidyl inositol turnover, the generation of cGMP has been described as an additional mechanism of transduction.

4.6 Differences in Receptor Activating Requirements as Evidence for Alternative Metabolic Pathways

Previous data established with PTH revealed that the in vitro stimulation of cAMP production required an intact amino terminus; amino-terminal deletions or modifications resulted in peptides with little or no adenylate cyclase-stimulating activity (JÜPPNER 1989). Some PTH molecules with amino-terminal truncations had inhibitor properties in vitro, although they failed to exhibit these antagonistic properties in vivo (SEGRE et al. 1985). Only the introduction of the further truncated [Tyr^{34}]bPTH(7-34) (HORIUCHI et al. 1983) allowed in vivo inhibition of PTH-induced phosphate transport and cAMP accumulation. Mobilization of calcium from calvarial bone, however, was induced by PTH(1-34) as well as truncated analogues (HERMANN-ERLEÉ et al. 1976). These observations thus raised the possibility that distinct receptors are activated by different peptides (LOWIK et al. 1985).

Recent data, using clonal or primary bone- and kidney-derived cell lines, have extended these observations, showing that PTH peptides with an intact amino terminus activate different second messengers while the truncated PTH(3-34) acts only in some cAMP-independent systems (COLE et al. 1987; STERN et al. 1988; MARTIN et al. 1989). YAMAGUCHI et al. (1987) showed that the PTH-mediated calcium influx into UMR106 cells may be subdivided into two different phases. The first rapid phase of calcium influx can be blocked by verapamil, lanthanum, cAMP, or phorbol esters, implying partial IP_3 dependency. After a transient decrease in ion influx, intracellular calcium increased cAMP-dependently, suggesting the activation of a different second messenger system. Furthermore, similar to the results of others, identical activation of cAMP production by either PTH or PTHrP was obtained in the osteoblastic cell line MC3T3-E1 (YAMADA et al. 1989). PTHrP failed to increase cytosolic calcium in this cell line (which would further support the two-receptor hypothesis), yet revealed the same responses as PTH in UMR106 cells (CIVITELLI et al. 1989).

Ligand-specific cellular responses were also observed in an opossum kidney cell line (OK), showing that the sodium-dependent phosphate transport is inhibited, presumably via IP_3 metabolism, by PTH(1-34) and to a lesser extent by PTH(3-34). The production of cAMP was only observed after challenge with PTH(1-34) (COLE et al. 1987; MARTIN et al. 1989). Protein kinase A (PK-A) activation, a cAMP-dependent mechanism, was detected with both PTH(1-34) and PTH(3-34) (MARTIN et al. 1989). Since the inhibition of phosphate uptake is observed without measurable cAMP production, PK-A activity, as the more sensitive indicator of cAMP generation, is thought to indicate that both peptides inhibit phosphate transport through the same second messenger system. This discrepancy in sequence requirements for ligand-mediated cell activation was even more pronounced in other systems, i.e., the PTH-induced glucose-6-phosphate dehydrogenase activity in guinea pig renal tubules (SAKAGUCHI et al. 1987), the PTH-induced thymidine incorporation into chicken chondrocytes (SCHLÜTER et al. 1989), and the vascular actions of PTH (PANG et al. 1981; HELWIG et al. 1987). The significant homology established for the carboxyl-terminal portions of all known PTH species (KHOSLA et al. 1988) and previous evidence for specific binding sites for a carboxyl-terminal PTH peptide (McKEE and MURRAY 1985; RAO and MURRAY 1985; DEMAY et al. 1985) also provide evidence for additional receptor(s) for the carboxyl-terminal portions of PTH. One biological event mediated through this pathway may be a PTH(53-84)-induced increase in alkaline phosphatase activity, recently observed in dexamethasone-treated bone-derived ROS 17/2.8 cells (MURRAY et al. 1989). Similar to these observations with PTH that raise the possibility of different receptor proteins for specific regions within the molecule, various ANP peptides appear not only to bind to particular receptors but also to initiate apparently different second messenger(s) and thus different biological events (TAKAYANAGI et al. 1987).

5 Physicochemical Characterization of Receptor Proteins

Hormone receptors are usually membrane proteins of very low abundance. In order to study their physicochemical properties, radioactive ligands are covalently attached to the protein sequence by homo- or heterobifunctional reagents. After solubilization of the intact cells or the membrane preparation, the proteins can be analyzed by chromatographic techniques or sodium dodecyl sulfate polyacrylamide gel electrophoresis followed by autoradiography (RUOHO et al. 1984; PILCH and CZECH 1984; SWEET and MURDOCK 1987).

The major advantage of affinity labeling techniques is their specificity since theoretically only the receptor protein is being labeled. The amino function of lysine residues and/or the amino-terminal amino acid can be derivatized on peptides or the receptor proteins. Thus all successful affinity cross-linking protocols for a variety of hormone receptors have employed bifunctional reagents which react with these amino groups. Commonly used cross-linking reagents include glutaraldehyde, dimethyl suberimidate, and disuccinimidyl suberate. These chemicals require appropriate geometrical spacing between free amino groups on the peptide ligand and its receptor, and may, therefore, fail under certain conditions. Photoaffinity labeling techniques are less selective, yet have the advantage that photoactivatable heterobifunctional reagents form highly reactive carbenes, which insert in multiple types of bonds within receptor protein.

However, due to the introduction of an additional group into the ligand, the resulting analogues were recently shown to have quite different biological properties. For example, derivatization of $[Nle^{8,18},Tyr^{34}]bPTH(1-34)$amide with the photoreactive, heterobifunctional reagent 4-fluoro-3-nitrophenylazide (FNPA) resulted in the generation of multiple reaction products, some of which were extensively evaluated with respect to their chemical and biological properties (SHIGENO et al. 1988 a, 1989). Derivatizing the peptide on Lys^{13} resulted only in a slight decrease in affinity for the receptor on ROS 17/2.8 cells and canine renal membranes. This peptide analogue was successfully used for the physicochemical characterization of the PTH receptor (SHIGENO et al. 1988 a, b). More pronounced changes were observed if the peptide was derivatized on either the amino terminus or $Lys^{26/27}$ (SHIGENO et al. 1989). Similar observations were also made for FNPA-derivatized $[Tyr^{36}]PTHrP(1-36)$amide analogues. Interestingly, the introduction of FNPA on either the amino terminus or Lys^{11} revealed no significant change in affinity for the receptor on ROS 17/2.8 cells, yet a more than 90% decrease for the analogue modified on Lys^{13}. The biological properties of these ligands were even more discrepant since the amino-terminally derivatized peptide had lost most of its adenylate cyclase-stimulating capacity, despite its high affinity for the receptor (JÜPPNER et al. 1990). These data demonstrate that the use of cross-linking reagents requires extensive evaluation of the various reaction products before being successfully employed for photoaffinity labeling of membrane receptors.

References

Abou-Samra AB, Jueppner H, Westerberg D, Potts, JT Jr, Segre GV (1989) Parathyroid hormone causes translocation of protein kinase-C from cytosol to membranes in rat osteosarcoma cells. Endocrinology 124:1107–1113

Ashkenazi A, Peralta EG, Winslow JW, Ramachanran J, Capon DJ (1989) Functionally distinct G proteins selectively couple different receptors to PI hydrolysis in the same cell. Cell 56:487–493

Baulieu EE, Mester J (1989) Steroid hormone receptors. In: DeGroot LJ (ed) Endocrinology, vol 1, 2nd edn. W. B. Saunders, Philadelphia, pp 16–39

Benovic JL, Mayor F Jr, Somers RL, Caron MG, Lefkowitz RJ (1986) Light-dependent phosphorylation of rhodopsin by β-adrenergic receptor kinase. Nature 321:869–872

Berridge MJ (1987) Inositol trisphosphate and diacylglycerol: two interacting second messengers. Ann Rev Biochem 56:159–193

Bonner TI, Buckley NJ, Young AC, Brann MR (1987) Identification of a family of muscarinic acetylcholine receptor genes. Science 237:527–532

Boutin JM, Jolicoeur C, Okamura H et al. (1988) Cloning and expression of the rat prolactin receptor, a member of the growth hormone/prolactin receptor family. Cell 53:69–77

Caulfield MP, Nutt RF, Levy JJ et al. (1989) Removal of partial agonism from 7-to-34 parathyroid hormone-related peptide (PTHrP) by substitution with amino acids from the parathyroid hormone sequence. J Bone Miner Res 4 (Suppl 1):896

Changeux JP, Giraudat J, Dennis M (1987) The nicotinic acetylcholine receptor: molecular architecture of a ligand-regulated ion channel. TIPS 8:459–465

Chinkers M, Garbers DL (1989) The protein kinase domain of the ANP receptor is required for signaling. Science 245:1392–1394

Chinkers M, Garbers DL, Chang MS, Lowe DG, Chin H, Goeddel DV, Schulz S (1989) A membrane form of guanylate cyclase is an atrial natriuretic peptide receptor. Nature 338:78–83

Civitelli R, Martin TJ, Fausto A, Gunsten SL, Hruska KA, Avioli LV (1989) Parathyroid hormone-related peptide transiently increases cytosolic calcium in osteoblast-like cells: comparison with parathyroid hormone. Endocrinology 125:1204–1210

Cockroft S (1987) Phosphoinositide phosphodiesterase: regulation by a novel guanine nucleotide binding protein, Gp. Trends Pharmacol Sci 12:75–78

Cole JA, Eber SL, Poelling RE, Thorne PK, Forte LR (1987) A dual mechanism for regulation of kidney phosphate transport by parathyroid hormone. Am J Physiol 253:E221–E227

Cunningham BC, Jhurani P, Ng P, Wells JA (1989) Receptor and antibody epitopes in human growth hormone identified by homolog-scanning mutagenesis. Science 243:1330–1336

Demay M, Mitchell J, Goltzman D (1985) Comparison of renal and osseous binding of parathyroid hormone and hormonal fragments. Am J Physiol 249:E437–E446

Dixon RA, Kobilka BK, Strader DJ et al. (1986) Cloning of the gene and cDNA for mammalian β-adrenergic receptor and homology with rhodopsin. Nature 321:75–79

Ebina Y, Ellis L, Jarnagin K et al. (1985) The human insulin receptor cDNA: the structural basis for hormone-activated transmembrane signalling. Cell 40:747–758

Gilman AG (1984) G proteins and dual control of adenylate cyclase. Cell 36:577–579

Gilman AG (1987) G proteins: transducers of receptor-generated signals. Annu Rev Biochem 56:615–649

Goldman ME, McKee RL, Caulfield MP et al. (1988) A new highly potent parathyroid hormone antagonist: [D-Trp^{12}, Tyr^{34}]bPTH-(7-34)NH$_2$. Endocrinology 123:2597–2599

Grunwald GB, Gierschik P, Nirenberg M, Spiegel A (1986) Detection of α-transducin in retinal rods but not cones. Science 231:856–859

Helwig JJ, Yang MCM, Bollack C, Judges C, Pang PKT (1987) Structure–activity relationship of parathyroid hormone: relative sensitivity of rabbit renal microvessel and tubule adenylate cyclases to oxidized PTH and PTH inhibitors. Europ J Pharmacol 140:247–257

Hermann-Erleé MPM, Heersche JNM, Hekkelman JW, Gaillard PJ, Tregear GW, Parsons JA, Potts JT Jr (1976) Effects on bone in vitro of bovine parathyroid hormone and synthetic fragments representing residues 1-34, 2-34 and 3-34. Endocr Res Commun 3:21–35

Horiuchi N, Holick MF, Potts JT Jr, Rosenblatt M (1983) A parathyroid hormone inhibitor in vivo: design and biological evaluation of an analogue. Science 220:1053–1055

Julius D, MacDermott AB, Axel R, Jessell TM (1988) Molecular characterization of a functional cDNA encoding the serotonin 1c receptor. Science 241:558–564

Jüppner H, Abou-Samra AB, Uneno S, Gu WX, Potts JT Jr, Segre GV (1988) The parathyroid hormone-like peptide associated with humoral hypercalcemia of malignancy and parathyroid hormone bind to the same receptor on the plasma membrane of ROS 17/2.8 cells. J Biol Chem 263:8557–8560

Jüppner H (1989) Parathyroid hormone: biological activity of synthetic and endogenous peptide fragments. In: Martinez J (ed) Peptide hormones as prohormones: processing, biological activity, pharmacology. Ellis Horwood, Chichester, England, pp 325–354

Jüppner H, Abou-Samra AB, Uneno S, Keutmann HT, Potts JT Jr, Segre GV (1989) Preparation and characterization of [N^{π}-(4-Azido-2-nitrophenyl)Ala^1,Tyr^{36}]-parathyroid hormone related peptide (1–36)amide: a high-affinity, partial agonist having high cross-linking efficiency with its receptor on ROS 17/2.8 cells. Biochemistry (in press)

Kahn CR (1976) Membrane receptors for hormones and neurotransmitters. J Cell Biol 70:261–286

Kahn CR (1989) Membrane receptors for peptide hormones. In: DeGroot LJ (ed) Endocrinology, vol 1, 2nd edition, W. B. Saunders, Philadelphia, pp 40–57

Katoh M, Raguet S, Zachwieja J, Djiane J, Kelly PA (1987) Hepatic prolactin receptors in the rat: characterization using monoclonal antireceptor antibodies. Endocrinology 120:739–749

Khosla S, Demay M, Pines M, Hurwitz S, Potts JT Jr, Kronenberg HM (1988) Nucleotide sequence of cloned cDNAs encoding chicken preproparathyroid hormone. J Bone Miner Res 3:689–698

Kobilka BK, Matsui H, Kobilka TS et al. (1987a) Cloning, sequencing, and expression of the gene coding for the human platelet α_2-adrenergic receptor. Science 238:650–656

Kobilka BK, Frielle T, Collins S et al. (1987b) An intronless gene encoding a potential member of the family of receptors coupled to guanine nucleotide regulatory proteins. Nature 329:75–79

Kris RM, Lax I, Gullick W, Waterfield MD, Ullrich A, Fridkin M, Schlessinger J (1985) Antibodies against a synthetic peptide as a probe for the kinase activity of the avian EGF receptor and v-erbB protein. Cell 40:619–625

Krupinski J, Coussen F, Bakalyar HA et al. (1989) Adenylyl cyclase amino acid sequence: possible channel- or transporter-like structure. Science 244:1558–1564

Kuno T, Andresen J, Kamisaki Y et al. (1986) Co-purification of an atrial natriuretic factor receptor and particulate guanylate cyclase from rat lung. J Biol Chem 261:5817–5823

Lefkowitz RJ, Roth J, Pricer W, Pastan I (1970) ACTH receptors in the adrenal: specific binding of ACTH-^{125}I and the relation to adenyl cyclase. Proc Natl Acad Sci USA 65:745–752

Lerea CL, Somer DE, Hurley JB, Klock IB, Bunt-Milan AH (1986) Identification of specific transducin alpha subunits in retinal rod and cone photoreceptors. Science 234:77–80

Leung DW, Spencer SA, Cachianes G et al. (1987) Growth hormone receptor and serum binding protein: purification, cloning and expression. Nature 330:537–543

Libert F, Parmentier M, Lefort A et al. (1989) Selective amplification and cloning of four new members of the G protein-coupled receptor family. Science 244:569–572

Lochrie MA, Hurley JB, Simon MI (1985) Sequence of the alpha subunit of photoreceptor G protein: homologies between transducin, ras, and elongation factors. Science 228:96–99

Löwik CW, van Leeuwen JP, van der Meer JM, van Zeeland JK, Scheven BA, Herrmann-Erleé MPM (1985) A two receptor model for the action of parathyroid hormone on osteoblasts: a role for intracellular free calcium and cAMP. Cell Calcium 6:311–326

Martin KJ, McConkey CL, Garcia JC, Montani D, Betts CR (1989) Protein kinase-A and the effects of parathyroid hormone on phosphate uptake in opossum kidney cells. Endocrinology 125:295–301

Masu Y, Nakayama K, Tamaki H, Harada Y, Kuno M, Nakanishi S (1987) cDNA cloning of bovine substance-K receptor through oocyte expression system. Nature 329:836–838

McFarland KC, Sprengel R, Phillips H et al. (1989) Lutropin-choriogonadotropin receptor: an unusual member of the G protein-coupled receptor family. Science 245:494–499

McIntosh CHS, Hesch RD (1975) Labelled antibody membrane assay for parathyroid hormone: a new approach to the measurement of receptor bound hormone. Biochem Biophys Res Comm 64:376–383

McKee MD, Murray TM (1985) Binding of intact parathyroid hormone to chicken renal membranes: evidence for a second binding site with carboxyl-terminal specificity. Endocrinology 117:1930–1939

Mittel CK, Murad F (1982) Guanylate cyclase: regulation of cyclic GMP metabolism. In: Nathason JA, Kebabian JW (eds) Handbook of experimental pharmacology, vol 58(1). Springer, Berlin Heidelberg New York, pp 225–260

Murad F (1986) Cyclic guanosine monophosphate as a mediator of vasodilation. J Clin Invest 78:1–5

Murray TM, Rao LG, Muzaffar SA, Ly H (1989) Human parathyroid hormone carboxylterminal peptide (53-84) stimulates alkaline phosphatase activity in dexamethasone-treated rat osteosarcoma cells in vitro. Endocrinology 124:1097–1099

Nussbaum SR, Rosenblatt M, Potts JT Jr (1980) Parathyroid hormone: renal receptor interactions. Demonstration of two receptor-binding domains. J Biol Chem 255:10183–10187

Okamura H, Zachwieja J, Raguet S, Kelly PA (1989) Characterization and application of monoclonal antibodies to the prolactin receptor. Endocrinology 124:2499–2508

Pang PKT, Yang MCM, Tenner TE, Chang JK, Shimizu M (1981) Hypotensive action of synthetic fragments of parathyroid hormone. J Pharmacol Exp Ther 216:567–571

Paul AK, Marala RB, Jaiswal RK, Sharma RK (1987) Coexistence of guanylate cyclase and atrial natriuretic factor receptor in a 180-kD protein. Science 235:1224–1226

Pilch PF, Czech MP (1984) Affinity cross-linking of peptide hormones and their receptors. In: Venter JC, Harrison LC (eds) Membranes, detergents, and receptor solubilization. Alan R. Liss, New York, pp 161–175

Rall TW, Sutherland EW, Berthet J (1957) The relationship of epinephrine and glucagon to liver phosphorylase. J Biol Chem 224:463–475

Ramachandran J, Ullrich A (1987) Hormonal regulation of protein tyrosine kinase activity. Trends Pharmol Sci 8:28–31

Rao LG, Murray TM (1985) Binding of intact parathyroid hormone to rat osteosarcoma cells: major contribution of binding sites for the carboxyl-terminal region of the hormone. Endocrinology 117:1632–1638

Rosenblatt M, Goltzman D, Keutmann HT, Tregear GW, Potts JT Jr (1976) Chemical and biological properties of synthetic, sulfur-free analogues of parathyroid hormone. J Biol Chem 251:159–164

Ruoho AE, Rashidbaigi A, Roeder PE (1984) Approaches to the identification of receptors utilizing photoaffinity labeling. In: Venter JC, Harrison LC (eds) Membranes, detergents, and receptor solubilization. Alan R. Liss, New York, pp 119–160

Sakaguchi K, Fukase M, Kobayashi I et al. (1987) Synthetic parathyroid hormone fragments shortened at the amino-terminus stimulate glucose-6-phosphate dehydrogenase activity in the distal renal tubule. J Bone Miner Res 2:83–90

Schimerlik MI (1989) Structure and regulation of muscarinic receptors. Annu Rev Physiol 51:217–227

Schlüter KD, Hellstern H, Wingender E, Mayer H (1989) The central part of parathyroid hormone stimulates thymidine incorporation of chondrocytes. J Biol Chem 264:11087–11092

Segre GV, Rosenblatt M, Reiner BL, Mahaffey JE, Potts JT Jr (1979) Characterization of parathyroid hormone receptors in canine renal cortical plasma membranes using a radioiodinated sulfur-free hormone analogue. J Biol Chem 254:6980–6986

Segre GV, Rosenblatt M, Tully GL, Laugharn J, Reit B, Potts JT Jr (1985) Evaluation of an in vitro parathyroid hormone antagonist in vivo in dogs. Endocrinology 116:1024–1029

Shigeno C, Hiraki Y, Westerberg DP, Potts JT Jr, Segre GV (1988a) Photoaffinity labeling of parathyroid hormone receptors in clonal rat osteosarcoma cells. J Biol Chem 263:3864–3871

Shigeno C, Hiraki Y, Westerberg DP, Potts JT Jr, Segre GV (1988b) Parathyroid hormone receptors are plasma membrane glycoproteins with asparagine-linked oligosaccharides. J Biol Chem 263:3872–3878

Shigeno C, Hiraki Y, Keutmann HT, Stern AM, Potts JT Jr, Segre GV (1989) Preparation of a photoreactive analog of parathyroid hormone [Nle^8, Lys(N-ε-4-azido-2-nitrophenyl)^{13}, Nle^{18}, Tyr^{34}] bovine parathyroid hormone (1-34)NH$_2$, a selective, high-affinity ligand for characterization of parathyroid hormone receptors. Anal Biochem 179:268–273

Sibley DR, Benovic JL, Caron MG, Lefkowitz RJ (1988) Phosphorylation of cell surface receptors: a mechanism for regulating signal transduction pathways. Endocr Rev 9:38–56

Singh S, Lowe DG, Thorpe DS et al. (1988) Membrane guanylate cyclase is a cell surface receptor with homology to protein kinases. Nature 334:708–712

Steinbach JH (1989) Structural and functional diversity in vertebrate skeletal muscle nicotinic acetylcholine receptors. Annu Rev Physiol 51:353–365

Stern PH, Stewart PJ, Stathopoulus V, Rappaport MS (1988) Parathyroid hormone and phosphatidylinositol. Proceedings of the First International Conference on New Actions of Parathyroid Hormone 1987. Kobe, Japan, pp 17–21

Strader CD, Sigal IS, Candelore MR, Rands E, Hill WS, Dixon RA (1988) Conserved aspartic acid residues 79 and 113 of the beta-adrenergic receptor have different roles in receptor function. J Biol Chem 263:10267–10271

Sweet F, Murdock GL (1987) Affinity labeling of hormone-specific proteins. Endocr Rev 8:154–184

Takayanagi R, Inagami T, Snajdar RM, Imada T, Tamura M, Misono KS (1987) Two distinct forms of receptors for atrial natriuretic factor in bovine adrenocortical cells. J Biol Chem 262:12104–12113

Ullrich A, Coussens L, Hayflick JS et al. (1984) Human epidermal growth factor receptor cDNA sequence and aberrant expression of the amplified gene in A431 epidermoid carcinoma cells. Nature 309:418–425

Ullrich A, Bell JR, Chen EY et al. (1985) Human insulin receptor and its relationship to the tyrosine kinase family of oncogenes. Nature 313:756–761

Ullrich A, Gray A, Tam AW et al. (1986) Insulin-like growth factor I receptor primary structure: comparison with insulin receptor suggests structural determinants that define functional specificity. EMBO J 5:2503–2512

Yamada H, Tsutsumi M, Fukase M et al. (1989) Effects of human PTH-related peptide and human PTH on cyclic AMP production and cytosolic free calcium in an osteoblastic cell clone. Bone Mineral 6:45–54

Yamaguchi DT, Hahn TJ, Iida-Klein A, Kleeman CR, Muallem S (1987) Parathyroid hormone-activated calcium channels in an osteoblast-like clonal osteosarcoma cell line. J Biol Chem 262:7711–7718

Yamamoto I, Shigeno C, Potts JT Jr, Segre GV (1988) Characterization and agonist-induced down-regulation of parathyroid hormone receptors in clonal rat osteosarcoma cells. Endocrinology 122:1208–1217

Yarden Y, Escobedo JA, Kuang WJ et al. (1986) Structure of the receptor for platelet-derived growth factor helps define a family of closely related growth factor receptors. Nature 323:226–232

Morphological Characterization of Cell Receptors

M. DIETEL

1 Introduction

The understanding of what receptors are and how they work began with the "receptor concept" proposed by EHRLICH and LANGLEY (1878). The coordination by receptors of thousands of interactions between living cells was later realized to be a fundamental mechanism of intercellular communication and a prerequisite for the functional integrity of multicellular organisms. Receptor proteins are essential (a) to transmit information from one cell to another, (b) to transduce signals from the cellular environment to intracellular targets and to nuclear DNA for regulating cell function, (c) to take up special substances in a highly controlled manner to maintain cell metabolism, and (d) to handle toxins and artificial drugs. These functions are accomplished by a specific set of proteins which can be localized in the cell membrane, the intracellular organelles, or the cytosol and/or the nucleus. A wide variety of biochemically very different signal molecules can bind to these receptor proteins, transporting specific information which in turn induces specific cellular reactions.

1.1 Function of Cell Receptors in Intercellular Communication

To fulfill the complex task of regulating diverse cellular activities, eukaryotic cells are thought to communicate in three different ways:

1. The *neural system* performs rapid long-distance cell-to-cell communication, innervating the voluntary and smooth muscle systems, blood and lymph vessels, most endocrine organs, the diffuse endocrine system, and many other target cells by specialized junctions (synapses). Nerve cells form extremely long cell protrusions (axons) to contact the selected target cells directly. At synapses the nerve cells secrete very short-range chemical mediators (catecholamines), called neurotransmitters, which bridge the small distance between the nerve end and the receptor of the target cell, e.g., muscle cell.
2. The *humoral or endocrine system* controls a great variety of processes and activities, such as embryological development, growth, maturation, cellular differentiation, and metabolic homeostasis. The vehicle of signal transportation is the blood and the extracellular fluid. The system is characterized by wide distribution of signals throughout the body. One has to imagine that nearly every cell of the body is surrounded by the same "mixture of signals" released from a great number of endocrine or endocrine-related cells into the circulation. The precision of this system is guaranteed by the specificity of cell receptors, which recognize only that one signal which fits to the related receptor. Signal transmission in the humoral system is slower (minutes) than in the neural system (seconds). Thus, the former is more likely to regulate and balance processes which have a longer turnover rate.
3. In addition there exist specialized cells that secrete chemical signals into the intercellular fluid, where they are rapidly detected and taken up by receptors of adjacent cells (paracrine) or by themselves (autocrine), providing a signaling system. The majority of these signals are peptide hormones or hormone-like substances that function as local chemical mediators and act only on cells in the immediate environment.

Besides these three established mechanisms there may be further forms of signal exchange between adjacent cells by communicating cell surface molecules, by transient gap junctions, or by other types of cell-cell channel, such as direct voltage-gated calcium channels.

In complex organisms all the functions of each cell have to be controlled continuously. This means that the controlling systems always have to "know" what all the other cells do; this includes, for example, various metabolic processes (calcium homeostasis, glucose level, etc.), growth regulation, e.g., of millions of blood cells developing from hematopoietic stem cells or of the rapidly growing and differentiating enterocytes of the small intestine, and so on. To harmonize all these complex functions it is necessary to establish a highly specialized multifunctional system using different types of receptors (see Sect. 1.2) and a great variety of different ligands (see Sect. 1.3).

This chapter will be confined to the indirect mechanisms of communication – those mediated by secreted chemical signals. The neural system using nerve cells with axons, synapses, and postsynaptic events will be discussed elsewhere in this volume.

1.2 Classification of Receptors

To identify the enormous number of different signals and substrates (ligands), receptor proteins possess high affinity and sensitivity for their specific ligand. There are only rare examples of cross-reactivity, and these often cause pathological conditions. There are different classes of specialized receptor located in the cell or organelle membrane, cytosol, or nucleus.

1. *Cell surface receptors* are (glyco)proteins which are localized in the plasmalemma. They predominantly bind large hydrophilic (lipophobic) ligands like peptide and protein hormones, certain drugs, prostaglandins, and related compounds. An exception are the catecholamines, which are small molecules, but which nonetheless bind to surface receptors. The reason for this is the lipophobicity of the catecholamines, which cannot pass the plasma membrane and thus act by binding to membrane receptors.

2. *Intracellular membrane receptors* are molecules integrated into membranes of cytoplasmic organelles, e.g., endoplasmic reticulum, mitochondria, and Golgi stacks. They are helpful in (a) transmitting and transforming signals on their way to the cytoplasmic target proteins and (b) recognizing intracellular signaling molecules which influence and control diverse cell functions (see Sect. 2.6).

3. *Cytosolic/nuclear receptors* are primarily localized in the cytosol. They bind small lipophilic molecules, such as steroid or thyroid hormones, which pass the cell membrane by passive diffusion and, after activation, move towards the nucleus where the ligand-receptor complex exerts its specific function by directly or indirectly modifying DNA transcription (see Sect. 3). A variety of lipophilic drugs also bind to intracytoplasmic binding sites.

1.3 Classification of Ligands

The number of different types of molecule interacting with cells via receptor proteins is extremely high. They fall into four broad categories: (a) large signal transducing peptides or (glyco)proteins, such as hormones and growth factors,

Table 1. Molecules and viruses that affect cells via surface receptors (*SR*) or cytosolic/nuclear receptors (*CR*)

Peptide hormones (SR)	*Serum transport proteins (SR)*
Insulin	Transferrin
Gastrin	Low-density lipoprotein
Glucagon	Yolk proteins
Growth hormone	Transcobalamin
Melanocyte-stimulating hormone	
Calcitonin	*Altered-serum proteins (SR)*
LH (HCG)	α_2-Macroglobulin-protease complexes
Prolactin	Acetylated LDL
Catecholamines	Thrombin
Parathyroid hormones	
Gastrointestinal hormones	*Viruses (SR)*
Adrenocorticotropic hormone	Rous sarcoma virus
	Semliki forest virus
Growth factors (SR)	Vesicular stomatitis virus
Epidermal growth factor	Adenovirus
Platelet-derived growth factor	Human immunodeficiency virus
Interferon	
Transforming growth factor α	*Antibodies (SR)*
Transforming growth factor β	IgE
Nerve growth factor	Polymeric IgA
	Maternal IgG
Neurotransmitters (SR)	IgG (Fc receptor)
Epinephrine	
Norepinephrine	*Specific carbohydrate determinants (SR)*
Dopamine	Lysosomal enzymes
	β-Galactosidase
Steroid hormones (CR)	Mannose glycoproteins
Cortisol	Asialoglycoproteins
Estrogen	
Progesteron	*Prostaglandins*
Androgens	
Testosterone	
Thyroid hormones (CR)	
Thyroxine (T_4)	
Triiodothyronine (T_3)	
Toxins and lectins (SR, CR)	
Diphtheria toxin	
Pseudomonas toxin	
Cholera toxin	
Ricin	
Wheat germ agglutinin	
Concanavalin A	

(b) small signal transducing lipophilic molecules, such as steroid or thyroid hormones, (c) transport proteins that carry, for example, cholesterol in the form of low density lipoproteins (see NIENDORF and BEISIEGEL in this volume) or iron as transferrin, α_2-macroglobulin complexes, or asialoglycoproteins, and (d) extracellular enzymes, the endocytosis of which is mediated by receptors in many cell types. Recently it could be shown by morphological methods that uptake of β-galactosidase in Chinese hamster ovary cells is mediated by receptors. This lysosomal enzyme is internalized through a concentrating receptor-mediated endocytosis involving clathrin-coated pits and receptorsomes (WILLINGHAM et al. 1981). In addition to these physiologically important molecules, there are two groups of nonphysiological substances which often use receptors to enter a cell, i.e., toxins and viruses. Table 1 gives a selection of the principal representatives of ligands with receptor-mediated uptake and function.

2 Cell Surface Receptors

Receptors are highly conserved cellular structures. Not one living eukaryotic cell exists without expression of membrane-bound and/or cytosolic receptors. This is true not only for multicellular but also for monocellular organisms. Bacteria, plasmodia, flagellates, and many others possess receptor proteins which function as a kind of "sense organ," retaining contact to the cell's environment. General information regarding biochemical and biokinetic properties is given in the reviews by BERRIDGE (1981), COOPER and HUNTER (1983), DIETEL (1987a) and PASTAN and WILLINGHAM (1985).

2.1 Structure and Function

The three major aspects of receptor function are (a) binding affinity, (b) binding specificity for the ligand, and (c) effector specificity (see Sect. 2.8) for the resulting cellular behavior. Surface receptors are transmembrane proteins which may contain carbohydrate or phospholipid constituents. Some receptor structures proposed for different types of ligand are shown in Fig. 1. Receptors may consist of one unique transmembrane protein (Fig. 1A), of three separated proteins which are structurally and functionally linked (Fig. 1B), of more complex helical structures, such as a bimolecular heterotetrameric form with two or four subunits covalently linked by interchain disulfide bonds (Fig. 1C), or of multiple membrane-spanning molecules (Fig. 1D). The number of receptors per ligand varies between 20000 and 1000000, corresponding to 0.01% – 0.2% of the total mass of membrane proteins (RUSSELL et al. 1983; TEITELBAUM and STREWLER 1984).

After binding, a signaling ligand usually appears to have no function other than having given the information to the cell [exception: transporting

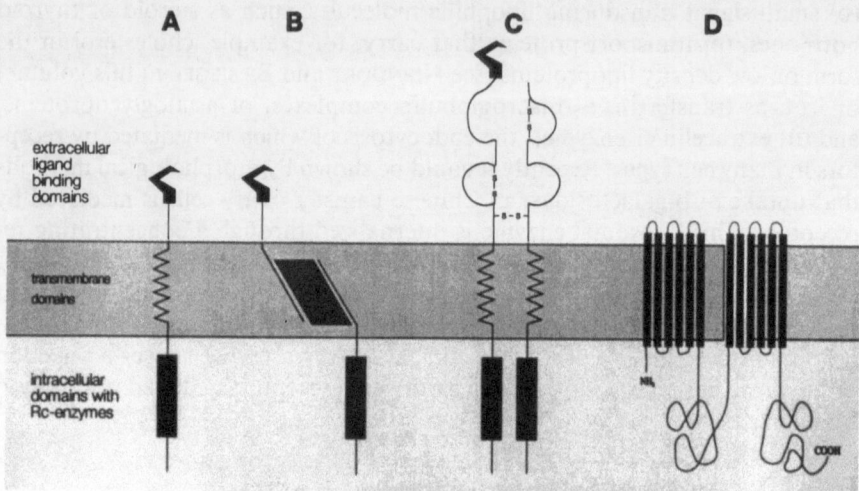

Fig. 1 A–D. Principal types of receptor structure. **A** a unique transmembrane protein; **B** three separated proteins structurally and functionally linked together; **C** a complex bimolecular helical molecule with subunits covalently linked by interchain disulfide bonds; **D** multiple membrane-spanning helical molecules connected by linear proteins. *Rc*, receptor

molecules (see Sect. 2.12) and internalized enzymes]. The ligand is not metabolized to useful products, is not itself an intermediate in cellular activity, and usually has no enzymatic properties. The ligand's only function is to change the property of a receptor which signals to the cell the presence of a specific product in the environment. Subsequent processes are described in the next section.

2.2 The Endocytotic Pathway and Intracellular Sorting

After ligand binding and signal induction one basic property of surface receptors is the ability to move along the plasmalemma, followed by clustering and internalization, called receptor-mediated endocytosis (Fig. 2). Most ligands bind tightly to their receptors. This prevents rebinding of dissociated ligands to the same or an adjacent receptor, which would initiate uncontrolled signaling. Receptor occupancy triggers the following process: After receptor binding

Fig. 2a, b. Receptor-mediated endocytosis. **a** Diagram showing ligand binding to surface receptor, ▶ lateral receptor movement, internalization by clathrin (*CL*)-coated pits (*CP*) and coated vesicle (*CV*), transport to the compartment of uncoupling of receptor and ligand (*CURL*), lysosomal degradation (*LYS*) of the ligand, and reintegration of the receptor protein into the plasmalemma. **b** Ultrastructural demonstration of ligand receptor processing: **b1** gold-labeled ligand binding to surface receptor with disintegration of the plasmalemma and condensation of cytoplasmic proteins, e.g., clathrin; **b2** formation of coated pits and **b3** coated vesicles followed by **b4** transport to the CURL (for further intracellular pathways, see Fig. 3)

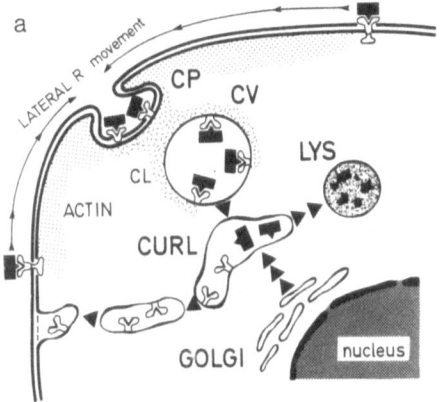

a

LATERAL R movement

CP

CV

LYS

CL

ACTIN

CURL

GOLGI

nucleus

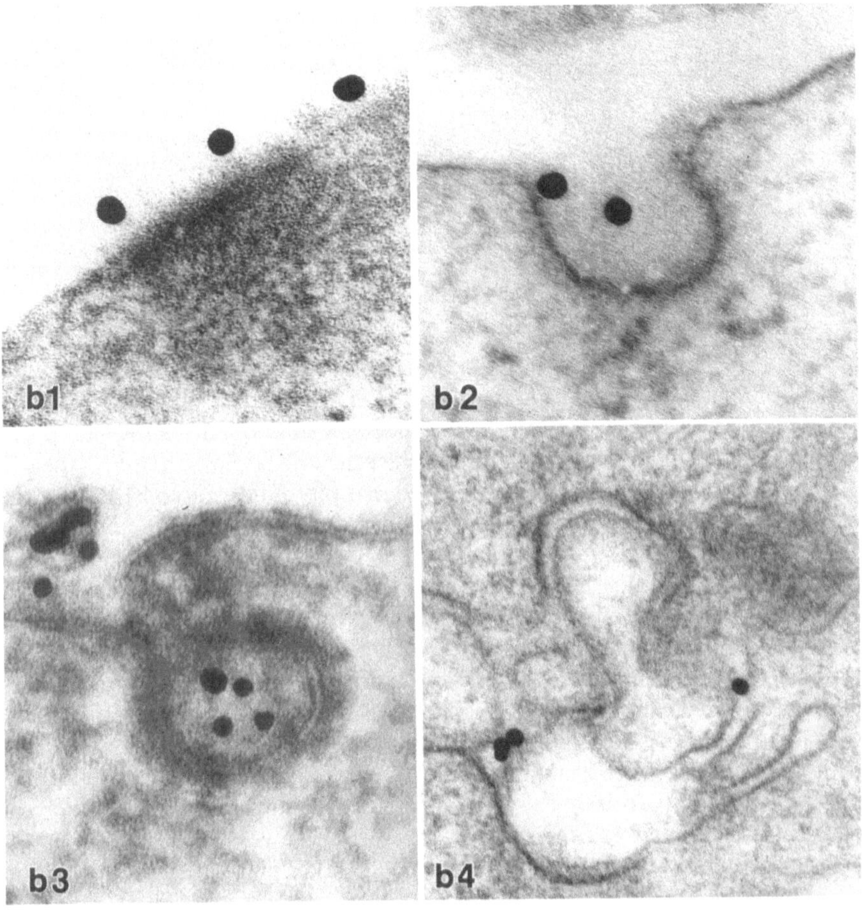

b1

b2

b3

b4

Table 2. Forms of sorting receptor-ligand systems according to cell type

	Cell type
1. Receptor recycled: ligand to lysosomes	
Mannose receptor	Macrophages
Galactose receptor	Hepatocytes
LDL receptor	Fibroblasts
Mannose 6-phosphate receptor	Fibroblasts
α_2-Macroglobulin receptor	Fibroblasts
Nerve growth factor	Ganglion cells
2. Receptor recycled: ligand recycled	
IgG (neonatal gut)	Rat ileum
Transferrin	Reticulocytes
	Fibroblasts
(Mannose)*	Macrophages
(Galactose)*	Hepatocytes
(LDL)*	Fibroblasts
IgG monomer	Macrophages
3. Receptor demise: ligand to lysosomes	
EGF	Fibroblasts
Insulin	Adipocytes
	Lymphocytes
Human choriogonadotropin	Follicular cells
IgG multimer	Macrophages
4. Receptor demise: ligand not to lysosomes	
IgA	Hepatocytes

the ligand-receptor complex (LRC) is concentrated at certain areas of the cell membrane by lateral migration called clustering or capping. The first morphological evidence that receptors are mobile was provided by the demonstration of labeled ligands which bind to the receptor and subsequently move along the membrane to accumulate in patches on the cell surface. Subsequently, the cell membrane shows invaginations, called coated pits. These dissociate from the plasmalemma, forming vesicles which are surrounded by a characteristic coat of the fibrous protein clathrin. These intracellular vesicles are called coated vesicles (Brown et al. 1983). The coated vesicles migrate to the inner part of the cell by saltatory movements along tracts of microtubules (Pastan and Willingham 1985). During this process the coated vesicles are freed from the clathrin coat, and are then called receptorsomes. These are transport vesicles that carry ligands and receptors from the cell surface to the cell interior. They measure about 2000 Å in diameter, but they can grow up to 6000 Å by fusing with one another. Receptorsomes have a low pH and do not contain significant amounts of hydrolytic enzymes. Therefore, ligands and receptors are not extensively degraded or otherwise modified in these vesicles.

Intracellular sorting (Courtoy et al. 1985) takes place after the formation of receptorsomes and is attributed to four basic mechanisms (Table 2, Fig. 3):

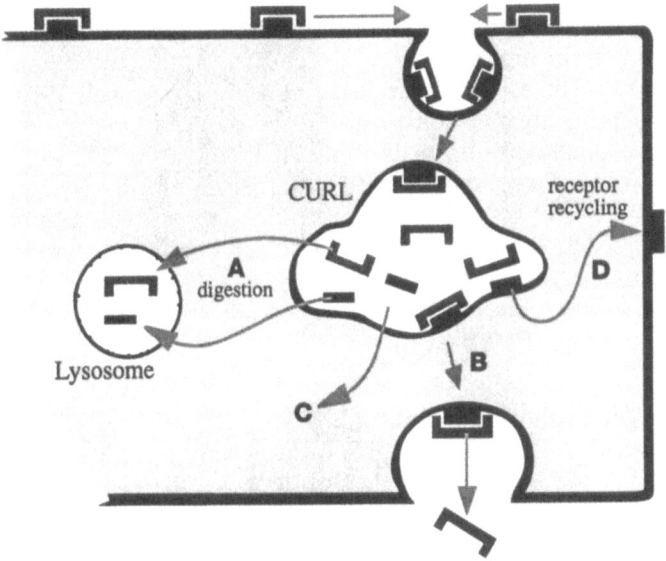

Fig. 3. Four basic pathways of intracellular sorting of receptor and ligand. After receptor-mediated endocytosis and transport to the compartment of uncoupling of receptor and ligand (*CURL*), the complex is (*A*) discharged into lysosomes for degradation of ligand and/or receptor, (*B*) moved arcoss the cell for excretion, (*C*) crossed through the endosomal membrane to the cytosol, or (*D*) the recepor recycled to the cell membrane

1. Receptorsomes fuse with tubules of the transgolgi region, which contains an acid pH. This represents the "compartment of uncoupling of receptor and ligand" (CURL). Inside the endocytotic vesicular compartment most ligands are dissociated from their receptors. Acid-sensitive ligands are then released in the fluid phase, discharged into lysosomes, and further degraded. The receptors are returned to the cell surface to be reutilized.

2. The endosomes avoid lysosomes and are addressed to either the basolateral or the apical cell surface, which respectively face the internal and external milieu.

3. Receptor and ligand are recycled to the plasmalemma, where the ligand is released into the environment while the receptor is reintegrated into the cell membrane; this occurs, for example, in the case of transport proteins.

4. The permeation pathway is defined by the penetration of some ligands through the endosomal or receptorsomal membrane to enter the cytosol. This is observed for toxins and viral genomes. Different pathways are used by different ligands in relationship to their special function (see following sections). Transport functions of vesicular structures are associated with tubular elements such as microtubules, some of which lay near the cell surface and probably comprise the endocytic and exocytic processes (see Sect. 2.5).

Several types of receptor, e.g., for epidermal growth factor, low density lipoproteins, and transferrin, are found to be concentrated at the surface even in an unoccupied state. This is called preclustering (BOONSTRA et al. 1985). These ligand-free receptors are constantly internalized via receptorsomes but subsequently are not ordinarily routed to lysosomes to be degraded but instead are returned to the cell surface to be reused. Internalization, intracellular movement, and reintegration of unoccupied receptors happens within minutes. Thus, receptor mediated endocytosis is a rapid and continuous process exchanging considerable areas of the plasmalemma.

2.3 Ligand Binding and Signal Transmission

The interaction of ligands and receptors is a highly specific, reversible, and concentration-dependent process. Comparable with the antigen-antibody reaction, special parts of the ligand molecule fit exactly to the extracellular binding region of the related receptor. Hormone affinity for receptors is due to noncovalent binding which involves electrostatic forces including hydrogen bonding, hydrophobic interactions, and van der Waals forces. The hydrophobic interactions drive the reaction, since the hydrophobic regions on the surface of the hormone shun the hydrophobic environment. Thus, these regions seek contact with similar hydrophobic regions of the protein receptor. In addition, electrostatic interactions occur and hydrogen bonds are formed between charged groups of the hormone and oppositely charged groups of the complementary side of the receptor. This interaction activates the receptor molecule with modification of intramembrane proteins and associated enzymes. Alternatively, cell surface receptors may open or close gated ion channels in the plasma membrane. Four different types of signal transducing by surface receptors are described (Fig. 4):

1. The adenylate cyclase-associated receptor contains three different domains separated from each other: (a) the hydrophilic extracellular domain for specific signal detection, (b) the lipophilic intramembrane regulatory unit or G protein mediating the signal across the plasmalemma, and (c) the hydrophilic domain with enzymatic activity, called receptor enzyme, located at the inner part of the plasmalemma and inducing further steps of target protein activation.

Receptor stimulation induces modification of the intramembrane G protein which is essential for further steps of signal transmission. This protein binds the energy-rich guanosine triphosphate (GTP) on its cytoplasm-attached domain (GTP-binding region) cleaves, and initiates subsequent activation of the adenylate cyclase (ROSENTHAL and SCHULTZ 1988). The activation of adenylate cyclase results in increased production of cyclic adenosine 3',5'-monophosphatase (cAMP), which stimulates a cytosolic protein kinase. This enzyme activates an effector protein inducing the cellular response by phosphorylation of proteins at specific serine and threonine residues (CATT et al. 1979; PASTAN and WILLINGHAM 1981). Additionally, the G protein is essential for the time course of receptor activation. Normally activation persists only for

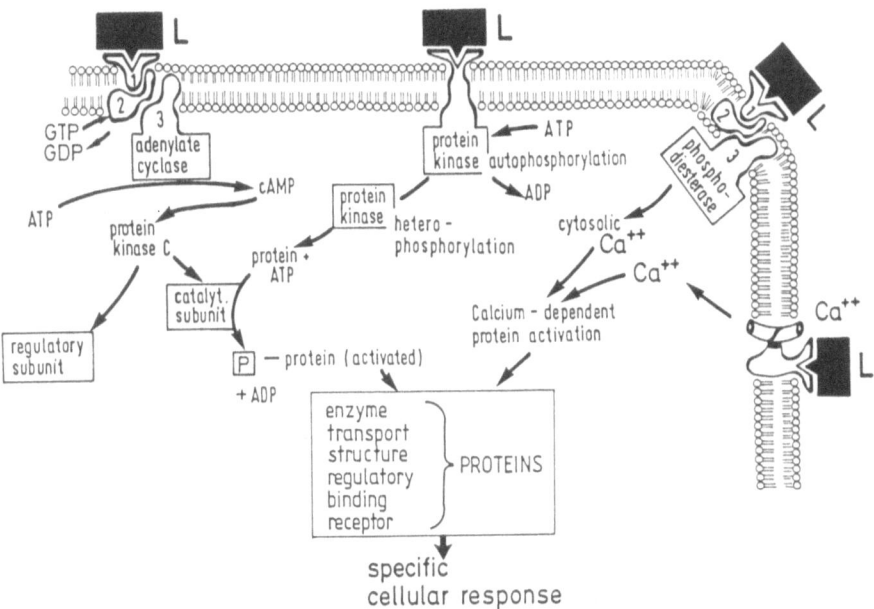

Fig. 4. Ligand-induced cell surface receptor activation, signal transmission, and subsequent intracellular processing. *Left:* ligand binding to an adenylate cyclase-associated receptor [*1*, extracellular domain; *2* intramembranous regulatory protein (G protein); *3*, intracellular enzymatically active domain], production of cAMP, and phosphorylation of cellular effector proteins. *Center:* ligand binding, to a protein kinase-associated receptor with autophosphorylation and heterophosphorylation. *Top right:* ligand receptor complex acting via a phosphodiesterase with elevation of the cytosolic calcium concentration. *Middle right:* ligand binding to a calcium-associated receptor with activation of gated calcium channels, elevation of intracellular calcium levels, and stimulation of calcium-dependent cellular events

seconds. This gives cells the possibility to respond rapidly to varying concentrations of the information-transporting substances, e.g., to pulsatile signals. The type of response is assured because the G protein is a GTPase that self-inactivates by hydrolyzing GTP to guanosine diphosphate (GDP) and inorganic phosphate. Thus, G proteins terminate adenylate cyclase activity by cycling between active and resting forms. It is likely that the hormone receptor, the regulatory unit, and the cyclase reach with each other through their mobility in the lipid-membrane matrix.

The protein kinase activated by cAMP has been shown to consist of two subunits – a regulatory subunit to which cAMP binds and a catalytic subunit that catalyzes phosphorylation of the substrate protein molecules. In the absence of cAMP the regulatory and catalytic subunits are inactive. The binding of cAMP alters a conformation of the regulatory subunit so that it dissociates from the complex and releases a catalytic subunit which is thereby activated. The effects of cAMP are usually transient and regulated by diphosphorylation which is catalyzed by phosphoproteinphosphatase, an enzyme itself regulated by cAMP.

2. A second type of receptor is characterized by unique amphophilic transmembrane protein. The receptor enzyme is a protein kinase, the activation of which induces autophosphorylation of the receptor protein followed by heterophosphorylation of cytosolic effector proteins (KASUGA et al. 1982; USHIRO and COHEN 1980). The heterophosphorylation then activates cellular target proteins. This system is especially involved in regulating cell growth via hormones and growth factors (DIETEL 1987b).

3. Some membrane receptors are associated with a membrane-bound phosphodiesterase that converts phosphatidyl inositol-4,5-diphosphate (PIP_2), a phospholipid confined to the inner leaflet of the plasma membrane, to inositol-1,4,5-triphosphate (IP_3) and diacylglycerol (BERRIDGE and IRVINE 1984). The phosphate diesterase and the cytosolic protein kinase interact with each other and thereby modulate their activity. Activation of the enzymes then leads to an increase in intracellular calcium and cytosolic pH. These events are especially involved in the control of cell proliferation (see Sect. 2.6).

4. The activation of some surface receptors opens voltage-dependent membrane calcium channels. Cells utilize the large calcium gradient across plasma membranes for transducing extracellular signals. The transient opening of such channels following receptor activation allows calcium to enter the cytosol, where it acts as second messenger. This is followed by an immediate increase in cytosolic calcium which activates a variety of enzymes (BISHOP 1985; CARAFOLI and PENESTON 1986). In some cases receptor activation first depolarizes plasma membrane and changes membrane potential and opens calcium channels. In other cases opening of the calcium channels in the plasmalemma occurs independently of changes in the membrane potential.

An important alternative mechanism for calcium-mediated signaling is release of free calcium from intracellular organelles, increasing the cytoplasmic calcium concentration and thereby activating target proteins. This can also occur during intracellular signaling.

2.4 Receptor Regulation

When cells are exposed to moderate or high concentrations of a ligand, the number and activity of surface receptors are decreased; this process is called down-regulation. Under conditions of ligand excess, it was measured that the rate of receptor synthesis is reduced while the rate of receptor degradation is enhanced. For example, if a tumorous endocrine gland secretes a hormone in unphysiologically high levels, the number of receptors at the target cells is decreased. Most cell types decrease surface receptors by accelerating the rate of receptor degradation. Only in response to chronic excess is the rate of synthesis decreased as well.

If there is reduced synthesis of a ligand a compensatory increase in receptor density on target cells has been shown, named up-regulation (CATT et al. 1979). In addition, severe short-term variations in the concentration of a ligand can generate modifications of the receptor affinity (desensitization) without changing the number of receptors. Obviously, receptor up-regulation is due to a decreased rate of receptor degradation instead of increased synthesis.

Exceptions to these rules are known for gonadotropins. The increased concentrations during the menstruation cycles are followed by an increased receptor density. In addition, gonadotropin-releasing hormone (GnRH), prolactin, and angiotensin initiate so-called reverse receptor regulation in accordance with the physiological function of these hormones.

2.5 Involvement of Cytoskeleton Elements in Receptor Action

As mentioned previously (see Sect. 2.2), receptor-mediated endocytosis induces clustering, internalization, and intracellular routing of organelles, i.e., there are actively directed movements of the ligand-receptor complex. Because of such observations the involvement of the cytoskeletal system in receptor distribution has frequently been considered. Microtubule and microfilament disrupting agents, however, were unable to resolve this issue, because the agents proved to be either ineffective or nonspecific. The directed intramembrane movement of receptors may be due to microfilament activation, in part achieved by intracellular contractile elements, such as certain subunits of actin.

2.6 Intracellular Membrane Receptors

In addition to the cell surface, the intracellular membrane-coated organelles also contain receptors which are involved in intracellular sorting of functionally active signal molecules. For example, lysosomal enzymes are synthesized in the endoplasmic reticulum and then selected by specific lysosome receptors which concentrate the lysosomal enzyme into the lysosomes. Using techniques of immunocytochemical localization it could be shown that phosphomannosyl, an essential residue for certain lysosomal enzymes, is taken up by the cell via surface receptors of the plasmalemma; once inside the cell it is delivered to the Golgi apparatus, which takes up the molecule via its membrane receptors. Further transport to the lysosomes is also receptor mediated.

By applying light and electron microscope autoradiography it was demonstrated that ^{125}I-insulin and ^{125}I-human growth hormone bind to the cisternal face of Golgi secretory vesicles and to microsomal structures, suggesting biologically important intracellular hormone receptors (BERGERON et al. 1978).

Thus, multiple steps of receptor-mediated transport and selection can be necessary to convey certain substances from the environment to certain intracellular targets (WILLINGHAM et al. 1983). In addition, delivery of proteins, mostly synthesized in the endoplasmic reticulum, from one organelle to another organelle of the cell, e.g., lysosomes, mitochondria, Golgi apparatus, and secretory granules, is regulated by organelle membrane receptors. The delivered proteins are called endogenous ligands.

2.7 Peptide Hormone Receptors

Almost all peptide hormones are water-soluble molecules which carry highly specialized information. They bind to membrane receptors which are most often adenylate cyclase associated (Fig. 4, Table 3a). The binding immediately induces a specific cellular response. This covers a wide variety of reactions, including metabolic changes, cell growth, differentiation, cellular movement, polarization, and special cellular functions.

Morphological receptor localization is important for the understanding of many physiological processes, such as calcium and glucose metabolism, thyroid and adrenal regulation, renal functions, and intracerebral interactions. Receptor visualization can be performed by introducing labeled (radioactive or biotinylated) ligands or by applying antibodies against the receptor protein or antibodies against the ligand after receptor binding. Only by morphological methods can one disclose the exact distribution of receptor-positive and -negative cells in certain tissues. Some examples are as follows:

Binding sites for biotinylated parathyroid hormone (PTH) could be detected (Fig. 5) in the proximal tubule, the thin limb of Henle's loop, the distal tubule, and the collecting duct (NIENDORF et al. 1987b; DIETEL et al. 1985a) as well as cultured kidney cells (NIENDORF et al. 1986). The staining intensity varied from cell to cell and between different parts of the nephron. Another example (Fig. 6) is provided by the morphological demonstration and quantification of thyrotropin (TSH)-binding sites in thyroid tissue (SCHRÖDER et al. 1986). The growth and function of thyroid glands are regulated by TSH. The initial step of TSH action on thyroid tissue is its binding to specific receptors located on the thyroid plasma membrane and activation of the adenylate

Table 3a. Hormones that stimulate adenylate cyclase (examples)

Adrenocorticotropic hormone	Luteinizing hormone-releasing hormone
Calcitonin	Lipotropin
Catecholamines (β-adrenergic)	Melanocyte-stimulating hormone
Chorionic gonadotropin	Nerve growth factor
Follicle-stimulating hormone	Parathyroid hormone
Glucagon	Prostaglandin E_1
Luteinizing hormone	Thyrotropin-stimulating hormone

Table 3b. Hormones and hormone-like agents that do not stimulate adenylate cyclase

Angiotensin	Insulin-like growth factors
Catecholamines (α-adrenergic)	Multiplication-stimulating activity
Chorionic somatomammotropin	Oxytocin
(placental lactogen)	Prolactin
Epidermal growth factor	Prostaglanding $F_{2\alpha}$
Fibroblast growth factor	Somatomedin
Growth hormone	Somatostatin
Insulin	

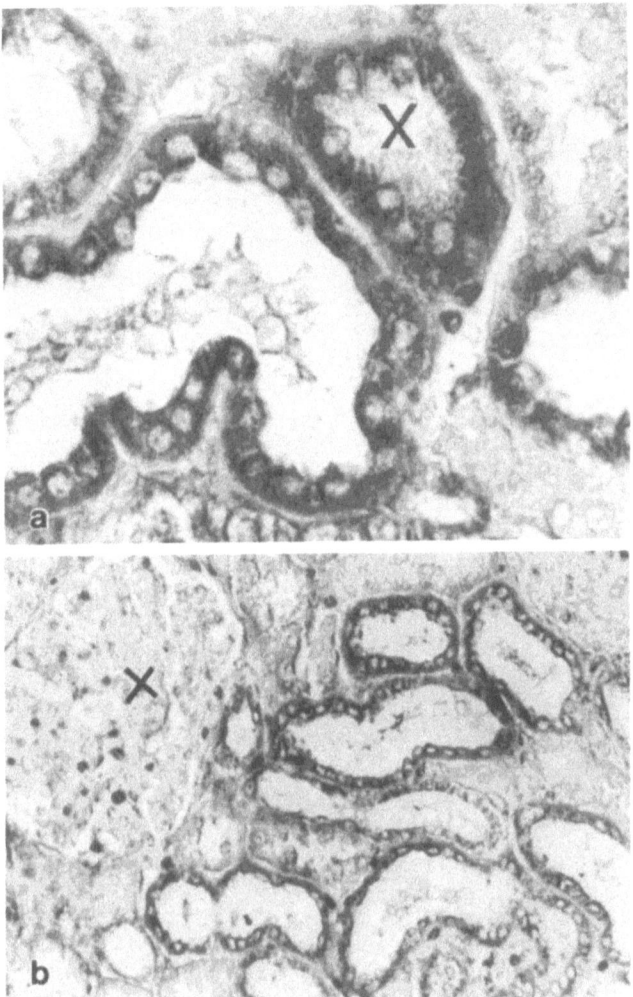

Fig. 5. Immunocytochemical demonstration of PTH receptor. Cells of proximal (X) and distal tubules as well as single cells within the collecting duct express PTH-binding immunoreactivity. No staining was seen in glomerula, the thin limb of Henle's loop, blood vessels, or connective tissue. (NIENDORF et al. 1987 b)

cyclase. Applying ^{125}I-TSH to thyroid sections obtained with a cryostat, the TSH binding could be localized to normactive thyrocytes as well as to adenomatous, hyperstimulated, or carcinomatous follicle cells in various degrees. This localization of TSH binding may play a role in unbalanced growth.

A second group of peptide hormone receptors exist which are not associated with adenylate cyclase but with guanylate cyclase, tyrosine kinase, protein kinase, and phosphodiesterase. Among these the insulin receptor is one of the

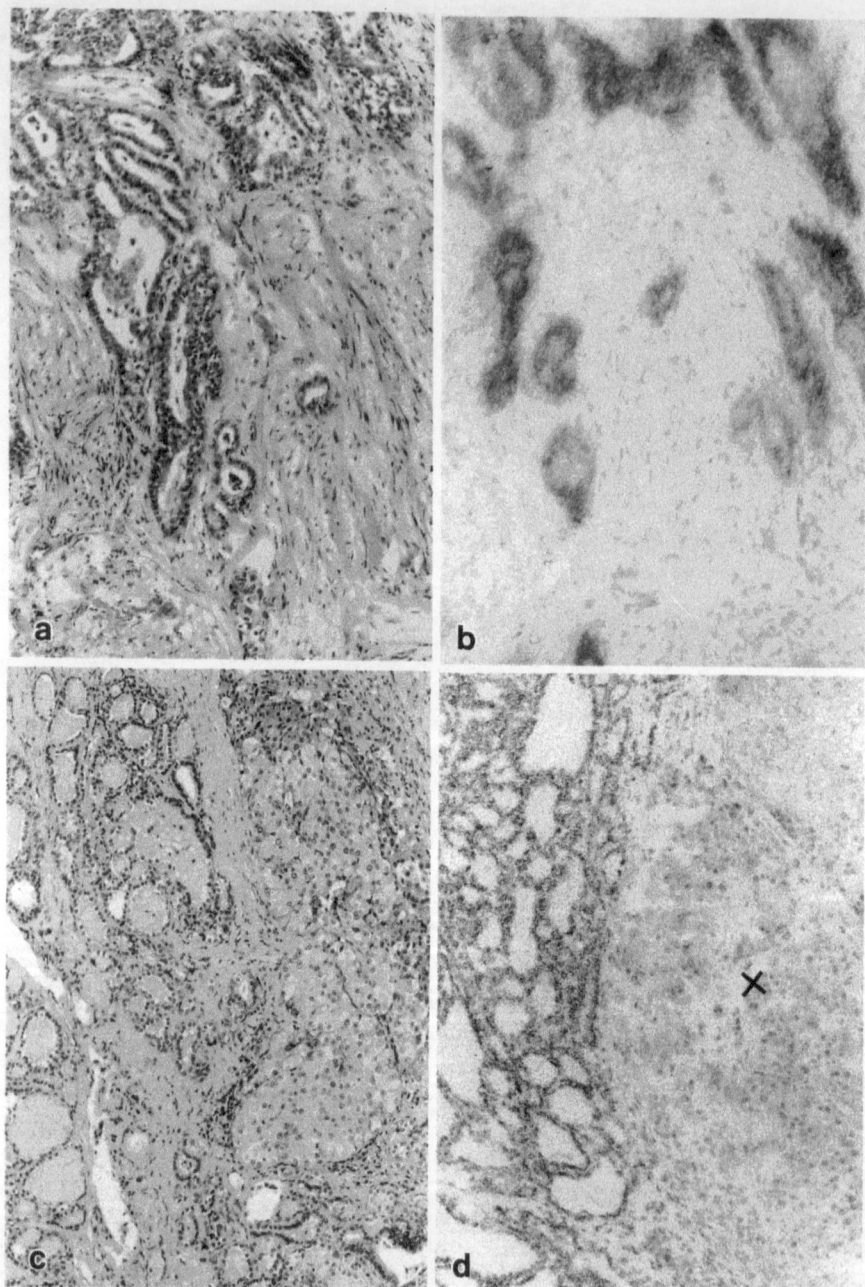

Fig. 6 a–d. Morphological visualization of TSH receptors in thyroid carcinoma **a** HE staining and **b** ^{125}I-thyrotropin autoradiography of follicular variant of papillary carcinoma. **c** HE staining and **d** ^{125}I-thyrotropin binding sites in poorly differentiated follicular carcinoma with intensive binding to normal thyroid tissue (*left*) and no binding to tumor cells (*X*). (Schröder et al. 1986)

most intensively studied (for review see SHECHTER 1985). Insulin acts on the majority of cells in the body. Its immediate action is to increase the rate of glucose uptake from the blood into the cell by increasing the activity of glucose permease in the plasma membrane. The insulin receptor can be detected on most cells biochemically by the criteria of saturability and specificity of binding and morphologically by detecting labeled insulin. The insulin receptor contains an insulin-activated tyrosine-specific protein kinase (for structural details see Sect. 2.12 and Fig. 1).

Dynamic processes of receptor activation, such as ligand binding in different conditions, lateral movement in the plasma membrane, capping, and internalization, mostly have been studied in time course experiments by morphological visualization of ligands using cell cultures. For these studies biotinylated ligands have often been used. CHILDS et al. (1987) were able to localize the corticotropin-releasing factor (CRF) receptor on pituitary cells by incubation with biotinylated CRF. They demonstrated that binding and internalization of the labeled CRF appeared in less than 5 min, while routing through the cell required a further 15–30 min. NIENDORF et al. (1987c) developed a new method for visualizing labeled hormones on living cells which enables the investigator to observe directly receptor mobility at the cell surface and during internalization (Fig. 7). Similar studies have been performed by DRAZNIN et al. (1985), who visualized somatostatin binding sites on the surface of primary cultures of anterior pituitary cells. For this purpose somatostatin was conjugated to gold particles, which allows the molecule to be followed during binding and internalization. Within the cells somatostatin is routed to either lysosomes or Golgi apparatus (see Sect. 2.2).

2.8 Structurally Similar Receptors Inducing Different Responses

Depending upon their functions, cells express a specific set of receptors. Different cells may have different sets of receptors for the same signal, each receptor inducing a different response. Alternatively, the same receptor may occur on various cells, but binding of a signal triggers different cellular responses. Clearly, cells respond in a variety of ways to the same ligand. These differentiated cellular responses are transduced by similar but not identical receptor proteins, introducing a tissue specificity for ligands by modulating the information provided by signaling substances.

Studies using photoaffinity labeling of pituitary and gonadal receptors for GnRH exhibited that the divergent action of GnRH in these tissues is expressed through receptors with similar structures. This finding is important since the gonadal action of GnRH is predominantly inhibitory, in contrast to the primary role of GnRH in stimulating luteotropin release in the pituitary. Thus, small differences in the molecular structure of the GnRH receptors can induce a very different cell response in pituitary and gonadal target cells.

Modification of receptor molecules with changes in the ligand's information is a unique finding in receptor physiology. Such modification is known to

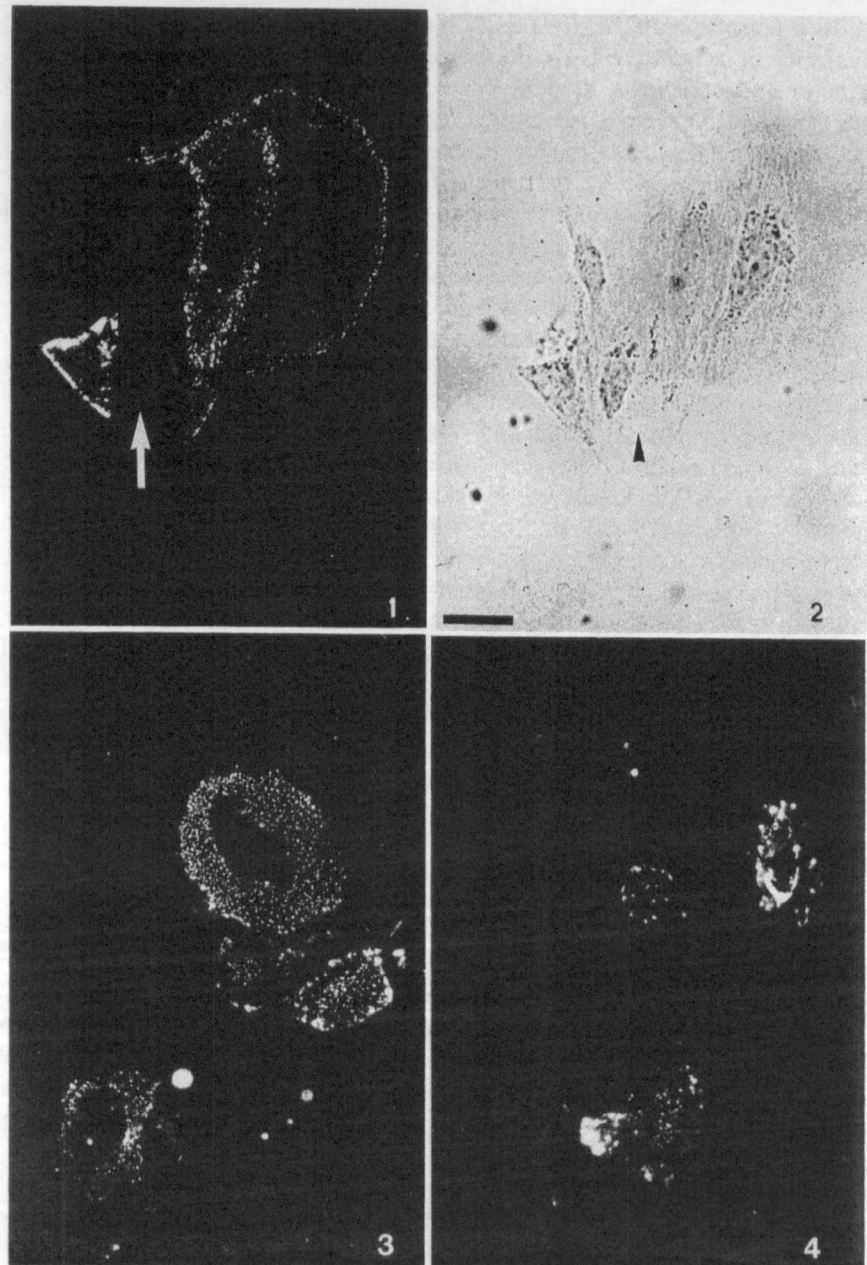

Fig. 7. Living cortical kidney cells with PTH-binding sites visualized by incubation with biotinyl-b-PTH. **1** As expected for a heterogeneous cell population, not all cells are stained; see *arrow* and compare to **2** phase contrast demonstration of the same cell group. **3** Dependent upon the incubation time, first a diffuse distribution pattern is shown, followed **4** by a clumpy concentration of PTH-binding sites representing receptor clustering. (Niendorf et al. 1987c)

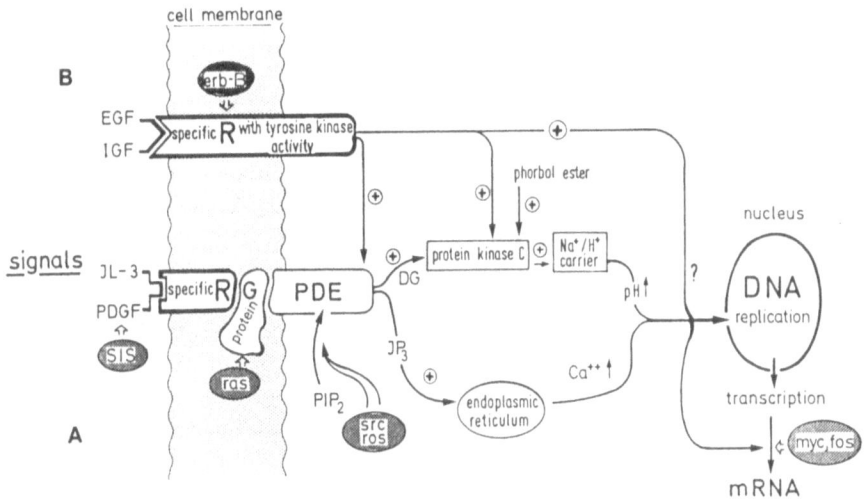

Fig. 8 A, B. Growth factor receptor-mediated cellular proliferation. Suggested interactions of growth factors, cell surface receptor, and subsequent signal transmission. In several steps oncogenes and oncogene products are involved. Two mechanisms are demonstrated: **A** Activation of the phosphodiesterase (*PDE*)-associated receptor converts phosphatidyl inositol diphosphate (*PIP_2*) to diacylglycerol (*DG*), elevating cytoplasmic pH, or to inositol triphosphate (*IP_3*), elevating cytosolic Ca^{2+} concentration. Both events are suspected to stimulate cell growth. **B** Activation of the trosine kinase-containing receptor appears to stimulate PDE, the cytoplasmic protein kinase C, and possibly directly the DNA replication

occur in the case of growth factor receptors, e.g., nerve growth factor (NGF) receptors (see Sect. 2.9): The NGF receptors have unique parts which are altered by association with accessory proteins. This slightly modifies the signal given by binding.

2.9 Growth Factor Receptors

The molecular structure of most growth factor receptors is somewhat different from that of hormone receptors in that they often consist of a single continuous transmembrane polypeptide chain (Fig. 1 A). This has an extracellular growth factor binding site, a transmembrane region, and an intracellular domain with enzyme activity, mostly tyrosine-specific protein kinase C (Table 3 B, Fig. 8). Autophosphorylation followed by heterophosphorylation is the means by which the signal is transduced (Fig. 4). A second means of signal transmission is known for platelet-derived growth factor (PDGF) and interleukin-3, namely via G protein and phosphodiesterase activation (Fig. 8).

Growth factors and consequently growth factor receptors are essential for proportional growth, maturation, and differentiation of tissues as well as for healing of tissue lesions. One example is NGF and its receptor, used here to explain the basic mechanisms important in the interaction of growth factors and related receptors. NGF is necessary for the survival and development of cells of the peripheral nervous systems and possibly also has some influence on the central nervous system. Three important functions of NGF are known in peripheral neurons: (a) ensuring the survival (not proliferation) of ganglionic cells, (b) inducing neurite outgrowth from peripheral ganglia, and (c) directing the growth of the axon following trauma which disturbs the normal connection between the axon and its target cell. All these functions are mediated by different classes of the NGF receptor (see Sect 2.8) at the surface of nerve cells (EVELETH 1988). Similar to other growth factor receptor proteins, the NGF receptor is a continuous protein with three domains, as explained above.

A further example of a protein kinase-associated growth factor receptor is that specific for epidermal growth factor (EGF). After activation of the EGF receptor the ligand-receptor complex is internalized and degraded completely (USHIRO and COHEN 1980) without separation of receptor and ligand (Table 2). EGF-producing cells and the related target cells have been located by immunocytochemistry in normal and tumorous cells. The biological effects are associated with activation of the receptor tyrosine kinase. The receptor is a 170 000-dalton integral membrane glycoprotein that undergoes autophosphorylation on tyrosine residues during EGF binding. Activation of the EGF receptor induces a broad range of mitogenic and nonmitogenic actions. It is responsible for normal development of many types of tissues in cells, e.g., the physiological proliferation and differentiation of epithelial mammary cells. Also macrostructural development may depend on the presence of EGF, e.g., the fusion of the palatal shelves is controlled by an appropriate concentration of EGF. The effects on proliferation are discussed later (see Sect. 2.10.1).

2.10 Cell Surface Receptors and Growth Modulation of Tumor Cells

The exact process by which cancer cells arise and develop into malignant disease is an exceptionally perplexing medical challenge. The concept that a single physiological or morphological aberration can define all cancer cells and distinguish them from their normal counterparts has been discarded largely as a consequence of molecular biological studies. The family of growth factors is thought to play a dominant role in the regulation of tumor growth, i.e., in growth promotion as well as growth inhibition. Both mechanisms are associated with synthesis of appropriate receptor proteins by the tumor cells.

2.10.1 Growth Factor Receptors and Growth Stimulation

Binding of stimulating growth factors to cell surface receptors is a prerequisite for the initiation and maintenance of biochemical events required for cellular proliferation (for details of signal transmission see Sect. 2.3 and Fig. 8). It is known that several growth factors, such as EGF, granulocyte and macrophage colony stimulating factor (GM-CSF), and interleukin 2 (IL-2) bind to both high and low affinity cell surface receptors. It appears that it is the small number of high affinity receptors which are involved in the generation and maintenance of the mitogenic signals and that only these should be considered when functional interactions are investigated.

It has been known for some time that particular peptides, e.g., EGF, not only bind to their receptors but down-regulate unoccupied receptors of other growth factors, such as CSF, PDGF, and transforming growth factor (TGF). This interreceptor modulation appears to be a particularly intriguing aspect of growth factor signal transduction. It has been shown that different growth-promoting substances interact using similar types of receptor to transduce the signal to the cells.

The influence of EGF receptor status on proliferation rates of tumor cells in a clinically important manner has been investigated in detail (CARPENTER and COHEN 1979; HUNTER 1984; SPORN and ROBERTS 1985). However, results are contradictory. Some authors have demonstrated a correlation between growth rate, depth of invasion, and density of EGF receptors in bladder and breast cancer (NEAL et al. 1985; SAINSBURY et al. 1985) and have thus claimed that EGF receptor status is of prognostic significance. In other studies, however, no correlation was shown between EGF receptor expression and growth fractions, grading, tumor diameter, and lymph node status, suggesting that EGF receptor status is irrelevant for tumor growth and other prognostic parameters (FITZPATRICK et al. 1984).

To address this problem further we investigated whether 15 cell lines of different epithelial malignant tumors responded to EGF and whether this was receptor dependent (unpublished data). The receptors were detected by immunocytochemistry using monoclonal antibodies against the receptor protein and by in situ hybridization using labeled cDNA specific for the mRNA of the EGF receptor (Fig. 9). Six cell lines increased cell growth due to addition of EGF. They all expressed the EGF receptor. Three lines contained EGF receptors without being stimulated by EGF, indicating "nonfunctioning" receptors or receptor-like proteins. The rest (n = 6) did not express EGF receptors and were not influenced by EGF application. The results demonstrate the limited relevance of EGF receptor detection in evaluating the growth-promoting activity of EGF, and in our experience suggestions regarding clinical approaches should be made with caution.

2.10.2 Growth Factor Receptors and Autocrine Growth Stimulation

It was recently shown that tumor cells synthesize and secrete growth factors, e.g., PDGF, bombesin, TGF-α, and EGF, and express the related receptor at their own surface. It has been proven that growth factor production by a cell can be responsible for the uncontrolled growth of that same cell. At least in some types of tumors this autostimulation is necessary to preserve cellular transformation and proliferation. The process is called autocrine stimulation.

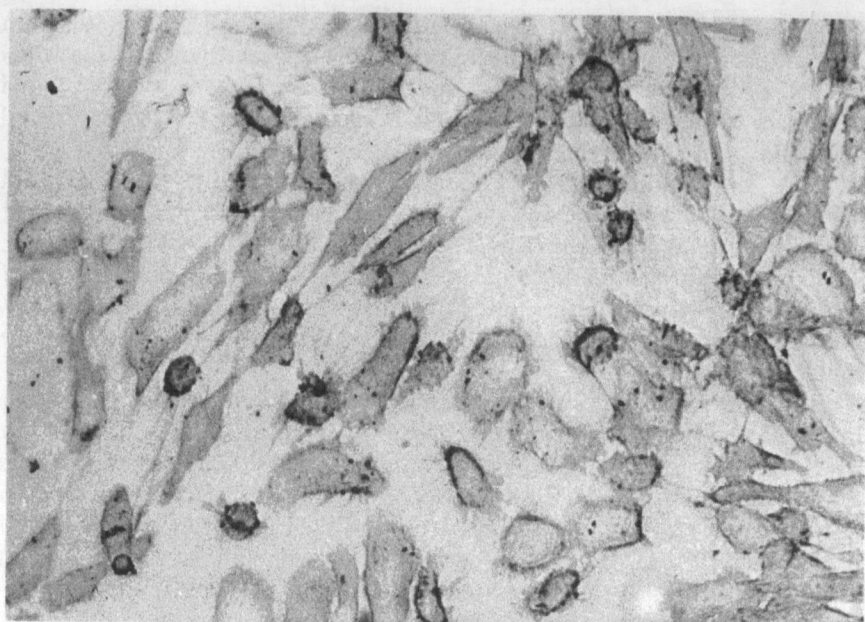

Fig. 9. EGF receptor on cultured carcinoma cells shown by immunocytochemistry using monoclonal antibodies against the receptor protein. Typical membrane-concentrated dot-like distribution

Why do cells use such a laborious way to preserve growth? Certain signals, predominantly peptides, obviously have to deliver their information via surface receptors in order to promote cell growth. There is no possibility for tumor cells to induce this method of growth stimulation via intracellular signaling, as discussed above for other signals (see Sect. 2.6). The signal has to come from the environment, to be taken up by surface receptors, and then to be transmitted to the nucleus by the mechanism described previously (see Sect. 2.3).

An example is a human small cell lung cancer (SCLC) cell line which was shown to secrete a bombesin-like peptide (Cuttitta et al. 1985). Bombesin, a tetradecapeptide, has potent pharmacological effects in animals and man, and activates protein kinase C. Evidence of autocrine growth stimulation was obtained from the application of a bombesin-specific monoclonal antibody directed against the receptor binding region. This blocked the receptor-mediated effects of bombesin on cell proliferation. A certain tumor specificity could be ascertained, since non-SCLC cell lines did not express bombesin receptors and, of course, did not react to the peptide.

Similar mechanisms with "neutralizing antibodies" may become important for the treatment of other tumors, such as the different forms of leukemia, gastrointestinal tumors, and breast cancer. Antibodies which neutralize PDGF and a CSF have been shown to inhibit transformed fibroblasts. Several carcinomas can produce growth factor-like proteins (tumor-derived growth factors, TDGF) capable of binding to receptors of other known growth factors. For example, TGF-α is able to bind to the receptor for EGF, promoting cell growth. The autostimulatory mechanisms of tumor growth appear to be complex (presumably with great individual variations) and are far from completely understood.

2.10.3 Growth Factor Receptors and Growth Inhibition

In addition to growth-promoting growth factors there exists a new class of proteins which inhibit cell growth, so-called negative growth factors, e.g., TGF-β. The negative growth factors induce growth arrest and therefore may antagonize the effect of stimulating growth factors.

Recent developments have made it clear that most epithelial cells and lymphocytes are strongly inhibited by TGF-β. Only fibroblasts increase proliferation in response to this factor. TGF-β has been isolated and purified from a variety of normal as well as neoplastic tissues (SPORN et al. 1987; MOSES et al. 1985). The receptor of TGF-β was recently identified and proven to be widely distributed among various normal tissues (KYPRIANOU and ISAACS 1988; TUCKER et al. 1984; FANGER et al. 1986) and human cancer cell lines (ARTEAGA et al. 1988; KNABBE et al. 1987). New therapeutic schemes could be developed which take advantage of negative growth factors.

Platelet-derived growth inhibitor (PDGI) has been partially purified and shown to inhibit endothelial proliferation (BROWN and CLEMMONS 1986). It blocks DNA synthesis and mitosis in endothelial cells. Although nothing is definitely known about its physiological function, it has been suggested that PDGI could take part in growth regulation of endothelial cells, especially after injury, and of malignant tumor cells.

In a human colon carcinoma cell line, a tumor inhibitory factor (TIF) was identified which reversibly prevents the anchorage-independent growth of several other colon carcinoma cell lines (LEVINE et al. 1985).

2.10.4 Interaction of Growth Factor Receptors and Oncogenes

There is a complex interaction between chromosomal translocation, mechanisms of molecular integration, control and regulation of gene activity, the synthesis of growth factors, and expression of related receptors. In normal cells with unaltered genetic constitution, growth regulation results from either cell contact or a low level of growth factor receptor. Receptor activation is communicated through the cytoplasm to the nucleus, thus regulating the synthesis of mRNA, essential for the unique proteins needed to initiate subsequent DNA synthesis and mitosis. Any genetic change that influences the regulation of mitosis may result in malignancy. However, as a rule one genetic change is insufficient to drive a cell toward malignancy. A so-called second hit at the genetic or the receptor level must be involved before immortalization is induced. Tumor promoters, for example, are known to induce mitogenic responses through signal transduction mechanisms necessary to allow the expression of genetic defects. This second hit also requires specific receptors in order to have an effect.

Proteins encoded by oncogenes have been shown to have receptor properties by their cellular localization, their enzymatic activity, and their (poly)nucleotide binding affinities. The most exciting outcome of these analyses is that several oncogene proteins can be identified as altered variants of physiological receptors that are involved in growth-regulating processes (Fig. 8). Most of these receptor proteins are truncated, mutated, or otherwise modified versions of normal membrane proteins. The molecular alterations can affect the localization and function of the encoded protein, causing altered, often hyperstimulated, activity.

In a number of tumors amplified EGF receptor genes were found, e.g., primary brain tumors (LIEBERMAN et al. 1985), gastric carcinoma (YAMAMOTO et al. 1986), several lines of squamous cell carcinomas (KAMATA et al. 1986), and the intensively studied human epidermoid carcinoma cell line A 431. The EGF receptor protein of A 431 cells is encoded by the activated erb-B oncogene (ULLRICH et al. 1984), which was originally described for the avian erythroblastosis virus (AEV). The encoded protein has a high degree of sequence homology with the EGF receptor and both v-erb-B and the EGF receptor map to the same region of human chromosome 7 (SPURR et al. 1984). The erb-B oncogene expresses a truncated extracellular domain different from that of the normal EGF receptor. A 431 cells exhibit an EGF-independent rapid growth which appears to be due to uncontrolled chronic autophosphorylation of the EGF receptor. This observation, first made by ULLRICH et al. (1984), suggests that loss of the ligand-binding site of a receptor, in this case the EGF receptor, is sufficient to activate its thyrosine kinase and induce malignant transformation. Indeed, AEV can easily transform fibroblasts by incorporation of the truncated EGF receptor protein. The identification of erb-B as the modified gene for the hyperactivated EGF receptor provides a model for growth factor receptor-mediated malignancies.

A variety of other oncogene-encoded proteins with a receptor-like structure possess thyrosine kinase activities similar to those associated with growth factor receptor. For example, the feline sarcoma virus contains an oncogene, v-fms, the product of which has been shown to be almost identical to the receptor of the growth factor CSF-1. The c-fms product was indistinguishable from the CSF-1 receptor and, vice versa, CSF-1 was found to bind to the c-fms protein (SHERR et al. 1985). Similar to the EGF receptor, the fms protein exhibits thyrosine kinase activity even in the absence of CSF-1. These data suggest that the fms protein is constitutively activated and that the molecular defect resulting in autophosphorylation is the basic mechanism through which uncontrolled cell proliferation is stimulated.

Special interest has been directed to the ras oncogene, which codes for guanine nucleotide (GDP and GTP)-binding proteins, suggesting a relation to the intramembrane G protein (see Sect. 2.3) of the receptor complex. Three ras genes (H-*ras*, K-*ras*, and N-*ras*) have been identified and their products localized to the cytoplasmic site of the plasma membrane, similar to the subcellular location of G proteins. Activation of the *ras*-oncogene product results in an overproduction of protein kinase C, with elevation of cytoplasmic pH and calcium (Fig. 8), thus triggering cell proliferation (HELMREICH 1986).

The *sis* oncogene codes for the synthesis of PDGF (DOWNWARD et al. 1984). PDGF acts via receptor and stimulates the production of the intracellular messengers diacylglycerol and inositol triphosphate (Fig. 8) in 3T3 cells. The gene or oncogen coding for the PDGF receptor is not known.

Simultaneous activation of two or more oncogenes may enhance the transforming potency of oncogenes (RAPP 1985). The combination of *ras* and *myc* transforms fibroblasts and bone marrow cells with a higher activity than does either of these oncogenes alone. This phenomenon is explained by the observation that the oncogene products may be involved in different steps of the complex mechanisms of receptor-mediated signal recognition and subsequent transmission to the nucleus followed by induction of DNA replication.

The oncogene *neu* was discovered after serial transfection of cells with neuroblastoma DNA. *neu* was found to have 50% sequence homology with the normal EGF receptor gene. The *neu* oncogene-encoded protein reacts with antibodies to the EGF receptor and has the characteristic thyrosine kinase activity.

The mechanisms so far described offer explanations as to how the activation of oncogenes can result in the uncontrolled growth of tumor cells. At the level of growth factors and their receptors, physiological development and malignant transformation may be more closely related than expected. Since the receptors are located in the cell membrane, they can be easily affected by systemically administered substances. This may result in new therapeutic strategies, e.g., use of GF antagonists, receptor-blocking drugs, and receptor antibodies. Also, the regulation of oncogene activity may be influenced by regulatory genes, so called anti-oncogenes (GREEN and WYKE 1985), which are involved in the switch on/switch off process of oncogenes.

2.11 Virus Receptors

In human oncogenesis several tumor types are known to be associated with viral infections. The tumor cells express receptors at their surface by which virions can enter (Fig. 10). Some of the tumor types are T-cell leukemia, Burkitt's lymphoma, nasopharyngeal carcinoma, liver carcinomas, and squamous cell carcinomas of the skin and the lower genitourinary, upper digestive, and respiratory tracts. General mechanisms of viral cell infection have been reviewed recently (BISHOP 1985; VARMUS 1985, THORLEY-LAWSON 1988). Many viruses enter cells by receptor-mediated endocytosis (DALES 1973; MARCH 1984; PASTAN and WILLINGHAM 1985). For example, it is known that Epstein-Barr virus (EBV) binds to any cell that carries a receptor specifically interacting with one of the viral envelope glycoproteins (GP350/220) (NE-

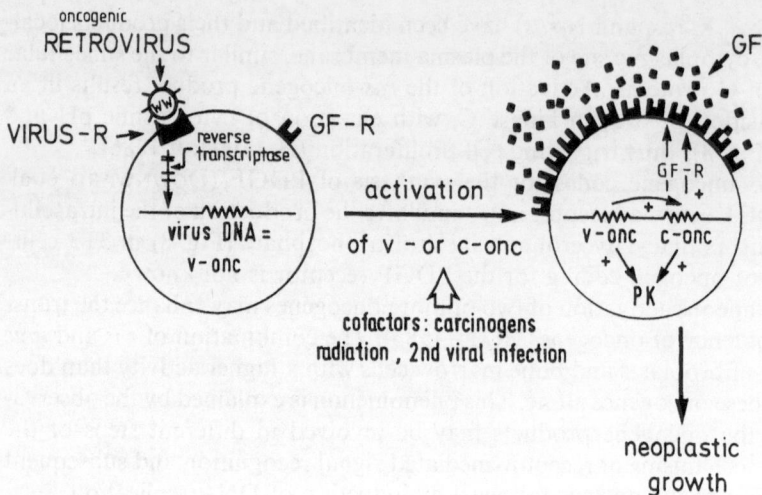

Fig. 10. Virus-induced cell transformation with involvement of cell surface receptor at two different sites. Virions enter the cell by binding to a special receptor (*R*); the virus (*v*) RNA is transcribed to v-DNA. Oncogenic v-DNA (*v-onc*) is integrated into cellular DNA (*left cell*). After activation by cofactors v-*onc* can activate cellular oncogenes (*c-onc*) that code for growth factor receptor (*right cell*). The elevated expression of growth factor receptor is responsible for neoplastic growth promotion

MEROW et al. 1987; TANNER et al. 1987). The virus-binding receptor is also the receptor for the C3d component of the complement system (a 140 000-dalton glycoprotein) designated CD21. EBV prefentially infects resting B cells and epithelial cells of the mouth mucosa which contain a high number of the mentioned receptors. Target cells are exposed to a high number of infectious viral variants which may become manifest as simple infection or may induce malignant dedifferentiation, particularly Burkitt's lymphoma and nasopharyngeal carcinoma (Table 4), usually by integrating viral genome into the DNA of the host cell. The primary EBV infection results in activated B cells which produce a 45 000-dalton glycoprotein identified as a low molecular weight B-cell growth factor. This CD23 protein is released in a high number of copies, being the "unphysiological" stimulus for the lymphoblasts which recognize the CD23 protein by membrane receptors. The general mechanisms are shown in Fig. 10.

The first virus with proven tumor-inducing potency was the Rous sarcoma virus, which produces malignant mesenchymal tumors in chickens. Its tumor-promoting potency was found to be based on the binding of the virus to a surface receptor with subsequent endocytosis, followed by the transcription of the viral RNA to DNA by means of reverse transcriptase and integration of the oncogenic virus DNA region (*v-src*-oncogene) into the host cell's DNA. The *src* oncogene product is a receptor-associated protein kinase, the activation of which induces mitosis (HUNTER and SEFTON 1980). Another virus, human T-cell leukemia virus type III (HTLV-III), has been reported to enter

Table 4. Some of the viruses, oncogenes, and receptor-related oncogene products involved in carcinogenesis

Onco-genes	Virus	Species of origin	Receptor-related *onc* proteins	Type of tumor
erb-B	Avian erythroblastosis	Fowl		Leukemia, sarcoma
src	Rous sarcoma	Fowl		Sarcoma
yes	Yamaguchi sarcoma	Fowl	Membrane or	Sarcoma
fps	Fujinama sarcoma	Fowl	cytoplasmic,	Sarcoma
fgr	Pasheed feline sarcoma	Cat	tyrosine-specific	Sarcoma
fms	McDonough feline sarcoma	Cat	protein kinase	Sarcoma
fes	Feline sarcoma	Cat		Sarcoma
ros	Rochester-2 sarcoma	Fowl		Sarcoma
abl	Abelson murine leukemia	Mouse		Leukemia
sis	Simian sarcoma	Monkey	PDGF homologues	Sarcoma
mos	Moloney murine sarcoma	Mouse	Phosphoprotein	Sarcoma
Ha-ras	Harvey murine sarcoma	Rat		Sarcoma
Ki-ras	Kirsten murine sarcoma	Rat	G-receptor protein	Sarcoma, human
N-ras	??			carcinoma, leukemia
myc	Myelocytomatosis MC 29	Fowl	Nucl. DNA receptor	Sarcoma, human carcinoma leukemia
B-lym	??	Fowl	Nucl. DNA receptor	Malignant lymphoma

T4 lymphocytes via receptor-mediated endocytosis (MCDOUGLAS et al. 1986). The HTLV-III receptor can be partly blocked by an anti-receptor antibody, reducing HTLV-III binding and thus internalization. Malignant transformation has also been shown for HTLV-I, which can initiate leukemia (98 WN1). The HTLV-I proviral genome is integrated into the T-cell DNA, inducing activation of the gene that codes for the IL-2 receptor (IL-2 is a T-cell growth factor). Then, the IL-2 receptor is expressed at the surface of the infected cell in high amounts. This was demonstrated in an HTLV-I-transformed T-cell line (POIESZ et al. 1980) with enhanced transduction of the IL-2-specific growth signal. It is not clear whether the IL-2 receptor-mediated signal is transduced by activating protein kinase C or whether other mechanisms are involved. In this respect, autointernalization of the IL-2 receptor with induction of cell growth in HTLV-1-transformed T-cell lines is discussed (SUGAMURA et al. 1986).

Morphological detection of the virus itself or of the receptor protein necessary for viral infection is possible by in situ hybridization and immunocytochemistry. This allows the investigator not only to detect infection itself but also to localize and attribute the infection to certain cell populations (LÖNING and MILDE 1987). A special advantage of in situ hybridization is the possibility of demonstrating virions in low productive virus infections and in states of immunosuppression when this is not possible with serological methods. Thus, in diagnostic work this technique will become a supplement to immunological assays in the near future.

2.12 Miscellaneous

The regulation of the highly complex hematopoietic system is mainly under the control of regulating proteins, such as IL-2 and other lymphokines. The lymphocytotrophic hormones give receptor-mediated signals to lymphocyte subpopulations, such as T cells, and other hematopoietic cells, to proliferate, to mature, to differentiate, and to perform specialized functions of the immune system (Smith 1988). The IL-2-specific receptor (see Sect. 2.10.1) has been discovered just recently and shows a complex bimolecular structure (Fig. 1) with two distinct polypeptide chains (α and β), each of which contains an IL-2-binding site (Tsudo et al. 1987). The α and the β chain are not connected via covalent disulfide bonds, as most other complex receptor proteins are, and thus represent a type of receptor presumably specific to hematopoietic cells. The unusual interaction of the α and the β chain may have importance in realizing the complex IL-2-molecule.

The whole immune system depends on precisely functioning surface receptors. A highly complex signal-detecting molecule is the T-cell receptor, which has to detect and distinguish simultaneously pathological antigens as well as nonpathological proteins belonging to the same organism, i.e., same immunological compartment. These proteins, which characterize cells as belonging or not belonging to a body, are termed major histo-compatibility antigens (MHC proteins). An extensive review is given by Marrack and Kappler (1987).

Transport protein receptors play a fundamental role in normal cell metabolism and in cell proliferation. During mitosis of rapidly growing tumor cells relatively large amounts of membrane have to be synthesized. This requires cholesterol usually taken up by the cells as low density lipoproteins (LDLs). This process requires highly active LDL receptors which can be demonstrated in tumor cell membranes and may contribute to new therapeutic approaches (see Niendorf and Beisiegel in this volume). Proliferating cells need large amounts of iron because it is a cofactor for the ribonucleotide reductase. For transport and internalization the iron molecule is coupled to the transport protein transferrin, forming di-ferro-transferrin. This complex is taken up by the cells via receptors (Hannover and Dickson 1985). To satisfy the iron requirement of growing cells, large amounts of the transferrin receptor have to be expressed in cell membranes (Musgrove et al. 1984).

Cell surface receptors for extracellular matrix components such as collagen IV, fibronectin, laminin, vinculin, and other glycoproteins involved in cellular attachment are necessary for adequate anchorage of differentiated cells in the extracellular matrix. In tumor cells, loss or modification of the receptor proteins has been advanced as a hypothesis which, with other factors, may be responsible for the different morphology and growth properties of tumor cells as well as for the occurrence of invasive growth and formation of metastases (Hynes 1985; Hynes and Yamada 1982; Pierschbacher and Rouslahti 1984).

Drug action on cells is often also transduced by more or less specific receptor or binding proteins. For example, in tumor therapy a membrane-bound receptor protein has recently received attention, the 170 000-dalton P-glycoprotein (Pgp), a drug-binding and transporting, channel-forming molecule with 12 membrane-spanning domains (Fig. 1). Pgp functions as an energy-dependent cytostatic drug efflux pump which contributes to the elimination of toxic drugs from tumor cells. By this mechanism resistance of tumor cells is induced to several structurally divergent antiproliferative substances; this process is called multidrug resistance. The cytostatic substances involved in multidrug resistance are all amphipathic lipophilic molecules. They bind to the Pgp receptor molecule by which they are transported out of the cell. More information on structure, function, and morphological distribution is given by BRADLEY et al. (1988), GOTTESMANN and PASTAN (1988), KARTNER and LING (1989), and DIETEL et al. (1990).

2.13 Receptor-Associated Diseases

Two subgroups of disorders related to the structure and/or function of cell receptors have to be distinguished:

1. Diseases with defective receptors
2. Diseases with cellular abnormalities mediated by normal receptors

The latter group has been discussed extensively in previous sections. Some examples of the former group are given in Table 5. In general, diseases with receptor abnormalities can be genetically determined or acquired (for review see DREYER and RÜDIGER, 1986 a, b).

Familial hypercholesterolemia is due to defective genes for LDL receptors which result in excessively high plasma LDL levels (up to 10 mg/ml in homozygotes). The abnormality can involve either a reduction in LDL receptors as shown using monoclonal antibodies (BEISIEGEL et al. 1981) or a structurally disturbed receptor protein. The resulting hyperlipidemia predisposes the patients to progressive premature atherosclerosis, often leading to death from cardiovascular disease during the fourth decade (TOLLESHAUG et al. 1984).

In pseudohypoparathyroidism an abnormality of the gene coding the PTH receptor or one of the intracellular effector proteins is suspected. There is either reduced binding of PTH to its target cells or normal binding but disturbed signal processing in the cell, resulting in severe hypocalcemia. In response to the low calcium level the parathyroid glands are hyperactive and patients exhibit excessively elevated PTH serum concentrations.

Acquired receptor defects are often caused by autoantibodies, e.g., against the receptor for insulin, TSH, acetylcholine, catecholamine, FSH, and transferrin (KAHN et al. 1976; CHIAUZZI et al. 1982; DRACHMANN 1978; LARRICK and HYMAN 1984; VENTER and FRASER 1980). Binding of the autoantibody to the receptor can destroy its integrity, causing insufficiency. If an antibody with so-called intrinsic activity is present (anti-idiotypic antibody) the receptor is overstimulated by antibody binding (MUTSCHLER 1986). Graves' disease is a

Table 5. Examples of disorders with abnormalities of cell receptors

Ligand	Type of receptor defect	Localization of the receptor defect	State of disease
Insulin	Deficiency or absence Autosomal recessive	All target cells	Insulin-resistant diabetes with acanthosis nigricans
	Disturbed affinity	Fat cells	Hereditary lipodystrophy
Parathyroid hormone	Insufficiency G protein Autosomal dominant or X-chromosomal	Bone cells Kidney cells	Pseudohypoparathyroidism (type I)
GH	Deficiency Autosomal recessive	All target cells	Pseudohyposomatotropism
	Augmentation	Tumor cells	Tumor promotion
Gonado-tropins	Deficiency, inherited	Leydig cells	Lack of testosterone with secondary feminization
	Deficiency, inherited	Granulosa cells	Primary amenorrhea
ADH	Insufficiency X-chromosomal	Distal renal tubule	Renal diabetes insipidus
ACTH	Receptor defect Autosomal recessive	Adrenal cortex	ACTH resistance
	Pathological stimulation of MSH receptor in ACTH hypersecretion	Melanocytes	Hyperpigmentation
TSH	Pathological hetero-stimulation by TSH receptor antibodies	Thyrocytes	Graves' disease
	Deficiency, inherited	Thyrocytes	Congenital hypothyroidism
T_3	Deficiency, insufficiency Autosomal recessive	Target cells	Hypothyroidism
Testosterone	Insufficiency inherited X-chromosomal	Target cells	Testicular feminization
	Hypersensitivity/ augmentation	Prostate (stroma, epithelial cells)	Benign hyperplasia
Acetylcholine	Auto-AB	Muscle cells	Myasthenia gravis
Progesterone	Deficiency, inherited	Endometrium	Pseudo corpus luteum insufficiency
$1,25\text{-}D_3$	Insufficiency, inherited	Osteoblasts	D_3-resistant rickets
GF	Augmentation	Tumor cells	Tumor promotion
LDLs	Deficiency, inherited	All target cells	Familial hypercholesterolemia
Transferrin	Auto-AB	All target cells	Hypochromic anemia

Abbreviations: auto-AB, autoantibodies; GH, growth hormone; ADH, antidiuretic hormone; ACTH, adrenocorticotropic hormone; MSH, melanocyte-stimulating hormone; TSH, thyrocyte-stimulating hormone; T_3, triiodthyronine; $1,25\text{-}D_3$, 1,25-dihydroxycholecalciferol; GF, growth factors (see text); LDLs, low density lipoproteins.

typical example, with anti-idiotypic antibodies inducing stimulation analogous to that produced by TSH (MEHDI and KRISS 1978).

3 Cytosolic/Nuclear Receptors

Cytosolic/nuclear receptors represent the recognition site for small ligands. They are localized predominantly in the nucleus and only to a limited extent in the cytosol. Most ligands are steroid hormones, such as male and female sex steroids, glucocorticoids, and cholecalciferols, as well as thyroid hormones. The mechanisms of cellular uptake and intracellular signal processing are outlined below.

3.1 Steroid Hormone Receptors

Steroid hormones are transported in the blood, mainly in a protein-linked form. The binding proteins include albumin and special testosterone-, estradiol-, androgen- or glucocorticosteroid-binding globulins, named steroid-binding globulins (SBGs). In the non-protein-linked form steroids are relatively small molecules which can easily enter the cell by diffusion or other (active) transport mechanisms not well understood (MOROZAVA et al. 1986).

3.1.1 Steroid Receptors and Mechanisms of Signal Transmission

Our current understanding of the sequence of events following the approach of steroids to a target cell has evolved from the "two-step-model" proposed by JENSEN et al. (1968) and GORSKI et al. (1968). Inside the cell steroids are bound to a cytosolic acceptor protein (Fig. 11, see L 2971 Fig. 1). Whether this binding is specific or whether the cytosolic acceptor proteins possess a general steroid-binding capacity has not yet been clarified. Once the binding has happened, the steroid-receptor complex is activated (transformation), as shown by an allosteric change of conformation of the coefficient of sedimentation from 4 S to 5 S. The activated complex moves to and accumulates in the nucleus (translocation) and binds to nuclear DNA acceptor sites. The exact nature of the binding sites is still under discussion. The proposals are: acidic proteins of the nonhistone class, specific DNA sequences, and the nuclear matrix. Apart from precise localization and chemical properties of the binding subunit, the steroid-receptor DNA interaction stimulates transcriptionally active regions of the DNA (BEATO 1987) which are upstream of the transcriptional starting sites in responsive genes.

By means of light microscopic immunocytochemistry with monoclonal antibodies (GREENE et al. 1980; ERIKSSON and GUSTAFSSON 1983), steroid receptors were localized almost exclusively in the nucleus (Fig. 12); only a small

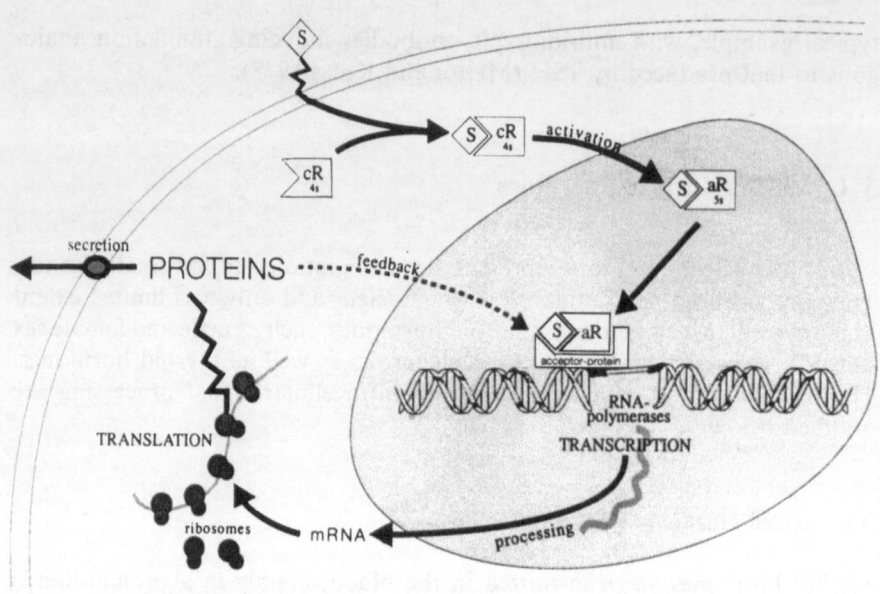

Fig. 11. The mechanisms of signal recognition and transmission and cellular events caused by binding steroid and thyroid hormones (for details see text). *S*, steroid hormone; *cR*, cytosolic receptor protein; *aR*, activated receptor protein

fraction was found to be present in the cytoplasm too. Electron microscopic immunocytochemistry with the protein A-gold detection technique (PERROT-APPLANAT et al. 1986) disclosed that the receptor protein was associated with the condensed chromatin of the nucleus and areas surrounding condensed chromatin. The nucleolus was not decorated. It is noteworthy that receptor distribution varied in response to steroid hormone application. When no or only a few steroid molecules were present, the receptor was randomly scattered over the condensed chromatin, but steroid administration induced receptor concentration at the border of condensed chromatin, the nucleoplasm, and dispersed euchromatin, which are known as the most active sites of gene transcription. The observed redistribution is compatible with hormone-activated DNA regions. Subsequently, the DNA cord is unwound, activating the polymerases I and II (GORSKI et al. 1968; JENSEN et al. 1982; BEATO 1987). The transcription of mRNA and subsequent translation are performed as in the case of protein synthesis.

The presence of low amounts of steroid receptor in the cytoplasm may account for some binding capacity and a biological function in the translational machinery (LIAO et al. 1983; THAMPAN 1985). Since some receptors are associated with membrane-bound and free ribosomes, a portion of the nonnuclear protein could represent newly synthesized receptor.

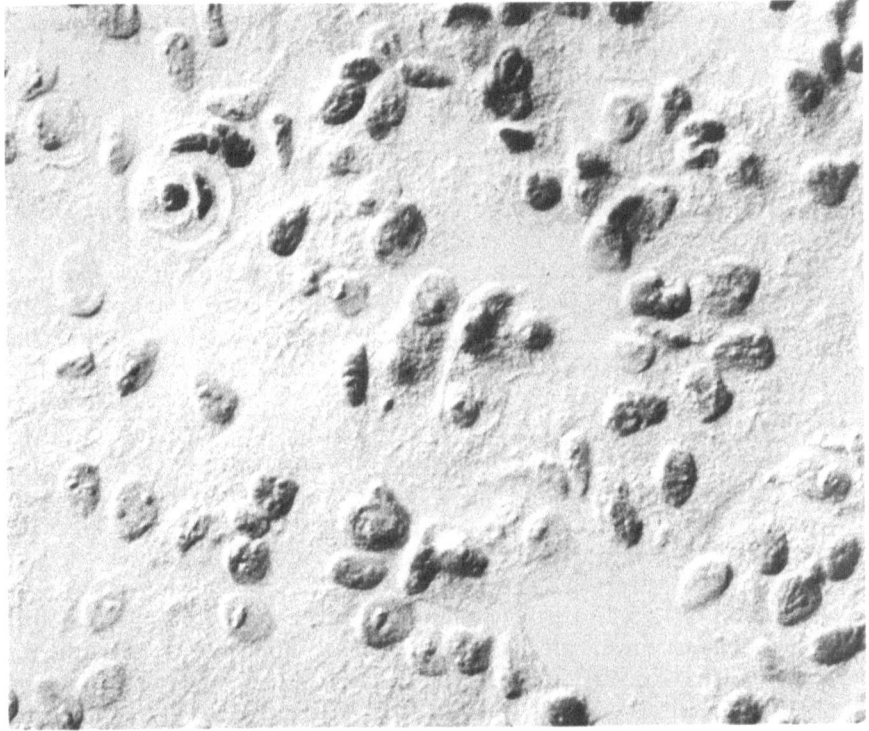

Fig. 12. Immunocytochemistry of steroid hormone receptor in the mammary carcinoma cell line MCF-7 M1 (see DIETEL et al. 1989) with exclusive localization in the nuclei. Although the cells have been cloned in earlier passages, the ER expression is heterogeneous, indicating genomic variability

3.1.2 Steroid Receptors and Growth Modulation

Receptor-mediated growth-enhancing effects on tumor cells have been investigated most intensively for the female sex steroid hormones. The stimulating activity of estrogen and progesterone in mammary and endometrial carcinomas and in benign and malignant hepatic neoplasms (NAGASUE et al. 1986; WANLESS and MEDLINE 1982) is generally accepted. In addition, other steroids have been suggested to play a role in modulating cellular growth activities and mitosis in other types of tissue: androgens in benign prostate hyperplasia (HABIB et al. 1986), prostate carcinomas (GELLER and ALBERT 1984; MARKLAND et al. 1978; NEUMANN et al. 1977), and both normal and malignant pancreatic tissue (CORBISHLEY et al. 1986), and glucocorticosteroids and 1,25 $(OH)_2$ vitamin D_3 (see below) in leukemia and several human malignancies. Recently, estrogen receptors have also been demonstrated in gastric cancer of the diffuse type (TOKUNAGA et al. 1986), in other gastrointestinal tumors (SICA et al. 1984), in several types of carcinoma (GREENWAY et al. 1981; KOBAYASHI et al. 1982), and in gynecological sarcomas (LANTTA et al. 1984). However, in these cases the growth-promoting potency of estrogen has not yet been evidenced.

The exact mechanism by which the binding of estrogen to its receptor triggers cell growth is unknown. The receptor activation induces synthesis of several gene products (see Sect. 3.1.1) which are involved in cell division. To obtain greater insight into this problem, GREEN et al. (1986) cloned and sequenced the complementary DNA (cDNA) of the estrogen receptor (ER) in the ER-positive breast cancer cell line MCF-7. Subsequently, it was possible to express the cDNA in HeLa cells and Chinese hamster ovary cells. The cDNA was found to code for a protein the same molecular mass and same estrogen-binding capacity as the ER of MCF-7 cells. It is of considerable interest that the cDNA sequence of the ER shares strong sequence homology with the v-*erb*A gene (GREEN et al. 1986) of the oncogenic AEV (see Sect. 2.11). Amino acid sequence comparisons revealed significant homology among the human ER, the human glucocorticoid receptors, and the v-*erb*A oncogene product. Although the *erb*A oncogene in itself is not carcinogenic, the *erb*A product may be involved in the receptor-mediated growth-promoting effect of estrogen in breast cancer.

The action of 1,25-dihydroxyvitamin D_3 is mediated by specific receptor. By immunohistochemistry (using a monoclonal antibody, PIKE 1984) the receptor protein has recently been visualized in normal kidney tissue (DIETEL and HARDERS in preparation). It was mainly located in the nuclei of the proximal tubule epithelium. Other studies found the 1,25-dihydroxyvitamin D_3 receptor protein in several tumor cell lines (EISMAN et al. 1983; NIENDORF et al. 1987 a). 1,25-Dihydroxyvitamin D_3 action on benign and malignant tumors and cell lines has been described for breast cancer (MCF-7 cell line), osteosarcoma, malignant melanoma, myeloid leukemia, colon carcinoma, parathyroid tumors, and gastrointestinal tumors (DIETEL et al. 1979; COLSTON et al. 1982; EISMAN et al. 1983; NIENDORF et al. 1987 a). The 1,25-dihydroxyvitamin D_3-dependent modulation of growth and differentiation of cultured tumor cells is accepted to be a receptor-mediated process and may therefore be used to influence tumor growth. However, cell culture experiments disclosed an uncertain reaction of tumor cells in culture, with both growth-inhibiting and growth-enhancing effects. It has to be concluded that in vitro 1,25-dihydroxyvitamin D_3 is not necessarily growth reducing and thus will not contribute to general antitumor strategies (NIENDORF et al. 1987 a).

3.1.3 Clinical Relevance of Steroid Receptor Determinations

Many excellent reviews have been published on the extraordinary clinical importance of estrogen and progesterone receptor determination in malignant mammary tumors (LIPPMAN 1978; WITTLIFF et al. 1977) and on the (less relevant) significance of androgen receptor determination in prostate cancer (MOBBS et al. 1980; TRACHTENBERG and WALSH 1982). Since a detailed description of the results would go beyond the scope of present paper, the discussion below is confined to the significance and problems of morphological receptor detection.

The development of a monoclonal antibody specific for the nuclear receptor protein of estrogen (GREENE et al. 1980) marked the starting point of a great number of studies on morphological determination of this receptor using the ER-immunocytochemical assay (ER-ICA). The ER antibody binds to a receptor domain not essential for steroid binding and thus it is possible to visualize ER independent of the hormonal status of the patient, the tumor tissue, or the cell preparation in vitro. Therefore, ER-ICA yields more information than the estrogen binding-dependent dextran-coated charcoal assay (DCC assay). Immunocytochemical receptor localization revealed a heterogeneous ER presence in breast cancer cells. Often not more than 50%–70% of the tumor cells express the receptor protein (Fig. 12), explaining the varying response to endocrine receptor-blocking treatment.

3.1.4 Steroid Receptors and Steroid Hormone Antagonists

To decrease the proliferation rate of steroid receptor-expressing tumors, steroid hormone antagonists are used. This mode of therapy is mainly applied in breast cancer, but it may also be helpful in other types of ER-expressing malignancies, such as cancer of the stomach, prostate, endometrium, and colon and several sarcomas (see above). For treatment of mammary carcinomas, among the antagonists employed are the antiestrogen tamoxifen (TAM) and more recently 3-hydroxytamoxifen (3-OH-TAM) (LÖSER et al. 1985; DIETEL et al. 1989), 4-OH-TAM (TERAKAWA et al. 1988), or others. They interfere (a) with estrogen-binding sites of the ER and (b) with ER binding sites at the specific DNA sequences (HORWITZ and McGUIRE 1978; ROCHEFORT and BOGNA 1981). Exact knowledge of the blocking action is lacking. One prerequisite for successful therapy with steroid antagonists is the presence of the appropriate receptor, which should be demonstrated both morphologically and biochemically. The dual evidence is important, since an antiproliferative effect of steroid antagonists on tumor cell growth depends (a) on the number of receptor molecules per tissue weight (determinable biochemically) and (b) on the number of receptor-positive cells per tumor specimen (determinable morphologically).

Immunohistochemical studies demonstrated that even tumors with a high amount of the receptor often contain receptor-negative cell clones (GREENE et al. 1980; KING and GREENE 1985). The heterogeneity can also be observed in ER-positive breast cancer cell lines, e.g., MCF-7 M1 (DIETEL et al. 1989). It is surprising that even after cloning of MCF-7 M1 cells it only takes several passages to again find cells with varying ER content (Fig. 12). It is not clear whether the different cells are mutants or whether they are only in different states of ER expression. Nonetheless, it seems to be a general process in tumor cells for cellular properties to be changed. This explains experimentally and clinically obtained data (Dietel et al. 1985b; Simon et al. 1984) proving lack of the expected inhibitory effect of steroid antagonists in steroid receptor-positive tumors and derived cell lines (see 3.1.5).

ER-positive malignant breast tumors primarily sensitive to endocrine therapy often develop resistance to antihormone application. The biochemical basis of this insensitivity is unknown. Experimental studies on antiestrogen-resistant ER-positive cells lines MCF-7^{R27} and LY 2 (Vignon et al. 1986; Bronzert et al. 1985, unpublished observation) showed that the nuclear ER protein remains present in the resistant cells. It is, however, unclear whether (a) the ER still has estrogen-binding activity, (b) the activation of ER-binding DNA sequences is altered, and (3) other transcriptional, translational, or posttranslational mechanisms are disturbed.

The interpretation of ER-ICA results is still a matter of controversy. There are many approaches to the "quantitation" of immunocytochemical staining using immunoreactive scores (Remmele et al. 1986). It is generally agreed that these scores have to include both reaction intensity and percentage of positive cells. The cut-off points between ER-positive and ER-negative tumors have to be defined in relation to the clinical outcome of antiestrogen therapy. Nonetheless, morphological ER determination has several advantages over biochemical procedures:

1. The tumor tissue to be removed surgically for diagnostic purpose can be very small (0.2–0.5 g) without influencing the result. This creates the opportunity to combine primary histological diagnosis and receptor analysis.
2. Even fine needle biopsies and cytological preparations can be used for receptor determination. This is especially relevant for the recognition of the receptor status of metastases. ER positivity of satellite tumors justifies endocrine therapy prior to other, more aggressive approaches.
3. Microtissue biopsies can be taken from several parts of the tumor and from metastases, giving an impression of the tumor's heterogeneity (see Sect. 3.1.4). This may be of importance in combined therapeutic approaches, e.g., antihormones in combination with cytostatics, specific antibodies, or other anticancer drugs.

As regards the clinical relevance of ER-ICA, there appears to be increasing evidence that morphological detection yields more relevant results than the biochemical DCC assay (Remmele et al. 1986). It has to be considered that in only 60%–70% of cases does the DCC assay correlate with the clinical response to endocrine therapy. Thus, a positive ER-ICA result should cause antiestrogen therapy to be initiated even if the DCC results indicate receptor negativity. The latter may be due to obscure hormonal status (pre- vs postmenopausal situation), to iatrogenic hormonal effects (often due to use of prolonged action drugs), or to a low number of tumor cells in comparison with the ER-negative stroma cells. Future investigations will reveal which methods show the best correlation with clinical follow-up.

Recently, endocrine treatment of prostate cancer has received increased attention (Markland et al. 1978; Geller and Albert 1984). Correlation of receptor status, tumor histology, grading and staging of the disease, and clinical follow-up is not evident and differs greatly from that of ER status and breast cancer characteristics. Whether the presence of androgen receptor reflects responsiveness of the tumor to antiandrogens and whether antihormone sensitivity is a general property of malignant prostate cells are still under investigation (TRACHTENBERG and WALSH 1982).

Clinical investigations exhibited analogous results in that long-term antiestrogen therapy (e.g., with TAM) often induced tumors with increasing resistance to endocrine measures. One reason may be the long half-life of TAM, which results in constantly elevated serum levels of the drug during clinical treatment. This in turn initiates down-regulation of receptor activity (see Sect. 2.4) with decreased responsiveness to TAM (HORWITZ and MCGUIRE 1978; NAWATA et al. 1981). This problem can be overcome by using 3-OH-TAM, which has a short half-life. Intermittent administration produces rises and falls in its serum concentration, which may slow down the process of acquiring resistance to antiestrogens (DIETEL et al. 1989).

3.1.5 Interaction of Steroid Receptors and Peptide Receptors

The complexity of receptor-mediated growth-controlling mechanisms is exemplified by the observation that application of progesterone to progesterone receptor (PgR)-positive breast cancer cell lines (MURPHY et al. 1986) induces an increase in the EGF receptor expression. EGF is a potent mitogen for many breast cell lines in vitro and presumably also for tumors in vivo (see Sects. 2.9, 2.10.1). Although the pathophysiological role of this interaction is still under discussion, modulation of the PgR with antiprogestins decreases the growth-promoting potency of EGF. A receptor-transmitted complementary influence of steroids and peptide hormones in growth promotion of tumors was reported by SIMON et al. (1984). In the breast carcinoma cell line EFM-19 prolactin stimulated cell growth. The cellular binding of prolactin was increased by physiological concentrations of 17β-estradiol or dihydrotestosterone, enhancing cell proliferation.

Autocrine growth stimulation in MCF-7 cells by an estrogen-regulated 52000-dalton glycoprotein was reported by VIGNON et al. (1986). Estrogen triggers synthesis of this protein through binding to the cytosolic ER. Subsequently, the protein is secreted and reinternalized through cell surface receptors.

TGF-β, a negative growth factor (see Sect. 2.10.3), inhibits cell proliferation of many epithelial cell lines (ARTEAGA et al. 1988), among them ER-negative breast cancer cells (KNABBE et al. 1987). TAM stimulates synthesis of TGF-β and presumably TGF-β receptors. The complex interacting mechanisms are exemplified in Fig. 13.

All the described interactions between steroid hormones, growth factors, and peptide hormones demonstrate a complex regulation of tumor cell growth that functions at the receptor level. Thus, agents directly interfering with

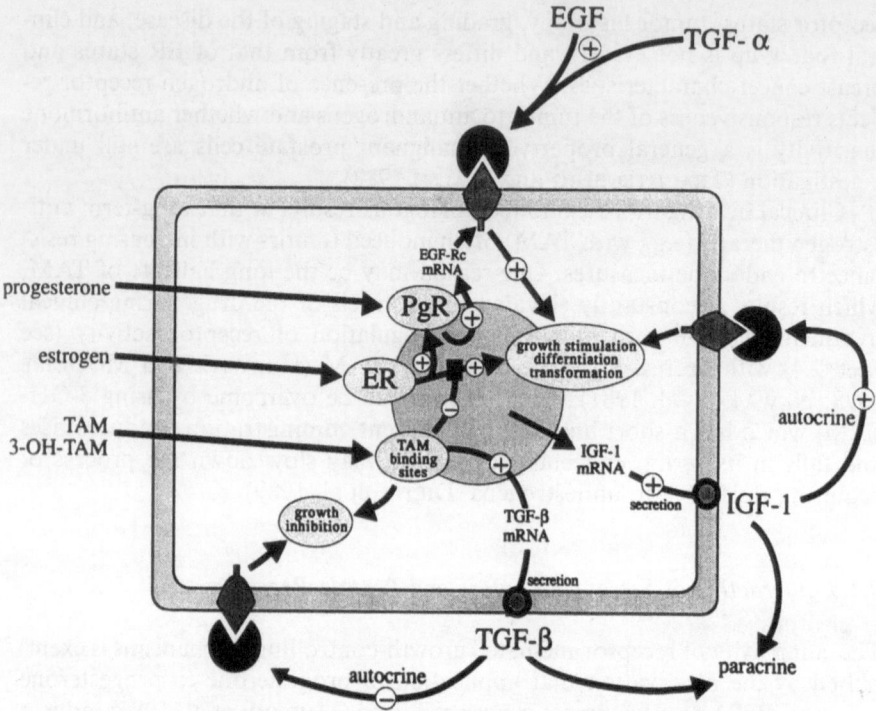

Fig. 13. Complex interactive network between steroid and peptide hormones and related receptors exemplified for the mammary carcinoma cell line MCF-7. Estrogen binding to ER stimulates PgR synthesis, which, after activation, enhances EGF receptor synthesis. Overexpressed EGF receptors are stimulated by EGF and TGF-α inducing cellular growth, differentiation, and/or transformation. During this process IGF-I synthesis, secretion, and receptor expression are elevated. TAM and 3-OH-TAM inhibit estrogen induced growth and in parallel stimulate production and secretion of the negative growth factor TGF-β- *PgR*, progesterone receptor; *ER*, estrogen receptor; *EGF*, epidermal growth factor; *TGF*, transforming growth factor; *IGF*, insulin-like growth factor; *TAM*, tamoxifen; +, stimulation; −, inhibition

receptors by modulating binding capacity, sensitivity, functional activity, etc. or antihormones preventing receptor activation may prove useful tools in cancer treatment.

3.2 Thyroid Hormone Receptors

Thyroid hormones are small molecules that enter the cell by diffusion, similar to steroid hormones. The first step in thyroid hormone action on target cells is the interaction of the hormone with the thyroid hormone receptor. Specific binding of T_3 has been found in subcellular fractions from cell membranes (PLIAM and GOLDFINE 1977), cytoplasm (BARSANO and DEGROOT 1983), mitochondria (STERLING et al. 1978), and nuclei (OPPENHEIMER 1979) of several

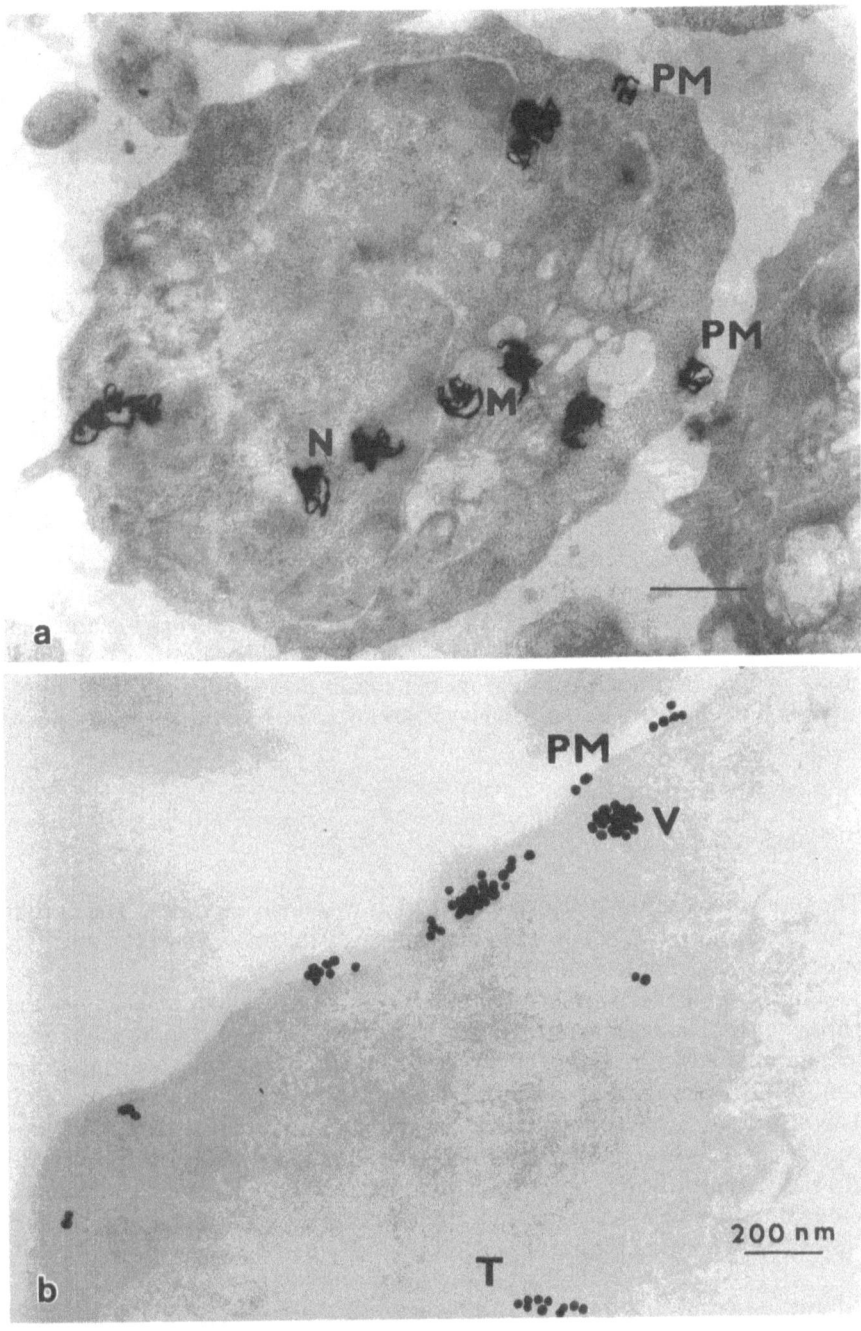

Fig. 14. a Granulocyte incubated with gold-labeled T_3 complexes, showing T_3 binding to the plasmalemma and intracellular organelle membranes. **b** The complex is internalized by vesicles (V) transported in connection with tubular structures (T). (KOSTROUCH et al. 1987)

types of tissue. Concerning signal transmitting functions, the predominant site of action is thought to be located (OPPENHEIMER et al. 1976) in the nucleus. Nuclear binding sites for the biologically active thyroid hormone T_3 have been located in nearly all cells of animal and human tissues (OPPENHEIMER et al. 1974). It is still not clear which proteins specifically bind T_3 and transfer the information to the DNA. Interaction of initiated T_3-receptor complex with DNA sequences induced activation of mRNA followed by multiple and in many respects undefined cellular responses. The differentiated effects of T_3 theoretically require several types of T_3 receptor to transmit the various metabolic effects, effect cellular differentiation, control balanced cell water and salt exchange, and so on. However, up to now the precise mechanism of T_3 and different receptor proteins has not been clarified.

To study T_3 binding sites living cells have been used, applying ultrastructural autoradiography with ^{125}I-labeled T_3 (KOSTROUCH et al. 1987). This technique has revealed binding of T_3 in many cellular compartments: mitochondria > plasma membrane > nucleus > nuclear envelope > other cytoplasmic structures. From competitive experiments it seems probable that the binding of T_3 to plasma membrane plays a major role in hormone internalization. This was visualized in target and producing cells using immunocytochemistry on ultrathin cryosections (Fig. 14) with colloidal gold stabilized T_3 complexes (GT_3A). Thus, thyroid hormone signal transmission does not function exclusively by cytosolic/nuclear receptors but rather also requires T_3 binding on plasma membranes and on other intracellular membranous structures of target cells.

4 Conclusions

The topic of receptor-mediated processes is extremely complex. The current paper gives only an overview of the main mechanisms of receptor functions and does not aspire to being exhaustive.

Precise understanding of receptor functions in cellular physiology and pathology is of increasing importance, since new therapeutic approaches interfering directly with receptor or postreceptor processes could become important. Proteins designed by molecular biologists may be of special importance. This is true for metabolic diseases, genetic abnormalities, several infections, and cancer. It is becoming increasingly clear that in the pathogenesis of many diseases a considerable number of different regulatory systems, previously thought to be rather independent of each other, are part of a complex network with extensive mutual influences.

Acknowledgements. The author thanks Gisela Broers for her excellent help in preparing the manuscript. The high quality technical work concerning own results was performed by Bjarne Lauritzen, Birgit Schaefer, Manuela Sieck, Nebahat Ayhan, and Maria Trapp. The work was supported by the Deutsche Forschungsgemeinschaft (SFB 232, Hamburg, Grant Di 276/1–3) and by the Hamburger Krebsgesellschaft, the Hamburger Stiftung zur Förderung der Krebsbekämpfung, and the Erich und Gertrud Roggenbuck Stiftung.

References

Arteaga CL, Tandon AK, von Hoff DD, Osborne CK (1988) Transforming growth factor β: potential autocrine growth inhibitor of estrogen receptor-negative human breast cancer cells. Cancer Res 48:3898–3904

Barsano CP, DeGroot LJ (1983) Nuclear-cytoplasmic interrelationships. In: Oppenheimer JH, Samuels HH (eds) Molecular basis of thyroid hormone action. Academic, New York, 139–178

Beato M (1987) Induction of transcription by steroid hormones. Biochim Biophys Acta 910:95–102

Beisiegel U, Schneider WJ, Goldstein JL, Anderson RGW, Brown MS (1981) Monoclonal antibodies to the low density lipoprotein receptors as probes for study of receptor-mediated endocytosis and the genetics of familial hypercholesterolemia. J Biol Chem 256:11923–11931

Berridge MJ (1981) Receptor and calcium signalling. In: Lamble JW (ed) Towards understanding of receptors. Elsevier/North Holland, Amsterdam, pp 122–131

Berridge MJ, Irvine RF (1984) Inositol triphosphate: a novel second messenger in cellular signal transduction. Nature 312:315–321

Bishop JM (1985) Viruses, genes and cancer. II. Retrovirus and cancer genes. Cancer 55:2329–2333

Boonstra J, van Mauric P, Defice LHK, de Laat SW, Lenissen JML, Verkleij AJ (1985) Visualization of epidermal growth factor receptor in cry-section of cultured A 431 cells by immuno-gold labeling. Eur J Cell Biol 36:209–216

Bradley G, Juranka PF, Ling V (1988) Mechanism of multidrug resistance. Biochim Biophys Acta 948:87–128

Brown MS, Anderson RGW, Goldstein JL (1983) Recycling receptors: the round trip itinerary of migrant membrane proteins. Cell 32:663–667

Brown MT, Clemmons DR (1986) Platelets contain a peptide inhibitor of endothelial cell replication and growth. Proc Natl Acad Sci USA 83:3321–3325

Carafoli E, Peneston JT (1986) The calcium signal. Sci Am (German edn) Jan 1:76–85

Carpenter G, Cohen S (1979) Epidermal growth factor. Ann Rev Biochem 48:193–216

Catt KJ, Harwood JP, Aguilera G, Dufau ML (1979) Hormonal regulation of peptide receptors and target cell responses. Nature 280:109–116

Chiauzzi V, Cigarraga S, Escobar ME, Rivarola MA, Charreau EH (1982) Inhibition of follicle-stimulating hormone receptors binding by circulating immunoglobulins. J Clin Endocrinol 54:1221–1227

Childs GV, Marchetti C, Brown AM (1987) Involvement of sodium channels and two types of calcium channels in the regulation of adrenocorticotropin release. Endocrinology 120:2059–2069

Clark JH, Guthrie S (1986) Subcellular localization of triphenylethylene antiestrogen binding sites (Tabs) in rat liver. J Steroid Biochem 25:635–639

Colston K, Colston MJ, Fieldsteel AH, Feldman D (1982) 1,25-Dihydroxyvitamin D3 receptors in human epithelial cancer cell lines. Cancer Res 42:856–859

Cooper JA, Hunter T (1983) Regulation of cell growth and transformation by tyrosine-specific protein kinase: the search for important cellular substrate proteins. Curr Top Microbiol Immunol 107:125–161

Corbishley TP, Iqbal J, Wilkinson ML, Williams R (1986) Androgen receptor in human normal and malignant pancreatic tissue and cell lines. Cancer 57:1992–1995

Courtoy PJ, Quintart J, Limet JN, De Roe C, Baudheim P (1985) Polymeric-IgA and galactose specific pathways in rat hepatocytes: evidence for intracellular ligand sorting. In: Pastan I, Willingham MC (eds) Endocytosis. Plenum, New York, pp 163–184

Cuttitta F, Camey DN, Mulshine J, Moody TW, Fedorko J, Fischler A, Minna JD (1985) Bombesin-like peptides can function as autocrine growth factors in human small-cell lung cancer. Nature 316:823–826

Dales S (1973) Early events in cell-animal virus interaction. Bacteriol Rev 37:103–135

Dietel M (1987a) Structure and function of cell receptors. In: Seifert G, Hübner K (eds) Pathology of cell receptors and tumor markers. Gustav Fischer, Stuttgart

Dietel M (1987b) What's new in receptor mediated growth promotion of normal and malignant cells? Pathol Res Pract 182:431–442

Dietel M, Dorn G, Montz R, Altenähr E (1979) Influence of vitamin D_3, 1,25-dihydroxyvitamin D_3, and 24,25-dihydroxyvitamin D_3 on parathyroid hormone secretion, adenosine 3',5'-monophosphate release, and ultrastructure of parathyroid glands in organ culture. Endocrinology 105:237-245

Dietel M, Arps H, Niendorf A (1985a) Visualization of PTH binding sites, CaBP, and tubulin in cell culture and intact tissue. In: Norman AW, Schäfer K (eds) Vitamin D. A chemical, biochemical and clinical update. Walter de Gruyter, Berlin, pp 1168-1169

Dietel M, Hölzel F, Arps H, Simon WE, Albrecht M (1985b) Mamma carcinomas in vitro: cytoskeleton, tumor markers, nuclear DNA, and rate of proliferation in relation to hormones and cytostatics. Verh Dtsch Ges Pathol 69:212-217

Dietel M, Löser R, Röhlke P et al. (1989) Effect of continuous vs. intermittent application of 3-OH-tamoxifen or tamoxifen on the proliferation of the human breast cancer cell line MCF-7M1. J Cancer Res Clin Oncol 115:36-40

Dietel M, Arps H, Lauritzen B, Niendorf A (1990) Membrane vesicle formation due to acquired mitoxantrone resistance in the human gastric carcinoma cell line EPG85-257. Cancer Res

Downward J, Yarden Y, Mayers E et al. (1984) Close similarity of epidermal growth factor receptor and v-erb-B oncogene protein sequences. Nature 307:521-527

Drachmann DM (1978) Myasthenia gravis. N Engl J Med 298:136

Draznin B, Sherman N, Sussman K, Dahl R, Vatter A (1985) Internalization and cellular processing of somatostatin in primary culture of rat anterior pituitary cells. Endocrinology 117:960-966

Dreyer M, Rüdiger HW (1986a) Erbliche Rezeptordefekte als Krankheitsursache. Dtsch Med Wochenschr 12:465-471

Dreyer M, Rüdiger HW (1986b) Erworbene Rezeptordefekte. Dtsch Med Wochenschr 11:427-433

Ehrlich P (1956-1960) In: Himmelweit F (ed) The collected papers of Paul Ehrlich, 3 vols. Pergamon, Oxford

Eisman JA, Frampton RJ, Sher E, Suva LJ, Martin TJ (1983) Presence and role of 1,25-dihydroxyvitamin D receptors in human cancer cells. In: Meyskens FL, Prasad KN (eds) Modulation and mediation of cancer by vitamins. Karger, Basel, pp 282-286

Eriksson H, Gustafsson JA (eds) (1983) Nobel symposium on steroid hormone receptors: structure and function. Elsevier Scientific, Amsterdam

Eveleth DD (1988) Nerve growth factor receptors: structure and function. In vitro Cellular Development Biol 24:1148-1153

Fanger BO, Wakefield LM, Sporn MB (1986) Structure and properties of the cellular receptor for transforming growth factor type β. Biochemistry 25:3083

Fitzpatrick SL, Brightwell J, Whittliff JL, Barrows GH, Schultz GS (1984) Epidermal growth factor binding by breast tumor biopsies and relationship to estrogen receptor and progestin receptor. Cancer Res 44:3448-3453

Geller J, Albert JD (1984) Antiandrogen and small dose of estrogen therapy as preferred treatment for prostate carcinoma. In: Hormones and cancer. Raven, New York

Gorski J, Toft D, Shyamala G, Smith D, Notides A (1968) Hormone receptors: studies on the interaction of estrogen with the uterus. Recent Prog Horm Res 24:45-80

Gottesman MM, Pastan I (1988) The multidrug transporter – a double-edged sword. J Biol Chem 263:12163

Green AR, Wyke JA (1985) Anti-oncogenes. A subset of regulatory genes involved in carcinogenesis. Lancet II:475-477

Green S, Walter P, Kumar V, Krust A, Bornert JM, Argos P, Chambon P (1986) Human oestrogen receptor cDNA: sequence, expression and homology to v-erb-A. Nature 320:134-139

Greene GL, Nolan C, Engler JP, Jensen EV (1980) Monoclonal antibodies to human estrogen receptor. Proc Natl Acad Sci USA 77:5115-5119

Greenway B, Iqbal MJ, Johnson PJ, Williams R (1981) Oestrogen receptor proteins in malignant and fetal pancreas. Br Med J 283:751-753

Grove RI, Pratt RM (1984) Influence of epidermal growth factor and cyclic AMP on growth and differentiation of palatal epithelial cells in culture. Dev Biol 106:427-437

Gullino PM (1981) Angiogenesis factor(s). In: Baserga R (ed) Tissue growth factors. Springer, Berlin Heidelberg New York (Handbook of Experimental Pharmacology, vol 57, pp 427-450)

Habib FK, Odoma S, Busuttil A, Chisholm GD (1986) Androgen receptors in cancer of the prostate. Correlation with the stage and grade of the tumor. Cancer 57:2351–2356

Hannover JA, Dickson RB (1985) Transferrin: receptor mediated endocytosis and iron delivery. In: Pastan I, Willingham MC (eds) Endocytosis. Plenum, New York, pp 131–162

Helmreich EJM (1986) Fortschritte der molekularen Endokrinologie. Klin Wochenschr 64:669–681

Hesch RD, Bodenstein H, Atkinson MJ (1984) Ectopic production and autocrine action of parathyroid hormone in acute leukemia. Serono Symp Publ 19

Horwitz KG, McGuire WL (1978) Nuclear mechanisms of estrogen action: effects of estradiol and antiestrogens on estrogen receptors and nuclear receptor processing. J Biol Cehm 253:8185–8191

Hunter T (1984) The proteins of oncogenes. Sci Am 251:70–79

Hunter T, Sefton BM (1980) Transforming gene product of Rous sarcoma virus phosphorylates tyrosine. Proc Natl Acad Sci USA 77:1311–1315

Hynes RO (1985) Molecular biology of fibronectin. Ann Rev Cell Biol 1:67–91

Hynes RO, Yamada KM (1982) Fibronectin: multifunctional modular glycoproteins. J Cell Biol 95:369–377

Jensen EV, Suzuki T, Kawashima T et al. (1968) A two-step mechanism for the interaction of estradiol with rat uterus. Proc Natl Acad Sci USA 59:632–639

Jensen EV, Greene GL, Closs LE, Desombre ER (1982) Receptors reconcidered; a 20-year perspective. In: Rec Prog Horm Res Academic Press, New York, London

Kahn CRJ, Flier JS, Bar S, Archer JA, Gorden P, Martin MM, Roth J (1976) The syndromes of insulin resistance and acanthosis nigricans. N Engl J Med 296:739–745

Kamata N, Chida K, Rikimaru K, Horikoshi M, Enomoto S, Kuroki T (1986) Growth-inhibitory effects of epidermal growth factor and overexpression of its receptors on human squamous cell carcinomas in culture. Cancer Res 46:1648–1653

Kartner N, Ling V (1989) Multidrug resistance in cancer. Sci Am March 1989 (German edn) p 26

Kasuga M, Karlsson FA, Kahn CR (1982) Insulin stimulates the phosphorylation of the 96000-dalton subunit of its own receptors. Science 215:185

King WJ, Greene GL (1985) Monoclonal antibodies localize oestrogen receptor in the nuclei of target cells. Nature 307:745–747

Knabbe C, Lippman ME, Wakefield LM, Flanders KC, Kasid A, Derynck R, Dickson RB (1987) Evidence that transforming growth factor-β is a hormonally regulated negative growth factor in human breast cancer cells. Cell 48:417–428

Kobayashi S, Mizuno T, Tobioka T (1982) Sex steroid receptors in diverse human tumors. Gann 73:439–445

Kostrouch Z, Felt V, Raska I, Nedvidková J, Holecková E (1987) Binding of (^{125}I) triiodothyronine to human peripheral leukocytes and its internalization. Experientia 42:1117–1118

Kyprianou N, Isaacs JT (1988) Identification of a cellular receptor for transforming growth factor-β in rat ventral prostate and its negative regulation by androgens. Endocrinology 123:2124–2131

Langley JN (1878) J Physiol 1:339–369

Lantta M, Kärkkäinen J, Wahlström T, Widholm O (1984) Estradiol and progesterone receptors in gynecologic sarcomas. Acta Obstet Gynecol Scand 63

Larrick JW, Hyman ES (1984) Acquired iron-deficiency anaemia caused by an antibody against the transferrin receptor. N Engl J Med 311:214–219

Levine A, McRae LJ, Hamilton DA, Brattain DE, Yeoman LC, Brattain MG (1985) Identification of endogenous inhibitory growth factors from a human colon carcinoma cell line. Cancer Res 45:2248–2254

Liao S, Chang C, Saltzman AG (1983) Androgen-receptor interaction – an overview. In: Eriksson H, Gustafsson IA (eds) Nobel symposium on steroid hormone receptors: structure and function. Elsevier Scientific, Amsterdam, pp 407–417

Liberman TA, Nusbaum HR, Razon N et al. (1985) Amplification, enhances expression and possible rearrangement of EGF receptor gene in primary human brain tumors of glial origin. Nature 313:144–147

Lippman ME (1978) Receptors in breast cancer. N Engl J Med 299:930–933

Löning T, Milde K (1987) Viral tumor markers. In: Seifert G (ed) Morphological tumor markers. Springer, Berlin Heidelberg New York (Current Topics in Pathology 77)

Löser R, Seibel K, Roos W, Eppenberger U (1985) In vivo and in vitro antiestrogenic action of 3-hypdroxytamoxifen, tamoxifen and 4-hydroxytamoxifen. Eur J Cancer Clin Oncol 21:985–990

March M (1984) The entry of enveloped viruses into cells by endocytosis. Biochem J 218:1–10

Markland FS, Chiopp RT, Cosgrove MD, Howard EB (1978) Characterization of steroid hormone receptors in the Dunning R-3327 rat prostatic adenocarcinoma. Cancer Res 38:2818–2826

Marrack P, Kappler J (1987) Der T-Zell-Rezeptor. In: Immunsystem – Abwehr und Selbsterkennung auf molekularem Niveau. Spektrum der Wissenschaft-Verlagsgesellschaft, Heidelberg, pp 98–108

McDouglas JS, Kennedy MS, Sligh JM, Cort SP, Mawle A, Nicholson JKA (1986) Binding of HTLV-III/LAV to T4$^+$ T cells by a complex of the 110K viral protein and the T4 molecule. Science 231:382–385

Mehdi SQ, Kriss JP (1978) Preparation of radiolabelled thyroid-stimulation immunoglobulins (TSH) by recombining TSH heavy chains with 1,25-I-labelled light chains. Direct evidence that the product binds to the membrane thyrotropin receptors and stimulates adenylate cyclase. Endocrinology 103:296–304

Mobbs BG, Johnson IE, Connolly JG (1980) The effect of therapy on the concentration and occupancy of androgen receptors in human prostatic cytosol. Prostate 1:37–51

Morozowa TM, Mitina RL, Nagibneva IN, Ozhogina ZB, Salganik RI (1986) The biochemical mechanisms of the stimulating action of estradiol on the growth of mammary tumours. Exp Clin Endocrinol 88:293–302

Moses HL, Tucker RF, Leof EB, Coffey JR, Halber J, Shipley GD (1985) Type beta transforming growth factor is a growth stimulator and a growth inhibitor. In: Feramisco J, Ozanne B, Stiles C (eds) Cancer cells, vol 3. Cold Spring Harbor Laboratory, Cold Spring Harbor, pp 65

Murphy LJ, Sutherland RL, Stead B, Murphy LC, Lazarus L (1986) Progestin regulation of epidermal growth factor receptor in human mammary carcinoma cells. Cancer Res 46:728–748

Musgrove E, Rugg C, Taylor I, Hedley D (1984) Transferrin receptor expression during exponential and plateau phase growth of human tumour cells in culture. J Cell Physiol 118:6–12

Mutschler E (1986) Arzneimittelwirkungen. Wissenschaftlicher Verlag (Berlin)

Nagasue N, Ito A, Yukaya H, Ogawa Y (1986) Estrogen receptors in hepatocellular carcinoma. In: Cancer 57:87–91

Nawata H, Bronzert D, Lippman ME (1981) Isolation and characterization of a tamoxifen-resistant cell line derived from MCF7 human breast cancer. J Biol Chem 256:5016–5021

Neal DE, Marsh C, Bennett MK, Abel PD, Hall RR, Sainsbury JRC, Harris AI (1985) Epidermal growth factor receptors in human bladder cancer: comparison of invasive and superficial tumors. Lancet II:366–368

Nemerow GR, Mold C, Schwend VK et al. (1987) Identification of gp350 as the viral glycoprotein mediating attachmet of Epstein-Barr virus (EBV) to the EBV/C3d receptor B cells: sequence homology of gp350 and C3 complement fragment C3d. J Virol 61:1416–1420

Neumann F, Graf KJ, Hason SH, Schenche B, Steinbeck H (1977) Central actions of antiandrogens. In: Martini C, Motta D (eds) Androgens and antiandrogens. Raven, New York, pp 163–171

Niendorf A, Dietel M, Arps H, Lloyd J, Childs GW (1986) Visualization of binding sites for bovine parathyroid hormone (PTH 1–84) on cultured kidney cells with a biotinyl-b-PTH (1–84) antagonist. J Cytochem Histochem 34:357–361

Niendorf A, Arps H, Dietel M (1987a) Effect of 1,25-dihydroxyvitamin D$_3$ on human cancer cells in vitro. J Steroid Biochem 27:825–828

Niendorf A, Arps H, Sieck M, Dietel M (1987b) Immunoreactivity of PTH-binding in intact bovine kidney tissue and cultured cortical kidney cells indicative for specific receptors. Acta Endocrinol [Suppl] (Copenh) 281:207–211

Niendorf A, Dietel M, Arps H, Childs GW (1987c) A novel method to demonstrate parathyroid hormone binding on unfixed living target cells in culture. J Cytochem Histochem 36:307–309

Oppenheimer JH (1979) Thyroid hormone action at the cellular level. Science 203:971–979

Oppenheimer JH, Schwartz HL, Surks MI (1974) Tissue differences in the concentrations of tri-iodothyronine nuclear binding sites in the rat liver, kidney, pituitary, heart, brain, spleen, and testes. Endocrinology 95:897

Oppenheimer JH, Schwartz HL, Surks MI, Koener D, Dillmann WH (1976) Nuclear receptors and initiation of thyroid hormone action. Recent Prog Horm Res 32:529

Pastan I, Willingham MC (1981) Journey to the center of the cell: role of the receptorsome. Science 214:504–509

Pastan I, Willingham MC (1985) Endocytosis. Plenum, New York, pp 1–40

Perrot-Applanat M, Groyer-Picard MT, Logeat F, Milgrom E (1986) Ultrastructural localization of the progesterone receptor by an immunogold method: effect of hormone administration. J Cell Biol 102:1191–1199

Pierschbacher MD, Ruoslahti E (1984) Cell attachment activity of fibronectin can be duplicated by small synthetic fragments of the molecule. Nature 309:30–33

Pike JW (1984) Monoclonal antibodies to chick intestinal receptors for 1,25-dihydroxyvitamin D_3. J Biol Chem 258:8554–8556

Pliam NB, Goldfine ID (1977) High affinity thyroid hormone sites on purified rat liver plasma membranes. Biochem Biophys Res Commun 79:166–173

Poiesz BJ, Ruscetti FW, Mier JW, Woods AM, Gallo RC (1980) T-cell lines established from human T-lymphocytic neoplasias by direct response to T-cell growth factors. Proc Natl Acad Sci USA 77:6815–6819

Rapp UR, Bonner TI, Moelling K, Jansen HW, Bister K, Ihle J (1985) Genes and gene products involved in growth regulation of tumor cells. In: Havemann K, Sorenson G, Gropp C (eds) Recent results in cancer research, vol 99. Springer, Berlin Heidelberg New York, pp 221–236

Remmele W, Hildebrand U, Hienz HA et al. (1986) Comparative histological, histochemical, im-munohistochemical and biochemical studies on oestrogen receptors, lectin receptors, and Barr bodies in human breast cancer. Virchows Arch [A] 409:127–147

Rochefort H, Bogna JL (1981) Differences between estrogen receptor and nuclear receptor process-ing. Nature 292:257–259

Rosenthal W, Schultz G (1988) Guaninnucleotid-bindende Proteine als membranäre Signaltrans-duktionskomponenten und Regulatoren enzymatischer Effektoren. Klin Wochenschr 66:511–523

Russell DW, Yamamoto T, Schneider WJ, Slaughter CJ, Brown MS, Goldstein JL (1983) cDNA cloning of the bovine low density lipoprotein receptor: feedback regulation of a receptor mRNA. Proc Natl Acad Sci USA 80:7501–7505

Sainsbury JRC, Farndon JR, Sherbet GV, Harris A (1985) Epidermal growth factor receptors and estrogen receptors in human breast cancer. Lancet II:364–366

Schröder S (1988) Pathologie und Klinik maligner Schilddrüsentumoren. Veröff Pathologie, vol 130. Gustav Fischer, Stuttgart

Schröder S, Müller-Gärtner HW, Schroiff R, Schmiegelow P, Niendorf A, Böcker W (1986) Mor-phological demonstration and quantification of TSH binding sites in neoplastic and non-neo-plastic thyroid tissues. Virchows Arch [A] 409:555–570

Shechter Y (1985) Studies on insulin receptors: implications for insulin action. In: The receptors II. Academic, New York, pp 221–244

Sherr CJ, Rettenmier CW, Sacca R, Roussel MF, Look AT, Stanley ER (1985) The c-*fms* proto-oncogene product is related to the receptor for the mononuclear phagocyte growth factor, SCF-1. Cell 41:665–676

Sica V, Nora E, Contieri E (1984) Estradiol and progesterone receptors in malignant gastrointesti-nal tumors. Cancer Res 44:4670–4674

Simon WE, Albrecht M, Tram G, Dietel M, Hölzel F (1984) In vitro growth promotion of human mammary carcinoma cells by steroid hormones, tamoxifen and prolactin. J Natl Cancer Inst 73:313–321

Smith KA (1988) Interleukin-2: Inception, impact, and implications. Science 240:1169–1176

Sporn MB, Roberts AB (1985) Autocrine growth factor and cancer. Nature 313:745–747

Sporn MB, Roberts AB, Wakefield LM, de Crombrugghe B (1987) Some recent advances in the chemistry and biology of transforming growth factor-beta. Cell 105:1039

Spurr NK, Solomon E, Janson M, Sheer D, Goddfellow PN, Bodmer WF, Vennstrom B (1984) Chromosomal localization of the human homologues to the oncogenes erbA and B. EMBO J 3:159–163

Sterling K, Lazarus JH, Milch PO, Sakurada T, Brenner MA (1978) Mitochondrial thyroid hormone receptor: localization and physiological significance. Science 201:1126–1129

Sugamura K, Fuji M, Ishii T, Hinuma Y (1986) Possible role of interleukin 2 receptor in oncogenesis of HTLV-I/ATLV. Cancer Rev 1:96–114

Tanner J, Weis J, Fearon D et al. (1987) Epstein-Barr virus gp350/220 binding to the B lymphocyte C3d receptor mediates adsorption, capping, and endocytosis. Cell 50:203–213

Teitelbaum AP, Strewler G (1984) Parathyroid hormone receptors coupled to cyclic adenosine monophosphate formation in an established renal cell line. Endocrinology 114:980–985

Terakawa N, Shimizu I, Ikegami H, Tanizawa O, Matsumoto K (1988) 4-Hydroxytamoxifen binds to estrogen receptors and inhibits the growth of human endometrial cancer cells in vitro. Cancer 61:1312–1315

Thampan RV (1985) The nuclear binding of estradiol stimulates ribonucleoprotein transport in the rat uterus. J Biol Chem 260:5420–5426

Thorley-Lawson DA (1988) Basic virological aspects of Epstein-Barr-virus infection. Semin Hematol 25:247–260

Tokunaga A, Nishi K, Matsukura N et al. (1986) Estrogen and progesterone receptors in gastric cancer. Cancer 57:1376–1379

Tolleshaug H, Hobgood KK, Brown MS, Goldstein JL (1984) The LDL receptor locus in familial hypercholesterolemia. Multiple mutations disrupt transport and processing of a membrane receptor. Cell 32:941–949

Trachtenberg J, Walsh PC (1982) Correlation of prostatic nuclear androgen receptor content with duration of response and survival following hormonal therapy in advanced therapy prostatic cancer. J Urol 127:466–471

Tucker RF, Branum EL, Shipley GD, Ryan RJ, Moses HL (1984) Specific binding to cultured cells of [^{125}I]labelled type β transforming growth factor from human platelets. Proc Natl Acad Sci USA 81:6757

Ullrich A, Coussens L, Hayflick JS et al. (1984) Human epidermal growth factor receptor cDNA sequence and aberrant expression of the amplified gene in A 431 epidermoid carcinoma cells. Nature 309:418–425

Ushiro H, Cohen S (1980) Identification of phosphotyrosine as a product of epidermal growth factor activated protein kinase in A 431 cell membranes. J Biol Chem 255:8263

Varmus HE (1985) Viruses, genes and cancer I. The discovery of cellular oncogenes and their role in neoplasia. Cancer 55:2324–2328

Venter JC, Fraser CM (1980) Autoantibodies to β$_2$-adrenergic receptors: a possible cause of adrenergic hyperresponsiveness in allergic rhinitis and asthma. Science 207:1361–1364

Vignon F, Capony F, Chambon M, Freiss G, Garcia M, Rochefort H (1986) Autocrine growth stimulation of the MCF-7 breast cancer cell by the estrogen-regulated 52K protein. Endocrinology 118:1537–1545

Wanless IR, Medline A (1982) Role of estrogens as promoters of hepatic neoplasia. Lab Invest 46:313–320

Willingham MC, Pastan IH, Sahagian GG, Jourdain GW, Neufeld EF (1981) A morphologic demonstration of the pathway of internalization of a lysosomal enzyme through the mannose-6-phosphate receptor in cultured CHO cells. Proc Natl Acad Sci USA 78:6967

Willingham MC, Pastan IH, Sahagian GG (1983) Ultrastructural immunocytochemical localization of the phosphomannosyl receptor in Chinese hamster ovary (CHO) cells. J Histochem Cytochem 31:1–11

Wittliff JL, Beatty BW, Baker DT, Savlov ED, Cooper RA (1977) Clinical significance of molecular forms of estrogen receptors in human breast cancer. Res Steroids 7:393–403

Yamamoto T, Kamata N, Kawano H et al. (1986) High incidence of amplification of the EGF receptor gene in human squamous carcinoma cell lines. Cancer Res 46:414–416

Molecular Biology of Receptors for Neuropeptide Hormones

D. Richter, W. Meyerhof, F. Buck, and S. D. Morley

1 Introduction – Neuropeptide Hormone Receptor Classes

The theoretical concept of cell membrane-associated molecules which act as specific "receptors" for ligands such as peptide hormones, neurotransmitters, and various growth factors is long established. However, it is only in the last 8–10 years that convincing evidence has emerged for the functional role of these receptor proteins in the transduction of external stimuli into intracellular signals.

In the case of neuropeptide hormone ligands, at least two major types of transduction system have been recognized (Fig. 1). One group of neuropeptide receptors shows structural and functional relationships to a class of transmembrane signaling molecules, termed the G protein-coupled receptors, which includes multiple receptors for catecholamines, serotonin, acetylcholine, light, and probably odorant substances. Neuropeptide ligand-binding representatives of this class include the substance K receptor (Masu et al. 1987) and the *mas* oncogene product (Young et al. 1986), which encodes a functional angiotensin receptor (Jackson et al. 1988). Such transmembrane signaling systems usually involve at least three distinct components which respond sequentially to the initial stimulus: the receptor itself, a specific guanine nucleotide

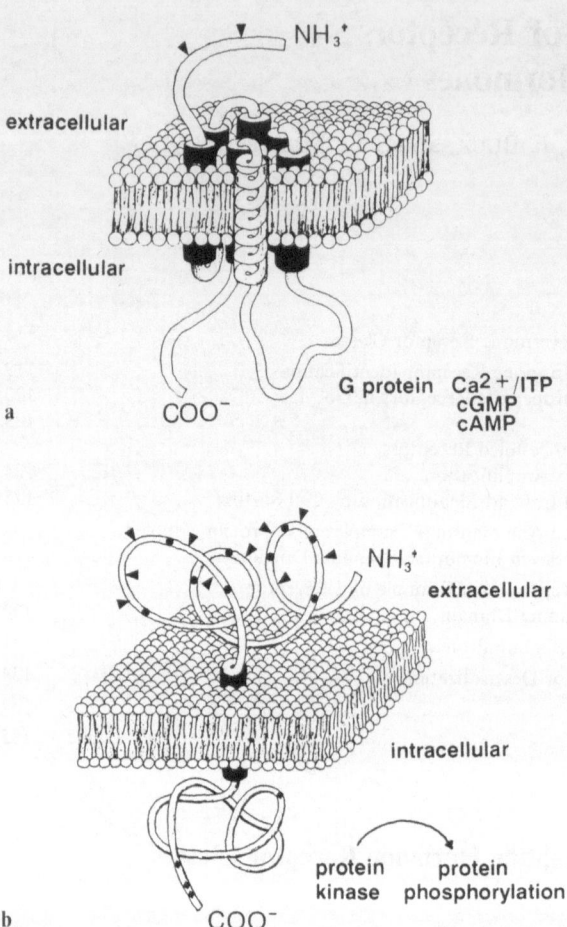

Fig. 1 a, b. Schematic representation of receptor superfamilies. **a** The G protein-coupled receptors are constituted by a single polypeptide chain which probably crosses the plasma membrane seven times, forming three intra- and three extracellular loops. In this diagram, one of the membrane-spanning domains is shown "cut-away" to reveal the predicted α-helical peptide structure of the receptor protein within the membrane. Potential sites for receptor protein glycosylation (*arrowheads*) and phosphorylation (*dots*) are shown. The second messenger systems to which each type of receptor is coupled are listed. **b** A receptor family, of which receptors for peptide growth factors, insulin, or the presently discussed ANP are typical, consists of one or more subunits, the polypeptide chain(s) of which spans the membrane lipid bilayer only once

regulatory protein or "G protein" (GILMAN 1987), and an effector protein. The receptor, consisting typically of a single polypeptide with seven membrane-spanning domains, interacts directly with the G protein located at the intracellular face of the membrane, which in turn modulates the activity of the specific effector protein. The latter may be an enzyme generating a second messenger molecule or protein(s) being capable of acting as ion channels or transporters (Ross 1989).

Recently, another class of neuropeptide receptors has been recognized that possess a single transmembrane-spanning domain; part of the cytoplasmic region of this receptor is an integral guanylate cyclase which is activated by atrial natriuretic peptide, thus regulating directly cyclic guanosine 3',5'-monophosphate (cGMP) second messenger production (LOWE et al. 1989). This class presently includes shared receptor subtypes for the homologous atrial and brain natriuretic peptides (see Sect. 4). Interestingly, these receptors possess significant structural similarities with a further group of receptors (e.g., for growth factors and insulin), the so-called protein tyrosine kinase receptors, which also contain single membrane-spanning domains but respond to ligand occupancy by activation of an integral protein tyrosine kinase (reviewed in YARDEN and ULLRICH 1988).

This review focuses primarily on G protein-coupled receptors for neuropeptide hormones and also the natriuretic peptide receptor family as an example of receptors coupled directly to cytoplasmic second messenger modulation. The term neuropeptide is used here in the general sense as referring to a class of bioactive peptides found in the central nervous system as well as in peripheral organs (KRIEGER 1983). Since little is known of the structure-function relationship of neuropeptide receptors, means by which such receptors may be functionally identified and analyzed and the use of such methods in molecular cloning are discussed. Two examples of previously cloned neuropeptide receptors – those for substance K and angiotensin – are compared and contrasted with other members of the G protein-coupled receptor family which respond to non-peptide ligands. The general mechanisms by which such receptors communicate, via G proteins, with the intracellular signaling pathways and the means by which such communication may be regulated are considered. Finally, some suggestions are presented as to how such a complex, multifunctional family of receptor proteins may have arisen during evolution.

2 Strategies for Neuropeptide Hormone Receptor Identification

The major difficulties associated with the study of neuropeptide hormone receptors have been their relative lack of abundance in the cell membrane and the technical problems associated with isolating them in a purified form. Recently a variety of innovative molecular approaches have enabled corresponding cDNA clones to be isolated without the need to embark on laborious protein purification procedures. The molecular cloning approach requires, however, that appropriate tools for identifying the cDNA encoding the respective receptor be available. A number of strategies addressing this problem have been developed which include functional expression of neuropeptide hormone receptors in frog oocytes, the polymerase chain reaction amplification of cDNAs encoding putative receptors, and expression of transfected receptor genes in cell cultures followed by identification using cell-sorting techniques.

2.1 Functional Expression of Neuropeptide Receptors in Oocytes

Xenopus oocytes have become a valuable tool for the analysis of neuropeptide receptors. Encouraged by the following list of observations, various investigators have attempted the functional identification in oocytes of neuropeptide receptors which, in their normal physiological environment, are thought to be linked to an elevation of the intracellular calcium concentration: The ability of oocytes

1. to translate efficiently exogenous mRNAs (GURDON et al. 1971)
2. to carry out a variety of posttranslational modifications (reviewed in COLMAN 1984)
3. to correctly assemble and implant complicated proteins, such as the heteropentameric nicotinic acetylcholine receptor, into their membranes (BARNARD et al. 1983)
4. the electrophysiological detection of membrane chloride channels activated by intracellular calcium (MILEDI 1982; BARISH 1983).

In order to approach the functional expression of receptors for certain neuropeptides, mRNA from receptor-containing tissues has been isolated and injected into *Xenopus* oocytes (RICHTER et al. 1988a, b). After an incubation period of 1–5 days to allow for expression of exogenous proteins, oocytes can be examined for ligand-induced responses employing various assays for the functional identification of neuropeptide receptors.

Electrophysiological Identification. Neuropeptide receptors implanted into the membranes of *Xenopus* oocytes can be identified by their ability to respond to ligand challenge by eliciting membrane current changes as recorded by the conventional two-electrode voltage-clamp technique (METHFESSEL et al. 1986). Thus in our hands many peptide receptors, including those for vasopressin, oxytocin, thyrotropin-releasing hormone, angiotensin, and bombesin, and more recently for vasotocin, isotocin, and bradykinin, have been functionally expressed and electrophysiologically detected in oocytes (MEYERHOF et al. 1988a; MORLEY et al. 1988; MEYERHOF and RICHTER 1989; Table 1). In all cases, voltage-clamped oocytes responded to ligand challenge in a specific and concentration-dependent manner by displaying rapidly inward chloride currents followed by slower oscillating inward currents. Importantly, responses could be reversibly suppressed by the introduction of appropriate antagonists into the oocyte bathing medium, demonstrating that at least some of the in vivo pharmacological properties of the various receptors are retained in the oocyte expression system. Most instructive is the case of vasopressin, which interacts with two pharmacologically distinct receptor types, namely a V_1 type present in liver and blood vessels and coupled to the inositol phosphate/calcium second messenger system (CREBA et al. 1983), and a V_2 type present in kidney and linked to cyclic adenosine 3',5'-monophosphate (cAMP) second messenger metabolism (FAHRENHOLZ et al. 1985). Injection of rat liver mRNA rendered oocytes responsive to vasopressin challenge, as demonstrated by the

Table 1. Neuropeptide receptor expressed in frog oocytes

Receptor	Source of mRNA	Remarks	References
V_1-vasopressin	Rat liver	Pharmacologically characterized	MEYERHOF et al. 1988a
Oxytocin	Bovine endometrium	Pharmacologically characterized	MORLEY et al. 1988
Vasotocin	White sucker hypothalamus	Inhibited by V_1-antagonist	MEYERHOF and MORLEY, unpublished
Isotocin	White sucker hypothalamus		MEYERHOF and MORLEY, unpublished
Bradykinin	White sucker hypothalamus		MEYERHOF and MORLEY, unpublished
TRH	Bovine pituitary gland, GH$_3$ pituitary cells	Accumulation of inositol phosphates reported; Inhibited by chlordiazepoxide $^{45}Ca^{2+}$ efflux, activates chloride conductance, PIP_2 cleavage; response dependent on internal but not external Ca^{2+}	MCINTOSH and CATT 1987; MEYERHOF et al. 1988a, ORON et al. 1987; MAHLMANN et al. 1989a; MAHLMANN et al. 1988b
	Rat brain		MEYERHOF et al. 1988b
Angiotensin	Bovine adrenal glands	Accumulation of inositol phosphates; Ca^{2+} mobilization	MCINTOSH and CATT 1987; SANDBERG et al. 1988
	Rat liver	Pharmacologically characterized	MEYERHOF et al. 1988a
	In vitro RNA	MAS oncogene product, A III > A II > A I	JACKSON et al. 1988a
CCK	Rat brain	CCK 1–8 used as agonist; chloride channel activation depends on internal, not external, Ca^{2+}	MORIARTY et al. 1988
	Pancreatic cell line	$^{45}CA^{2+}$ efflux measured	WILLIAMS et al. 1988
Bombesin	Rat brain	Chloride channel activity induced by intracellular, not extracellular, Ca^{2+}	MORIARTY et al. 1988
	Rat brain	Various peptides of the bombesin family tested; response inhibited by antagonists	MEYERHOF et al. 1988b
Substance P	Rat stomach, Rat brain, Rat small intestine	Homologous desensitization, chloride current, smooth current with unknown ionic composition; Activates chloride channels	HARADA et al. 1987; PARKER et al. 1986; AOSHIMA et al. 1987
Substance K	Rat stomach, Rat brain, In vitro RNA	Substance K > neuromedin K > substance P	HARADA et al. 1987; MASU et al. 1987
Neurotensin	Rat brain	Activation of chloride channels and of a smooth current of unknown ionic composition	PARKER et al. 1986; HIRONO et al. 1987
GnRH	Rat anterior pituitary gland	Identified with superactive agonist buserelin; response blocked by Mg^{2+}; response alternates with that of TRH	EIDNE et al. 1988

measurement of inward membrane currents, showing that functional vaso-pressin receptors were being expressed. Importantly, this response could be completely suppressed by a V_1 receptor-specific antagonist, lending weight to the idea that liver mRNA encodes V_1-type vasopressin receptors. In contrast, no signals were seen from oocytes injected with bovine kidney mRNA, consis-tent with the observation that receptors which are normally coupled to cAMP as second messenger are unable to activate the Ca^{2+} mobilizing pathway in *Xenopus* oocytes.

The fact that all receptors listed in Table 1 respond to ligand challenge with depolarizing responses in oocytes and that they, in their normal physiological environment, are coupled to an elevation of the intracellular calcium concen-tration via phosphoinositide hydrolysis, suggests that they activate a common transmembrane signaling pathway in oocytes. It seems plausible that this signal transduction pathway is, for the most part, identical to that used by the oocyte's endogenous muscarinic acetylcholine receptor (Kusano et al. 1977; Parker et al. 1987). The principal steps in this signal transduction cascade include the coupling of the expressed receptors to the G protein(s) of the oocytes that, when activated by the receptor-ligand complex, cause in turn the intracellular mobilization of endogenous calcium to membrane chloride chan-nels via the inositol 1,4,5-trisphosphate (IP_3) second messenger system (Ber-ridge and Irvine 1984; Oron et al. 1985; Mahlmann et al. 1989a; Parker and Miledi 1986; Gillo et al. 1987).

The above observations raise the question of the physiological role of this signal transduction pathway in oocytes. Obviously, the fate of an oocyte is to mature in an egg which awaits fertilization by a sperm. Recent evidence sug-gests that the inositol phosphate/calcium signal transmission pathway is used by the sperm in egg activation and that this process is mediated by a specific "sperm receptor protein" (Kline et al. 1988). Immature oocytes can be inject-ed with mRNA or in vitro RNA from cloned M1 muscarinic acetylcholine receptor for example, and then stimulated to develop to the mature egg stage by progesterone treatment. When such eggs are exposed to appropriate pep-tide ligands, occupancy of the exogenously added receptors results in mem-brane activation potentials indistinguishable from those mediated by sperm fusion and phenomena characteristic of the fertilization process and onset of development (Kline et al. 1988). It seems, therefore, that under these condi-tions exogenously added neurotransmitter receptors, by activating the endoge-nous inositol phosphate/calcium signaling pathway, can substitute for the sperm receptor in initiating the oocyte developmental program.

Alternative Methods. As an alternative to the electrophysiological method described above, receptors may be identified by the analysis of ligand-initiated phospholipid hydrolysis. In this approach oocytes prelabeled with tritiated myo-inositol are injected with receptor-encoding mRNA. The expressed re-ceptor (e.g., for angiotensin II) can then be identified by a ligand-dependent accumulation of radioactive inositol phosphates, determined in the presence of lithium ions, which inhibit the breakdown and metabolism of IP_3 (MCINTOSH and CATT 1987).

A further receptor identification method is based on the ability of the photoprotein aequorin to elicit calcium-dependent changes in light emission. Aequorin is coinjected with the receptor-encoding mRNA; a ligand-mediated increase in light emission detected by liquid scintillation counting indicates the presence of exogenous receptors (SANDBERG et al. 1988). Finally, the measurement of ligand-induced $^{45}Ca^{2+}$ efflux has been used for the identification of receptors for cholecystokinin (CCK), angiotensin II, and vasopressin (V_1 type) (WILLIAMS et al. 1988).

2.2 Expression Cloning

A fundamentally new cloning strategy has been developed involving the synthesis of in vitro mRNA from a cDNA library and its expression in oocytes (NOMA et al. 1986). Briefly, cDNA derived from receptor-encoding tissues is inserted into an expression vector containing a prokaryotic RNA polymerase promoter in front of the cloning site. In vitro synthesized RNA generated from pools of clones can then be injected into *Xenopus* oocytes and assayed for receptor expression under voltage-clamp conditions. Fractionation of positive clone pools should result in the isolation of a single clone encoding a peptide receptor. It should thus be possible to clone any receptor which is coupled to phosphoinositol turnover leading to intracellular mobilization of Ca^{2+} ions (RICHTER 1988). Indeed, using such a strategy, two receptors have been cloned to date, namely those for substance K (MASU et al. 1987) and the serotonin 1c receptor (JULIUS et al. 1988). The application of this approach may be limited by the relative concentration of the specific receptor-encoding mRNA within the tissue source and the efficiency with which it is reverse-transcribed into "full-size" cDNA. Other important parameters are likely to be mRNA stability in the oocyte, its translational efficiency, stability of the translated protein and the efficiency with which it is inserted into the plasma membrane, and finally the ability of a receptor protein to couple effectively to the endogenous signal transduction pathway of the oocyte. Further, the importance of such factors is likely to differ between particular receptors.

The oocyte system may in addition be used to identify putative receptor cDNA clones isolated by other means (RICHTER 1988), such being the case for the *mas* oncogene (JACKSON et al. 1988). The human *mas* oncogene was originally detected by its ability to render NIH 3T3 cells tumorigenic in nude mice (YOUNG et al. 1986). Transcripts of its rat homologue are confined to certain neuronal tissues and are not detected in other tissues (YOUNG et al. 1988). Its predicted amino acid sequence displays a putative seven transmembrane topology characteristic of the G protein-coupled receptors. When compared to this class of receptors, it became apparent that the highest degree of sequence relationship existed with the substance K receptor, suggesting that the *mas* gene product might also be a peptide receptor. In vitro synthesized RNA

rendered oocytes responsive to angiotensins, in the rank order AIII
> AII > AI, by showing dose-dependent oscillating membrane currents under
voltage-clamp conditions (JACKSON et al. 1988). Furthermore, a neuroblas-
toma-glioma hybrid cell line (NG 115-4OIL), when transiently transfected
with appropriate DNA constructs, also acquired angiotensin responsiveness.
These data suggest that the *mas* proto-oncogene product is a brain-specific
angiotensin receptor (YOUNG et al. 1988; JACKSON et al. 1988) and further
demonstrate that such receptor candidates can be identified in oocytes.

2.3 Cloning of Putative G Protein-Coupled Receptors by Polymerase Chain Reaction Amplification

The recently developed "polymerase chain reaction" (PCR) allows the rapid
synthesis of large amounts of a desired DNA sequence from a complex substrate, such as genomic or cDNA, in which the target sequence may be present
at very low concentrations (SAIKI et al. 1988; MARX 1988). This approach can
be applied to the cloning of neuropeptide receptors by exploiting the known
similarities between the family of G protein-coupled receptors to identify new
family members. Advantage is taken of the conserved nucleic acid sequences
corresponding to the third and sixth membrane-spanning domains to synthesize complementary PCR primers, which can also include restriction site recognition sequences at their 5' ends. Double-stranded cDNA derived from reverse
transcription of mRNA from a suitable receptor-encoding tissue is heat-denatured and the separated chains are allowed to anneal to the primers, which thus
define the ends of the sequence to be amplified. Synthesis of two new complementary chains is initiated by the thermostable "Taq" DNA polymerase, after
the completion of which, the reaction mixture is heat-denatured to provide a
doubled number of templates for further rounds of primer annealing and
extension. Thirty to fifty-five repetitions, which each cycle giving a doubling
of the DNA, yield a large amount of the sequence framed by the primers, while
molecules not containing primer-complementary sequences remain unampli-
fied. The PCR product is then restricted by the enzymes specified in the 5'
sequences of the PCR primers to yield "sticky" ends which facilitate direction-
al cloning into a suitable plasmid or phage vector for sequencing. Sequences
so obtained are then computer-translated and compared with the amino acid
sequences bounded by the third and sixth membrane-spanning domains of
known G protein-coupled receptors, in order to identify conserved sequences
and structures. Such a fragment of a putative new G protein-coupled receptor
can then be used as a probe to recover full-length clones from cDNA libraries.

A recent application of this approach (LIBERT et al. 1989) has yielded three
known receptors identified by comparison with existing cDNA sequences, thus
demonstrating the facility of the method, and in addition clones for four new
putative receptors. Such novel sequences can be further analyzed by pharmacological and expression studies in *Xenopus* oocytes and mammalian cell systems as described in Sect. 2.1, to complete their identification.

2.4 Receptor Cloning in Cell Cultures and Identification by Cell Sorting

As an alternative to cDNA cloning methods, a strategy based on gene purification methods is currently being developed for receptor cloning (Fig. 2). High molecular weight genomic DNA from an appropriate source is cotransfected with the plasmid ZipneoSV (CEPKO et al. 1984) carrying the gene for aminoglycoside phosphotransferase (APH) into NIH 3T3 cells. The APH gene mediates resistance to aminoglycoside antibiotics, such as G418, which inhibit protein synthesis in eukaryotes. Cotransfected cells which have taken up the biochemical marker as well as up to 1% of genomic donor DNA are selected and grown in the presence of G418.

The specific selection of cells expressing functional receptors is carried out using a cell sorter. This method is based on the measurement of receptor-regulated intracellular Ca^{2+} concentrations. Cells are loaded with the membrane-permeable Ca^{2+} chelator indo-1-acetoxymethylester (GRYNKIEWICZ et al. 1985). After intracellular ester hydrolysis, the dye is retained in the cell and emits, when excited with UV light, green (lambda max = 482 nm) and blue (lambda max = 398 nm) fluorescence light. A large shift in fluorescence emission maxima occurs upon chelation of Ca^{2+}. Indo-1-acetoxymethylester-loaded cells are exposed to Ca^{2+}-mobilizing ligands and analyzed using the cell sorter. Cells displaying a changed ratio of fluorescence emission are sorted out, subcultured, and analyzed again. When receptor-expressing cells have been enriched after several cycles of selection, high molecular weight DNA is prepared. Since in the host genome the biochemical marker is usually linked to the donor DNA (PERUCHO et al. 1980), the DNA isolated from primary cell lines can be transfected into host cells without necessitating cotransfection with ZipneoSV DNA. The entire protocol can then be conducted a second and a third time with the intention of isolating clonal cell lines expressing transfected receptor-encoding donor DNA. An alternative method for detecting peptide hormone receptor-expressing transfected clonal cell lines utilizes replica filter screening with iodinated receptor ligand of transfected cells grown on filter supports (MACHIDA et al. 1989). Finally, in both cases, genomic DNA libraries in phage lambda can be constructed from such receptor-expressing, clonal cell lines and analyzed using radioactive probes either for donor-specific repetitive DNA sequences or for the cotransfected plasmid DNA sequences. Identified clones can then be characterized by standard molecular biology techniques.

3 Receptors for Substance K and Angiotensin – Examples of G Protein-Coupled Neuropeptide Receptors with Seven Membrane-Spanning Domains

Prominent among the isolation of new members of the G protein-coupled receptor family has been the cloning of the receptors for neuropeptides of the tachykinin family, substance K (MASU 1987) and substance P (OHKUBO, per-

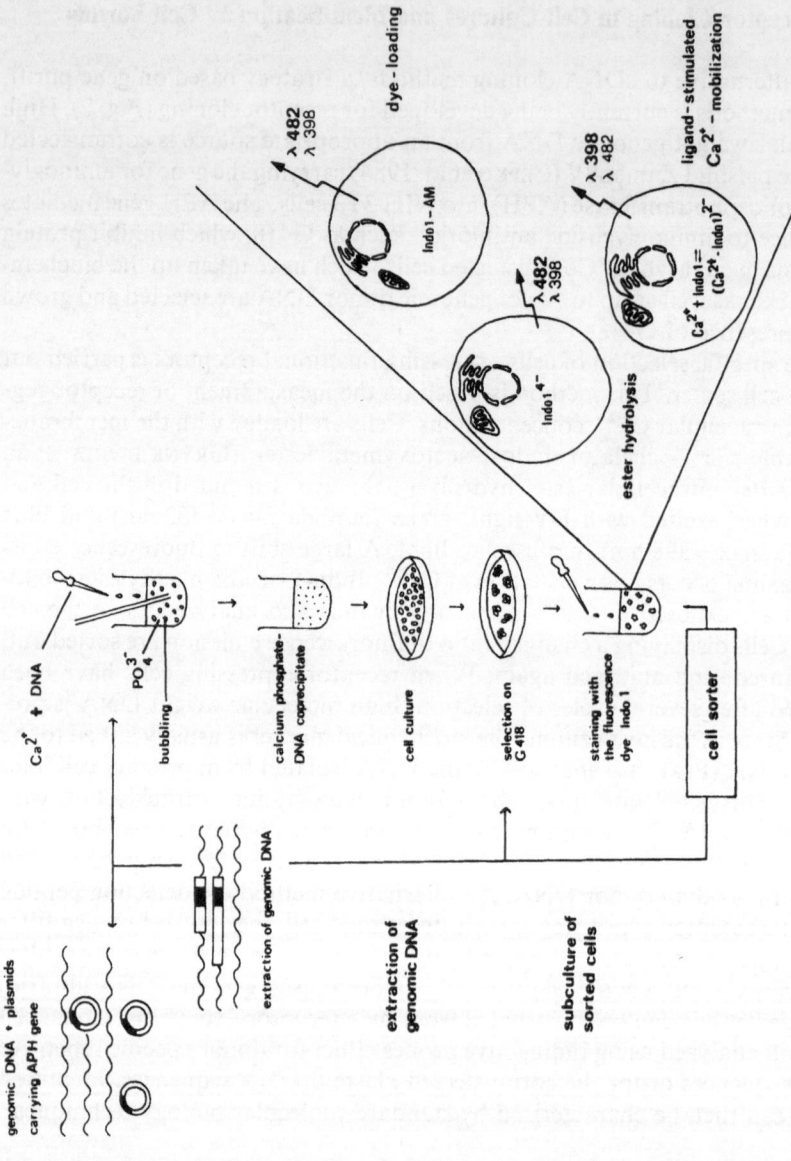

Fig. 2. Schematic representation of the cloning strategy based on receptor gene purification by DNA transfection and cell sorting

sonal communication), and the recognition that the *mas* oncogene (YOUNG et al. 1986) encodes a functional angiotensin receptor (JACKSON et al. 1988, and see Sect. 2.1). These successes have confirmed the widely held view that receptors for neuropeptide hormones would show not only functional but also structural homologies to rhodopsin (FINDLAY and PAPPIN 1986) and its related family of receptors (DOHLMAN et al. 1987). The key to this work was the fact that both of these receptor types stimulate phosphoinositide turnover and, indirectly, Ca^{2+}-dependent chloride channel opening, and their ability to be functionally expressed in oocytes of the frog *Xenopus laevis*.

Cloning of receptors for the neuropeptides substance K and angiotensin reveals that they have essentially similar structures to rhodopsin and the G protein-coupled neurotransmitter receptors. Perhaps the most striking feature of the deduced structures of all the G protein-coupled receptors is the presence of seven stretches of 20–28 predominantly hydrophobic amino acids which, by analogy with rhodopsin, are predicted to form seven transmembrane-spanning domains (DOHLMAN et al. 1987), pointing to an essentially conserved role, perhaps in ligand binding and the receptor signal transduction mechanism. The membrane-spanning domains are separated by segments of hydrophilic amino acids that form loops projecting either extra- or intracellularly (Fig. 1). The predicted G protein-coupled receptor sequences also closely resemble each other and rhodopsin in that they lack N-terminal signal sequences and contain potential N-linked glycosylation sites close to the N-terminus and multiple serines and threonines in the C-terminal cytoplasmic domain, which could be substrates for phosphorylation by cAMP-dependent protein kinase, by protein kinase C, or by receptor-specific kinases.

The *mas* oncogene/angiotensin receptor shows some sequence divergence from other members of the family (Fig. 3), including the substitution of some amino acids that are otherwise present in all other G protein-coupled receptors examined to date. However, the significance of these changes is hard to assess in the absence of further examples of neuropeptide receptors and since the receptor is obviously functional. Apart from this, the only respect in which the substance K and angiotensin receptors differ substantially from the majority of the G protein-coupled receptors is that they both possess very short third cytoplasmic "variable" loops between the putative fifth and sixth membrane-spanning domains, which results in their predicted amino acid sequences (384 and 325 amino acids respectively) being 100 or more residues shorter than those for the various muscarinic acetylcholine, catecholamine, and serotonin receptors (see for example JULIUS et al. 1988). In this respect, they most resemble rhodopsin itself, which has a sequence of 348 amino acids, the lack of an extensive third cytoplasmic variable domain presumably reflecting aspects of their interactions with specific ligands and in particular appropriate G proteins of the signal transduction pathway.

A survey of all the currently cloned G protein-coupled receptors suggests that those which inhibit the adenylate cyclase system tend to have larger third cytoplasmic domains than those which act to stimulate phosphoinositide turnover. However, this rule is not without exceptions, and the precise nature

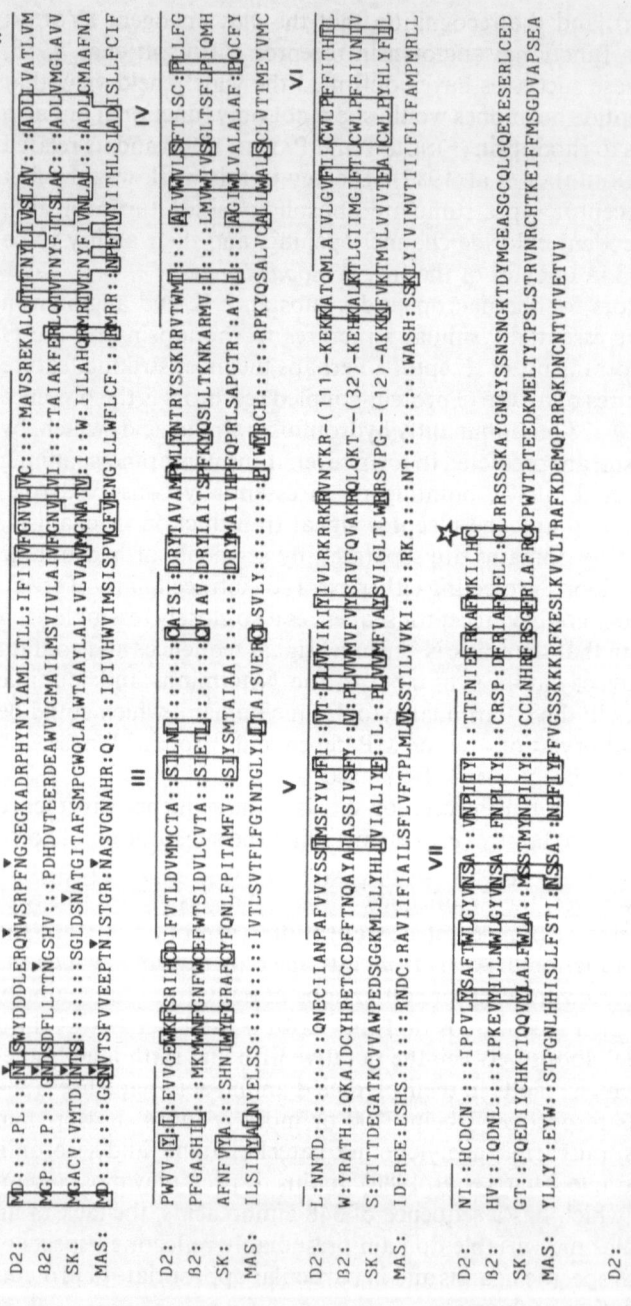

Fig. 3. Amino acid sequence comparison of G protein-coupled receptors. The amino acid sequences of dopamine (*D2*) (Bunzow et al. 1988), β₂-adrenergic (*β₂*) (Dixon et al. 1986), substance K (*SK*) (Masu et al. 1987), and angiotensin (*MAS*) (Young et al. 1986; Jackson et al. 1988) receptors are aligned to show maximal homology. *Boxed residues* are conserved in at least three of the sequences shown. *Triangles* denote potential N-linked glycosylation sites. The *star* marks cysteine residues which are conserved in many G protein-coupled receptors and have recently been demonstrated to be a site for palmitoylation in the human β₂-adrenergic receptor (O-Dowd et al. 1989), thereby attaching the polypeptide chain to the plasma membrane and forming a fourth intracellular loop. The seven membrane-spanning domains are indicated by *lines* and *Roman numerals*

of the G protein-binding domain of each receptor must reflect features not only of the signaling system to which the receptor is coupled, but particularly the specific G protein with which it interacts.

The mechanisms by which receptors of the G protein-coupled family interact with their respective ligands have been studied by constructing chimeric receptors by combining segments of different cloned receptor cDNAs and expressing them in the *Xenopus* oocyte system. Most of the available data have been obtained for the α_2- and β_2-adrenergic receptors (KOBILKA et al. 1988). Based on chimeras of the α_2- and β_2-adrenergic receptors, it has been shown that the specificity for coupling to the stimulatory G protein lies between the amino terminus of the fifth and the C-terminus of the sixth hydrophobic domain, a region that includes the "highly variable" third cytoplasmic loop. The same study revealed that the major determinants of the α_2- and β_2-adrenergic receptor agonist and antagonist ligand-binding specificity are contained within the seventh membrane-spanning domain (KOBILKA et al. 1988). Other work has indicated the importance for ligand binding of a conserved aspartate residue present within the second putative transmembrane domain (FINDLAY and PAPPIN 1986; JULIUS et al. 1988; and references therein). The results from the chimeric receptor experiments do not contradict these conclusions, but rather show that more than one part of the receptor molecule is involved in determining ligand binding and specificity. Thus replacement of the seventh hydrophobic domain of the β_2-adrenergic receptor with that from its α_2-couterpart results in a predominantly α_2-type ligand-binding specificity, although with certain small pharmacological differences presumably resulting from the heterologous environment.

Ligand-binding and signal transduction experiments in frog oocytes have recently been carried out using a chimeric receptor that consisted of the N-terminal part of the β_2-adrenergic receptor including the first and second transmembrane-spanning domain and the serotonin HT2 receptor moiety comprising the remaining five transmembrane domains and the intracellular C-terminus. The preliminary data suggest that binding of [^{3}H]-spiperone, a serotonin antagonist, is not impaired; in contrast ligand-induced membrane currents are inhibited. This may indicate either that binding sites for the ligand and antagonist are not identical and/or that the N-terminal part of the serotonin receptor is essential for triggering the signal transduction pathway (F. BUCK, unpublished data).

The utility of using such molecular approaches to dissect receptor function lies in the fact that, in contrast to standard mutagenic strategies where the endpoint is often loss of function, it enables the acquisition of function to be correlated with the introduction of specific protein sequences. Furthermore, it is tempting to extrapolate such conclusions concerning the G protein and ligand-binding specificity of the α_2- and β_2-adrenergic receptors to other receptors of this class. Clearly there will be no shortage of candidates for further analysis by the chimeric receptor approach as more cloned G protein-coupled receptor sequences become available.

4 Atrial Natriuretic Peptide Receptor – An Example of a Receptor with a Single Membrane-Spanning Domain

Atrial natriuretic peptide (ANP), a 24–28 amino acid endocrine hormone secreted primarily from the adult atrial myocytes, but also found in other peripheral tissues and brain, plays a central role in the regulation of diuresis, natriuresis, and vasorelaxation (FLYNN and DAVIES 1985; BAXTER et al. 1988). At least two apparent ANP receptor subtypes have been recognized in mammals (GARBERS 1989). These comprise a low molecular weight form (ANP-C receptor), perhaps responsible for the clearance of ANP from the circulation, and a high molecular weight receptor (ANP-A) which has been shown to be coupled to ANP-induced elevations of intracellular cGMP (reviewed in GARBERS 1989). Biochemical evidence as well as recent success in cloning and sequencing the ANP-A receptor (CHINKERS et al. 1989) has shown that the membrane-associated form of guanylate cyclase and the ANP receptor are one and the same molecule.

The cDNA for the mammalian guanylate cyclase ANP receptor was first isolated by cross-hybridization with a cDNA clone encoding the membrane-associated form of guanylate cyclase from sea urchin (SINGH et al. 1988), the homology at the cDNA level reflecting the known immunological cross-reactivity between these two proteins. Various biochemical data had already identified this sea urchin guanylate cyclase as a cell surface receptor on spermatozoa for the chemotactic peptide "Resact" (reviewed in GARBERS 1989). Both cDNAs predict proteins with cytoplasmic and extracellular domains linked by a single membrane-spanning domain, thus resembling the protein tyrosine kinase family of receptors in overall structure. Remarkably, this structural similarity is reflected at the amino acid sequence level, insofar as a region conserved in all protein kinases studied to date and thought to represent the ATP-binding domain is also conserved in the mammalian and sea urchin forms of membrane-associated guanylate cyclase, although its function here is not clear.

The membrane-associated guanylate cyclase differs crucially from the tyrosine kinase receptors and the G protein-coupled receptors, however, in that the binding of an extracellular agonist, ANP, is transduced directly into the formation of an intracellular second messenger, cGMP, without, on the one hand, the intervention of an enzyme cascade or, on the other, the intermediate of G proteins. As such the ANP-A receptor represents a new paradigm for second messenger signal transduction (LOWE et al. 1989) and may represent the archetypal member of a new diverse family of cell surface receptors for peptide ligands. In this respect, the membrane-associated ANP-A receptor and a cytoplasmic form of guanylate cyclase regulated by nitrovasodilators contain at least one homologous domain. Further, the ANP-C receptor for which a cDNA clone has recently been isolated (FULLER et al. 1988) and the ANP-A receptor share 34% amino acid homology in their extracellular domains.

Finally, the so-called brain natriuretic peptide (BNP) (SUDOH et al. 1988), which shows a similar tissue distribution and spectrum of activities to ANP

(MINAMINO et al. 1988), shows considerable cross-reactivity with the guanylate cyclase ANP-A receptor, but also binds preferentially to a distinct receptor type within the brain and peripheral tissues, the so-called ANP-B receptor (CHANG et al. 1989). Nevertheless, the close functional analogy between these two peptides and the structural and functional homologies between the two receptor subtypes suggest that the receptors for BNP fall into the same general class as those for ANP.

5 Signal Transduction by GTP-Binding Proteins

Many receptors for neurotransmitters are coupled by a GTP-binding protein (G protein) to the enzymatic system producing the second messenger (ROSS 1989). The introduction of mediators (the G proteins) between receptor and enzyme allows for additional signal attenuation, while slowing down the signal response. So far, two major second messenger signaling systems are known: (a) cAMP generated by the action of a G protein-dependent adenylate cyclase and (b) a system triggered by the G protein-dependent activation of phospholipase C, resulting in the generation of IP_3 and diacylglycerol from phosphatidyl inositol-4,5-diphosphate (PIP_2) and a subsequent rise in the intracellular Ca^{2+} concentration (BERRIDGE 1985, 1987). Most of the information on signal transduction via G proteins is derived from studies of the acetylcholine and catecholamine receptors, but presently available evidence points to a similar coupling via G proteins of peptide hormone receptors to their respective second messenger signaling systems.

In recent years, a rapidly growing number of GTP-binding proteins have been isolated which, on the basis of certain common features, can be considered as members of a family of "G proteins" responsible for receptor-effector coupling (GILMAN 1987; LOCHRIE and SIMON 1988). G proteins have been found to be involved in the stimulatory (G_s) or inhibitory (G_i) coupling of receptors to adenylate cyclase, to the cGMP phosphodiesterase (transducin), to phospholipase C, and to ion channels (ROSS 1989). In addition, a number of G proteins have recently been described for which a function remains to be found (STERNWEIS and ROBISHAW 1984; KATADA et al. 1987; DICKEY et al. 1987; IYENGART et al. 1987; WALDO et al. 1987; FONG et al. 1988; BOKOCH et al. 1988). All G proteins isolated thus far are heterotrimers, which differ in their large subunit (α), while the smaller β- and γ-subunits of different G proteins are highly homologous (LOCHRIE and SIMON 1988) and have often been shown to be functionally interchangeable.

The first members of the G protein family to be identified, which are now most thoroughly characterized both structurally and functionally, were the G proteins responsible for the coupling of receptors to adenylate cyclase (NORTHUP 1985). Adenylate cyclase-stimulating (G_s) and inhibiting (G_i) G proteins can be distinguished. In terms of G_s receptor specificity, guanosine triphosphate (GTP) binding, GTPase activity, and interaction with adenylate

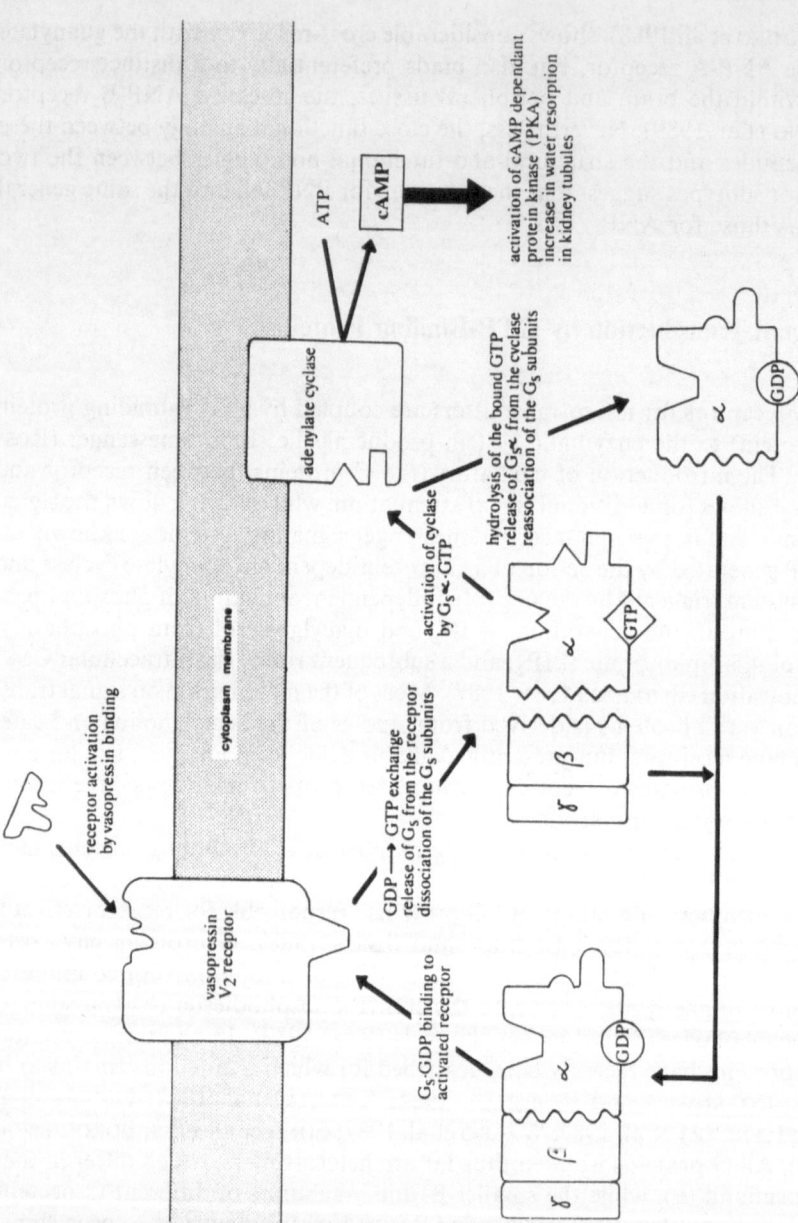

Fig. 4. G protein-mediated signal transduction: vasopressin-induced stimulation of cAMP formation in kidney tubules

cyclase all have been demonstrated to be localized within the α-subunit, while the hydrophobic βγ subunits, in addition to anchoring the G protein to the membrane, seem to have a regulatory function (CASEY and GILMAN 1988).

The mechanism of signal transduction by G_s is schematically shown in Fig. 4 for the kidney vasopressin receptor system. The free G protein contains tightly bound guanosine diphosphate (GDP), which is released upon binding to the vasopressin-activated receptor. The receptor-G protein complex is relatively stable in the absence of GTP. In the presence of GTP, however, GTP is bound to the α-subunit, resulting in the dissociation of the GTP-G protein complex from the receptor and of the $G_α$-subunit from the $G_{βγ}$ subunits; the $G_{βγ}$ subunits are tightly associated and cannot be separated from each other under nondenaturating conditions. In the case of G_s, the GTP-$G_α$ complex is then able to bind to and activate the adenylate cyclase. It was therefore formerly often referred to as the "regulatory subunit" of the adenylate cyclase. Deactivation of the adenylate cyclase is achieved by hydrolysis of the bound GTP through the intrinsic GTPase activity of $G_α$. The GDP-$G_α$ complex dissociates from the adenylate cyclase and reassociates with the $G_{βγ}$-subunits to form the free GDP-$G_{αβγ}$ protein complex (CASEY and GILMAN 1988).

In contrast to the V_2 vasopressin receptor, $α_2$-adrenergic, opiate, and M2 muscarinic receptors all mediate inhibition of adenylyl cyclase via G_i (inhibitory) protein coupling (ROSS 1989). Briefly, receptor activation and G protein binding causes βγ subunit release from G_i and inhibition of adenylyl cyclase by chelation of $G_{αs}$ (stimulatory)-subunits which would otherwise be bound to the active enzyme. The various steps in this mechanism are accompanied by GTP hydrolysis as described above. A complex interplay between the various G proteins of the adenylyl cyclase system apparently ensures that hormonal inhibition of adenylyl cyclase is the only such inhibition commonly observed (ROSS 1989).

Less well understood is the process by which receptors are linked to phospholipase C, which generates IP_3 and diacylglycerol from PIP_2, resulting in the release of Ca^{2+} from intracellular stores such as the endoplasmic reticulum. The structural homology between the receptors acting on adenylate cyclase and those activating phospholipase C, as well as the GTP requirement of the hormone-stimulated phospholipase C activation (LITOSCH et al. 1985) and the hormone-independent stimulation by $GTP_γS$ and GDP(NH)P (COCKCROFT and GOMPERTS 1985), clearly indicates the involvement of G proteins (CASEY and GILMAN 1988). However, the putative G protein responsible for the signal transduction from the respective receptors to the phospholipase C has so far not been identified, even though the isolation of an intact V_1 vasopressin receptor-G protein complex has already been described (FITZGERALD et al. 1986). Studies of the influence of cholera and pertussis toxin treatment on the receptor-phospholipase C coupling have indicated the involvement of different G proteins in different cell types. In certain cell types, phospholipase C activation by hormones known to stimulate phosphoinositide turnover in these particular cells was inhibited by treatment with either cholera and/or pertussis toxin, while in other cells phospholipase C activation was unaffected by toxin treatment (LO and HUGHES 1987; GIERSCHIK and JACOBS 1987; BIFFEN and MARTIN 1987).

6 Signal Attenuation by Receptor Desensitization

The signaling pathways described above cannot be considered as independent and unidirectional. Rather, they are part of an intricate regulatory network with numerous cross-connectivities and feedback controls. The most obvious of these is receptor "desensitization," i.e., the loss of signal intensity upon prolonged exposure of the cell to a particular hormone. Most of the knowledge on the mechanisms of desensitization is derived from experiments with rhodopsin, adrenergic, and acetylcholine receptors, but it seems probable that receptor desensitization is achieved by similar mechanisms in all G protein-coupled receptors, including those for neuropeptide hormones. The effect is most probably mediated by the phosphorylation of the receptor, by the action either of a protein kinase specific for the particular systems (homologous desensitization) or of a kinase with a broader substrate specificity which is not necessarily activated by the particular system (heterologous desensitization) (SIBLEY et al. 1985, 1987; ATTRAMADAL et al. 1988). Potential phosphorylation sites have been found clustered in the C-terminal part of all G protein-coupled receptors, and in the case of adrenergic receptors, modification of these sites by site-directed mutagenesis or the deletion of C-terminal sequences results in receptors which are no longer susceptible to desensitization (LEFKOWITZ and CARON 1988).

The β-adrenergic receptors have been shown directly to undergo homologous as well as heterologous desensitization by phosphorylation within their C-terminal regions. The homologous desensitization is due to the action of a specific β-adrenergic receptor kinase, which is not cAMP dependent, but preferentially phosphorylates β-adrenergic receptors in their ligand-occupied state (BENOVIC et al. 1986). Heterologous phosphorylation leading to desensitization of adrenergic receptors is observed with both cAMP-dependent kinase (BENOVIC et al. 1985) and protein kinase C (BOUVIER et al. 1987).

7 Genes and Evolution

A most remarkable feature of the presumed seven membrane-spanning topology of the G protein-coupled receptor family is the stability of this structure throughout evolution, even in members of this class that show little or no significant amino acid homology. Thus such an organization of α-helices within the cell membrane has been inferred not only for neurotransmitter and neuropeptide receptors and the light-sensitive pigment rhodopsin mentioned previously, but also for the enzyme HMG-CoA reductase, the α and a mating factor receptors of *Saccharomyces cerevisiae,* and bacteriorhodopsin, the purple membrane protein of *Halobacterium halobium* (reviewed in DOHLMAN et al. 1987).

Presumably this particular membrane organization confers special properties in terms of ligand binding and transmission of an extracellular signal across the plasma membrane, by means of conformational change, to the coupling proteins on the cytoplasmic surface. It seems probable that once such a mechanisms for generalized signal transduction to the intracellular space has evolved, it will be conserved with only minor alterations subsequently to accommodate different ligand and second messenger coupling specificities. Indeed, the close resemblance between family members argues for a common origin in the same ancestral gene. This might have been assembled, in the current view, from DNA segments with particular functions from preexisting genes, these segments being separated by noncoding introns in the new gene (GILBERT 1985). Certainly, the positions of introns in the genes encoding rhodopsin and the more recently cloned 5HT1c serotonin, dopamine, and substance K receptors are consistent with this notion, although it should be noted that the splicing pattern of the former differs from the latter three (MAELICKE 1989). It was something of a surprise, therefore, when cloning studies revealed that genes for such as the various muscarinic acetylcholine and catecholamine receptors and also the G-21 gene and the *mas* oncogene, which encode respectively a serotonin 5HT-IA subtype (FARGIN et al. 1988) and an angiotensin receptor (JACKSON et al. 1988), all lacked introns in their coding regions.

Various mechanisms exist whereby genes may either gain or lose sequence information. However, the finding that the gene for the β_2-adrenergic receptor is flanked by direct repeats (KOBILKA et al. 1987) suggests that it might have originated as a consequence of a reverse transcription event whereby a DNA copy of a mature spliced receptor-encoding mRNA was inserted back into the genome to yield a new intronless gene. It seems improbable that each intronless receptor gene could have originated from independent reverse transcription events, so they could all be descended from a single ancestral intronless gene.

Thus a picture begins to emerge whereby the ancestral G protein-coupled receptor gene might have arisen by exon shuffling, after which many variants were produced by simple modification of the basic functional structure. At some point, the lines leading to the rhodopsins and somewhat later to the intronless receptor genes would have branched off to produce discrete subfamilies of receptors. This model predicts that individual receptors within a subfamily should show greater similarity to each other than to members of the other subfamilies. Indeed, this appears to be the case for the rhodopsin, muscarinic acetylcholine, and catecholamine receptors. However, there are both intronless and intron-containing genes encoding serotonin receptor subtypes. Clearly, a good deal more information is required before one will be able to draw any firm conclusions about the evolutionary history of this gene family.

References

Aoshima H, Iio H, Anan M, Ishii H, Kobayashi S (1987) Induction of muscarinic acetylcholine, serotonin and substance P receptors in *Xenopus* oocytes injected with mRNA prepared from the small intestine of rats. Mol Brain Res 2:15–20

Attramadal H, Eikvar L, Hansson V (1988) Mechanisms of glucagon-induced homologous and heterologous desensitization of adenylate cyclase in membranes and whole Sertoli cells of the rat. Endocrinology 123:1060–1068

Barish ME (1983) A transient calcium-dependent chloride current in the immature *Xenopus* oocyte. J Physiol 342:309–325

Barnard EA, Beeson D, Bilbe G et al. (1983) Acetylcholine and GABA receptors: subunits of central and peripheral receptors and their encoding nucleic acids. CSH Symp on Quant Biology, The Cold Spring Harbor Laboratory, Cold Spring Harbor, vol XLVIII, pp 109–124

Baxter JD, Lewicki JA, Gardner DG (1988) Atrial natriuretic peptide. Biotechnology 6:529–546

Benovic JL, Pike LJ, Cerione RA et al. (1985) Phosphorylation of the mammalian β-adrenergic receptor by cyclic AMP-dependent protein kinase. J Biol Chem 260:7094–7101

Benovic JL, Strasser RH, Caron MG, Lefkowitz RJ (1986) β-adrenergic receptor kinase: identification of a novel protein kinase that phosphorylates the agonist-occupied form of the receptor. Proc Natl Acad Sci USA 83:2797–2801

Berridge MJ (1985) Inositol trisphosphate and diacylglycerol as intracellular second messengers. In: Poste G, Crooke ST (eds) Mechanisms of receptor regulation. Plenum Press, New York, pp 111–130

Berridge MJ (1987) Inositol trisphosphate and diacylglycerol: two interacting second messengers. Annu Rev Biochem 56:159–193

Berridge MJ, Irvine RF (1984) Inositol trisphosphate, a novel second messenger in cellular signal transduction. Nature 312:315–321

Biffen M, Martin BR (1987) Polyphosphoinositide labeling in rat liver plasma membranes is reduced by preincubation with cholera toxin. J Biol Chem 262:7744–7750

Bokoch GM, Parkos CA, Mumby SM (1988) Purification and characterization of the 22000-dalton GTP-binding protein substrate for ADP-ribosylation by botulinum toxin, G_{22K}. J Biol Chem 263:16744–16749

Bouvier M, Leeb-Lundberg LMF, Benovic JL, Caron MG, Lefkowitz RJ (1987) Regulation of adrenergic receptor function by phosphorylation. J Biol Chem 262:3106–3113

Bunzow JR, van Tol HHM, Grandy DK, Albert P, Salon J, Christie M, Machida CA, Neve KA, Civelli O (1988) Cloning, expression of a rat D2 dopamine receptor cDNA. Nature 336:783–787

Casey PJ, Gilman AG (1988) G protein involvement in receptor-effector coupling. J Biol Chem 263:2577–2580

Cepko CL, Roberts BE, Mulligan RC (1984) Construction and applications of a highly transmissible murine retrovirus shuttle vector. Cell 37:1053–1062

Chang M-S, Lowe DG, Lewis M, Hellmiss R, Chen E, Goeddel DV (1989) Differential activation by atrial and brain natriuretic peptides of two different receptor guanylate cyclases. Nature 341:68–72

Chinkers M, Garbers DL, Chang M-S, Lowe DG, Chin H, Goeddel DV, Schulz S (1989) A membrane form of guanylate cyclase is an atrial natriuretic peptide receptor. Nature 338:78–83

Cockcroft S, Gomperts BD (1985) Role of guanine nucleotide binding protein in the activation of polyphosphoinositide phosphodiesterase. Nature 314:534–536

Colman A (1984) Translation of eukaryotic messenger RNA in *Xenopus* oocytes. In: Hames BD, Higgins SJ (eds) Transcription and translation, a practical approach. IRL, Oxford, pp 271–302

Creba JA, Downes CP, Hawkins PT, Brewster G, Michell RH, Kirk CJ (1983) Rapid breakdown of phosphatidylinositol 4-phosphate and phosphatidyl-inositol 4,5-biphosphate in rat hepatocytes stimulated by vasopressin and other Ca^{2+}-mobilizing hormones. Biochem J 212:733–747

Dickey BF, Pyun AY, Williamson KC, Navarro J (1987) Identification and purification of a novel G protein from neutrophils. FEBS Lett 219:289–292

Dixon RAF, Kobilka BK, Strader DJ et al. (1986) Cloning of the gene and cDNA for mammalian β-adrenergic receptor and homology with rhodopsin. Nature 321:75–79

Dohlman HG, Caron MG, Lefkowitz RJ (1987) A family of receptors coupled to guanine nucleotide regulatory proteins. Biochemistry 26:2657–2664

Eidne KA, McNiven AI, Taylor PL, Plant S, House CR, Lincoln DW, Yoshida S (1988) Functional expression of rat pituitary gonadotrophin-releasing hormone receptors in Xenopus oocytes. J Mol Endocr 1:R9–R12

Fahrenholz F, Boer F, Crause P, Toth MV (1985) Photoaffinity labelling of the renal V2 vasopressin receptor. Identification and enrichment of a vasopressin-binding subunit. Eur J Biochem 152:589–595

Fargin A, Raymond JR, Lohse MJ, Kobilka BK, Caron MG, Lefkowitz RJ (1988) The genomic clone G-21 which resembles a β-adrenergic receptor sequence encodes the 5-HT$_{1A}$ receptor. Nature 335:358–360

Findlay J, Pappin DJC (1986) The opsin family of proteins. Biochem J 238:625–642

Fitzgerald TJ, Uhing RJ, Exton JH (1986) Solubilization of the vasopressin receptor from rat liver plasma membranes. J Biol Chem 261:16871–16877

Flynn TG, Davis PL (1985) The biochemistry and molecular biology of atrial natriuretic factor. Biochem J 232:313–321

Fong HKW, Yoshimoto KK, Eversole-Cire P, Simon MI (1988) Identification of a GTP-binding protein α subunit that lacks an apparent ADP-ribosylation site for pertussis toxin. Proc Natl Acad Sci USA 85:3066–3070

Fuller F, Porter JG, Arfsten AE et al. (1988) Atrial natriuretic peptide clearance receptor. Complete sequence and functional expression of cDNA clones. J Biol Chem 263:9395–9401

Garbers DL (1989) Guanylate cyclase, a cell surface receptor. J Biol Chem 264:9103–9106

Gierschik P, Jacobs KH (1987) Receptor-mediated ADP-ribosylation of a phospholipase C-stimulating G protein. FEBS Lett 224:219–223

Gilbert W (1985) Genes-in-pieces revisited. Science 228:823–824

Gillo B, Lass Y, Nadler E, Oron Y (1987) The involvement of inositol 1,4,5-triphosphate and calcium in the two component response to acetylcholine in Xenopus oocytes. J Physiol 392:349–361

Gilman AG (1987) G proteins: transducers of receptor-generated signals. Ann Rev Biochem 56:615–649

Grynkiewicz G, Poenie M, Tsien RY (1985) A new generation of Ca^{2+} indicators with greatly improved fluorescence properties. J Biol Chem 260:3440–3450

Gurdon JB, Lane CD, Woodland HR, Marbaix G (1971) Use of frog eggs and oocytes for the study of messenger RNA and its translation in living cells. Nature 233:177–182

Harada Y, Takahashi T, Kuno M, Nakayama K, Masu Y, Nakanishi S (1987) Expression of two different tachykinin receptors in Xenopus oocytes by exogenous mRNAs. J Neurosci 7:3265–3273

Hirono C, Ito J, Sugiyama H (1987) Neurotensin and acetylcholine evoke common responses in frog oocytes injected with rat brain messenger ribonucleic acid. J Physiol 382:523–535

Iyengart R, Rich KA, Herberg JT, Grenet D, Mumby S, Codina J (1987) Identification of a new GTP-binding protein. J Biol Chem 262:9239–9245

Jackson TR, Blair LAC, Marshall M, Goedert M, Hanley MR (1988) The mas oncogene encodes an angiotensin receptor. Nature 335:437–440

Julius D, MacDermott AB, Axel R, Jessell T (1988) Molecular characterization of a functional cDNA encoding the serotonin 1c receptor. Science 241:558–564

Katada T, Oinuma M, Kusakabe K, Ui M (1987) A new GTP-binding protein in brain tissues serving as the specific substrate of islet-activating protein, pertussis toxin. FEBS Lett 213:353–358

Kline D, Simoncini L, Mandel G, Maue RA, Kado RT, Jaffe LA (1988) Fertilization events induced by neurotransmitters after injection of mRNA in Xenopus eggs. Science 241:464–467

Kobilka BK, Frielle T, Dohlman H et al. (1987) Delineation of the intronless nature of the genes for the human and hamster β$_2$-adrenergic receptor and their putative promoter regions. J Biol Chem 262:7321–7327

Kobilka BK, Kobilka TS, Daniel K, Regen JW, Caron MG, Lefkowitz RJ (1988) Chimeric α_2-, β_2-adrenergic receptors: delineation of domains involved in effector coupling and ligand binding specificity. Science 248:1310–1316

Krieger D (1983) Brain peptides: What, where and why? Science 222:975–985

Kusano K, Miledi R, Stinnakre J (1977) Acetylcholine receptors in the oocyte membrane. Nature 270:739–741

Lefkowitz RJ, Caron MG (1988) Adrenergic receptors. J Biol Chem 263:4993–4996

Libert F, Parmentier M, Lefort A et al. (1989) Selective amplification and cloning of four new members of the G protein-coupled receptor family. Science 244:569–572

Litosch I, Wallis C, Fain JN (1985) 5-hydroxytryptamine stimulates inositol phosphate production in a cell-free system from blowfly salivary glands. J Biol Chem 260:5464–5471

Lo WWY, Hughes J (1987) Receptor-phosphoinositidase C coupling. FEBS Lett 224:1–3

Lochrie MA, Simon MI (1988) G protein multiplicity in eukaryotic signal transduction systems. Biochemistry 27:4957–4965

Lowe DG, Chang M-S, Hellmiss R, Chen E, Singh S, Garbers DL, Goeddel DV (1989) Human atrial natriuretic peptide receptor defines a new paradigm for second messenger signal transduction. EMBO J 8:1377–1384

Machida CA, Bunzow J, Hanneman E, Grandy D, Civelli O (1989) Replica filter screening technique to detect transfected cells expressing β_2-adrenergic receptor. DNA 8:447–455

Maelicke A (1989) Cloning of a rat D_2-dopamine receptor. Trends Biochem Sci 14:41–42

Mahlmann S, Meyerhof W, Schwarz JR (1989a) Different roles of IP4 and IP3 in the signal pathway coupled to the TRH receptor in microinjected Xenopus oocytes. FEBS Lett 249:108–112

Mahlmann S, Schwarz JR, Meyerhof W (1989b) Modulation of neuropeptide-induced membrane currents by protein kinase C in Xenopus oocytes injected with GH_3 pituitary cell poly(A)+ RNA. J Neuroendocrinol 1:65–69

Marx JL (1988) Multiplying genes by leaps and bounds. Science 240:1408–1410

Masu Y, Nakayama K, Tamaki H, Harada Y, Motoy K, Nakanishi S (1987) cDNA cloning of bovine substance-K receptor through oocyte expression system. Nature 329:836–838

McIntosh RP, Catt KJ (1987) Coupling of inositol phospholipid hydrolysis to peptide hormone receptors expressed from adrenal and pituitary mRNA in Xenopus laevis oocytes. Proc Natl Acad Sci USA 84:9045–9048

Methfessel C, Witzemann V, Takahashi T, Mishina M, Numa S, Sakmann B (1986) Patch clamp measurements on Xenpus laevis oocytes: currents through endogenous channels and implanted acetylcholine receptor and sodium channels. Pflugers Arch 407:577–588

Meyerhof W, Richter D (1989) Characterization of neuropeptide-induced membrane chloride currents in Xenopus oocytes primed with exogenous mRNA. J Protein Chem 8:365–368

Meyerhof W, Morley S, Schwarz J, Richter D (1988a) Receptors for neuropeptides are induced by exogenous poly(A)+ RNA in oocytes from Xenopus laevis. Proc Natl Acad Sci USA 85:714–717

Meyerhof W, Morley SD, Richter D (1988b) Expression and electrophysiological identification of the receptor for bombesin and gastrin-releasing peptide in Xenopus laevis oocytes injected with polyA+ RNA from rat brain. FEBS Lett 239:109–112

Miledi R (1982) A calcium-dependent transient outward current in Xenopus laevis oocytes. Proc R Soc Lond [Biol] 215:491–497

Minamino N, Aburaya M, Ueda S, Kangawa K, Matsuo H (1988) The presence of brain natriuretic peptide of 12000 daltons in porcine heart. Biochem Biophys Res Commun 155:740–746

Moriarty TM, Gillo B, Sealfon S, Roberts JL, Blitzer RD, Landau EM (1988) Functional expression of brain cholecystokinin and bombesin receptors in Xenopus oocytes. Mol Brain Res 4:75–79

Morley SD, Meyerhof W, Schwarz J, Richter D (1988) Functional expression of the oxytocin receptor in Xenopus laevis oocytes primed with mRNA from bovine endometrium. J Mol Endocrinol 1:77–81

Noma Y, Sideras P, Naito T et al. (1986) Cloning of a cDNA encoding the murine IgG1 induction factor by a novel strategy using SP6 promoter. Nature 319:640–646

Northup JK (1985) Overview of the guanine nucleotide regulatory protein systems, N_s and N_i, which regulate adenylate cyclase activity in plasma membranes. In: Cohen P, Housley MD (eds) Molecular mechanisms of transmembrane signalling. Elsevier, Amsterdam, pp 91–116

O'Dowd BF, Hnatowich M, Caron MG, Lefkowitz RJ, Bouvier M (1989) Palmitoylation of the human β2-adrenergic receptor. J Biol Chem 264:7564–7569

Oron Y, Dascal N, Nadler E, Lupu M (1985) Inositol 1,4,5-trisphosphate mimics muscarinic response in Xenopus oocytes. Nature 313:141–143

Oron Y, Gillo B, Straub RE, Gershengorn MC (1987) Mechanism of membrane electrical response to thyrotropin-releasing hormone in Xenopus oocytes injected with GH3 pituitary cell messenger ribonucleic acid. Mol Endocrinol 1:918–925

Parker I, Miledi R (1986) Changes in intracellular calcium and in membrane currents evoked by injection of inositol trisphosphate into Xenopus oocytes. Proc R Soc Lond [Biol] 228:307–315

Parker I, Sumikawa K, Miledi R (1986) Neurotensin and substance P receptors expressed in Xenopus oocytes by messenger RNA from rat brain. Proc R Soc Lond [Biol] 229:151–159

Parker I, Sumikawa K, Miledi R (1987) Activation of a common effector system by different brain neurotransmitter receptors in Xenopus oocytes. Proc R Soc Lond [Biol] 231:37–45

Perucho M, Hanahan D, Wigler M (1980) Genetic and physical linkage of exogenous sequences in transformed cells. Cell 22:309–317

Richter D (1988) Molecular events in expression of vasopressin and oxytocin and their cognate receptors. Am J Physiol 255:F207–F219

Richter D, Morley SD, Schwarz J, Meyerhof W (1988a) Characterising neuropeptide signaling pathways. In: Thorn NA, Treiman M, Petersen OH (eds) Molecular mechanisms in secretion. Alfred Benzon Symposium 25. Munksgaard, Copenhagen, pp 544–553

Richter D, Meyerhof W, Morley SD, Mohr E, Fehr S, Schmale H (1988b) Molecular biology of brain peptides and their cognate receptors. In: Kleinkauf, von Döhren, Jaenicke (eds) The roots of modern biochemistry. Walter de Gruyter, Berlin, pp 305–321

Ross EM (1989) Signal sorting and amplification through G protein-coupled receptors. Neuron 3:141–152

Saiki RK, Gelfand DH, Stoffel S et al. (1988) Primer-directed enzymatic amplification of DNA with a thermostable DNA polymerase. Science 239:487–491

Sandberg K, Markwick AJ, Trinh DP, Catt KJ (1988) Calcium mobilization by angiotensin II and neurotransmitter receptors expressed in Xenopus laevis oocytes. FEBS Lett 241:177–180

Sibley DR, Nambi P, Lefkowitz RJ (1985) Molecular mechanisms of hormone receptor desensitization. In: Cohen P, Houslay MD (eds) Molecular mechanisms of transmembrane signalling. Elsevier, Amsterdam, pp 359–374

Sibley DR, Benovic JL, Caron MG, Lefkowitz RJ (1987) Regulation of transmembrane signaling by receptor phosphorylation. Cell 48:913–922

Singh S, Lowe DG, Thorpe DS et al. (1988) Membrane guanylate cyclase is a cell-surface receptor with homology to protein kinases. Nature 334:708–712

Sternweis PC, Robishaw JD (1984) Isolation of two proteins with high affinity for guanine nucleotides from membranes of bovine brain. J Biol Chem 259:13806–13813

Sudoh T, Kangawa K, Minamino N, Matsuo H (1988) A new natriuretic peptide in porcine brain. Nature 332:78–81

Waldo GL, Evans T, Fraser ED, Northup JK, Martin MW, Harden TK (1987) Identification and purification from bovine brain of a guanine-nucleotide-binding protein distinct from Gs, Gi and G0. Biochem J 246:431–439

Williams JA, McChesney DJ, Calayag MC, Lingappa VR, Logsdon CD (1988) Expression of receptors for cholecystokinin and other Ca2+mobilizing hormones in Xenopus oocytes. Proc Natl Acad Sci USA 85:4939–4943

Yarden Y, Ullrich A (1988) Molecular analysis of signal transduction by growth factors. Biochemistry 27:3113–3119

Young D, Waitches G, Birchmeier C, Fasano O, Wigler M (1986) Isolation and characterization of a new cellular oncogene encoding a protein with multiple potential transmembrane domains. Cell 45:711–719

Young D, O'Neill K, Jessell T, Wigler M (1988) Characterization of the rat mas oncogene and its high-level expression in the hippocampus and cerebral cortex of rat brain. Proc Natl Acad Sci USA 85:5339–5342

Cell Surface Receptors

Peptide Hormone Receptors

J. Lloyd, R. Tibolt, and G. V. Childs

1 Introduction

A large number of peptide hormones have been identified and are known to elicit a wide variety of biological responses from target cells throughout the body. These hormones exert their effects by interaction with receptors which are integral components of the cell surface membrane. The observation that many of the peptide hormones affect only specific cell types led to the concept that these cells must have specific receptor sites for each hormone.

Hormone-induced biological responses are the result of a complex series of biochemical events initiated by the binding of the ligand to its specific receptor. Furthermore, the response of a cell to hormonal stimulation is dependent on both the concentration of the hormone and its receptor, and the affinity of the receptor for that hormone. The formation of the "hormone-receptor complex" stimulates one of several second messengers, which then activate protein kinases, resulting in the phosphorylation of enzymes involved in the biosynthetic and metabolic pathways of the target cell (reviewed in Roth and Grunfeld 1985). Consequently, regulation of the cell surface receptors plays a critical role in determining the effects of peptide hormones (Catt et al. 1979).

The study of hormone receptors has been aided greatly by the development of radioreceptor assays and affinity cytochemical techniques which permit their measurement and localization. Radioreceptor assays measure the binding of ligands to their receptors and can be used to determine the level of receptors during different physiological states and following experimental manipulation (Clayton and Catt 1981). Affinity cytochemistry is a useful technique for the localization of membrane receptors (Bayer et al. 1979). It involves conjugation of a biologically active ligand to a probe which, following incubation with the tissue of interest, can be visualized by various staining methodologies. A variety of probes have been used to visualize receptors at both the light and electron microscopic levels. These include ferritin (Anderson et al. 1977; HOPKINS and GREGORY 1977), rhodamine (NAOR et al. 1981), peroxidase (WILLINGHAM and PASTAN 1980), and radioactive labels (CARPENTER and COHEN 1976; CARPENTIER et al. 1978).

The advantage of this method is that it provides a means whereby one can measure quantitatively receptor binding to specific target cells while maintaining the anatomical integrity of the plasma membranes. This is especially important in tissues comprising heterogeneous cell populations. For example, in the anterior pituitary, hypothalamic releasing hormones bind specifically to only one cell type which may represent as little as 10% of the total cell number.

In this chapter we will discuss our recent studies on the characterization of specific receptors on anterior pituitary gonadotropes and corticotropes. Affinity cytochemical techniques were used to localize potent biotinylated analogues of gonadotropin-releasing hormone (GnRH), corticotropin-releasing hormone (CRH), and arginine vasopressin (AVP).

2 GnRH Studies

The hypothalamic decapeptide GnRH stimulates the release of luteinizing hormone (LH) and follicle-stimulating hormone (FSH) from the anterior pituitary in a calcium-dependent manner (MARIAN and CONN 1979). The precise mechanism of action of GnRH is an area of active investigation. In addition to the calcium requirement, GnRH appears to operate through a complex series of interactions involving phosphatidylinositol turnover and protein kinase C activation (reviewed in CONN 1986; HUCKLE and CONN 1988). GnRH-mediated gonadotropin release is dependent on specific binding to high affinity receptors on pituitary gonadotropes (CLAYTON and CATT 1981). At present little information is available concerning the biochemical structure of the GnRH receptor. It is a glycoprotein containing sialic acid residues (HAZUM 1982), and photoaffinity labeling has identified a single component with a molecular weight of 60000 (HAZUM and KEINAN 1982). More recent studies solubilized GnRH receptors with a zwitterionic detergent, CHAPS (WINIGER et al. 1983), and suggested that the receptor is larger than originally thought. Radiation inactivation studies suggest a functional molecular weight of 136000 (CONN and VENTER 1985) and solubilization has revealed two components of 59000 and 57000 daltons respectively (HAZUM et al. 1986). It is thought that the additional component is probably the carbohydrate moiety of the glycoprotein.

Pituitary GnRH receptor levels change throughout the estrous cycle of the rat (SAVOY-MOORE et al. 1980; CLAYTON et al. 1980). Receptor levels were low during estrus and metestrus but gradually increased on diestrus II. Peak levels were maintained until late in the afternoon of proestrus, when they fell just prior to the preovulatory gonadotropin surge. These studies employed radioreceptor binding assays for determination of receptor concentrations. They did not, however, distinguish whether the observed changes were due to the number of receptors per cell or the number of GnRH-receptive cells. In 1983, our laboratory reported the localization of a biotinylated GnRH analogue (bio-GnRH) on the surface of dispersed anterior pituitary monolayer cultures (CHILDS et al.). The bio-GnRH was localized using the avidin–biotin peroxidase (ABC) technique and the black peroxidase substrate, nickel-intensified diaminobenzidine. The percentage of labeled cells (13%–16%) agreed with that obtained from labeling of living monolayer cultures with fluorescein-labeled GnRH analogues (NAOR et al. 1981), and they matched the percentage of total gonadotropes reported in dual immunocytochemical stains for both gonadotropins (CHILDS 1983). Using the affinity cytochemical technique we have been able to quantify changes in the number of GnRH receptors and to determine whether the reported fluctuations during the cycle were reflected by changes in the number of cells that bound the ligand (LLOYD and CHILDS 1988).

Figure 1 shows that the magnitude and timing of the changes in the percentage of GnRH-receptive cells compared favorably with results from radioreceptor studies. There was a two- to threefold increase in receptive cells from diestrus I (D-I; 5.2% $\pm$ 0.5%) to proestrous morning (PRO-AM; 20.2% $\pm$ 4%), followed by an acute drop on proestrous afternoon (PRO-PM; 7.4% $\pm$ 0.8%). Figure 2 shows an example of bio-GnRH visualized on the surface of a cell from a proestrous monolayer. Localization of the bio-GnRH with the ABC technique required that the cells be fixed prior to staining. Therefore, in a parallel study the biotinylated ligand was visualized on living cells using an avidin-fluorescein conjugate. Counts of living stained cells produced a similar change in the number of receptive cells (Table 1). There was a

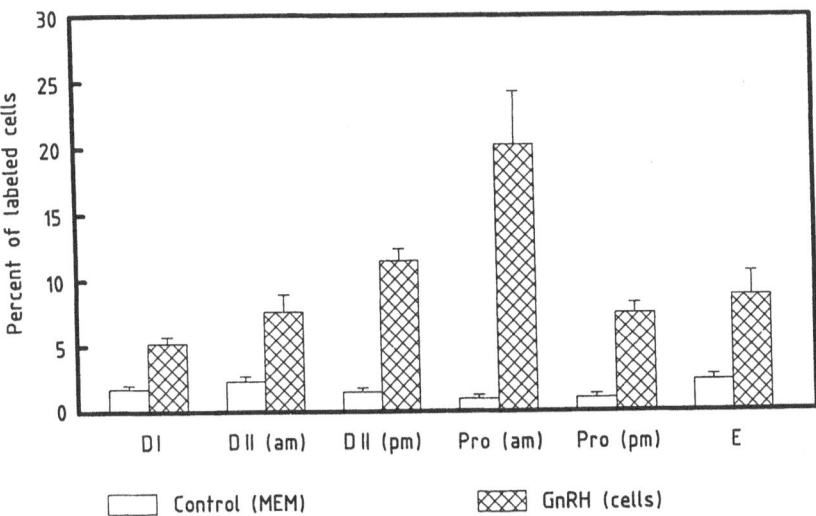

Fig. 1. Bio-GnRH was localized on 1-day monolayer cultures using the ABC technique. This figure shows the percentage ($\pm$ SE) of labeled cells identified at specific stages of the estrous cycle. All values were significantly greater than vehicle (MEM)-treated controls. (LLOYD and CHILDS 1988)

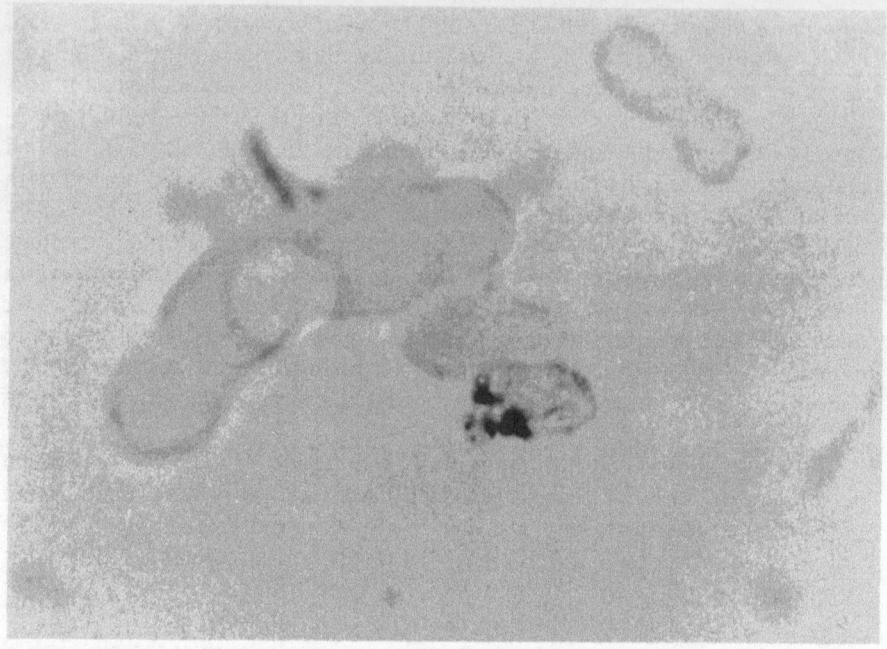

Fig. 2. Application of the ABC stain to proestrous monolayer culture following 3-min bio-GnRH exposure. × 960

Table 1. Localization of bio-GnRH on living cells with an avidin–fluorescein conjugate (LLOYD and CHILDS 1988)

Group	% labeled cells
Vehicle (MEM)	1.0 ± 0.5
Estrus	7.4 ± 0.9[a]
Diestrus II	10.8 ± 0.5[a]
Proestrus	16.2 ± 1.0[a]

[a] Significantly greater than vehicle-treated (MEM) control.

two-fold increase in GnRH-bound cells between the morning of estrus and proestrus ($7.4\% \pm 0.9\%$ vs $16.2\% \pm 1\%$). The percentages did not match exactly those from the fixed cultures, but the differences are probably due to differences in the rates of internalization and loss of fluorescence in acidic vesicles. These data indicate that changes in GnRH receptors during the rat estrous cycle reflect, at least in part, changes in the number of receptive cells.

Once the changes in GnRH-receptive cells during the cycle had been established, it was of interest to investigate possible mediators of the phenomenon. It is well established that estrogen feedback affects gonadotropin secretion at both the hypothalamic and pituitary levels (FERLAND et al. 1975). During the preovulatory period, the anterior pituitary responsiveness to GnRH stimulation increases dramatically and, by the afternoon of proestrus, is 50 times greater than the basal response at estrus (FINK et al. 1975). Increased serum estradiol levels have been implicated in this phenomenon (AIYER and FINK 1974). In addition, the positive correlation between rising GnRH receptor and estradiol levels (SAVOY-MOORE et al. 1980) suggests that perhaps the steroid up-regulates GnRH receptors, enabling the anterior pituitary to respond to the low endogenous levels of GnRH in the portal circulation.

In light of these data, we investigated the effects of estradiol benzoate (EB) administered in vitro on the percentage of GnRH-receptive cells during the cycle (LLOYD and CHILDS 1988). The results of these studies are shown in Fig. 3. Monolayer cultures from rats at specific stages of the cycle were pre-treated for 24 h with EB (10^{-9}–10^{-12} M) prior to a 3-min bio-GnRH stimulation and fixation. EB had a biphasic effect that was dependent on the stage of the cycle at which the cells were taken. In cultured cells taken from animals on the morning of diestrus I and diestrus II, pretreatment with physiological concentrations of EB increased the percentage of GnRH-labeled cells to levels normally observed on the afternoon of diestrus II (11.0% ± 1% and 12.0% ± 1% vs 11.5% ± 0.9%). Effectively, in cells taken from these rats, which had been exposed to low endogenous levels of the steroid, EB pretreatment mimicked the endogenous rise of estradiol normally encountered on the afternoon of diestrus II. In contrast, in cultures of cells taken on the morning of proestrus there was a dose-dependent decrease in the number of GnRH-bound cells.

These data suggest that estradiol is capable of recruiting GnRH-receptive cells prior to the proestrous gonadotropin surge and provide a cytological

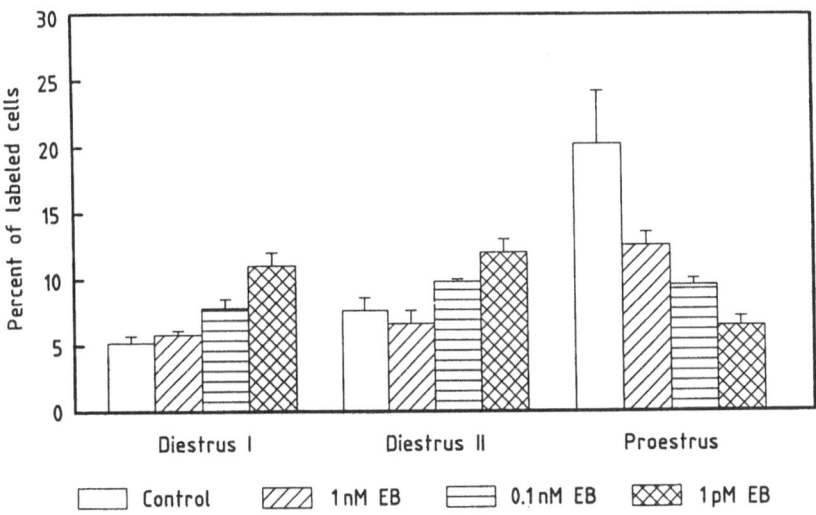

Fig. 3. The percentage of bio-GnRH-labeled cells following 24 h pretreatment of cultures with physiological concentrations of EB (modified from LLOYD and CHILDS 1988)

basis for changes in GnRH receptors and pituitary responsiveness to the decapeptide during the estrous cycle. However, it is interesting to note that in EB-treated diestrous cultures, the percentage of labeled cells never reached the peak levels observed on the morning of proestrus. This indicates that other factors may be involved in the up-regulatory mechanism. A possible candidate is progesterone, which may be involved in initiating the rise in GnRH receptors (SAVOY-MOORE et al. 1980). Estradiol has been shown to increase the number of anterior pituitary GnRH receptors in rats (MENON et al. 1985) and ewes (GREGG and NETT 1989). A similar effect of the steroid has been reported for TRH receptors (DELEAN et al. 1977). However, this study is the first to demonstrate a steroid-induced increase in the number of receptive cells. The mechanism by which estradiol up-regulates GnRH receptors remains unclear. MENON's work indicates the possible involvement of new protein synthesis, but whether this represents synthesis of new GnRH receptors has yet to be determined (MENON et al. 1985).

The EB-induced decline in GnRH-receptive cells in cultures taken on the morning of proestrus is difficult to explain since endogenous steroid levels are high at this time, and there is no loss of GnRH receptors until later in the afternoon. However, exposure of cells taken at 12.30 p.m. proestrus to estradiol resulted in a 50% reduction in their response to GnRH (TURGEON and WARING 1981), a result that would be expected given a loss of GnRH receptors. Several studies have reported steroid-induced loss of GnRH receptors. GIGUERE et al. (1981) found that 48 h preincubation of pituitary cells with the testosterone metabolite 5α-dihydrotestosterone (DHT) resulted in a 40% decrease in both GnRH receptors and GnRH-stimulated LH release. Similarly, both estradiol and testosterone decreased GnRH receptor binding in steers (ZOLMAN 1983).

In addition to the steroid effect on GnRH-receptive cells during the cycle, we have also investigated the relationship in male rats. These studies combined affinity cytochemical localization of bio-GnRH with immunocytochemical stains for either LHβ or FSHβ (TIBOLT and CHILDS 1985). DHT pretreatment decreased the percentage of GnRH-bound cells relative to vehicle-treated controls (9.2% ± 0.4% vs 15.8% ± 0.6%), thus agreeing with the results from radioreceptor studies. This negative effect was most pronounced in the LH gonadotrope population, a result corroborated by the fact that in DHT-treated cultures, the magnitude of the reduction in GnRH-induced LH release was far greater than that for FSH release. Interestingly, pretreatment of cultures with corticosterone, which also suppressed GnRH-induced gonadotropin release, had no effect on the percentage of labeled cells. This suggested that adrenal steroid modulation of gonadotropin release is independent of the GnRH receptor. Thus, while generally the number of GnRH-receptive cells is a good indicator of gonadotrope responsiveness to the releasing hormone, it is important to realize that postreceptor interactions are probably also involved.

3 CRH and AVP Studies

The hypothalamic peptide, corticotropin-releasing hormone (CRH), appears to be a major physiological regulator of ACTH secretion from the anterior pituitary (VALE 1981; RIVIER et al. 1983). In addition to CRH, several other neuropeptides have been identified as secretagogues, including arginine vasopressin (AVP) and angiotensin II (A-II) (BENY and BAERTSCHI 1981; GILLIES et al. 1982; SPINEDI and NEGRO-VILAR 1983). All these peptides stimulate ACTH release to varying degrees in a calcium-dependent manner (GIGUERE et al. 1982; AGUILERA et al. 1983; ABOU-SAMRA et al. 1987a). However, in spite of this requirement for calcium, these peptides do not all activate the same second messenger system. CRH acts by increasing adenylate cyclase activity and stimulating cAMP-dependent protein kinases (AGUILERA et al. 1983; HOLMES et al. 1984), whereas AVP and A-II action involves phospholipid turnover and protein kinase C activation (ABOU-SAMRA et al. 1986). While AVP alone is a relatively weak secretagogue, it has been reported to be a highly effective potentiator of CRH-mediated ACTH release in vivo (RIVIER and VALE 1983) and in vitro (ANTONI et al. 1983). This effect of AVP is complex and is thought to involve both protein kinase C and enhancement of CRH-induced cAMP accumulation (ABOU-SAMRA et al. 1987b).

Corticotropin-releasing hormone exerts its effects by binding to specific high affinity receptors on the surface of anterior pituitary corticotropes (WYNN et al. 1983). Corticotrope plasma membranes also contain AVP receptors which are distinct from the CRH-binding sites (ANTONI et al. 1984; BAERTSCHI and FRIEDLI 1985). In this laboratory affinity cytochemical techniques were used to localize biotinylated analogues of CRH and AVP on dispersed pituitary monolayers (WESTLUND et al. 1984; CHILDS et al. 1986, 1987a, b). This simple technique permitted investigation of how AVP, glucocorticoids, second messenger activation, and membrane currents affect CRH binding at the single cell level.

The use of affinity cytochemistry to study physiological regulation of peptide hormone receptors depends on ligands which retain full biological activity. In an earlier study from this laboratory, WESTLUND et al. (1984) reported that biotinylation of CRH decreased binding activity slightly. This was probably due to interference with the critical lysine residue at position 36 near the C-terminal. Therefore, a bio-CRH analogue was synthesized by AGUILERA and MORELL in which the biotin molecule was attached to the N-terminal serine residue (CHILDS et al. 1986). Dose-response curves and specificity tests revealed no loss in potency or biological activity when compared to the native hormone. When applied to mixed pituitary cells, the bio-CRH stained 10% – 11% of the cells in both fixed (localized with ABC) and living (localized with avidin-fluorescein) cultures. Competition with AVP, A-II, or somatostatin caused no reduction in labeling of cells. In combination with immunocytochemical labels for ACTH or β-endorphin, bio-CRH bound 90% of the corticotropes (WESTLUND et al. 1984).

Following validation of the technique, the effect of steroid pretreatment on bio-CRH binding was investigated (CHILDS et al. 1986). Glucocorticoids are known to feedback and suppress ACTH release (ABOU-SAMRA et al. 1986). Furthermore, CRH receptor binding capacity declines following adrenalectomy (WYNN et al. 1985), probably as a result of receptor down-regulation. This effect was prevented by dexamethasone and it was postulated that this might be due to suppression of the postadrenalectomy rise in CRH release from the hypothalamus. However, there remained the possibility of a pituitary effect. Pretreatment of pituitary cells for 1 h with corticosterone or dexamethasone resulted in a 50%–60% reduction in the number of bio-CRH-stained cells accompanied by a decline in basal and CRH-stimulated ACTH release. As mentioned earlier, this phenomenon was also observed in gonadotropes by demonstrating a loss of GnRH-receptive cells in response to steroid pretreatment (TIBOLT and CHILDS 1985; LLOYD and CHILDS 1988). In corticotropes, however, the loss of receptive cells after 1 h pretreatment was not paralleled by a receptor decline evidenced by radioreceptor assay. Electron microscopic study of bio-CRH internalization provided a possible explanation of this discrepancy. Bio-CRH was localized on a subpopulation of cytoplasmic granules (Fig. 4), suggesting a means whereby receptors can be recycled to the cell surface. Perhaps short-term glucocorticoid exposure slows down this process and causes a reduction in available receptors for binding to bio-CRH.

Since CRH-stimulated ACTH release is calcium dependent, it was of interest to determine the nature of any calcium membrane currents and their involvement in the CRH–corticotrope interaction. As mentioned previously, a major obstacle in studying specific pituitary cell types is their dilution in a heterogeneous cell population. This problem was especially debilitating in the case of corticotropes, which comprise a mere 10% of anterior pituitary cells. The ability to visualize bio-CRH (or bio-AVP) on living corticotropes with the avidin-fluorescein stain suggested a possible means to obviate this problem. Since CRH binds specifically to corticotropes, living cells "tagged" with the avidin-fluorescein label might be used for subsequent electrophysiological detection of membrane channels. However, the procedure would only be useful provided that the biotinylated ligand did not alter corticotrope responsiveness for a prolonged period. Therefore, initial studies were designed to test the feasibility of this idea (CHILDS et al. 1987a).

Both bio-CRH and bio-AVP bound 8%–10% of anterior pituitary cells. They did not compete with one another for receptor sites. This was in agreement with previous work indicating the presence of separate receptors for the two peptides (ANTONI et al. 1984; BAERTSCHI and FRIEDLI 1985). An example of bio-CRH localization with avidin-fluorescein is shown in Fig. 5. During the labeling procedure the two ligands induced a 4- to 18-fold increase in ACTH secretion (Table 2). Basal secretion was restored after only 1 h or 24 h after labeling was the same as the initial stimulation, indicating that the biotinylated ligands did not desensitize corticotropes. Consequently, in ensuing studies on bio-CRH-labeled corticotropes, MARCHETTI et al. (1987) detected the presence of a tetrodotoxin (TTX)-sensitive sodium current and two calcium currents,

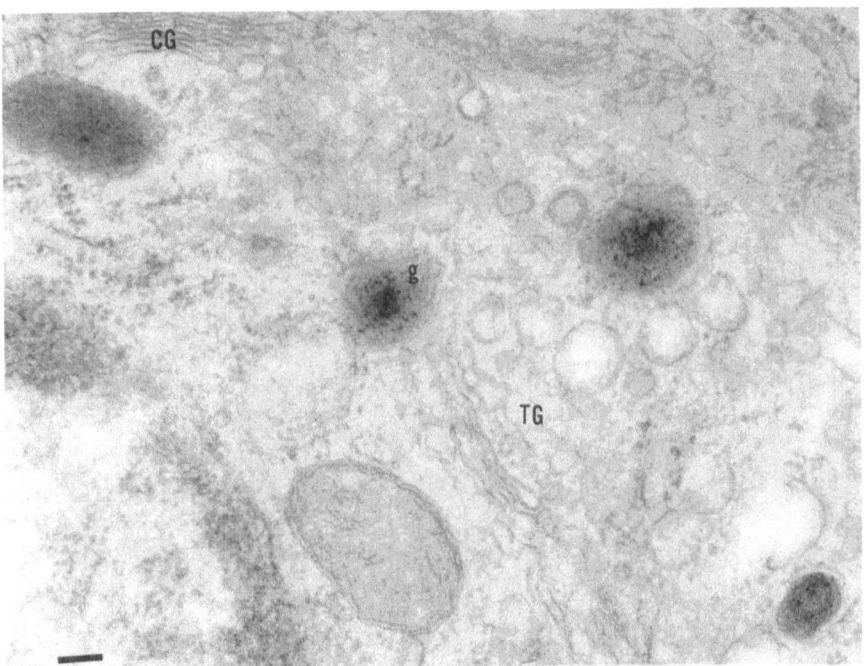

Fig. 4. Label for bio-CRH in granules developing in the Golgi complex region (*TG*). Label is also evident on mature granules. (CHILDS et al. 1986) × 84 000, *bar* = 0.1 μm

Table 2. Effect of avidin–fluorescein stain on stimulated ACTH release (CHILDS et al. 1987 a)

Treatment	Rate of ACTH released (pg/min)	
	Bio-CRH 5 min	Bio-AVP 10 min
Stimulus during stain	183 ± 48	41 ± 8.6
Basal levels 60 min after stain	7.6 ± 3	9 ± 4
Stimulus 60 min after stain (acute)	116 ± 36	36 ± 14

one transient and the other long-lasting. Parallel studies using bio-GnRH-labeled gonadotropes were less successful due to desensitization of these cells by the ligand.

Given the fact that corticotrope membranes possess calcium currents and that calcium channel blockade attenuated CRH-mediated ACTH release (AGUILERA et al. 1986), the next step was to ask the question, how does modulation of these channels affect CRH receptor binding? Blockade of calcium channels with the dihydropyridine antagonist, nimodipine, resulted in up to a 74% decrease in the percentage of CRH-bound cells in parallel with a decrease in the rate of CRH-mediated ACTH secretion (CHILDS et al. 1987b). Similar

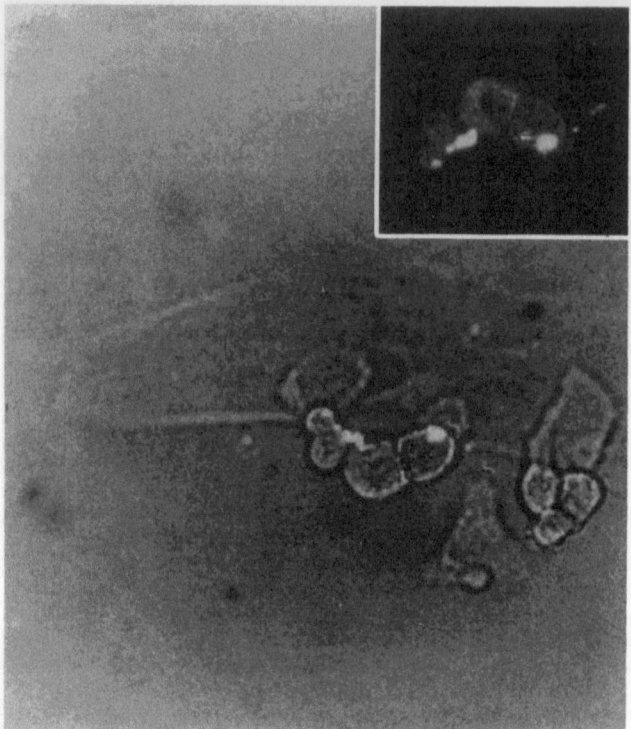

Fig. 5. Living mixed pituitary cell cultures were treated with 0.1 nM bio-CRH and then labeled with avidin–fluorescein after 1 min. Label is in patches on the surface of corticotropes. (CHILDS et al. 1987a) ×960

effects were observed with $CdCl_2$, which blocks low and high threshold calcium currents, and TTX blockade of sodium channels, although both were effective only at higher concentrations. Conversely, pretreatment with either the dihydropyridine agonist Bay K 8644 or AVP caused a 30% increase in CRH-bound cells. These studies indicated a relationship between the activity of membrane currents and CRH receptivity.

Subsequent studies of this system determined the possible involvement of second messenger systems in the enhancement of CRH binding (CHILDS and UNABIA 1989). Two lines of evidence from previous studies had suggested an AVP-induced recruitment of CRH-bound cells. Firstly, as shown above, AVP pretreatment for 1 h increased the percentage of bio-CRH-labeled corticotropes (CHILDS et al. 1987b). Secondly, in studies of ACTH secretion with the reverse hemolytic plaque assay, only 7% of mixed pituitary cells secreted ACTH under basal conditions, while immunocytochemical stains for ACTH demonstrated that 10% of the population were corticotropes (CHILDS and BURKE 1987). However, the percentage of ACTH-secreting (plaque-forming) cells could be increased to 10% by exposure to either high concentrations of

CRH or low concentrations of CRH in combination with AVP. This indicated that as many as one-third of corticotropes are held in reserve during basal conditions, and that perhaps these cells required the activation of two second messenger systems in order to realize their secretory potential. Thus, the effect of AVP and its second messengers on CRH binding was investigated.

These studies benefitted from the availability of enriched corticotrope cell fractions (CHILDS et al. 1988). Counterflow centrifugation of monodispersed pituitary cells produced partially enriched fractions. However, following CRH-induced corticotrope hypertrophy and reelutriation, we were able to collect fractions with greater than 90% enrichment as assessed by immunocytochemical labeling for ACTH and β-endorphin. This advance in purification technique made studies of corticotropes far easier than in the past because they were no longer diluted by other pituitary cell types.

Enriched corticotropes were pretreated with either AVP, Bay K 8644, or the protein kinase C activator, 12-O-tetradecanoyl-phorbol-13-acetate (TPA)

Table 3. Spectrophotometric analysis of CRH binding to individual corticotropes (area of label/cell)

Treatment	Area of label for bio-CRH (μm^2)	Average cell area (μm^2)
Medium alone	7 ± 1.0	145 ± 11
10 nM AVP	11 ± 0.8[a]	146 ± 29
10 ng/ml TPA	9 ± 1.0[a]	128 ± 23
10 nM Bay K 8644	11.5 ± 0.9[a]	161 $\pm$ 9

[a] Significantly different from control (medium alone).

prior to localization of bio-CRH with the ABC technique (CHILDS and UNABIA 1989). The area and density of the label were analyzed with the Cue 3 image analysis system, which detects label by its color (wavelength). Treatment with either 10 nM AVP or either of the activating drugs caused a 30% increase in the area of label for bio-CRH (Table 3). This increase occurred in the absence of any change in the average cell area. Furthermore, analysis of the density of the label revealed increases in saturation of the color (13%) and gray level density (18% lower gray level), and a decrease in brightness (23%). In conjunction with the increase in the area of the label, these data suggest an overall increase in CRH binding. In mixed pituitary cultures, treatment with AVP, Bay K 8644, or TPA increased the percentage of CRH-bound cells by 30% – 40%, and the inhibitory effect on binding by calcium blockade with nimodipine was prevented by coincubation with TPA (Tables 4, 5). These results suggest that AVP potentiation of CRH-mediated ACTH release is due, at least in part, to enhancement of CRH binding by mechanisms involving activation of calcium channels and second messengers.

Table 4. Effect of protein kinase C activation on CRH-bound corticotropes

Pretreatment	% bio-CRH
Medium alone	8.6 ± 0.9
TPA (10 ng/ml)	11.3 ± 0.3[a]
TPA (100 ng/ml)	13.2 ± 0.1[a]
Bay K (10 nM)	12.0 ± 0.4[a]
Bay K (10 nM) + TPA (100 ng/ml)	13.7 ± 1.4[a]

[a] Significantly different from control (medium alone).

Table 5. Effect of a voltage-dependent calcium channel antagonist on enhancement of CRH binding by TPA (CHILDS and UNABIA 1989)

Treatment	% bio-CRH
Medium alone (vehicle)	9.0 ± 0.9
Nimodipine alone	4.3 ± 1.2[a]
TPA (100 ng/ml) + 10 nM nimodipine	8.4 ± 1.9

[a] Significantly different from vehicle-treated, $P < 0.05$.

4 Conclusion

Recent studies of peptide hormone receptors serve to underline the importance of these structures in endocrine physiology. The regulation of these receptors is critical in our understanding of how cells respond to a multitude of biological signals. The work presented in this chapter shows that many factors act in concert to regulate cell receptivity. These include up- and down-regulation by steroids, activation of membrane currents, and involvement of second messenger systems. Future electron microscopic studies are needed to elucidate how these factors might work to alter receptor processing during the estrous cycle and other physiological states. Furthermore, cytochemical and cytophysiological studies are needed to characterize reserve cell populations during recruitment of bio-CRH-labeled cells. In conjunction with biochemical characterization and isolation of receptor structure, these methods will enable us to understand better the regulation of hormonal function.

References

Abou-Samra AB, Catt KJ, Aguilera G (1986) Involvement of protein kinase C in the regulation of adrenocorticotropin release from rat anterior pituitary cells. Endocrinology 118:212–217
Abou-Samra AB, Catt KJ, Aguilera G (1987a) Calcium-dependent control of corticotropin release in rat anterior pituitary cell cultures. Endocrinology 121:965–971

Abou-Samra AB, Harwood JP, Manganiello VC, Catt KJ, Aguilera G (1987b) Phorbol 12-myristate 13-acetate and vasopressin potentiate the effect of corticotropin-releasing factor on cyclic AMP production in rat anterior pituitary cells. J Biol Chem 262:1129–1136

Aguilera G, Harwood JP, Wilson JX, Morell J, Brown JH, Catt KJ (1983) Mechanisms of corticotropin-releasing factor and other regulators of corticotropin release in rat pituitary cells. J Biol Chem 258:8039–8045

Aguilera G, Wynn PC, Harwood JP, Hauger RL, Millan MA, Grewe C, Catt KJ (1986) Receptor-mediated actions of corticotropin-releasing factor in pituitary gland and nervous system. Neuroendocrinology 43:79–86

Aiyer MS, Fink G (1974) The role of sex steroid hormones in modulating the responsiveness of the anterior pituitary gland to luteinizing hormone-releasing factor in the female rat. J Endocrinol 62:553–574

Anderson RG, Brown MS, Goldstein JL (1977) Role of the coated endocytotic vesicle in the uptake of receptor-bound low density lipoprotein in human fibroblasts. Cell 10:351–364

Antoni FA, Holmes MC, Jones MT (1983) Oxytocin as well as vasopressin potentiate ovine CRF in vitro. Peptides 4:411–415

Antoni FA, Holmes MC, Makara GB, Karteszi M, Laszlo FA (1984) Evidence that the effects of arginine-8-vasopressin (AVP) on pituitary corticotropin (ACTH) release are mediated by a novel type of receptor. Peptides 5:519–522

Baertschi AJ, Friedli M (1985) A novel type of vasopressin receptor on anterior pituitary corticotrophs? Endocrinology 116:499–502

Bayer EA, Skutelsky E, Wilchek M (1979) The avidin-biotin complex in affinity cytochemistry. Methods Enzymol 62:308–315

Beny JL, Baertschi AJ (1981) Corticotropin-releasing factors (CRF) secreted from the rat median eminence in vitro in the presence or absence of ascorbic acid: quantitative role of vasopressin and catecholamines. Endocrinology 109:813–817

Carpenter G, Cohen S (1976) ^{125}I-labeled human epidermal growth factor. Binding, internalization, and degradation in human fibroblasts. J Cell Biol 71:159–171

Carpentier J, Gorden P, Amherdt M, Van Obberghen E, Kahn CR, Orci L (1978) ^{125}I-insulin binding to cultured human lymphocytes. J Clin Invest 61:1056–1070

Catt KJ, Harwood JP, Aguilera G, Dufau ML (1979) Hormonal regulation of peptide receptors and target cell responses. Nature 280:109–116

Childs GV (1983) Application of dual pre-embedding stains for gonadotropins to pituitary cell monolayers with avidin-biotin (ABC) and peroxidase-antiperoxidase (PAP) complexes: light microscopic studies. Stain Technol 58:281–289

Childs GV, Burke JA (1987) Use of the reverse hemolytic plaque assay to study the regulation of anterior lobe adrenocorticotropin (ACTH) secretion by ACTH-releasing factor, arginine vasopressin, angiotensin II and glucocorticoids. Endocrinology 120:439–444

Childs GV, Unabia G (1989) Activation of protein kinase C and L calcium channels enhances binding of biotinylated corticotropin-releasing hormone by anterior pituitary corticotropes. Mol Endocrinol 3:117–126

Childs GV, Naor Z, Hazum E, Tibolt RE, Westlund KN, Hancock MB (1983) Localization of biotinylated gonadotropin releasing hormone on pituitary monolayer cells with avidin-biotin-peroxidase complexes. J Histochem Cytochem 31:1422–1425

Childs GV, Morell JL, Niendorf A, Aguilera G (1986) Cytochemical studies of corticotropin-releasing factor (CRF) receptors in anterior lobe corticotropes: binding, glucocorticoid regulation, and endocytosis of [biotinyl-ser[1]] CRF. Endocrinology 119:2129–2142

Childs GV, Unabia G, Burke JA, Marchetti C (1987a) Secretion from corticotropes after avidin-fluorescein stains for biotinylated ligands (CRF or AVP). Am J Physiol 252:E347–E356

Childs GV, Marchetti C, Brown AM (1987b) Involvement of sodium channels and two types of calcium channels in the regulation of adrenocorticotropin release. Endocrinology 120:2059–2069

Childs GV, Lloyd JM, Unabia G, Rougeau D (1988) Enrichment of corticotropes by counterflow centrifugation. Endocrinology 123:2885–2895

Clayton RN, Catt KJ (1981) Gonadotropin-releasing hormone receptors: characterization, physiological regulation and relationship to reproductive function. Endocr Rev 2:186–209

Clayton RN, Solano AR, Garcia-Vela A, Dufau ML, Catt KJ (1980) Regulation of pituitary receptors for gonadotropin releasing hormone during the rat estrous cycle. Endocrinology 107:699–706

Conn PM (1986) The molecular basis of gonadotropin-releasing hormone action. Endocr Rev 7:3–10

Conn PM, Venter JC (1985) Radiation inactivation (target cell analysis) of the gonadotropin-releasing hormone receptor: evidence for a high molecular weight complex. Endocrinology 116:1324–1326

DeLean A, Ferland L, Drouin J, Kelly PA, Labrie F (1977) Modulation of pituitary thyrotropin releasing hormone receptor levels by estrogens and thyroid hormones. Endocrinology 100:1496–1503

Ferland L, Drouin J, Labrie F (1975) Role of sex steroids on LH and FSH secretion in the rat. In: Labrie F, Meites J, Pelletier G (eds) Current topics in molecular endocrinology. Plenum, New York, pp 191–209

Fink G, Aiyer MS, Jamieson MG, Chiappa SA (1975) Factors modulating the responsiveness of the anterior pituitary gland in the rat with special reference to gonadotrophin releasing hormone (GnRH). In: Motta M, Crosignani PG, Martini L (eds) Hypothalamic hormones. Academic, London, pp 139–160

Giguere V, Lefebvre F, Labrie F (1981) Androgens decrease LHRH binding sites in rat anterior pituitary cells in culture. Endocrinology 108:350–352

Giguere V, Lefevre G, Labrie F (1982) Site of calcium requirement for stimulation of ACTH release in rat anterior pituitary cells in culture by synthetic ovine corticotropin-releasing factor. Life Sci 31:3057–3062

Gillies GE, Linton EA, Lowry PJ (1982) Corticotropin releasing activity of the new CRF is potentiated several times by vasopressin. Nature 299:355–357

Gregg DW, Nett TM (1989) Direct effects of estradiol-17β on the number of gonadotropin-releasing hormone receptors in ovine pituitary. Biol Reprod 40:288–293

Hazum E (1982) GnRH-receptor of rat pituitary is a glycoprotein: differential effects of neuroamidase and lectins on agonists and antagonists binding. Mol Cell Endocrinol 26:217–222

Hazum E, Keinan D (1982) Photoaffinity labeling of pituitary gonadotropin-releasing hormone receptors during the rat estrous cycle. Biochem Biophys Res Commun 107:695–698

Hazum E, Schvartz I, Waksman Y, Keinan D (1986) Solubilization and purification of rat pituitary gonadotropin releasing hormone receptor. J Biol Chem 261:13043–13048

Holmes MC, Antoni FA, Szentendrei T (1984) Pituitary receptors for corticotropin-releasing factor: no effect of vasopressin on binding or activation of adenylate cyclase. Neuroendocrinology 39:162–169

Hopkins CR, Gregory H (1977) Topographical localization of the receptors for luteinizing hormone–releasing hormone on the surface of dissociated pituitary cells. J Cell Biol 75:528–540

Huckle WR, Conn PM (1988) Molecular mechanism of gonadotropin releasing hormone action. II. The effector system. Endocr Rev 9:387–395

Lloyd JM, Childs GV (1988) Changes in the number of GnRH-receptive cells during the rat estrous cycle: biphasic effects of estradiol. Neuroendocrinology 48:138–146

Marchetti C, Childs GV, Brown AM (1987) Membrane currents of identified isolated rat corticotropes and gonadotropes. Am J Physiol 252:E340–E346

Marian J, Conn PM (1979) Gonadotropin releasing hormone stimulation of cultured pituitary cells requires calcium. Mol Pharmacol 16:196–201

Menon M, Peegel H, Katta V (1985) Estradiol potentiation of gonadotropin-releasing hormone responsiveness in the anterior pituitary is mediated by an increase in gonadotropin-releasing hormone receptors. Am J Obstet Gynecol 151:534–540

Naor Z, Atlas D, Clayton RN, Forman DS, Amsterdam A, Catt KJ (1981) Interaction of fluorescent gonadotropin-releasing hormone with receptors in cultured pituitary cells. J Biol Chem 256:3049–3052

Rivier C, Vale W (1983) Interaction of corticotropin-releasing factor and arginine vasopressin on adreno-corticotropin secretion in vivo. Endocrinology 113:937–942

Rivier J, Spiess J, Vale W (1983) Characterization of rat hypothalamic corticotropin-releasing factor. Proc Natl Acad Sci USA 80:4851–4855

Roth J, Grunfeld C (1985) Mechanism of action of peptide hormones and catecholamines. In: Wilson JD, Foster DW (eds) Textbook of endocrinology, 7th edn. WB Saunders, Philadelphia, pp 76–122

Savoy-Moore RT, Schwartz NB, Duncan JA, Marshall JC (1980) Pituitary gonadotropin releasing hormone receptors during the rat estrous cycle. Science 209:942–944

Spinedi E, Negro-Vilar A (1983) Angiotensin II and ACTH release: site of action and potency relative to corticotropin releasing factor and vasopressin. Neuroendocrinology 37:446–453

Tibolt RE, Childs GV (1985) Cytochemical and cytophysiological studies of gonadotropin-releasing hormone (GnRH) target cells in the male rat pituitary: differential effects of androgens and corticosterone on GnRH binding and gonadotropin release. Endocrinology 117:396–404

Turgeon JL, Waring DW (1981) Acute progesterone and 17β-estradiol modulation of luteinizing hormone secretion by pituitaries in cycling rats superfused in vitro. Endocrinology 108:413–419

Vale W, Spiess J, Rivier C, Rivier J (1981) Characterization of a 41-residue ovine hypothalamic peptide that stimulates secretion of corticotropin and β-endorphin. Science 213:1394–1397

Westlund KN, Wynn PC, Chmielowicz S, Childs GV (1984) Characterization of a potent biotin-conjugated CRF analog and the response of anterior pituitary corticotropes. Peptides 5:627–631

Willingham MC, Pastan IH (1980) The receptosome: an intermediate organelle of receptor-mediated endocytosis in cultured fibroblasts. Cell 21:67–77

Winiger BP, Birabeau MA, Lang U, Capponi AM, Sizonenko PC, Aubert ML (1983) Solubilization of pituitary GnRH binding sites by means of a zwitterionic detergent. Mol Cell Endocrinol 31:77–91

Wynn PC, Aguilera G, Morell J, Catt KJ (1983) Properties and regulation of high-affinity pituitary receptors for corticotropin-releasing factor. Biochem Biophys Res Commun 110:602–605

Wynn PC, Harwood JP, Catt KJ, Aguilera G (1985) Regulation of corticotropin-releasing factor (CRF) receptors in the rat pituitary gland: effects of adrenalectomy on CRF receptors and corticotroph responses. Endocrinology 116:1653–1659

Zolman JC (1983) Peptide–receptor protein relationships: steroid feedback in GnRH stimulation of the anterior pituitary. J Steroid Biochem 18:297–301

Growth Factor Receptors

I. DAMJANOV

1 Introduction

Proliferation of normal cells is regulated through the complex interaction of each cell with neighboring homologous and heterologous cells and extracellular tissue components forming its microenvironment. In addition to the cell–cell and cell–extracellular matrix contact, the proliferation of cells is regulated by a variety of soluble growth factors acting as endocrine, paracrine, or autocrine stimulators of cell division (SPORN and ROBERTS 1988). These growth factors, defined as polypeptides that stimulate cell proliferation through binding to specific high affinity cell membrane receptors (GOUSTIN et al. 1986), have pleiotropic effects which are often diametrically opposite from one another in different cell systems. Thus, a growth factor that is highly mitogenic for one cell type may act as an inhibitor of proliferation of other cells and vice versa. The transforming growth factor beta (TGF-β) stimulates the growth of nonneoplastic fibroblasts in soft agar, while inhibiting the growth of many normal and neoplastic cells (SPORN and ROBERTS 1985). On the other hand a polypeptide that induces differentiation and halts the proliferation of leukemia cells promotes the growth of undifferentiated embryonic cells (WILLIAMS et al. 1988).

The response of the cell to exogenous stimuli activates not only internal metabolic changes but also provides a feedback linking the stimulator and the responder cells into complex interactive loops designed to maintain the integrity of the tissues, organs and the whole mammalian body. Since the breakdown of these autoregulatory mechanisms leads to malignancy, it is obvious that the study of growth factors and corresponding cellular receptors relates directly to the core problems of oncology (ROBERTS et al. 1988).

Recent developments that have demonstrated marked homology between polypeptide hormones and their receptors on the one hand, and viral and cytoplasmic oncogenes and proto-oncogenes on the other, have highlighted the pivotal role of growth factors and cell surface molecules in the pathogenesis of neoplasia even more. The numerous discoveries made over the last 10 years in the field of growth factors, cellular growth factor receptors, and related oncogenes have been extensively reviewed (HELDIN and WESTERMARK 1984; GOUSTIN et al. 1986; SALOMON and PERROTEAU 1986; FOLKMAN and KLAGSBRUN 1987; CARPENTER 1987; DEFIZE et al. 1987; ADAMSON 1987; DEUEL, 1987; DERYNCK 1988; SIBLEY et al. 1988; FISHER et al. 1989). Hence, the present article will only summarize some of the background information about the principal growth factors and their cognate receptors and will review some of the recent research data of interest to the anatomical pathologist.

2 Growth Factors

The list of endogenous factors regulating the proliferation of mammalian cells has been expanding rapidly owing to the advances in laboratory technology as well as the development of computerized information storage and retrieval methodology that have made possible the comparison of data gathered in various ostensibly unrelated systems. This multidisciplinary approach has led to the discovery of polypeptide superfamilies to which various growth factors and their receptors can be assigned, the correspondence between growth factors/receptors and various viral oncogenes, and the linkage of growth factors/receptors to other functionally unrelated proteins that show structural homology. An overview of the superfamilies encompassing growth hormones and their receptors is beyond the scope of this article and I will concentrate only on several of the best characterized growth factors and their receptors: epidermal growth factor, platelet-derived growth factor, nerve growth factor, insulin-like growth factors, fibroblast growth factors, and the transforming growth factors.

2.1 Epidermal Growth Factor

Epidermal growth factor (EGF) is a 6-kilodalton (kDa) polypeptide composed of 53 amino acids (COHEN 1987). Originally isolated from the murine salivary glands (COHEN 1962), it was later found in many normal and pathologically altered mammalian tissues (COHEN 1987). The proven and hypothetical functions of EGF could be classified as endocrine, paracrine, and exocrine (KASSELBERG et al. 1985). Some of the well-known functions of the EGF are summarized in Table 1.

Table 1. Important biological functions of EGF (COHEN 1987; READ 1987)

Cell proliferation
Eyelid opening (rodent pups)
Eruption of teeth
Maturation of intestinal epithelium
Inhibition of pepsin secretion
Wound healing
Smooth muscle cell contraction

Epidermal growth factor is present in blood, cerebrospinal fluid, milk, seminal fluid, saliva, and gastric and pancreatic juices (CARPENTER and COHEN 1979; CARPENTER 1985; COHEN 1987). The urinary growth factor known as *urogastrone* was proven to be identical to EGF (reviewed by READ 1987). Body fluids, secretions, and serum contain in various proportions either the 6-kDa EGF, the 9-kDa pre-EGF, a 25-kDa secretory complex, or a 75-kDa complex composed of two molecules of EGF and two carrier proteins (reviewed by COHEN 1987). The exact physiological roles of the circulating and/or secreted EGF in various body compartments remain incompletely understood.

Immunohistochemical studies with antibodies to the EGF have confirmed the biochemical data and have shown that the salivary glands are the major site of synthesis of this polypeptide growth factor (HEITZ et al. 1978). However, EGF is not restricted to the salivary glands and it could be immunohistochemically demonstrated in Brunner's glands, gastric and pyloric glands of the stomach, renal medullary tubules, pancreas, anterior pituitary, sweat glands, and adrenal medullary cells (KASSELBERG et al. 1985; SALIDO et al. 1986; POULSEN et al. 1986). Lactating breast glands and some of the primordial follicles in the ovary of the newborn also contain EGF (KASSELBERG et al. 1985), indicating that the expression of EGF may be developmentally and hormonally regulated. Some tissues, like the distal tubule of the kidney, synthesize the prepro-epidermal growth factor but do not form the completely processed EGF (RALL et al. 1985). The function of the incompletely processed EGF is also unknown.

Epidermal growth factor could play a critical role in carcinogenesis (STOSCHECK and KING 1986a). Thus, it can cause malignant transformation of nonneoplastic cells under certain circumstances (VELU et al. 1987); it promotes viral and chemical transformation in several model systems; and it shares several properties with the viral oncogenes, and among others mediates phosphorylation of crucial cytoplasmic proteins (STOSCHEK and KING 1986 b). EGF induces rapidly c-*fos* and c-*myc* proto-oncogenes (BRAVO et al. 1985).

As expected from histochemical data on normal tissues EGF is found in tumors of salivary glands (MORI et al. 1987; TSIKUTANI et al. 1987) and Brunner gland tumors (RUFENACHT et al. 1986). Immunoreactive EGF may be found in the cytoplasm of some human tumors (SATO et al. 1985; FOWLER et al. 1988), although as stated by GOUSTIN et al. (1986), all claims to this effect should be supported by proof that the immunoreactive substance is EGF and not the closely related TGF-α.

The 53 amino acid EGF polypeptide shows a 30% structural similarity with the 50 amino acid long transforming factor alpha (TGF-α) (DERYNCK et al. 1984, 1987). Both growth factors show remarkable spacing of six cysteines and similar disulfide bridges. Most importantly, both factors bind to the same cell surface receptor, although TFG is only half as efficient as EGF. TGF-α is synthesized by many carcinomas, and since many of these express the EGF receptor, it has been postulated that the binding of the TGF-α to the EGF receptor may play an autocrine function in promoting the autonomous growth of these cell lines. TGF-α is also secreted by normal keratinocytes in the skin (COFFEY et al. 1987), activated macrophages, pituitary cells, and some neural cell in the brain (WILCOX and DERYNCK 1988 a). Synthesis of TGF-α is especially prominent in the placenta and several fetal tissues, such as the developing kidney, the nasopharyngeal pouch, and the otic vesicles (WILCOX and DERYNCK 1988 a, b). In the pregnant rat uterus TGF-α is secreted by the decidua (HAN et al. 1987), and its production peaks shortly after implantation of the embryo. All this suggests that TGF-α, like EGF, participates in regulating normal cell proliferation, as well as in mediating the neoplastic cell growth (DERYNCK 1988).

As reported by COFFEY et al. (1987), TGF-α has a remarkable ability to autocatalyze its own synthesis even in normal cells. Since many tumor cells coexpress TGF-α with TGF-β and the EGF receptor (DERYNCK et al. 1987), it seems that TGF-α could represent a "neoplastic form" of EGF, or a factor that is more suitable for autocrine stimulation of cells. Analogously, high levels of TGF-α in embryonic cells (DEFIZE et al. 1987) suggest that TGF-α could represent an "embryonic" equivalent of EGF. TGF-α may be involved in nonneoplastic hyperproliferative states, as evidenced by its overexpression in psoriasis (ELDER et al. 1989).

Epidermal growth factor has considerable structural similarity to several other polypeptides that belong to the same superfamily. This family includes, among others, viral growth factors such as vaccinia growth factor (STROOBANT et al. 1985), myxoma growth factor, Shope fibroma growth factor (CHANG et al. 1987), and several mammalian proteins such as tissue type plasminogen activator, clotting factors X and XI, the low density lipoprotein receptor, human proteoglycan core protein, cytotactin, and a new bifunctional cell growth modulator named amphiregulin (SHOYAB et al. 1989). Due to the structural similarity with EGF, some of these polypeptides also bind to the EGF receptor.

2.2 Platelet-Derived Growth Factor

Platelet-derived growth factor (PDGF) is one of the potent mitogenic polypeptides in human serum (reviewed by Ross and Vogel 1978, Ross et al. 1986). The human PDGF consists of two chains, A and B, linked together with disulfide bonds into AA or BB homodimers and AB heterodimers (Johnsson et al. 1982). Human PDGF derived from platelets is mostly an AB dimer (Hammacher et al. 1988). The gene for the 16-kDa human A polypeptide has been localized to chromosome 7, whereas the gene for the 16-kDa B chain is on chromosome 22 (Betsholtz et al. 1986; Rao et al. 1986; Deuel 1987). The A and B chains show 56% homology. The B chain is identical to the polypeptide encoded by the simian sarcoma virus oncogene or its cellular counterpart c-*sis* (Johnsson et al. 1984).

Human PDGF in the serum is mostly derived from the alpha granules of platelets (Ross et al. 1986), although it has been shown that endothelial cells (Kavanaugh et al. 1988) and macrophages (Shimokado et al. 1985) also have the capacity to secrete this growth factor. The placenta is a major source of PDGF (Goustin et al. 1985). Early mouse embryos (Rappolee et al. 1988) and the aortic smooth muscle cells, but not those of the adult rats, also express PDGF (Seifert et al. 1984), suggesting that PDGF may play a role in development. The coexpression of the PDGF and the PDGF receptor on human placental cytotrophoblastic cells suggests that PDGF may be an autocrine factor in the placenta and account for the invasive nature of the trophoblastic cells (Goustin et al. 1985).

The structural identity between the B chain of PDGF and the c-*sis* protooncogene (Rao et al. 1986) indicates that PDGF may play a crucial role in simian sarcoma virus-induced malignant transformation of infected cells. The function of PDGF in human tumors is less clear (Heldin et al. 1987). Several malignant cell lines secrete PDGF-like growth factors, which most likely act through the autocrine stimulation loop promoting cell growth in vitro (Betsholtz et al. 1986). Thus, PDGF or PDGF-like factors have been identified in mouse and human embryonal carcinoma cell lines (Rizzino et al. 1988; Rizzino and Bowen-Pope 1988; Weima et al. 1988), osteosarcoma, breast carcinoma, lung cancer, melanoma, and glioma (Pantazis et al. 1985; Westermark et al. 1986; Betsholtz et al. 1986; Perez et al. 1987; Nister et al. 1988 a, b). Curiously enough, although these tumor cells express mRNA for both chains of PDGF, the growth factor produced is mostly composed of homodimers of PDGF A chains (Nister et al. 1988 a), known to be less mitogenic than the B chain (Nister et al. 1988 b). The possible autocrine function of tumor-derived PDGF has been postulated, but it is not fully understood because many of the tumor cell lines producing PDGF do not express the cognate receptor (Westermark et al. 1986). The role of tumor-derived PDGF in promoting the growth of the stromal cells in the host (PEREZ et al. 1987) deserves further scrutiny.

Platelet-derived growth factor is involved in several aspects of acute inflammatory response, wound healing, and scar formation (DEUEL and SENIOR 1987). Pulmonary fibrosis may evolve due to the action of PDGF from alveolar macrophages (MARTINET et al. 1987). It has been postulated that PDGF mediates atherosclerosis (WILCOX et al. 1988), and many other conditions such as glomerulonephritis, myelofibrosis, and keloid formation (WILLIAMS 1989).

2.3 Nerve Growth Factor

Nerve growth factor (NGF) is the first well-characterized polypeptide growth factor discovered by Rita LEVI-MONTALCINI, who received the Nobel prize for her pioneering work in this field (LEVI-MONTALCINI 1987). NGF is a complex protein composed of an α-, a β-, and a γ-subunit (reviewed by BRADSHAW 1978; GREENE and SHOOTER 1980). The β-subunit, a 26-kDa dimer composed of two identical chains, represents the active form of NGF. The gene for human NGF has been localized to chromosome 1 and the pathway of synthesis from the initial gene-encoded polypeptide of 307 amino acids to the 118 acid mature NGF subunit protein formed by cleavage has been mostly elucidated (LEVI-MONTALCINI 1987).

Nerve growth factor was first purified from snake venom and mouse salivary glands. However, it should be remembered that the original experiments that led to the identification of this nerve growth-promoting substance were performed with a mouse sarcoma implanted into the chick embryo (reviewed by LEVI-MONTALCINI 1987) and that many tumor cell lines secrete NGF (BRADSHAW 1978). Thus, it appears that all cells in the body synthesize NGF in small amounts, although it is not known for what purposes.

The function of NGF is considered to be primarily trophic, as evidenced by its stimulation of the neurite outgrowth and the maintenance of cholinergic brain neurons (DREYFUS 1989).

Numerous studies with clonal pheochromocytoma cells (PC 12), a well-characterized target for NGF (GREENE and TISCHLER 1976), show that NGF affects the growth, metabolic functions, and internal structure of these cells (e.g. VAN HOOF et al. 1989). However, as pointed out by LEVI-MONTALCINI (1987), the facts that it is present in snake venom, that the concentration of NFG is tenfold higher in salivary glands of males than female mice, and that it is discharged into blood in large amounts in male mice fighting each other, suggest that NGF could have a general role regulating responses of the body to stress and the "the defense or offense mechanisms." A possible role for NGF in reproduction was suggested by the high concentration of NGF in male genital organs (HARPER and THOENEN 1980).

2.4 Insulin-Like Growth Factors

Insulin-like growth factors (IGF) I and II are serum growth factors related to insulin (BLUNDELL and HUMBEL 1980), also known as somatomedin C and A respectively. These growth factors are widely distributed in many tissues (e.g., ARON et al. 1989). They bind to specific cell surface receptors and also compete for some other growth factor receptors (reviewed by ROSENFELD and HINTZ 1986). In the serum, cerebrospinal fluid and amniotic fluid IGFs circulate bound to specific carrier proteins (ROSENFELD et al. 1989); however, their precise physiological function in the body is not known. High levels of IGF-II have been found in the cerebrospinal fluid and it appears that this factor is secreted from the choroid plexus and leptomeningeal cells (HASELBACHER et al. 1985; STYLIANOPOULOU et al. 1988). DAUGHADAY and ROTWEIN (1989) have reviewed the current molecular biology data on both IGF-I and IGF-II.

2.5 Fibroblast Growth Factors

Heparin-binding proteins extracted from the bovine brain and pituitary are mitogenic to fibroblasts and are known under the name of fibroblast growth factors (FGFs) (reviewed by GOSPODAROWICZ et al. 1986, 1987). An acidic 17.4-kDa form and a basic form that could be resolved into several species have been isolated. The FGFs are potent mitogens and stimulators of angiogenesis (FOLKMAN and KLAGSBRUN 1987). Three different oncogenes, *int*-2, *hst,* and a bladder oncogene, are closely related to the basic FGF, which itself seems to be a potential oncogene (TAIRA et al. 1987; ROGELJ et al. 1988). Both the acidic and the basic FGF have been sequenced (ABRAHAM et al. 1986; GIMENEZ-GALLEGO et al. 1985). Some human tumor cells, like chondrosarcoma and hepatoma, produce basic FGF-like polypeptides (KLAGSBRUN et al. 1986). Sex hormones modulate the synthesis of the basic FGF in human endometrial adenocarcinoma cells (PRESTA 1988), indicating that this growth factor is linked into regulatory circuits with other hormones.

2.6 Transforming Growth Factors

Transforming growth factor alpha (TGF-α) is structurally related to EGF and it binds to the EGF receptor. It is unrelated and quite distinct from the transforming growth factor beta (TGF-β). TGF-β is a 25-kDa homodimer that transforms normal cells and stimulates them to form colonies in soft agar (reviewed by SPORN et al. 1986; ROBERTS et al. 1988). Two molecular forms of TGF-β have been identified – named β1 and β2 (reviewed by BARNARD et al. 1988). Each of these polypeptides consist of 112 amino acids. Eighty of these amino acids including all nine cysteines, are conserved and identical in both TGF-β1 and TGF-β2 (CHEIFETZ et al. 1987). The exact physiological function of TGF-β is not known, although the response seems to be cell specific

(BARNARD et al. 1988). TGF-β is secreted by a number of normal and neoplastic cells and even during embryonic development (HEINE et al. 1987). TGF-β belongs to a superfamily of polypeptide hormones which includes mullerian inhibitory substance, inhibins, and Vg-1, a protein from *Xenopus levis* that promotes mesodermal induction (MASSAGUÉ 1987). An autocrine function has been postulated for this growth factor in embryonal carcinoma cells (WEIMA et al. 1988) and breast carcinoma cells (ARTEAGA et al. 1988), some of which coexpress TGF-β with its cognate receptor and TGF-α.

3 Growth Factor Receptors

The effects of soluble growth factors on cells would not be possible without the cell surface receptors, which represent the crucial ligand and make the transmittal of the specific messages across the cell membrane to the effector system in the cell cytoplasm and the nucleus possible (reviewed by ROZENGURT 1986; SIBLEY et al. 1988; DIETEL et al. 1989 and this volume). A series of receptors for various endocrine, paracrine, and autocrine growth factors have been identified, and knowledge about their role in normal and neoplastic growth is accumulating at an exponential rate.

3.1 Epidermal Growth Factor Receptor

Epidermal growth factor receptor (EGFr) is a well-characterized transmembrane glycoprotein whose structure and function have been reviewed extensively (CARPENTER 1985, 1987). Originally isolated from the squamous cell carcinoma A-431, which overexpresses this glycoprotein (WRANN and FOX 1979), it was subsequently found in normal liver (COHEN et al. 1982a), in developing mouse tissues (ADAMSON and MEEK 1984), and in many tissues of the adult organism (DAMJANOV et al. 1986). The number of EFGrs varies from one cell type to another (Table 2). It seems to be more prominent in embryonic tissues and the proliferative cell compartments of adult epithelial tissues

Table 2. Number of EGFrs on various cells

Cell type	Number of receptor cells	Reference
A-431 squamous cell carcinoma (human)	1 – 4 000 000	WRAN and FOX 1979
Hepatocytes (rat)	600 000	DUNN et al. 1986
Breast epithelium (human)	100 000	ZAJCHOWSKI et al. 1988
Fibroblasts (human)	40 000 – 100 000	CARPENTER and COHEN 1979
Aortic smooth muscle cells (rat)	14 000	NANNEY et al. 1988
Hematopoietic cells	0	CARPENTER 1987

(GREEN et al. 1983; GUSTERSON et al. 1984; DAMJANOV et al. 1986), and its expression is age dependent (OLIVER 1988). EGFr is present in the placenta (MAGID et al. 1986) and its expression increases during pregnancy (CHEN et al. 1988). However, it is also present on nonproliferating mesenchymal cells, such as aortic smooth muscle, although in considerably smaller numbers than on epithelial cells (NANNEY et al. 1988). Analogues of human EFGr have been identified in other mammals and lower animals, including *Drosophila* (SCHEJTER and SHILO 1989). The gene for human EGFr was localized to the p14–p12 region of chromosome 7 (CARPENTER 1987).

The entire EGFr has a molecular weight of 170 kDa, which includes a 62 amino acid long extracellular domain that has two cysteine-rich regions, a short membrane spanning domain, and an intracytoplasmic part that includes a stretch of 250 amino acids homologous to the *src* family of tyrosine kinase (DOWNWARD et al. 1984a).

The kinase activity of EGFr is crucial for most of the receptor functions, including transduction of the signal, trafficking of the receptor, and initiation of DNA synthesis (YARDEN and ULLRICH 1988a, b; SCHLESSINGER 1988a, b). Several N-linked oligosaccharide chains, accounting for 30 kD of the total molecular weight of EGFr, are linked to the extracellular domain (CUMMINGS et al. 1985). Mannose phosphate has been reported in side chains (TODDERUD and CARPENTER 1988). Carbohydrate side chains show homology to blood group A (CHILDS et al. 1984) and react with antibodies to blood group antigens.

The receptor is synthesized in the rough endoplasmic reticulum. Glycosylation, which is apparently cotranslational, is essential for the intracellular trafficking of the receptor, whereas posttranslational processing of the extracellular domain is important for the acquisition of ligand-binding capacity (CARLIN and KNOWLES 1986). EGF binding to the receptor may induce reversible aggregation of the receptor (YARDEN and SCHLESSINGER 1987), allosteric changes, or oligomerization (SCHLESSINGER 1988a). Oligomerization of the receptor is, however, not an absolute prerequisite for its activation (NORTHWOOD and DAVIS 1988).

Clustering of ligand–receptor complexes into clathrin-coated pits is followed by rapid degradation of both molecules in the lysosomes (DUNN et al. 1986). In contrast to the rapid sequence of events that follows binding of the ligand, the initiation of DNA synthesis is a relatively late event and it occurs only if the receptor is occupied by the ligand for 8 h. EFGr is shed into the medium from the surface of the A-431 cells (COHEN et al. 1982b), but this seems to be of little if any physiological significance and has not been detected in other cell cultures (CARPENTER 1987).

Epidermal growth factor receptor shows homology to the *erb*B oncogene of the avian erythroblastosis virus (ULLRICH et al. 1984) and belongs to the family of cysteine-rich protein tyrosine kinases, which includes the insulin receptor and the receptor for IGF-I (YARDEN and ULLRICH 1988). The NGF receptor and IL-2 receptor have a cysteine-rich extracellular domain but have no protein kinase activity. PDGF and CSF-1 receptor have a similar structure to EGF but have no cysteine-rich regions and a different cytoplasmic domain. The v-*erb*B lacks the extracellular domain of the EGFr and a portion of the C-terminal tyrosine autophosphorylation sites, thus representing a truncated form of the human receptor (CARPENTER 1987).

The cellular equivalent of the v-*erb*B is also called c-*erb*B-1 to distinguish it from the c-*erb*B-2 or HER2, a completely different gene located on human chromosome 17 (COUSSENS et al. 1985; FUKUSHIGE et al. 1986). The c-*erb*B-2 gene encodes a protein with 50% homology to the EGFr (BARGMANN et al. 1986a, b). It is analogous to the activated *neu* oncogene in mouse-rat neuroblastoma induced with ethyl nitrosourea. Rat *neu* gene encodes a 180-kDa protein that has no binding affinity for the EGF. The ligand for the *neu* or c-*erb*B-2 has not yet been identified although a protein secreted by *ras*-transformed fibroblasts is a strong candidate (YARDEN and WEINBERG 1989). LEE et al. (1989) have transformed NIH 3T3 cells by transfecting them with human c-*erb*B-2 and have obtained evidence suggestive of autocrine production of an endogenous ligand that could thus autostimulate the receptor-containing cells.

Human EGFr binds not only EGF but also TGF-α, which may be the major mediator of cell proliferation in tumors, in reactive processes such as psoriasis (ELDER et al. 1989), and even in wound healing (SCHULTZ et al. 1987). Murine EGF binds to human EGFr and to chicken EGFr, suggesting that there is considerable evolutionary conservation of both ligands. However, the affinity of the mammalian EGF for the mammalian receptor is approximately 100 times higher than its affinity for the avian receptor (LAX et al. 1989). Vaccinia virus growth factor and Shope fibroma virus, which share only one-fifth of the amino acid identity with either EGF or TGF-α, have a similar affinity for the EGFr as the two mammalian ligands, and induce a set of cellular events similar to that induced by the natural ligands (BLOMQUIST et al. 1984; CHANG et al. 1987). A 10-kDa gene product of human adenovirus that resembles the putative transmembrane domain of EGFr at the cytoplasmic face is also able to bind to EGFr and to downregulate it by forming hetero-oligomers that are internalized and degraded (CARLIN et al. 1989).

Autophosphorylation of mammalian EGFr can be accomplished with soluble EGF and TGF-α, but also with uncleaved cell membrane-bound pro-TGF (WONG et al. 1989) or the nonsecreted growth factor (BRACHMANN et al. 1989). Apparently EGFr could be activated by cell-to-cell contact and even by cells that do not secrete the fully assembled growth factors. Furthermore, since many cells coexpress EGFr with TGF-α, one could hypothesize that these cells would not necessarily have to secrete the growth factor into the pericellular milieu but could autostimulate themselves in an autocrine mode or through cell-to-cell contact stimulate adjacent cells in a paracrine mode. The effect of TGF-α derived from activated macrophages (MADTES et al. 1988) on tumor cells remains to be explored further. Antibodies to EGF linked to an immunotoxin may inhibit the growth of cell lines overexpressing the EGFr (OZAWA et al. 1989), indicating that this receptor indeed plays a pivotal role in regulating tumor cell growth.

Epidermal growth factor receptor expression depends on many factors, including TGF-β (THOMPSON et al. 1988). EGFr is not expressed on undifferentiated mouse embryonal carcinoma cells, but it does appear after these cells have been induced to differentiate (ADAMSON and HOGAN 1984). Interestingly, EGFr is amplified on drug-resistant neuroblastoma cells (MEYERS et al. 1988). It is not known why some tumor cells produce an altered EGFr (STECK et al. 1988). Obviously, the intricacies of EGFr regulation are far from clear.

3.1.1 Epidermal Growth Factor Receptor in Human Tumors

Aberrant overexpression of EGFr, especially if accompanied by synthesis of the homologous ligands, such as TGF-α, could provide the cells with proliferative advantage. As predicted, the human gene for EGFr was found to be actively transcribed or overexpressed in many tumors in·vivo and many cell lines permanently propagated in vitro.

Central nervous system tumors, such as gliomas (LIEBERMANN et al. 1984, 1985; WONG et al. 1987; REIFENBERGER et al. 1989), meningiomas (LIEBERMANN et al. 1984; WESTPHAL and HERMANN 1986; WEISMAN et al. 1987; REIFENBERGER et al. 1989), medulloblastomas, and anaplastic neurinoma (REIFENBERGER et al. 1989) express EGFr. The expression of the EGFr correlates with the histological level of anaplasia of gliomas, and several glioblastoma-derived cell lines established in vitro overexpress the receptor (GEROSA et al. 1989). Since the gene for the human EGFr is located on chromosome 7 and abnormalities of chromosome 7 are common in gliomas (HENN et al. 1986), the overexpression the receptor may be related to the chromosomal derangements.

Some glioma cells show selective amplification of the cytoplasmic domain of EGFr (MALDEN et al. 1988). The lack of EGFr in highly malignant medulloblastomas and primitive neuroectodermal tumors of the brain (REIFENBERGER et al. 1989) and the presence of the receptor on most benign meningiomas, however, indicate that it would be premature to predict the malignancy of all intracranial tumors on the basis of their EGFr status. It is nevertheless encouraging that the antibodies to EGFr localize in gliomas xenografted to nude mice (TAKAHASHI et al. 1987) and could thus be used in the diagnosis and treatment of human brain tumors as well.

Respiratory tract tumors and cell lines derived from them express EGFr. These include squamous cell carcinomas of the nasopharynx (EISBRUCH et al. 1987; MAXWELL et al. 1989), and lung (BERGER et al. 1987 b; VEALE et al. 1987, 1989; HAEDER et al. 1988). Since some lung carcinoma cells coexpress TGF-α (SIEGFRIED and OWENS 1988; IMANISHI et al. 1989), it is most plausible that EGFr expressed on these tumors is involved in autocrine stimulation of tumor cell proliferation. The fact that the externally added TGF-α inhibits the proliferation of human adenocarcinoma cell lines capable of producing TGF-α provides additional support for the existence of an autocrine pathway of receptor stimulation in these cells (IMANISHI et al. 1989).

Gastrointestinal tract tumors, such as esophageal carcinoma (OZAWA et al. 1987; HOLLSTEIN et al. 1988), gastric carcinoma (YASUI et al. 1988a, b; BENNETT et al. 1989), and colon carcinoma (YASUI et al. 1988b) express EGFr. Concurrent expression of EGFr and TGF-α was observed more often in tumors than in nonmalignant tissue of the stomach (BENNETT et al. 1989) and only in one of 14 normal samples. However, gastritis alters EGFr expression and the normal esophagus contains both TGF-α and EGFr. Hence, the significance of EGFr on neoplastic and nonneoplastic gastrointestinal tissues cannot be interpreted unequivocally.

Breast tumors commonly overexpress EGFr (FITZPATRICK et al. 1984; SAINSBURY et al. 1985, 1987, 1988; SKOOG et al. 1986; WRBA et al. 1988; CAPPELLETTI et al. 1988). The consensus of these studies is that the EGFr measurements provide useful clinical data (NICHOLSON et al. 1988a, b) and that there is an inverse relationship between the EGFr and steroid receptors. Similar inverse correlation has been repeatedly reported from in vitro studies on breast carcinoma cell lines (SARUP et al. 1988; CORMIER et al. 1989). Approximately 20% of lobular and 35% of ductal carcinomas are EGFr-positive (SAINSBURY et al. 1988). With some exceptions (WRBA et al. 1988) the existing data indicate a positive correlation between the expression of EGFr, the presence of lymph node metastases (BATTAGLIA et al. 1988), the increased risk of early recurrence of the tumor, and poor prognosis (SAINSBURY et al. 1988). Coexpression of growth factors and oncogenes (ZAJCHOWSKI et al. 1988) or EGFr and TGF-α in the same cells (VALVERIUS et al. 1989) is consistent with autocrine stimulation of tumor cell growth in the breast.

Urinary tract tumors have been reported to express EGFr (BERGER et al. 1987a). In this respect it is noteworthy that the urine contains urogastrone, i.e., human EGF that could act on the EGFr of the normal urinary bladder cells. Since the EGF in urine amplifies some cell functions, like the well-known tumor promoters (YURA et al. 1989), the expression of EGFr on various preneoplastic and neoplastic lesions of the urinary bladder deserves closer scrutiny. Increased excretion of TGF-α in urine of patients with breast cancer (STROMBERG et al. 1987) is another factor whose effects on the urinary bladder have not fully been investigated. Enhanced coexpression of EGFr and c-*myc* oncogene has been reported in renal carcinomas (YAO et al. 1988).

Female reproductive tract tumors express EGFr in variable amounts (GULLICK et al. 1986). Interestingly, leiomyomas have fewer EGFrs than normal myometrium (FAYED et al. 1989). However, more systematic studies are needed before any conclusion is reached.

Endocrine tumors have not been studied in great detail for EGFr expression and the data are fragmentary. MIYAMOTO et al. (1988) have reported that thyroid cancer cells have fewer EGFrs than normal thyroid.

The significance of EGFr in tumors has not been established unequivocally, although there is considerable evidence that the overexpression of the receptor portends an unfavorable prognosis at least in some tumors, such as breast carcinoma. In view of the fact that there are reliable radioimmunoassays (GULLICK et al. 1984) and that many monoclonal antibodies against EGFr have been produced (catalogued by DEFIZE et al. 1987), one can only hope that more definitive studies will be forthcoming.

3.1.2 C-erbB-2 in Human Tumors

The c-*erb*B-2 or HER2 gene was identified as the human equivalent of the rat *neu* transforming gene expressed on rat neuroblastoma (PADHY et al. 1982). In contrast to rat *neu*, the human gene is a potent oncogene when overexpressed in NIH 3T3 cells (DIFIORE et al. 1987). Although the gene product of *erb*B-2 shows some similarity to EGFr (also known as *erb*B-1), the two genes are distinct (SCHECHTER et al. 1985) and located on different chromosomes (FUKUSHIGE et al. 1986). The 185-kDa *erb*B-2 protein does not bind EGF (SCHECHTER et al. 1984).

Interest in the expression of *erb*B-2 has grown since SLAMON et al. (1987) reported that the gene is amplified in breast cancer cells and claimed that the receptor expression has prognostic implications in breast cancer patients. These claims have been confirmed by some (VAN DE VIJVER et al. 1987, 1988; VENTER et al. 1987; ZHOU et al. 1987; BERGER et al. 1988) and disputed by other scientists (GUSTERSON et al. 1988).

Using antibodies developed by GULLICK et al. (1987), VENTER et al. (1987) showed that the *erb*B-2-encoded protein can be quantified in immunoblots or in histological sections. WRIGHT et al. (1989) demonstrated c-*erb*B-2 immuno-histochemically in 58% of breast tumors and found that the receptor status has prognostic significance. The prognostic value of *erb*B-2 amplification has been reconfirmed by SLAMON et al. (1989), who used several methods to assess the level of gene expression in breast tumors. GUSTERSON et al. (1988), however, did not find any correlation between the c-*erb*B-2 overexpression in tumor cells and the extent of tumor spread to lymph nodes. Peculiarly, more in situ ductal carcinomas (44%) than infiltrating duct carcinomas (16%) were found to overexpress c-*erb*-2, indicating that this protein might indeed be related to malignant transformation. The significance of overexpression of c-*erb*B-2 in some intraductal carcinomas remains to be determined, especially since it may be mediated by several mechanisms (KRAUS et al. 1987).

Data on the expression of c-*erb*B-2 in other human tumors are limited. It has been reported that this gene is not expressed in esophageal carcinoma (HOLLSTEIN et al. 1988). Ovarian carcinomas overexpress *erb*B-2 proto-onco-gene in a manner similar to breast carcinomas (SLAMON et al. 1989).

3.2 Platelet-Derived Growth Factor Receptor

The cell surface receptor for the platelet-derived growth factor (PDGFr) is a 185-kDa glycoprotein that undergoes autophosphorylation upon binding of PDGF to the cell surface of fibroblasts (EK and HELDIN 1982). The literature on PDGF has been reviewed by DEUEL (1987) and WILLIAMS (1989).

Platelet-derived growth factor receptor is expressed on many cells, such as vascular smooth muscle and glial cells, but is not found on endothelial and most hematopoietic cells. Like the other receptors of the protein tyrosine kinase group, it has intrinsic tyrosine kinase activity, and undergoes autophosphorylation upon binding with the ligand (WILLIAMS 1989). On the basis of its amino acid sequence and structure, PDGF belongs to the family that includes the colony-stimulating factor 1 (CSF-1), v-*fms*, and c-*kit* proto-oncogene. It has an extracellular, transmembrane, and intracytoplasmic domain. The extracellular domain of PDGFr has five immunoglobulin-like segments, which contain regularly spaced cysteine residues and are stabilized by disulfide bonds (WILLIAMS 1989).

Binding of PDGF to the 185-kDa receptor involves primarily the BB form of the growth factor (WILLIAMS 1989). HELDIN et al. (1988) found that PDGF BB binds to two components of 160 and 175-kDa. The AA form of PDGF binds predominantly to a 125-kDa and to a lesser extent to the 160-kDa component. Binding of the growth factor results in a set of early responses and ultimately in DNA synthesis which is predominantly mediated by the PDGF-B receptor (HELDIN et al. 1988). PDGF increases the expression of c-*myc* and c-*fos* proto-oncogenes (WILLIAMS 1989).

Simian sarcoma virus-derived oncogene v-*sis* binds to PDGFr and only receptor-positive cells can be transformed with this oncogene (DEUEL 1987). PDGFr have been demonstrated on the surface of osteosarcoma cells and glioma cells (BETSHOLTZ et al. 1986), some of which also secrete PDGF (HELDIN et al. 1986).

The role of PDGFr in regulating tumor cell proliferation is unclear, especially since the antibodies to PDGF do not inhibit cell proliferation. Autocrine stimulation of the receptor by the ligand produced by the same tumor cell, without actual secretion of the growth factor, is one of the possible explanations of these findings (SITARAS et al. 1988).

It may be that PDGFr-positive cells down-regulate the receptor in response to PDGF and other growth factors and even as a result of increased cell density in vitro (RIZZINO et al. 1988). It is feasible that similar down-regulation could occur in vivo or in response to endogenously synthesized PDGF. Epithelial tumor cells like prostatic or breast carcinoma cells (ROZENGURT et al. 1985; PEREZ et al. 1987) do not express PDGFr but synthesize PDGF, which may promote the proliferation of stromal fibroblasts and contribute to the desmoplastic reaction of the host.

3.3 Nerve Growth Factor Receptor

The NGF binds with high affinity to cholinergic neurons (reviewed by DREY-FUS 1989), especially during development, and to sensory peripheral neurons and sympathetic neurons (RICHARDSON et al. 1986). Glia cells and meninges also have NGF receptor (NGFr) (DREYFUS 1989). The adrenal medullary cells, paraganglia, and Schwann cells (TANIUCHI et al. 1988) and even some nonneural cells, like mast cells, may respond to the growth factor (LEVI-MONTALCINI 1987). Immunohistochemically, NGFr can be detected in human peripheral nerves and perivascular cells (PEROSIO and BROOKS 1989).

Rat pheochromocytoma cell line PC 12 (GREENE and TISCHLER 1976) undergoes differentiation and ceases to proliferate under the influence of the growth factor. This effect of NGF is presumably mediated through the receptor. Several species of NGFr have been isolated and a gene coding a 90-kDa NGFr from the PC 12 cells has been cloned (LEVI et al. 1985). The gene for the human NGF is localized on chromosome 17 (HUEBNER et al. 1986).

Nerve growth factor receptor is expressed on neuroblastoma and melanoma cells (FABRICANT et al. 1977; ROSS et al. 1984; GROB et al. 1985) and in some nonneural cell lines such as colorectal carcinoma (RAKOWICZ-SZULCZYNSKA et al. 1988) and soft tissue sarcomas (PEROSIO and BROOKS 1989). The significance of the NGFr in neoplasia and the relationship between the cell membrane and the nuclear receptor (RAKOWICZ-SZULCZYNSKA et al. 1988) needs to be elucidated.

The nature of proteins in the range from 35 to 230 kDa that are immunoprecipitated with the antibody to the NGFr remains poorly understood.

Cholinergic neurons of the rat brain lose NGFr during postnatal development. However, these nerves can be stimulated to reexpress NGFr, suggesting that NGF responsiveness can be reestablished following nerve cell injury (GAGE et al. 1989). Since NGF is released at the site of neural injury it is possible that NGFr plays a crucial role in the response of the nerve cells to injury. The hypothetical involvement of NGFr in various degenerative brain diseases deserves additional studies (DREYFUS 1989).

3.4 Insulin-Like Growth Factor Receptors

Insulin growth factors (IGFs) have specific receptors (ROSENFELD and HINTZ 1986) distinct from those for insulin. IGF-I, also known as somatomedin C, binds with high affinity to a single gene-encoded protein derived from a 180-kDa precursor. Like the insulin receptor, the processed IGF-I receptor consists of an α- and a β-subunit. The 130-kDa α-subunit binds the IGF-I and the 98-kDa subunit has tyrosine kinase activity and is capable of autophosphorylation. The binding of IGF-I to normal cells may be modulated by soluble 34-kDa IGF-I binding protein secreted by several tissues including the human decidua (PEKONEN et al. 1988 b).

IGF-II or somatomedin A binds to a single polypeptide receptor of 250 kDa that is different from the receptor for IGF-I and insulin and has no tyrosine kinase activity (ROSENFELD and HINTZ 1986).

Insulin-like growth factor receptors have been reported on rat and human neuroblastoma cells (RECIO-PINTO et al. 1984; MATTSSON et al. 1986; STURM et al. 1989) and breast carcinoma cells (DE LEON et al. 1988). Almost all breast carcinomas studied by PEYRAT et al. (1988a) expressed amounts of increased receptor for IGF-I, in contrast to benign tumors or normal breast tissues (PEYRAT et al. 1988b). The expression of the receptor for IGF-I in breast tissue correlates directly with the expression of steroid receptors in human breast cancer tissue (PEKONEN et al. 1988a; PEYRAT et al. 1988a).

3.5 Fibroblast Growth Factor Receptors

The acidic and basic FGFs, also known as heparin-binding polypeptides, induce proliferation of fibroblasts, endothelial cells, neuroectodermal cells (GOSPODAROWICZ et al. 1987), and also some epithelial cells such as those from the rat prostate (MCKEEHAN et al. 1987). Due to the high affinity for heparin, FGFs are mostly bound in tissues to the extracellular matrix.

The cellular receptor for the acidic FGF is expressed on the surface of human hepatoma cells (KAN et al. 1988), from which it has been isolated as a 130-kDa protein. The role of the FGF receptors in tumorigenesis is not known. In view of the fact that the oncogene isolated from Kaposi's sarcoma seems to be closely related to FGF (DELLI BOVI et al. 1987), and since endothelial cells have receptors for FGF (KLAGSBRUN et al. 1986), it is plausible that these receptors account for the vascularity of the lesions in this disease.

3.6 Transforming Growth Factor Beta Receptors

The three forms of TGF-β, i.e., the two homodimeric forms of TGF-β1 and TGF-β2 and the heterodimer TGF-β1.2, bind with varying affinity to three receptors (MASSAGUÉ et al. 1987; MASSAGUÉ 1987). The molecular weights of these receptors are 65 kDa, 95 kDa, and 280–330 kDa respectively. The lower molecular weight receptors have higher affinity for TGF-β1 than for TGF-β2. The high molecular weight receptor (the only one that binds all three forms of TGF equally well) carries heparan and chondroitin sulfate glycosaminoglycan chains. The binding site for TGF-β resides in the 100- to 120-kD core polypeptide (CHEIFETZ et al. 1988). All three receptors are usually coexpressed and have been found on most of the cells so far surveyed, the exceptions being retinoblastoma and pheochromocytoma cells (KIMCHI et al. 1988).

The exact role of TGF-β receptors in tumorigenesis is not known. In view of the fact that some human carcinoma cell lines express the receptor and also secrete the TGF-β (ARTEAGA et al. 1988) it is feasible that the growth factor and its receptor are linked into an autocrine loop. However, since TGF-β inhibits the growth of most epithelial tumor cells it is also possible that the cells that are unresponsive have escaped from the effects of the TGF-β by having altered the receptor, having lost it, or having a receptor that cannot transmit the signals from the bound ligand (ROBERTS et al. 1988).

4 Conclusion

A large number of growth factors and growth factor receptors have been identified during the period from the discovery of nerve growth factor to the discovery of the most recent members of the receptor family, such as amphiregulin (SHOYAB et al. 1989). Many of these growth-regulating molecules have been fully characterized, but for others the data are still fragmentary. There is considerable evidence that measurements of receptors on tumors could be of clinical significance. The clinicopathological correlations for EGFr and erbB-2 appear to be most encouraging. However, there is still insufficient information about most of the other growth factor receptors. Expression of growth factor receptors in inflammatory, metabolic, and degenerative disorders has not been studied in greater detail, but even so there is no doubt that these pleiotropic factors play an important role in almost all pathological processes.

Acknowledgements. The original work referred to in this review was supported by the W. W. Smith Charitable Trust, Rosemont, Pennsylvania and PHS Grant AA-07186.

References

Abraham JA, Mergia A, Whang JL, Tumolo A, Friedman J, Hyerrild KA, Gospodarowicz D, Fiddes JC (1986) Nucleotide sequence of a bovine clone encoding the angiogenic protein, basic fibroblast growth factor. Science (Wash) 223:545–548

Adamson ED (1987) Oncogenes in development. Development 99:449–471

Adamson ED, Hogan BLM (1984) Expression of EGF receptor and transferrin by F9 and PC13 teratocarcinoma cells. Differentiation 27:152–157

Adamson ED, Meek K (1984) The ontogeny of epidermal growth factor receptors during mouse development. Dev Biol 103:67–71

Aron DC, Rosenzweig JL, Abboud HE (1989) Synthesis and binding of insulin-like growth factor I by human glomerular mesangial cells. J Clin Endocrin Metab 68:585–591

Arteaga CL, Tandon AK, von Hoff DD, Osborne CK (1988) Transforming growth factor β: potential autocrine growth inhibitor of estrogen receptor-negative human breast cancer cells. Cancer Res 48:3898–3904

Bargmann CI, Hung M-C, Weinberg RA (1986a) The *neu* concogene encodes an epidermal growth factor-related protein. Nature (Lond) 319:226–230

Bargmann CI, Huang M-C, Weinberg RA (1986b) Multiple independent activations of the *neu* oncogene by a point mutation altering to transmembrane domain of p185. Cell 45:649–657

Barnard J, Bascom CS, Lyons RM, Sipes NJ, Moses HL (1988) Transforming growth factor β in the control of epidermal proliferation. Am J Med Sci 296:159–163

Battaglia F, Scambia G, Rossi S, Benedetti Panici P, Bellantone R, Polizzi G, Querzoli P, Negrini R, Iacobelli S, Crucitti F, Mancuso S (1988) Epidermal growth factor receptor in human breast cancer: correlation with steroid hormone receptors and auxiliary lymph node involvement. Eur J Cancer Clin Oncol 24:1685–1690

Bennett C, Paterson IM, Corbishley CM, Luqmani YA (1989) Expression of growth factor and epidermal growth factor receptor encoded transcripts in human gastric tissues. Cancer Res 49:2104–2111

Berger MS, Greenfield C, Gullick WJ, Haley J, Downward J, Neal DE, Harris AL, Waterfield MD (1987a) Evaluation of epidermal growth factor receptors in bladder tumours. Br J Cancer 56:533–537

Berger MS, Gullick WJ, Greenfield C, Evans S, Addis BJ, Waterfield MD (1987b) Epidermal growth factor receptors in lung tumors. J Pathol 152:297–307

Berger MS, Gottfried WL, Saurer S, Gullick WJ, Waterfield MD, Groner B, Hynes NE (1988) Correlation of c-erbB-2 gene amplification and protein expression in human breast carcinoma with nodal status and nuclear grading. Cancer Res 48:1238–1243

Betsholtz C, Johnsson A, Heldin CH, Westermark B, Lind P, Urdea MS, Eddy R, Shows TB, Philpott K, Mellor A, Knott TJ, Scott J (1986) cDNA sequence and chromosomal localization of human platelet-derived growth factor A-chain and its expression in tumor cell lines. Nature (Lond) 320:695–699

Blomquist MC, Hunt LT, Barker WC (1984) Vaccinia virus 19-kilodalton protein: relationship to several mammalian proteins, including two growth factors. Proc Natl Acad Sci USA 81:7363–7367

Blundell TL, Humbel RE (1980) Hormone families: pancreatic hormones and homologous growth factors. Nature (Lond) 287:781–787

Brachmann R, Lindquist PB, Nagashima M, Kohr W, Lipari T, Napier M, Derynck R (1989) Transmembrane TGF-α precursors activate EGF/TGF-α receptors. Cell 56:691–700

Bradshaw RA (1978) Nerve growth factor. Annu Rev Biochem 47:191–216

Bravo R, Burckhardt J, Curran T, Muller R (1985) Stimulation and inhibition of growth by EGF in different A431 cell clones is accompanied by the rapid induction of c-fos and c-myc proto-oncogenes. EMBO J 4:1193–1197

Cappelletti V, Brivio M, Miodini P, Granata G, Coradini D, DiFronzo G (1988) Simultaneous estimation of epidermal growth factor receptors and steroid receptors in a series of 136 resectable primary breast tumors. Tumor Biol 9:200–211

Carlin CR, Knowles BB (1986) Biosynthesis and glycosylation of the epidermal growth factor receptor in human tumor-derived cell lines A431 and Hep3B. Mol Cell Biol 6:257–264

Carlin CR, Tollefson AE, Brady HA, Hoffman BL, Wold WSM (1989) Epidermal growth factor receptor is down-regulated by a 104,000 MW protein encoded by the E3 region of adenovirus. Cell 57:135–144

Carpenter G (1985) Epidermal growth factor: Biology and receptor metabolism. J Cell Sci Suppl 3:1–9

Carpenter G (1987) Receptors for epidermal growth factor and other polypeptide mitogens. Annu Rev Biochem 56:881–914

Carpenter G, Cohen S (1979) Epidermal growth factor. Annu Rev Biochem 48:193–216

Chang W, Upton C, Hu S-L, Purchio AF, McFadden G (1987) The genome of Shope fibroma virus, a tumorigenic poxvirus, contains a growth factor gene with sequence similarity to those encoding epidermal growth factor and transforming growth factor alpha. Mol Cell Biol 7:535–540

Cheifetz S, Weatherbee JA, Tsang ML-S, Anderson JK, Mole JE, Lucas R, Massagué J (1987) The transforming growth factor-beta system, a complex pattern of cross-reactive ligands and receptors. Cell 48:409–415

Cheifetz S, Andres JL, Massagué J (1988) The transforming growth factor-β receptor type III is a membrane proteoglycan. J Biol Chem 263:16984–16991

Chen C-F, Kurachi H, Fujita Y, Terakawa N, Miyake A, Tanizawa O (1988) Changes in epidermal growth factor receptor and its messenger ribonucleic acid levels in human placenta and isolated trophoblast cells during pregnancy. J Clin Endocr 67:1171–1177

Childs RA, Gregoriou M, Scudder P, Thorpe SJ, Rees AR, Feizi T (1984) Blood group-active carbohydrate side chains on the receptor for epidermal growth factor of A431 cells. EMBO J 3:2227–2233

Coffey RJ, Derynck R, Wilcox JN, Bringman TS, Goustin AS, Moses HL, Pittelkow MR (1987) Production and autoinduction of transforming growth factor-α in human keratinocytes. Nature (Lond) 328:817–820

Cohen S (1962) Isolation of a mouse submaxillary gland protein accelerating incisor eruption and eyelid opening in the newborn animal. J Biol Chem 237:1555–1562

Cohen S (1987) Epidermal growth factor. In Vitro Cell Develop Biol 23:239–250

Cohen S, Fava RA, Sawyer ST (1982a) Purification and characterization of epidermal growth factor receptor/protein kinase from normal mouse liver. Proc Natl Acad Sci USA 79:6237–6241

Cohen S, Ushiro H, Stoscheck CM, Chinkers M (1982b) A native 170,000 epidermal growth factor receptor-kinase complex from shed plasma membrane vesicles. J Biol Chem 257:1523–1531

Cormier EM, Wolf MF, Jordan VC (1989) Decrease in estradiol-stimulated progesterone receptor production in MCF-7 cells by epidermal growth factor and possible clinical implication for paracrine-regulated breast cancer growth. Cancer Res 49:576–580

Coussens L, Yang-Feng TL, Liao Y-C, Chen E, Gray A, McGrath J, Seedburg PH, Liberman TA, Schlessinger J, Francke U, Levinson A, Ullrich A (1985) Tyrosine kinase receptor with extensive homology to EGF receptor shares chromosomal location with neu oncogene. Science (Wash) 230:1132–1139

Cummings RD, Soderquist AM, Carpenter G (1985) The oligosaccharide moieties of the epidermal growth factor receptor in A-431 cells. Presence of complex-type N-linked chains that contain terminal N-acetylgalactosamine residues. J Biol Chem 260:11944–11952

Damjanov I, Mildner B, Knowles BB (1986) Immunohistochemical localization of the epidermal growth factor receptor in normal human tissues. Lab Invest 55:588–592

Daughaday WH, Rotwein P (1989) Insulin-like growth factors I and II. Peptide messenger ribonucleic acid and gene structures, serum, and tissue concentrations. Endocr Rev 10:68–91

Defize LHK, Mummery CL, Molenaar WH, deLaat SW (1987) Antireceptor antibodies in the study of EGF–receptor interaction. Cell Differentiation 20:87–102

De Leon DD, Bakker B, Wilson DM, Hintz RL, Rosenfeld RG (1988) Demonstration of insulin-like growth factor (IGF-I and -II) receptors and binding protein in human breast cancer cell lines. Biochem Biophys Res Commun 152:398–405

Delli Bovi P, Curatolo A, Kern FG, Greco A, Ittman M, Basilico C (1987) An oncogene isolated by transfection of Kaposi sarcoma DNA encodes a growth factor that is a member of the FGF family. Cell 50:729–737

Derynck R (1988) Transforming growth factor. Cell 54:593–595

Derynck R, Roberts AB, Winkler MA, Chen EY, Goeddel DV (1984) Human transforming growth factor-α: precursor structure and expression in E. coli. Cell 38:287–297

Derynck R, Goeddel DV, Ullrich A, Gutterman JU, Williams RD, Bringman TS, Berger WH (1987) Synthesis of messenger RNAs for transforming growth factors α and β and the epidermal growth factor receptor by human tumors. Cancer Res 47:707–712

Deuel TF (1987) Polypeptide growth factors: Roles in normal and abnormal cell growth. Ann Rev Cell Biol 3:443–492

Deuel TF, Senior RM (1987) Growth factors in fibrotic diseases. New Engl J Med 317:236–237

Dietel M, Kostrouch Z, Courtoy PJ, Boonstra J, Toth J (1989) What's new in the importance of receptors in pathology? Path Res Pract 184:116–127

DiFiore PP, Pierce JH, Kraus MH, Segatto O, Richter King C, Aaronson SA (1987) erbB-2 is a potent oncogene when overexpressed in NIH/3T3 cells. Science (Wash) 237:178–182

Downward J, Parker P, Waterfield MD (1984a) Autophosphorylation sites on the epidermal growth factor receptor. Nature 311:483–485

Downward J, Yarden Y, Mayes E, Scrace G, Totty N, Stockwell P, Ullrich A (1984b) Close similarity of EGF receptor and v-erb-B-oncogene protein sequences. Nature (Lond) 307:521–527

Dreyfus CF (1989) Effects of nerve growth factor on cholinergic brain neurons. Trends Pharmacol Sci 10:145–148

Dunn WA, Connolly TP, Hubbard AL (1986) Receptor-mediated endocytosis of epidermal growth factor by rat hepatocytes: receptor pathway. J Cell Biol 102:24–36

Eisbruch A, Blick M, Lee JS, Sacks PG, Guttermann J (1987) Analysis of the epidermal growth factor receptor gene in fresh human head and neck tumors. Cancer Res 47:3603–3605

Ek B, Heldin C-H (1982) Characterization of a tyrosine-specific kinase activity in human fibroblast membranes stimulated by platelet-derived growth factor. J Biol Chem 257:10486–10492

Elder JT, Fisher GJ, Lindquist PB, Bennett GL, Pittelkow MR, Coffey RJ Jr, Ellingsworth L, Derynck R, Voorhees JJ (1989) Overexpression of transforming growth factor a in psoriatic epidermis. Science (Wash) 243:811–814

Fabricant RN, DeLarco JE, Todaro GJ (1977) Nerve growth factor receptors on human melanoma cells in culture. Proc Natl Acad Sci USA 74:565–569

Fayed YM, Tsibris JCM, Langenberg PW, Robertson AL Jr (1989) Human uterine leiomyoma cells: binding and growth responses to epidermal growth factor, platelet-derived growth factor and insulin. Lab Invest 60:30–37

Fisher RJ, Bader JP, Papas TS (1989) Oncogenes and the mitogenic signal pathway. In: Important Advances in Oncology 1989, DeVita VT, Hollman S, Rosenberg SA (Eds) pp 3–27. Lippincott, Philadelphia

Fitzpatrick SL, Brightwell J, Wittliff JL, Barrows GH, Shultz GS (1984) Epidermal growth factor binding by breast tumor biopsies and relationship to estrogen receptor and progestin receptor levels. Cancer Res 44:3448–3453

Folkman J, Klagsbrun M (1987) Angiogenic factors. Science 235:442–448

Fowler JE, Lau JLT, Ghosh L, Mills SE, Mounzer A (1988) Epidermal growth factor and prostatic carcinoma: an immunohistochemical study. J Urol 133:857–1110

Fukushige S-I, Matsubara K-I, Yoshida M, and 5 others (1986) Localisation of a novel v-erbB-related gene, c-erbB-2 on human chromosome 17 and its amplification in a gastric cancer cell line. Mol Cell Biol 5:1442–1446

Gage FH, Batchelor P, Chen KS, Chin D, Higgins GA, Koh S, Deputy S, Rosenberg MB, Fisher W, Bjorklund A (1989) NGF receptor reexpression and NGF mediated cholinergic neuronal hypertrophy in the damaged adult neostriatum. Neuron 2:1177–1184

Gerosa MA, Talarico D, Fognani C, Raimondi E, Colombata M, Tridente G, DeCarli L, Della Valle G (1989) Overexpression of N-ras oncogene and epidermal growth factor receptor gene in human glioblastomas. J Natl Cancer Inst 81:63–67

Gimenez-Gallego G, Rodkey K, Bennett C, Rios-Candelore MR, DiSalvo J, Thomas K (1985) Brain-derived acidic fibroblast growth factor: complete amino acid-sequence and homologies. Science (Wash) 230:1385–1388

Gospodarowicz D, Neufeld G, Schweigerer L (1986) Fibroblast growth factor. Mol Cell Endocr 46:187–204

Gospodarowicz D, Ferrara N, Schweigerer L, Neufeld G (1987) Structural characterization and biological functions of fibroblast growth factor. Endocr Rev 8:95–114

Goustin AS, Betsholtz C, Pfeifer-Ohlsson S, Persson H, Rydnert J, Bywater M, Holmgren G, Heldin C-H, Westermark B, Ohlsson R (1985) Co-expression of the sis and myc proto-oncogenes in human placenta suggest autocrine control of trophoblast growth. Cell 41:301–312

Goustin AS, Leof EB, Shipley GD, Moses HL (1986) Growth factors and cancer. Cancer Res 46:1015–1029

Green MR, Baskeetter DA, Couchman JR, Rees DA (1983) Distribution and number of EGF receptors in skin is related to epithelial cell growth. Dev Biol 100:506–512

Greene LA, Shooter EM (1980) The nerve growth factor: biochemistry, synthesis and mechanism of action. Annu Rev Neurosci 3:353–402

Greene LA, Tischler AS (1976) Establishment of a noradrenergic clonal line of rat adrenal pheochromocytoma cells which responds to nerve growth factor. Proc Natl Acad Sci USA 73:2424–2428

Grob PM, Ross AH, Koprowski H, Bothwell M (1985) Characterization of the human melanoma nerve growth factor receptor. J Biol Chem 260:8044–8049

Gullick WJ, Julian D, Downward H, Marsden JJ, Waterfield MD (1984) A radioimmunoassay for human epidermal growth factor receptor. Anal Biochem 141:253–261

Gullick WJ, Berger MS, Bennett PLP, Rothbard JB, Waterfield D (1986) Expression of epidermal growth factor receptors on human cervical, ovarian, and vulval carcinomas. Cancer Res 46:285–292

Gullick WJ, Berger MS, Bennett PLP, Rothbard JB, Waterfield MD (1987) Expression of the c-erb-B-2 protein in normal and transformed cells. Int J Cancer 40:246–254

Gusterson A, Cowley G, Smith JA, Ozanne B (1984) Cellular localization of human EGF receptors. Cell Biol Int Rep 8:649–658

Gusterson BA, Gullick WJ, Venter DJ, Poweles TJ, Elliott C, Ashley S, Tidy A, Harrison S (1987) Immunohistochemical localization of c-*erb*B-2 in breast carcinoma. Mol Cell Probes 1:383–391

Gusterson BA, Machin LG, Gullick WJ, Gibbs NM, Powles TJ, Elliott C, Ashley S, Monaghan P, Harrison S (1988a) c-*erb*B-2 expression in benign and malignant breast disease. Brit J Cancer 58:453–457

Gusterson BA, Machin LG, Gullick WJ, Gibbs NM, Powles TL, Price P, McKinna A, Harrison S (1988b) Immunohistochemical distribution of c-*erb*B-2 in infiltrating and in situ breast cancer, Int J Cancer 42:642–845.

Haeder M, Rotsch M, Bepler G, Henning C, Havemann K, Heimann B, Moelling K (1988) Epidermal growth factor receptor expression in human lung cancer cell lines. Cancer Res 48:1132–1136

Hammacher A, Hellman U, Johnson A, Ostman A, Gunnarsson K, Westermark B, Wasteson A, Heldin CH (1988) A major part of PDGF purified from human platelets is a heterodimer of one A and one B chain. J Biol Chem 263:6493–6498

Han VKM, Hunter ES III, Pratt RM, Zendegui JG, Lee DC (1987) Expression of transforming growth factor alpha mRNA during development occurs predominantly in the maternal decidua. Mol Cell Biol 7:2335–2343

Harper GP, Thoenen H (1980) The distribution of nerve growth factor in the male sex organs of mammals. J Neurochem 34:893–903

Haselbacher G, Schwab ME, Pasi A, Humbel RE (1985) Insulin-like growth factor II (IGF II) in human brain: regional distribution of IGF II and of higher molecular mass forms. Proc Natl Acad Sci USA 82:2153–2157

Heine UI, Munoz EF, Flanders KC, Ellingsworth LR, Lam H-YP, Thompson NL, Roberts AB, Sporn MS (1987) Role of transforming growth factor-β in the development of the mouse embryo. J Cell Biol 105:2861–2876

Heitz PU, Kasper M, Van Noorden S, Polak JM, Gregory H, Pearse AGE (1978) Immunohistochemical localisation of urogastrone to human duodenal and submandibular glands. Gut 19:408–413

Heldin C-H, Westermark B (1984) Growth factors: mechanism of action and relation to oncogenes. Cell 37:9–20

Heldin C-H, Johnsson A, Wennergren S, Wernstedt C, Betsholtz C, Westermark B (1986) A human osteosarcoma cell line secretes a growth factor structurally related to a homodimer of PDGF A chains. Nature (Lond) 319:511–514

Heldin C-H, Betsholtz C, Claesson-Welsch L, Westermark B (1987) Subversion of growth regulatory pathways in malignant transformaton. Biochim Biophys Acta Cancer Rev 907:219–244

Heldin C-H, Backstrom G, Ostman A, Hammacher A, Ronnstrand L, Rubin K, Nister M, Westermark B (1988) Binding of different dimeric forms of PDGF to human fibroblasts: evidence for two separate receptor types. EMBO J 7:1387–1393

Henn W, Blin N, Zang KD (1986) Polysomy of chromosome 7 is correlated with overexpression of the *erb*B oncogene in human glioblastoma cell lines. Hum Genet 74:104–106

Hoffman GE, Rao CV, Barrows GH, Schultz GS, Sanfilippo JS (1984) Binding sites for epidermal growth factor in human uterine tissues and leiomyomas. J Clin Endocr Metab 58:880

Hollstein MC, Smits AM, Galiana C, Yamasaki H, Bos JL, Mandard A, Partensky C, Montesano R (1988) Amplification of epidermal growth factor receptor gene but no evidence for *ras* mutations in primary human esophageal cancers. Cancer Res 48:5119–5123

Huebner K, Isobe M, Chao M, Bothwell M, Ross AH, Finan J, Hoxie JA, Sehgal A, Buck GR, Lanahan A, Nowell PC, Koprowski H, Croce CM (1986) The nerve growth factor receptor is at human chromosome region 17q12-17q22, distal to the chromosome 17 breakpoint in acute leukemias. Proc Natl Acad Sci USA 83:1403–1407

Imanishi K, Yamaguchi K, Kuranami M, Kyo E, Hozumi T, Abe K (1989) Inhibition of growth of human lung adenocarcinoma cell lines by anti-transforming growth factor-α monoclonal antibody. J Natl Cancer Inst 81:220–223

Johnsson A, Heldin C-H, Westermark B, Wasteson A (1982) Platelet-derived growth factor: Identification of constituent polypeptide chains. Biochem Biophys Res Commun 104:66–74
Johnsson A, Heldin C-H, Wasteson A, Westermark B, Deuel TF, Huang JS, Seeburg PH, Gray A,
   Ullrich A, Scrace G, Stroobant P, Waterfield MD (1984) The c-sis gene encodes a precursor of
   the B chain of platelet-derived growth factor. EMBO J 3:921–928
Kan M, DiSorbo D, Hou J, Hoshi H, Mansson P-E, McKeehan WL (1988) High and low affinity
   binding of heparin-binding growth factor to a 130-kDa receptor correlates with stimulation
   and inhibition of growth of a differentiated human hepatoma cell. J Biol Chem 263:11 306–
   11 313
Kasselberg AG, Orth DN, Gray ME, Stahlman MT (1985) Immunocytochemical localization of
   human epidermal growth factor/urogastrone in several human tissues. J Histochem Cytochem
   33:315–322
Kavanaugh WM, Harsh GR IV, Starksen NF, Rocco CM, Williams LT (1988) Transcriptional
   regulation of the A and B chain genes of platelet-derived growth factor in microvascular
   endothelial cells. J Biol Chem 263:8470–8472
Kimchi A, Wang X-F, Weinberg RA, Cheifetz S, Massagué J (1988) Absence of TGF-β receptors
   and growth inhibitory responses in retinoblastoma cells. Science (Wash) 240:196–199
Klagsbrun M, Sasse J, Sullivan R, Smith JA (1986) Human tumor cells synthesize an endothelial
   cell growth factor that is structurally related to basic fibroblast growth factor. Proc Natl Acad
   Sci USA 83:2448–2451
Kraus MH, Papescu NC, Amsbaugh SC, King CR (1987) Overexpression of the EGF-receptor-re-
   lated protooncogene erb-B-2 in human mammary tumor cell lines by different molecular
   mechanisms. EMBO J 6:605–610
Lax I, Bellot F, Howk R, Ullrich A, Givol D, Schlessinger J (1989) Functional analysis of the ligand
   binding site of EGF-receptor utilizing chimeric chicken/human receptor molecules. EMBO J
   8:421–427
Lee J, Dull TJ, Lax I, Schlessinger J, Ullrich A (1989) HER2 cytoplasmic domain generates normal
   mitogenic and transforming signals in a chimeric receptor. EMBO J 8:167–173
Levi A, Eldrige JD, Paterson BM (1985) Molecular cloning of a gene sequence regulated by nerve
   growth factor. Science (Wash) 223:393–395
Levi-Montalcini R (1987) The nerve growth factor 35 years later. Science (Wash) 237:1154–1162
Liebermann TA, Razon N, Bartal AD, Yarden Y, Schlessinger J, Soreq H (1984) Expression of
   epidermal growth factor receptors in human brain tumors. Cancer Res 44:753–760
Liebermann TA, Nusbaum HR, Razon N, Kris R, Lax I, Soreq H, Whittle N, Waterfield MD,
   Ullrich A, Schlessinger J (1985) Amplication, enhanced expression and possible rearrangement
   of EGF receptor gene in primary human brain tumours of glial origin. Nature (Lond) 313:144–
   147
Madtes DK, Raines EW, Sakariassen KS, Assoian RK, Sporn MB, Bell GI, Ross R (1988) Induction of transforming growth factor-α in activated human alveolar macrophages. Cell 53:285–
   293
Magid M, Nanney LB, Stoscheck CM, King LE (1986) Epidermal growth factor binding and
   receptor distribution in term human placenta. Placenta 6:519–526
Malden LT, Novak U, Kaye AH, Burgess AW (1988) Selective amplification of the cytoplasmic
   domain of the epidermal growth factor receptor gene in glioblastoma multiforme. Cancer Res
   48:2711–2714
Martinet Y, Rom WN, Grotendorst GR, Martin GR, Crystal RG (1987) Exaggerated spontaneous
   release of platelet-derived growth factor by alveolar macrophages from patients with idopathic
   pulmonary fibrosis. N Engl J Med 317:202–209
Massagué J (1987) The TGF-β family of growth and differentiation factors. Cell 49:437–438
Massagué J, Cheifetz S, Ignotz RA, Boyd FT (1987) Multiple type-β transforming growth factors
   and their receptors. J Cell Physiol suppl 5:43–47
Mattsson MEK, Engerg G, Ruusala A-I, Hall K, Pahlman S (1986) Mitogenic response of human
   SH-SY5Y neuroblastoma cells to insulin-like growth factor I and II is dependent on the stage
   of differentiation. J Cell Biol 102:1949–1954
Maxwell SA, Sacks PG, Gutterman JU, Gallick GE (1989) Epidermal growth factor receptor
   protein-tyrosine kinase activity in human cell lines established from squamous carcinomas of
   the head and neck. Cancer Res 49:1137–1141

McKeehan WL, Adams PS, Fast D (1987) Different hormonal requirements for androgen-independent growth of normal and two epithelial cells from rat prostate. In Vitro Cell Dev Biol 23:147–156

Meyers MB, Shen WPV, Spengler BA, Ciccarone V, O'Brien JP, Donner DB, Furth ME, Biedler JL (1988) Increased epidermal growth factor receptor in multi-drug-resistant human neuroblastoma cells. J Cell Biochem 38:87–97

Miyamoto M, Sugawa H, Mori T, Hase K, Kuma T, Imura H (1988) Epidermal growth factor receptors on cultured neoplastic human thyroid cells and effects of epidermal growth factor and thyroid-stimulating hormone on their growth. Cancer Res 48:3652–3656

Mori M, Naito R, Tsikutani K, Okada Y, Hayashi T, Kato K (1987) Immunohistochemical distribution of human epidermal growth factor in salivary gland tumors. Virchows Arch A Pathol Anat 411:499–507

Nanney LB, Stoscheck CM, King LE (1988) Characterization of binding and receptors for epidermal growth factor in smooth muscle. Cell Tissue Res 254:125–132

Nicholson S, Halcrow P, Sainsbury JRC, Angus B, Chambers P, Farndon Jr, Harris AL (1988a) Epidermal growth factor receptor (EGFr) status associated with failure of primary endocrine therapy in elderly postmenopausal patients with breast cancer. Br J Cancer 58:810–814

Nicholson S, Sainsbury JRC, Needham GK, Chambers P, Farndon JR, Harris AL (1988b) Quantitative assays of epidermal growth factor receptor in human breast cancer: Cut off points of clinical relevance. Int J Cancer 42:36

Nister M, Hammacher A, Mellstrom K, Siegbahn A, Ronnstrand L, Westermark B, Heldin C-H (1988a) A glioma-derived PDGF A chain homodimer has different functional activities than a PDGF AB heterodimer purified from human platelets. Cell

Nister M, Libermann TA, Betsholtz C, Pettersson M, Claesson-Welsh L, Heldin C-H, Schlessinger J, Westermark B (1988b) Expression of messenger RNAs for platelet-derived growth factor and transforming growth factor alpha and their receptors in human malignant glioma cell lines. Cancer Res 48:3910–3918

Northwood IC, Davis RJ (1988) Activation of the epidermal growth factor receptor tyrosinase protein kinase in the absence of receptor oligomerization. J Biol Chem 263:7450–7453

Oliver AM (1988) Epidermal growth factor receptor expression in human fetal tissues is age-dependent. Br J Cancer 58:461–463

Ozawa S, Ueda M, Ando N, Abe O, Shimizu N (1987) High incidence of EGF receptor hyperproduction in esophageal squamous cell carcinomas. Int J Cancer 39:333–337

Ozawa S, Ueda M, Ando N, Abe O, Minoshima S, Shimizu N (1989) Selective killing of squamous carcinoma cells by an immunotoxin that recognizes the EGF receptor. Int J Cancer 43:152–157

Padhy LC, Shih C, Cowing D, Finkelstein R, Weinberg RA (1982) Identification of a phosphoprotein specifically induced by the transforming DNA of rat neuroblastomas. Cell 28:865–871

Pantazis P, Pelicci PG, Dalla-Favera R, Antoniades HN (1985) Synthesis and secretion of proteins resembling platelet-derived growth factor by human glioblastoma and fibrosarcoma cells in culture. Proc Natl Acad Sci USA 82:2404–2408

Pekonen F, Partanen S, Makinen T, Rutanen EM (1988a) Receptors for epidermal growth factor and insulin-like growth factor and their relation to steroid receptors in human breast cancer. Cancer Res 48:1343–1347

Pekonen F, Suikkari A-M, Makinen T, Rutanen E-M (1988b) Different insulin-like growth factor binding species in human placenta and decidua. J Clin Endocr Metab 67:1250–1258

Perez R, Betsholtz C, Westermark B, Heldin C-H (1987) Frequent expression of growth factors for mesenchymal cells in human mammary carcinoma cell lines. Cancer Res 47:3425–3429

Perosio RM, Brooks JJ (1989) Expression of growth factor receptors in soft tissue tumors. Implications for the autocrine hypothesis. Lab Invest 60:245–253

Peyrat J-P, Bonneterre J, Beuscart R, Djiane J, Demaille A (1988a) Insulin-like growth factor 1 receptors in human breast cancer and their relation to estradiol and progesterone receptors. Cancer Res 48:6429–6433

Peyrat J-P, Bonneterre J, Laurent JC, Amrani S, Leroy-Martin B, Vilain MO, Delobelle A, Demaille A (1988b) Presence and characterization of insulin-like growth factor I receptors in benign breast diseases. Eur J Cancer Clin Oncol 24:1425–1431

Poulsen SS, Nexo E, Skov Olsen P, Hess J, Kirkegaard P (1986) Immunohistochemical localization of epidermal growth factor in rat and man. Histochemistry 85:389–394

Presta M (1988) Sex hormones modulate the synthesis of basic fibroblast growth factor in human endometrial adenocarcinoma cells: implication for the neovascularization of normal and neoplastic endometrium. J Cell Physiol 137:593–597

Rakowicz-Szulczynska EM, Herlyn M, Koprowski H (1988) Nerve growth factor receptors in chromatin of melanoma cells, proliferating melanocytes, and colorectal carcinoma cells in vitro. Cancer Res 48:7200–7206

Rall LB, Penschow JD, Niall HD, Coghlan JP (1985) Mouse prepro-epidermal growth factor synthesis by the kidney and other tissues. Nature (Lond) 313:228–231

Rao CD, Igarashi H, Chiu I-M, Robbins KC, Aaronson SA (1986) Structure and sequence of the human c-sis/platelet derived growth factor 2 (SIS/PDGF 2) transcriptional unit. Proc Natl Acad Sci USA 83:2392–2396

Rappolee DA, Brenner CA, Schultz R, Mark D, Werb Z (1988) Developmental expression of PDGF, TGF-α and TGF-β genes in preimplantation mouse embryos. Science (Wash) 241:1823–1825

Read AE (1987) What happened to urogastrone? Quart J Med 62:1–6

Recio-Pinto E, Lang FF, Ishii DN (1984) Insulin and insulin-like growth factor II permit nerve growth factor binding and the neurite formation response in cultured human neuroblastoma cells. Proc Natl Acad Sci USA 81:2562–2566

Reifenberger G, Prior R, Deckert M, Wechsler W (1989) Epidermal growth factor receptor expression and growth faction in human tumours of the nervous system. Virchows Arch A Pathol Anat 414:147–155

Richardson PM, Verge Issa VMK, Riopelle RJ (1986) Distribution of neuronal receptors for nerve growth factor in the rat. J Neurosci 6:2312–2321

Rizzino A, Bowen-Pope DF (1988) Production of PDGF-like growth factors by embryonal carcinoma cells and binding of PDGF to their endoderm-like differentiated cells. Dev Biol 110:15–22

Rizzino A, Kazakoff P, Ruff E, Kuszynski C, Nebelsick J (1980) Regulatory effects of cell density on the binding of transforming growth factor B, epidermal growth factor, platelet derived growth factor and fibroblast growth factor. Cancer Res 48:4266–4271

Rizzino A, Kazakoff P, Ruff E, Kuszynski C, Nebelsick J (1988) Regulatory effects of cell density on the binding of transforming growth factor beta, epidermal growth factor, platelet-derived growth factor, and fibroblast growth factor. Cancer Res 38:4266–4271

Rizzino A, Kuszynski C, Ruff E, Tiesman J (1988) Production and utilization of growth factors related to fibroblast growth factor by embryonal carcinoma cells and their differentiated cells. Dev Biol 129:61–71

Roberts AB, Thompson NL, Heine U, Flanders C, Sporn MB (1988) Transforming growth factor-β: possible roles in carcinogenesis. Br J Cancer 58:594–600

Rogelj S, Weinberg RA, Fanning P, Klagsbrun M (1988) Basic fibroblast growth factor fused to a signal peptide transforms cells. Nature (Lond) 331:173–175

Rosenfeld RG, Hintz RL (1986) Somatomedin receptors: structure, function and regulation. In: Conn PM (ed) The Receptors, Vol 3, pp 281–239. New York: Academic Press

Rosenfeld RG, Pham H, Conover CA, Hintz RL, Baxter RC (1989) Structural and immunological comparison of insulin-like growth factor binding proteins of cerebrospinal and amniotic fluids. J Endocr Metabol 68:638–646

Ross AH, Grob P, Bothwell M, Elder DE, Ernst CS, Marano N, Ghrist BFD, Slemp CC, Herlyn M, Atkinson B, Koprowski H (1984) Characterization of nerve growth factor receptor in neural crest tumors using monoclonal antibodies. Proc Natl Acad Sci USA 81:6685–6689

Ross R, Vogel A (1978) The platelet-derived growth factor. Cell 14:203–210

Ross R, Rains EW, Bowen-Pope DF (1986) The biology of platelet-derived growth factor. Cell 46:155–169

Rozengurt E (1986) Early signals in the mitogenic response. Science (Wash) 234:161–166

Rozengurt E, Sinnett-Smith J, Taylor-Papadimitriou J (1985) Production of PDGF-like growth factor by breast cancer cell lines. Int J Cancer 36:247–252

Rufenacht H, Kasper M, Heitz Ph U, Streule K, Harder F (1986) "Brunneroma"-hamartoma or tumor? Pathol Res Pract 181:107–109

Sainsbury JRC, Malcolm AJ, Appleton DR, Farndon JR, Harris AL (1985) Presence of epidermal growth factor receptor as an indicator of poor prognosis in patients with breast cancer. J Clin Pathol 38:1225–1228

Sainsbury JRC, Needham GK, Malcolm A, Farndon JR, Harris AL (1987) Epidermal growth factor receptor status as predictor of early recurrence of and death from breast cancer. Lancet i:1398–1402

Sainsbury JRC, Nicholson S, Angus B, Farndon JR, Malcolm AJ, Harris AL (1988) Epidermal growth factor receptor status of histological sub-types of breast cancer. Br J Cancer 58:458–460

Salido EC, Barajas L, Lechago J, Laborde NP, Fisher DA (1986) Immunocytochemical localization of epidermal growth factor in mouse kidney. J Histochem Cytochem 34:1155–1160

Salomon DS, Perroteau I (1986) Growth factors in cancer and their relationship to oncogenes. Cancer Invest 4:43–60

Sarup JC, Rao KVS, Fox CF (1988) Decreased progesterone binding and attenuated progesterone action in cultured human breast carcinoma cells treated with epidermal growth factor. Cancer Res 48:5071–5078

Sato M, Yoshida H, Hayashi Y, Miyakami K, Bando T, Yanagawa T, Yura Y, Asuma M, Ueno A (1985) Expression of epidermal growth factor and transforming factor-β in a human salivary gland adenocarcinoma cell line. Cancer Res 45:6160–6167

Schechter AL, Stern DF, Vaidyanathan L, Decker SJ, Drebin JA, Greene MI, Weinberg RA (1984) The *neu* oncogene: an *erb*-B related gene encoding a 185,000-M$_r$ tumor antigen. Nature (Lond) 312:513–516

Schechter AL, Hung MC, Vaidyanathan L, Weinberg RA, Yang-Feng TL, Francke U, Ulrich A, Coussens L (1985) The *neu* gene: an *erb*-B homologous gene distinct from and unlinked to the gene encoding the EGF-receptor. Science (Wash) 229:976–978

Schejter ED, Shilo B-Z (1989) The drosophila EGF receptor homolog (DER) gene is allelic to *faint little ball*, a locus essential for embryonic development. Cell 56:1093–1104

Schlessinger J (1988 a) Signal transduction by allosteric receptor oligomerization. Trends Biochem Sci 13:443–447

Schlessinger J (1988 b) The epidermal growth factor receptor as a multifunctional allosteric protein. Biochemistry 27:3119–3123

Schultz GS, White M, Mitchell R, Brown G, Lynch J, Twardzik DR, Todaro GJ (1987) Epithelial wound healing enhanced by transforming growth factor-alpha and vaccinia growth factor. Science (Wash) 235:350–352

Seifert RA, Schwartz SM, Bowen-Pope DF (1984) Developmentally regulated production of platelet-derived growth factor-like molecules. Nature (Lond) 311:669–671

Shimokado K, Raines EW, Madtes DK, Barrett TB, Benditt EP, Ross R (1985) A significant part of macrophage-derived growth factor consists of at least two forms of PDGF. Cell 43:277–286

Shoyab M, Plowman GD, McDonald VL, Bradley JG, Todaro GJ (1989) Structure and function of human amphiregulin: a member of the epidermal growth factor family. Science (Wash) 243:1074–1076

Sibley DR, Benovic JL, Caron MG, Lefkowitz RJ (1988) Phosphorylation of cell surface receptors: a mechanism for regulating signal transduction pathways. Endocr Rev 9:38–62

Siegfried JM, Owens SE (1988) Response of primary human lung carcinomas to autocrine growth factors produced by a lung carcinoma cell line. Cancer Res 48:4976–4981

Sitaras NM, Sariban E, Bravo M, Pantazis P, Antoniades HN (1988) Constitutive production of platelet-derived growth factor-like protein by human prostate carcinoma cell lines. Cancer Res 48:1930–1935

Skoog L, Macias A, Azavedo E, Lombardero J, Klinterberg C (1986) Receptors for EGF and oestradiol and thymidine kinase activity in different histological subgroups of human mammary carcinomas. Br J Cancer 54:271

Slamon DJ, Clark GM, Wong SG, Levin WJ, Ullrich A, McGuire WL (1987) Human breast cancer: Correlation of relapse and survival with amplication of the HER-2/*neu* oncogene. Science (Wash) 235:177–182

Slamon DJ, Godolphin W, Jones LA, Holt JA, Wong SG, Keith DE, Levin WJ, Stuart SG, Udove
    J, Ullrich A, Press MF (1989) Studies of the HER-2/*neu* proto-oncogene in human breast and
    ovarian cancer. Science (Wash) 244:707–712
Sporn MB, Roberts AB (1988) Peptide growth factors are multifunctional. Nature (Lond) 332:217–
    219
Sporn MB, Roberts AB, Wakefield LM, Assoian RK (1986) Transforming growth factor-β: biolog-
    ical function and chemical structure. Science (Wash) 233:532–534
Steck PA, Lee P, Hung M-C, Yung WKA (1988) Expression of an altered epidermal growth factor
    receptor by human glioblastoma cells. Cancer Res 48:5433–5439
Stoscheck CM, King LE (1986a) The role of epidermal growth factor in carcinogenesis. Cancer Res
    46:1030–1037
Stoscheck CM, King LE (1986b) Functional and structural characteristics of EGF and its receptor
    and their relationship to transforming proteins. J Cell Biochem 31:135–152
Stromberg K, Hudgins WR, Orth D (1987) Urinary TGFs in neoplasia: immunoreactive TGF-α in
    the urine of patients with disseminated breast carcinoma. Biochem Biophys Res Commun
    144:1059–1068
Stroobant P, Rice AP, Gullick WJ, Cheng DJ, Kert IM, Waterfield MD (1985) Purification and
    characterization of vaccinia virus growth factor. Cell 42:383–393
Sturm MA, Conover CA, Pham H, Rosenfeld RG (1989) Insulin-like growth factor receptors and
    binding protein in rat neuroblastoma cells. Endocrinology 124:388–396
Stylianopoulou F, Herbert J, Soares MB, Efstratiadis A (1988) Expression of the insulin-like growth
    factor II gene in the choroid plexus and the leptomeninges of the adult rat central nervous
    system. Proc Natl Acad Sci USA 85:141–145
Taira M, Yoshida T, Miyagawa K, Sakamoto H, Terada M, Sigimura T (1987) cDNA sequence of
    human transforming gene *hst* and identification of the coding sequence required for transform-
    ing activity. Proc Natl Acad Sci USA 84:2980–2984
Takahashi H, Herlyn D, Atkinson B, Powe J, Rodeck U, Alavi A, Bruce DA, Koprowski H (1987)
    Radioimmunodetection of human glioma xenografts by monoclonal antibody to epidermal
    growth factor receptor. Cancer Res 47:3847–3850
Taniuchi M, Clark HB, Schweitzer JB, Johnson EM Jr (1988) Expression of nerve growth factor
    receptors by Schwann cells of axotomized peripheral nerves: ultrastructural location, suppres-
    sion by axonal contact, and binding properties. J Neurosci 8:664–681
Thompson KL, Assoian R, Rosner MR (1988) Transforming growth factor-β increases transcrip-
    tion of the genes encoding the epidermal growth factor receptor and fibronectin in normal rat
    kidney fibroblast. J Biol Chem 263:19519–19524
Todderud G, Carpenter G (1988) Presence of mannose phosphate on the epidermal growth factor
    receptor in A-431 cells. J Biol Chem 263:17893–17896
Tsikutani K, Tatemoto Y, Noda Y, Mori M, Hayashi T, Kato K (1987) Immunohistochemical
    detection of human epidermal growth factor in submandibular glands and their tumors using
    a polyclonal antiserum and a monoclonal antibody. Histochemistry 87:293–300
Ullrich A, Coussens L, Hayflick JS, Dull TJ, Gray A, Tam AW, Lee J, Yarden Y, Liberman TA,
    Schlessinger J, Downward J, Mayes EL, Whittle N, Waterfield MD, Seeburg PH (1984) Human
    epidermal growth factor receptor cDNA sequence and aberrant expression of the amplified
    gene in A431 epidermal carcinoma cells. Nature (Lond) 309:418–425
Ullrich A, Gray A, Tam AW, Yang-Feng T, Tsubokawa M, Collins C, Henzel W, LeBon T, Kathuria
    S, Chen E, Jacobs S, Francke U, Ramachandran J, Fujita-Yamaguchi Y (1986) Insulin-like
    growth factor I receptor primary structure: Comparison with insulin receptor suggests structur-
    al determinants that define functional specificity. EMBO J 5:2503–2512
Valverius EM, Bates SE, Stampfer MR, Clark R, McCormick F, Salomon DS, Lippman ME,
    Dixon RB (1989) Transforming growth factor-α production and EGF receptor expression in
    normal and oncogene transformed human mammary epithelial cells. Mol Endocr 3:203–214
Van De Vijver M, van de Bersselaar R, Devilee P, Cornelisse C, Peterse J, Nusse R (1987) Ampli-
    fication of the *neu* (c-*erb*B-2) oncogene in human mammary tumors is relatively frequent and
    is often accompanied by amplification of the linked c-*erb*A oncogene. Mol Cell Biol 7:2019–
    2023

Van De Vijver MJ, Mool WJ, Wisman P, Peterse JL, Nusse R (1988) Immunohistochemical detection of the *neu* protein in tissue sections on human breast tumors with amplified neu DNA. Oncogene 2:175–178

Van Hoof COM, Holthuis JCM, Destreicher AB, Boonstra J, DeGraan PNE, Gispen WH (1989) Nerve growth factor-induced changes in the intracellular localization of the protein kinase C substrate B-50 in pheochromocytoma PC12 cells. J Cell Biol 108:1115–1126

Veale D, Ashcroft T, Marsh C, Gibson GJ, Harris AL (1987) Epidermal growth factor receptors in non-small cell lung cancer. Br J Cancer 55:513–516

Veale D, Kerr N, Gibson GJ, Harris AL (1989) Characterization of epidermal growth factor receptor in primary human non-small cell lung cancer. Cancer Res 49:1313–1317

Velu TJ, Beguinot L, Vass WV, Willingham MC, Merlino GT, Pastan I, Lowy DR (1987) Epidermal growth factor-dependent transformation by a human EGF receptor proto-oncogene. Science (Wash) 238:1408–1410

Venter DJ, Tuzi NL, Kumar S, Gullick WJ (1987) Overexpression of the c-*erb*B-2 onco-protein in human breast carcinomas: immunohistological assessment correlates with gene amplification. Lancet II:69–72

Weima SM, van Rooijen MA, Mummery CL, Feijen A, Kruijer W, deLaat SW, van Zoelen EJJ (1988) Differentially regulated production of platelet-derived growth factor and transforming factor beta by a human teratocarcinoma cell line. Differentiation 38:203–210

Weisman AS, Raguet SS, Kelly PA (1987) Characterization of the epidermal growth factor receptor in human meningioma. Cancer Res 47:2172–2176

Westermark B, Johnsson A, Paulsson Y, Betsholtz C, Heldin C-H, Herlyn M, Rodeck U, Koprowski H (1986) Human melanoma cell lines of primary and metastatic origin express the genes encoding the constituent chains of PDGF and produce a PDGF-like growth factor. Proc Natl Acad Sci USA 83:7197–7200

Westphal M, Hermann HD (1986) Epidermal growth factor-receptors on cultured human meningioma cells. Acta Neurochir (Wien) 83:62–66

Wilcox JN, Derynck R (1988a) Localization of cells synthesizing transforming growth factor-alpha mRNA in the mouse brain. J Neurosci 8:1901–1904

Wilcox JN, Derynck R (1988b) Developmental expression of transforming growth factors alpha and beta in mouse fetus. Mol Cell Biol 8:3415–3422

Wilcox N, Smith KM, Williams LT, Schwartz SM, Gordon D (1988) Platelet-derived growth factor mRNA detection in human atherosclerotic plaques by in situ hybridization. J Clin Invest 82:1134–1143

Williams LT (1989) Signal transduction by the platelet-derived growth factor receptor. Science (Wash) 243:1564–1570

Williams RL, Hilton DJ, Pease S, Willson TA, Stewart CL, Gearing DP, Wagner EF, Metcalf D, Nicola NA, Gough NM (1988) Myeloid leukemia inhibitory factor maintains the developmental potential of embryonic stem cells. Nature (Lond) 336:684–688

Wong AJ, Bigner DD, Kinzler KW, Hamilton SR, Vogelstein B (1987) Increased expression of the epidermal growth factor receptor gene in malignant gliomas is invariably associated with gene amplification. Proc Natl Acad Sci USA 84:6899–6903

Wong S, Winchell LF, McCune BK, Earp HS, Teixido J, Massagué J, Herman B, Lee DC (1989) The TGF-α precursor expressed in the cell surface finds to the EGF receptor on adjacent cells, leading to signal transduction. Cell 56:459–506

Wrann MM, Fox CF (1979) Identification of epidermal growth factor receptors in a hyperproducing epidermal carcinoma cell line. J Biol Chem 254:8083–8086

Wrba F, Reiner A, Ritzinger E, Holzner JH (1988) Expression of epidermal growth factor receptors on breast carcinomas in relation to growth fractions, oestrogen receptor status and morphological criteria. Path Res Pract 183:25–29

Wright C, Angus B, Nicholson S, Sainsbury JRC, Cairns J, Gullick WJ, Kelly P, Harris AL, Horne CHW (1989) Expression of c-*erb*B-2 oncoprotein: a prognostic indicator in human breast cancer. Cancer Res 49:2087–2090

Yao M, Shuin T, Misaki H, Kubota Y (1988) Enhanced expression of c-*myc* and epidermal growth factor receptor (c-*erb*B-1) genes in primary human renal cancer. Cancer Res 48:6753–6757

Yarden Y, Schlessinger J (1987) Epidermal growth factor induces rapid, reversible aggregation of the purified epidermal growth factor receptor. Biochemistry 26:1443–1451

Yarden Y, Ullrich A (1988a) Molecular analysis of signal transduction by growth factors. Biochemistry 27:3113–3119

Yarden Y, Ullrich A (1988b) Growth factor receptor tyrosine kinases. Annu Rev Biochem 57:443–478

Yarden Y, Weinberg RA (1989) Experimental approaches to hypothetical hormones: detection of a candidate ligand of the *neu* proto-oncogene. Proc Natl Acad Sci USA 86:3179–3183

Yasui W, Hata K, Yokozaki H, Nakatani H, Ochiai A, Ito H, Tahara E (1988a) Interaction between epidermal growth factor and its receptor in progression of human gastric carcinoma. Int J Cancer 41:211–217

Yasui W, Sumiyoshi H, Hata J, Kameda T, Ochiai A, Ito H, Tahara E (1988b) Expression of epidermal growth factor receptor in human gastric and colonic carcinomas. Cancer Res 48:137–141

Yura Y, Hayashi O, Kelly M, Oyasu R (1989) Identification of epidermal growth factor as a component of the rat urinary bladder tumor-enhancing urinary fractions. Cancer Res 49:1548–1553

Zajchowski D, Band V, Pauzie N, Tager A, Stampfer M, Sager R (1988) Expression of growth factors and oncogenes in normal and tumor-derived human mammary epithelial cells. Cancer Res 48:7041–7047

Zhou D, Battifora H, Yokota J, Yamamoto T, Cline MJ (1987) Association of multiple copies of the c-*erb*B-2 oncogene with spread of breast cancer. Cancer Res 47:6123–6125

Low-Density Lipoprotein Receptors

A. NIENDORF and U. BEISIEGEL

1 Introduction

The origins of the receptor theory are based on the work of Paul Ehrlich (1854–1915) and John Newport Langley (1852–1925) (PARASCANDOLA 1981), who were the first to postulate that specific cellular reactions to drugs must be based on so-called receptive substances. Receptors are proteins that specifically bind and take up hormones, growth factors, transport proteins, viruses, and toxins. The understanding of the physiological effects that are caused by certain ligands depends to a considerable extent on knowledge of where the target cells are located and how they interact with their ligands. In Chap. 4 of this book, M. DIETEL reviews the different locations of cell receptors with special reference to morphological and biochemical aspects. This chapter focuses on specific receptors for low-density lipoproteins (LDL).

LDL is a transport protein that delivers cholesterol to the liver as well as to the cells of other organs. It is composed of a polar coat consisting of unesterified cholesterol and phospholipids and a core of cholesteryl esters which are packed into the center of the particle. The hydrophilic shell including apoprotein B (apoB) mediates the solubility of the hydrophobic core. ApoB also recognizes the receptor, as will be discussed below.

Atherosclerosis is closely related to serum cholesterol levels. It is well known that especially high levels of LDL cholesterol are correlated with early onset of the development of severe atherosclerotic lesions. This is of special interest in respect of coronary heart disease and myocardial infarction, which together represent the most common cause of death in Europe and the United States. The mechanisms leading to foam cell formation and to advanced atheromatous lesions are discussed in Sect. 5.

The term "tumor-associated hypocholesterolemia" derives from the observation that low serum cholesterol levels are positively correlated with cancer incidence. This observation could lead to the fatal conclusion that low cholesterol levels might be a contributing factor in the development of cancer. It could then be concluded that the current public health message, which recommends that individuals with elevated cholesterol levels should seek to lower them (Study Group, European Atherosclerosis Society 1987), in fact risks increasing the number of cancer deaths. On the basis of epidemiological studies it is most likely that tumor-associated hypocholesterolemia represents a phenomenon which is caused by preclinical and advanced cancer. This raises the question of whether LDL receptors are present in tumor cells. Their relevance will be discussed in Sect. 6.

As a basis for further discussion, Section 2 provides a review of data on the LDL receptor protein with special reference to biochemical aspects, including cellular mechanisms to regulate cholesterol homeostasis. However, this review is by no means intended to give a comprehensive account of lipoprotein metabolism. The data are condensed in such a way that the essential role of a specific and controlled LDL metabolism and, furthermore, the morphological observations that have fundamentally contributed to the concept of the process termed receptor-mediated endocytosis can be easily understood. The detection of the receptor pathway of LDL particles has clarified why LDL is recognized by target cells and how it is delivered to its final destination (i.e., the lysosome within the target cell). These studies are reviewed in Sects. 3 and 4.

2 Biochemistry of LDL Receptors

2.1 Lipoproteins and Lipid Metabolism

Cholesterol is needed for membrane synthesis, bile acids, and synthesis of steroid hormones. It is transported in the plasma, mainly incorporated into LDL particles, and is delivered to target cells via specific receptors. LDL represents one of six different lipoproteins, the others being chylomicrons, chylomicron remnants, very low-density lipoproteins (VLDL), intermediate-density lipoproteins (IDL), and high-density lipoproteins (HDL). These particles, which are composed of different proportions of cholesteryl esters and triglycerides and possess different apoproteins, transport lipids from the site of synthesis to the site of catabolism (HAVEL et al. 1980; BROWN and GOLDSTEIN 1981; GOLDSTEIN et al. 1983). In brief, an exogenous pathway which delivers dietary fat and an endogenous pathway which delivers fat derived mainly from the liver can be distinguished. Triglyceride-rich chylomicrons are derived from the intestine, while VLDL is synthesized in the liver. Once these particles pass by the endothelium, an enzyme called lipoprotein lipase (for review see ECKEL 1989) hydrolizes triglycerides and thereby creates particles (chylomicron remnants from chylomicrons and IDL from VLDL) that are now smaller in diameter and relatively cholesterol enriched. Both can be taken up via a receptor-mediated process by the liver. However, most of the IDL remains in the circulation; the remaining triglycerides are removed and LDL particles are generated. The LDL is taken up via the LDL receptor in all peripheral cells. Every cell is able to meet its cholesterol need by this process. The main LDL catabolism takes place in the liver, where cholesterol is used for VLDL synthesis and is converted to bile acids. Via bile acid secretion cholesterol is catabolized. This process is illustrated in Fig. 1. It is obvious that the liver, being the main site of synthesis of lipoproteins as well as the most important catabolic unit, is extremely important in the regulation of cholesterol metabolism.

2.2 The LDL Receptor Protein

In 1974 it was found that normal cells express a specific receptor site for high affinity binding of LDL, while cells of a patient with homozygous familial hypercholesterolemia (FH) did not show such a binding site (GOLDSTEIN and BROWN 1974). Because of the severity of this disease due to a defect in the LDL receptor, with very early onset of coronary heart disease, it became obvious that the LDL receptor must have an important function in cholesterol metabolism. The group around GOLDSTEIN and BROWN therefore tried to identify and to isolate the receptor protein. The solubilization of the receptor was achieved with the detergent octyl-glucoside (SCHNEIDER et al. 1979). This detergent was the only one which preserved the binding activity after solubilization of the receptor and thereby allowed the detection of the receptor throughout the purification procedure.

Exogenous Pathway

Endogenous Pathway

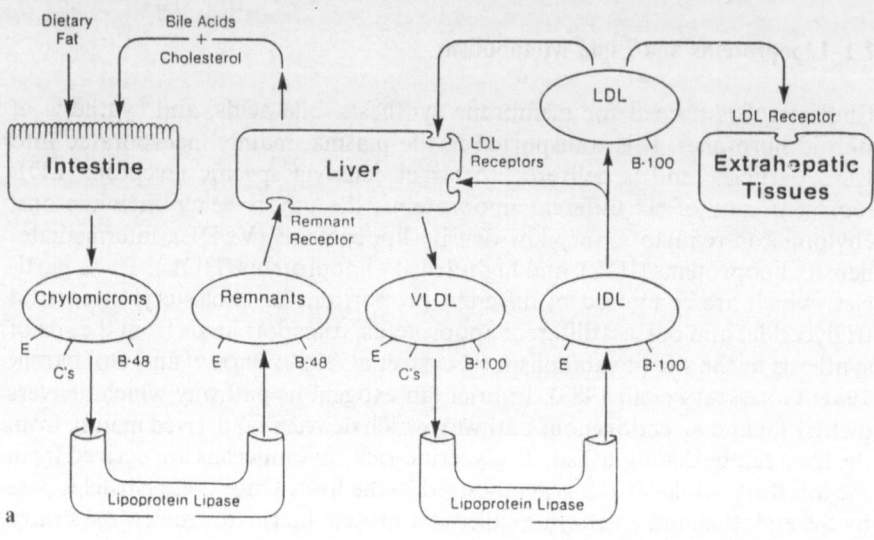

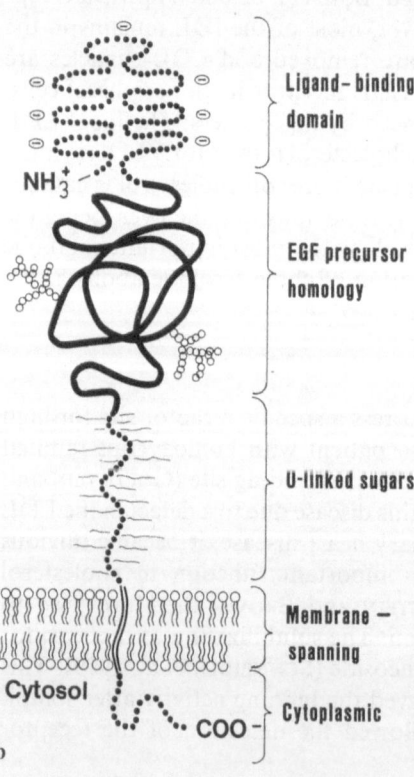

Fig. 1 a–c. The exogenous and endogenous pathways of lipoproteins, the structure–function relationship of the LDL receptor protein, and the regulatory mechanisms at the cellular level. **a** The metabolism of chylomicrons and VLDL (modified from GOLDSTEIN et al. 1983). **b** Model of the LDL receptor protein including the five important structure domains of the receptor. It is known that mutations in the different domains of the receptor cause different defects in the receptor function: (1) Defects in the domain with O-and N-linked sugar might influence the growth of the precursor protein to the mature receptor protein. (2) Mutation in the binding domain will produce a nonfunctional reaction on the cell surface. (3) Disturbances in the cytoplasmic tail cause the so-called internalization defect, where binding but not internalization can be measured. In addition to all these mutations, the total absence of the LDL receptor protein has been observed in FH patients. (Modified from BROWN and GOLDSTEIN 1986. **c** The cellular LDL receptor pathway, showing the three main regulatory consequences of the delivery of free cholesterol from the lysosomes into the cell: the HMG CoA reductase synthesis will be inhibited, the ACAT will be stimulated to produce cholesteryl esters for storage, and the synthesis of LDL receptors will be inhibited (modified from BROWN and GOLDSTEIN 1978)

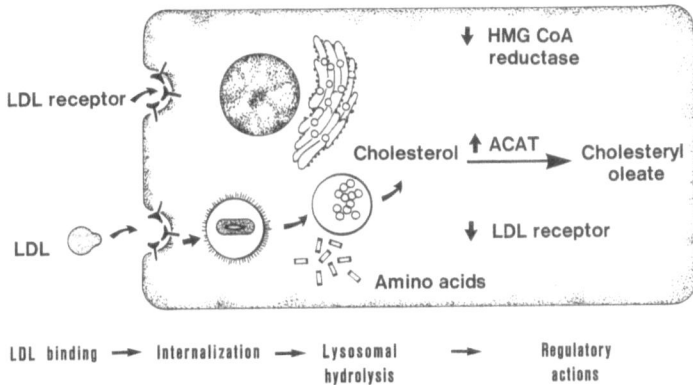

LDL binding ⟶ Internalization ⟶ Lysosomal ⟶ Regulatory
hydrolysis actions

Fig. 1 c

As a tool for further structural studies, monoclonal antibodies against this receptor were prepared (BEISIEGEL et al. 1981 b). The purification of the 164 000-dalton glycoprotein was achieved by affinity chromatography (SCHNEIDER et al. 1982). The receptor is synthesized as a 120 000-dalton precursor and in post-translational modifications the protein matures to the 164 000-dalton membrane protein.

Receptor-defective fibroblasts from patients with FH allowed a deeper insight in the structure-function relationship of the LDL receptor, and it was found that multiple mutations can disrupt the binding or uptake in the receptor pathway (TOLLESHAUG et al. 1983).

The cDNA cloning of the bovine LDL receptor guided the way to the LDL receptor gene (RUSSEL et al. 1983). Today we know that the receptor is a multidomain protein with five distinctive domains, as shown in Fig. 1 b (GOLDSTEIN and BROWN 1986). After the protein structure had been described, it was shown that the site for sterol-dependent suppression of the receptor synthesis is located in the LDL receptor gene promoter (DAWSON et al. 1988).

2.3 LDL Receptor Ligands

Not only is the structure of the LDL receptor known; the structural features of its main ligand, apoB, have also been verified in recent years. ApoB is a glycoprotein with a molecular weight of 514000. The complete nucleic acid and amino acid sequence is known and the binding domain has been determined (LAW et al. 1986; KNOTT et al. 1986; YANG et al. 1986). Studies with monoclonal antibodies and heparin-binding studies have reinforced the evidence that the region around the T3/T2 junction is involved in binding to the LDL receptor. The LDL receptor is also called apoB,E-receptor, since it is known that apoE can be bound with high affinity. As on apoB, the LDL receptor-binding site on apoE, a second ligand of the LDL receptor, coincides with the heparin-binding site. Heparin can therefore displace both ligands from the LDL receptor (WEISGRABER and RALL 1987). ApoE can be found in VLDL and IDL, as well as in chylomicron remnants; however, it is not involved in the LDL clearance, since the LDL particle contains one molecule of apoB only and no other apoproteins.

2.4 Biochemical Localization of LDL Receptors

The first evidence for the LDL receptor derived from binding studies on fibroblasts with ^{125}I-LDL (GOLDSTEIN and BROWN 1974). Later plasma membrane preparations were used to perform binding experiments and high affinity binding could be demonstrated. The binding was Ca^{2+} dependent. Membrane preparations from patients with FH did not express high affinity binding (BASU et al. 1978).

For further characterization of the receptor it was important not to lose the physiological binding activity, which is the only evidence for the presence of the receptor in the isolation steps. Use of the detergent octyl-glucoside turned out to preserve the binding activity and in the filter assays the receptor could be followed over the whole purification procedure. The receptor was packed in liposomes and incubated with ^{125}I-LDL; the free LDL was then separated on filters, where the receptor-ligand complex could be measured by its radioactivity. In order to follow the LDL receptor pathway in the cell with biochemical methods it is possible to separate intracellular vesicles with density gradients. Subsequently the LDL receptor can be analyzed in the different fractions. In comparison to plasma membranes, a receptor enrichment can be measured in endosomes after stimulation of the LDL receptor-mediated endocytosis of lipoproteins (JAECKLE et al. 1989).

2.5 Regulatory Steps Mediated Through Binding of LDL to the Receptor

By 1973 the feedback suppression of HMG CoA reductase by LDL had been recognized (GOLDSTEIN and BROWN 1973). In FH patients a complete absence of this mechanism was found, which led to overproduction of cholesterol. HMG CoA reductase is the key enzyme in cholesterol biosynthesis. A general scheme for the regulation of cholesterol metabolism in mammalian cells was described by BROWN and GOLDSTEIN (1978); a modified version is shown in Fig. 1 c. The free cholesterol released into the cell by the degradation of LDL affects the cellular cholesterol metabolism in three main ways: the HMG CoA reductase is inhibited, the LDL receptor synthesis is decreased, and the ACAT activity is increased. The ACAT esterifies the cholesterol to be stored in cholesteryl ester droplets. The regulatory effect of free cholesterol in the cell prevents overloading of the cell with cholesterol by decreased uptake via receptors and decreased synthesis (BROWN and GOLDSTEIN 1986).

The mechanism of cholesterol-induced suppression of HMG CoA reductase and LDL receptor synthesis is today known to involve an eight-base pair sequence in the promoter of the genes of both proteins (DAWSON et al. 1988; OSBORNE et al. 1988). Moreover, an oxysterol protein has been described which might mediate the cholesterol effect at the genes (DAWSON et al. 1989).

3 Morphology of LDL Receptors

GOLDSTEIN and colleagues defined the concept of receptor-mediated endocytosis by criteria which depend essentially on morphological observations (GOLDSTEIN et al. 1979a):

1. Receptors are trapped spontaneously in coated pits or are retained in these areas after binding of the ligand (the LDL receptor is most likely spontaneously arrested in coated pits).
2. Internalization of the ligand is effectively coupled to binding with a half-time in the range of 10 min.
3. Upon so-called uncoupling and sorting of internalized receptors and ligands, the receptor protein is recycled and reinserted into the plasma membrane, where it can be utilized again.
4. The internalized ligand ends up being digested in lysosomes.

Binding experiments with cultured cells with LDL-ferritin reveal typical distribution patterns which represent different stages of the receptor-mediated internalization process. The receptor protein is randomly distributed on the plasma membrane before it clusters in coated pits (see below). Thus, at the light microscopic level initial stages of ligand-receptor interactions reveal a typically fine punctuated and linearly orientated distribution. After endocytosis has occurred, aggregates are seen in the perinuclear region, which is an area that corresponds to the trans Golgi region and where lysosomes normally reside (Fig. 2a, b). Throughout this section these typical morphological patterns will be referred to, which were only understood after ultrastructural investigations had been performed. Biochemical as well as morphological data have indicated that, unlike the ligand, which has to be degraded in order to be catabolized, the receptor protein is separated before the formation of lysosomes within the internalization process and reintegrated into the plasma membrane. This mechanism is termed recycling of receptors (BROWN et al. 1982, 1983) and represents an economical usage of a protein that can be reutilized and therefore must not be digested. One single receptor protein is calculated to perform approximately 150 round trips in its 30-h life span (BROWN et al. 1983).

The pathway of receptor-mediated endocytosis is not unique for the uptake of LDL even though it is probably best studied in this system. Other ligands such as epidermal growth factor (EGF) and α_2-macroglobulin also are first bound in coated pits before they enter the cell. The joint pathway of receptor-

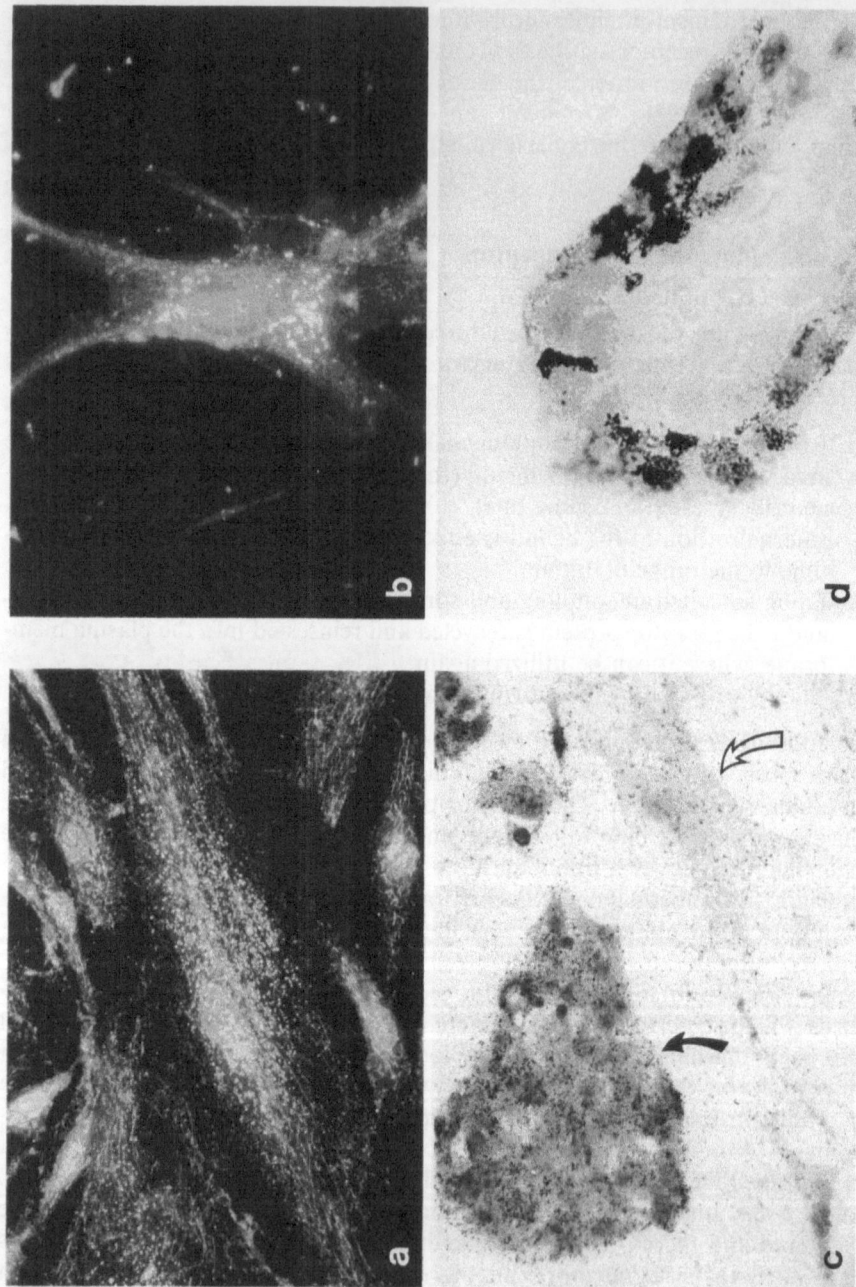

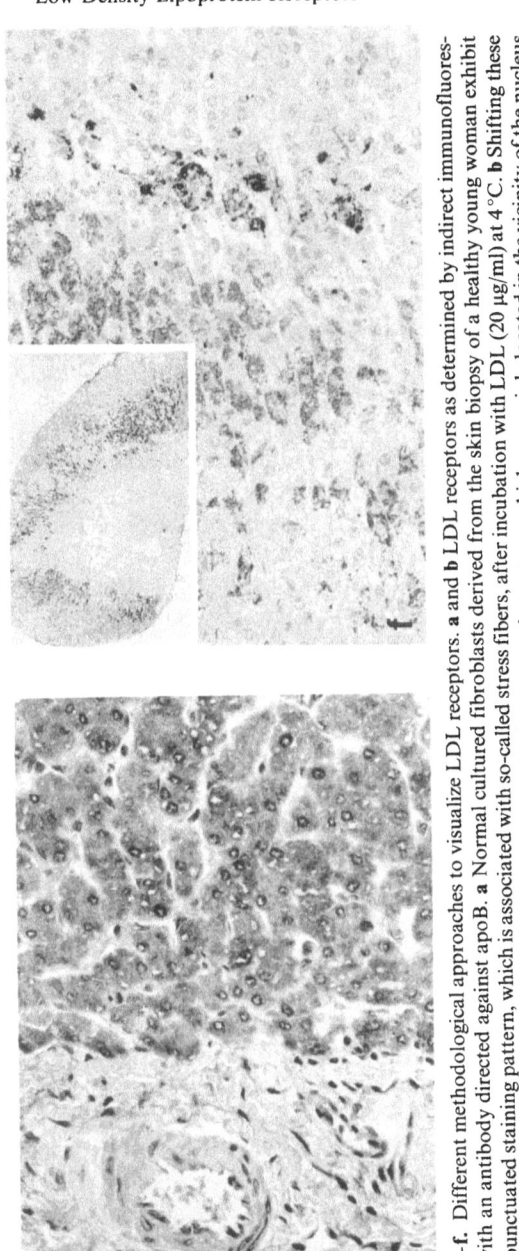

Fig 2a–f. Different methodological approaches to visualize LDL receptors. **a** and **b** LDL receptors as determined by indirect immunofluorescence with an antibody directed against apoB. **a** Normal cultured fibroblasts derived from the skin biopsy of a healthy young woman exhibit a fine, punctuated staining pattern, which is associated with so-called stress fibers, after incubation with LDL (20 µg/ml) at 4 °C. **b** Shifting these cells to 37 °C for 10 min leads to an accumulation of now enlarged punctuated structures which are mainly located in the vicinity of the nucleus and represent the different vesicles of the endocytotic pathway. This change in staining pattern represents the process of internalization. **c** ^{125}I-LDL binding to HepG2 cells that have been co-cultured with LDL receptor-negative fibroblasts. These cells have been derived from a homozygous FH patient. Autoradiography reveals a concentration of silver grains over the nest of HepG2 cells (*dark arrow*), whereas the fibroblasts (*light arrow*) do not express more than background staining. Coculturing of two cell types, one of which does not express the receptor protein in question, is an excellent test of the specificity of the staining technique. This method allows use of a ligand which is generally accepted to perform biochemical experiments. However, the resolution is poor. **d** LDL-gold binding to cultured colon carcinoma cells at the light microscopic level. These cells (D216/86) were exposed to LDL-gold (20 µg/ml) at 4 °C, fixed, and stained by the silver enhancement technique. Only cells in the periphery of this nest of cells exhibit a signal for LDS binding. This result is in accordance with data from others who stress the dependence of LDL receptor expression on the cell density. **e** Human liver stained with an antibody (MAB C7) directed against the LDL receptor protein by use of the ABC method. Note that only hepatocytes are stained. Neither Kupffer cells nor fibroblasts show any staining. **f** In vivo uptake of LDL-gold in the cortex of the mouse adrenal gland. Again the receptor-bound LDL-gold has been visualized by the silver enhancement technique. The most dense staining can be observed over the zona fasciculata. The zona reticularis displays moderate staining. No reaction product is visible over the zona glomerulosa or over the medulla. Single cells within a given region can be identified as being extremely active in terms of DL uptake and can be distinguished from adjacent cells that do not express such activity. The *inset* gives an overview of the whole section. These data are in accordance with biochemical observations and underline the need of sterol-synthesizing cells for cholesterol. (Original magnification: **a** and **b**, × 360; **c**, × 400; **d**, × 1000; **e** and **f**, × 400; *inset in* **f**, × 25)

ligand complexes into the cell and their sorting within the cell, which leads to delivery of the ligand into lysosomes and to recycling of the receptor back to the plasma membrane, will be discussed in detail below. Special attention will be paid to morphological findings that have played a key role in the present understanding of this complex process.

4 Methodological Aspects of Morphological Receptor Detection

There are three different experimental designs for the localization of receptors in intact tissues or in cultured cells. Firstly, the ligand can be labeled using either fluorescent markers such as fluorescent cholesteryl ester derivatives and FITC or radioactive markers such as ^{3}H and ^{125}I at the light microscopic level. Electron-dense materials such as ferritin and colloidal gold are suitable markers for electron microscopy. Secondly, the ligand can be bound to target cells and then detected by immunocytochemical methods. In both cases it is concluded that the receptor must be present where the ligand is found, which is in fact only true in the early stages of receptor-mediated endocytosis, as discussed in Sect. 4.3. However, it is generally accepted that in heterogeneous cell populations target cells can be identified by this approach. If competition with an excess of unlabeled ligand reveals negativity for the stain, it can be concluded that specific binding sites have been detected. The third possibility is a direct approach in which antibodies directed against the receptor protein are used to visualize it by immunocytochemical methods.

R. G. W. Anderson, who pioneered the field of the mechanism of receptor-mediated endocytosis by investigation of LDL uptake in cultured fibroblasts, reviewed the topic of LDL receptor visualization (Anderson 1986). In that paper he pointed out that light microscopy should be applied to establish optimal experimental conditions and that electron microscopy should be applied where quantitative statements on receptor distribution are desired. All the different approaches for receptor localization on cultured cells were discussed. The paper can therefore be considered as an essential reference work for those morphologists who intend to perform receptor studies on their own. Some more general considerations concerning the experimental setup for LDL receptor visualization are given below.

In order to receive an optical signal of maximal intensity, an attempt should be made to induce a maximum receptor number by up-regulating their expression on the cell surface. This is generally achieved by preincubation in a culture medium lacking LDL 48 h prior to the binding experiments. The incubation temperature is also important inasmuch as endocytosis does not take place at 4°C. At this temperature a maximum number of receptors can be occupied without initiation of subsequent catabolic steps. If tracking of the internalization pathway is to be performed, the culture medium should be warmed to 37°C in order to promote the passage of the labeled receptor into the subsequent intracellular compartments. Anderson has called this approach "inducing a wave of endocytosis."

4.1 Visualization of LDL Receptors

4.1.1 LDL Receptors Visualized
by Antibodies Directed Against the LDL Receptor

BEISIEGEL and co-workers were the first to obtain antibodies against the LDL receptor protein (BEISIEGEL et al. 1981 a, b). These studies revealed a structural similarity of the LDL receptor in different organs from a variety of species. Both polyclonal and monoclonal receptor antibodies were able to inhibit LDL binding to specific receptors. Furthermore, the uptake and degradation kinetics of LDL and the monoclonal antibody were similar. When the monoclonal antibody was incubated with target cells after LDL had already bound to the receptor, it was still able to detect this protein. This property made it a promising tool in investigating the different fates of ligand and receptor during the internalization process. Again with these antibodies, the typical punctuated and linearly orientated distribution pattern was observed (ANDERSON et al. 1982). Immunological localization of LDL receptors may require permeabilization of the cell membrane in order to detect those receptors that were on their way into or out of the cell at the time of fixation in addition to those receptor proteins already located in the plasma membrane at that point in time.

4.1.2 LDL Receptors Visualized by Indirect Immunofluorescence

KRUTH et al. (1979) determined the intensity of staining for LDL receptors by indirect immunofluorescence using a methodological approach that was adopted from BROWN and colleagues (BROWN et al. 1976) and was again published in 1980 (ANDERSON et al.). They focused special attention on the effect of cell density and receptor number, which seem to display an inverse correlation. This can be explained by the fact that cholesterol is an important membrane component which is primarily needed in proliferating systems (see also Fig. 2 a, b).

4.1.3 LDL Receptors Visualized by Fluorochrome-Labeled LDL

BARAK and WEBB (1981) demonstrated LDL uptake in cultured cells by use of a fluorescent LDL derivative, dil (3)-LDL. This method was first established by KREIGER and co-workers (KREIGER et al. 1978, 1979). The method of BARAK and WEBB is of special interest since their way of preparing a fluorescent LDL probe reveals a ligand that yields morphological results which match those obtained in previous studies and since it can be shown that this preparation retains physiological activity, as determined by its potency in suppressing HMG CoA reductase activity. Knowledge about the biological value of a ligand which has been designed for morphological studies is a prerequisite in estimating the value of the data obtained by the use of such a tool. The same group determined the mobility of receptor-ligand complexes by the use of this method and were able to demonstrate that the internalization defect of a certain patient with FH was not due to decreased mobility within the cell membrane (Barak and Webb 1982). In addition these investigators quantified the LDL receptor number by digital intensified video optical microscopy (Gross and Webb 1986).

4.1.4 LDL Receptors Visualized by Gold-Labeled LDL

A fascinating technique of LDL receptor visualization was introduced when Handley et al. (1981 a) described a method whereby LDL was coupled to colloidal gold particles. These particles are extremely suitable for electron microscopy. Due to negative charges on their surface they are bound to the LDL. One LDL-gold complex is formed by approximately eight LDL molecules located at the periphery with the gold grain in the center. Experiments on the ultrastructural level yielded results that were in accordance with those published previously. This method acquired particular importance after it was biochemically shown that LDL-gold complexes bind to LDL receptors on nitrocellulose paper (Roach et al. 1987). The advantage of gold as a marker for electron microscopy lies in the fact that gold is much more easily distinguishable than, for instance, ferritin or peroxidase reaction products. The journey of LDL-gold complexes to the inside of the cell is observable until isolated gold grains are finally found in lysosomes. At the beginning of the internalization process, which is at that point when the receptor-ligand complexes are still located in coated pits, there is a distinguishable distance between the gold grain and the plasma membrane. This measures approximately 20 nm and represents the diameter of the LDL molecule. Thus indirect evidence can be gained as to the sites at which the receptor-ligand complex is still intact. Robenek et al. (1982, 1983) applied gold-labeled LDL in combination with freeze-fracture, deep etching, and surface replication techniques to study the reinsertion of LDL receptors into the plasma membrane. Contrary to results obtained by Goldstein and co-workers they found the distribution of the LDL receptor in cultured human fibroblasts to be not random but rather patchy. They stated that the coated pit is formed at the site where receptors are inserted into the plasma membrane and doubt whether there is lateral movement of receptor proteins over longer distances. However, the vast majority of studies, including those with ligands other than LDL, have favored a model in which lateral movement of receptor proteins occurs.

4.2 Receptor-Mediated Endocytosis of LDL in Cultured Cells

As outlined above, the vast majority of morphological receptor studies have been performed on cultured fibroblasts. Being interested in the mechanisms which lead to the failure of LDL uptake in patients with FH, most investigators have first defined the system under normal circumstances by investigation of normal fibroblasts and then performed parallel experiments with mutant cells derived from FH patients. As early as 1976, ANDERSON et al. demonstrated in a quantitative ultrastructural study that 70% of LDL-ferritin is located in distinctly indented regions that cover only 2% of the cell surface. They were termed "coated regions" due to a fuzzy electron-dense material which formed a coat over the plasma membrane at the site of these indentations. Interestingly, these coated areas were also present in fibroblasts derived from a patient with FH but did not contain the LDL-ferritin, a finding that was confirmed in 1977 (ANDERSON et al. 1977b). Later the same group demonstrated by immunohistochemical techniques that antibodies directed against the material which represents the coat protein were arranged in a pattern corresponding to the distribution of LDL-ferritin (ANDERSON et al. 1978). It was suggested that this coat consists of a protein named clathrin, which was first described by PEARSE (1975, 1976). The exact route that receptor-ligand complexes take when travelling to the inside of the cell was published in 1977 (GOLDSTEIN et al. 1977; ANDERSON et al. 1977a). At the light microscopic level it was observed that at an incubation temperature of 4°C the autoradiographic grain pattern of ^{125}I-LDL was seen in the peripheral margins of the cell membrane and furthermore in the vicinity of the nucleus. When cells were moved to 37°C (induction of endocytosis), perinuclear aggregation and a concomitant loss of peripheral staining were revealed. If electron microscopic experiments were then performed using LDL-ferritin, prechilled cells initially bound the ligand, as already mentioned, in coated regions, and delivered it to distinguishable compartments within the cell after warming to 37°C. The coated regions indented more and more deeply into the cell until they finally pinched off from the plasma membrane, forming so-called coated vesicles. These vesicles lost their coat and as they increased in size and content as well as in the number of ferritin molecules, regions of increased density appeared within these compartments. It has been suggested by the current authors that the formation of secondary lysosomes, which could be explained by the fusion of primary lysosomes with endocytotic vesicles, is responsible for this phenomenon. Figure 4 exemplifies this pathway of receptor-mediated endocytosis in cultured mammary carcinoma cells.

4.3 Uncoupling and Sorting of Receptor and Ligand

The exact route of LDL receptor and LDL trafficking has been demonstrated in an elegant study by PATHAK et al. (1988). These investigators performed a double-staining electron microscopic technique for the ligand as well as for the receptor protein. Gold-labeled LDL in normal fibroblasts could be found to reside at the same locus as the ligand, at first in coated pits and later in coated vesicles and irregularly shaped multivesicular bodies. At the periphery of multivesicular bodies aggregates of receptor protein were observed that sometimes seemed to bud off again in small vesicular structures that were suggested to represent the compartment where receptors are recycled back to the cell membrane. However, there is a considerable amount of semantic and morphological confusion about the different compartments that are described when receptor-mediated endocytosis is the issue. PATHAK and colleagues stated that the multivesicular bodies resemble compartments that have previously been named endosomes (HELENIUS et al. 1983), receptosomes (PASTAN and WILLINGHAM 1981), or the compartment of uncoupling of receptor and ligand (GEUZE et al. 1983). PAAVOLA et al. (1985) demonstrated the presence of gold-labeled LDL in lysosomes as determined by parallel staining for acid phosphatase, a marker that is considered to identify these compartments. There is no doubt that LDL particles are finally delivered to lysosomes, but the functional meaning of different morphologically distinguishable stages en route to this destination has not, so far, been determined with any certainty.

4.4 Investigation of LDL Receptors in Cultured Cells Other Than Fibroblasts

LDL receptors have been visualized in a large number of cells that have been derived from different organs. The special need of cholesterol for steroid hormone synthesis has been underlined by studies by PAAVOLA et al. (1985) and TÓTH et al. (1988). Endothelial cells represent an important barrier for plasma proteins, separating the subendothelial tissue from those substances present in the bloodstream. These cells do possess LDL receptors in vitro when cultured at high density such that they are contact inhibited (SANAN et al. 1985, 1987). MOMMAAS-KIENHUIS et al. (1985a) visualized LDL receptors in dependence on LDL depletion. VASILE et al. (1983), NISTOR and SIMIONESCU (1986), and SNELTING-HAVINGA et al. (1989) demonstrated both transcytosis and, though to a much lesser extent, receptor-mediated endocytosis of LDL by endothelial cells. These results can best be interpreted as indicating that endocytosis serves to supply the endothelial cells' need for cholesterol while transcytosis supplies and possibly controls the delivery of LDL to subendothelial cells. Interestingly, transcytotic promotion of native LDL parallels that of reductively methylated LDL. The way in which the LDL particle, modified or native, interferes with the integrity of the arterial wall in the sense of a "primary injury" or chronic accumulation of cholesterol has not really been clarified by these studies. Further experiments to investigate the mode of LDL-cholesterol accumulation on the basis of chronically elevated LDL serum levels could be helpful in elucidating this process.

As outlined below, there is increasing evidence that more than one LDL receptor exists. The question of whether modified LDL is taken up in a pathway similar to that which has been established in fibroblasts as well as many other cultured cells has been investigated by MOMMAAS-KIENHUIS (1985b) and FUKUDA et al. (1986). Both groups found that there is no difference at the morphological level between the receptor-mediated endocytosis of native and acetylated LDL in cultured cells.

Cells of other organs, such as placental cells (MALASSINÉ et al. 1987) and keratinocytes (WILLIAMS et al. 1987), have also been shown to possess LDL receptors by means of morphological receptor staining. In these studies the focus was on the expression of LDL receptors as a function of growth and differentiation.

4.5 Localization of LDL Receptors in Intact Tissue

In a recent study TOMPKINS et al. (1988) visualized the LDL receptor-mediated uptake of iodinated LDL in the squirrel monkey. This study is of special interest because quantitative autoradiography provides information about the extent of LDL uptake not only in a given organ but also by distinct target cells. If only single cell types incorporate the ligand via specific receptors, the signal that arises from this binding will be too weak to be detected in tissue homogenates. By use of quantitative autoradiography it could be shown, for instance, that in the testis, which is an organ that does not exhibit a particularly high LDL uptake when whole-sample gamma counting is performed, the Leydig cells bound significantly higher amounts of iodinated LDL than other cellular components of the organ.

In the liver, which is the most important organ in terms of quantitative LDL catabolism, LDL binding and uptake takes place in hepatocytes, Kupffer cells, and endothelial cells (HANDLEY et al. 1981 b). However, there is still some controversy about the biological significance of LDL receptors in nonparenchymal cells. PITAS et al. (1985) state that specific sinusoidal endothelial cells are responsible for the removal of modified lipoproteins from the circulation in vivo. If this is true, in those morphological studies which employ labeled ligands it should be borne in mind that the uptake of oxidatively modified LDL, the generation of which cannot be prevented under any experimental condition, interferes with the uptake of native LDL. As outlined above and below, both pathways involve coated pits, but since they function via different receptors it is impossible to know which type of LDL is detected if the same label is used. Simple evidence about a certain pathway does not yield any information about this question. In immunocytochemical studies performed in our laboratory on human liver sections, exclusively hepatocytes expressed detectable amounts of immunoreactivity for the LDL receptor protein (Fig. 2e). An example of in vivo LDL uptake in the adrenal gland of the mouse is given in Fig. 2f.

5 Atherosclerosis and LDL Receptors

The World Health Organization (1958) defines atherosclerosis as "a variable combination of changes of the intima of arteries (as distinguished from arterioles) consisting of the focal accumulation of lipids, carbohydrates, blood and blood products, fibrous tissue and calcium deposits, and associated with medial changes."

As mentioned earlier, atherosclerosis, leading to myocardial infarction and stroke, is the most common cause of death in western Europe and North America. Its pathogenesis is closely linked to hypercholesterolemia as one of the major risk factors for myocardial infarction. Since the beginning of 1975 cellular cholesterol metabolism has been quite well understood. A series of experiments performed predominantly in the laboratory of GOLDSTEIN and BROWN revealed the specific receptor-mediated uptake of cholesterol-rich LDL particles. The detection of the LDL receptor pathway and especially the detection of its absence in patients suffering from the homozygous form of FH explained how and why LDL accumulates in the plasma, building up atherogenic levels. However, these investigations did not explain how high plasma cholesterol triggers the formation of atherosclerotic lesions. In this regard the understanding of modified LDL and the detection of a receptor that binds oxidized, i.e., modified LDL (see below) elucidated the mechanism by which LDL affects the arterial wall. Thus the association of high plasma cholesterol and atherosclerosis has to be seen as involving two components: (a) cholesterol homeostasis, which is regulated predominantly in the liver, and (b) injury of the arterial wall, where modified LDL seems to play a key role.

5.1 Pathogenesis of Atherosclerosis

The arterial wall consists of three layers, the intima, media, and adventitia. Depending on the type of artery, elastic or muscular, elastic laminae or smooth muscle cells represent the major component of the media. The intima in newborns consists of an endothelial layer and an underlying thin layer of loose connective tissue. With increasing age the intima becomes thicker. Glycosaminoglycans, some smooth muscle cells, and macrophages accumulate. This phenomenon, which is termed "intimal thickening," is generally regarded as representing a physiological process.

Ross (1986) has reviewed the pathogenesis of atherosclerosis and proposed a "revised response to injury hypothesis" which is based on a number of in vitro and in vivo data in animals and humans. He points out that the intimal smooth muscle cell proliferation represents the key event in the development of advanced atherosclerotic lesions. Special emphasis is put on the involvement of growth factors and chemotactic factors which trigger the replication and migration of cellular components forming atherosclerotic lesions. In brief, on the basis of elevated cholesterol levels, and certainly in conjunction with high blood pressure, cigarette smoking, and diabetes mellitus, a primary injury of the endothelial layer which can be denuding, as in the vicinity of fatty streaks (DAVIES et al. 1976), or nondenuding is postulated. Subsequently circulating monocytes become attached to the endothelium, invade the intimal layer, become macrophages, and begin to take up lipids. These macrophages then show the appearance of foam cells which build up the earliest atheromatous lesion, namely the fatty streak. Platelets that clump at locations where a denuding injury of the endothelial layer has occurred, in conjunction with macrophages and endothelial cells, are capable of secreting a number of growth factors. The growth factor-dependent invasion of medial smooth muscle cells which can secrete large amounts of intercellular matrix then leads to the formation of so-called fibrous plaques. If these plaques undergo further modification, such as ulceration and mural thrombosis, and if necrotic areas develop, complicated lesions form.

The literature does not provide generally accepted and consistent terminology concerning the morphological classification of atheromatous or atherosclerotic lesions. STARY (1989) has systematically investigated the early events in the development of atherosclerosis in subjects who died between full-term birth and the age of 29 years and classified the lesions into isolated macrophage foam cells, fatty streak, preatheroma, atheroma, and fibroatheroma. Definite morphological criteria are given that identify each of these classes, which take pathophysiological concepts of atherogenesis into consideration. It would certainly be helpful if a consensus were to be achieved on which classification to use, as this would enable findings from different investigators to be compared more easily.

5.2 Receptors Recognizing Modified LDL

As outlined in the introduction to Sect. 5, it is not the LDL receptor which mediates the accumulation of lipids in the arterial wall. This conclusion is based on the fact that those individuals who have the highest serum cholesterol levels due to an absolute defect in LDL receptor-dependent lipoprotein catabolism (i.e., FH patients) are the candidates for the most severe accumulation of lipids in the arterial wall.

A pathway first described by GOLDSTEIN et al. (1979c) was shown to function via so-called scavenger receptors. These receptors, which are present on monocyte/macrophages, Kupffer cells, and specialized endothelial cells, take up chemically modified LDL such as acetylated LDL. Native LDL does not compete for these binding sites. It could be shown that macrophages can only be overloaded with cholesterol via this scavenger pathway. Clearly, native LDL binding to the B/E receptor, which is expressed in only small numbers on macrophages, does not lead to the formation of foam cells in vitro. The results of a number of in vitro studies indicate that endothelial cells and macrophages themselves are able to modify LDL in vitro such that it is recognized by a receptor mechanism other than the LDL receptor. Modification of LDL in vivo (acetylated LDL does not exist in vivo) is most likely to be mediated by oxidation occurring in extravascular compartments (for review see STEINBERG et al. 1989).

Consequences of LDL oxidation are (a) chemotactic recruitment of macrophages (adhesion and invasion of monocytes), (b) uptake by macrophages which is several times more efficient than that of native LDL (formation of foam cells), (c) immobilization of these cells (formation of fatty streaks), and (d) a toxic effect on the endothelial layer (injury *and* initiation of a vicious circle) (Steinberg et al. 1989). These data are based primarily on in vitro experiments. The initiating step is the peroxidation of polyunsaturated fatty acids leading to the formation of a number of aldehydes, including malondialdehyde (for review see Esterbauer et al. 1988). Subsequently malondialdehyde is coupled covalently to apoB in a way that this, now modified, LDL particle will not be recognized by the LDL (B/E) receptor. In addition to the polyunsaturated fatty acids, LDL cholesterol is oxidized, rendering it cytotoxic.

Recently Sparrow et al. (1989) demonstrated with binding experiments that two forms of modified LDL (i.e., acetylated LDL and oxidized LDL) bind specifically to two forms of LDL receptor present on macrophages. Displacement in competition experiments revealed that oxidized LDL binds not only to the receptor which recognizes acetylated LDL but also to a distinct receptor that has no affinity for this type of chemically modified LDL. These experiments underline the significance of lipid peroxidation and support the concept of "oxidized LDL receptor"-mediated lipoprotein uptake. Thus the scavenger pathway, previously viewed as a single entity, in fact involves at least two distinct types of receptor.

The vast majority of these experiments were performed in vitro, and the in vivo relevance of lipid peroxidation has been questioned. Haberland et al. (1988) were able to demonstrate in an immunohistochemical study that malondialdehyde is colocalized with apoB preferentially in atheromatous lesions in an animal model of FH, namely the WHHL (Watanabe heritable hyperlipidemic) rabbit. This finding provided direct evidence for the existence in vivo of protein modified by a physiological end product of lipid peroxidation within arterial lesions. Further evidence underlining the participation of modified LDL in the generation of atherosclerotic lesions of the arterial wall was provided by the studies of Kita et al. (1987) and Carew et al. (1987), who demonstrated that substances which are transported incorporated into the LDL particle and which prevent its oxidation, such as probucol, are able to prevent the progression of atherosclerosis in WHHL rabbits independently of the lipid-lowering effect of the substance. Figure 3 shows a part of an advanced lesion of the aorta of a WHHL rabbit in which two types of cell with lipid inclusions, one of which exhibits the typical shape of a foam cell, can be seen. These cells represent the target cells for modified lipoproteins. However, the oxidized LDL receptor has not been isolated and therefore not been demonstrated by immunocytochemical studies in intact tissue sections of atherosclerotic lesions.

On the basis of these data some questions can be answered. As we all know, individuals with comparable high serum cholesterol levels do not necessarily develop the same degree of atherosclerosis, inasmuch as they might differ as regards potency of oxidative lipoprotein modification. If the LDL particle per se does not lead to foam cell formation and subsequent development of atherosclerotic lesions, it would make sense not only to lower plasma cholesterol but also to protect the remaining LDL particles from oxidation.

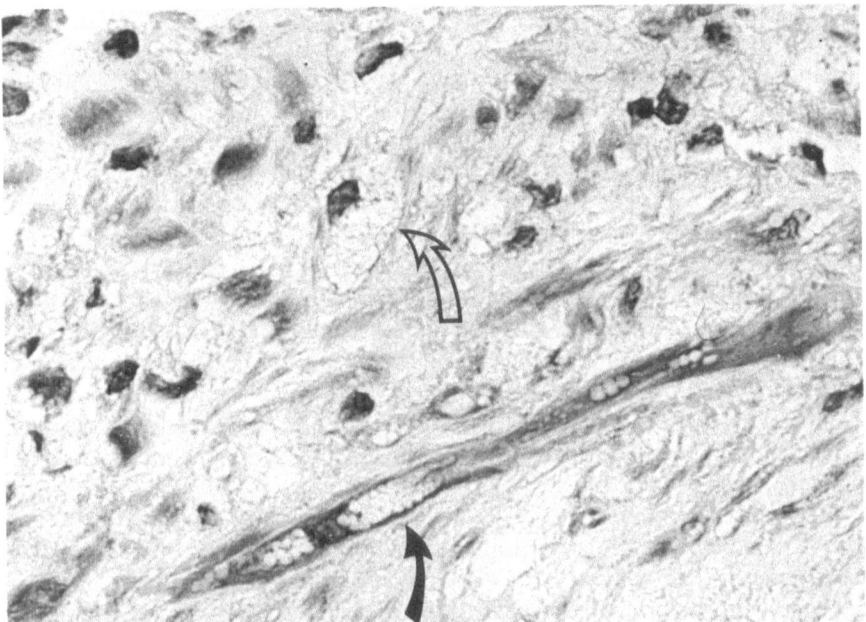

Fig. 3. Demonstration of foam cell formation in an area of a complicated lesion of the aorta from a WHHL rabbit. Two types of cell are revealed, one spear-shaped cell with multiple droplike lipid inclusions (*dark arrow*) and multiple foam cells with ballooned cytoplasm caused by aggregation of amorphous lipid material (*light arrow*). These cells contribute to the formation of atherosclerotic lesions by accumulation of oxidized LDL, which they take up via the scavenger receptor pathway. This receptor has so far not been visualized by methods other than ligand binding. (Masson-Goldner stain, original magnification × 1000)

6 LDL Receptors and Cancer

The literature provides a number of contradictory data concerning the association of plasma cholesterol levels and cancer mortality. A negative as well as a positive correlation has been reported. The studies in question are listed in the next section. If the sequence is "tumor – low cholesterol" and not "low cholesterol – tumor," a working hypothesis can be proposed that regards the presence of LDL receptors in tumor cells as a prerequisite to these cells being able to take up exogenous cholesterol. Thus the tumor could represent an LDL-cholesterol consuming tissue and therefore contribute to an augmented LDL metabolism. Low serum cholesterol levels in cancer patients could then be interpreted as a state of "additional," receptor-mediated, LDL consumption in tumor tissue. The rationale for this hypothesis is the need of proliferating cells to gain cholesterol as an essential membrane component.

6.1 Epidemiology

An inverse correlation between serum cholesterol and cancer mortality in epidemiological studies has been reported by KARK et al. (1980), BEAGLEHOLE et al. (1980), KAGAN et al. (1981), GARCIA-PALMIERI et al. (1981), WILLIAMS et al. (1981), MORRIS et al. (1983), GINSBERG et al. (1982), ALEXOPOULOS et al. (1987), SCHATZKIN et al. (1988), and VENTKATANARAYANAN et al. (1988). ALEXOPOULOS and co-workers made an exception for breast cancer, in which increased cholesterol levels were observed. ROSE and SHIPLEY (1980) called the inverse correlation between plasma cholesterol and cancer mortality "a source of error" since this inverse correlation was confined to the first 2 years of follow-up. The excess of cancer deaths in the lowest cholesterol groups was also reported to attenuate over time in two other big studies (International Collaborative Group 1982; SHERWIN et al. 1987). It is concluded from these studies that low cholesterol levels might be indicative of preclinical as well as clinical cancer.

A positive correlation between cholesterol and cancer has been reported in men (statistically significant) and women (statistically not significant) for rectal cancer (TÖRNBERG et al. 1986). In women with breast cancer augmentation of serum cholesterol levels has been reported to indicate a progression of the disease (ZIELINSKI et al. 1988). These data are difficult to interpret, given that RUDLING et al. (1986) in contrast found that LDL receptor content shows an inverse correlation with survival time of women with breast cancer. Even though cholesterol levels were not determined in this study, the authors surmise that their findings suggest increased cholesterol consumption in breast cancer tissue. A site-specific analysis of cancer incidence could reveal two effects that mask each other. First, a causative mechanism where increased fat uptake leads to hypercholesterolemia and subsequently to the creation of carcinogenic compounds. And on the other hand the development, by way of the mechanisms discussed elsewhere in this chapter, of tumor-associated hypocholesterolemia once a tumor has already been established.

In patients with acute myelogenous leukemia, VITOLS et al. (1985) demonstrated that low cholesterol levels were inversely correlated with the high affinity degradation of LDL per leukemic cell multiplied by the white blood cell count. In another study of plasma cholesterol levels in acute myelogenous leukemia patients (REVERTER et al. 1988), high leukocyte counts were associated with hypocholesterolemia at the time of diagnosis, while there was a significant increase in plasma cholesterol at the time of remission.

FEINLEIB (1983), who reviewed this issue, concludes that there is no ready explanation for the divergence of results. In our opinion, investigation of both LDL-cholesterol uptake and cholesterol synthesis from acetate in the tumor cell itself, correlated with individual cholesterol levels in cancer patients taking tumor size and rate of growth into consideration, should help to resolve the controversies surrounding the phenomenon of tumor-associated hypocholesterolemia. Especially information about the LDL receptor content of tumors would help in understanding when tumor-associated hypocholesterolemia will occur and when not.

6.2 Biochemical Demonstration of LDL Receptors in Tumors

LDL receptors have been determined by biochemical methods in a variety of tumors derived from different organs, including thyroid, kidney, gastric and colon carcinoma (1986), mammary carcinoma (RUDLING et al. 1986), brain tumors (RUDLING 1986), and leukemic cells (VITOLS et al. 1984, 1985). In an animal model of renal cell carcinoma (CLAYMAN et al. 1986a) it was observed that LDL receptor activity and disappeared entirely or to a great extent (CLAYMAN et al. 1986b). In both studies by CLAYMAN et al., a fivefold increase in endogenous cholesterol synthesis was determined. Thus, tumors whose normal tissue counterparts express LDL receptors do not necessarily need to follow that pathway.

6.3 Morphological Demonstration of LDL Receptors in Tumor Cells and Tissues

Only a few studies have examined the morphology of receptor-mediated endocytosis in cultured tumor cells. HESZ et al. (1987) investigated the uptake of apoB- and apoE-containing lipoproteins into HepG2 cells and concluded that there was little if any qualitative difference in binding, internalization, and intracellular processing. Both ligands bound to receptors initially located in clathrin-coated pits, with subsequent accumulation in lysosomes. VERMEER et al. (1985) demonstrated LDL-gold binding and internalization to be excessive when squamous carcinoma cells were cultured at low density, and in contrast found internalization to be defective when these cells were cultured at high density. They concluded that in squamous carcinoma cells the LDL receptor-mediated endocytosis is dependent on terminal differentiation. In another study VERMEER et al. (1986) presented two cell lines, one of which (squamous carcinoma, derived from the tongue) internalized LDL-gold complexes excessively whereas the other (SV40-transformed keratinocytes) failed to take up the ligand.

We have demonstrated the presence of LDL receptors on cultured tumor cells as well as on intact tumor tissue. A number of different methodological approaches, discussed in Sect. 3, have been employed. Colon carcinoma cells were incubated with an LDL-gold preparation (for methodology see HANDLEY et al. 1981a) for a constant length of time (2 h at 4 °C) and subsequently stimulated at 37 °C to induce internalization. The receptor-bound and/or internalized LDL-gold complexes were visualized by application of a silver-enhancement technique whose constituents are commercially available (Janssen, Belgium). Figure 2d shows that the highest receptor content is expressed in the periphery of a cell complex that has been grown on glass slides to perform these binding studies. This observation underlines the relevance of LDL receptor expression as a function of cell density. This technique, which was developed by C. RÖCKEN, represents a valuable method for morphological receptor demonstration and in addition serves as an excellent precontrol for receptor visualization at the electron microscopic level.

Using ^{125}I-labeled LDL, Keilhau was able to show the expression of LDL receptors on HepG2 cells in culture by means of autoradiography. In these experiments HepG2 cells were cocultured with receptor-deficient fibroblasts that were derived from the skin biopsy of a homozygous subject suffering from type IIa hyperlipidemia. Figure 2c reveals a specific grain pattern which is restricted to the HepG2 cells and clearly not present on the fibroblasts, which have been shown in previous biochemical binding studies not to bind or internalize LDL by means of receptor-mediated endocytosis. HepG2 cells can be distinguished by their characteristic polygonal to oval shape and frequently include some vacuoles in the cytoplasm, in contrast to skin fibroblasts. In general, coculturing of cells which are expected either to express or not to express the receptor in question is an excellent internal control that reveals the specificity of the staining technique.

At the electron microscopic level we were able to demonstrate the consecutive events representing the process termed receptor-mediated endocytosis of LDL in cultured mammary carcinoma cells (MDA-231) (Niendorf et al. 1989b). These cells show receptor-mediated uptake via coated pits, coated vesicles, and endosomes of different sizes and final delivery to lysosome-like compartments of different density. A morphometric analysis of receptor-mediated uptake of gold-labeled LDL particles in these mammary carcinoma cells as compared with normal human fibroblasts reveals an efficiency in the tumor cells that exceeds by several times the uptake in fibroblasts (Niendorf and Keilhau in preparation). Figure 4 shows LDL-connected gold grains in the different membrane-associated locations or intracellular compartments. LDL-gold binding to cultured tumor cells has furthermore been demonstrated on cells that have been derived from gastric, pancreatic, and colon carcinoma (Niendorf et al. 1988a, 1989). In conclusion, these experiments demonstrate that cultured tumor cells possess the ability to bind and internalize LDL particles using the same pathway as was established by Goldstein et al. (1977). However, as Anderson et al. (1981) demonstrated, this need not necessarily be the case, as shown in cultured A-431 cells, which express an internalization defect.

In intact tumor tissue the LDL receptor protein has been demonstrated to be present in a series of epithelial and mesenchymal tumors derived from different organs (Niendorf et al., in preparation). In this study we also investigated the distribution of apoB. In 62% of a series comprising a total sample of 50 cases, LDL receptor and apoB were found to be present. A high degree of congruence between the distribution patterns of the two proteins was observed. This means that the ligand is located where the receptors are present and vice versa. A grading of staining intensity performed in the same study indicated that differences exist in the extent of expression of the receptor protein. Whether these differences represent different regulatory states of receptor synthesis remains to be investigated. Figure 5 shows the immunoreactivity for the LDL receptor and for apoB in an adenocarcinoma of the colon. The congruent staining pattern for both proteins is exemplified. It is not easy to differentiate between background staining and a true reaction when looking

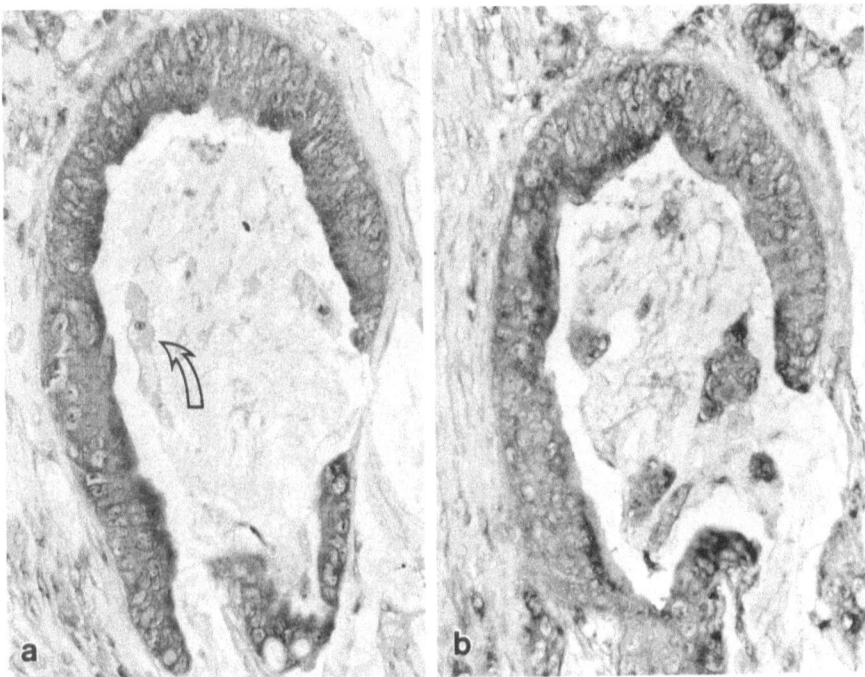

Fig. 4a, b. Immunohistochemical evidence for the presence of LDL receptor in tumor tissue.
a Demonstration of the LDL receptor protein by use of the monoclonal antibody C7. A relatively
strong staining (comparable with the intensity obtained in hepatocytes) can be observed in an
adenocarcinoma of the colon. Stromal cells exhibit only slight reactivity. Note the macrophages in
the center (*light arrow*), which do not express a detectable amount of LDL receptor protein. b A
consecutive section of the same specimen stained for apoB by the use of a polyclonal antibody. The
staining pattern is almost congruent. Cells corresponding to those represented in a are stained. One
exception has to be noted: the macrophages which do not express LDL receptor activity do take up
LDL (center of the tubule). In conclusion, these figures demonstrate that the ligand (LDL) is found
predominantly in cells which possess the receptor (B/E receptor). (Original magnification: a and b,
× 1000)

for LDL receptors and apoB in solid tumors. For instance, stromal fibroblasts
frequently express a considerable intensity of immunostaining for the LDL
receptor. This can easily be explained since these cells proliferate just as well
as the tumor does. Staining for both proteins (LDL receptor and apoB) can be
demonstrable in the tumor with varying intensity and is frequently observed
in surrounding tissues, such as smooth muscle tissue, liver, and the adrenal
gland. As explained in Sect. 2, the LDL receptor could be demonstrated to a
varying degree by biochemical methods in almost all tissues so far investigated.
It is rather the intensity of staining in comparison with surrounding cells that
engages the attention of the morphologist. ApoB shows up with serum levels
in the range of 100 mg/dl. This explains why the lumina of blood vessels and
necrotic areas almost uniformly express a relatively high degree of immunoreactivity for this protein.

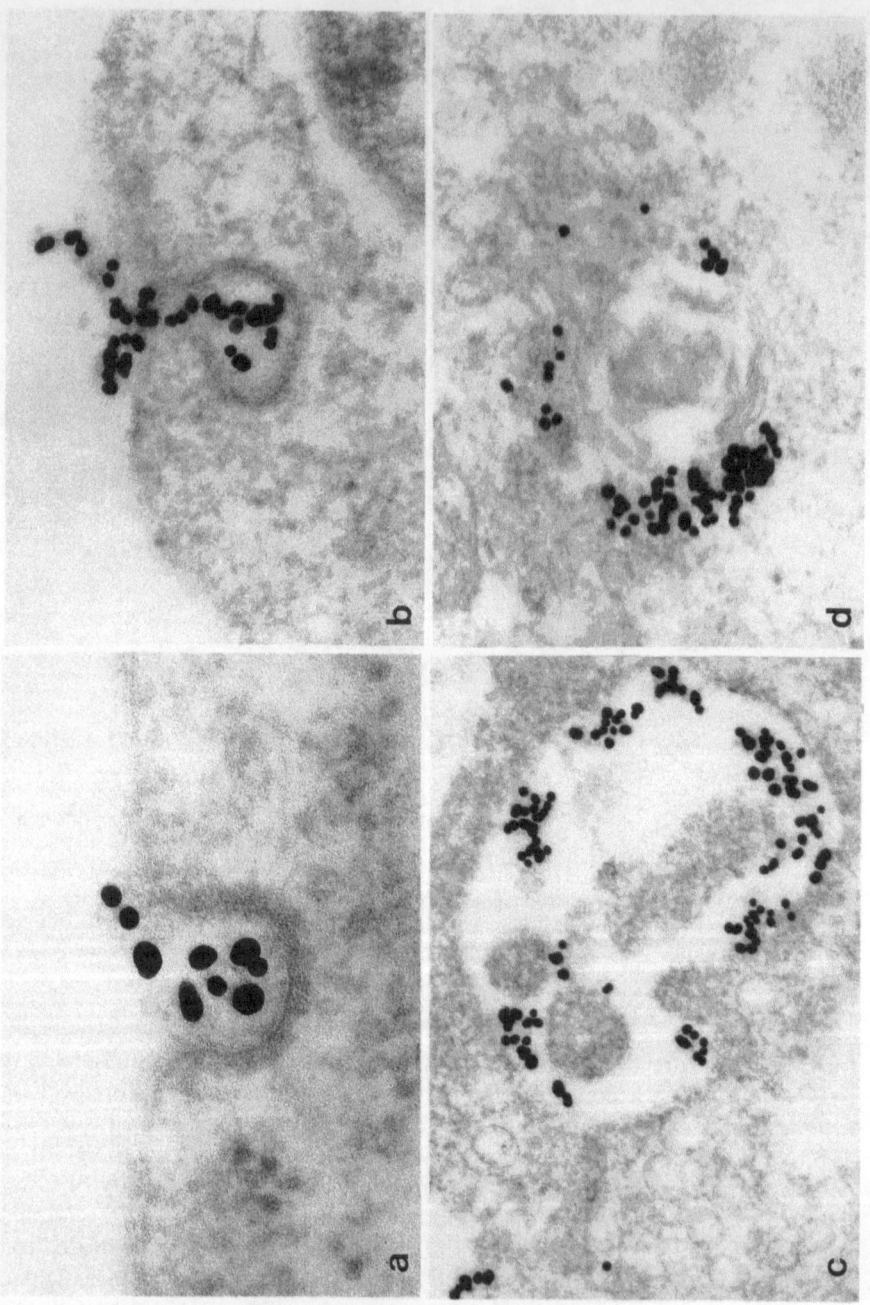

◀ **Fig. 5a–d.** Visualization of the receptor-mediated endocytosis of LDL-gold conjugates. At the beginning of the internalization process LDL is bound to so-called coated pits (**a**) which invaginate (**b**) until they pinch off from the plasma membrane. The electron-dense material covering the membrane at the site of these invaginations represents the protein clathrin. Later the gold grains are detected in lysosome-like compartments (**c** and **d**). The LDL particle is only visible in **a** through **c** covering the gold grain, which is located in the center of such a complex. (Original magnification: **a** and **b**, × 30 000; **c** and **d**, × 20 000)

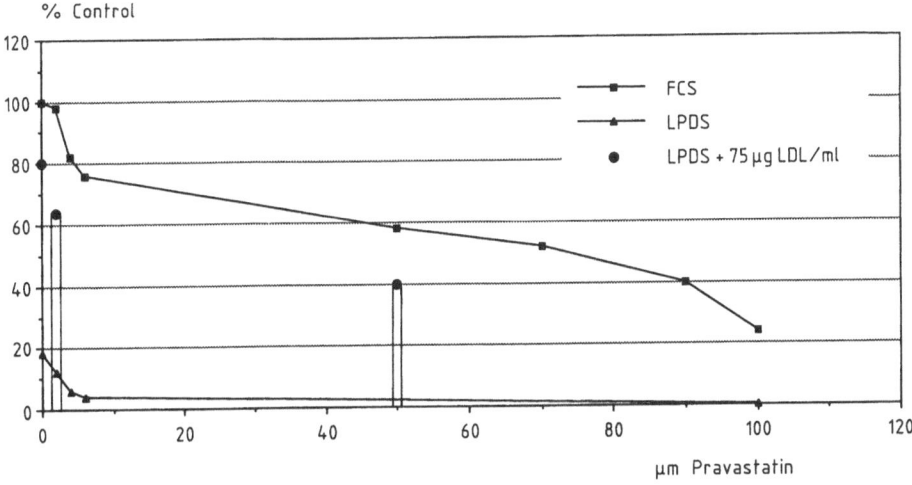

Fig. 6. Demonstration of tumor cell growth (D198/84; cell line originally established from a colon carcinoma) as dependent on cholesterol supply. Cell growth is indicated as percent of control on the *abscissa*. Increasing concentrations of the HMG CoA reductase inhibitor Pravastatin (CS-514) are indicated on the *ordinate*. Cells were grown in 10% Leibowitz L15 cell culture medium. They were counted after a 4-day incubation period in a hemocytometer. The *upper curve* represents cell growth in FCS (fetal calf serum) which had been dialyzed against the same buffer as LPDS (lipoprotein-deficient medium). Cell growth in LPDS is shown in the *lower curve*. The *bars* represent incubation conditions where the cells were grown in LPDS which had been resupplemented with LDL at a concentration of 75 µg/ml. Depletion from LDL-cholesterol slows the growth rate of this cell line to 18% of control. Cell growth is inhibited by Pravastatin in a dose-dependent manner under both conditions (incubation in FCS and in LPDS). However, the difference in growth rate can be largely compensated by addition of LDL to LPDS culture medium. Pravastatin-induced inhibition of growth rate in FCS is probably due to decreased synthesis of metabolites of the polyisoprenoid pathway other than cholesterol (a finding which has only been observed in this particular extremely rapidly proliferating cell line). In conclusion, these data underline the essential significance of cholesterol for membrane synthesis in proliferating cell systems

6.4 Cell Kinetic Studies

GAL and co-workers investigated cholesterol metabolism in gynecological cancer cell lines and found that these cells exhibit the same regulatory mechanisms as are present in nonneoplastic cells (which need not be true for all cancer cells) (GAL et al. 1981 a, b). In brief, this group found a decrease in LDL metabolism and HMG CoA reductase activity and in contrast an increase in the rate of cholesteryl ester formation in correlation with increasing cell density. This is consistent with the concept that cholesterol metabolism is regulated by the requirements of the cells for membrane synthesis.

The growth inhibition in LDL-deficient states was first described by GOLDSTEIN et al. (1979 b) in Chinese hamster ovary cells and later by ROTHENEDER and KOSTNER (1989) in mammary carcinoma cells. The concept that LDL-cholesterol is needed to establish optimal growth was supported by both studies, which, however, disagreed as to whether HDL is able to restore growth under lipoprotein-depleted conditions. GOLDSTEIN et al. demonstrated furthermore that in an LDL-depleted medium growth can be arrested by inhibition of HMG CoA reductase with compactin.

In our laboratories we have investigated growth patterns of different gastrointestinal tumor cell lines (NIENDORF et al. 1988, 1989 a). In a cell line derived from a human colon carcinoma, growth could be inhibited to 18% [lipoprotein-deficient serum (LPDS) vs normal fetal calf serum (FCS)], to 24% [FCS + 100 µM Pravastatin (CS-514; kindly provided by Dr. H.-M. Müller, Squibb-von Heyden, Munich)], and to 0% (LPDS + 100 µM Pravastatin). The results of these experiments are given in detail in Fig. 6. These studies underline the essential role of cholesterol and other metabolites of the polyisoprenoid pathway for the growth characteristics of malignant cells.

6.5 Clinical Implications of Receptor-Mediated Uptake of LDL in Tumor Tissues

From the available data the following conclusions can be drawn:

1. It is far from clear whether all those cancer cells which possess LDL receptors regulate their cholesterol metabolism in the same way as has been established for cultured fibroblasts.
2. HMG CoA reductase inhibitors are most likely not suitable to serve as antineoplastic drugs by direct inhibition of cholesterol synthesis in the tumor itself, as has been proposed by MALTESE et al. (1985). This would probably lead to severe side-effects since the mechanism should (in our opinion) be basically growth inhibiting and not directly cytotoxic. An approach based on long-term treatment would necessarily also affect any other organ with the same intensity. Drug safety experiments with lovastin, pravastatin, and simvastatin show a very high LD_{50} dose for all three substances, but such studies have not been designed to elucidate the benefit of cancer treatment.

3. It has been suggested that the delivery of cytotoxic compounds incorporated into LDL might be of benefit to cancer patients (RUDLING et al. 1983; FIRESTONE et al. 1984; VITOLS et al. 1985). A selective enrichment could be obtained by smuggling a "Trojan horse" into the cancer cell. In our opinion LDL receptors are so widespread in almost all tissues of the body that this approach represents an area bombing rather than a concentrated shot at a defined target!
4. Iatrogenically induced hypocholesterolemia seems to be one possibility of growth inhibition in malignant tumors. Excellent candidates should be tumors that are growing rapidly and possess little if any HMG CoA reductase activity.

Even if all the above-mentioned mechanisms should fail to yield relevant insights, the investigation of LDL receptors in cancer cells, paying particular attention to the other parameters of cholesterol metabolism, will certainly help to shed some light on the so far poorly understood pathogenesis of tumor-associated hypocholesterolemia.

References

Alexopoulos CG, Blatsios B, Avgerinos A (1987) Serum lipids and lipoprotein disorders in cancer patients. Cancer 60:3065–3070

Anderson RGW (1986) Methods for visualization of the LDL pathway in cultured fibroblasts. Methods Enzymol 129:201–216

Anderson RGW, Goldstein JL, Brown MS (1976) Localization of low density lipoprotein receptors on plasma membranes of normal human fibroblasts and their absence in cells from a familial hypercholesterolemia homozygote. Proc Natl Acad Sci USA 73:2434–2438

Anderson RGW, Brown MS, Goldstein JL (1977a) Role of the coated endocytic vesicle in the uptake of receptor-bound low density lipoprotein in human fibroblasts. Cell 10:351–364

Anderson RGW, Goldstein JL, Brown MS (1977b) A mutation that impairs the ability of lipoprotein receptors to localize in coated pits on the cell surface of human fibroblasts. Nature 270:695–699

Anderson RGW, Vasile E, Mello RJ, Brown MS, Goldstein JL (1978) Immunocytochemical visualization of coated pits and vesicles in human fibroblasts: relation to low density lipoprotein receptor distribution. Cell 15:919–933

Anderson RGW, Goldstein JL, Brown MS (1980) Fluorescence visualization of receptor-bound low density lipoprotein in human fibroblasts. J Recept Res 1:17–39

Anderson RGW, Brown MS, Goldstein JL (1981) Inefficient internalization of receptor-bound low density lipoprotein in human carcinoma A-431 cells. J Cell Biol 88:441–452

Anderson RGW, Brown MS, Beisiegel U, Goldstein JL (1982) Surface distribution and recycling of the low density lipoprotein receptor as visualized with antireceptor antibodies. J Cell Biol 93:523–531

Balasubramaniam S, Goldstein JL, Brown MS (1977) Regulation of cholesterol synthesis in rat adrenal gland through coordinate control of 3-hydroxy-3-methylglutaryl coenzyme A synthase and reductase activities. Proc Natl Acad Sci USA 74:1421–1425

Barak LS, Webb WW (1981) Fluorescent low density lipoprotein for observation of dynamics of individual receptor complexes on cultured human fibroblasts. J Cell Biol 90:595–604

Barak LS, Webb WW (1982) Diffusion of low density lipoprotein-receptor complex on human fibroblasts. J Cell Biol 95:846–852

Basu SK, Goldstein JL, Brown MS (1978) Characterization of the low density lipoprotein receptor in membranes prepared from human fibroblasts. J Biol Chem 253:3852–3856

Beaglehole R, Foulkes MA, Prior IAM, Eyles EF (1980) Cholesterol and mortality in New Zealand Maoris. Br Med J II:285–287

Beisiegel U, Kita T, Anderson RGW, Schneider WJ, Brown MS, Goldstein JL (1981a) Immunologic cross-reactivity of the low density lipoprotein receptor from bovine adrenal cortex, human fibroblasts, canine liver and adrenal gland, and rat liver. J Biol Chem 256:4071–4078

Beisiegel U, Schneider WJ, Goldstein JL, Anderson RGW, Brown MS (1981b) Monoclonal antibodies to the low density lipoprotein receptor as probes for study of receptor mediated endocytosis and the genetics of familial hypercholesterolemia. J Biol Chem 256:11923–11931

Brown MS, Goldstein JL (1981) Lowering plasma cholesterol by raising LDL receptors. N Engl J Med 305:515–517

Brown MS, Goldstein JL (1986) A receptor mediated pathway for cholesterol homeostasis. Science 232:34–47

Brown MS, Faust JR, Goldstein JL (1975) Role of the low density lipoprotein receptor in regulating the content of free and esterified cholesterol in human fibroblasts. J Clin Invest 55:783–793

Brown MS, Ho YK, Goldstein JL (1976) The low-density lipoprotein pathway in human fibroblasts: relation between cell surface receptor binding and endocytosis of low-density lipoprotein. Ann NY Acad Sci USA 68:244–247

Brown MS, Anderson RGW, Basu SK, Goldstein JL (1982) Recycling of cell-surface receptors: observations from the LDL receptor system. Cold Spring Harbor Symp Quant Biol XLVI:713–721

Brown MS, Anderson RGW, Goldstein JL (1983) Recycling receptors: the round-trip itinerary of migrant membrane proteins. Cell 32:663–667

Carew TE, Schwenke DC, Steinberg D (1987) Antiatherogenic effect of probucol unrelated to its hypocholesterolemic effect: evidence that antioxidants in vivo can selectively inhibit low density lipoprotein degradation in macrophage-rich fatty streaks and slow the progression of atherosclerosis in the Watanabe heritable hyperlipidemic rabbit. Proc Natl Acad Sci USA 84:7725–7729

Clayman RV, Bilhartz LE, Buja ML, Spady DK, Dietschy JM (1986a) Renal cell carcinoma in the Wistar rat: a model for studying the mechanism of cholesterol acquisition by a tumor in vivo. Cancer Res 46:2958–2963

Clayman RV, Bilhartz LE, Spady DK, Buja ML, Dietschy JM (1986b) Low density lipoprotein receptor activity is lost in vivo in malignant transformed renal tissue. FEBS Lett 196:87–90

Davies PF, Reidy MA, Goode TB, Bowyer TE (1976) Scanning electron microscopy in the evaluation of endothelial integrity of the fatty lesion in atherosclerosis. Atherosclerosis 25:125–130

Dawson P, Hofman SL, Westhuyzen DR, Südhoff TC, Brown MS, Goldstein JL (1988) Sterol dependent repression of low density lipoprotein receptor promoter mediated by 16-base pair sequence adjacent to binding site for transcription factor sp1. J Biol Chem 263:3372–3379

Dawson P, Westhuyzen DR, Goldstein JL, Brown MS (1989) Purification of oxysterol binding protein from hamster liver cytosol. J Biol Chem 264:9046–9052

Eckel RH (1989) Lipoprotein lipase: a multifunctional enzyme relevant to common metabolic disease. N Engl J Med 320:1060–1068

Esterbauer H, Quehenberger O, Jürgens G (1988) Effect of peroxidative conditions on human plasma low density lipoproteins. In: Nigam SK, Brien DCH, Slater TF (eds) Eicosanoids, lipidperoxidation and cancer. Springer, Berlin Heidelberg New York, pp 203–213

Feinleib M (1983) Review of the epidemiological evidence for a possible relationship between hypocholesterolemia and cancer. Cancer Res 43:2503–2507

Firestone RA, Pisano JM, Falck JR, McPhaul MM, Krieger M (1984) Selective delivery of cytotoxic compounds to cells by the LDL pathway. J Med Chem 27:1037–1043

Fukuda S, Horiuchi S, Tomita K, Murakami M, Morino Y, Takahashi K (1986) Acetylated low-density lipoprotein is endocytosed through coated pits by rat peritoneal macrophages. Virchows Arch [B] 52:1–13

Gal D, MacDonald PC, Porter JC, Simpson ER (1981a) Cholesterol metabolism in cancer cells in monolayer culture. III. Low-density lipoprotein metabolism. Int J Cancer 28:315–319

Gal D, MacDonald PC, Porter JC, Smith JW, Simpson ER (1981b) Effect of cell density and confluency on cholesterol metabolism in cancer cells in monolayer culture. Cancer Res 41:473–477

Garcia-Palmieri MR, Sorlie PD, Costas R, Havlik RJ (1981) An apparent inverse relationship between serum cholesterol and cancer mortality in Puerto Rico. Am J Epidemiol 114:29–40

Geuze HJ, Slot JW, Strous GJAM, Peppard J, von Figura K, Hasilik A, Schwartz AL (1983) Intracellular site of asialoglycoprotein receptor-ligand uncoupling. Double label immunoelectronmicroscopy during receptor mediated endocytosis. Cell 35:277–287

Ginsberg H, Gilbert HS, Gibson JC, Le NA, Brown WV (1982) Increased low-density lipoprotein catabolism in myeloproliferative disorders. Ann Intern Med 96:311–316

Goldstein JL, Brown MS (1973) Familial hypercholesterolemia: identification of a defect in the regulation of 3-hydroxy-3-methylglutaryl coenzyme A reductase activity associated with overproduction of cholesterol. Proc Natl Acad Sci USA 70:2804–2808

Goldstein JL, Brown MS, Anderson RGW (1977) The low density lipoprotein pathway in human fibroblasts. In: Brinkley BR, Porter KR (eds) International cell biology. Rockefeller University Press, New York, pp 639–648

Goldstein JL, Anderson RGW, Brown MS (1979a) Coated pits, coated vesicles, and receptor-mediated endocytosis. Nature 279:679–685

Goldstein JL, Helgeson JAS, Brown MS (1979b) Inhibition of cholesterol synthesis with compactin renders growth of cultured cells dependent on the low density lipoprotein receptor. J Biol Chem 254:5403–5409

Goldstein JL, Ho YK, Basu SK, Brown MS (1979c) Binding site on macrophages that mediates uptake and degradation of acetylated low density lipoprotein, producing massive cholesterol deposition. Proc Natl Acad Sci USA 76:333–337

Goldstein JL, Kita T, Brown MS (1983) Defective lipoprotein receptors and atherosclerosis: lessons from an animal counterpart of familial hypercholesterolemia. N Engl J Med 309:288–295

Gross D, Webb WW (1986) Molecular counting of low-density lipoprotein particles as individuals and small clusters on cell surfaces. Biophys J 49:901–911

Haberland ME, Fong D, Cheng L (1988) Malondialdehyde-altered atheroma of Watanabe heritable hyperlipidemic rabbits. Science 241:215–218

Handley DA, Arbeeny CM, Witte LD, Chien S (1981a) Colloidal gold-low density lipoprotein conjugates as membrane receptor probes. Proc Natl Acad Sci USA 78:368–371

Handley DA, Arbeeny CM, Eder HA, Chien S (1981b) Hepatic binding and internalization of low density lipoprotein-gold conjugates in rats treated with 17α-ethinylestradiol. J Cell Biol 90:778–787

Havel RJ, Goldstein JL, Brown MS (1980) Lipoproteins and lipid transport. In: Bondy PK, Rosenberg LE (eds) Metabolic control and disease, 8th edn. WB Saunders, Philadelphia, pp 393–494

Helenius A, Mellman I, Wall D, Hubbard A (1983) Endosomes. Trends Biochem Sci 8:245–250

Hesz A, Ingolic E, Krempler F, Kostner GM (1987) The existence of B/E and E receptors on HEP-G2 cells: a study using colloidal gold- and ^{125}I-labeled lipoproteins. Exp Mol Pathol 46:372–382

International Collaborative Group (1982) Circulating cholesterol level and risk of death from cancer in men aged 40 to 69 years. JAMA 248:2853–2859

Jaeckle S, Brady SE, Havel RJ (1989) Membrane liver binding sites for plasma lipoproteins on endosomes from rat liver. Proc Natl Acad Sci USA 86:1880–1884

Kagan A, McGee DL, Yano K, Rhoads GG, Nomura A (1981) Serum cholesterol and mortality in a Japanese-American population – the Honolulu heart program. Am J Epidemiol 114:11–20

Kark JD, Smith AH, Hames CG (1980) The relationship of serum cholesterol to the incidence of cancer in Evans County, Georgia. J Chronic Dis 33:311–322

Kita T, Nagano Y, Yokode M et al. (1987) Probucol prevents the progression of atherosclerosis in Watanabe heritable hyperlipidemic rabbit, an animal model for familial hypercholesterolemia. Proc Natl Acad Sci USA 84:5928–5931

Knott TJ, Pease RJ, Powell LM et al. (1986) Complete protein sequence and identification of structural domains of human apo-lipoprotein B. Nature 323:734–738

Kovanen PT, Faust JR, Brown MS, Goldstein JL (1979a) Low density lipoprotein receptors in bovine adrenal cortex. I. Receptor mediated uptake of low density lipoprotein and utilization of its cholesterol for steroid synthesis in cultured adrenocortical cells. Endocrinology 104:599–609

Kovanen PT, Basu SK, Goldstein JL, Brown MS (1979 b) Low density lipoprotein receptors in bovine adrenal cortex. II. Low density lipoprotein binding to membranes prepared from fresh tissue. Endocrinology 104:610–616

Kreiger M, Smith LC, Anderson RG et al. (1979) Reconstituted low density lipoprotein: a vehicle for the delivery of hydrophobic fluorescent probes to cells. J Supramol Struct 10:467–478

Kreiger M, Brown MS, Faust JR, Goldstein JL (1978) Replacement of endogenous cholesteryl esters of low density lipoprotein with exogenous cholesteryl linoleate. J Biol Chem 129:4093–4101

Kruth HS, Avigan J, Gamble W, Vaughan M (1979) Effect of cell density on binding and uptake of low density lipoprotein by human fibroblasts. J Cell Biol 83:588–594

Law SW, Grant SM, Higuchi K, Hospattankar A, Lackner K, Lee N, Brewer HB (1986) Human liver apolipoprotein B-100 cDNA: complete nucleic acid and derived amino acid sequence. Proc Natl Acad Sci USA 83:8142–8146

Malassiné A, Besse C, Roche A, Alsat E, Rebourcet R, Mondon F, Cedard L (1987) Ultrastructural visualization of the internalization of low density lipoprotein by human placental cells. Histochemistry 87:457–464

Mommaas-Kienhuis AM, Krijbolder LH, van Hinsbergh VWM, Daems WT, Vermeer BJ (1985a) Visualizing of binding and receptor-mediated uptake of low density lipoproteins by human endothelial cells. Eur J Cell Biol 36:201–208

Mommaas-Kienhuis AM, van der Schroeff JG, Wijsman MC, Daems WT, Vermeer BJ (1985b) Conjugates of colloidal gold with native and acetylated low density lipoproteins for ultrastructural investigations on receptor-mediated endocytosis by cultured human monocyte-derived macrophages. Histochemistry 83:29–35

Morris DL, Borhani NO, Fitzsimons E et al. (1983) Serum cholesterol and cancer in the hypertension detection and follow-up program. Cancer 52:1754–1759

Niendorf A, Röcken C, Hupfeld S et al. (1988) Morphologische und funktionelle Bedeutung von LDL-Rezeptoren an malignen Tumoren. Verh Dtsch Ges Pathol 72:427

Niendorf A, Hupfeld S, Keilhau A, Peters A, Arps H, Beisiegel U, Dietel M (1989a) HMG-CoA-Reductase-Inhibition hemmt das Wachstum von kultivierten Coloncarcinomzellen. Verh Dtsch Ges Pathol 73:459

Niendorf A, Peters A, Keilhau A, Arps H, Behnke B, Beisiegel U, Dietel M (1989b) LDL deficiency and suppression of HMG-CoA-reductase activity cause inhibition of proliferation rates in mammary carcinoma cells in vitro. Proceedings of the 80th annual meeting of the American Association for Cancer Research 30:80 (abstr)

Nistor A, Simionescu M (1986) Uptake of low density lipoproteins by the hamster lung – interactions with capillary endothelium. Am Rev Respir Dis 134:1266–1272

Osborne TF, Gil G, Goldstein JL, Brown MS (1988) Operator constitutive mutation of 3-hydroxy-3-methylglutaryl coenzyme A reductase promoter abolishes protein binding sterol regulator element. J Biol Chem 263:3380–3387

Paavola LG, Strauss JF, Boyd CO, Nestler JE (1985) Uptake of gold- and [^{3}H]cholesteryl linoleate-labeled human low density lipoprotein by cultured rat granulosa cells: cellular mechanisms involved in lipoprotein metabolism and their importance to steroidogenesis. J Cell Biol 100:1235–1247

Parascandola J (1981) Origins of the receptor theory. In: Lamble W (ed) Towards understanding receptors. Elsevier/North-Holland Biomedical Press, Amsterdam, pp 1–7

Pastan IH, Willingham MC (1981) Journey to the center of the cell: role of the receptosome. Science 214:504–509

Pathak RK, Merkle RK, Cummings RD, Goldstein JL, Brown MS, Anderson RGW (1988) Immunocytochemical localization of mutant low density lipoprotein receptors that fail to reach the Golgi complex. J Cell Biol 106:1831–1841

Pearse BMF (1975) Coated vesicles from pig brain: purification and biochemical characterization. J Mol Biol 97:93–98

Pearse BMF (1976) Clathrin: a unique protein associated with intracellular transfer of membrane by coated vesicles. Proc Natl Acad Sci USA 73:1255–1259

Pitas RE, Boyles J, Mahley RW, Montgomery Bissell D (1985) Uptake of chemically modified low density lipoproteins in vivo is mediated by specific endothelial cells. J Cell Biol 100:103–117

Reverter JC, Sierra J, Marti-Tutusaus, Montserrrat E, Granena A, Rozman C (1988) Hypocholesterolemia in acute myelogenous leukemia. Eur J Haematol 41:317–320

Roach PD, Zollinger M, Noel SP (1987) Detection of the low density lipoprotein (LDL) receptor on nitrocellulose paper with colloidal gold-LDL conjugates. J Lipid Res 28:1515–1521

Robenek A, Hesz A (1983) Dynamics of low-density lipoprotein receptors in the plasma membrane of cultured human skin fibroblasts as visualized by colloidal gold in conjunction with surface replicas. Eur J Cell Biol 31:275–282

Robenek H, Rassat J, Hesz A, Grünwald J (1982) A correlative study on the topographical distribution of the receptors for low density lipoprotein (LDL) conjugated to colloidal gold in cultured human skin fibroblasts employing thin section, freeze-fracture, deep-etching, and surface replication techniques. Eur J Cell Biol 27:242–250

Robenek H, Hesz A, Rassat J (1983) Variability of the topography of low-density lipoprotein (LDL) receptors in the plasma membrane of cultured human skin fibroblasts as revealed by gold-LDL conjugates in conjunction with the surface replication technique. J Ultrastruct Res 82:143–155

Rose G, Shipley MJ (1980) Plasma lipids and mortality: a source of error. Lancet I:523–526

Ross R (1986) The pathogenesis of atherosclerosis – an update. N Engl J Med 314:488–500

Rotheneder M, Kostner GM (1989) Effects of low- and high-density lipoproteins on the proliferation of human breast cancer cells in vitro: differences between hormone-dependent and hormone-independent cell lines. Int J Cancer 43:875–879

Rudling M (1986) Low density lipoprotein receptors in normal and malignant tissues: physiological aspects and clinical significance. Akademisk Avhandling, Department of Pharmacology, Karolinska Institute, Stockholm, Sweden

Rudling MJ, Collins VP, Peterson CO (1983) Delivery of aclacinomycin A to human glioma cells in vitro by the low-density lipoprotein pathway. Cancer Res 43:4600–4605

Rudling MJ, Stahle L, Peterson CO, Skoog L (1986) Content of low density lipoprotein receptors in breast cancer tissue related to survival of patients. Br Med J 292:580–582

Russel DW, Yamamoto T, Schneider WJ, Slaughter CJ, Brown MS, Goldstein JL (1983) cDNA cloning of the bovine low density lipoprotein receptor: feedback regulation of a receptor mRNA. Proc Natl Acad Sci USA 50:7501–7505

Sanan DA, Strümpfer AEM, van der Westhuyzen DR, Coetzee GA (1985) Native and acetylated low density lipoprotein metabolism in proliferating and quiescent bovine endothelial cells in culture. Eur J Cell Biol 36:81–90

Sanan DA, Van der Westhuyzen DR, Gevers W, Coetzee GA (1987) The surface distribution of low density lipoprotein receptors on cultured fibroblasts and endothelial cells – ultrastructural evidence for dispersed receptors. Histochemistry 86:517–523

Schatzkin A, Hoover RN, Taylor PR et al. (1988) Site-specific analysis of total serum cholesterol and incident cancer in the national health and nutrition examination survey. I. Epidemiologic follow-up study. Cancer Res 48:452–458

Schneider WJ, Basu SK, McPhaul MJ, Goldstein JL, Brown MS (1979) Solubilization of the low density lipoprotein receptor. Proc Natl Acad Sci USA 76:5577–5581

Schneider WJ, Goldstein JL, Brown MS (1980) Partial purification and characterization of the low density lipoprotein receptor from bovine adrenal cortex. J Biol Chem 255:11442–11447

Schneider WJ, Beisiegel U, Goldstein JL, Brown MS (1982) Purification of the low density lipoprotein receptor, an acidic glycoprotein of 164000 molecular weight. J Biol Chem 257:2664–2673

Sherwin RW, Wentworth DN, Cutler JA, Hulley SB, Kuller LH, Stamler J (1987) Serum cholesterol levels and cancer mortality in 361662 men screened for the multiple risk factor intervention trial. JAMA 257:943–948

Snelting-Havinga I, Mommaas M, Van Hinsbergh VWM, Daha MR, Daems WT, Vermeer BJ (1989) Immunoelectron microscopic visualization of the transcytosis of low density lipoproteins in perfused rat arteries. Eur J Cell Biol 48:27–36

Sparrow CP, Parthasarathy S, Steinberg D (1989) A macrophage receptor that recognizes oxidized low density lipoprotein but not acetylated low density lipoprotein. J Biol Chem 264:2599–2604

Stary H (1989) Evolution and progression of atherosclerotic lesions in coronary arteries of children and young adults. Arteriosclerosis 9:119–132

Steinberg D, Parthasarathy S, Carew T, Khoo JC, Witzum JL (1989) Beyond cholesterol modifica-
    tion of low-density lipoprotein that increase its atherogenicity. N Engl J Med 320:915–924
Study Group, European Atherosclerosis Society (1987) Strategies for the prevention of coronary
    heart disease: a policy statement of the European Atherosclerosis Society. Eur Heart J 8:77–88
Südhof TC, Goldstein JL, Brown MS, Russel DW (1985) The ldl receptor gene: a mosaic of exons
    shared with different proteins. Science 228:815–822
Tolleshaug H, Hobgood KK, Brown MS, Goldstein JL (1983) The LDL receptor locus in familial
    hypercholesterolemia: multiple mutations disrupt transport and processing of a membrane
    receptor. Cell 32:941–951
Tompkins RG, Schnitzer JJ, Yarmush ML, Colton CK, Smith KA (1988) Measurement of ^{125}I-low
    density lipoprotein uptake in selected tissues of the squirrel monkey by quantitative auto-
    radiography. Am J Pathol 132:526–542
Törnberg SA, Holm LE, Carstensen JM, Eklund GA, Odont D (1986) Risks of cancer of the colon
    and rectum in relation to serum cholesterol and beta-lipoprotein. N Engl J Med 315:1629–
    1633
Tóth IE, Szabó D, Szalay KS, Gyévai A, Szollár LG, Gláz E (1988) Colloidal gold-labeled lipo-
    protein binding and internalization in adrenocortical cells in vitro. Clin Biochem 21:101–105
Vasile E, Simionescu M, Simionescu N (1983) Visualization of the binding, endocytosis, and
    transcytosis of low-density lipoprotein in the arterial endothelium in situ. J Cell Biol 96:1677–
    1689
Ventkatanarayanan S, Nagarajan B (1988) Association between tumour status and serum lipo-
    protein cholesterol in hemopoietic malignancy. Biochem Int 17:499–507
Vermeer BJ, Wijsman MC, Mommaas-Kienhuis AM, Ponec M, Havekes L (1985) Modulation of
    low density lipoprotein receptor activity in squamous carcinoma cells by variation in cell
    density. Eur J Cell Biol 38:353–360
Vermeer BJ, Wijsman MC, Mommaas-Kienhuis AM, Ponec M (1986) Binding and internalization
    of low-density lipoproteins in SCC25 cells and SV40 transformed keratinocytes. A morpholog-
    ic study. J Invest Dermatol 86:195–200
Vitols S, Gahrton G, Peterson C (1984) Significance of the low density lipoprotein (LDL) receptor
    pathway for the in vitro accumulation of AD-32 incorporated into LDL in normal and
    leukemic white blood cells. Cancer Treat Rep 68:515–520
Vitols SG, Masquelier M, Peterson CO (1985) Selective uptake of a toxic lipophilic anthracycline
    derivative by the low-density lipoprotein receptor pathway in cultured fibroblasts. J Med Chem
    28:451–454
Weisgraber KH, Rall SC Jr (1987) Human apolipoprotein B-100 heparin-binding sites. J Biol Chem
    262:11 097–11 103
Williams ML, Mommaas-Kienhuis AM, Rutherford SL, Grayson S, Vermeer BJ, Elias PM (1987)
    Free sterol metabolism and low density lipoprotein receptor expression as differentiation
    markers of cultured human keratinocytes. J Cell Physiol 132:428–440
Williams RR, Sorlie PD, Feinleib M, McNamara PM, Kannel WB, Dawber TR (1981) Cancer
    incidence by levels of cholesterol. JAMA 245:247–252
World Health Organization (1958) Classification of atherosclerotic lesions; report of a study group.
    World Health Organization Technical Report Series No. 143 3–20
Yang CY, Chen SH, Gianturco SH et al. (1986) Sequence, structure, receptor binding domains and
    internal repeats of human apolipoprotein B-100. Nature 323:738–740
Zielinski CC, Stuller I, Müller C (1988) Increased serum concentrations of cholesterol and triglyc-
    erides in the progression of breast cancer. J Cancer Res Clin Oncol 114:514–518

Neurotransmitter Receptors in Human Brain Diseases

A. Probst, G. Mengod, and J. M. Palacios

1 Introduction

Progress in our understanding of how the brain works is intimately related to the development of new techniques allowing for a more refined analysis of the brain machinery at the microscopic and molecular levels. The broad applications, in the last three decades, of biochemical techniques to the study of nervous tissues has greatly increased our understanding of the chemical make-up of the brain. Because of the high regional and cellular complexity of the brain and the low resolution of the classical biochemical approaches, histochemical methods have been devised that permit more accurate description of molecular events at a cellular or subcellular level. These new methods have taken advantage of the recent development of antibodies, radioactive markers, and molecular probes for brain antigens or nucleic acid sequences.

As many of the brain molecules do not significantly alter their properties after death for some hours, they are still accessible to analysis, using biochemical and histochemical procedures (Bird and Iversen 1982; Palacios et al. 1986).

The systematic study of samples from diseased human brains and their comparison with samples of control populations has led to the discovery of selective biochemical deficits in a number of neurodegenerative diseases. Classical examples are the dopamine deficit in Parkinson's disease (Ehringer and Hornykiewicz 1960) and the cholinergic deficit of Alzheimer's disease (Bowen et al. 1976; Davies and Maloney 1976).

The broad issue that we are concerned with in this review is the visualization of neurotransmitter receptor expression and distribution in nervous tissues and more specifically in diseases of the human brain. The main focus will be on the analysis of neurotransmitter receptor distribution and regulation as a consequence of specific neuronal degeneration encountered in specific neurodegenerative diseases. We will review how the autoradiographic techniques for the detection of receptors are applied to their analysis in human postmortem materials and how the information obtained to date can help in understanding the basic mechanisms involved in neurotransmitter actions and in the neuropathology and neuropharmacology of the human brain.

2 Neurotransmitter Receptors: Description and Molecular Biology Approaches

Neurons are highly and differently specialized cells which interact with each other in complex networks. Each neuron communicates with many different target cells by releasing chemical messengers, called neurotransmitters. At the synaptic junctions, however, chemical messengers delivered by neurons have no direct action on the lipid bilayer of the postsynaptic membrane. In order to be effective they must interact with special transmembrane protein complexes having specific recognition sites for the messenger molecules: the receptors. Since essentially all information processing in the brain involves specific receptor sites for neurotransmitter substances, one of the main tasks of neuroscience is the molecular characterization and the cellular localization of receptors.

A major function of a neurotransmitter receptor is to mediate the physiological effect of chemical messengers into the interior of the target cells. The postsynaptic effect of a chemical messenger is not specifically characteristic of the transmitter as a chemical but results from its interaction with specific receptors. This means that for one particular transmitter, for instance acetylcholine, it is the receptor which will determine the type of postsynaptic effect on the receptive cell, for instance inhibitory or excitatory. Each neuron has a multiplicity of receptors on its cell membrane and it is through the repertoire of receptors that neuronal versatility is developed (McGeer et al. 1987).

Considerable progress has been made in the last few years toward elucidation of the molecular chemistry of neurotransmitter receptors and of their mechanism of action. Not only can we now study and measure receptors using biochemical techniques and determine the second messenger mechanism involved in many receptor types; we can also visualize these receptors at the microscopic and in some cases at the submicroscopic level.

2.1 General Classes of Neurotransmitter Receptor

Neurotransmitter receptors can be divided into two general classes based on the mechanism by which they transmit information in the target cell. One class mediates changes in membrane conductance by having within their structure the ability to conduct ions across the plasma membrane. The other class mediates changes in the metabolic machinery of postsynaptic cells by coupling to a family of signal-transducing proteins (second messengers) located on the cytoplasmic surface of the plasma membrane (SCHWARZ 1985) (G protein-coupled receptors).

2.1.1 Receptors Containing Ion Channels (Ligand-Gated Channels)

In this type of receptor, binding of the external chemical messenger produces a conformational change in which channels are opened to accommodate ions of suitable charge and diameter (McGEER et al. 1987). The paradigmatic example is the nicotinic cholinergic receptor that open channels to Na^+, K^+, and ions of similar size. The $GABA_A$ and glycine receptors are both in themselves Cl^- ion channels and constitute further examples of receptor complexes with an ionotropic function. Nicotinic receptors are unique among neurotransmitter receptors in that they are present at relatively high concentrations in some specific tissues. Their presence as a major component of total membrane protein in electroplax organs of *Torpedo californica* (electric fish) made it possible to isolate and purify these receptors to homogeneity and to characterize their molecular structure. We now know that this ion channel is a multimetric, intrinsic membrane glycoprotein that traverses the postsynaptic membrane. We also know from biophysical and biochemical studies that the receptor is formed from five subunits with stoichiometry α_2, β, γ, δ, with β-subunits represented twice in the complex. Two molecules of the transmitter can bind to this multimetric complex: each acetylcholine molecule binds on one of the two α-subunits of the receptor. Increased permeability to ions is postulated to be the result of a cooperative rearrangement of the subunits, resulting from the binding of the α-subunit with the acetylcholine molecules.

γ-Aminobutyric acid (GABA) receptors, which are widely distributed in the mammalian brain and spinal cord, can be pharmacologically distinguished as $GABA_A$ receptors, selectively stimulated by muscimol, and $GABA_B$, selectively stimulated by baclofen. The $GABA_A$ receptor channels stabilize the cell resting potential by increasing the membrane conductance of Cl^- during the activation of excitatory receptors and are therefore inhibitory. In contrast to $GABA_A$ receptors, $GABA_B$ receptors do not contain an integral ion channel but are coupled to Ca^{2+} or K^+ channels involving the activation of guanine nucleotide binding protein in the cell membrane.

The $GABA_A$ receptor Cl^- channel complex is composed of two distinct subunits, α and β, of similar molecular weight (53 000 and 57 000 respectively) with presumptive subunit structure $\alpha_2\beta_2$ (MAMALAKI et al. 1987). The binding site for GABA, which is located on the β-subunit, is recognized by structural analogues of GABA, such as muscimol, whereas benzodiazepines bind to the α-subunit.

The Cl^- ion channels of the glycine receptor, like the $GABA_A$ receptor, consist of two subunits. The α-subunit binds glycine but the function of the β-subunit remains unknown. It has been suggested that the β-subunit is required for function since the α-subunit has been proved to be nonfunctional.

In recent years, dramatic insights have been obtained into the molecular biology of the ion channel receptors, particularly of the nicotinic receptor. Using chemical methods, RAFTERY et al. (1980) determined the amino acid sequence for each of the subunits of the nicotinic receptor. Short pieces of synthetic DNA were then generated from the known amino acid sequence, and these probes used to screen as cDNA library of the electric fish electroplax organs. Hybridizing clones were then isolated and sequenced. The deduced protein sequences were found to have several stretches of hydrophobic, membrane-spanning domains, suggesting that each subunit spans the membrane several times. Furthermore, sequence homologies between the four subunits suggest an evolutionary relationship between the units.

Several polypeptides from other ligand-gated channels of the central nervous system have been sequenced: four subunits of the $GABA_A$ receptor channel (SCHOFIELD et al. 1987, 1989; PRITCHETT et al. 1989; LOLAIT et al. 1989) and one subunit from the glycine receptor channel (GRENNINGLOH et al. 1987). They share about 50% homology at the amino acid level. Furthermore, strong homologies in protein sequences have been found between nicotinic, $GABA_A$, and glycine receptors, particularly in the putative transmembrane domains. This has led to the suggestion that all the ion channels may have arisen from an unknown ancestor gene.

2.1.2 G Protein-Coupled Receptors

Not all receptors mediate changes in ionic conductance directly. There is an important group of more complex receptor types which modify the cell's biochemistry through a cascade of events involving several distinct proteins, among them intracellular second messenger systems. Second messengers are defined as molecules that serve as functional links between receptors of external chemical messengers and effector mechanisms in the target cell (metabolic processes, genes, ion channels). The best understood of the enzymatic second messenger system is the type that uses a cyclic nucleotide as the intracellular signal molecule. Most prevalent is cyclic adenosine 3′,5′-monophosphate (cAMP), which is synthesized by the enzyme adenylate cyclase and is degraded by a phosphodiesterase to the inactive AMP. The steps leading to cAMP synthesis include binding of a first chemical messenger (transmitter or hormone) to a receptor protein, causing allosteric change which activates the linking protein (called G protein, N protein, or transducing protein) to bind preferentially guanosine triphosphate (GTP) instead of guanosine diphosphate (GDP). When this occurs, the G protein moves laterally in the membrane to associate with the adenylate cyclase molecule (SHEPHERD 1988), which is then activated or inhibited, depending on whether stimulatory or inhibitory ligand has bound to the receptor protein. The cAMP produced binds to the regulatory subunit of protein kinase A, releasing its catalytic unit to phosphorylate specific target proteins, which, in turn, will be involved in a variety of metabolic and physiological cell processes. The cAMP system appears to be particularly important in nerve cells and, in fact, adenylate cyclase activity is very high in the brain.

A variety of neurotransmitters or putative neurotransmitters are able to stimulate the adenylate cyclase system. They include dopamine (D_1), epinephrine (at β-adrenoceptors), serotonin, histamine (H_2), adenosine, and various peptides.

Another cyclic nucleotide that acts as a second messenger is cyclic guanosine 3',5'-monophosphatase (cGMP). The chemistry of cGMP is parallel to that of cAMP in a number of ways. Guanyl cyclase is localized primarily to the postsynaptic component of neurons and is also found in glial cells. cGMP is present in the brain at concentrations far below those of cAMP, except in the cerebellum, where the concentrations are approximately equal. cGMP may be involved in the action of acetylcholine. Cholinergic agents increase cGMP, while atropine blocks such increases.

In addition to acting on adenylate cyclase, G-binding proteins can also act on other molecules, like those of ion channels themselves. For instance, G protein-stimulated opening of K^+ channels, coupled in some cases to inhibition of adenylate cyclase, has been reported in the brain. A common mechanism at many muscarinic cholinergic receptors is for the receptor to activate a G protein which modulates a K^+ channel. The G protein may also modulate adenylate cyclase production of cAMP, which in turn modulates Ca^{2+} channels, usually in a way that complements the effects on K^+ channels (SHEPHERD 1988).

Polyphosphoinositides are yet another class of second messengers (for review see MCGEER et al. 1987). Catabolism of this membrane-bound material is stimulated when a cell surface receptor is activated, inducing a complex sequence of events whose result is a release of phosphate groups for activating protein kinase and mobilizing intracellular Ca^{2+}. As for the cAMP system, a G protein is involved as the coupling agent. Glutamate/aspartate, acetylcholine, dopamine, and norepinephrine are among the neurotransmitters that have been reported to act on phosphoinositide systems.

Another important second messenger is Ca^{2+}. This ion also brings about protein phosphorylation through Ca^{2+}-dependent protein kinases.

The mentioned second messenger systems do not achieve their regulatory effects independently of each other; rather they have been shown to interact and influence each other in several key reactions (SCHWARZ 1985).

Different G protein-coupled receptor subtypes have been "classically" identified by their pharmacological differences, either in their response to drugs in vivo or to ligand binding in vitro. For example, muscarinic receptors have been classified into pharmacological subtypes termed M_1 and M_2 based on the different affinity for the antagonist pirenzepine. The M_2 subtype can be further subdivided into "M_2 cardiac" and "M_2 glandular" according to the relative affinity for the antagonists hexahydrosiladifenidol and AF-DX 116. Equivalent tools have been used to define the subtypes of other G protein-coupled receptor.

During the past 3 years, molecular cloning techniques have revealed the primary amino acid sequences of several receptors coupled to guanine nucleotide regulatory proteins (G proteins), showing that all of them are highly related proteins but coded by different genes. Thus, five different genes have been identified that encode muscarinic receptor subtypes m1, m2, m3, m4, and m5 (Bonner et al. 1987, 1988; Peralta et al. 1987). Also the adrenergic receptors, subtypes $\alpha_2 A$ (Kobilka et al. 1987a), $\alpha_2 B$ (Regan et al. 1988), β_1 (Frielle et al. 1987), and β_2 (Dixon et al. 1986; Kobilka et al. 1987b) are encoded by separate genes. Some serotonin receptor subtype genes have also been cloned: 5-HT_{1A} (Kobilka et al. 1987c; Fargin et al. 1988), 5-HT_{1C} (Lübbert et al. 1987; Julius et al. 1988), and 5-HT_2 (Pritchett et al. 1988). Dopamine D_2 subtype has also been cloned and sequenced (Bunzow et al. 1988), as has substance K receptor (Masu et al. 1987).

Hydropathicity profiles for each G protein coupled receptor have shown seven hydrophobic regions of 20–25 amino acids, which are potentially membrane spanning, a long C-terminal hydrophilic sequence, a shorter N-terminal hydrophilic sequence, and a long loop between transmembrane segments V and VI. The high degree of amino acid homology and the similarities in the general structure have led to the classification of membrane receptors modulating cellular function via G proteins in the same multigene family (Dohlman et al. 1987).

3 Imaging Receptors in the Brain

One of the most basic questions that we have to ask a neuropathologist interested in brain receptors concerns the localization of these receptors in normal and pathological brain tissues. Different techniques have been devised in order to "image" or localize receptors in the brain. The following main technical approaches will be described in this section:

1. Receptor radiohistochemistry
2. Receptor immunohistochemistry
3. In situ Hybridization histochemistry using molecular probes for detection of receptor mRNA in brain sections.

The last two approaches have as yet found only limited application within human neuropathology. However, as they represent important tools for determining the cellular and subcellular localization of receptors and for our understanding of receptor gene expression in brain tissues, a short account of these techniques and of their results is included.

3.1 Receptor Radiohistochemistry

The localization of receptors can be ascertained by manually dissecting out brain regions of interest and labeling receptors in membrane preparations obtained from these regions by means of simple binding techniques with radioactive ligands. Such techniques have been widely used for the study of neurotransmitter receptors in human and animal brain tissues (BIRD and IVERSEN 1982). Ligand-binding techniques have also been used successfully to characterize ion channels modulated by neurotransmitters as well as second messenger systems such as cAMP and the phosphoinositide cycle.

However, it is clear that in an organ with such highly diversified architecture as the brain, techniques that allow "imaging" of receptors at a microscopic level possess important advantages over techniques using homogenized tissues. Among these techniques are the microscopic autoradiographic methods, which provide resolution in the micrometer range and allow accurate quantitative high resolution assessment of the distribution of binding sites for neurotransmitters throughout the brain.

Initial studies utilized in vivo autoradiography in which the radiolabeled drug was administered intravenously and the brain rapidly removed and sectioned for autoradiography. In recent years the concept underlying in vivo receptor labeling has been applied in humans in positron emission tomography (PET) imaging of receptors.

An autoradiographic technique using in vitro incubation of labeled ligand with brain sections was subsequently developed (YOUNG and KUHAR 1979) (in vitro labeling autoradiography). With this technique human tissue can be used, which would not be possible with other approaches. The technique also allows much better control of receptor labeling conditions and permits circumvention of the blood–brain barrier. Furthermore, the distribution of different receptors can be examined in adjacent sections, i.e., in the same anatomical field. However, the main advantage of the in vitro labeling technique over purely biochemical studies is that it allows for correlations between changes in specific neuronal populations and changes in specific receptors. This means that the ligand binding can be directly correlated with histopathology in the same or in immediately consecutive sections.

Technical Aspects. A number of characteristics of the patients are collected, including age, gender, interval between death and freezing of tissues, cause of death, and, when possible, clinical records and drug history. Brains obtained at autopsy are promptly dissected at room temperature. Half of the brain is immersed in formaldehyde (4%) in phosphate buffer saline (pH 7.2) and further processed for routine neuropathological examination. The remainder of the CNS is cut into 4 mm thick tissue blocks which are numbered and frozen at −80 °C on aluminum foil. Usually 30–40 blocks are taken from each brain, including several neocortical areas, the hippocampus and other limbic structures, the basal ganglia, the diencephalon, the cerebellum, and different levels of the brain stem and, if possible, of the spinal cord.

The tissue samples are then brought to –20 °C and mounted onto microtome chucks. 10 μm thick sections are cut using a microtome cryostat, thaw mounted onto gelatin-coated microscope glass slides, and stored at –20 °C until used.

Slide-mounted tissue sections are then incubated with different radiolabeled ligands (see Table 1) at concentrations roughly equivalent to their known KD values. Incubations are carried out at room temperature or at 4 °C for different amounts of time depending on the time at which steady state binding of the investigated receptor occurs. This will be preceded by a preincubation step if the binding of the ligand proved to be enhanced by this procedure. Buffers (mostly isotonic) selected for incubation (and washing) are similar to those used in previous biochemical studies of the ligand.

The sections are then washed in ice-cold buffer, dipped in cold distilled water, and then dried rapidly in a cold airstream. The wash time is selected in such a way that a workable specific/nonspecific ratio is obtained while most specific binding is maintained intact. Nonspecific binding is determined by incubating consecutive sections with the labeled ligand in the presence of unlabeled (cold) ligand at concentrations in μM range, and is defined as the binding not displaced at these concentrations of the cold ligand. Autoradiographs are generated by apposing slide-mounted tissue sections with labeled receptors to an emulsion. An emulsion-coated glass or plastic coverslip or a tritium-sensitive sheet-film such as an Ultrofilm (LKB Sweden) may be employed) (Unnerstall et al. 1982; Palacios et al. 1981). For later quantification of binding sites on sheet-films, standards made of brain gray or white matter or plastic polymers containing known amounts of tritium or ^{121}I are exposed along with the labeled tissues.

The emulsion on the coverslip or the X-ray film is then developed after an adequate exposure time, the length of which depends on the amounts of radioactivity in the tissue. Examination of autoradiographic images on emulsion-coated coverslips is facilitated by dark-field microscopy, where the silver grains appear white on a dark background. The resolution achieved with this technique is better than with film autoradiography because the crystal diameter of the emulsion is smaller than that used in tritium-sensitive films. Furthermore, tissue sections are attached permanently to the autoradiographic image, making the identification of discretely labeled regions more accurate. However, quantification of binding sites is generally more time consuming when using nuclear emulsions, as it can only be achieved by counting the number of silver grains.

Despite the aforementioned advantages of the emulsion coverslip method over film autoradiography, we have used the latter method more systematically for investigations of postmortem brain material. The main reasons for this are that film-based autoradiography is comparatively easy to perform and that it allows the simultaneous investigation of a great number of labeled slides from many different brain regions on one film sheet. This is achieved by placing the labeled slides in a side-by-side array in apposition to the film (Palacios et al. 1981). Another reason is that in sheet-film-based receptor autoradiography, gray levels associated with receptor localization can be readily converted into optical densities and translated into amounts of receptors per mg protein or tissue equivalent thanks to the use of standards and computer-assisted image analysis systems. This is achieved by digitizing the autoradiographic image into a matrix of 512×512 picture units using a TV camera connected to a microscope. The autoradiographic image is displayed at the same time on the screen of a TV monitor, where the areas and nuclei of interest are drawn with a light-sensitive pen.

Once the autoradiographic procedure is completed, the labeled tissue slices can be stained for histopathological examinations. Furthermore, by combining receptor autoradiography with immunohistochemistry, it is theoretically possible to examine at the same time, in the same or in consecutive sections, alterations at both pre- and postsynaptic levels. A further development would consist in hybridization experiments with molecular probes for receptor mRNAs combined with immunocytochemistry or receptor autoradiography on consecutive sections.

3.2 Receptor Immunohistochemistry

An alternative approach for mapping receptors in nervous tissues, namely the immunohistochemical approach, has been developed in recent years. This new development was prompted by some limitations of the radiohistochemical technique for receptor mapping. These are mainly the radiation scatter and the diffusion of the radioligand which limit the resolution of the radiohistochemical approach such that receptor sites cannot be attributed to defined subcellular structures (RICHARDS et al. 1986a, b). With the immunohistochemical technique, receptor antigenic sites are visualized by non-diffusible markers such as peroxidase-coupled (LENTZ and CHESTER 1977) or colloidal gold-coupled secondary antibodies (NGHIÊM et al. 1983; SEALOCK et al. 1984). This procedure allows the ulstrastructural localization of binding sites, such as those for benzodiazepine, acetylcholine, and glycine, on synaptic membranes. Monoclonal antibodies against these receptors have been raised by hybridoma techniques using the isolated and purified receptor as an immunogen.

Taking advantage of the isolation/purification of the $GABA_A$/benzodiazepine receptor complex from bovine cerebral cortex (SIGEL et al. 1983; SCHOCH and MÖHLER 1983), RICHARDS et al. (1984) were able to raise monoclonal antibodies for this complex. To immunize mice, they used a receptor preparation containing high and low affinity binding sites for GABA in addition to the neuronal binding site for benzodiazepine. The epitopes recognized by the monoclonal antibodies were localized on a protein of 50000 or 55000 daltons, which correspond to the known subunits of the receptor complex (RICHARDS et al. 1986a). Using these antibodies they were able to localize immunoreactive, presumably GABAceptive, neuronal cell bodies and neuronal processes in animal and human brain tissues with a resolution far superior to that obtained with radiohistochemistry. Furthermore, using high reso-

Table 1. Receptors, subtypes, and ligands

Receptors	Subtypes	Ligands
Amines		
Muscarinic–cholinergic	M_1	[N-methyl-^{3}H]Scopolamine
		[N-methyl-^{3}H]Pirenzepine
		[^{3}H]Quinuclidinyl benzilate
		[^{3}H]Acetylcholine
	M_2	[N-methyl-^{3}H]Scopolamine
		[methyl-^{3}H]Oxotremorine M
Nicotinic		[^{3}H]Nicotine
		[^{3}H]-α-Bungarotoxin
		[^{3}H]Acetylcholine
Serotonin	5-HT_{1A}	5-[1,2^{3}H(N)]Hydroxytryptamine
		8-hydroxy-[^{3}H]DPAT
		[^{125}I]-5-MeO-DPAT
		[N-methyl-^{3}H]Lysergic acid diethylamide
	5-HT_{1C}	5-[1,2^{3}H(N)]Hydroxytryptamine
		[N-methyl-^{3}H]Lysergic acid diethylamide
		[^{125}I]Lysergic acid diethylamide
		[^{3}H]Mesulergine
	5-HT_{1D}	5-[1,2^{3}H(N)]Hydroxytryptamine
	5-HT_2	[^{3}H]Ketanserin
		[^{3}H]Spiperone
		[N-methyl-^{3}H]Lysergic acid diethylamide
	5-HT_3	[^{3}H]ICS 205-930 [(3 α-tropanyl)-1H--indole-3-carboxylic acid ester
		[^{3}H]Zacopride
Dopamine	D_1	[N-methyl-^{3}H]SCH 23390
		[^{125}I]SCH 23982
	D_2	[^{3}H]CV 205-502
		[^{125}I]Iodosulpride
		[^{3}H]Spiperone
Epinephrine	α_1	[^{125}I]BE 2254
		[7-methoxy-^{3}H]Prazosin
	α_2	[^{3}H]Para-aminoclonidine
		[^{3}H]Rauwolscine
		[^{3}H]Idazoxan
	β_1	[^{125}I]Iodocyanopindolol
	β_2	[^{125}I]Iodocyanopindolol
Histamine	H_1	[^{125}I]Iodobolpyramine
		[^{3}H]Mepyramine
	H_2	[^{125}I]Iodoaminopotentidine
	H_3	[^{3}H]-R-α-Methylhistamine

Table 1. (Continued)

Receptors	Subtypes	Ligands
Amino acids		
γ-Amino	A	[³H]Flunitrazepam
butyric acid		[³H]Muscimol
	B	[³H]GABA
Glycine		[³H]Strychnine
Glutamate		[³H]Glutamate
Adenosine	A_1	[³H]8-cyclopentyl-1,3-dipropylxanthine
		[³H]N^6-cyclohexyladenosine
		[¹²⁵I]Hydroxyphenyl-isopropyladenosine
	A_2	[carboxyethyl-³H(N)]CGS 21680
Neuropeptides		
Opiates	μγκ	[³H]-(−)-bremazocine
	μ	[¹²⁵I]FK-33824
	μ	[³H]Dihydromorphine
Substance P		[¹²⁵I]BH-SP
Cholecystokinin (CCK-8)		[¹²⁵I]BH-CCK-8
Thyrotropin-releasing hormone		[³H]-Me-TRH
Neurotensin		[¹²⁵I-Tyr3]Neurotensin
		[³H]Neurotensin
Vasopressin		[³H]Vasopressin
Oxytocin		[¹²⁵I]Oxytocin antagonist
		[³H]Oxytocin
Somatostatin		[¹²⁵I]204-090
		[¹²⁵I]SS-28
Vasoactive intestinal		M-[¹²⁵I]Vasoactive intestinal
peptide		peptide
Galanin		[¹²⁵I]Galanin

lution immunohistochemistry it was possible to visualize subcellular sites of benzodiazepine interaction. Selective staining of pre- and postsynaptic membranes of axosomatic and axodendritic contacts was obtained in the rat substantia nigra (RICHARDS et al. 1986 a). This is an indication that at least in this localization, the benzodiazepine receptor ligands affect not only the postsynaptic GABA receptor but also presynaptic, probably GABA autoreceptors.

Antibodies raised against purified nicotinic acetylcholine receptors (NChR) have been important probes in the study of these receptors. Particularly monoclonal antibodies to both muscles and fish electric organs which bind to neuronal receptors and do not recognize BTX sites (Whiting and Lindstrom 1987) have proved very useful in the study of NChR in the mammalian CNS. Classical immunoperoxidase (Deutch et al. 1987) and radioimmunohistochemistry (Swanson et al. 1987) have been used to visualize receptors in the rodent brain. Two ^{125}I-labeled monoclonal antibodies to purified chicken and rat brain NChR were shown to label sites in all major subdivisions of the CNS with a distribution similar to that found with ^{3}H-nicotine as ligand but different from that of ^{125}I-BTX binding sites. To our knowledge no immunohistochemical study has been published with antibodies against the muscarinic cholinergic receptor (MChR).

In two recent studies, the distribution of the glycine receptor immunoreactivity was characterized in some areas of animal brains, using monoclonal antibodies raised against the purified glycine receptor (Pfeiffer et al. 1984). Antibodies included the antiserum 2 b, which recognize a 48 000-dalton (48-K) protein, one of the three proteins of the glycine receptor. In one study on rat spinal cord (Basbaum 1988), the pattern of glycine receptor immunoreactivity was comparable to that seen with binding of the tritiated glycine receptor antagonist strychnine to sections of rat and human spinal cord (Zarbin et al. 1981; Probst et al. 1986). However, in contrast to high levels of strychnine binding reported in the substantia gelatinosa, there was almost no glycine receptor immunoreactivity in corresponding layers of the dorsal horn or in the trigeminal nucleus caudalis. Such a discrepancy is difficult to explain since the 48-K protein recognized by the antiserum 2 b contains the strychnine binding site. In a second study on the ultrastructural localization of the glycine receptor in the ventral cochlear nucleus of the guinea pig (Wenthold et al. 1988), immunoreactivity was seen to be essentially restricted to postsynaptic membranes on dendrites and cell bodies of cochlear neurons.

3.3 In situ Hybridization Histochemistry using DNA Probes

DNA probes complementary to specific nucleic acid sequences of neurotransmitter receptors can be hybridized to RNA from tissue samples by the Northern blot technique. They also can be hybridized to tissue sections, a procedure referred to as in situ hybridization histochemistry.

Northern blot analysis has been done for many of the cloned receptors, indicating in most cases that there is tissue specificity in the distribution of their mRNAs. In situ hybridization histochemistry in the case of the NChR has shown the presence of their mRNAs in different regions of the mammalian central nervous system (Goldman et al. 1986, 1987), agreeing in general with the distribution of the receptors shown by autoradiography of bound nicotine (Clarke et al. 1985) or by immunohistochemistry (Swanson et al. 1987).

The areas of the brain containing mRNAs coding for the α- and β-subunits (Sequier et al. 1988) and the γ2-subunit (Pritchett et al. 1989) of the GABA$_A$ receptor were also mapped by in situ hybridization, presenting, in general, a

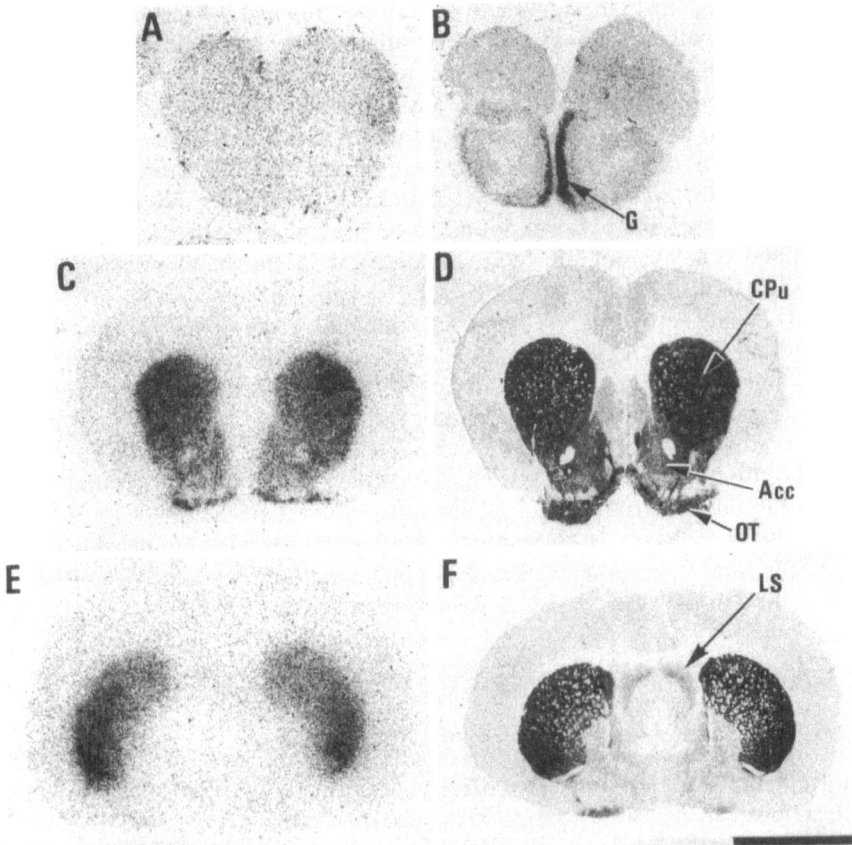

Fig. 1A–F. Distribution of the D_2 receptor mRNA in the rat brain. Comparison with the distribution of D_2 receptor binding. On the *left* (**A, C, E**) are autoradiographs from sections hybridized with a ^{32}P-labeled oligomer complementary to the rat D_2 mRNA. On the *right* (**B, D, F**) are autoradiographs showing the distribution of the binding of ^{3}H-SDZ 205-502 to D_2 receptor at levels similar to those shown for the in situ hybridization histochemistry. Coronal brain sections are presented in a rostrocaudal progression. *Dark areas* are those rich in hybridization or binding signal. The nonspecific binding was homogeneous and comparable to that seen in white matter areas. Note that the distribution of D_2 binding sites and of D_2 receptor mRNA is similar, except in the olfactory bulb, where no significant hybridization signal could be detected. *Acc*, nucleus accumbens; *CPu*, caudate–putamen nucleus; *G*, glomerular layer of the olfactory bulb; *LS*, lateral septum; *OT*, olfactory tubercle. *Bar* = 5 mm. (Reproduced with permission from Mengod et al. 1989)

pattern of distribution very similar to that obtained by immunohistochemistry and ligand-binding autoradiography (SEQUIER et al. 1988).

The mRNAs coding for the muscarinic receptor subtypes m1, m2, m3, m4, and m5 have been found to show a differential localization in the central nervous system (BUCKLEY et al. 1988; VILARÓ et al. 1990a). The pyramidal cell layer of the hippocampus, as well as the granule cell layer of the dentate gyrus, the olfactory bulb, the amygdala, the olfactory tubercle, and the piriform cortex, showed the highest levels of m1 mRNA. m3 and m4 mRNAs were also present in the olfactory bulb and the pyramidal layer of the hippocampus, but at lower levels in the dentate gyrus. m3 mRNA was also found in the cerebral cortex and in a number of thalamic and brain stem nuclei, while m4 mRNA predominated in the neostriatum (BUCKLEY et al. 1988; VILARÓ et al. 1990 b). m2 mRNA was detected in the medial septum, the diagonal band of Broca, the olfactory bulb, and pontine nuclei (BUCKLEY et al. 1988; VILARO et al., in preparation). m5 mRNA was found to be present, at very low levels, in the pyramidal cell layer of the hippocampus and in the substantia nigra pars compacta (VILARO et al. 1990 a).

5-HT$_{1C}$ receptor mRNA has been mapped in the epithelial cells of the choroid plexus, throughout the limbic system, in both catecholaminergic and serotoninergic neurons, and in the hypothalamus and subthalamus, showing a good correlation with receptor binding (HOFFMAN and MEZEY 1989; Mengod et al. 1990).

The mRNA coding for the D$_2$ receptor subtype has been found to be present in the neostriatum and nucleus accumbens and the olfactory tubercle (MENGOD et al. 1989). In these areas, the distribution of hybridization signal was comparable to that of the D$_2$ receptor binding sites as visualized by autoradiography using ^{3}H-SDZ 205-502 as a ligand (Fig. 1).

4 Receptor Radiohistochemistry in Human Neuropathology

The application of receptor autoradiography to the study of human post-mortem brain materials from different diseases has revealed that receptor alterations can be found as a consequence of disease.

In vitro light microscopic autoradiographic studies of neurotransmitter receptors have been carried out in a variety of human brain disease in our as well as in other laboratories. As most of this work has been done in neurodegenerative diseases, we will focus on characteristic receptor changes encountered in some of these conditions and discuss receptor changes in relation to known transmitter alterations and/or degeneration of specific neuronal systems.

4.1 Amyotrophic Lateral Sclerosis

The essential histological change in amyotrophic lateral sclerosis (ALS) is a loss of large motor cells in cerebral cortex, brain stem, and spinal cord. This is most easily observed in the anterior horns of the lumbar and cervical enlargement and in the hypoglossal nuclei. Remaining neurons are shrunken or rounded up, disclosing abnormally pale cytoplasm suggesting that affected cells do not die suddenly but undergo progressive and eventually lethal sickness.

A number of abnormalities of transmitters and of transmitter-related markers have been documented in post-mortem studies of spinal cords from

ALS cases. One of the most consistent changes consists in a profound reduction in the activity of choline acetyltransferase (CAT) not only within the motor neuron area but also in the dorsal horn (GILLBERG et al. 1982).

In vitro receptor autoradiographic studies have been used for localizing and quantitating neuroreceptors within sections of spinal cord from nonneurological controls and from ALS cases (WHITEHOUSE et al. 1983; GILLBERG et al. 1984; GILLBERG and AQUILONIUS 1985; MANAKER et al. 1985, 1988). ^{3}H-Quinuclidinylbenzilate or ^{3}H-N-methylscopolamine (NMS) has been used as a radioligand for the MChR. In control spinal cord, the highest densities of MChR are found in lamina IX and in the substantia gelatinosa (lamina II–III) (WHITEHOUSE et al. 1983; GILLBERG et al. 1984; SCATTON et al. 1984). In ALS, the muscarinic binding sites are markedly reduced in lamina IX. Furthermore, cell counts and quantification of receptors reveal a high degree of correlation between reduction in MChR and the degree of motor neuron loss in the anterior horn (WHITEHOUSE et al. 1983; GILLBERG et al. 1984; MANAKER et al. 1988). Although this suggests that a majority of these receptors are located on motor neuron somata, we do not know to what extent they are also located on collaterals and on Renshaw cells. Interestingly, MChR losses were also found to be slightly reduced in the substantia gelatinosa (WHITEHOUSE et al. 1983; GILLBERG et al. 1984; MANAKER et al. 1988).

Selective modification of the M_1 and "non-M_1" subtypes of the MChR in ALS has also been investigated. Carbachol has been used to displace ^{3}H-NMS from the non-M_1 sites in the incubation medium. Using this compound, a marked decrease in ^{3}H-NMS binding was observed in all areas of the spinal cord, indicating that high affinity agonist sites, or non-M_1 sites, constitute a majority of binding sites in the spinal cord and are the sites most involved in ALS (WHITEHOUSE et al. 1983).

Autoradiographic studies of other spinal receptor binding sites, namely ^{3}H-α-bungarotoxin, opioid, benzodiazepines, glycine (WHITEHOUSE et al. 1983; GILLBERG et al. 1984), β-adrenergic, and norepinephrine (MANAKER et al. 1988), have shown unchanged (β-adrenergic, norepinephrine) or slightly reduced (benzodiazepines, glycine) densities in ALS.

Substance P-like immunoreactivity (SPLI) has been measured in tissue parts dissected from the normal human spinal cord and the highest values were found in the dorsal horn at all segment levels (GILLBERG et al. 1982). Conflicting results have been obtained concerning SPLI in spinal cords from ALS cases. DIETL et al. (1989), using radioimmunohistochemical methods, found apparently unaffected SPLI even in the anterior horn of such patients. However, PATTEN and CROFT (1984) and SCHOENEN et al. 1985), using immunohistochemical methods, found severe loss of spinal SP-containing fibers in ALS.

The distribution of substance P receptors has been examined by autoradiography in the spinal cord of ALS and control patients (DIETL et al. 1989). In normal control spinal cords the substance P receptors were found to be evenly distributed throughout the different parts of the spinal cord gray matter. In ALS, a marked reduction in substance P receptors was measured in the ventral horns in the areas corresponding to motor neurons (Fig. 2).

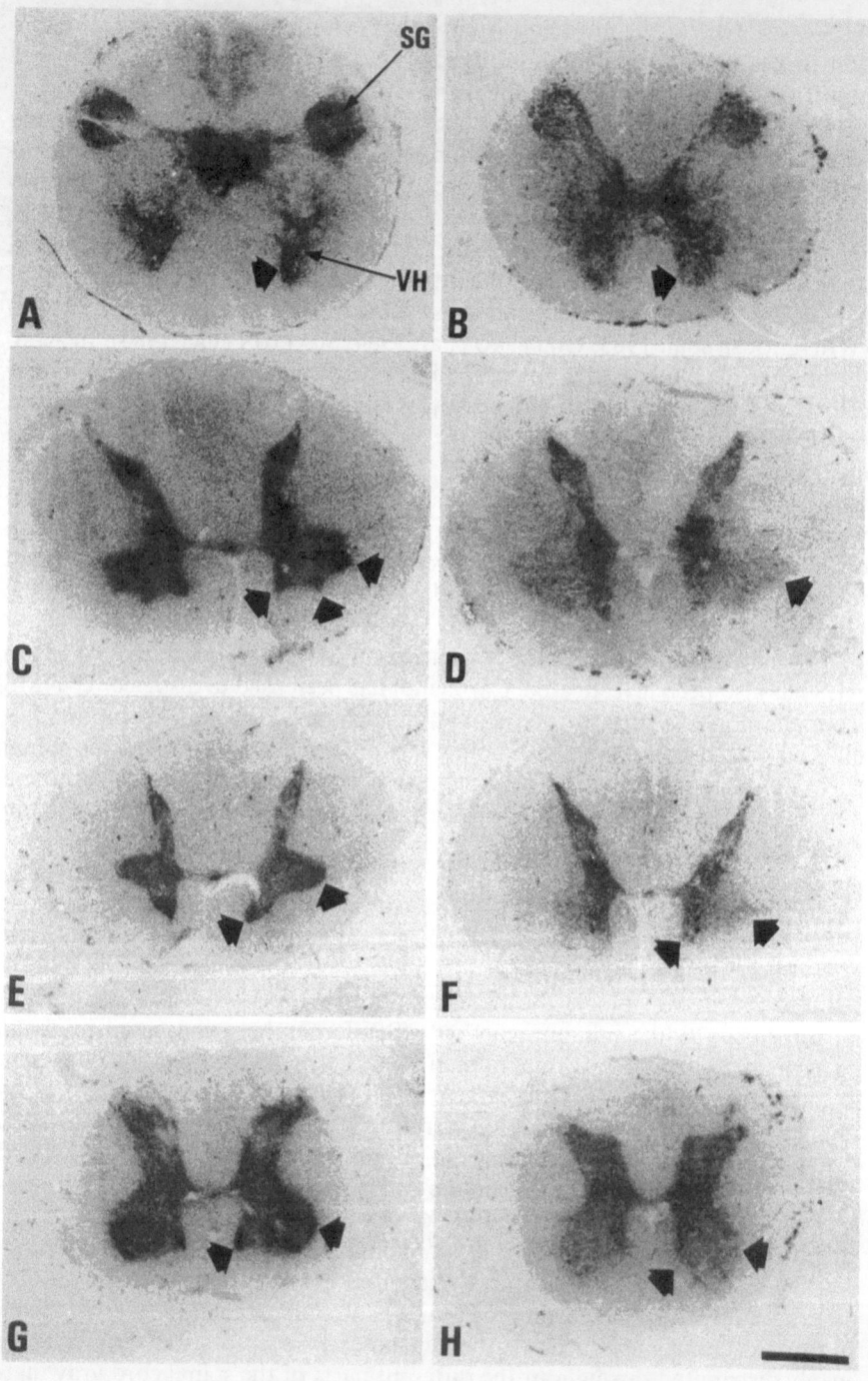

In 1983, ENGEL and associates reported that infusion of thyrotropin-releasing hormone (TRH) improved strength and lessened spasticity in patients with ALS. In addition to effects on the pituitary gland, TRH may act directly on the nervous system by being released from the hypothalamus or produced in the brain outside of the hypothalamus. In fact, TRH has long been thought to play a role in spinal cord function. Immunocytochemical studies have demonstrated localization of TRH immunoreactive fibers and terminals throughout the spinal gray matter, with particular concentrations of terminals adjacent to motor neurons (LECHAN et al. 1984; APPEL et al. 1987). Some investigators have found reduced TRH levels in the anterior horn of ALS cases (MITSUMA et al. 1984), whereas others have not found any changes in spinal TRH levels in ALS (JACKSON et al. 1986).

Thyrotropin-releasing hormone receptors have been described in the spinal cord of many species, including man (MANAKER et al. 1985). TRH receptors appear to be particularly enriched in lamina II and moderately concentrated in lamina IX (MANAKER et al. 1985). As shown by recent quantitative autoradiographic study, TRH receptors in lamina IX are reduced by almost 90% and the corresponding reduction in lamina II is about 50% (MANAKER et al. 1988). As for MChR, such a decrease may result from the loss of motor neurons and is at least compatible with a postsynaptic localization of TRH receptors on these cells. Alternatively, many of these receptors may be located on terminals innervating motor neurons.

An interesting finding of the same research group consisted in an increase of up to 140% in 5-HT_{1A} receptor densities in lamina IX of spinal cord in ALS patients, using $^3\text{H-8-OHDPAT}$ as a ligand. It has been suggested that this increase is due to an up-regulation of 5-HT receptors in response to some serotonin depletion in the cord (MANAKER et al. 1988). Unfortunately, no data are available on 5-HT content of the spinal cord in ALS. Since 5-HT and TRH are known to be colocalized in raphe neurons projecting to the spinal cord (JOHANNSON et al. 1981), coincidental depletion of 5-HT and TRH could occur in ALS cord due to some dysfunction of this descending pathway. This would imply that the pathophysiological process in ALS involves not only motor neurons with their postsynaptic receptors but also presynaptic terminals bearing TRH receptors. Certainly more investigations are needed in order to clarify the way 5-HT and TRH modify motor neuron activity and to obtain more precise information on the localization of their respective receptors in the normal cord as well as in ALS.

◄ Fig. 2 A–H. Distribution of substance P receptors in several levels of human spinal cord of a control case (*left column*) and a patient dying from ALS (*right column*). A and B, the first cervical segment; C and D the seventh cervical segment; E and F, the first thoracic segment; G and H, the third lumbar segment. Note the high densities of substance P receptors in the substantia gelatinosa (*SG*) and motor nuclei (*arrows*) of the ventral horn (*VH*) in the control case (A, C, E, G) while an important decrease in these receptors is observed in the ventral horn of the ALS case (B, D, F, H) (*arrows*). *Bar* = 5 mm. (Reproduced with permission from Dietl et al. 1989)

4.2 Olivopontocerebellar Atrophy

The term olivopontocerebellar atrophy (OPCA) was introduced by Déjerine and Thomas (1900) to identify a group of clinically and genetically heterogenous syndromes, whose only common feature was a loss of neurons in the ventral portion of the pons, inferior olives, and cerebellar cortex (Petito et al. 1973). Together with Shy-Drager syndrome and striatonigral degeneration, OPCA is included under the general heading of multiple system atrophy (Berciano 1988; Oppenheimer 1984).

Morphologically, OPCA is characterized by gross shrinkage of the ventral aspect of the pons, middle cerebellar peduncles, cerebellar cortex, and inferior olives. Histologically, neuronal losses are found in the Purkinje cell layer of the cerebellum, with resulting empty basket cell formations, in pontine nuclei, and in the inferior olives, with loss of olivocerebellar projection fibers. Granule cells of cerebellar cortex are much less affected than the Purkinje cells, and the dentate nuclei only present gliosis owing to the degeneration of incoming Purkinje axons. The lesions are not confined to the cerebellar system but may extend to the long tracts of the spinal cord, to anterior horns, or to basal ganglia.

Reduced levels of GABA have been described in the cerebellar cortex and dentate nucleus of some patients with OPCA (Perry et al. 1981). GABA is almost certainly the neurotransmitter of Purkinje cells, which project to the dentate nucleus, and of inhibitory interneurons of the cerebellar cortex, including basket, stellate, and Golgi cells (Storm-Mathisen 1976). Thus a decrease in GABA could imply loss of either inhibitory interneurons or of Purkinje cells. The latter seems more likely since a more marked decrease in GABA was observed in the dentate nucleus than in the cortex in most OPCA cases (Perry et al. 1981).

In normal human cerebellar cortex both granule cell and molecular layers are richly endowed with benzodiazepine receptors. In contrast, only low levels of receptors are found in the dentate nucleus (Zezula et al. 1988). In an autoradiographic study of benzodiazepine receptors in postmortem cerebellar tissues from OPCA patients, a 150% increase in receptor density was found in the dentate nucleus (Whitehouse et al. 1986 b). Although GABA levels measured in homogenized cerebellar cortex of OPCA patients were not significantly lower than in controls, an increased number of GABA receptors in the dentate nucleus was interpreted as a result of denervation supersensitivity following loss of GABA-ergic projections from dysfunctioning Purkinje cells. In the same investigation high affinity GABA receptors (recognized by the ligand muscimol) were found to be decreased in number in the granule cell layer.

In another study on OPCA patients from five different pedigrees (Kish et al. 1983), Purkinje cell loss, as judged from subnormal dentate nucleus GABA values, was associated with an increase in GABA receptor binding in homogenates of cerebellar cortex. This increase reached 60% of control values in one pedigree, lesser increases being observed in the other pedigrees. This result is clearly at variance with that reported by Whitehouse et al. (1986 b). One explanation for the discrepancy in the reported results could lie in the different nature of the OPCA material and especially in differences in the degree of Purkinje cell and granule cell loss in the examined patient cohorts. KISH et al. (1983) suggested that increased binding of GABA receptors in their cases might represent a form of denervation supersensitivity affecting neuronally located GABA sites. This could be induced by dysfunction or loss of GABA-ergic cells in cerebellar cortex, like Purkinje, Golgi, or stellate cells, with a compensatory increase in the affinity or density of GABA receptors located on preserved but denervated cells. The observation that mild to moderate Purkinje cell loss was associated with the most marked increase in GABA receptor binding is compatible with such an assumption. A similar mechanism may explain the increased density of cerebellar cortical benzodiazepine receptors at early stages of canine inherited ataxia, a disorder of Gordon setters characterized by progressive degeneration of Purkinje and granule cells (TRONCOSO et al. 1984).

Marked reductions in the cerebellar excitatory amino acids aspartate and glutamate have been reported in some OPCA pedigrees by PERRY et al. (1981). These deficiencies are thought to result from a degeneration of incoming climbing and mossy fibers, although loss of granule cells might also be a contributory factor because glutamic acid is probably the excitatory neurotransmitter of these neurons (see PERRY et al. 1981).

TSIOTOS et al. (1989) recently described decreased ^{3}H-glutamate-binding sites in homogenized cerebellar specimens from OPCA cases. KISH et al. (1989), using a similar procedure, found decreased ^{3}H-inositol-1,4,5-triphosphate (IP$_3$) binding in OPCA cerebellar cortex. Biochemical evidence suggests that IP$_3$ may be a second messenger in glutamatergic and aspartatergic neurotransmission (reviewed in KISH et al. 1989). Since the Purkinje cell dendrites receive neuronal input from granule cells and climbing fibers, using glutamate and aspartate as neurotransmitters, the reduced IP$_3$ binding in OPCA may reflect a loss of Purkinje cells containing these amino acid receptors linked to the phosphatidylinositide second messenger system (KISH et al. 1989).

More precise information on the anatomical distribution of excitatory amino acid receptors (N-methyl-D-aspartate, quisqualate, and kainate) in human cerebellum is expected to accrue from autoradiographic investigations on control and OPCA cerebellar specimens.

In the next three disorders to be discussed in this review (progressive supranuclear palsy, Parkinson's disease, and Huntington's disease) major pathological changes are found in the basal ganglia, although other brain structures are affected as well. Some introductory remarks on basal ganglia may therefore be appropriate at this point.

The basal ganglia constitute the so-called extrapyramidal system and comprise a group of gray matter structures situated deep in the brain and including the caudate nucleus, the putamen, both segments (internal and external) of the globus pallidus, the subthalamic nucleus, and the substantia nigra (pars compacta and pars reticulata). It is generally believed that these structures are mainly involved in modulating and facilitating various motor and cognitive programs.

Recent progress in our understanding of the basal ganglia has shown that these areas present a very rich subnuclear organization. The main cellular types and some of their chemical characteristics are also well documented. The basal ganglia provide a good example where the combination of receptor imaging with the study of pathology can help to elucidate the chemical organization of a brain nucleus (GRAYBIEL and RAGSDALE 1983).

We have used receptor autoradiography to examine the distribution of a number of receptor sites in the normal and pathological basal ganglia. The distribution and density of the expression of some neuropeptides were examined in parallel by in situ hybridization techniques using synthetic oligonucleotides as probes (MENGOD and PALACIOS 1990).

Receptors for neurotransmitters such as acetylcholine, dopamine, serotonin, GABA, and some peptide neurotransmitters such as enkephalin, somatostatin, cholecystokinin (CCK), neurotensin, and substance P are found to be more or less enriched in the human basal ganglia. In general these receptors present a heterogeneous distribution, with areas enriched relative to the background. The distribution of these high density patches is different for the different receptors examined. Overlapping patches of high density of GABA/benzodiazepine and neurotensin receptors are predominant in the ventral aspects of the anterior putamen. These patches do not correspond to the patches of high μ-opioid receptor binding. When compared with the distribution of AChE activity, it appears that benzodiazepine receptors are localized in areas of high AChE activity while enkephalin receptors are enriched in areas with low AChE activity, the so-called striosomes (GRAYBIEL and RAGSDALE 1983). Substance P receptors again present a different distribution, with high receptor densities concentrated in small areas more or less uniformly distributed throughout the striatum.

A heterogeneous distribution of the expression of the mRNA for several neuropeptides can also be observed. Enkephalin mRNA presents a patchy distribution overlapping with the distribution of AChE activity. In contrast, in several striatal areas, cells expressing somatostatin mRNA are organized around the patches of high levels of enkephalin expression. This distribution agrees well with that found in the cat striatum using immunohistochemical techniques (CHESSELET and GRAYBIEL 1986). CCK mRNA is not expressed in the striatum, indicating that striatal CCK immunoreactivity is localized to processes from cells extrinsic to this nucleus.

4.3 Progressive Supranuclear Palsy

Progressive supranuclear palsy (PSP) is a fatal neurological disorder in adults which combines the clinical features of a supranuclear ophthalmoplegia, pseudobulbar palsy, dysarthria, dystonic rigidity of the neck and upper trunk, and dementia (STEELE et al. 1964; STEELE 1972). The histological appearance of the brain consists in loss of neurons associated with gliosis and neurofibrillary degeneration in the subthalamic nucleus, nucleus basalis of Meynert, globus pallidus, pretectal area and tectum of the midbrain, substantia nigra, locus coeruleus, dentate nucleus, and brain stem nuclei (STEELE et al. 1964; PROBST and DUFRESNE 1975). Over the past few years PSP has become a routine diagnostic consideration for neurologists confronted with atypical parkinsonism (GOLBE and DAVIS 1988). Developments in understanding PSP have been slow, largely because of the condition's rarity and the difficulty in separating it clinically from Parkinson's disease.

In the last few years neuronal changes of PSP have been investigated by immunohistochemistry using various panels of antibodies against cytoskeletal proteins (BANCHER et al. 1987; PROBST et al. 1988a; GALLOWAY 1988). These studies have indicated some overlap between subcellular changes in PSP and Alzheimer's disease. In particular, tangles in both PSP and Alzheimer's disease stain with antibodies for abnormally phosphorylated tau protein (POLLOCK et al. 1986; BANCHER et al. 1987; PROBST et al. 1988a).

Recent neuropathological studies have extended our knowledge about lesional distribution in PSP. In particular, losses of nerve cells and neurofibrillary degeneration of neurons have been described in the pedunculopontine nucleus (ZWEIG et al. 1987; HIRSCH et al. 1987). This nucleus is considered to be the caudal part of the "central" cholinergic system, which includes basal forebrain nuclei, and most of its cells are considered to be cholinergic. Alterations of this cell population in PSP could contribute to disorders of movement, of the sleep–wake cycle (PERRET and JOUVET 1980; GROSS et al. 1978), and of cognition that are known to be characteristic of PSP.

Among the different neuronal systems affected in PSP, the dopaminergic nigrostriatal system has been that most thoroughly investigated until now. This may be explained by clinical evidence for extrapyramidal disorders with parkinsonian features and by morphological and biochemical evidence indicating involvement of the dopaminergic nigrostriatal system in PSP.

Two studies (KISH et al. 1985; RUBERG et al. 1985) revealed a marked (80%–90%) decrease in dopamine in the caudate nucleus and putamen, thus confirming the results of a previous study by JELLINGER et al. (1980). However, unlike in Parkinson's disease, dopamine concentrations have been found to be normal in the nucleus accumbens and frontal cortex, indicating that the mesolimbic and mesocortical dopaminergic neurons do not degenerate in PSP (KISH et al. 1985; RUBERG et al. 1985).

As in Parkinson's disease, striatal homovanillic acid was much less decreased (45%–50%) than dopamine, suggesting a presynaptic compensatory mechanism with increased turnover of dopamine. Decreased dopamine content in striatal nuclei is easily explained by the loss of pigmented neurons in the pars compacta in PSP. Loss of nigrostriatal dopaminergic projections to the striatum can also be evidenced by the loss of dopamine uptake sites in the target nuclei of these projections, as visualized autoradiographically using the uptake blocker ^{3}H-mazindol (Fig. 3).

Decreased densities of D_2 type dopamine receptors, measured by the binding of ^{3}H-spiperone, have been reported in the caudate nucleus and the puta-

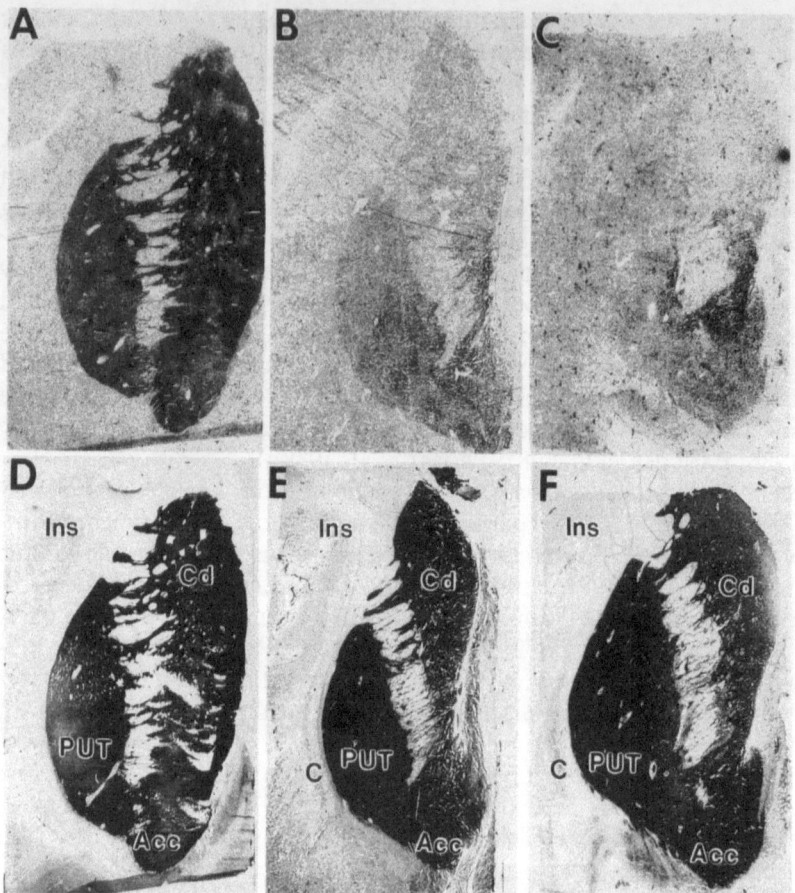

Fig. 3. A–C Autoradiographs illustrating the distribution of dopamine uptake sites as labeled with ³H-mazindol. D–F Photographs from AChE-stained sections adjacent to those shown in A–C. A and D are from control cases, B and E are from PSP cases, and C and F are from cases of Parkinson's disease. *Cd,* caudate nucleus, head; *PUT;* putamen; *Acc,* nucleus accumbens; *C,* claustrum; *Ins,* cortex of insula

men in PSP (PIEROT et al. 1988; RUBERG et al. 1985). The mapping of dopamine-receptors in vivo by PET with labeled bromospiperone injected intravenously in patients with evidence of PSP has also confirmed the decreased number of spiperone binding sites in this disorder (BARON et al. 1985).

Decreased choline acetyltransferase (CAT) activity in several brain regions, among them the neostriatum, was found to be an additional feature in PSP by RUBERG et al. (1985), who also found both CAT activity and ³H-spiperone binding to be decreased in parallel fashion in the striatum. Thus, the loss of ³H-spiperone binding was best explained by the loss of cholinergic interneurons of the striatum, which are thought to be the target cells of the nigrostriatal system.

Decreased ^{3}H-spiperone binding is unlikely to result from the degeneration of dopaminergic nigrostriatal fibers since no similar decrease is observed in Parkinson's disease (BOKOBZA et al. 1984). Furthermore, in Parkinson's disease and in animals with lesions of the dopaminergic nigrostriatal pathway, dopaminergic denervation rather results in a hypersensitivity response of D_2 receptors (SAVASTA et al. 1987).

Examining several cases of PSP by autoradiographic methods we have confirmed this decrease in ^{3}H-spiperone binding sites in the neostriatum. However, using other radiolabeled ligands (^{125}I-sulpride, ^{3}H-SDZ 205-502) of D_2 receptors in the same cases we were unable to find any alteration in the density of the receptors. The reason for such differential involvement of dopamine receptors in the striatum of PSP is not known. One explanation would be that different D_2 receptor subtypes are localized on different striatal cell populations.

Only minor morphological changes, if any, have been reported in the striatum in PSP. However, more severe neuronal changes than previously accepted were documented in the neostriatum by application of the Gallyas silver iodide technique and by protein tau immunocytochemistry (PROBST et al. 1988a). It remains to be established whether neurons with neurofibrillary changes and accumulation of abnormal tau in the striatum correspond to a population of dopaminoreceptive neurons, including cholinergic interneurons. In fact a significant negative correlation was found between the number of ^{3}H-spiperone binding sites and the number of neurofibrillary tangles in the neostriatum of PSP patients (RUBERG et al. 1985), suggesting that D_2 receptors could be localized on neurons which are at risk in this condition.

We have already mentioned the reduced CAT activity in the neostriatum of PSP patients. Recent investigations have also shown cortical cholinergic deficiency in this disease, although of lesser extent than in Alzheimer's disease. When compared with levels in control brains, CAT levels were decreased by 20%, 35%, and 40%, respectively, in the frontal, cingulate cortex, and hippocampus (RUBERG et al. 1985). The decrease in enzymatic activity was more dramatic in the substantia innominata. These results suggest a loss of innominatocortical and of septohippocampal cholinergic projections in PSP, although the very severe enzymatic change in the substantia innominata may reflect additional loss of other cholinergic afferents to this nucleus (AGID et al. 1986). It has been suggested that cortical cholinergic deficiency in PSP, although of moderate degree, may account for the relatively mild memory impairment observed in patients (AGID et al. 1986).

Conflicting results have been obtained concerning MChR in the brain of PSP patients. Whereas the number of ^{3}H-QNB binding sites has been found to be unchanged throughout the brain by one group (RUBERG et al. 1985), we have found significant decreases in MChR in the caudate nucleus and putamen using ^{3}H-NMS as a ligand. Increased numbers of ^{3}H-NMS binding sites have been found in the medial segment of the globus pallidus in the same cases.

It is clear that more detailed investigations of MChR in PSP cases are needed. Furthermore, the lack of cellular resolution of the autoradiographic technique does not allow for the identification of the striatal cells expressing MChR as well as dopamine receptors. Further studies using in situ hybridization histochemistry or immunohistochemistry, once antibodies against these receptors are available, will be necessary to clarify this issue.

4.4 Parkinson's Disease

Clinically Parkinson's disease (PD) is characterized by motor disabilities, such as akinesia, tremor, and rigidity and by mental disturbances, including dementia and depression. Neuropathologically, PD is characterized by progressive loss of neuromelanin-containing dopaminergic neurons in the substantia nigra pars compacta, with consequent degeneration of the nigrostriatal axonal projections. This degeneration is believed to be responsible for the akinesia and rigidity (Hornykiewicz 1978). However, neuronal losses are not restricted to dopaminergic nuclei and areas, since significant cell losses can be found in noradrenergic locus coeruleus, in cholinergic nucleus basalis of Meynert, and in vagal nuclei.

The discovery by Ehringer and Hornykiewicz (1960) of a marked deficit of dopamine in the putamen of patients with PD led to the hypothesis that dopaminergic mechanisms are involved in this condition. In parkinsonian brains, dopamine concentrations are found to be decreased not only in the target nuclei of the nigrostriatal projections but also in the substantia nigra itself. Furthermore, depletion of dopaminergic cells and decreased TH activity have also been found in the ventral tegmental nuclei of the midbrain, suggesting additional involvement of the mesocorticolimbic dopaminergic system (Bogerts et al. 1983; Javoy-Agid and Agid 1980; Uhl et al. 1985). Involvement of this projection system has been thought responsible for cognitive impairment and for certain aspects of akinesia found in PD (Javoy-Agid and Agid 1980; Uhl et al. 1985).

Alongside alterations in dopamine levels, various other neurotransmitter systems have been found to be altered in PD. As a consequence of the loss of noradrenergic cells in the locus coeruleus, there is a decreased content of norepinephrine in several areas of parkinsonian brains (Scatton et al. 1983). However, normal norepinephrine content in the hypothalamus and nucleus paranigralis as well as normal levels of dopamine-β-hydroxylase activity in A1 and A2 areas of the medulla oblongata suggests that noradrenergic innervation is partially maintained in PD (Javoy-Agid et al. 1982). A dementia-related decrease in the CAT level has also been observed in the neocortex and substantia innominata of patients with PD (Ruberg et al. 1982; Dubois et al. 1983), suggesting that cholinergic neurons of the substantia innominata may degenerate in some patients with PD and that subcortical cholinergic dysfunction may be responsible for intellectual impairment in these patients. Furthermore, levels of serotonin have been found to be markedly depleted in the cerebral cortex in PD patients (D'Amato et al. 1987) and in most components of the basal ganglia (Bernheimer et al. 1961).

A number of peptides are also altered in PD (AGID and JAVOY-AGID 1985). For instance, reduced concentrations of substance P (TENOVUO et al. 1984) and met-enkephalin (LLORENS-CORTES et al. 1984) have been measured in substantia nigra. Interestingly, the levels of the peptide enkephalin have been found to be decreased in the striatum in PD (AGID and JAVOY-AGID 1985). When basal ganglia from patients with PD are hybridized with the enkephalin probe, a pronounced decrease in hybridization signal is consistently observed in all cases when compared with control brains (MENGOD and PALACIOS 1989) (Figs. 4, 5). These results stand in contrast to those obtained with 6-hydroxydopamine (6-OHDA) lesioned rats, where a marked increase in the level of enkephalin mRNA was observed on the lesioned side (YOUNG et al. 1986; SAVASTA et al. 1989). It may be assumed that such differences may relay on different regulatory mechanisms of striatal peptide expression by mesencephalic dopamine neurons.

Alterations in neurotransmitter receptors in PD have been studied extensively using membrane binding assays (BOKOBZA et al. 1984; CASH et al. 1987; GUTTMANN et al. 1986; LEE et al. 1978; PIMOULE et al. 1985; RAISMAN et al. 1985). The detailed anatomical investigation of receptor changes by radiohistochemical methods is, however, limited. Because the main biochemical lesion in PD is a decrease in dopamine, we examined the distribution of D_1 and D_2 receptors in patients with PD and in controls (Fig. 6) using the D_1 antagonist

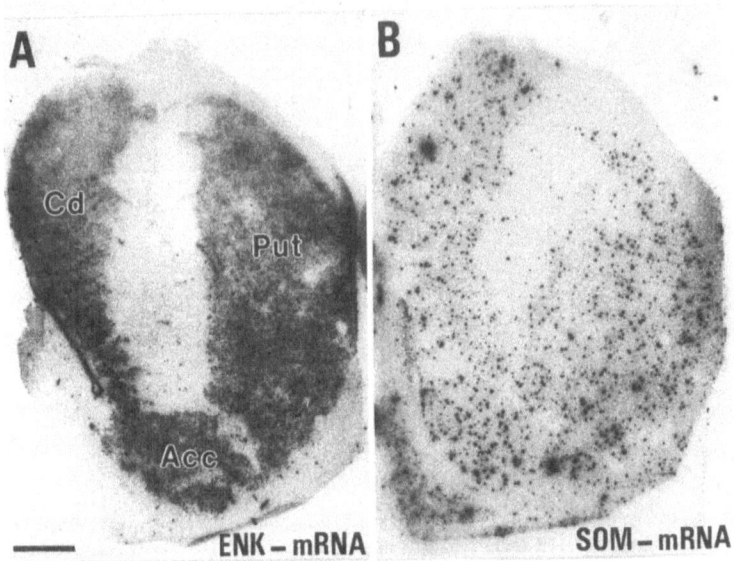

Fig. 4A, B. Visualization of mRNAs for two neuropeptides in the normal human basal ganglia by in situ hybridization histochemistry. The photographs are from film autoradiographs obtained by hybridizing sections from the same control case with ^{32}P-labeled oligonucleotides complementary to selective sequences from the mRNAs coding for **A** enkephalin (*ENK*) and **B** somatostatin (*SOM*). *Dark areas* are rich in the corresponding mRNA. *Cd*, caudate nucleus; *Put*, putamen; *Acc*, nucleus accumbens. *Bar* = 5 mm

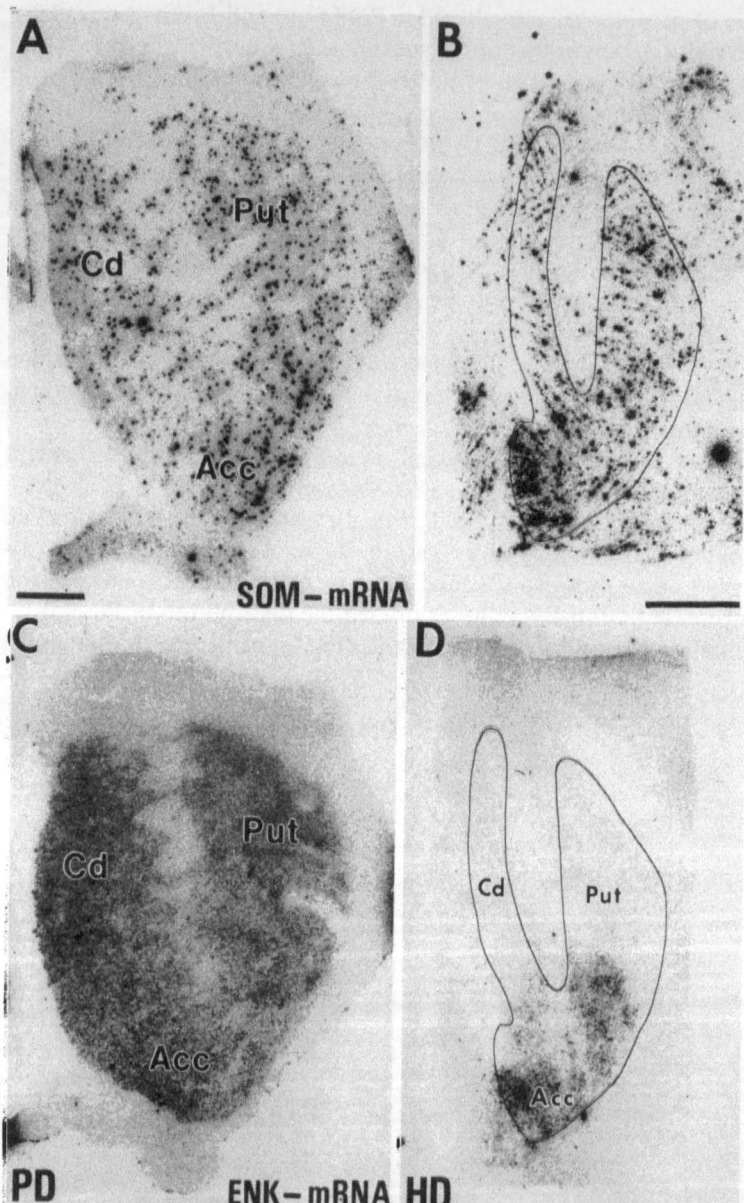

Fig. 5 A–D. Alterations in the levels of neuropeptide mRNAs in Parkinson's disease (*PD*) (**A** and **C**) and Huntington's chorea (*HD*) (**B** and **D**). Levels of somatostatin mRNA were preserved in PD (**A**) and HD (**B**) while those of enkephalin mRNA were decreased in both PD (**C**) and HD (**D**). Because of the severe shrinkage of the basal ganglia in HD, the boundaries of these nuclei have been drawn over the autoradiographs. For a control see Fig. 4. Abbreviations and scale bar are as in Fig. 4. (Reproduced with permission from Mengod and Palacios 1990)

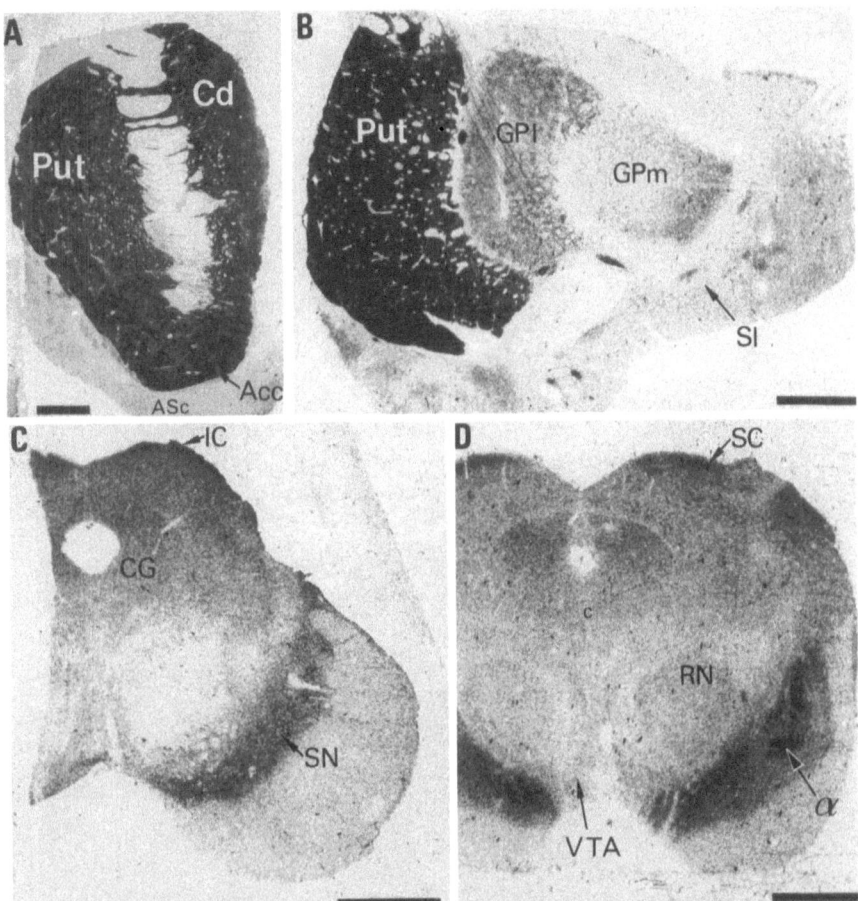

Fig. 6A–D. Distribution of ³H-CV 205-502 binding sites (D₂ receptors) in the human basal ganglia. **A–C** Autoradiographs obtained from sections of an adult human brain. High densities of D₂ receptors are observed in the putamen (*Put*), caudate nucleus (*Cd*), and nucleus accumbens (*Acc*). The globus pallidus pars lateralis (*GPl*) and the substantia nigra (*SNC* and *SNR*, pars compacta and reticulata, respectively) exhibit lower densities of receptor sites. In the latter, binding is localized mainly to the pars compacta, the highest levels being found in the lateral part of layer α. Other regions showing some ³H-CV 205-502 specific binding are the globus pallidus pars medialis (*GPm*), the substantia innominata (*SI*), and the central gray (*CG*). Note the low levels of binding in the ventral tegmental area (*VTA*). **D** was obtained from the superior half of the midbrain of an infant. The density of D₂ receptors in the substantia nigra is somewhat higher than in the adult case shown in **C**. *Bar* = 5 mm. (Reproduced with permission from Camps et al. 1989, Neuroscience 28:275–290)

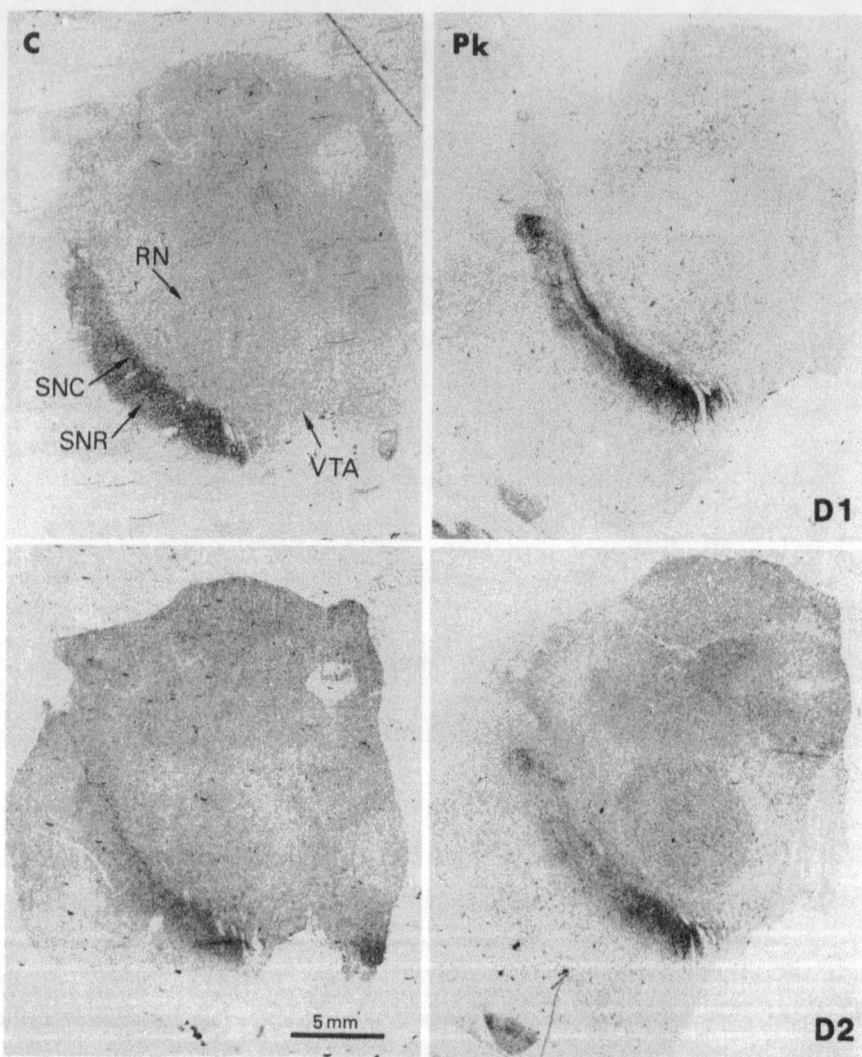

Fig. 7. Localization of D_1 (*above*) and D_2 (*below*) receptors as labeled with ^{3}H-SCH 23390 and ^{3}H-CV 205-502, respectively, in the substantia nigra [both pars compacta (*SNC*) and reticulata (*SNR*)] in a control case (*C*) and in Parkinson's disease (*PK*). Note that in Parkinson's disease D_1 and D_2 receptor densities in the substantia nigra do not differ from control levels. *VTA*, ventral tegmental area; *RN*, red nucleus. (Reproduced with permission from Palacios et al. 1989, Neurosurg. Rev. 12:11–20)

^{3}H-SCH 23390 and the D_2 agonist ^{3}H-CV 205-502 as ligands. In the population of patients examined, we did not detect any significant alteration in the number of D_1 or D_2 receptors in the areas examined, including the caudate nucleus, putamen, pallidum, and substantia nigra (Fig. 7) (CORTÉS et al. 1989; PALACIOS et al. 1988a, b). Our data confirm the lack of modification of both D_1 and D_2 receptors in the striatum of parkinsonian patients treated with

L-dopa, already described by several authors (BOKOBZA et al. 1984; GUTTMANN et al. 1986; CASH et al. 1987). On the other hand, they indicate that there are no modifications of D_1 and D_2 receptors in untreated patients.

The stability of dopamine receptor density in the substandia nigra of PD patients is rather surprising since dopamine receptors, at least the D_2 subtypes, have been shown to be significantly decreased in rats after stereotaxic injection of the dopaminergic neurotoxin 6-OHDA in the pars compacta of this nucleus (FILLOUX et al. 1987) and in monkeys after administration of 1-methyl-4-phenyl-1,2,3,5-tetra-hydropyridine (BURNS et al. 1983).

These experiments suggest that in experimental animals, D_2 receptors are located on dendrites and cell bodies of dopaminergic neurons of the substantia nigra. Absence of any alteration of D_2 receptors in human substantia nigra in spite of severe depletion of dopaminergic cells in this nucleus at least suggests species differences in the regulation or localization of this receptor, which clearly deserve further investigation.

Receptors for other classical neurotransmitters, such as the GABA-benzodiazepine or the serotonin-1 receptor binding sites, have also been found to be unchanged in the substantia nigra and in other brain areas, including the basal ganglia (PALACIOS et al. 1988 b; WAEBER and PALACIOS 1989). MChR examined in the substantia nigra as well as in the globus pallidus were, however, slightly increased in some cases (PALACIOS et al. 1989 b) (Fig. 8).

In contrast to the absence of changes or only limited modifications of classical neurotransmitter receptors in the substantia nigra in PD, we (PALACIOS et al. 1987) and others (UHL et al. 1984) have found that some neuropeptide receptors are markedly depressed in this nucleus. In normal human brain neurotensin receptor autoradiography reveals dense binding in the pars compacta and in the adjacent ventral tegmental area. In substantia nigra from patients with PD, receptor binding is only about one-third of control values (UHL et al. 1984; PALACIOS et al. 1987) (Fig. 8). Such a decrease could conceivably be due to the loss of dopaminergic cells in the pars compacta. Nigral neurotensin receptor depletion in the 6-OHDA-lesioned substantia nigra in rats (PALACIOS and KUHAR 1981) supports the assertion that in rats and man, at least the bulk of nigral neurotensin receptors are localized upon dopaminergic cells.

Of the neuropeptide receptors examined in the striatum, we found a decrease in the levels of both somatostatin and neurotensin receptors. However, the levels of expression of both peptides remain unaltered in parkinsonian brains (AGID and JAVOY-AGID 1985; GRAYBIEL and RAGSDALE 1983). No explanation is available for the down-regulation of somatostatin receptors in the parkinsonian brain. Somatostatin-like immunoreactivity has been detected in medium sized aspiny neurons of the mammalian striatum, which appear to be intrinsically organized in this nucleus (see references in GRAYBIEL and RAGSDALE 1983). However, very little is known about the cellular localization of the somatostatin receptors in the striatum. As will be seen in the next section, the decrease in somatostatin receptors in Huntington's chorea suggests that these receptors are expressed by cells intrinsic to the human striatum.

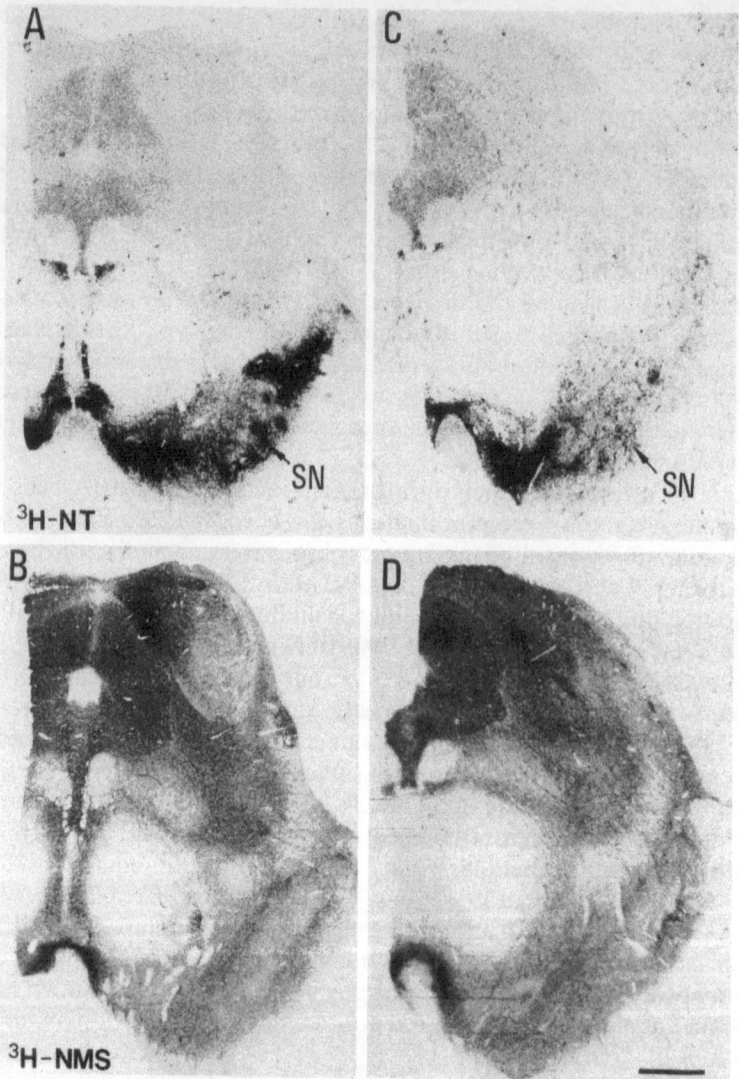

Fig. 8A–D. Receptor changes in Parkinson's disease. **A** and **B** are autoradiographs from midbrain sections from a control patient illustrating the localization of ^{3}H-neurotensin binding sites (**A**) and MChR (**B**). **C** and **D** are sections from a parkinsonian brain labeled with the same ligands showing a decrease in ^{3}H-neurotensin binding (**C**) but unchanged or slightly increased densities of MChR (**D**) in the substantia nigra (*SN*). *Bar* = 3 mm

Neurotensin receptors, as suggested by experimental evidence in animals (Quirion et al. 1985), appear to be partially presynaptically localized on striatal terminals of dopaminergic neurons. Thus, decreased numbers of striatal neurotensin receptors in PD could be a consequence of striatal dopaminergic denervation.

4.5 Huntington's Disease

Huntington's disease (HD), a disorder of midlife onset, is characterized by progressive involuntary choreiform movements, personality changes, and dementia. This disease occurs in approximately 5–10 per 100000 people and is known to be inherited as an autosomal dominant disorder with complete penetrance. Statistically, 50% of the children of heterozygotes for the HD gene will themselves inherit the gene, and therefore the disease (PENNEY and YOUNG 1988). Epidemiological data support the contention that the gene defect of HD originates from a common source and that new mutations are rare if not nonexistent (MYERS and MARTIN 1982).

The basal ganglia incur most of the pathological changes in HD (BRUYN 1968). The most severely affected parts are the caudate nucleus and the putamen. Furthermore, there is a dorsal–ventral gradient of the effects of the disease, such that the ventral putamen and nucleus accumbens are less affected than the dorsal putamen and caudate nucleus (VONSATTEL et al. 1985). HD also produces atrophy and gliosis of the pallidal structures and of the pars reticularis, but there is no loss of neurons in these regions. In fact, the density of neurons increases in the globus pallidus (LANGE et al. 1976). It can therefore be assumed that these changes are secondary to changes in the neostriatum and to the resulting loss of striatopallidal and striatonigral projections. Diffuse cerebral cortical neuronal loss is also reported to occur, particularly in the occipital lobe and frontobasal cortex (LANGE et al. 1981).

In HD some neurons of the caudate nucleus and putamen are selectively affected (KOWALL et al. 1987; FERRANTE et al. 1985, 1987), namely the spiny projecting output neurons, whereas aspiny interneurons such as those that contain the peptides somatostatin and neuropeptide Y (NPY) and the enzyme NADPH diaphorase are spared. The concentrations of a number of neurotransmitters which are known to be expressed by cells projecting outside of the striatum are decreased, e.g., acetylcholine, GABA, substance P, and enkephalins (see PENNEY and YOUNG 1988). Using in situ hybridization techniques, we have observed a marked decrease in the levels of preproenkephalin mRNA in both the caudate nucleus and the putamen of several HD cases (Fig. 5), but only minor changes in the nucleus accumbens (MENGOD and PALACIOS 1990).

Other markers, like serotonin, dopamine, and cholecystokinin, appear to be unchanged in the striatum of HD patients (BERNHEIMER and HORNYKIEWICZ 1973; BEAL and MARTIN 1986). Some neurotransmitters have been reported to be spared or even increased in the striatum of affected patients. These substances include somatostatin and NPY (which are localized in the same striatal interneurons), TRH, and neurotensin (SPINDEL et al. 1981; ARONIN et al. 1983; NEMEROFF et al. 1983; BEAL et al. 1984; MARSHALL and LANDIS 1985). Using in situ hybridization techniques, we observed an apparent increase of the hybridization signal for both somatostatin (Fig. 5) and NPY mRNAs in the striatum of HD patients (PALACIOS et al. 1989a; MENGOD and PALACIOS 1990). This increase may be explained by the marked shrinkage of the striatum due to losses of caudate nucleus and putamen neurons. Thus it may be assumed that the amount of somatostatin and NPY remains constant in the striatum as a whole.

Many receptors have been found to be altered in the striatum of the victims of HD. For instance, GABA and benzodiazepine, dopamine, and kainic and glutamic acid receptors have been found to be decreased in number (see Penney and Young 1988). A marked decrease in serotonin-1 receptor subtypes (most of them belonging to the serotonin-1D class) has been reported in the basal ganglia. This, together with unchanged serotonin-1D sites in Parkinson's disease, suggests that serotonin-1D receptors are expressed by striatal output cells which degenerate in HD (Waeber and Palacios 1989). In general, alterations of receptors were more marked in areas showing more severe morphological changes, i.e., in the anterior and dorsal aspects of the caudate nucleus and of the anterior putamen, than in the ventral parts of the striatum or in the nucleus accumbens.

There is evidence that the disease discriminates between the D_1 and D_2 receptors (Joyce et al. 1988): Whereas the striatum of HD cases shows a reduction in the density of D_1 (^{3}H-SCH 23 390) and D_2 (^{3}H-spiroperidol) receptors, the magnitude of D_2 receptor loss (28%) does not match that of D_1 receptor loss (65%). These results are compatible with the proposed localization of D_2 receptors on aspiny striatal interneurons, which are preserved in HD (Ferrante et al. 1985, 1987), and of the D_1 subtype on the spiny output neurons, which are depleted in HD (Joyce et al. 1988).

Further, increased numbers of GABA and benzodiazepine receptors have been found in the pars reticulata and in the lateral and medial globus pallidus, presumably reflecting receptor supersensitivity of the denervated nigral and pallidal GABAceptive neurons (Penney and Young 1982; Walker et al. 1984; Whitehouse et al. 1985). Interestingly, GABA and benzodiazepine receptors have been found to be decreased in the striatum and increased in the globus pallidus in early cases of HD where cell loss and atrophy could not yet be observed in the striatum (grade 0 of Vonsattel et al. 1985) (Walker et al. 1984).

Biochemical as well as autoradiographic studies in HD have shown a marked loss (over 50%) of muscarinic binding in the caudate nucleus and putamen (Hiley and Bird 1974; Wastek et al. 1976; Penney and Young 1982; Whitehouse et al. 1985; Joyce et al. 1988). We have examined several cases of HD belonging to grades 2 and 3 according to the morphological classification of Vonsattel et al. (1985) (Palacios et al. 1989b). We found reductions in MChR binding sites in the head of the caudate nucleus and in the anterior putamen, more so in the dorsal part. The reductions were found to be similar in grades 2 and 3. In contrast, MChR binding in the pallidum was not altered in grade 2 but was significantly decreased in grade 3. Our results suggest that at least one portion of the MChR is localized on striatal cells which degenerate in HD. However, as the loss of MChR is not in proportion with the degeneration of nearly 95% of the striatal neurons in grade 3 (Vonsattel et al. 1985), we have to admit that another portion of these receptors is located on the cells reported to be spared in HD, namely somatostatinergic, NPY-ergic, and NADPH diaphorase-positive cells (KOWALL et al. 1987).

Opiate receptor loss has also been noted in the striatum and lateral pallidum of HD patients (PENNEY et al. 1984). There is a dense projection of enkephalinergic neurons from the putamen to the lateral pallidum (YOUNG and PENNEY 1984). Thus the loss of opiate receptors in the lateral pallidum is consistent with their presynaptic localization on terminals of striatopallidal neurons.

Finally, receptors for several neuropeptides such as somatostatin, CCK (Fig. 9), and enkephalin have been found to be decreased in HD while substance P (Fig. 9) and neurotensin receptors were present in normal concentrations (MENGOD and PALACIOS 1990).

As in Parkinson's disease, alterations of receptors in HD were not related to the modifications in the levels of the corresponding endogenous peptides. While the concentrations of neurotensin and CCK were normal, that of somatostatin was increased, and that of substance P, decreased. Neurotensin receptors were little affected, CCK receptors were markedly decreased, somatostatin receptors were markedly decreased, and substance P receptors were not altered. This clearly questions the role of the endogenous neurotransmitter in the regulation of receptor expression in disease states of the human brain. More important appears to be the status of the cell bearing the receptor. If the receptor is localized to a cell which degenerates in this disease, the receptor will be lost in spite of the preservation of the transmitter. What physiological role the transmitter could have in the absence of the receptor is difficult to understand. In the opposite situation, the presence of receptors could still serve a purpose at least from the therapeutic point of view. A clear example of the therapeutic usefulness of such denervated receptors is provided by the use of dopamine receptor agonists in the treatment of Parkinson's disease. Similarly, drugs acting on some of the surviving receptors could be useful in the treatment of Parkinson's disease and HD.

Imaging studies of the human basal ganglia at both pre- and postsynaptic levels have revealed a number of interesting features of these brain nuclei and of their modification in disease. First, we have seen that receptors and transmitters are heterogeneously distributed in these nuclei. The correlation between the distribution of the transmitter and that of the receptor is not always good. Furthermore, alterations in neurotransmitter levels are not mirrored by receptor modifications. Further experiments are necessary to clarify the mechanisms involved in the regulation of the expression of both neurotransmitters and receptors in the human basal ganglia. Of interest will be the ability to visualize the expression of the message for these receptors by using in situ hybridization techniques. On the other hand, because neurodegenerative diseases are known to be progressive, it is important to develop methods to visualize the evolution of these biochemical parameters with the progress of the disease. For this purpose, the development of PET probes for the visualization of receptor and transmitter expression is necessary. The application of these invasive and non-invasive techniques to the basal ganglia is of enormous

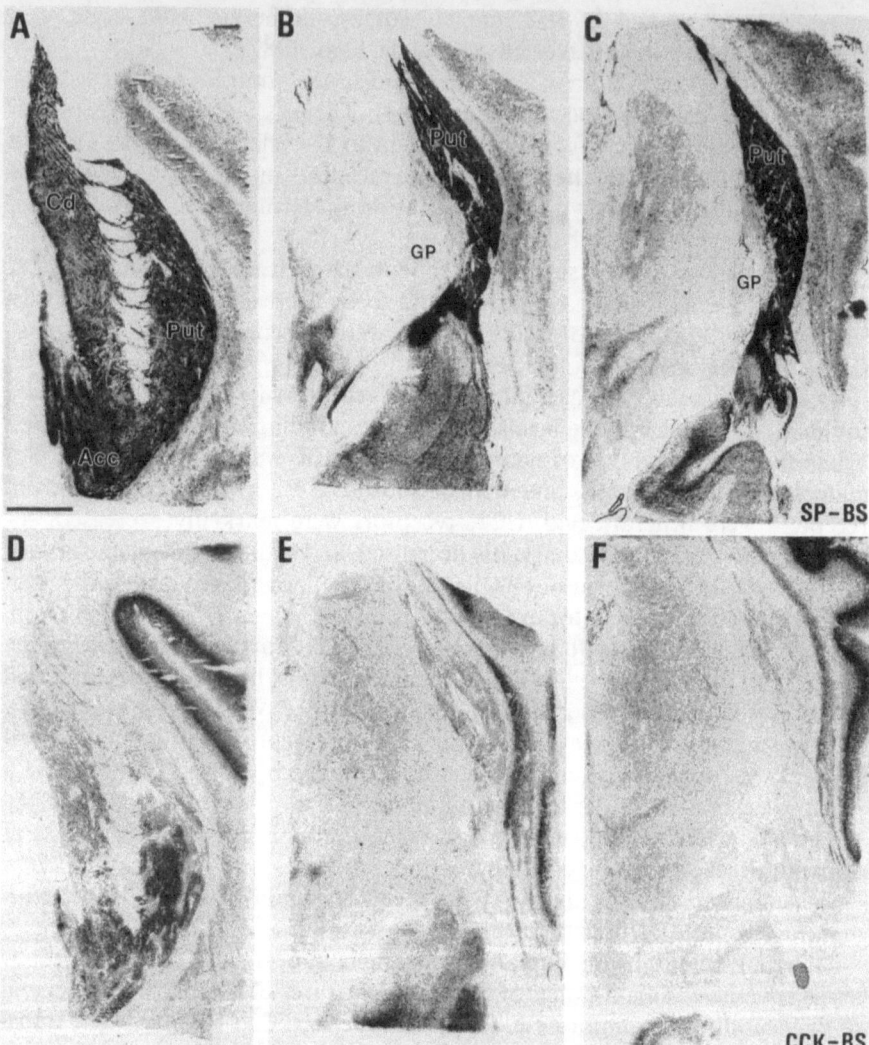

Fig. 9A–F. Alterations of neuropeptide receptor binding in the basal ganglia in a case of Huntington's chorea. Substance P-binding sites (*SP-BS*) are visualized in **A–C** at three different levels of the basal ganglia and show densities comparable to those seen in control cases (compare with Fig. 12A). In contrast, cholecystokinin-binding sites (*CCK-BS*), shown in **D–F**, are markedly reduced. The CCK-binding sites were, however, well preserved in the anterior and ventral striatum and in the nucleus accumbens (**D**). Here the binding sites had densities similar to control values. *GP,* globus pallidus; other abbreviations as defined previously. *Bar* = 5 mm. (Reproduced with permission from Palacios et al. 1989, Neurosurg. Rev. 12:11–20)

importance for the understanding of diseases of the brain and the development or rational therapeutic approaches.

4.6 Alzheimer's Disease

Alzheimer-type dementia (ATD) is one of the most frequent neurodegenerative diseases affecting the brain in old age and has emerged in the past decade as a major public health problem, mainly because of the continuous growth of the elderly population.

Alzheimer-type dementia is characterized clinically by a decline of memory function and progressive dementia and histologically by the presence of great numbers of neuritic or senile plaques and neurofibrillary tangles throughout the neocortex, hippocampus, and amygdaloid nucleus. These changes involve dysfunction and eventual death of several specific neuronal populations throughout the cortex and many subcortical nuclei.

Considerable efforts have been made in recent years to attack ATD using immunocytochemical tools and powerful molecular biological techniques mainly in order to characterize proteins implicated in the pathogenesis of this disease (see for instance ANDERTON et al. 1988; TANZI et al. 1989).

Another main research trend concerns alterations of neurotransmitters and their receptors in ATD. These studies have improved our understanding of the selectivity of the pathological process for certain populations of neurons (BEAL et al. 1987). The observation 13 years ago that enzymes involved in acetylcholine synthesis show reduced activity in brains of patients with ATD (BOWEN et al. 1976; DAVIES and MALONEY 1976) triggered a surge of interest in the biology of acetylcholine in the mammalian brain. However, since then it has become evident that the neurochemical pathology in ATD is not restricted to cholinergic pathways but extends to a variety of classical neurotransmitters and to a wide range of peptides throughout the brain. Only selected aspects of neurochemical pathology in ATD will be reviewed in this section; among them, the cholinergic alterations will be discussed more extensively.

In ATD, dysfunction or death of neurons in the medial septum, the nucleus of the diagonal band of Broca, and the nucleus basalis of Meynert (which are all components of the basal forebrain cholinergic system) (HEDREEN et al. 1984; MESULAM et al. 1984) is the likely substrate for the loss of presynaptic cholinergic markers in the telencephalon. Cholinergic loss is considered to be one major causative factor in the development of dementia in ATD. Loss of acetylcholinesterase (AChE) probably reflects the loss of cholinergic neurons, although this enzyme is also associated with noncholinergic, noradrenergic, and dopaminergic neurons. Histochemical demonstration of axons with AChE activity has shown a severe reduction in their number in the cortex of patients with ATD (ARMSTRONG et al. 1986; COLLERTON 1986). Furthermore, distorted cholinergic neurites have been demonstrated using CAT immunohistochemistry (ARMSTRONG et al. 1986) and a correlation has been established between the number of senile plaques in the cortex, mental test score, and

alteration of cholinergic innervation in ATD (Perry et al. 1978). In addition
to loss of cholinergic cells, neuronal pathology has been amply confirmed in
the noradrenergic locus coeruleus and in serotoninergic raphe nuclei (Forno
1978; Bondareff et al. 1982; Curcio and Kemper 1984). However, some but
not all patients with ATD and loss of neurons in the locus coeruleus had
reduced levels of norepinephrine and its metabolites in the brain (D'Amato
et al. 1987; Scatton et al. 1983). Generally, biochemical presynaptic markers
have been found to be much less affected than expected from cell loss in the
locus coeruleus. However, in neurosurgical cortex samples, a 90% reduction
in norepinephrine levels and a striking reduction in the numbers of norepinephrine reactive fibers were found in one study (Berger et al. 1976). In
contrast, evidence for involvement of 5-HT presynaptic markers (5-HT and
5-HIAA) in brain tissues of patients with ATD has been more compelling,
mainly in early onset ATD (Palmer et al. 1987). Recent data have also indicated a loss of dopaminergic cells in the ventral tegmental area in ATD (Mann
et al. 1987) and a significant decrease of dopamine has been detected in cortical
biopsy samples in presenile ATD (Berger et al. 1980). The amino acid transmitters glutamate and GABA have been reported to be markedly reduced by
some investigators in ATD (Arai et al. 1984; Sasaki et al. 1986; Perry T. L.
et al. 1987) but have been found to be normal by others (Perry E. K. et al.
1984) (for a review see Rossor and Iversen 1986). Some neuropeptides have
also been found to be altered in ATD, among them somatostatin (Davies et al.
1980), corticotropin releasing factor (De Souza et al. 1986) and galanin
(Chan-Palay 1988) (for a review on neuropeptides in ATD see Beal et al.
1987).

Loss of presynaptic cholinergic markers in brain tissues of patients with
ATD prompted intensive research on cholinergic receptors, both NChR and
MChR. Contrary to the converging data on the loss of presynaptic cholinergic
markers, reports on the state of MChR receptors in ATD have been inconsis-
tent. However, a majority of authors have reported normal levels of MChR
and a normal distribution pattern even in severely affected brain tissues. Furthermore, no obvious correlation could be found between the amount of
receptors and the density of senile plaques (Probst et al. 1988 b). Receptor
density over the plaques was comparable to that in the surrounding neuropil
(Palacios 1982; Lang and Henke 1983). In a more detailed study, Probst
et al. (1988 b) examined the correlation between MChR density in the CA 1
sector of the hippocampus and the number of pyramidal neurons per mm^2 in
this sector in a large cohort of ATD patients. The pyramidal cell density was
significantly decreased in ATD patients, although there were large variations
among the cases. In contrast, the densities of MChR were not significantly
decreased (Fig. 10 A, B) and the ratio of M_1 and "non-M_1" sites was unchanged. The average number of MChR per CA 1 pyramidal neuron was
significantly increased in ATD, suggesting a possible up-regulatory mechanism for MChR in the remaining hippocampal neurons. However, a marked
decrease in the concentration of MChR was found in a few severely demented
patients with more marked losses of pyramidal cells and numerous extracellu-

lar remnants of neurofibrillary tangles (ghost tangles) in the CA 1 sector (Fig. 10 C). These results suggest that compensatory mechanisms are no longer possible in cases with very severe destruction of hippocampal neurons. An incidental finding in the hippocampus of old patients consisted in circumscribed glial scars with loss of pyramidal cells, due to old ischemic injury to the hippocampus. Almost total loss of MChr and of other neurotransmitter receptors was found in such lesions (Fig. 11).

MASH et al. (1985) found a selective reduction of M_2 ("non-M_1") receptors in the hippocampus in ATD. This has been thought to favor a presynaptic localization of the M_2 subtype of MChR on the terminals of cholinergic neurons. AUBERT et al. (1989), using ^{3}H-acetylcholine to label M_2 receptors, found a reduction of M_2 in ATD. However, it should be mentioned that the density of M_2 receptors is extremely low compared with the total density of MChR. M_2 binding sites account for about 30 fmol/mg tissue, while the total density of MChR is about 500 fmol/mg tissue in the human hippocampus.

Although MChR are the predominant cholinergic receptor type in the CNS, some central cholinergic receptors have nicotinic pharmacological and physiological properties (MORLEY et al. 1979). Using quantitative autoradiography, ^{3}H-acetylcholine, and ^{3}H-nicotine, WHITEHOUSE et al. (1986a) demonstrated that the number of cortical NChR was significantly reduced in ATD, thus contrasting with the unaltered densities of MChR mentioned by most investigators. Moreover, they found that this reduction correlated with the magnitude of reduction in CAT in the same brain areas. The cellular location of NChR in cortex is not yet known. Parallel reductions in NChR and in CAT activity in ATD suggest that these receptor sites could be presynaptic, which would be consistent with the fact that nicotine can stimulate the release of acetylcholine from cholinergic terminals in the cerebral cortex (ROWELL and WINKLER 1984). Alternatively, NChR could be located postsynaptically, with the loss of these receptors being due to dysfunction or loss of neurons expressing them (WHITEHOUSE et al. 1986a). Clearly, the cellular localization of NChR in the brain, particularly in cortex and hippocampus, has to be clarified using techniques with higher resolution than receptor autoradiography. Concerning the hippocampus, there is evidence from in situ hybridization studies in the rat that local neurons contain transcripts of several NChR genes, but until now interpretation has proven difficult due to low levels of gene expression and high levels of nonspecific hybridization (WADA et al. 1988; BOULTER et al. 1986; GOLDMAN et al. 1986, 1987).

Alterations of receptors for other neurotransmitters have been observed in ATD. Of particular interest is the finding of a decreased number of somatostatin (BEAL and MARTIN 1985) and serotonin receptors (CROSS et al. 1984). This effect is probably related to the loss not only of presynaptic neurons but also of the target neurons which bear the receptor for these neurotransmitters. Greater reduction of ^{3}H-serotonin binding (5-HT$_1$ receptors) was found in ATD patients with a younger age of onset (CROSS et al. 1984) which is consistent with other studies reporting more extensive dysfunction in monoaminergic cell groups in younger patients (BONDAREFF et al. 1982; PALMER et al. 1987).

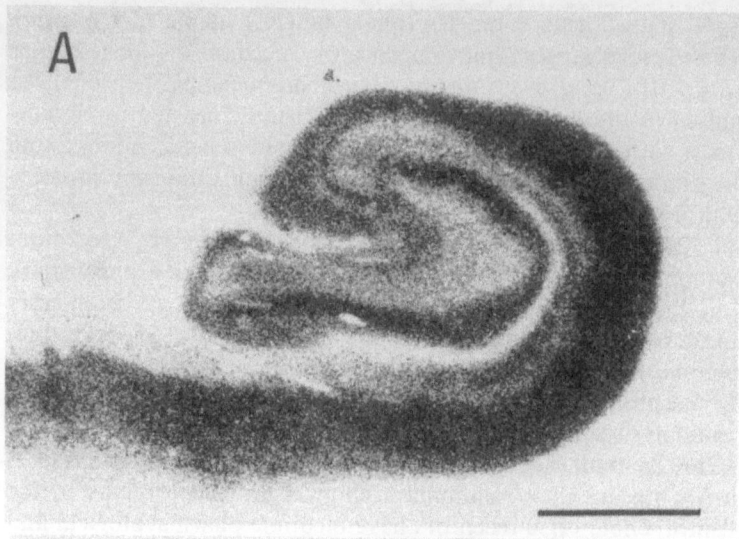

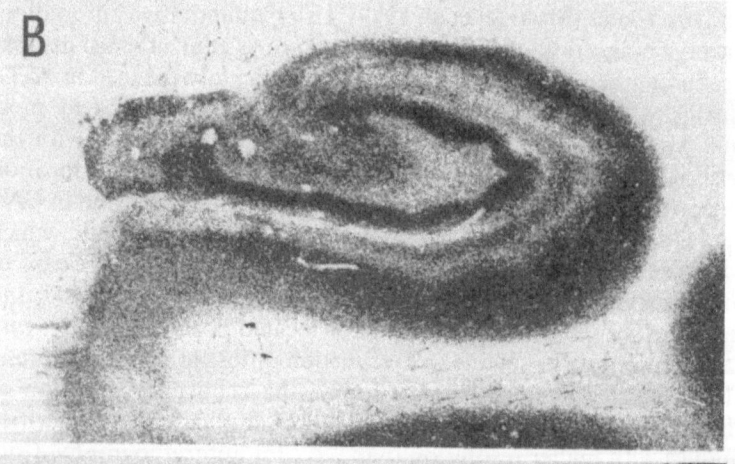

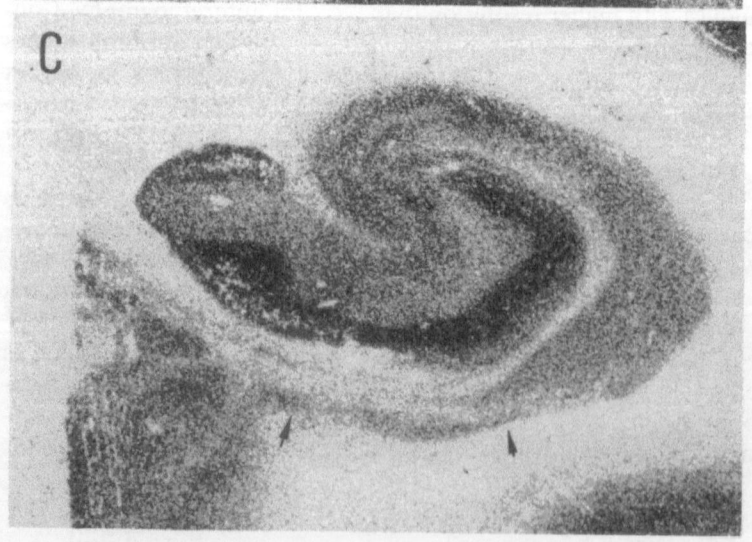

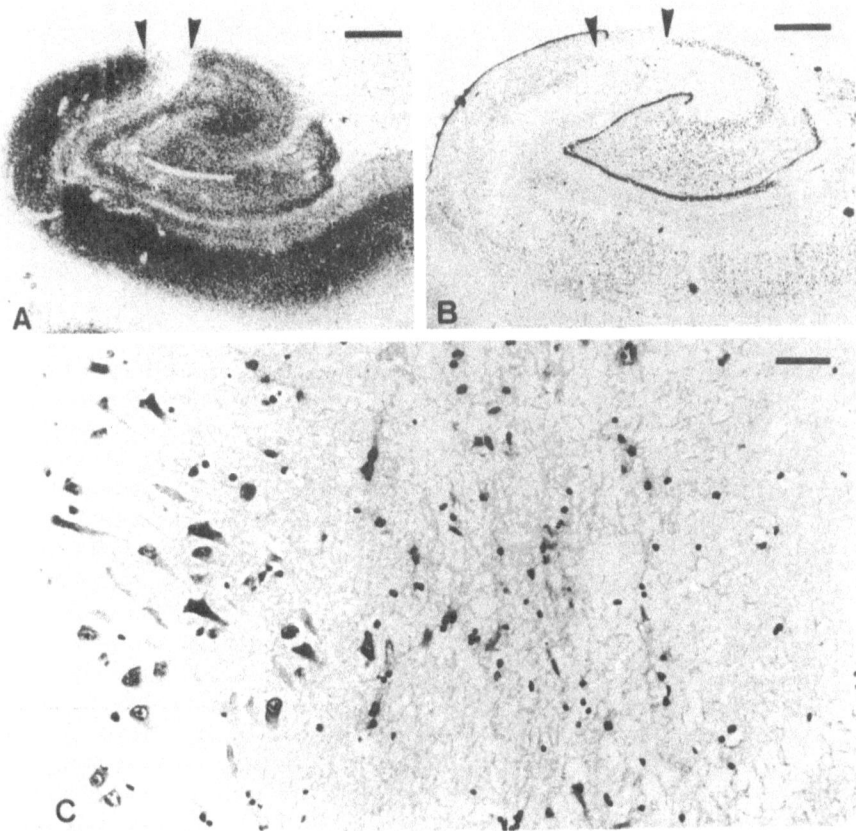

Fig. 11 A–C. Effects of a vascular lesion on MChR binding in the human hippocampus. **A** Loss of ³H-NMS binding over a small glial scar due to focal ischemia (*arrowheads*) in the CA 1 field. Loss of pyramidal neurons in the lesion is demonstrated by the cresyl violet stain in **B**. **C** illustrates the disappearance of neurons and the increase in glial cells in the lesion. *Bars:* **A** and **B** = 2 mm; **C** = 0.2 mm

◀ **Fig. 10 A–C.** Photomicrographs of autoradiographs showing the distribution of ³H-NMS-labeled MChR in the hippocampus of a control (**A**) and of two patients with Alzheimer's disease (**B, C**). MChR densities comparable to control values can be observed in the Alzheimer case illustrated in **B**. This case showed a slightly decreased number of pyramidal cells in the CA 1 sector of the hippocampus, and a medium amount of neurofibrillary tangles and senile plaques. In contrast, the Alzheimer case shown in **C** had a marked decrease in MChR in the CA 1 sector, corresponding to severe loss of pyramidal cells with numerous extracellular so-called ghost neurofibrillary tangles. *Bar* = 2 mm

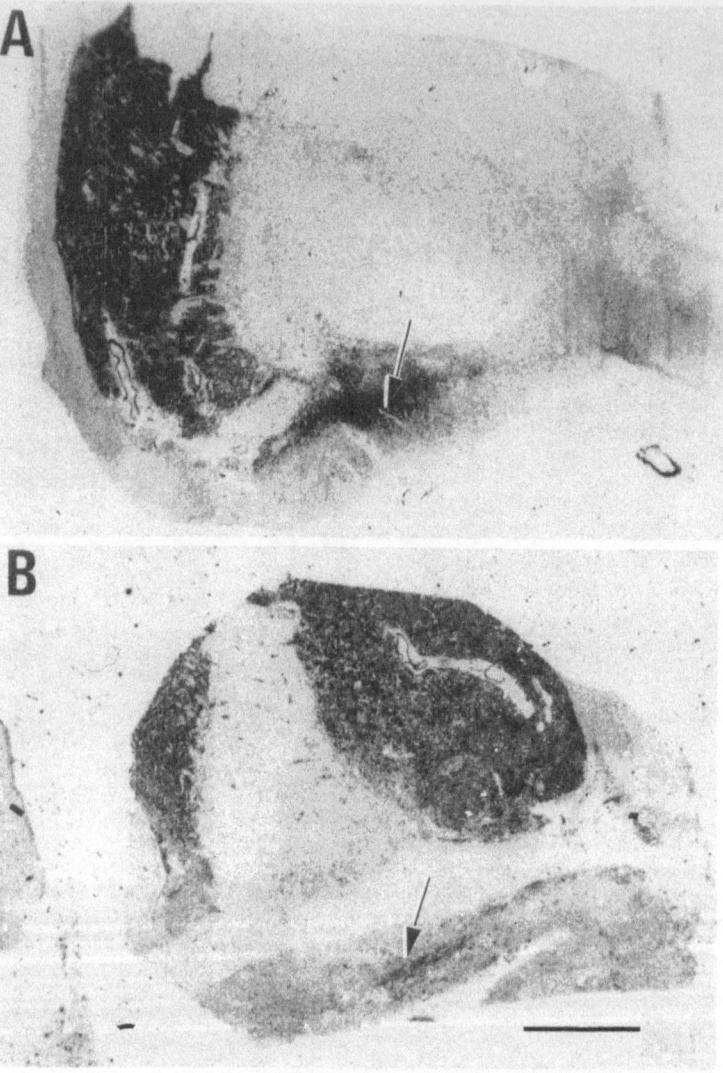

Fig. 12 A, B. Substance P receptors in the nucleus basalis of Meynert (*arrow*) in a control case (**A**) and in a patient with Alzheimer's disease (**B**). Note the loss of substance P receptors in Alzheimer's disease. *Bar* = 5 mm. (Courtesy of M. Dietl)

However, decreases in ^{3}H-ketanserin binding (5-HT$_2$ receptors) were more severe and constant in all age groups and were considered to be more specific to ATD (CROSS et al. 1986).

Another interesting finding in ATD concerns substance P receptors. High densities of these receptors have been found in the nucleus basalis of Meynert in control patients. In ATD, substance P receptors were found to be significantly decreased in this nucleus (Fig. 12) (DIETL et al. 1986). As in anterior horns in amyotrophic lateral sclerosis, the decrease in substance P receptors in the nucleus basalis in ATD seems to correlate with the local loss of neurons. Thus it appears likely that substance P receptors are located on neurons which are at risk in both conditions.

There have been only a few autoradiographic studies on excitatory amino acid binding sites in ATD. In one report (GEDDES et al. 1985) expansion of kainic acid receptor distribution was observed in the dentate gyrus molecular layer of hippocampal samples obtained postmortem from ATD patients. A similar expansion of AChE activity was observed in the same area, suggesting that a sprouting of glutamate-sensitive neurons (probably of septal origin) had occurred in this area, as a response to dentate gyrus denervation.

In contrast, significant reductions of glutamate receptors, particularly of NMDA subtype, have been found in the neocortex of ATD patients, when compared with age-matched controls and patients with Huntington's disease (GREENAMYRE et al. 1985). Glutamate receptors are concentrated on the spinous processes of distal dendrites of pyramidal neurons. It has been speculated (MARAGOS et al. 1987) that, early in ATD disease, glutamatergic inputs to cortical association areas and hippocampal pyramidal neurons become hyperactive and that damage to dendritic appendages ensues, with loss of glutamate receptors and eventually death of the neurons. On the other hand, excitatory amino acid neurotransmitter glutamate has been implicated in learning and memory (LYNCH and BAUDRY 1984) and NMDA receptors, a subtype of glutamate receptors, appear to be particularly important in these processes. Thus, impaired glutamate receptor function may contribute to learning and memory impairments in ATD (MARAGOS et al. 1987).

5 Concluding Remarks

We have seen that neurodegenerative diseases are characterized by significant changes in neurotransmitter receptors as well as by well-defined morphological changes and alterations of specific neurotransmitter systems.

One of the main conclusion of the cited studies is that neurotransmitter receptors, because of their association with specific neuronal populations, can be used as neuronal markers. Thus, death of a neuronal population will be mirrored by a corresponding loss of its specific receptors. Good examples of receptor decreases following cell losses have been mentioned in the spinal cord of patients with amyotrophic lateral sclerosis (WHITEHOUSE et al. 1983), in the hippocampus of patients with severe Alzheimer-type disease (PROBST et al. 1988 b), and in the striatum of patients with Huntington's disease (PENNEY and YOUNG 1982).

While local receptor losses associated with neuronal death are probably not surprising, receptor autoradiography also reveals that receptor changes may be sensitive markers for dysfunction or early degeneration of nerve cells in the brain (WHITEHOUSE 1985). One example is the decrease in MChR and GABA-related receptors in the striatum of patients with early Huntington's disease, in a stage at which cell loss and atrophy cannot yet be detected in this nucleus (WALKER et al. 1984). Alterations of neuronal receptors also occur when there are reactive and potentially reversible changes of neurons, e.g., as a response to axotomy. Hypoglossal axotomy has been shown to induce marked but reversible decreases in MChR in motor cells of the hypoglossal nucleus in rat (ROTTER et al. 1979), and facial nerve transection induces similar MChR decreases in neurons of the facial nucleus (HOOVER and HANCOCK 1985).

Other examples have been shown where neuronal loss not only results in receptor modification at the local level but also in distant areas. This is illustrated by the increased numbers of GABA and benzodiazepine receptors in the globus pallidus and pars reticularis in Huntington's disease, presumably reflecting receptor supersensitivity (PENNEY and YOUNG 1982).

Other situations have been mentioned in this review where receptor autoradiography has provided evidence of neuronal circuit dysfunction in the absence of neuronal alterations visible using other tissue-staining techniques. Examples are changes in MChR in the dorsal horn of the spinal cord in patients with amyotrophic lateral sclerosis (WHITEHOUSE et al. 1983).

A reverse situation may occur where receptor numbers are unchanged in spite of an obvious decrease in neurons considered to bear these receptors. This is the case for substantia nigra in Parkinson's disease, where a decrease of dopamine receptors was expected from animal models but in fact did not occur.

We have seen that there is no good correlation between the distribution of the transmitter and that of the receptor. For instance, alterations in neurotransmitter levels are not always paralleled by compensatory receptor modifications. Both a decrease in the transmitter concentration without modification of the receptor and a modification of the receptor in the presence of unchanged or even increased transmitter levels have been reported. These discrepancies have been commented upon in the present review, particularly in the context of basal ganglia disorders and may have several practical consequences, mainly in the therapeutic approach to these diseases. We do not yet understand the meaning of such a lack of correlation between the distribution of receptors and that of the transmitters, and clearly much remains to be learned regarding the pre- or postsynaptic localization of receptors and their modulation mechanisms in order to improve our anatomical interpretation of receptor alterations.

Although important information has already been obtained on the complex interrelationships between receptor alterations, as shown by autoradiographic techniques, alterations of chemical messengers, and neurohistopathological data, our understanding of these interrelationships is still very limited. New insights into the mechanisms of receptor regulation in human brain diseases will be obtained by the development of new ligands for the labeling of important functional sites, including not only receptor sites but also enzymes, transport sites, and ionic channels. The cloning of genes for neurotransmitter receptors has allowed prediction of the primary structure of these proteins. Antibodies can now be raised against synthetic peptides made from selected regions of these receptors and used to examine further the localization of these receptors at the subcellular level. This approach, combined with in situ hybridization techniques, will hopefully provide additional insights into the regulatory mechanisms controlling receptor synthesis, transport, and degradation. New approaches will also combine the mentioned techniques with immunocytochemical and histochemical staining in order to improve anatomical resolution and better characterize the relationships between receptors and cellular alterations.

In addition, animal models (some of which have been mentioned in this review) are available for the study of human neurodegenerative disorders. Models using experimental interventions (such as lesions or drugs) as well as natural models can be applied to the study of receptor modulation in restricted neuronal pathways or systems and receptor changes as a function of disease progression.

Although autoradiographic receptor studies on postmortem tissue samples provide a static view of receptor modifications in brain diseases, a more dynamic approach is to be expected from in vivo application of positron emission tomography (PET). Information generated by autoradiographic experiments on postmortem human brain samples, when applied to PET, could help in the early diagnosis of some degenerative diseases and in the monitoring of the therapeutic effects of drugs.

References

Agid Y, Javoy-Agid F (1985) Peptides and Parkinson's disease. Trends Neurosci 1:30–35

Agid Y, Javoy-Agid F, Ruberg M et al. (1986) Progressive supranuclear palsy: anatomical and biochemical considerations. Adv Neurol 45:191–206

Anderton BH, Brion JP, Flament-Durand J et al. (1988) Structure and chemistry of Alzheimer neurofibrillary tangles. In: Ulrich J (ed) Interdisciplinary topics in Gerontology 25. Karger, Basel, pp 106–118

Appel NM, Wessendorf MW, Elde RP (1987) Thyrotropin-releasing hormone in spinal cord: coexistence with serotonin and with substance P in fibers and terminals apposing identified preganglionic-sympathetic neurons. Brain Res 415:137–143

Arai H, Kobayashi K, Ichimiya Y (1984) A preliminary study of free amino acids in the postmortem temporal cortex from Alzheimer-type dementia patients. Neurobiol Aging 5:319–321

Armstrong DM, Bruce G, Hersch LB, Terry RD (1986) Choline acetyltransferase immunoreactivity in neuritic plaques of Alzheimer brain. Neurosci Lett 71:229–234

Aronin N, Cooper PE, Lorenz LJ, Bird ED, Sagar SM, Martin JB (1983) Somatostatin is increased in the basal ganglia in Huntington's disease. Ann Neurol 13:519–526

Aubert I, Aranjo DM, Gauthier S, Quirion R (1989) Muscarinic receptor alterations in human cognitive disorders. Abstr. of the 4th International Symposium on Subtypes of Muscarinic Receptors. Trends Pharmacol Sci (in press)

Bancher C, Lassmann H, Budka H, Grundke-Iqbal I, Iqbal K, Wiche G, Seitelberger F (1987) Neurofibrillary tangles in Alzheimer's disease and progressive supranuclear palsy: antigenic similarities and differences. Acta Neuropathol 74:39–46

Baron JC, Mazière B, Loc'h C, Sgouropoulos P, Bonnet AM, Agid Y (1985) Progressive supranuclear palsy: loss of striatal dopamine receptors demonstrated in vivo by positron tomography.
Lancet I:1163-1164

Basbaum AI (1988) Distribution of glycine receptors' immunoreactivity in the spinal cord of the rat:
cytochemical evidence for a differential glycinergic control of lamina I and V nociceptive
neurons. J Comp Neurol 278:330-336

Beal MF, Martin JB (1985) Reduced numbers of somatostatin receptors in the cerebral cortex in
Alzheimer's disease. Science 229:239-253

Beal MF, Martin JB (1986) Neuropeptides and neurological disease. Ann Neurol 20:547-565

Beal MF, Bird ED, Llanglais PJ, Martin JB (1984) Somatostatin is increased in the nucleus accumbens in Huntington's disease. Neurology 34:663-666

Beal MF, Kowall NW, Mazurek F (1987) Neuropeptides in Alzheimer's disease. In: Wurtman RJ,
Corkin SH, Crowdon JH (eds) Alzheimer's disease: advances in basic research and therapies.
Center for Brain Sciences and Metabolism Charitable Trust, Cambridge, MA, pp 151-168

Berciano J (1988) Olivopontocerebellar atrophy. In: Jankovic J, Tolosa E (eds) Parkinson's disease
and movement disorders. Urban and Schwarzenberg, Munich, pp 131-151

Berger B, Escourolle R, Moyne MA (1976) Axones catécholaminergiques du cortex cérébral hu-
main. Observation en histofluorescence, de biopsies cérébrales dont 2 cas de maladie
d'Alzheimer. Revue Neurol (Paris) 132:183-194

Berger B, Tassin JP, Rancurel G, Blanc G (1980) Catecholaminergic innervation of the human
cerebral cortex in presenile and senile dementia. Histochemical and biochemical studies. In:
Usdin E, Sourkes TL, Youdim MBH (eds) Enzymes and neurotransmitters in mental disease.
Wiley, Chichester, pp 317-328

Bernheimer H, Hornykiewicz O (1973) Brain amines in Huntington's chorea. Adv Neurol 1:1872-
1972

Bernheimer H, Birkmayer W, Hornykiewicz O (1961) Verteilung des 5-hydroxytryptamins (Serotonin) im Gehirn des Menschen und sein Verhalten bei Patienten mit Parkinson-Syndrom. Klin
Wochenschr 39:1056-1059

Bird ED, Iversen LL (1982) Human brain postmortem studies of neurotransmitter and related
markers. In: Lajtha A (ed) Handbook of neurochemistry, vol 2. Plenum, London, pp 225-251

Bogerts B, Häntsch J, Herzer M (1983) A morphometric study of the dopamine-containing cell
groups in the mesencephalon of normals, Parkinson patients and schizophrenics. Biol Psychiatry 18:951-969

Bokobza B, Ruberg M, Scatton B, Javoy-Agid F, Agid Y (1984) 3H-spiperone binding, dopamine
and HVA concentrations in Parkinson's disease and supranuclear palsy. Eur J Pharmacol
99:167-175

Bondareff W, Mountjoy CQ, Roth M (1982) Loss of neurons of origin of the adrenergic projection
to cerebral cortex (nucleus locus coeruleus) in senile dementia. Neurology 32:164-168

Bonner TI, Buckley NJ, Young AC, Brann MR (1987) Identification of a family of muscarinic
acetylcholine receptor genes. Science 237:527-532

Bonner TI, Young AC, Brann M, Buckley NJ (1988) Cloning and expression of the human and rat
m5 muscarinic acetylcholine receptor genes. Neuron 1:403-410

Boulter J, Evans K, Goldman D, Martin G, Treco D, Heinemann S, Patrick Y (1986) Isolation of
a cDNA clone coding for possible neural nicotinic acetylcholine receptor alpha-subunit. Nature 319:368-374

Bowen DM, Smith CB, White P, Davison AN (1976) Neurotransmitter-related enzymes and indices
of hypoxia in senile dementia and other abiotrophies. Brain 99:459-496

Bruyn GW (1968) Huntington's chorea: historical, clinical and laboratory synopsis. In: Vinken P,
Bruyn GW (eds) Handbook of clinical neurology, vol 6. North Holland, Amsterdam, 298-378

Buckley NJ, Bonner TI, Brann MR (1988) Localization of a family of muscarinic receptor mRNAs
in rat brain. J Neurosci 8:4646-4652

Bunzow Y, van Tol HHM, Grandy DK et al. (1988) Cloning and expression of a rat D2 dopamine
receptor cDNA. Nature 336:783-787

Burns RS, Chiueh CC, Markey SP, Ebert MH, Jacobowitz DM, Kopin IJ (1983) A primate model
for parkinsonism: selective destruction of dopaminergic neurons in the pars compacta of the
substantia nigra by N-methyl-4-phenyl-1,1,3,6-tetrahydropyridine. Proc Natl Acad Sci USA
80:4546-4550

Cash R, Raisman R, Ploska A, Agid Y (1987) Dopamine D1 receptor and cyclic AMP-dependent phosphorylation in Parkinson's disease. J Neurochem 49:1075–1083

Chan-Palay V (1988) Galanin hyperinnervates surviving neurons of the human basal nucleus of Meynert in dementia of Alzheimer's and Parkinson's disease: a hypothesis for the role of galanin in accentuating cholinergic dysfunction in dementia. J Comp Neurol 273:543–557

Chesselet MF, Graybiel AM (1986) Striatal neurons expressing somatostatin-like immunoreactivity: evidence for a peptidergic interneuronal system in the cat. Neuroscience 17:547–572

Clarke PBS, Schwartz RD, Paul SM, Pert CB, Pert A (1985) Nicotinic binding in rat brain: autoradiographic comparison of ^{3}H-acetylcholine, ^{3}H-nicotine and ^{125}I-alpha-bungarotoxin. J Neurosci 5:1307–1315

Collerton D (1986) Cholinergic function and intellectual decline in Alzheimer's disease. Neuroscience 19:1–28

Cortés R, Camps M, Gueye B, Probst A, Palacios JM (1989) Dopamine receptors in human brain: autoradiographic distribution of D1 and D2 sites in Parkinson syndrome of different etiology. Brain Res 483:30–38

Cross AJ, Crow TJ, Ferrier IN, Johnson JA, Bloom SR, Corsellis JAN (1984) Serotonin receptor changes in dementia of the Alzheimer type. J Neurochem 43:1574–1581

Cross AJ, Cross TJ, Ferrier IN, Johnson JA (1986) The selectivity of the reduction of serotonin S$_2$ receptors in Alzheimer-type dementia. Neurobiol Aging 7:3–7

Curcio CA, Kemper T (1984) Nucleus raphe dorsalis in dementia of the Alzheimer type: neurofibrillary changes and neuronal packing density. J Neuropathol Exp Neurol 43:359–368

D'Amato RJ, Zweig RM, Whitehouse PJ et al. (1987) Aminergic systems in Alzheimer's disease. Ann Neurol 22:229–236

Davies P, Maloney AJF (1976) Selective loss of central cholinergic neurons in Alzheimer senile dementia. Nature 288:279–280

Davies P, Katzman R, Terry RD (1980) Reduced somatostatin-like immunoreactivity in cerebral cortex from cases of Alzheimer disease and Alzheimer senile dementia. Nature 288:279–280

Déjerine J, Thomas A (1900) L'atrophie olivo-pontocérébelleuse. Nouv Iconogr Salpêt 13:330–370

De Souza EB, Whitehouse PJ, Kuhar MJ, Price DL, Vale WW (1986) Reciprocal changes in corticotropin-releasing factor (CRF)-like immunoreactivity and CRF receptors in cerebral cortex of Alzheimer's disease. Nature 319:593–595

Deutch AY, Holliday J, Roth RH, Chan LLY, Hawrot E (1987) Immunohistochemical localization of a neuronal nicotinic acetylcholine receptor in mammalian brain. Proc Natl Acad Sci USA 84:8697–8701

Dietl MM, Probst A, Palacios JM (1986) Mapping of substance P receptors in the human brain: high densities in the substantia innominata and effect of senile dementia. Soc Neurosci Abstr 12:831

Dietl MM, Sanchez M, Probst A, Palacios JM (1989) Substance P receptors in the human spinal cord: decrease in amyotrophic lateral sclerosis. Brain Res 483:39–49

Dixon RAF, Kobilka BK, Strader DJ et al. (1986) Cloning of the gene and cDNA for mammalian beta-adrenergic receptor and homology with rhodopsin. Nature 321:75–79

Dohlman HG, Caron MG, Lefkowitz RJ (1987) A family of receptors coupled to guanine nucleotide regulatory proteins. Biochemistry 26:2657–2664

Dubois B, Ruberg M, Javoy-Agid F, Ploska A, Agid Y (1983) A subcortico-cortical cholinergic system is affected in Parkinson's disease. Brain Res 288:213–218

Ehringer H, Hornykiewicz O (1960) Verteilung von Noradrenalin und Dopamin (3-Hydroxytryptamin) im Gehirn des Menschen und ihr Verhalten bei Erkrankungen des extrapyramidalen Systems. Klin Wochenschr 38:1236–1239

Engel WK, Siddique T, Nicoloff JT (1983) Effect on weakness and spasticity in amyotrophic lateral sclerosis of thyrotropin-releasing hormone. Lancet II:73–75

Fargin A, Raymond JR, Lohse MJ, Kobilka BK, Caron MG, Lefkowitz RJ (1988) The genomic clone G-21 which resembles a beta-adrenergic receptor sequence encodes the 5-HT 1A receptor. Nature 335:358–360

Ferrante RJ, Kowall NW, Beal MF, Richardson EP, Bird ED, Martin JB (1985) Selective sparing of a class of striatal neurons in Huntington's disease. Science 230:561–563

Ferrante RJ, Beal MF, Kowall NW, Richardson EP, Martin JB (1987) Sparing of acetylcholinesterase-containing striatal neurons in Huntington's disease. Brain Res 411:162–166

Filloux F, Wamsley JK, Dawson TN (1987) Dopamine D2 auto- and postsynaptic receptors in the nigrostriatal pathway system of the rat brain: localization by quantitative autoradiography with ^{3}H-sulpiride. Eur J Pharmacol 138:61–68

Forno LS (1978) The locus coeruleus in Alzheimer's disease. J Neuropathol Exp Neurol 37:614

Frielle T, Collins S, Daniel KW, Caron MG, Lefkowitz RJ, Kobilka BK (1987) Cloning of the cDNA for the human beta-adrenergic receptors. Proc Natl Acad Sci USA 84:7920–7924

Galloway PG (1988) Antigenic characteristics of neurofibrillary tangles in progressive supranuclear palsy. Neurosci Lett 91:148–153

Geddes JW, Monaghan DT, Cotman CW, Lott IT, Kim RC, Chui HC (1985) Plasticity of hippocampal circuitry in Alzheimer's disease. Science 230:1179–1181

Gillberg PG, Aquilonius SM (1985) Cholinergic, opioid and glycine receptor binding sites localized in human spinal cord by in vitro autoradiography. Acta Neurol Scand 72:299–306

Gillberg PG, Aquilonius SM, Eckernäs SA, Lundqvist A, Windblad B (1982) Choline acetyltransferase and substance P in the human spinal cord: changes in amyotrophic lateral sclerosis. Brain Res 250:394–397

Gillberg PG, Nordberg A, Aquilonius SM (1984) Muscarinic binding sites in small homogenates and in autoradiographic sections from spinal cord of rat and man. Brain Res 300:327–333

Golbe LI, Davis PH (1988) Progressive supranuclear palsy. Recent advances. In: Jankovic J, Tolosa E (eds) Parkinson's disease and movement disorders. Urban and Schwarzenberg, Baltimore, pp 121–130

Goldman D, Simmon D, Swanson LW, Patrick J, Heinemann S (1986) Mapping of brain areas expressing RNA homologous to two different acetylcholine receptor alpha-subunit cDNAs. Proc Natl Acad Sci USA 83:4076–4080

Goldman D, Deneris E, Luyten W, Kochhar A, Patrick J, Heinemann S (1987) Members of a nicotinic acetylcholine receptor gene family are expressed in different regions of the mammalian central nervous system. Cell 48:965–973

Graybiel AM, Ragsdale CW Jr (1983) Biochemical anatomy of the striatum. In: Emson PC (ed) Chemical neuroanatomy. Raven, New York, 427–504

Greenamyre JT, Penney JB, D'Amato CJ, Hicks SP, Shoulson I (1985) Alterations in L-glutamate binding in Alzheimer's and Huntington's disease. Science 227:1496–1499

Grenningloh G, Rienitz A, Schmitt B et al. (1987) The strychnine-binding subunit of the glycine receptor shows homology with nicotinic acetylcholine receptors. Nature 328:215–220

Gross RA, Spehlmann R, Daniels JC (1978) Sleep disturbances in progressive supranuclear palsy. Electroencephalogr Clin Neurophysiol 45:16–25

Guttmann M, Seeman P, Reynolds GP, Riederer P, Jellinger K, Tourtellotte WW (1986) Dopamine D2 receptor density remains constant in treated Parkinson's disease. Ann Neurol 19:487–492

Hedreen JC, Struble RG, Whitehouse PJ, Price DL (1984) Topography of the magnocellular basal forebrain system in human brain. J Neuropath Exp Neurol 43:1–21

Hiley CR, Bird ED (1974) Decreased muscarinic receptor concentration in post-mortem brain in Huntington's chorea. Brain Res 80:355–358

Hirsch EC, Graybiel AM, Duyckaerts, Javoy-Agid F (1987) Neuronal loss in the pedunculopontine tegmental nucleus in Parkinson disease and in progressive supranuclear palsy. Proc Natl Acad Sci USA 84:5976–5980

Hoffman BJ, Mezey E (1989) Distribution of serotonin 5-HT 1C receptor mRNA in adult rat brain. FEBS Lett 247:453–462

Hoover DB, Hancock JC (1985) Effect of facial nerve transection on acetylcholinesterase, choline acetyltransferase and ^{3}H-quinuclidinyl benzilate binding in rat facial nuclei. Neuroscience 15:481–487

Hornykiewicz O (1978) Historical aspects and frontiers of Parkinson's disease research. Adv Exp Med Biol 90:1–20

Jackson IMD, Adelman LS, Munsat TL, Forte S, Lechan RM (1986) Amyotrophic lateral sclerosis: thyrotropin-releasing hormone and histidyl proline diketopiperazine in the spinal cord and cerebrospinal fluid. Neurology 36:1218–1223

Javoy-Agid F, Agid Y (1980) Is the mesocortical dopaminergic system involved in Parkinson disease? Neurology 30:1326–1330

Javoy-Agid F, Ruberg M, Taquet H et al. (1982) Biochemical neuropathology and Parkinson disease. Advances in neurology: proceedings of the VII International Symposium on Parkinson disease. Raven, New York, pp 189–198

Jellinger K, Riederer P, Tomonaga M (1980) Progressive supranuclear palsy: clinicopathological and biochemical studies. J Neural Transm 16 [Suppl]: 111–128

Johansson O, Hökfelt T, Pernow B (1981) Immunocytochemical support for three putative transmitters in one neuron: coexistence of 5-hydroxytryptamine, substance P and thyrotropin releasing hormone like immunoreactivity in medullary neurons projecting to the spinal cord. Neuroscience 6: 1857–1881

Joyce JN, Lexow N, Bird E, Winokur A (1988) Organization of dopamine D1 and D2 receptors in human striatum: receptor autoradiographic studies in Huntington's disease and schizophrenia. Synapse 2: 546–557

Julius D, MacDermott AB, Axel R, Jessel T (1988) Molecular characterization of a functional cDNA encoding the serotonin 1c receptor. Science 241: 558–564

Kish SJ, Perry TL, Hornykiewicz O (1983) Increased GABA receptor binding in dominantly inherited cerebellar ataxias. Brain Res 269: 370–373

Kish SJ, Chang LJ, Mirchandani L, Shannak K, Hornykiewicz O (1985) Progressive supranuclear palsy: relationship between extrapryamidal disturbances, dementia and brain neurotransmitter markers. Ann Neurol 18: 530–536

Kish SJ, Li PP, Robitaille Y, Currier R, Gilbert J, Schut L, Warsh JJ (1989) Cerebellar ^{3}H-inositol-1,4,5-triphosphatase binding is markedly decreased in human olivopontocerebellar atrophy. Brain Res 489: 373–376

Kobilka BK, Maysui H, Kobilka TS et al. (1987a) Cloning, sequencing and expression of the gene coding for the human platelet alpha-2 adrenergic receptor. Science 238: 650–656

Kobilka BK, Dixon RAF, Frielle T et al. (1987b) cDNA for the human beta-2 adrenergic receptor: a protein with multiple membrane spanning domains and encoded by a gene whose chromosomal location is shared with that of the receptor for platelet-derived growth factor. Proc Natl Acad Sci USA 84: 46–50

Kobilka BK, Frielle T, Collins S et al. (1987c) An intronless gene encoding a potential member of the family of receptors coupled to guanine nucleotide regulatory proteins. Nature 329: 75–79

Kowall NW, Ferrante RS, Martin JB (1987) Patterns of cell loss in Huntington's disease. Trends Neurosci 10: 24–29

Lang W, Henke H (1983) Cholinergic receptor binding and autoradiography in brains of non-neurological and senile dementia of Alzheimer type patients. Brain Res 267: 271–280

Lange H, Thorner G, Hopf A, Schroder KF (1976) Morphometric studies of the neuropathological changes in choreatic disease. J Neurol Sci 28: 401–425

Lange HW (1981) Quantitative changes of telencephalon, diencephalon and mesencephalon in Huntington's chorea, postencephalitic and idiopathic parkinsonism. Verh Anat Ges 75: 923–925

Lechan RM, Snapper SB, Jacobson S, Jackson IMD (1984) The distribution of thyrotropin-releasing hormone (TRH) in the rhesus monkey spinal cord. Peptides 5 [Suppl 1]: 186–194

Lee T, Seeman P, Rajput A, Farley JJ, Hornykiewicz O (1978) Receptor basis for dopmainergic supersensitivity in Parkinson's disease. Nature 278: 59–61

Lentz TL, Chester J (1977) Localization of acetylcholine receptors in central synapses. J Cell Biol 75: 258–267

Lindstrom JM (1986) Probing nicotinic acetylcholine receptors with monoclonal antibodies. Trends Neurosci 9: 401–407

Llorens-Cortes C, Javoy-Agid F, Taquet H, Schwarz JC (1984) Enkephalinergic markers in substantia nigra and caudate nucleus from Parkinsonian subjects. J Neurochem 43: 874–877

Lolait SJ, O'Caroll AM, Kusano K, Müller JM, Brownstein MJ, Mahan LC (1989) Cloning and expression of a novel rat GABA A receptor. FEBS Lett 246: 145–148

Lübbert H, Hoffman BJ, Snutch TP et al. (1987) cDNA cloning of a serotonin 5-HT 1C receptor by electrophysiological assays of mRNA injected Xenopus oocytes. Proc Natl Acad Sci USA 84: 4332–4336

Lynch G, Baudry M (1984) The biochemistry of memory: a new and specific hypothesis. Science 224: 1057–1063

Mamalaki C, Stephenson FA, Barnard EA (1987) The GABA A benzodiazepine receptor is a heterotetramer of homologous alpha and beta subunits. EMBO J 6: 561–565

Manaker S, Winokur A, Rhodes H, Rainbow TC (1985) Autoradiographic localization of thyrotropin-releasing hormone (THR) receptors in human spinal cord. Neurology 35: 328–332

Manaker S, Caine SB, Winokur A (1988) Alterations in receptors for thyrotropin-releasing hormone, serotonin, and acetylcholine in amyotrophic lateral sclerosis. Neurology 38: 1464–1474

Mann DMA, Yates PO, Marcyniuk B (1987) Dopaminergic neurotransmitter systems in Alzheimer's disease and Down's syndrome at middle age. J Neurol Neurosurg Psychiatry 50: 341–344

Maragos WF, Greenamyre JT, Penney JB Jr, Young AB (1987) Glutamate dysfunction in Alzheimer's disease. A hypothesis. Trends Neurosci 10: 65–68

Marshall PE, Landis D (1985) Huntington's disease is accompanied by changes in the distribution of somatostatin containing neuronal processes. Brain Res 329: 71–82

Mash DC, Flynn DD, Potter LT (1985) Loss of M2 muscarinic receptors in the cererbral cortex in Alzheimer's disease and experimental cholinergic denervation. Science 228: 1115–1117

Masu Y, Nakayama K, Tamaki H, Harada Y, Kuno M, Nakanishi S (1987) cDNA cloning of bovine substance K receptor through oocyte expression system. Nature 329: 836–838

McGeer PL, Eccles Sir JC, McGeer EG (1987) Molecular neurobiology of the mammalian brain. Plenum, New York

Mengod G, Palacios JM (1990) Molecular neuropathology: the study of transmitter receptor expression in human postmortem materials by in situ hybridization and receptor autoradiography. Neuropsychopharmacology

Mengod G, Martinez-Mir MI, Vilaró MT, Palacios JM (1989) Localization of the mRNA for the dopamine D2 receptor in the rat brain by in situ hybridization histochemistry. Proc Natl Acad Sci USA 86: 8560–8564

Mengod G, Nguyen H, Le H, Waeber C, Lübbert H, Palacios JM (1990) The distribution and cellular localization of the serotonin 1 C receptor mRNA in the rodent brain examined by in situ hybridization histochemistry. Comparison with receptor binding distribution. Neuroscience 35: 577–591

Mesulam MM, Mufson EJ, Levey AI, Wainer BH (1984) Atlas of cholinergic neurons in the forebrain and upper brainstem of the macaque based on monoclonal choline acetyltransferase immunohistochemistry and acetylcholinesterase histochemistry. Neuroscience 12: 669–686

Mitsuma T, Nogimori T, Adachi K, Mukoyama M, Ando K (1984) Concentrations of immunoreactive thyrotropin-releasing hormone in spinal cord of patients with amyotrophic lateral sclerosis. Am J Med Sci 287: 34–36

Morley BJ, Kemp GE, Salvaterra P (1979) Alpha-bungarotoxin binding sites in the CNS. Life Sci 24: 859–872

Myers RH, Martin JB (1982) Huntington's disease. Semin Neurol 2: 365–372

Nemeroff CB, Youngblood WW, Manberg PJ, Prange AJ, Kizer JS (1983) Regional brain concentration of neuropeptides in Huntington's chorea and schizophrenia. Science 221: 972–975

Nghiêm HO, Cartaud J, Dubreuil C, Kordeli C, Buttin G, Changeux JP (1983) Production and characterization of a monoclonal antibody directed against the 43 000-dalton V1 polypeptide from torpedo marmorata electric organ. Proc Natl Acad Sci USA 80: 6403–6407

Oppenheimer DR (1984) Diseases of the basal ganglia, cerebellum and motor neurons. In: Adam JH, Corsellis JAN, Duchen LW (eds) Greenfields neuropathology. Edward Arnold, London, pp 700–747

Palacios JM (1982) Autoradiographic localization of muscarinic cholinergic receptors in the hippocampus of patients with senile dementia. Brain Res 23: 173–175

Palacios JM, Kuhar MJ (1981) Neurotensin receptors are located on dopamine-containing neurons in rat midbrain. Nature 294: 273–285

Palacios JM, Niehoff DL, Kuhar MJ (1981) Receptor autoradiography with tritium-sensitive film: potential for computerized densitometry. Neurosci Lett 24: 111–116

Palacios JM, Probst A, Cortés R (1986) Mapping receptors in the human brain. Trends Neurosci 9: 284–289

Palacios JM, Cortés R, Probst A, Dietl M (1987) Autoradiographic mapping of neurotransmitter receptors in normal and pathological human brain. In: Tucek S (ed) Synaptic transmitters and receptors. Academia, Prague, pp 71–79

Palacios JM, Camps M, Cortés R., Probst A (1988 a) Mapping dopamine receptors in the human brain. J Neural Transm 27:227–235

Palacios JM, Cortés R, Dietl M, Probst A (1988 b) Receptors in human brain disease: a use for receptor autoradiography in neuropathology. J Recept Res 8:509–520

Palacios JM, Mengod G, Savasta M (1989 a) Mapping of receptor and transmitters expression in the human brain. In: Ottoson D (ed) Visualization of brain function. Macmillan, London

Palacios JM, Mengod G, Vilaró MT, Wiederhold KH, Boddeke H, Alvarez FJ, Chinaglia G, Probst A (1989 b) Cholinergic receptors in the rat and human brain. Microscopic visualization. Prog Brain Res

Palmer AM, Wilcock GK, Esiri MM, Francis PT, Bowen DM (1987) Monoamine innervation of the frontal and temporal lobes in Alzheimer's disease. Brain Res 401:231–238

Patten BM, Croft S (1984) Spinal cord substance P in amyotrophic lateral sclerosis. In: Clifford Rose F (ed) Research progress in motor neuron disease. Pitman, Bath, pp 283–289

Penney JB, Young AB (1982) Quantitative autoradiography of neurotransmitter receptors in Huntington's disease. Neurology 32:1391–1395

Penney JB, Young AB (1988) Huntington's disease. In Jankovic F, Tolosa E (eds) Parkinson's disease and movement disorders. Urban and Schwarzenberg, Baltimore, pp 167–178

Penney JB, Young AB, Walker FO, Shoulson I (1984) Quantitative autoradiography of opiate receptors in Huntington's disease. Neurology 34 [Suppl 1]:153

Peralta EG, Ashkenazi A, Winslow JW, Smith DH, Ramachandran J, Capon DJ (1987) Distinct primary structures, ligand-binding properties and tissue-specific expression of four human muscarinic acetylcholine receptors. EMBO J 6:3923–3929

Perret JL, Jouvet M (1980) Etude du sommeil dans la paralysie supranucléaire progressive. Electroencephalogr Clin Neurophysiol 49:323–329

Perry EK, Tomlinson BE, Blessed G, Bergmann K, Gibson PH, Perry RH (1978) Correlation of cholinergic abnormalities with senile plaques and mental testscores in senile dementia. Br Med J II:1427–1429

Perry EK, Atack JR, Perry RH et al. (1984) Intralaminar neurochemical distributions in human midtemporal cortex: comparison between Alzheimer's disease and the normal. J Neurochem 42:1402–1410

Perry TL, Kish SJ, Hansen S, Currier RD (1981) Neurotransmitter amino acids in dominantly inherited cerebellar disorders. Neurology 31:237–242

Perry TL, Young VW, Bergeron C, Hansen S, Jones K (1987) Amino acids, glutathione and glutathione transferase activity in the brains of patients with Alzheimer's disease. Ann Neurol 21:331–336

Petito CK, Hart MN, Porro RS, Earle KM (1973) Ultrastructural studies of olivopontocerebellar atrophy. J Neuropathol Exp Neurol 32:503–552

Pfeiffer F, Simler R, Greeningloh G, Betz H (1984) Monoclonal antibodies and peptide mapping reveal structural similarities between the subunits of the glycine receptors of rat spinal cord. Proc Natl Acad Sci USA 81:7224–7227

Pierot L, Desnas C, Blair J et al. (1988) D1 and D2 type dopamine receptors in patients with Parkinson's disease and progressive supranuclear palsy. J Neurol Sci 86:291–306

Pimoule C, Schoemaker H, Reynolds GP, Langer SZ (1985) ^{3}H-SCH 23 390 labeled D1 dopamine receptors are unchanged in schizophrenia and Parkinson's disease. Eur J Pharmacol 114:235–237

Pollock NJ, Mirra SS, Binder LI, Hansen LA, Wood JG (1986) Filamentous aggregates in Pick's disease, progressive supranuclear palsy and Alzheimer's disease share antigenic determinants with microtubule-associated protein, tau. Lancet II:1211

Pritchett DB, Bach AWJ, Wozny M, Talebo O, Di Toso R, Shih JC, Seeburg PH (1988) Structure and functional expression of cloned rat serotonin 5HT-2 receptor. EMBO J 7:4135–4140

Pritchett DB, Sontheimer H, Shivers B, Ymer S, Kettenmann H, Schofield PR, Seeburg PH (1989) Importance of a novel GABA A receptor subunit for benzodiazepine pharmacology. Nature 38:582–585

Probst A, Dufresne JJ (1975) Paralysie supranucléaire progressive (ou Dystonie oculo-facio-cervicale). Arch Suisses Neurol Neurochir Psychiatr 116:107–134

Probst A, Cortés R, Palacios JM (1986) The distribution of glycine receptors in the human brain. A light microscopic autoradiographic study using ^{3}H-strychnine. Neuroscience 17:11–35

Probst A, Langui D, Lautenschlager C, Ulrich J, Brion JP, Anderton HB (1988a) Progressive supranuclear palsy: extensive neuropil threads in addition to neurofibrillary tangles. Acta Neuropathol 77:61–68

Probst A, Cortés R, Ulrich J, Palacios JM (1988b) Differential modification of muscarinic cholinergic receptors in the hippocampus of patients with Alzheimer's disease: an autoradiographic study. Brain Res 450:190–201

Quirion R, Chieueh CC, Everist HD, Pert A (1985) Comparative localization of neurotensin receptors on nigrostriatal and mesolimbic dopaminergic terminals. Brain Res 327:385–389

Raftery MA, Hunkapiller MW, Strader CD, Hood LE (1980) Acetylcholine receptor: complex of homologous subunits. Science 208:1454–1457

Raisman R, Cash R, Ruberg M, Javoy-Agid F, Agid Y (1985) Binding of ^{3}H-SCH 23 390 to D1 receptors in the putamen of control and parkinsonian subjects. Eur J Pharmacol 113:467–468

Regan JW, Kobilka TS, Yang Feng TL, Caron MG, Lefkowitz RJ, Kobilka BK (1988) Cloning and expression of a human kidney cDNA for an alpha-2 adrenergic receptor subtype. Proc Natl Acad Sci USA 85:6301–6305

Richards G, Möhler H, Schoch P, Häring P, Takacs B, Stahli C (1984) The visualization of neuronal benzodiazepine receptors in the brain by autoradiography and immunohistochemistry. J Recept Res 4:657–669

Richards G, Schoch P, Möhler H, Haefely W (1986a) Benzodiazepine receptors resolved. Experientia 42:121–126

Richards G, Möhler H, Haefely W (1986b) Mapping benzodiazepine receptors in the CNS by radiohistochemistry and immunohistochemistry. In: Panula P, Päivärinta Svinita S (eds) Neurohistochemistry: modern methods and applications. Alan R. Liss, New York, pp 629–677

Rinne JO, Rinne JK, Laakso K, Lönnberg P, Rinne UK (1985) Dopamine D1 receptors in the parkinsonian brain. Brain Res 359:306–310

Rossor M, Iversen LL (1986) Non-cholinergic neurotransmitter abnormalities in Alzheimer's disease. Br Med Bull 42:70–74

Rotter A, Birdsall NJM, Burgen ASV, Field PM, Smolen A, Raisman G (1979) Muscarinic receptors in the central nervous system of the rat. IV. A comparison of the effects of axotomy and deafferentation on the binding of ^{3}H-propybenzilylcholine mustard and associated synaptic changes in the hypoglossal and pontine nuclei. Brain Res Rev 1:207–224

Rowell PP, Winkler DL (1984) Nicotinic stimulation of ^{3}H-acetylcholine release from mouse cerebral synaptosomes. J Neurochem 43:1593–1598

Ruberg M, Ploska A, Javoy-Agid F, Agid Y (1982) Muscarinic binding and choline acetyltransferase activity in Parkinsonian subjects with reference to dementia. Brain Res 232:129–139

Ruberg M, Javoy-Agid F, Hirsch E et al. (1985) Dopaminergic and cholinergic lesions in progressive supranuclear palsy. Ann Neurol 18:523–529

Sasaki H, Muramato O, Kanazawa I, Arai H, Kosaka K, Tizuka R (1986) Regional distribution of amino acid transmitters in postmortem brains of presenile and senile dementia of Alzheimer type. Ann Neurol 19:263–269

Savasta M, Dubois A, Feuerstein C, Manier M, Scatton B (1987) Denervation supersensitivity of striatal D2-dopamine receptors is restricted to the ventro- and dorsolateral regions of the striatum. Neurosci Lett 74:180–186

Savasta M, Ruberte E, Palacios JM, Mengod G (1989) The colocalization of cholecystokinin and tyrosine hydroxylase mRNAs in mesencephalic dopaminergic neurons in the rat brain examined by in situ hybridization. Neuroscience 29:363–369

Scatton B, Javoy-Agid F, Rouquier L, Dubois B, Agid Y (1983) Reduction of cortical dopamine, noradrenaline, serotonin and their metabolites in Parkinson's disease. Brain Res 275:321–328

Scatton B, Dubois A, Javoy-Agid F, Camus A (1984) Autoradiographic localization of muscarinic cholinergic receptors at various levels of the human spinal cord. Neurosci Lett 49:239–245

Schoch P, Möhler H (1983) Purified benzodiazepine receptor retains modulation by GABA. Eur J Pharmacol 95:323–324

Schoenen J, Reznik M, Delwaide PJ, Vanderhaeghen JJ (1985) Etude immunocytochimique de la distribution spinale de la substance P, des encéphalines, de la cholécystokinine et de la sérotonine dans la sclérose latérale amyotrophique. CR Soc Biol (Paris) 179:528–534

Schofield PR, Darlison MG, Fujita N et al. (1987) Sequence and functional expression of the GABA A receptor shows a ligand-gated receptor super family. Nature 328:221–227

Schofield PR, Pritchett DB, Sontheimer H, Kettenmann H, Sehburg PH (1989) Sequence and expression of human GABA A receptor alpha 1 and beta 1 subunits. FEBS Lett 244:361–364

Schwarz JH (1985) Molecular aspects of postsynaptic receptors. In: Kandel EM, Schwarz JH (eds) Principles of neural science, 2nd. edn Elsevier, Amsterdam, pp 159–168

Sealock P, Barnaby EW, Froehner SC (1984) Ultrastructural localization of the M_r 43 000 protein and the acetylcholine receptor in torpedo postsynaptic membranes using monoclonal antibodies. J Cell Biol 98:2239–2244

Sequier JM, Richards JG, Malherbe P, Price GW, Mathews S, Möhler H (1988) Mapping of brain areas containing RNA homologous to cDNAs encoding the alpha and beta subunits of the rat GABA A alpha-aminobutyrate receptor. Proc Natl Acad Sci USA 85:7815–7819

Shepherd GM (1988) Neurobiology, 2nd edn. Oxford University Press, Oxford

Sigel E, Stephenson A, Mamalaki C, Barnard EA (1983) A gamma-amino-butyric acid/benzodiazepine receptor complex of bovine cerebral cortex: purification and partial characterization. J Biol Chem 258:6965–6971

Spindel ER, Wurtman RJ, Bird ED (1981) Increased TRH content in the basal ganglia in Huntington's disease. N Engl J Med 303:1235–1236

Steele JC (1972) Progressive supranuclear palsy. Brain 95:693–704

Steele JC, Richardson JC, Olszewski J (1964) Progressive supranuclear palsy: a heterogeneous degeneration involving the brain stem, basal ganglia and cerebellum, with vertical gaze and pseudobulbar palsy, nuchal dystonia and dementia. Arch Neurol 10:333–359

Storm-Mathisen J (1976) Distribution of the components of the GABA system in neuronal tissue: cerebellum and hippocampus effects of axotomy. In: Robert E, Chase T, Tower DB (eds) GABA in nervous system function. Raven, New York, pp 149–168

Swanson LW, Simmons DM, Whiting PJ, Lindstrom J (1987) Immunohistochemical localization of neuronal nicotinic receptors in the rodent central nervous system. J Neurosci 7:3334–3342

Tanzi RE, St. George-Hyslop PH, Gusella F (1989) Molecular genetic approaches to Alzheimer's disease. Trends Neurosci 12:152–157

Tenovuo O, Rinne UK, Viljanen MK (1984) Substance P immunoreactivity in the post mortem Parkinsonian brain. Brain Res 303:113–116

Troncoso JC, Cork L, Whitehouse PJ, Kuhar MJ, Price DL (1984) Canine inherited ataxia: neurotransmitter receptors in the cerebellum. Ann Neurol 16:135

Tsiotos P, Plaitakis A, Mitsacos A, Voukelatou G, Michalodimitrakis M, Kouvelas ED (1989) Glutamate binding sites of normal and atrophic human cerebellum. Brain Res 481:87–96

Uhl GR, Whitehouse PJ, Price DL, Tourtellotte WW, Kuhar MJ (1984) Parkinson's disease: depletion of substantia nigra neurotensin receptors. Brain Res 308:186–190

Uhl GR, Hedreen JC, Price DL (1985) Parkinson's disease: loss of neurons from the ventral tegmental area contralateral to therapeutic surgical lesions. Neurology 35:1215–1218

Uhl GR, Hackney GO, Torchia M et al. (1986) Parkinson's disease: nigral receptor changes support peptidergic role in nigrostriatal modulation. Ann Neurol 20:194–203

Unnerstall JR, Niehoff DL, Kuhar MJ, Palacios JM (1982) Quantitative receptor autoradiography using ^3H ultrafilm: application to multiple benzodiazepine receptors. J Neurosci Methods 6:59–73

Vilaró MT, Palacios JM, Mengod G (1990a) Localization of m5 muscarinic receptor mRNA in rat brain examined by in situ hybridization histochemistry. Neurosci. Lett. 114:154–159

Vilaró MT, Wiederhold K-H, Palacios JM, Mengod G (1990b) Muscarinic cholinergic receptors in the rat caudate-putamen and olfactory tubercle belong predominantly to the m4 class: in situ and receptor autoradiography evidence. Neuroscience, in press.

Vonsattel JP, Myers RH, Stevens TJ, Ferrante RJ, Bird ED, Richardson EP (1985) Neuropathological classification of Huntington's disease. J Neuropathol Exp Neurol 44:559–577

Wada K, Ballivet M, Boulter J et al. (1988) Function and expression of a new pharmacological subtype of brain nicotonic acetylcholine receptor. Science 240:330–334

Waeber C, Palacios JM (1989) Serotonin-1 receptor binding sites in the human basal ganglia are decreased in Huntington's chorea but not in Parkinson's disease: a quantitative in vitro autoradiographic study. Neuroscience

Walker FO, Young AB, Penney JB, Dovorini-Zis K, Shoulson I (1984) Benzodiazepine and GABA receptors in early Huntington's disease. Neurology 34:1237–1240

Wastek GJ, Stern LZ, Johnson PC, Yamamura HI (1976) Huntington's disease: regional alteration in muscarinic cholinergic receptor binding in human brain. Life Sci 19:1033–1040

Wenthold RJ, Parakkal MH, Oberdorfer MD, Altschuler RA (1988) Glycine receptor immunoreactivity in the ventral cochlear nucleus of the guinea pig. J Comp Neurol 276:423–435

Whitehouse PJ (1985) Receptor autoradiography: applications in neuropathology. Trends Neurosci 8:434–437

Whitehouse PJ, Wamsley JK, Zarbin MA, Price DL, Tourtellotte WW, Kuhar MJ (1983) Amyotrophic lateral sclerosis: alterations in neurotransmitter receptors. Ann Neurol 14:8–16

Whitehouse PJ, Trifiletti RR, Jones BE, Jolstein S, Price DL, Snyder SH, Kuhar MJ (1985) Neurotransmitter receptor alterations in Huntington's disease: autoradiographic and homogenate studies with special reference to benzodiazepine receptor complex. Ann Neurol 18:202–210

Whitehouse PJ, Martino AM, Marcus K, Price DL, Kellar KJ (1986a) Reductions in ^{3}H-acetylcholine nicotinic binding sites in cortex in Alzheimer's disease: an autoradiographic study. Neurology 36:270

Whitehouse PJ, Muramoto O, Troncoso JC, Kanazawa I (1986b) Neurotransmitter receptors in olivopontocerebellar atrophy: an autoradiographic study. Neurology 36:193–197

Whiting P, Lindstrom J (1987) Purification and characterization of a nicotinic acetylcholine receptor from rat brain. Proc Natl Acad Sci USA 84:595–599

Young AB, Penney JB (1984) Neurochemical anatomy of movement disorders. Neurol Clin 2:417–433

Young WS III, Kuhar MJ (1979) Autoradiographic localization of benzodiazepine receptors in brains of humans and animals. Nature 280:393–395

Young WS III, Bonner TI, Brann MR (1986) Mesencephalic dopamine neurons regulate the expression of neuropeptide mRNAs in the rat forebrain. Proc Natl Acad Sci USA 83:9827–9831

Zarbin MA, Wamsley JK, Kuhar MJ (1981) Glycine receptor: light microscopic autoradiographic localization with strychnine. J Neurosci 1:532–547

Zezula J, Cortés R, Probst A, Palacios JM (1988) Benzodiazepine receptor sites in the human brain: autoradiographic mapping. Neuroscience 25:771–795

Zweig RM, Whitehouse PJ, Casanova MF, Walker LC, Jankel WR, Price DL (1987) Loss of pedunculopontine neurons in progressive supranuclear palsy. Ann Neurol 22:18–25

Lectin Receptors *

M. VIERBUCHEN

---

* Dedicated to Prof. Dr. med. R. Fischer on the occasion of his 60th birthday.

1 Introduction

The term lectin (from Latin = to pick out, to choose) was coined by BOYD and SHAPELEIGH (1954) to denote a heterogeneous class of (glyco)proteins, primarily of plant origin, possessing different biological activities such as hemagglutination (phytohemagglutinins), blastic transformation of lymphocytes (mitogens), and toxic effects on vertebrate cells (toxins). These proteins vary markedly in their physicochemical properties; for instance while the potato lectin has a carbohydrate content as high as 50%, other lectins like concanavalin A or the peanut agglutinin, are devoid of carbohydrates. The molecular weights of lectins vary from 11 000 for the blood group B specific lectin from *Streptomyces* species to 335 000 for the lectin from the horseshoe crab. Despite their wide range of physicochemical properties and biological activities, lectins share a common feature which is responsible for their various biological, biochemical and immunochemical activities: Lectins bind with high affinity and specificity to mono- and oligosaccharides of complex carbohydrates in solutions, cell surfaces, subcellular organelles, and tissue sections.

Lectins have been operationally defined as carbohydrate-binding proteins of nonimmune origin that agglutinate cells or precipitate polysaccharides and glyocproteins (GOLDSTEIN et al. 1980, SHARON and LIS 1972). Such a definition implies that lectins contain at least two carbohydrate-binding sites (PORETZ et al. 1974). However, this definition has been revised, and the definition of lectins as "carbohydrate binding proteins other than an enzyme or an antibody" is now considered more appropriate (BARONDES 1988). Lectins have been used for the elucidation of the determinants of blood group antigens. Furthermore, the discovery that some lectins have mitogenic properties was the starting point for the modern concept of the role of lymphocytes in our immune system.

The first lectins were isolated from plants and invertebrates, and it was originally thought that lectins occur only in these species. For a long time lectins were isolated from the different sources, manipulated, and used as tools for the characterization of carbohydrates. However, little was known of their physiological function. The subsequent discovery of many lectins and agglutinins in bacteria, in viruses, and especially in vertebrates led to increasing interest in the significance and physiological function of lectins in general, and especially endogenous carbohydrate-binding proteins in vertebrates. This class of proteins has become the focus of intense interest, since lectins are involved in such important biological activities as cell adhesion and cell–cell interaction, as well as in the organotropism of metastasis and infectious diseases. In this article the component carrying the carbohydrate-binding site will be called a lectin (from plants, invertebrates, bacteria, viruses, and vertebrates). The component to which the lectin binds will be called a lectin receptor, although in many cases it is not known whether this binding triggers further events in the target or whether the lectin binding itself is the only result.

However, endogenous lectins can also act as receptor molecules mediating the transport of glycoconjugates. In both processes, lectins play an essential role: in the first case carbohydrates act as receptors for lectins (glycoconjugate binding sites for plant or bacterial lectins), while in the second case mainly endogenous (vertebrate) lectins act as receptors for carbohydrates (asialoglycoprotein- or mannose-6-phosphate receptors). This discovery has implications beyond the development of "new tools," for example in carbohydrate histochemistry.

In brief, the aim of this review is to discuss some areas of research on lectins as histochemical markers as well as their physiological functions and their involvement in pathological processes. To denote this scientific field the term *lectinology* should be used.

2 Structure of Glycoprotein Glycans

In biological materials the sugar units of complex carbohydrates represent the binding sites (ligands) for the lectins (lectin receptors). The complex carbohydrates are a family of macromolecules (Fig. 1) including glycolipids, proteoglycans, bacterial polysaccharides, and glycoproteins. Glycoproteins and glycolipids often contain identical carbohydrate sequences. Both compounds result from the association through covalent linkages of certain carbohydrates called glycans.

Carbohydrate chains on glycoproteins are commonly divided into different types depending on the linkage between the protein backbone and the carbohydrate moiety. This leads to the definition of two classes of glycoprotein, N-glycosylproteins and O-glycosylproteins (Table 1).

Glycan structures may be divided into families in which the structures are very similar and contain common oligosaccharide sequences, whether they originate from vertebrates, plants, microorganisms, or viruses.

The carbohydrate moieties on N- and O-glycosylglycans derive from the substitution of oligosaccharide structures common to all glycans of a given

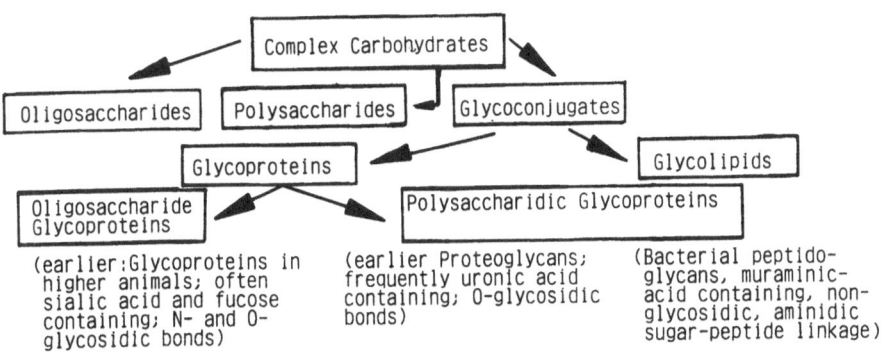

Fig. 1. Relationship and nomenclature of complex carbohydrates (according to SCHAUER et al. 1988)

Table 1. Protein-carbohydrate linkages in glycoproteins

Type of glycoprotein	Amino acids and sugars engaged in the linkages	
	Amino acid	Sugar
N-Glycosylproteins		
Membrane type	Asn	GlcNAc
Serumglycoprotein type	Asn	GlcNac
O-Glycosylproteins		
Mucin type	Thr/Ser	GalNAc
Proteoglycan type	Ser	Xyl
Collagen type	OH-Lys	Gal
Extensin type	OH-Pro	Ara

Asn, asparagine; *Thr*, threonine; *Ser*, serine; *OH-Lys*, hydroxyllysine; *OH-Pro*, hydroxylproline; *GlcNAc*, N-Acetylglucosamine; *GalNAc*, N-Acetylgalactosamine; *Gal*, galactose; *Xyl*, xylose; *Ara*, arabinose.

family. These nonspecific and invariant structures are conjugated to the peptide chain and constitute the most internal part of glycans, namely the inner core. Oligosaccharide units attached to N-glycosylproteins as a rule contain a common inner core consisting of a branched pentasaccharide (Fig. 2). The peripheral mannose residues of the oligosaccharide of Fig. 2 are always substituted, and in accordance with the substitution, three types N-glycosylprotein are derived from the core structure. Substitution may be due to the addition of mannosyl residues in the form of long and short chains that are usually branched (Fig. 3). Glycans of this type (high mannose type or oligomannosidic type) are already found during early stages in evolution and are therefore widely distributed in plants and animals. The structure is common to many glycoproteins with different functions, and thus does not represent any structural specificity.

Concerning the biosynthesis of N-linked glycoproteins (MONTREUIL 1980, 1982; LENNARTZ 1980, SHARON and LIS 1981, 1982; BERGER et al. 1982; HUGHES 1983), they could be regarded as precursor substances in the formation of complex-type glycans. However, this cannot be considered their only role. There is an increasing body of evidence to show that oligomannosaccharide structures act as recognition signals for infectious agents (microbial lectins) or lymphokines. In contrast to the oligomannosidic type of glycans, the

```
Man α (1-3)
   |
   ManB(1-4)GlcNAcB(1-4)GlcNAcB-1-Asn
   |
Man α (1-6)
```

Fig. 2. Structure of the inner core pentasaccharide of glycans in N-glycosylproteins

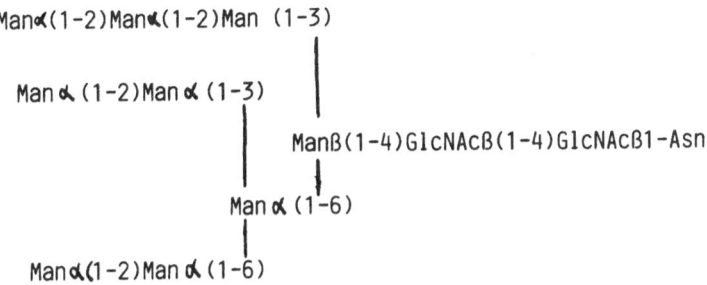

Manα(1-2)Manα(1-2)Man (1-3)

    Man α (1-2)Man α (1-3)

            ManΒ(1-4)GlcNAcΒ(1-4)GlcNAcΒ1-Asn

        Man α (1-6)

  Manα(1-2)Man α (1-6)

Fig. 3. Glycan structure of the oligomannosidic (high mannose) type present in human immunoglobulin D (MONTREUIL 1980, 1982)

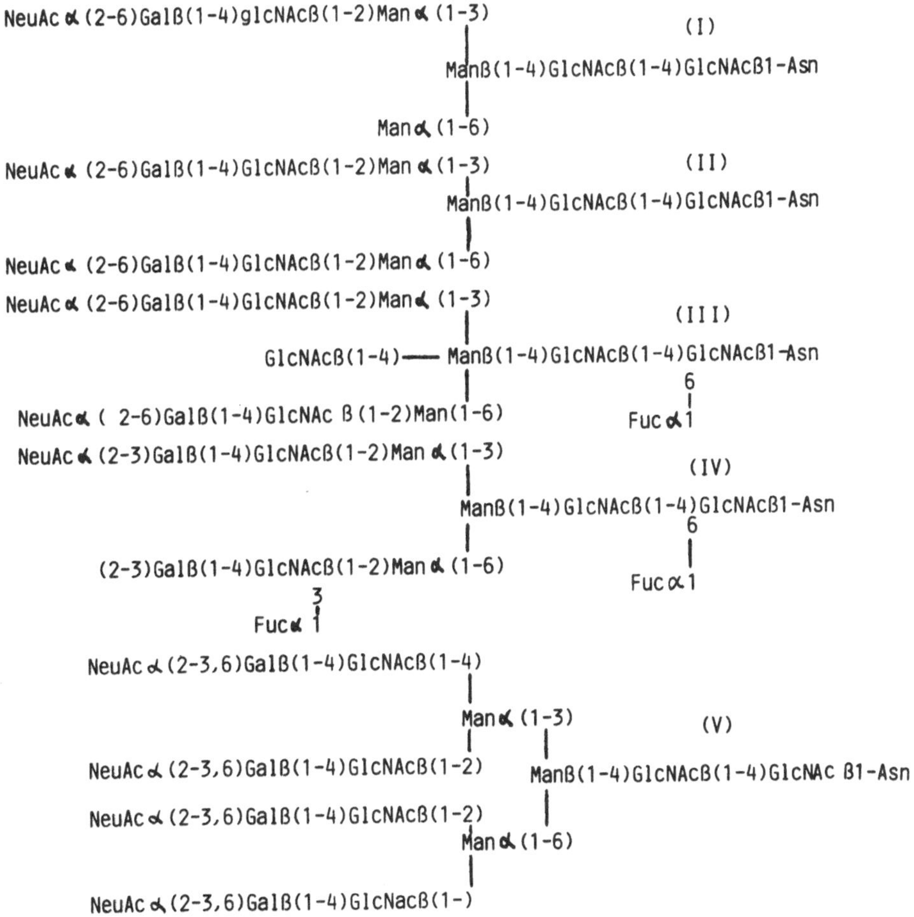

NeuAc α (2-6)GalΒ(1-4)glcNAcΒ(1-2)Man α (1-3)   (I)

                     ManΒ(1-4)GlcNAcΒ(1-4)GlcNAcΒ1-Asn

                 Manα (1-6)

NeuAcα (2-6)GalΒ(1-4)GlcNAcΒ(1-2)Man α (1-3)   (II)

                     ManΒ(1-4)GlcNAcΒ(1-4)GlcNAcΒ1-Asn

NeuAc α (2-6)GalΒ(1-4)GlcNAcΒ(1-2)Manα (1-6)

NeuAc α (2-6)GalΒ(1-4)GlcNAcΒ(1-2)Manα (1-3)   (III)

              GlcNAcΒ(1-4)—— ManΒ(1-4)GlcNAcΒ(1-4)GlcNAcΒ1-Asn

                                          6

NeuAcα (2-6)GalΒ(1-4)GlcNAc Β (1-2)Man(1-6)   Fuc α1

NeuAc α (2-3)GalΒ(1-4)GlcNAcΒ(1-2)Man α(1-3)   (IV)

                     ManΒ(1-4)GlcNAcΒ(1-4)GlcNAcΒ1-Asn

                                        6

   (2-3)GalΒ(1-4)GlcNAcΒ(1-2)Man α (1-6)

              3   Fuc α 1

           Fuc α 1

NeuAc α (2-3,6)GalΒ(1-4)GlcNAcΒ(1-4)

                       Man α (1-3)   (V)

NeuAc α (2-3,6)GalΒ(1-4)GlcNAcΒ(1-2)   ManΒ(1-4)GlcNAcΒ(1-4)GlcNAc Β1-Asn

NeuAc α (2-3,6)GalΒ(1-4)GlcNAcΒ(1-2)

                       Man α (1-6)

NeuAc α (2-3,6)GalΒ(1-4)GlcNacΒ(1-)

Fig. 4. Glycan structures of the complex type: I, monoantennary glycan of secretory component of human milk; II, biantennary glycan of human serum transferrin; III, monofucosylated and "bisected" biantennary glycan of human IgG; IV, difucosylated biantennary glycan of human lactotransferrin; V, tetraantennary glycan of humanα -1-acid glycoprotein (MONTREUIL 1980, 1982)

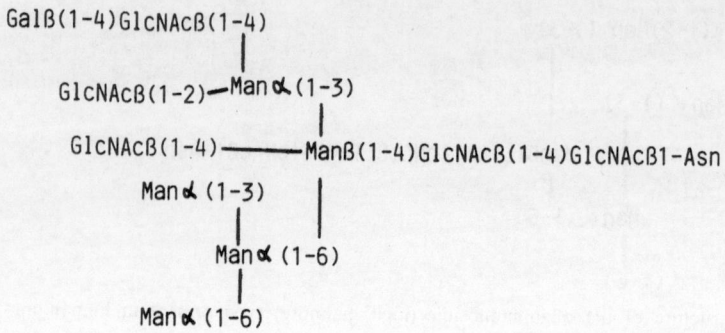

Fig. 5. Glycan structure of the hybrid type present in hen ovalbumin (MONTREUIL 1980, 1982)

carbohydrate moieties found in the complex-type oligosaccharides show a wide structural diversity which favors the hypothesis that these saccharide sequences act as recognition signals. This type of glycans contains two, three, or more branches on a pentasaccharide core (Fig. 4) and N-acetyllactosamine residues in the outer part of the oligo-saccharides.

In hybrid-type glycoproteins the glycan contains more than three mannose residues, presenting structures of the oligomannosidic and the N-acetyllac-tosaminic type (Fig. 5). These structures are considered as intermediates in the biosynthesis of glycans of the complex type.

In contrast to N-glycosidically linked carbohydrate sequences, few generalizations can be made about O-glycosidically linked glycans. The sugar residues vary from a single sugar residue as found in collagen, to branched oligosaccharides (called megalosaccharides) of up to 16–18 monosaccharide residues, as proposed for blood group substances. Perhaps the only common feature is the presence of the disaccharide β-D-Galβ(1-3)GalNAc linked O-glycosidically to serine or threonine (Fig. 6) of the protein backbone. The structures of the oligosaccharides of glycoproteins that have become known during recent years show several features. Despite an astronomical number of oligosaccharides that can be formed from monosaccharide constituents, there appear to be certain limitations due to very restrictive laws that reduce the number of structures actually formed by living organisms. The basis of this rests in the specificity of glycosyltransferases and in the conservative evolution of these enzymes. In fact the possibilities of substitution of a given monosaccharide are limited to one, two, or three well-defined monosaccharides, generally conjugated by a unique type of glycosidic linkage.

Substitution of the core structure is as follows:

1. Monosaccharide residues of the inner core Galβ(1-3)GalNAc- of O-glyco-sylproteins:
    a) GalNAC residue by siliac acid residue or by GlcNAc in the C-6 position
    b) Gal residue by sialic acid residue or by GalNAc residue in the C-3 position, or by α-L-Fuc in the C-2 position

2. Monosaccharides of the inner core of N-glycosylproteins (Fig. 7):
   a) GlcNAc-1 residue by α-L-Fuc in the C-6 position
   b) GlcNAc-5 by α-L-Fuc in the C 3 position, preventing the presence of sialic acid on Gal-6/Gal-6' due to the fact that glycosylation by sialyl- and fucosyl-transferases is exclusive
   c) Gal6/6' by NeuAc in the C-3 and C-6 positions, by GlcNAc in the C-3 position, and by α-L-Fuc in the C-2 position

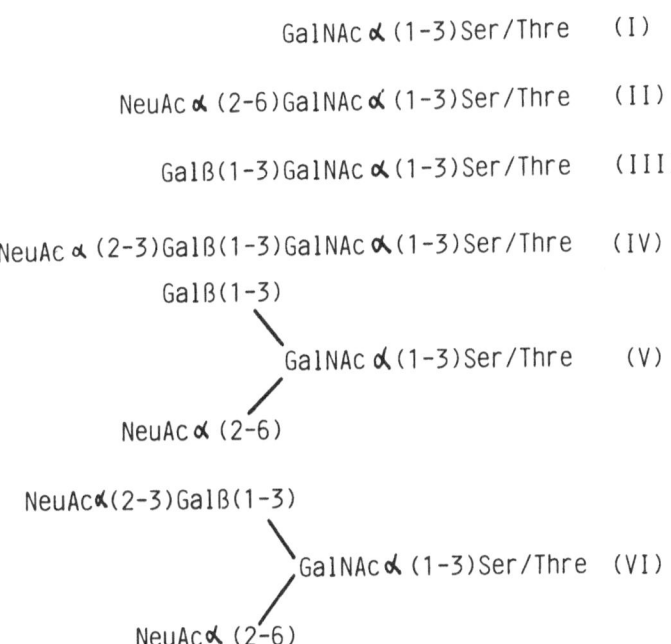

$$\text{GalNAc } \alpha \text{ (1-3)Ser/Thre} \quad (I)$$

$$\text{NeuAc } \alpha \text{ (2-6)GalNAc } \alpha \text{ (1-3)Ser/Thre} \quad (II)$$

$$\text{Gal}\beta\text{(1-3)GalNAc } \alpha \text{ (1-3)Ser/Thre} \quad (III)$$

$$\text{NeuAc } \alpha \text{ (2-3)Gal}\beta\text{(1-3)GalNAc } \alpha\text{(1-3)Ser/Thre} \quad (IV)$$

$$\text{Gal}\beta\text{(1-3)}$$
$$\searrow$$
$$\text{GalNAc } \alpha \text{ (1-3)Ser/Thre} \quad (V)$$
$$\nearrow$$
$$\text{NeuAc } \alpha \text{ (2-6)}$$

$$\text{NeuAc}\alpha\text{(2-3)Gal}\beta\text{(1-3)}$$
$$\searrow$$
$$\text{GalNAc } \alpha \text{ (1-3)Ser/Thre} \quad (VI)$$
$$\nearrow$$
$$\text{NeuAc}\alpha \text{ (2-6)}$$

Fig. 6. Glycan structures in O-glycosidically linked oligosaccharides: I, Tn antigen of erythrocytes; II, glycoprotein human erythrocyte; III, human serum immunoglobulin A1, TF-reactive erythrocytes; IV, human glycophorin N blood group substance; V, submaxillary mucins; VI, human glycophorin and blood group substance M

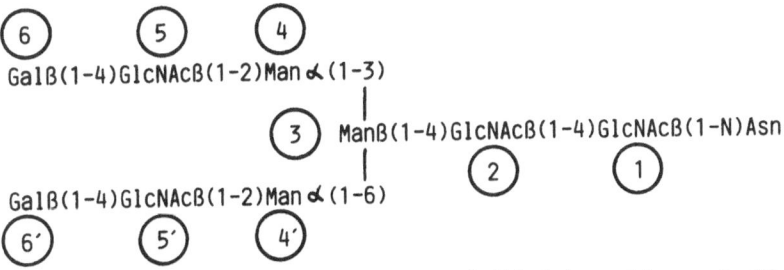

⑥   ⑤   ④
Galβ(1-4)GlcNAcβ(1-2)Man α(1-3)
                          |
          ③ Manβ(1-4)GlcNAcβ(1-4)GlcNAcβ(1-N)Asn
                          |
                          ②   ①
Galβ(1-4)GlcNAcβ(1-2)Man α(1-6)
⑥'   ⑤'   ④'

Fig. 7. Schematic representation of an N-glycosidically linked glycan of the complex (N-acetyllactosamine) type with some rules relative to the substitution of the basic biantennary structure

Given these rules, the particular importance of two monosaccharides, namely sialic acid and L-fucose, is evident. Their systematic external position designates them as important structures in biological systems. Two aspects of the function of sialic acid can be distinguished: the sugar may act as a receptor for soluble proteins (hormones, lectins, antibodies, toxins, or microbial lectins); on the other hand the sugar may mask other molecular or cellular recognition sites, preventing or reducing access to these structures by other receptor or effector systems (SCHAUER 1988). The structural variety of the glycan part ist increased by a certain freedom of motion or rotation which allows different spatial conformational changes of the oligosaccharide chains (DOUY and GAL-LOT 1980, GERVAIS and GALLOT 1982, LI et al. 1983, CARRER and BRISSON 1984). In the different spatial conformations the sugars of the core region can be either masked or exposed. Interaction between the oligosaccharide chain and the protein backbone is important for the spatial conformation. Sialic acid residues which can form noncovalent linkages with basic amino acids of the protein backbone are thought to be of special significance for the oligosaccharide–protein interaction.

This dynamic conception of the structure of glycans represents a prerequisite for the understanding of how oligosaccharides exert biological effects (Fig. 8).

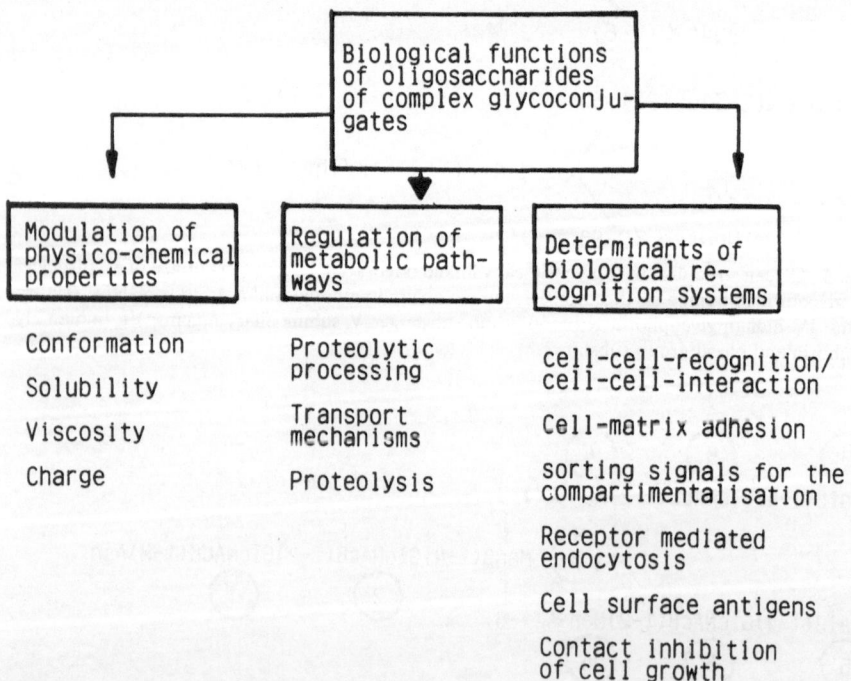

Fig. 8. Schematic representation of biological functions of complex glycoconjugates (carbohydrates)

3 Nature, Detection, and Action of Endogenous (Vertebrate) Lectins

Beside performing unspecific functions (Fig. 8), the oligosaccharides of complex carbohydrates represent – similar to the antigens in antigen–antibody reactions or the substrate in enzyme–substrate interaction – specific signals which are recognized and bound by endogenous carbohydrate-binding proteins in vertebrates, or endogenous (vertebrate) lectins. Whereas the structures of glycoconjugates have been widely documented, the characterization of endogenous lectins has only recently been initiated. Stimulated by the concept that vertebrate lectins are of special importance as components of multifunctional biological systems involved in cell recognition and related subjects, an increasing number of lectin activities have been isolated from different vertebrate tissues. On this basis, lectins can be divided into two groups according to their occurrence, function, and physico-chemical properties (ASHWELL and HARFORD 1982, BARONDES 1981):

1. Soluble lectins
2. Membrane lectins

Soluble lectins may play a role during secretion activities and in the organization of the extracellular matrix, while membrane lectins may be involved in such important biological activities as cell–cell interaction, cell–cell recognition, and cell adhesion.

As summarized in Table 2, vertebrate lectins can be detected by different approaches using biological, biochemical, immunochemical, and histochemical methods (STOCKERT et al. 1976, STAHL et al. 1976, 1978; ACHORD et al. 1977, 1978; PRIEELS et al. 1978, BARONDES 1978, KIEDA et al. 1978, KAWASAKI and ASHWELL 1977, GRABEL et al. 1979, GABIUS et al. 1987, VIERBUCHEN et al. 1988a, 1989a).

3.1 Soluble Lectins

Many vertebrate tissues contain materials which can agglutinate red blood cells and which have been shown to exhibit mitogenic properties. Therefore, these materials, which have recently been identified in *aqueous* extracts of solid tissues, can be called soluble lectins. An increasing number of tissues have been shown to possess soluble lectins which react with β-galactosides (Table 3). These carbohydrate-binding proteins have a dimeric structure in common and require a reducing substance to maintain their binding activity.

Another group of β-galactoside binding proteins has been isolated as monomers, such as the chicken lactose-binding lectin II, which possesses different immunological and physicochemical properties from chicken lactose-binding lectin I.

A monomeric β-galactoside-binding lectin has also been isolated from rabbit bone marrow.

Table 2. Methods for the detection, characterization, and localization of lectin activities in vertebrate tissues

Technique	Identification and charactization of lectin activity
Biochemical approach	Solubilization and purification of the lectin by affinity chromatography; glycoconjugate binding to the purified lectin; precipitation and precipitation inhibition test with appropriate glycoconjugates
Immunochemical/ serological approach	Agglutination of red blood cells; hemagglutination inhibition by competitive carbohydrates; blocking of the lectin recognition site(s) by monoclonal antibodies
Biological approaches in vitro	Pinocytosis of glycoconjugates and its inhibition with carbohydrates
Biological approaches in vivo	Clearance of labeled glycoproteins and neoglycoproteins and its inhibition by carbohydrates; effect of enzymatic and chemical alterations on the clearance of glycoconjugates
Cytological approaches	Mixed cell agglutination, inhibition of aggregation by inhibiting sugars: embryonic cells, tumor cells, sponge cells; microbe-host interaction, tumor cell-target cell interaction
Cytochemical approach (localization of the lectin in situ)	Using an antilectin antibody in a conventional immunohistochemical technique Glycosylated markers: gold- and ferritin-labeled neoglycoproteins for the ultrastructural demonstration of the carbohydrate-binding-activity; fluorescein-, biotin-, or peroxidase-labeled (neo)glycoproteins (fluorescence and light microscopy)

Table 3. Soluble β-galactoside-binding proteins (Barondes 1984)

Designation	Occurrence	Ligand
Calf β-Galactoside	Heart, spleen, muscle	β-Galactoside
Chicken lactose lectin I (CLLI)	Embryonic muscle, adult liver	β-Galactoside
Chicken lactose lectin II	Intestinal	β-Galactoside
Rat β-Galactoside lectin	Lung	β-Galactoside
Human β-Galactoside lectin	Lung	β-Galactoside
Thrombolectin	Snake venom	β-Galactoside
Anquilla rostrata eel lectin	Serum	L-Fucose
Xenopus laevis	Embryo egg	β-Galactoside
Erythroid developmental lectin	Rabbit bone marrow	β-Galactoside
Rana catesbiana	Egg	β-Galactoside
Rana japonica lectin	Egg	Sialic acid-rich glycoprotein
Chicken heparin lectin	Embryonic muscle, adult liver	Heparin, N-acetyl-glucosamine
Rat heparin lectin	Lung	Heparin, N-acetyl-glucosamine

A distinct property shared by many vertebrate lectins is that they are maximally synthesized during specific stages of the embryonic development of the tissues (BARONDES 1984). For example, chicken lactose-binding lectin I synthesis is maximal at the stage of muscle differentiation, while chicken lactose-binding lectin II synthesis is much higher in adult liver tissue than in the corresponding embryonic tissues. Erythroid developmental agglutinin is found in erythroblasts but does not occur in mature erythrocytes.

3.2 Membrane Lectins

This group of vertebrate lectins is believed to be inserted into the membrane, although in many cases this has not been rigorously established. An increasing number of membranous vertebrate lectins (Table 4) have been detected by

Table 4. Vertebrate membrane lectins

Species	Main occurrence	Carbohydrate specificity
Rat	Liver	
	Hepatocytes	D-Galactose, N-acetylgalactosamine
	Kupffer cells	Mannose, N-acetylglucosamine
	Brain	D-Galactose, lactose
	Muscle (myoblasts)	D-Galactose
Mouse	Spleen	D-Mannose, D-galactose, N-acetylglucosamine, N-acetylgalactosamine
	Thymus	D-galactose, lactose, N-acetylgalactosamine, N-acetylglucosamine
	Brain	N-Acetylgalactosamine, N-acetylglucosamine
	Hepatocytes	D-Galactose, L-fucose, N-acetylgalactosamine
	Kupffer cells	N-Acetylglucosamine, D-mannose
	Stomach	
	Surface and foveolar epithelium	N-Acetylgalactosamine, N-Acetylglucosamine
	Parietal cells	D-Galactose
	Lung	D-Galactose, N-acetylglucosamine, N-Acetylgalactosamine, L-fucose, D-mannose
	Kidney	D-Galactose, L-fucose, D-mannose, N-Acetylgalactosamine, N-acetylglucosamine
	Macrophages	D-Galactose, D-mannose, N-acetylglucosamine
	Cartilage	D-Glucosamine, 2-deoxy-D-glucose
Man	Kidney (collecting ducts)	D-Galactose, D-mannose, N-acetylgalactosamine
	Urothelium	N-Acetylgalactosamine, D-galactose, D-mannose
	Platelets	Hexosamine, basic amino acids
	Placenta	Lactose, asialofetuin, D-mannose, melibiose, L-fucose
	Fibroblasts	Mannose-6-phosphate, L-fucose, N-acetylgalactosamine

various methods in the rat (MORELL et al. 1971, GARTNER and PODLESKI 1976, STOCKERT et al. 1976, SIMPSON et al. 1977, ACHORD et al. 1978, STEER et al. 1978, STAHL et al. 1978, KOLB et al. 1978), mouse (CHANY-FOURNIER et al. 1978, PRIEELS et al. 1978, KIEDA et al. 1978, VIERBUCHEN et al. 1988 a), and man (BOWLES and HANKE 1977, GARTNER et al. 1978, GABIUS et al. 1987, VIERBUCHEN et al. 1989 a).

3.3 Endogenous Lectins – Intracellular Pathways and Mechanisms of Receptor Action

Vesicular intracellular transport of many macromolecules is mediated by specific membrane-bound receptors. These receptors and their ligands traverse various routes, depending on the subcellular source and the destination of the ligand.

3.3.1 The Hepatic Asialoglycoprotein Receptor (Hepatic Galactose-Binding Lectin)

Ashwell, Morrel and co-workers discovered and characterized the first mammalian lectin, a hepatic protein that specifically recognizes glycoproteins containing glycan parts exposing terminal galactose or N-acetylgalactosamine groups (ASHWELL and MORELL 1974). Many desialylated plasma glycoproteins are bound by this hepatic lectin (or galactose receptor), which then mediates their endocytosis and subsequent degradation (Fig. 9). Binding of the ligand requires calcium ions and the receptor complex rapidly dissociates below pH 6 (ASHWELL and MORELL 1974, MORELL et al. 1971). Internalization of galactosyl ligands is thought to occur through coated pits and coated vesicles (WALL et al. 1980, WALL and HUBBARD 1981). Within minutes after endocytosis the ligand appears in vesicles (HELENIUS et al. 1983). During the following minutes the ligand moves from the peripheral cytoplasm to the lysosomal-Golgi region and is degraded in lysosomes (WALL et al. 1980). Dissociation and segregation of the lectin occur intracellularly (BRIDGES et al. 1982, SIMMONS and SCHWARTZ 1984) in a prelysomal tubulovesicular compartment, termed the compartment of uncoupling of the receptor and ligand (CURL) or endosome.

3.3.2 The Mannose-6-phosphate Binding Receptor

In many respects lysosomal enzymes can be regarded as secretory proteins that are synthesized in the endoplasmic reticulum (ER). Within the lumen of the ER signal sequences are cleaved and polypeptides are glycosylated at asparagine residues analogous to those of other glycoproteins. However, in contrast to other secretory glycoproteins, lysosomal enzymes are modified by addition of the mannose-6-phosphate recognition marker to their N-linked

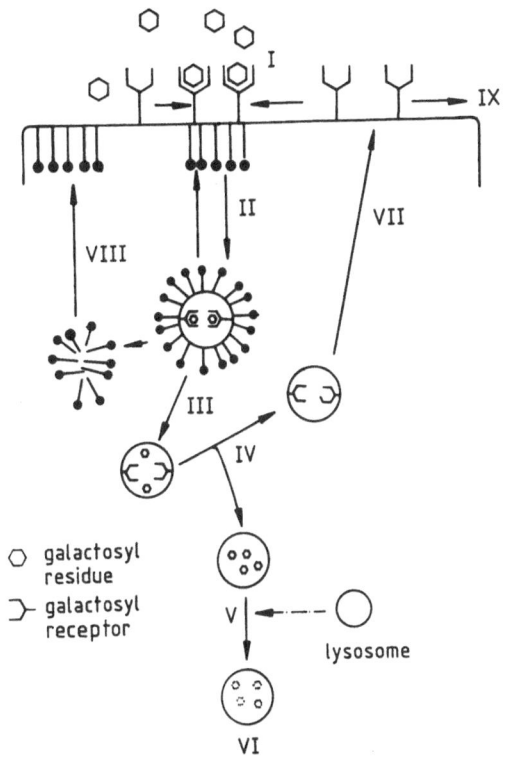

Fig. 9. Schematic representation of the function of the hepatic asialoglycoprotein (galactose/N-acetylgalactosamine) receptor system: The receptor-mediated endocytosis represents a series of events that include: I, ligand binding; II receptor-ligand internalization (invagination of the membrane; presence of cytoplasmic coat = coated membrane) and formation of receptor containing vesicles; III, delivery of the vesicular contents to specific intracellular compartments; IV: uncoupling of the receptor and the ligand; segregation of free ligand and receptor into different subcellular compartments; V, transport of the ligand to lysosomes or other compartments for degradation; VI, ligand degradation; VII, receptor processing and recycling to the plasma membrane, presumably via the tubular CURL elements; VIII, reincorporation of the receptor into coated pits; integration of the coated pit cycle with the receptor cycle; IX, possible partition of receptor complexes into different receptor cycles

glycan part (TABAS and KORNFELD 1980, HASILIK et al. 1980, GOLDBERG and KORNFELD 1981). Although lysosomal and secretory glycoproteins are initially identical, only lysosomal enzymes are effectively phosphorylated in the 6-hydroxy position of one or more mannose residues. The mannose-6-phosphate receptor (MPR) participates in the transport of newly synthesized acid hydrolases to lysosomes (SLY and FISHER 1982, KORNFELD 1986, v. FIGURA and HASILIK 1986).

The lysosomal enzymes exit the secretory route complexed either to a 270000-dalton-cation independent or a 47000 dalton cation-dependent MPR

(HOFLACK and KORNFELD 1985, STEIN et al. 1987a). Both receptor pathways act independently and can partly substitute for one another's function (STEIN et al. 1987b). The purified MPR is a glycoprotein (NEUFELD and ASHWELL 1980, SAHAGIAN et al. 1982); extensive disulfide cross-linking appears to be a characteristic shared with other receptors.

The lysosomal enzyme binding site has been shown to be located at luminal surface of the vesicle, and a 12 000-dalton part of the C-terminal end is exposed to the vesicle's cytoplasmic surface. A similar presentation has been reported for the MPR present in the microsomal fraction obtained from human fibroblasts (v. FIGURA et al. 1985). Previously, it has been suggested that the lysosomal enzymes leave the biosynthetic route by a distinctive class of vesicles developing from the trans Golgi region (GEUZE et al. 1988). These vesicles probably mediate enzyme transport to endosomes. After dissociation the MPR escapes from the endosomes via the tubular extensions of the trans Golgi region (GEUZE et al. 1988), whereas the ligand is delivered to the lysosomes (SAHAGIAN and NEUFELD 1983, GEUZE et al. 1985, GRIFFITHS et al. 1988). In normal human fibroblasts, binding of lysosomal enzymes to the receptor and subsequent delivery of the enzymes to the lysosomes occurs predominantly within the cells (v. FIGURA and WEBER 1978, SLY and STAHL 1978, HASILIK and NEUFELD 1980). Apparently, some of the complexes are transported to the plasma membrane (GEUZE et al. 1988). This population of the receptor appears to be in equilibrium with the total MPR population (v. FIGURA et al. 1984). Recently it has been demonstrated that the 270 000 dalton cation-independent receptor is identical to the insulin-like growth factor II receptor (KIESS et al. 1988).

3.3.3 Mannose and N-Acetylglucosamine Binding Proteins

In the course of studies on the clearance of glycosidases by rat liver it was found that the uptake of these glycoproteins could be inhibited by agalacto-orosomucoid, which is terminated by N-acetylglucosamine as well as by yeast mannan. As summarized in Table 5, this uptake system, reacting with both N-acetyl-glucosamine and mannose, was shown to be present in various mammalian (KAWASAKI et al. 1980, KOZUTSUMI et al. 1980, SHEPHERD et al. 1981, KAWASAKI 1982, ASHWELL and HARFORD 1982, KAWASAKI et al. 1983, MORI et al. 1983) and avian tissues (KUHLENSCHMIDT and LEE 1984, WANG et al. 1985). Following intravenous injection of glycoproteins terminated with mannose and/or N-acetylglucosamine, significant amounts of the proteins are concentrated in the cells of the mononuclear phagocytic system other than the sinusoidal cells of the liver (ACHORD et al. 1978, SCHLESINGER et al. 1980, KAWASAKI et al. 1980). A mannose/N-Acetylglucosamine receptor protein has been isolated from lymph nodes. Similar receptors have been found on alveolar macrophages, peritoneal macrophages, and macrophages derived from the bone marrow (EZEKOWITZ et al. 1981, STAHL and GORDON 1982, SHEPHERD et al. 1981). Finally, cytochemical studies in the author's laboratory using

Table 5. Occurrence of the mannose-binding proteins

Species/tissue	Sugar-binding specificity
Avian	
Chicken liver	GlcNAc > Man = Fuc
Serum	Man = L-Fuc > GlcNAc > ManNAc
Mammalian	
Lung macrophages	Man = Fuc > GlcNAc
Liver	
a) Nonparenchymal cells	ManNAc = Man-L-Fuc > GlcNAc
b) Parenchymal cells	ManNAc-Man-L-Fuc > GlcNAc
Lymphoid tissue	ManNAc > GlcNAc = Man > ManNH$_2$
Serum	ManNAc = Man = GlcNAc

ManNAc, N-Acetylmannosamine; *Fuc*, fucose; *GlcNac*, N-Acetylglucosamine; *ManNH$_2$*, mannosamin.

biotinylated bovine serum albumin covalently linked to mannose or N-acetylglucosamine have demonstrated the occurrence of this receptor system in the various phagocytic cell types of the spleen, lung, and liver (Fig. 10). Beside the mannose/N-acetylglucosamine receptor of the reticuloendothelial system, a mannose/N-acetylglucosamine-binding protein has been described in the hepatocytes which shares some physicochemical properties with corresponding protein of the reticuloendothelial cells. However, an antibody raised against this protein did not affect the pinocytosis of agalacto-orosomucoid by hepatic reticulo-endothelial cells (MAYNARD and BAENZIGER 1982). The lectin did not require a detergent for solubilization, indicating that this lectin may belong to the group of soluble proteins rather than to the family of membrane lectins, and might be related to mannose-binding serum lectins.

4 Lectin Involvement in the Immune System

4.1 Significance of Lectins in Lymphocyte Recirculation

Lymphocytes represent an extraordinarily mobile population of cells (FORD and SIMMONDS 1972). Most of the small lymphocytes in lymphoid organs are migratory cells that have recently arrived from the blood and will shortly return to the blood via collecting lymphatics. The rate of recirculation varies with the class of lymphocytes as well as with their stage of differentiation (FORD and SIMMONDS 1972, BRAHIM and OSMOND 1970, CANTOR and WEISSMAN 1976, FORD 1975).

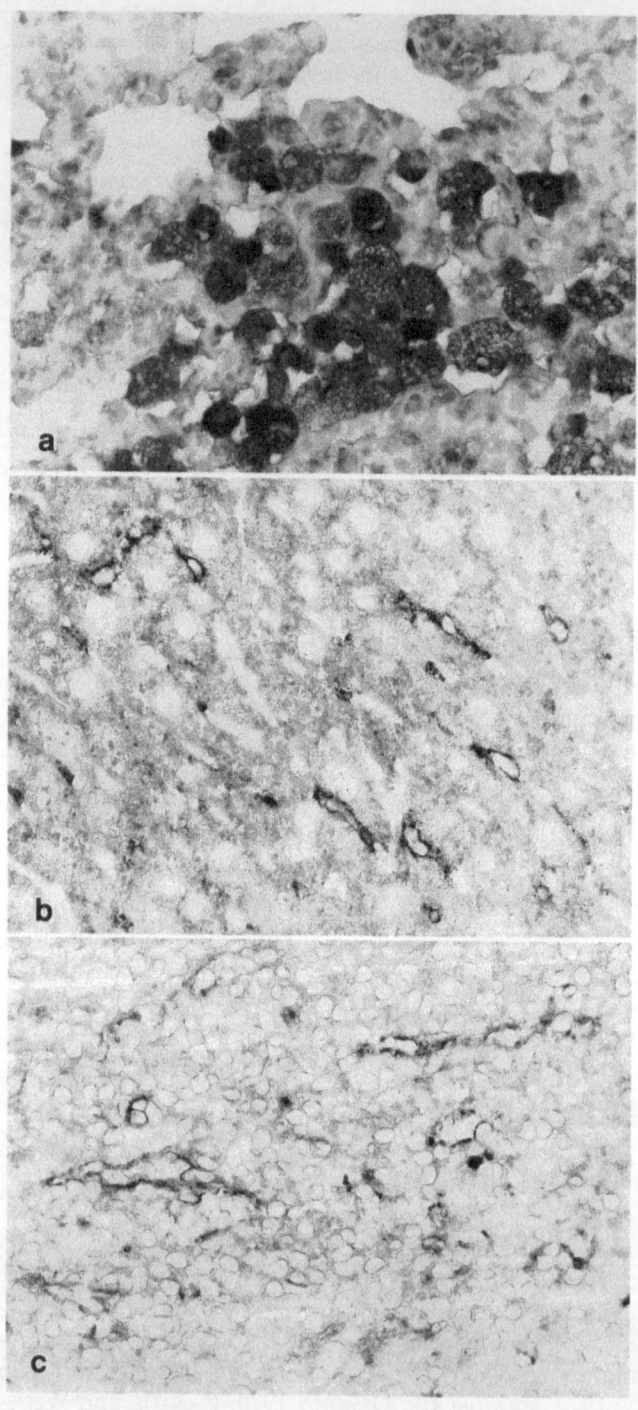

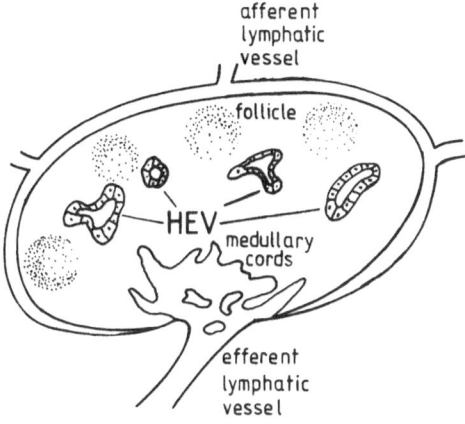

afferent
lymphatic
vessel

follicle

HEV

medullary
cords

efferent
lymphatic
vessel

Fig. 11. Architecture of a lymph node. Lymphocytes reach the lymphatic tissue through afferent lymphatic vessels. T and B lymphocytes bind to the luminal surface of the HEVs and then pass between the endothelial cells to enter the corresponding T- and B-cell compartments of the lymph node

The anatomy of the lymphocytic recirculation system has been elucidated only recently. In the lymph node lymphocytes migrate from the blood into the lymphatic tissue across a specialized vascular network, the postcapillary high endothelial venules (HEVs) (GOWANS 1957, 1959, GOWANS and KNIGHT 1964, MARCHESI and GOWANS 1964). This process helps to distribute normal lymphocytes throughout the body.

Figure 11 shows schematically the architecture of a lymph node and the location of HEVs. Circulating B and T lymphocytes first bind to the luminal surface of HEVs and then migrate between adjacent endothelial cells, crossing a thin basal membrane and entering the (para)cortex of the lymph node. The fact that relatively few blood granulocytes and monocytes adhere and traverse HEVs suggests that the lymphocyte-HEV attachment must involve highly selective recognition mechanisms (GOWANS and KNIGHT 1964, MARCHESI and GOWANS 1964, GOLDSCHNEIDER and McGREGOR 1968). The various lymphocytes show a high degree of specificity for the binding to the HEVs. Mouse T lymphocytes bind about 1.5 times better than mouse B-cells to peripheral lymph node HEVs, whereas B lymphocytes bind several times better to the HEVs of Peyer's patches. Furthermore, malignant lymphomas have been identified whose cells show a preference either for the HEVs of the peripheral lymph nodes or the HEVs of Peyer's patches (BUTCHER et al. 1980).

◀ **Fig. 10a–c.** Cytochemical demonstration of endogenous mannose-binding activities in the cells of the reticulohistiocytic system of the mouse. **a** Intense mannose-binding activity of the alveolar macrophages. **b** Kupffer cells in the liver exhibiting strong mannose-binding activities contrasting with the weak reactivity of the hepatocytes. **c** Distinct expression of mannose-binding activities in the sinusoidal cells of the spleen; note weak cell surface reactivity of some lymphocytes Mannose-BSA-biotin streptavidin technique. **a** hematoxylin counterstaining of cell nuclei; **b** and **c** without hematoxylin counterstaining of cell nuclei, snap frozen tissue sections

These observations suggest at least two very selective recognition systems for the homing of lymphocytes. The implication of carbohydrates and carbohydrate binding proteins (lectins) in the initial step of adhesion and migration as well as the recirculation of lymphocytes arose from in vitro and in vivo studies of lymphocyte binding to the HEVs (STAMPER and WOODRUFF 1976, WOODRUFF et al. 1977; BUTCHER et al. 1979, STEVENS and BUTCHER 1982). The inhibitory effect of sugars and the modulation of lymphocyte binding to HEVs under different incubation conditions suggested that both a lectin–carbohydrate interaction and ionic forces are involved in the attachment of lymphocytes to HEVs. Inhibition studies using a broad library of carbohydrates, structurally defined oligosaccharides, and glycoproteins indicated that the carbohydrate specificity of the lymphocyte lectin is for a phosphomannosyl determinant (mannose-6-phosphate); (STOOLMAN and ROSEN 1983, STOOLMAN et al. 1984). This concept was further supported by the effect of periodate treatment of tissue sections, which oxidizes carbohydrates and thereby inhibits the adhesion of lymphocytes to HEV's (STOOLMAN and ROSEN 1983, STOOLMAN et al. 1984). However, the specific contribution of carbohydrates and corresponding binding proteins in the different lymphoid tissues remains to be determined by further studies. In this context the possibility that endogenous lectins of HEVs also contribute to the homing of lymphocytes has to be elucidated. From this starting point, there may exist several possible carbohydrate recognition systems for the specific attachment of lymphocytes to HEVs (Fig. 12). Furthermore, there are reports on the occurrence of different endogenous carbohydrate-binding proteins on lymphocytes (KIEDA et al. 1978, APGAR and CRESSWELL 1982, BARZILAY et al. 1982, KOLB 1982, KIEDA and MONSIGNY 1983). Using neoglycoproteins it has been shown that the attachment of splenocytes to HEVs of mesenteric lymph nodes is inhibited by β-galactosides but not by other sugars (KIEDA and MONSIGNY 1983). High endothelial venules serve as the portal of lymphocyte entry into lymph nodes, including Peyer's patches, tonsils, adenoids, and the specialized submucosal lymphoid nodules in the respiratory and gastrointestinal tracts known as bronchus-associated lymphoid tissue (BALT) and gut-associated lymphoid tissue (GALT), respectively. However, the thymus, bone marrow, and the spleen do not possess HEVs. Therefore one may speculate that in these organs other carbohydrate recognition systems are involved in the homing of lymphocytes.

In the thymus subcortical lymphocytes express endogenous galactose-binding activities (Fig. 13) while the medullary thymocytes lack this endogenous lectin activity (VIERBUCHEN, unpublished results). Therefore, it is conceivable that the carbohydrate specificity of lymphocyte surface lectins varies with the source of lymphocytes and/or with the lymphoid target tissue. Further elucidation of the mechanisms for the interaction between lymphocytes and HEVs in lymphoid tissues may provide greater understanding of lymphocytic migration into lymphoid and nonlymphoid tissues. In addition, such studies may yield further insights into the mechanism by which malignant lymphomas metastasize from the organ from which they arise.

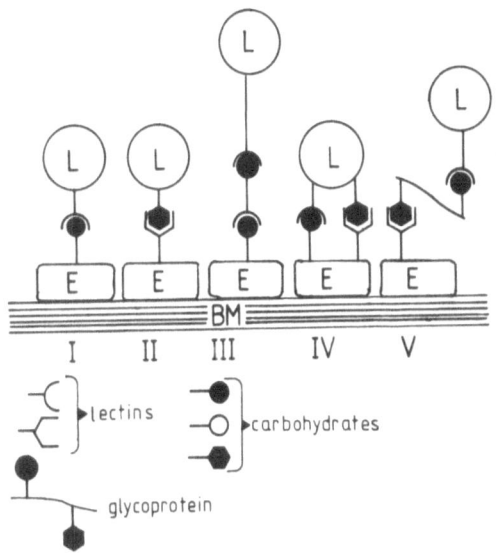

Fig. 12. Schematic representation of hypothetical models of lymphocyte-HEV binding. I, Phosphomannosyl lectin of the lymphocyte directly mediates the attachment. II, Endogenous lectins on the surface of the HEVs might bind to complementary carbohydrates of the lymphocyte surface. III, The phosphomannosyl lectin serves as a baseplate molecule (ligatin-like molecule) for other carbohydrate-binding proteins of the lymphocyte. IV, Endothelial and lymphocyte lectins are both involved in the specific adhesion. V, A glycoprotein acts as a bridging molecule and carries carbohydrates which are recognized by the lymphocyte and the endothelial cell lectin

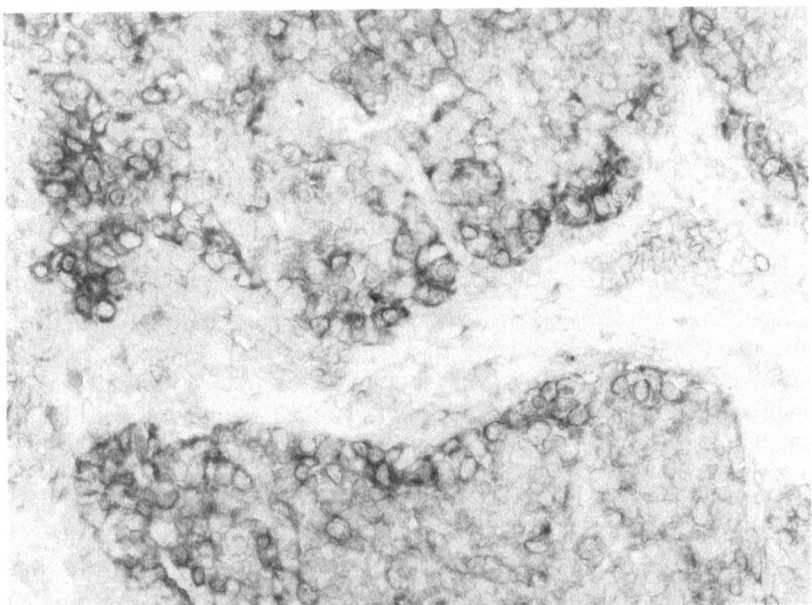

Fig. 13. Expression of endogenous galactose-binding proteins in cortical thymocytes. Galactose-BSA-biotin-streptavidin technique, without hematoxylin counterstaining of cell nuclei

The demonstration that malignant lymphomas may recapitulate the adhesive and migratory characteristics of their normal counterparts raises the provocative possibility that a selective adhesive interaction between circulating malignant lymphoid cells and the microvasculature may facilitate and determine the spread of malignant lymphomas.

4.2 Role of Lectins and Lectin Receptors in Immune Function and Immune Recognition

In the immune system, cellular recognition is a fundamental prerequisite for an effective and specific immunereaction. The various cell elements of the B- and T-cell systems represent a cornerstone of this complex recognition and effector system.

There is now increasing evidence that interactions between lectin like-molecules and their corresponding carbohydrates in a key–lock manner is one basis for immune recognition (BABA et al. 1979, MUCHMORE and BLAESE 1980, HIGGINS et al. 1978, TOMASKA and PARISH 1981, WEIR 1980, BLAESE et al. 1983, KASZENOWSKI and KARMER 1981) and nonimmune phagocytosis (SHARON 1984). Binding of lectins to cells of the immune system can result in dramatic effects on cellular function. Since NOWELL's observation (1960) that phytoghemagglutinin (PHA) stimulates mitotic division and proliferation of lymphocytes, the interaction of lectins with their ligands has become an area of intense investigation. From the standpoint of phylogenetics, the mononuclear phagocytes are probably the earliest cells to demonstrate a specialized host defense function. Using primitive mononuclear phagocytes, their ability to kill a variety of foreign targets in vivo and in vitro (BLAESE et al. 1983) has been demonstrated. These studies revealed that carbohydrates can inhibit the toxicity in a specific manner. Some carbohydrates showed inhibitory acitivity against some targets but not against others. Within the population of lymphoid cells in mammals, there exists another subpopulation of cells with the ability to lyse a variety of target cells without prior sensitization (HERBERMAN et al. 1975, HERBERMAN and HOLDEN 1978, KIESSLING and WIGZELL 1979, STUTMAN et al. 1980, FORBES et al. 1981). These cells have been called natural killer (NK) cells on the basis of their in vitro killing capacities and have been distinguished from conventional T and B lymphocytes.

Although NK cells exhibit a broad spectrum of reactivities, there also exists a distinct degree of selectivity, since some target cells are highly sensitive to cytolysis while others are not (RODER et al. 1981). In order to elucidate the nature of the target recognition system of NK cells, several carbohydrates have been used to block the NK cell-dependent cytolysis of the target.

Murine NK cell-mediated cytotoxicity could be inhibited by a number of sugars, such as mannose, galactose, and fructose (STUTMAN et al. 1980), suggesting that a lectin-like recognition system may be involved in this process. It has been demonstrated that mannose-6-phosphate, fructose-6-phosphate, and fructose-1-phosphate can inhibit human NK cell cytotoxicity (FORBES et al. 1981).

Further experiments have shown a mannose-6-phosphate receptor on the target cell. However, it could be demonstrated that the mannose-6-phosphate binding activity is involved in the uptake of lytic molecules which contain the receptor (lysosomal hydrolytic enzymes), rather than in the immediate interaction of the NK cell with the target.

4.3 Cell–Cell Interaction and Specific Carbohydrate Structures as Mediators of Natural Immunomodulation

The antigenic lymphocyte proliferation response to recall antigens using human peripheral blood monocytes as responders was examined in the presence of a great a number of carbohydrates (MUCHMORE and BLAESE 1980). These studies revealed three types of reaction toward the different sugars:

1. Sugars like cellobiose, lactose, maltose, and trehalose, which strongly inhibited cytotoxic reactions, had no effect on antigen-induced proliferation
2. Sugars that inhibited antigen-induced proliferation had no effect on PHA-induced proliferation.
3. Sugars like rhamnose and N-acetylgalactosamine blocked the elicitation of a second response only when they were present at the beginning of the assay.

An interesting explanation has been offered (PARISH et al. 1978, McKENZIE et al. 1977), namely that the sugars could be acting by blocking immune response (Ir) gene-restricted cellular interactions. These (I a) antigens express carbohydrate-defined antigenic determinants (PARISH et al. 1977) and have been implicated in many immunological phenomena, such as macrophage–T cell interactions (SCHWARTZ et al. 1978, ERB et al. 1980), T-cell help (HODES et al. 1980), and T-cell suppression (THEZE et al. 1977, TADA and OKUMURA 1979).

The observation that certain monosaccharides selectively inhibit the secondary immunoglobulin G response (TOMASKA and PARISH 1981), that the specificities of the inhibiting sugars differ depending on the strain of the mouse, and that the differences can be mapped to the I region of the MHC (TOMASKA and PARISH 1982) are consistent with the concept that carbohydrate-defined I a antigens play a role in communication between lymphoid cells and that their blockade can lead to loss of communication and immune function.

Most interesting are findings that suggest a homology in the primary structure of recognition molecules, including immunoglobulins, T-cell receptors, MHC products, and some lectins (WARR et al. 1983, VASTA et al. 1984).

Several immunomodulators bind specific target cells such as macrophages, cytotoxic cells, NK cells, and B cells. Recently, increasing evidence has been produced that the activities of these immunomodulators are specifically inhibited by simple glycolipids, glycopeptides, and simple sugars. Therefore, the interaction between some lymphokines and their target cells appears to involve lectin-like reactions. Table 6 summarizes the different sugar sensitivities of

Table 6. Carbohydrate specificities of lymphokines and natural immunoregulatory factors

Immunomodulator	Secretor cells	Target cells	Inhibitory sugars
Migration inhibitory factor	T	Macrophages	L-Rhamnose, L-fucose
Leukocyte inhibitory factor	T	Leukocytes	N-Acetylgalactosamine
Interferon	T	Cytotoxic	Gangliosides
Interferon	T_8	Cytotoxic	D-Mannose
Suppressor factor	T_8	B cells	L-Rhamnose
		N cytotoxic	D-Mannose, D-glucose
		N killer	D-Mannose, D-galactose, D-glucose, N-acetylgalactosamine
Soluble mediator	T_8	N killer	D-mannose, D-galactose, N-acetylgalactosamine
Suppressor factor	mononuclear cells	B cells	L-Rhamnose
		T cells	N-Acetylglucosamine
Lymphotoxin	T_k	L cells	Glycopeptides
Helper factor	T_h	B cells	N-Acetylglucosamine

T_8, T-suppressor; T_h, T-helper; T_k, T-killer.

lymphokines and immunoregulating factors with respect to their origin, target cells, and inhibitory carbohydrates (Vengris et al. 1976, Amsden et al. 1978, Kobayashi et al. 1978, Poste et al. 1979a, 1979b; Stutman et al. 1980, Kaszinovski and Karmer 1981, Wright and Bonavida 1981, Greene et al. 1981, Fleisher et al. 1981, Tomaska and Parish 1981, Kieda et al. 1982, Liu et al. 1980, 1982a, 1982b).

According to these findings one must conclude that some important functions of the immunocompetent cells, including proliferation and differentiation, and of the effector molecules of lymphocytes and macrophages may be induced and mediated by carbohydrate recognition proteins (lectins). However, little is known about the natural function of these endogenous carbohydrate-binding activities. Recently it has been shown that rIL-2 binds to glycoproteins which contain high mannose oligosaccharides and that this binding is inhibited by mannose glycopeptides, indicating that rIL-2 is a lectin (Sherblom et al. 1989). The carbohydrate-binding site is different from the cell surface receptor site, and might function in acidic microenvironments. It has been suggested that the lectin activity of IL-2 plays a role in the clearing, intracellular routing, and regulation of biological activity of this lymphokine. These observations cast an entirely new light on the importance of carbohydrate recognition systems in physiological regulation of the human immune response. However, much more information is needed to answer the question of how cells of the immune system interact (talk) with each other.

5 Lectin Receptors and Receptors for Lectins – Their Role in Metastasis

Metastasis, the process by which tumor cells from a primary site spread to near and distant secondary sites via the circulation, is the most important event in the pathogenesis of cancer and it accounts for most cancer deaths. Metastatic cells are characterized by quantitative alterations of the cell surface and other properties that confer upon these cells their ability to invade, disseminate, implant, survive, and grow at secondary sites. Metastasis is also determined by a variety of host factors that prevent, allow, or even stimulate metastatic progress. Studies of neoplastic spread have revealed that it is a complex process involving several sequential steps:

1. Following tumor growth, invasion of primary tumor cells into the surrounding tissues and blood vessels
2. Detachment of single tumor cells or tumor cell complexes from intravasated tumors, resulting in their dissemination, upon which they can adhere specifically or nonspecifically in the vessels of the target
3. Extravasation from the blood or lymph vessels into the tissue of the target organ, and proliferation to form secondary colonies

The mechanisms involved in the initial process leading to the metastasis (dissociation of tumor cells, invasion of blood or lymph vessels, and dissemination of tumor cells via the circulation) have been thoroughly investigated. The results of many studies indicate that the ability of tumor cells to form metastases is determined by a number of intrinsic properties, including growth rate, size, deformability, motility, antigenicity, susceptibility to host immune mechanisms, and secretion of angiogenesis factor as well as degradative enzymes (FIDLER 1984, KIERAN and LONGNECKER 1983, NICOLSON 1982, 1984; SCHIRRMACHER 1985, THORGEIRSON et al. 1984, WEISS and WARD 1983). An increasing number of studies have suggested that the sugar units of the complex carbohydrates are involved in cell recognition and cell adhesion (ROSEMAN 1970, OPPENHEIMER 1975). Many lines of evidence support the concept that tumor cell surface carbohydrates play an important role in metastasis (NICOLSON 1982, 1984).

Specific carbohydrate structures apparently are important for the recognition and adhesion of tumor cells to components of the extracellular matrix such as collagen IV and fibronectin (DENNIS et al. 1984, KENNEDY et al. 1983), to other cells of the tumor stroma (ASHWELL 1977, SIMPSON et al. 1978), and also to distinct organ cells such as hepatocytes (SCHIRRMACHER et al. 1980, BEUTH and UHLENBRUCK 1985, BEUTH et al. 1987). Clinical and experimental studies on the distribution of metastases to explain the occurrence or lack of organ-specific metastasis resulted in two opposing hypotheses:

a) Ewing's mechanical/anatomical hypothesis (EWING 1928)
b) Paget's seed and soil hypothesis (PAGET 1889)

In many cancers where tumor emboli are released into the lymph or blood vessels resulting in their limited distribution to regional sites (i.e., lymph nodes), metastasis is apparently solely due to circulatory patterns and mechanical lodgment. In contrast to such nonspecific lodgment of tumor cells in the microcirculation (for a review of this subject see WEISS 1977, 1983) there is increasing evidence of organ-specific metastasis, especially in the development of distant metastases (Table 7). Even neoplasms that are considered to be disseminated soon after their inception and are capable of being transported throughout the body, such as leukemias and lymphomas, reveal some preferences regarding organ localization and growth (Table 8). From these clinical observations it is reasonable to conclude that selective cell adhesion mechanisms do play an important role in tumor metastasis, although many factors are not exactly defined (BROSS andf BLUMENSON 1976, NICOLSON and POSTE 1983, LOTAN 1987).

The known tumor cell components that are thought to mediate tumor–host cell interactions include glycosyltransferases (SCHWARTZ et al. 1983, PODOLSKY et al. 1983), cell adhesion molecules (BRACKENBURY 1985), glycolipids (CHERESH 1986), fetal recognition systems (KAHAN 1979, McGUIRE et al. 1984), and lymphoid cell adhesion systems (GALLATIN et al. 1986, MENTZER et al. 1986).

Table 7. Clinical examples of human neoplasms not esplainable by circulating patterns and mechanical lodgment of blood-borne tumor cells

Type of tumor/common primary site	Common distant site(s) of metastases
Breast carcinoma (female breast)	Bone, brain
Neuroblastoma (adrenal gland)	Liver, bone
Prostatic adenocarcinoma	Bone
Small cell carcinoma (lung)	Bone, brain, liver
Renal cell carcinoma (kidney)	Bone, liver, thyroid gland
Follicular carcinoma (thyroid gland)	Bone
Adenocarcinoma (stomach)	Ovary (Krukenberg tumors)

Table 8. Frequency of organ infiltration (colonization) of some human leukemias (modified according to Kamenov et al. 1984)

Tumor type	Frequency of organ involvement
Acute lymphocytic T-cell leukemia	Bone marrow = lymph node > spleen = liver = brain
Acute lymphocytic B-cell leukemia	Bone marrow > lymph node = spleen = brain > liver
B-cell lymphocytic non-Hodgkin's lymphoma	Lymph node > bone marrow > liver
Hairy cell leukemia	Spleen > bone marrow > liver

5.1 Tumor Cell Lectins

The complex carbohydrates in normal and neoplastically transformed tissues carry biological information relevant as reconition structures which contribute to the sociological behavior of cells. Therefore, analysis of the corresponding endogenous natural *receptors* (lectins) for the carbohydrates of the glycan chains of complex glycoconjugates in normal and tumor cells has gained increasing importance for establishing a recognition system based on specific carbohydrate–binding protein interaction. As well as in normal tissues, endogenous lectins have been evidenced in an increasing number of human and animal tumor cells (tumor cell lectins) by their agglutinating ability, by binding of labeled (neo-)–glycoproteins, by their purification, or by immunochemical and immunohistochemical techniques. Tumor cell lectins have been detected in epithelial neoplasms and sarcomas (ROCHE et al. 1983, GABIUS et al. 1984, 1985 b, 1986 a, 1986 b, 1986 c; RAZ and LOTAN 1981, CRITTENDEN et al. 1982, JIANG et al. 1983), teratocarcinomas (GRABEL et al. 1979, 1981, CARROL et al. 1982, GRABEL 1984, GABIUS et al. 1984, 1985 a); T and B lymphoblastoid cells (APGAR and CRESSWELL 1982, CARDING et al. 1985), chronic lymphocytic leukemia and hairy cell leukemia (CARDING et al. 1985), and Hodgkin's lymphoma cells (PAIETTA et al. 1986).

The sugar specificities of the endogenous carbohydrate-binding proteins detected by the various methods cited above include galactose, mannose, fucose, fucoidin, heparin, glucuronic acid, and heparan sulfate. However, in many of the studies the exact nature and location of the tumor cell lectin has not been determined.

Figure 14 illustrates the different expression of endogenous carbohydrate-binding proteins in normal tissue and neoplastically transformed tissue. The proximal convoluted tubules of the kidney, which are thought to be the site of origin of renal adenocarcinoma of the clear cell type, do not express galactoside-binding proteins. In contrast, renal adenocarcinoma of the clear cell type exhibits endogenous carbohydrate-binding proteins for Gal-BSA (Vierbuchen, unpublished results).

The involvement of the endogenous tumor cell lectins in recognition and intercellular adhesion was demonstrated by the formation of rosettes between tumor cells (homotypic adhesion or homotypic aggregation) or tumor cells and red blood cells (heterotypic adhesion or heterotypic aggregation). The various forms of tumor cell aggregation could be inhibited by appropriate sugars and glycoconjugates containing the corresponding carbohydrates (GRABEL et al. 1981, GRAUPNER et al. 1984, MEROMSKY et al. 1986).

5.2 Endogenous Host Cell and Tumor Cell Lectins as Possible Mediators of Metastasis

Based on studies of experimental tumor systems and human neoplasms, various mechanisms have been suggested for the involvement of endogenous

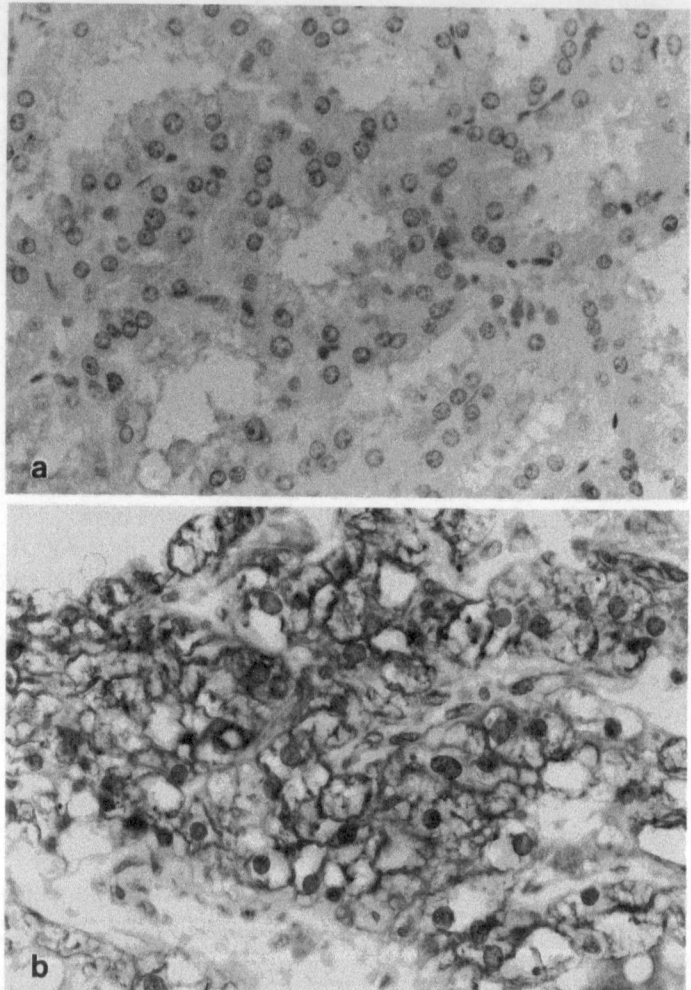

Fig. 14a, b. Expression of endogenous galactose-binding proteins in renal adenocarcinoma and in the proximal tubules of nonneoplastic renal tissue. **a** The nonneoplastic tubules lack endogenous galactose-binding activity, contrasting with **b** the expression of endogenous galactose-binding lectins in the renal (clear cell) adenocarcinoma. Tissue specimens were obtained from the same patient. Galactose BSA-biotin-streptavidin technique, hematoxylin counterstaining of cell nuclei

lectins and corresponding lectin receptors in the organotropism of tumor cell metastasis.

With respect to liver metastasis, UHLENBRUCK and co-workers (UHLEN-BRUCK et al. 1986, BEUTH et al. 1986, 1987) have proposed a hypothesis based on the function of the hepatic asialoglycoprotein receptor, which removes altered glycoproteins from the circulation. The basis of this concept lies in the tumor-associated alteration of the terminal position of the glycoconjugate (removal of sialic acid residue, and thereby exposing the subterminal galactose units). The altered glycoproteins are then accessible to the galactose-binding protein of the hepatocytes.

Subcutaneously inoculated fibrosarcoma cells in Balb/c mice resulted only in the formation of lung nodules. However, enzymatic removal of sialic acid residues by neuraminidase, which exposes subterminal galactose, resulted in the formation of *liver* and lung metastasis. These experiments clearly demonstrated that alteration of the glycocalyx modifies the organotropism of metastasis in this experimental tumor type. Further experiments showed that the liver metastasis of desialylated tumor cells could be blocked by the simultaneous administration of galactose or galactose-containing glycoconjugates but not by other carbohydrates (BEUTH et al. 1987). From these results the authors concluded that the liver metastasis of desialylated tumor cells is mediated by the galactose-binding lectin of the hepatocytes. From these findings it is reasonable to conclude that:

1. Desialylation of tumor cells modifies the organotropism of metastasis and specifically enhances the formation of liver metastases

2. The inhibition of the formation of liver metastases by galactose or galactose-containing glycoconjugates supports the concept that the β-galactoside binding protein of the liver is involved in the homing of tumor cells

3. The formation of lung metastases is influenced neither by desialylation nor by administration of galactose, indicating that in organs other than the liver, lectins with different carbohydrate specificities play an important role in the manifestation of metastatic islets.

Metastases are common in colorectal cancer patients and are the major cause of death. Thus the 5-year survival rate of patients operated on for local cancer has been reported to be more than 80% which contrasts markedly with the 5-year survival rate of 5% in patients having distant metastases at the time of operation (EISENBERG et al. 1982).

Alteration in the expression of complex carbohydrates plays an important role in metastasis of colorectal carcinomas to the liver. Such altered carbohydrates can act as ligands for endogenous lectins of the liver. In this context it has been shown that in comparison to nonneoplastic tissues, tumor tissues often display incomplete synthesis or altered expression of the glycan chains of complex carbohydrates. In colorectal carcinomas one of these carbohydrate chains is the so-called Thomsen-Friedenreich antigen, which is represented by the disaccharide sequence Galβ1-3GalNAc. Histochemically this antigen can be demonstrated by the use of the lectin from *Arachis hypogaea* (peanut agglutinin, PNA). Although the antigen is not exclusive for malignancy, there is t a correlation between the expression of this carbohydrate structure and the tumor stage of colorectal carcinomas (VIERBUCHEN 1987). In the cascade of colorectal liver metastasis this distinct carbohydrate chain may represent a ligand for the galactose-binding liver lectin and thereby may modulate the organotropism of distant metastasis in this tumor type. This hypothesis is supported by findings on the metastatic capacity of murine colon carcinoma cells (YEATMAN et al. 1989). In these experiments, liver metastasis was found to be proportional to the degree of tumor cell surface galactose expression, indicating that tumor galactose expression and recognition by the hepatic galactose lectin may be important components of a specific mechanism for the development of liver metastases from colorectal carcinomas.

Whereas structural alterations of glycoconjugates upon malignant transformation have been widely documented, the biochemical characterization of the endogenous carbohydrate-binding proteins (lectins) has only recently been investigated. There is increasing evidence that during the course of the metastatic cascade (Fig. 15), tumor cell lectins play a functional role in homotypic (tumor cell–tumor cell) and heterotypic (tumor cell-organ cell; tumor cell–extracellular matrix) tumor cell interaction (LOTAN and RAZ 1983, MEROMSKY et al. 1986, RAZ et al. 1984, 1986, 1987).

Homotypic tumor cell interaction may play a role in tumor growth, because it has been shown that certain vertebrate lectins possess mitogenic properties (Novogradsky and Ashwell 1977, Lipsick et al. 1980). Furthermore, the occurrence of tumor cells with different phenotypic properties in a tumor population, with qualitative and quantitative changes in tumor cell glycoconjugates as well as changes in the expression of the corresponding carbohydrate-binding proteins (tumor cell lectins), can influence the release of certain tumor cells from the primary site (Fig. 15).

During intravascular transport tumor cell lectins may induce embolization by homotypic and heterotypic interactions with host cell lectins, including lymphocytes (KIEDA et al. 1979, DECKER 1980) and platelets (GARTNER et al. 1978, JAFFE et al. 1982). Furthermore, tumor cell lectins can contribute to the selectivity in adhesive contacts between invading tumor cells and the target and thereby may be involved in the organotropism of metastasis (Fig. 15).

A quantitative estimation of cell surface lectin expression of tumor cell lines revealed that tumor cell subpopulations (clones) which exhibited a higher density of cell surface lectins possessed a greater metastatic capacity than tumor cells with a low cell surface lectin density (RAZ et al. 1984, 1987). The concept that tumor cell lectins are involved in tumor cell attachment to the target has been supported by different inhibition experiments. Blocking of the carbohydrate-binding activity by either asialofetuin or monoclonal antibodies raised against the lectins inhibited homotypic tumor cell interaction and the binding of tumor cells to tissue culture dishes (MEROMSKY et al. 1986).

Beside the tumor cell lectins, endogenous carbohydrate-binding proteins of the target may also be involved in the initial tumor cell interaction. Experiments on the adhesion of Lewis lung carcinomas to pulmonary cells revealed a more complex binding mechanism. Neither the blocking of tumor cell lectins by neoglycoproteins nor the blocking of organ cell lectins resulted in a complete inhibition of tumor cell binding to the target cells. However, simultaneous blocking of lung cell and tumor cell lectins resulted in an effective inhibition of tumor cell adhesion (KIEDA and MONSIGNY 1986). Given these

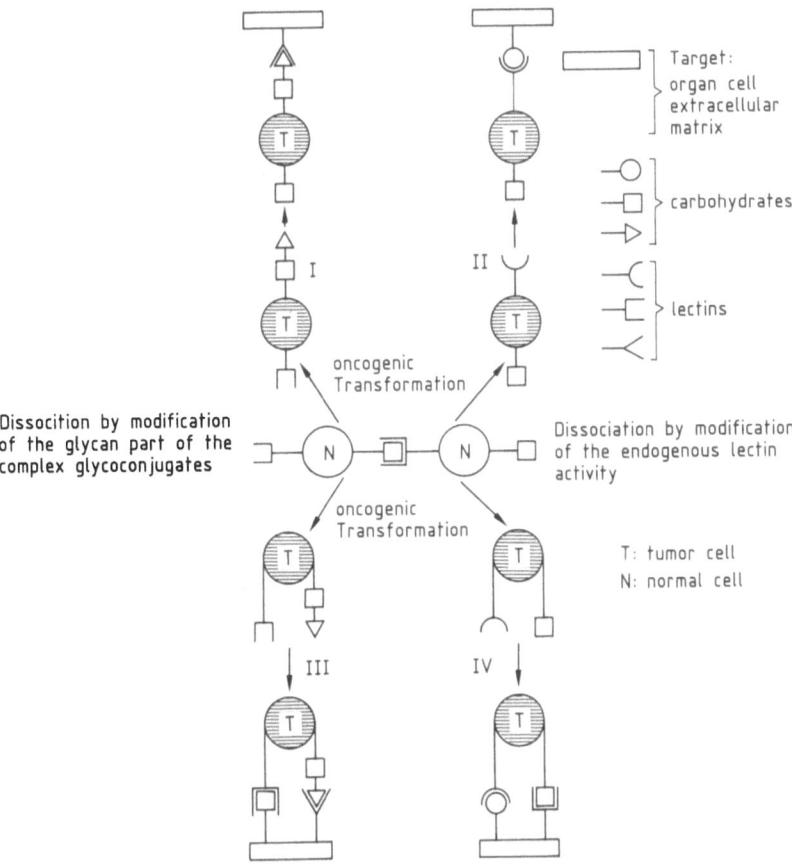

Fig. 15. Schematic representation of cell dissociation during oncogenesis and different metastatic pathways with respect to the modification of carbohydrates and carbohydrate-lectin interactions:

a) Monophasic recognition system

I, Binding of the metastatic tumor cell to the target via an altered carbohydrate of the tumor cell to the target cell-lectin

II, Binding of the metastatic tumor cell to the target by tumor cell-lectin carbohydrate interaction.

b) Biphasic recognition system

III and IV, Binding of the metastatic tumor cell is accomplished by the simultaneous interaction of the tumor cell and organ cell lectins with the corresponding carbohydrates

experimental data and the broad spectrum of organ and tumor cell lectins, one may speculate that the individual distribution of lectins in the different tumor types and the organ-typical expression of endogenous lectins together with the modification of the complex carbohydrates during oncogenic transformation may be responsible for the organotropism of metastatic spread (Fig. 15).

6 General Aspects of, and Tentative Views on, the Function of Vertebrate Lectins

Although the hepatic galactose/N-acetylgalactosamine-binding lectin was discovered in 1968, its physiological function remains unknown. Some suggestions as to the role of the hepatic receptor include:

1. *Clearance of desialylated glycoproteins.* However, there is evidence that under normal conditions glycoproteins are not desialylated in vivo and are not removed from the circulation by this receptor.
2. *Cell adhesion.* Several reports have shown the involvement of the heaptic receptor in the homing of metastatic tumor cells. Therefore, a possible role of this lectin in cell recognition phenomena must be considered.
3. *Clearance of antibodies.* Formation of immunoglobulin-antigen complexes alters the conformation of immunoglobulin G, exposing a galactose residue and thereby inducing the clearance of circulating immunecomplexes by the liver.

Beside the hepatic galactose-binding lectin, further endogenous carbohydrate-binding proteins with different sugar-binding specificities have been detected in vertebrate tissues. A certain number of mannose-binding proteins are now

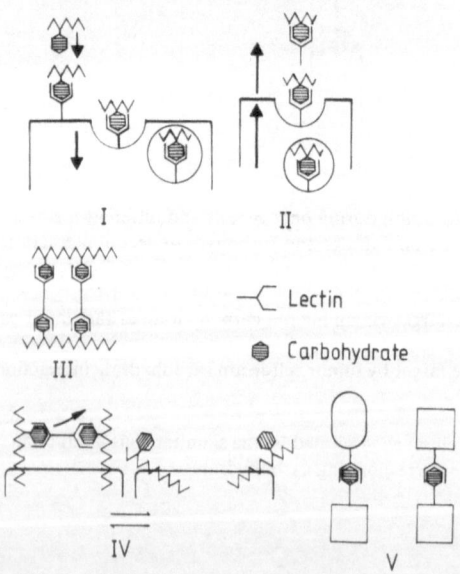

Fig. 16. Possible functions of endogenous carbohydrate-binding proteins in vertebrates: I, endocytosis of (altered) complex carbohydrates; II, as vehicles involved in the transport and secretion of complex carbohydrates; III, organization (bridging) of the complex carbohydrates of the extracellular matrix; IV, organization and modification of the glycocalyx (coupling and uncoupling of carbohydrates of the cell surface); V, mediation of cell-cell interaction (homotypic: interaction between identical cells, heterotypic: interaction between different cells, e.g., normal cell-tumor cell interaction, target cell-bacterium interaction)

reported to be present in vertebrate tissue at different cellular sites. It is reasonable to conclude that they have diverse structural functions. Mannose binding-proteins may be involved in endocytosis, as shown by the binding activity of the cells of the reticulohistiocytic system. Furthermore, mannose-binding proteins may be involved in the transport of high mannose type glycoproteins.

Immuno- and cytochemical experiments have provided the most clues about the function of endogenous carbohydrate binding-proteins. Figure 16 schematically summarizes the putative functions of endogenous lectins which have been deduced from the results of cytochemical studies on their occurrence in different tissues and organs.

7 Microbial Lectins – Mediators of Infection

The adherence of pathogenic agents to structures of the target cell (membrane) represents one of the most important early events in their interaction with the host, leading to multiplication in situ (colonization) and infection. This process provides the infectious agent with several advantages, such as resistance to the rinsing and cleaning action by body fluids, evasion of host defenses, and improved acquisition of nutrients from its environment. The specific adhesion to certain tissues may result in the organ tropism of microbial agents. The term "adhesion" describes a specific interaction of recognition proteins (adhesins) on the microbial cell surface with distinct structures of the target cell membrane. During recent years considerable evidence has accumulated showing that such adhesion is accomplished by interaction of a microbial adhesin with the carbohydrate moiety of complex carbohydrates (glycoproteins, glycolipids) of the target cell. This specific adhesion may be responsible for the organotropism of infectious agents. Since the receptors for the sugar-binding proteins are also often expressed on erythrocytes, hemagglutination is frequently induced by adhesive microbes. Thus, adhesins may be termed hemagglutinins (the more general term "lectins" is also used).

7.1 Receptors for Viral Hemagglutinins (Lectins)

The ability of viruses to agglutinate red blood cells was established decades ago. This important property afforded a simple and rapid method of quantifying and, in conjunction with the appropriate serological test (hemagglutination inhibition, hemadsorption), of typing viruses as etiological agents of infectious diseases. Structures involved in the virus attachment are:

1. Virus attachment proteins (VAP's)
2. Cellular receptor units – cellular molecules recognizing one VAP
3. Cellular receptor sites – cellular structures containing one or more cellular receptor unit which can effectively bind one virion

Removal of neuraminic acid by enzymatic treatment, or chemical modification of neuraminic acid residues by periodate oxidation, or by methylation of the carboxyl groups destroys the binding elements for ortho- and paramyxoviruses (SCHEID 1981).

Numerous investigators contributed to the eventual realization that sialic acid residues are required (Table 9) for the attachment of myxoviruses (LAVER and VALENTINE 1969, KATHAN et al. 1961, TIFFANY and BLOUGH 1971, HUANG et al. 1973, SUTTAJIT and WINZLER 1971, VARGHESE et al. 1983, ROGERS and PAULSON 1983, GETHING and SAMBROOK 1982), paramyxoviruses (PORTNER 1981, SCHEID and CHOPPIN 1974, MARKWELL and PAULSON 1980, FIDGEN 1975, HOLMGREN et al. 1980, HAYWOOD and BOYER 1982, MARKWELL et al. 1981, 1984, HSU et al. 1979), papovaviruses (ROGERS and PAULSON 1983, GETHING and SAMBROOK 1982, NOZIMA et al. 1968, YASUI et al. 1968, BOLEN et al. 1979, FRIED et al. 1981), and picornaviruses (BURNESS 1981, BURNESS and PARDOE 1981, PARDOE and BURNESS 1981). In the typical case the viral hemagglutinin is a component of the virus which is coded by the viral genome. Both nonenveloped and enveloped viruses display hemagglutinating ability. In nonenveloped viruses the VAP is a component of the capsid; in viruses that form a double-shelled particle it appears to be a component of the outer capsid (BOULANGER and LONBERG-HOLM 1981). In enveloped viruses it is a component of projections from the surface of the viral envelope called spikes or peplomers (FIELDS 1985).

In the case of the Sendai virus it has been shown that host cells containing the NeuAcα2-3Gal sequence are susceptible to infection. Although the se-

Table 9. Receptors for viral attachment proteins and viral agglutinins (lectins)

Viruses	Target tissue	Virus attachment protein	Carbohydrate specificity
Myxoviruses			
Influenza A	Respiratory	HA-spike, NA-spike	Sialic acid
Influenza B	tract	HA-spike	Sialic acid
Influenza C		gp-spike	O-Acetyl-Sialic acid
Papovaviruses			
Japanese encephalitis virus	Nervous system	–	Mannose
Polyoma viruses	Oncogenic	Capsid protein	Sialic acid
Paramyxovirus			
Sendai virus	Respiratory tract	HN-spike	Neu5Acα2-3Gal Neu5Acα2-8 NeuAcα2-3Gal
Picornavirus			
Cardiovirus (EMC)	Nervous system	Capsid protein	Sialic acid

gp, glycoprotein; *HA*, hemagglutination; *NeuAc*, N-acetylneuraminic acid, *Gal*, galactose; *NA*, neuraminidase.

quence NeuAcα2-3Gal is recognized by the virus carbohydrate binding protein, the related sequence NeuAcα2-8NeuAcα2-3Gal represents the receptor because the virus binds to it with a higher affinity. Upon testing the virus adhesion and infection in an actual host system, it was shown that the ganglioside GT1b, which contains the NeuAcα2-3Gal sequence and for which the virus has a moderate affinity, is sufficient for infection (MARKWELL et al. 1981). However, high affinity binding of the virus has been shown to the ganglioside GQ1b. The demonstration of the endogenous occurrence of this type of ganglioside in the target cells and its function as a receptor completed the first definition of a receptor for a lectin of a mammalian virus (MARKWELL et al. 1984).

Although several families have a requirement for the same sugar, i.e., sialic acid, the attachment of a virus is regarded as a very specific phenomenon. Therefore, it has been suggested that individual viruses have preferences for different forms of sialic acid. However, there are also experimental data which suggest only an indirect role of sialic acid in the attachment of the viruses (BURNESS 1981). This hypothesis is based on several findings. For example, on the one hand free sialic acid is not able to inhibit the attachment of viruses, and on the other hand extensive neuraminidase treatment does not completely inhibit the binding of the virus.

However, the mechanisms described for the interaction of virus attachment proteins with the corresponding carbohydrates on the host cells did not take into consideration the fact that many mammalian cells also contain endogenous carbohydrate-binding proteins (endogenous lectins). These endogenous lectins can likewise be involved in virus attachment by binding to carbohydrates of the virus. In most studies, binding of the virus has been attributed to the VAPs while the target cells have been considered to represent the passive glycoconjugate acceptor presenting the appropriate carbohydrates for the VAPs. Using a mutant of Sendai virus which was lacking the VAP, binding and infection by this virus were observed in hepatoma cell lines which expressed endogenous galactose-binding activity (asialoglycoproteinreceptor). Consequently binding of the virus could be inhibited by blocking the endogenous lectin of the target cell with galactose or N-acetyl-galactosamine (MARKWELL et al. 1985). As proposed for the mechanism of bacterial adhesion (VIERBUCHEN et al. 1989 a), these results open up the possibility that binding and infection by carbohydrate-containing viruses may be mediated by a dual recognition system involving the VAPs of the virus and the endogenous lectins of the host cells as well as the corresponding carbohydrates on the surface of the reacting cells. Although VAPs have some properties in common with other lectins of plant or microbial origin, they possess special features like high multivalency, an ability to exist as a mixed population, and the presence of receptor-destroying enzymes (RDEs). The presence of virus-associated enzymatic activity which destroys the receptor required by the virus for entry into the target cell is an intriguing feature of myxo- and paramyxoviruses. Several functions have been proposed for the viral sialidases (SCHEID 1981), including the inoculation of the virus into the target cell, the exit of the progeny virus from the infected cell, and the release of inoculating viruses from nonproductive binding to sialoglycoconjugates. Furthermore, it has been emphasized that at low temperature RDEs can become pseudolectins. At these temperatures the enzymatic activity is inhibited while enzyme-substrate and enzyme-product complexes can still be formed. In this context one can speculate that lectins might be enzymes that have lost their enzymatic activity during the isolation and purification procedures but have retained their substrate-binding capacity. With respect to the great number of different chemical forms of naturally occurring sialic acid types, some lectin activities might represent an enzyme with a very high substrate specificity and consequently some sugar types may act as nonfunctional substrate analogs. Finally, it might be possible that beneath the active enzymatic center there is a lectin-like binding site essential for forming the reactive enzyme-substrate complex.

7.2 Bacterial Lectins and Carbohydrate-Binding Bacterial Toxins

7.2.1 Bacterial Lectins

Bacterial attachment to target cells may be due to spcific or nonspecific binding mechanisms. Nonspecific binding of bacteria to target cells involves electrostatic forces, hydrophilic-, and hydrophobic forces as well as high polymer weight charged mediated adherence and aggregation. As already mentioned, the term adhesion describes a specific interaction of recognition proteins (adhesins) on the bacterial cell surface with the corresponding sugars of glycoconjugates (glycoproteins or glycolipids) of the target. Since the receptors for bacterial adhesins are found on erythrocytes, hemagglutination is often used as an in vitro assay for adhesion. Thus adhesins may also be termed hemagglutinins or bacterial lectins. The ability of bacteria to agglutinate erythrocytes was recognized at the beginning of this century (Guyot 1908). Binding of the bacteria is mediated by proteinaceous appendages on the bacterial surface (Duguid et al. 1955, Rivier and Darekar 1975, Salit and Gotschlich 1977a, 1977b) which carry the lectin activity for mannose. The sugar specificity of the E. coli lectin has been further supported by Ofek and co-workers (1977, 1978) who showed that the epithelial binding of the bacteria is associated with fimbriae and is sensitive to inhibition with mannose. It is now generally recognized that a large number of bacteria possess carbohydrate-binding proteins which mediate the specific binding (adhesion) to the carbohydrate receptor of the target cell (Mirelman and Ofek 1986). Most of these bacteria belong to the Enterobacteriae family. Based on their carbohydrate specificity, two types of fimbria can be distinguished, i.e., mannose specific and host specific. The mannose-specific fimbriae are found on saprophytic and pathogenic E. coli, while the host-specific fimbriae are restricted to pathogenic bacteria. To distinguish the mannose-specific fimbriae from other pili they were called type 1 fimbriae or common fimbriae. Hemagglutination was the first manifestation of the adhesive properties of E. coli.

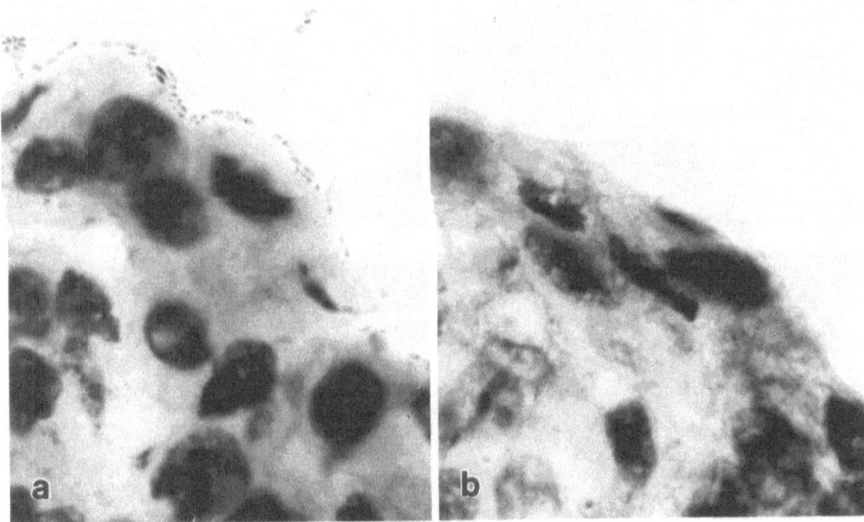

Fig. 17. a Binding of uropathogenic *E. coli* bacteria to the surface of the urothelium of the renal pelvis. **b** Inhibition of bacterial adhesion in the presence of 0.01 M mannose

The specificity of this interaction has been studied in in vitro experiments using whole bacteria and various types of cells. The binding reactions can be blocked by D-mannose. Detailed characterization of the mannose combining site was achieved by quantitative examination of the inhibitory effects of a large number of simple sugars and complex carbohydrates of mannose on the bacteria-induced agglutination of yeast (FIRON et al. 1984). These studies revealed that the combining site of the lectin accommodates best to branched mannosyl oligosaccharides.

In general, it is reasonable to conclude that bacteria with mannose specific lectins bind structures found in short oligomannose sequences of N-linked glycoproteins and not to glycolipids which do not contain this sugar. Using human urothelium, binding of the bacteria was reduced when *E. coli* was incubated in the presence of mannose (Fig. 17). However, detailed studies revealed that the urothelium also possesses endogenous carbohydrate-binding proteins (Fig. 18) which can also contribute to the adhesion of the bacteria (VIERBUCHEN et al. 1989b). Therefore a dual recognition mechanism for the specific adhesion of bacteria has been proposed which is accomplished by the simultaneous interaction of the bacterial and host cell lectins with the corresponding sugars of the bacterium and target cell, respectively (Fig. 19). NeuNAcα-(2-3,6)-β-Gal is recognized by S-specific adhesins of *E. coli* causing sepsis and neonatal meningitis (KORHONEN et al. 1984). β-GlcNAc-(1-3)-β-GlcNAc represents the combining site for G-specific adhesins of urotoxogenic bacteria. Other uropathogenic strains of *E. coli* with so-called X fimbriae bind to α 2-6 and α 2-3 neuraminic acid residues of glycophorin and other sialoglycoproteins (LINDAHL et al. 1988). Also *Campylobacter pyloris* which is asso-

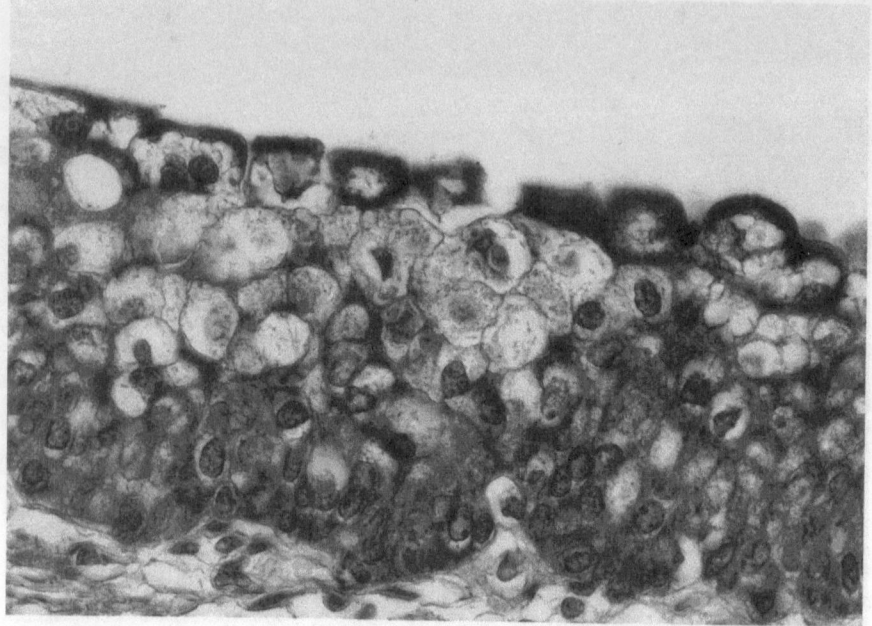

Fig. 18. Endogenous mannose-binding proteins concentrated along the luminal surface of the urothelium of the renal pelvis. Mannose-BSA-biotin-streptavidin technique, hematoxylin counterstaining of cell nuclei

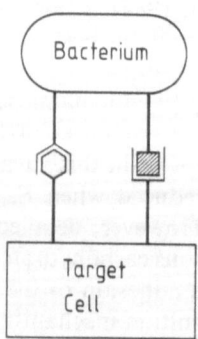

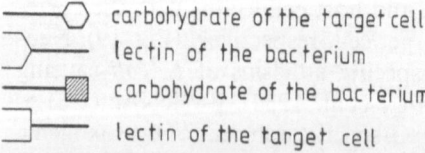

Fig. 19. Schematic representation of a dual recognition mechanism for bacterial adhesion mediated by the combined action of the bacterial and target cell lectins with the corresponding sugars of the bacterium and the target cell

ciated with human gastritis and pyloric ulcers, commonly expresses a sialic acid-specific cell surface lectin probably of nonfimbrial nature (WADSTRÖM 1988). Many carbohydrates which are recognized by bacterial lectins (especially mannose-specific lectins) are expressed on the glycan part of glycoproteins. Beside these lectin activities, bacterial lectins have been found in mannose-resistant strains which bind to the carbohydrate part of glycolipids of the globoside type (BOCK et al. 1985). Carbohydrate receptor of this bacterial lectin has been elucidated as Gal (1-4)Gal which is the receptor of the P-specific adhesins of urotoxogenic bacteria. This receptor for bacterial lectins is also found in the blood group P determinant. Therefore, from the genetic point of view the blood group systems may be an important element of the natural resistance systems, whether the individual possesses the blood group or not.

It has been known for some years that many bacteria and viruses produce and secrete glycosidases, especially neuraminidase. Immunochemical studies have shown that this enzyme removes terminal sialic acid from the glycan part of the complex carbohydrates. However, there is little information on the role of this viral and bacterial enzyme in the pathogenesis of infectious disease. Bacterial and viral neuraminidase can uncover sialic acid-masked Galβ(1-3)GalNAc residues which serve as receptors for certain *Actinomyces* strains or for certain strains of *Pseudomonas aeruginosa* (UHLENBRUCK 1987). On the other hand, neuraminidase destroys human blood group substance M, which specifically binds the pyelonephritic *E. coli* strain 1H 11165.

In this context, neuraminidase of viral or bacterial origin might be an important factor for the mechanism of co- and superinfection (UHLENBRUCK 1987). It is well known that influenzal pneumonia is often followed or superimposed by bacterial infection. Therefore one may speculate that the viral neuraminidase liberates cryptic carbohydrate structures which are recognized by bacterial lectins.

From lectin-histochemical studies it has been deduced that pneumococcal neuraminidase may liberate the cryptic, i.e., sialic acid-masked Thomsen-Friedenreich antigen (VIERBUCHEN and KLEIN 1983). By this mechanism an immunereaction is induced because all sera contain natural antibodies to this cryptic antigen. On the other hand, it has been proposed that neuraminidase may unmask the carbohydrate-binding site for bacterial lectins or bacterial toxins (VIERBUCHEN 1987) by removing terminal sialic acid, thereby enhancing bacterial and/or bacterial toxin binding. Much of the current knowledge on microbial lectins and agglutinins reflects some molecular aspects of the role of these molecules in pathogen–host interaction. Further studies on the combined action of microbial enzymes and microbial lectins on the target cell will reveal new aspects on the molecular mechanisms of infectious diseases.

7.2.2 Carbohydrate-Binding Bacterial Toxins

As described in the proceeding section, the adhesion of bacteria represents the initial and most important step in infectious disease, beeing followed by a cascade of further biochemical events leading to the pathological sequelae. In this context adhesion may be of special importance for toxin-producing bacteria, bringing the toxin into close contact with the toxin-binding cell surface receptor.

One of the earliest successes of micobiology was the successful prevention of bacterial diseases like diphtheria and tetanus by immunization with toxoid vaccines. A number of bacterial toxins have been shown to bind to the glycoproteins or glycolipids of the surface of the target cells. It is most likely that the carbohydrate parts of these complex carbohydrates are the receptos for the specific interaction between the toxin and the host cell. With respect to the carbohydrate specificity, certain bacterial toxins show some characteristics of (incomplete, monovalent) lectins.

A number of bacterial toxins have been shown to consist of two major components, each possessing distinct biochemical and biological properties (LEVINE et al. 1983, MIDDLEBROOK and DORLAND 1984, EIDELS et al. 1984).

These peptide chains are referred as the A (toxophoric) and the B (haptophoric) chain respectively. Both peptide chains, which are linked together by disulfide bonds, represent the holotoxin.

Recently a third domain has been postulated (KEUSCH 1981). As schematically illustrated in Fig. 20, this domain is responsible for the entry (entry domain, E-domain) of the toxin into the cell. In the absence of the B chain, the A chain is inactive to intact cells but can exert its effects in a cell-free system. In general, activation of the A chain requires limited proteolysis and often reduction of the disulfide bond.

Carbohydrate-binding bacterial toxins have been identified in various species; examples are the cholera toxin (FISHMAN et al. 1976, MOSS et al. 1976, CRITCHLEY et al. 1981, 1982, MORITA et al. 1980), the *E. coli* heat-labile toxin (MOSS et al. 1979, 1981; DONITA et al. 1982), the diphteria toxin (Draper et al. 1978, MEKADA et al. 1979), the pseudomonas toxin (MIDDLEBROOK et al. 1978,

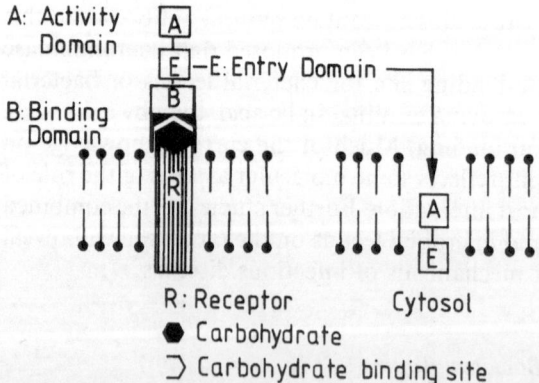

Fig. 20. Schematic representation of the interaction of carbohydrate-binding bacterial toxins with the target cell receptor. Modified according to the three domain-model of toxin receptor structure (KEUSCH 1981)

VASIL and IGLEWSKI 1978), the shigella toxin (KEUSCH and JACEWICZ 1977), the tetanus toxin (LEE et al. 1978), and the botulin toxin (KITAMURA et al. 1980).

Although all toxins are primarily bound via a complex glycoconjugate to the target, they exhibit different biological effects on the target cell. Some of these biological and biochemical characteristics as well as the carbohydrate nature of the receptor for these bacterial toxins are summarized in Table 10.

7.3 Receptors for Protozoan Lectins

Specific recognition systems are responsible for the selectivity with which the parasite binds to the host cell. For example the binding of *Giardia lamblia* is confined to the enterocytes of the proximal small intestine. *Leishmania* parasites infect macrophages exclusively. The binding of the parasites can be inhibited by carbohydrates, leading to the conclusion that the attachment of these parasites is mediated by lectin-like activities (PEREIRA 1986). However, there are no experimental data which further support the concept of a lectin-mediated recognition system for *Leishmania* or *Trypanosoma cruzi* infection (PEREIRA 1986).

In contrast, lectins from *Entamoeba histolytica* (TAKEUCHI and PHILLIPS 1975, GUERRANT et al. 1981, JARUMILINTA and MAEGRAITH 1969, RAVDIN and GUERRANT 1981, BRACHA et al. 1982, BRACHA and MIRELMAN 1983, 1984), *Giardia lamblia* (BOAZ et al. 1986), and *Plasmodium falciparum* (MILLER et al. 1977, PERKINS 1981, WEISS et al. 1981, DEAS and LEE 1981, HOWARD et al. 1982, PASVOL et al. 1982, JUNGERY et al. 1983, SCHULMAN et al. 1983, FACER 1983) have been detected.

Some data from the literature on the sugar receptors, the localizations of the lectins, and their target cells are summarized in Table 11.

Only the trophozoites of *Entamoeba histolytica* affect human tissues. The trophozoites may be confined to the intestinal lumen without giving rise to clinical disease or may penetrate the mucosa and destroy it because of the proteolytic enzymes they produce. Virulent strains of *Entamoeba histolytica* contain at least two distinct lectins. One appears to be a membrane-bound protein, while the other is loosely bound to the membrane and easily removed. The membrane-bound lectin has hemagglutinating activity and is inhibited by micromolar concentrations of oligosaccharides containing GLcNAc (KOBILER and MIRELMAN 1980, 1981). This lectin may be responsible for the attachment of the trophozoites to cultured human intestinal epithelial cells (KOBILER and MIRELMAN 1981). The soluble lectin is inhibited by micromolar concentrations of GalNAC and asialofetuin (RAVDIN and GUERRANT 1981, RAVDIN et al. 1985). The GalNAc-inhibitable lectin may play a role in contact-mediated cytolysis (RAVDIN and GUERRANT 1981, RAVDIN 1986). Recently a 220 000-dalton protein with lectin-like properties has been isolated from *Entamoeba histolytica*. The lectin is inhibited by micromolar concentrations of hyaluronic acid, chitin, chitin-derived products, and antibodies directed against the lectin.

Table 10. Bacterial toxins – Carbohydrate nature of bacterial toxin-binding sites (receptors) and their biological effects and biochemical activities

Microbial toxin	Biochemical/biological activities of the toxin	Chemical structures involved in the binding of the toxin
Protein synthesis inhibitors		
Diphtheria toxin	Interaction with cytosolic polypeptide elongation factor 2 (EF2) by transfer of one molecule of ADP-ribose from NADH to one molecule of EF2	Glycoproteins, N-acetylglucosamine
Pseudomonas exotoxin A	NAD-dependent ADP ribosylation for EF2	Glycoproteins, sugar structures unknown
Shigella toxin	Acts catalytically; irreversibly inactivates ribosomes through action on the 60s ribosomal subunit. Affects peptide chain elongation	Glycoproteins, glycolipids, β1-4 linked, N-acetylglucosamine
Nucleotide cyclase activating toxins		
Cholera toxin	ADP-ribosylation transferase and NAD-glycohydrolase activity, resulting in ADP ribosylation and prolonged activation of adenylate cyclase activity of the cell. Elevation of intracellular cAMP causes a decrease in the neutral absorption of sodium and chloride in the villus tip cell and an increase in the active transport into the intestinal lumen	Ganglioside G_{M1} $Gal\beta(1\text{-}3)GalNAc\beta(1\text{-}4)Gal\beta(1\text{-}4)Glc\text{-}Cer$ $\qquad\qquad\qquad\quad 3$ $\qquad\qquad\qquad\quad |$ $\qquad\qquad\qquad NeuAc\alpha2$
E. coli heat labile toxin	Identical to cholera toxin in structure and function	Ganglioside G_{M1}

Neurotoxins		
Tetanus toxin	Mechanism of action unknown; inhibition of choline uptake and release of acetylcholine and norepinephrine	1) Ganglioside G_{D1b}

Galβ(1-3)GalNAcβ(1-4)Galβ(1-4)Glc-Cer
                  3
                  |
            NeuAcα2
                  8
                  |
                  NeuAcα2

2) Ganglioside G_{T1b}

Galβ(1-3)GalNAcβ(1-4)Galβ(1-4)Glc-Cer
 3   3
 | |
NeuAcα2   NeuAcα2
                      8
                      |
                    NeuAcα2

3) Glycoprotein-binding sites

| Botulinus toxin | Blocking of acetylcholine from cholinergic nerve endings | Ganglioside G_{T1b} |

Cer, ceramide; *Glc*, glucose; *Gal*, galactose; *GalNAc*, N-acetylgalactosamine; *NeuAC*, N-acetylneuraminic acid.

Table 11. Sugar receptors for protozoan lectins

Protozoan	Target cell/ target tissue	Carbohydrate specificity	Localization of the lectin
Entamoeba histolytica	Large intestine	(GlcNAc)$_{2-3}$ GalNAc	Membrane Soluble, loosely bound to the membrane
Giardia lamblia	Small intestine	Asialofetuin GalNAc	Membrane Soluble
Plasmodium falciparum	Erythrocytes	GlcNAc-albumin Glycophorin Sialic acid	Membrane

GlcNac, N-acetylglucosamine; *GalNAc*, N-acetylgalactosamin.

It is assumed that this lectin could be one of the putative receptor molecules involved in cell and matrix attachment (ROSALES-ENCINA et al. 1987). In conclusion, *Entamoeba histolytica* contains two lectins of different carbohydrate specificity which are involved in adhesion to the epithelium, colonization, and cytolytic activities as well as the metastatic spread of the parasite. *Giardia lamblia* is one of the most widespread of intestinal protozoa. The infecting organism attaches to the lining of the small intestine and in heavy infections causes an inflammatory reaction and interference with the absorption of foodstuffs.

A membrane-bound lectin has been demonstrated in *Giardia lamblia* by hemagglutination inhibition tests. Most interestingly, the lectin is activated by proteolytic action consistent with a limited proteolysis (BOAZ et al. 1986). *Giardia lamblia* colonizes the proximal small intestine, whose fluid is rich in proteolytic enzymes of the pancreas. The proteolytic enzymes of the proximal intestinal fluid may activate the protozoan lectin, thereby enhancing the lectin-mediated attachment of *Giardia lamblia* to the enterocytes.

Malaria parasites are obligate intracellular parasites and ability to invade erythrocytes is crucial to their survival (HADLEY et al. 1986). In *Plasmodium falciparum* infection there are many lines of evidence that at least some stages of recognition and attachment occurring prior to the invasion of the target cell involve lectin-like activities.

Several areas of the glycophorin molecule play a coordinated and synergistic role in the merozoite association with the red blood cells, as indicated by the following findings:

1. Red blood cells deficient in one or more glycophorins are more resistant to merozoite invasion than the corresponding normal cells.
2. Glycophorin in solution can also block invasion.
3. Specific fragments of glycophorin can inhibit invasion.
4. Monoclonal antibodies to specific epitopes on glycophorin A can block invasion.

Sialic acid on the erythrocyte membrane is required for optimal invasion. Beside erythrocyte glycophorins, it has been shown that N-acetylglucosamine both in solution and inserted into liposomes can block the invasion of erythrocytes by *Plasmodium falciparum*.

Blocking of invasion was particularly effective when the sugar was coupled to bovine serum albumin.

Studies on the invasion of $M^k M^k$ erythrocytes that lack glycophorin A and B and of enzyme-treated erythrocytes by a sialic acid-dependent parasite and a relatively sialic acid-independent parasite revealed differences in the importance of sialic acid, the peptide backbone of glycophorin, and other erythrocyte ligands for the invasion of erythrocytes by *Plasmodium falciparum* (HADLEY et al. 1987). It is suggested that the ligand on glycophorin is primarily sialic acid and may not require the protein backbone. However, the receptor heterogeneity includes different requirements for sialic acids (MITCHELL et al. 1986) and the requirement for a trypsin-sensitive receptor.

Beside the receptor which interacts with sialic acid, there is a second receptor which interacts with a neuraminidase-insensitive, trypsin-sensitive ligand that is located on a molecule other than glycophorin A or B (HADLEY et al. 1987). In addition to the receptors for the glycophorins, this glycophorin-independent receptor needs to be identified if receptors are to be used as effective immunogens (HADLEY et al. 1987).

8 Lectins as Histochemical Tools for the In Situ Localizations of Carbohydrates

8.1 Carbohydrate Specificities of Lectins

Mammalian cells invariably contain an array of glycosylated moieties, both inside the cell and on the cell surface. Although the glycan parts of complex carbohydrates have been analyzed by biochemical techniques, until recently only relatively unspecific staining techniques have been available for the demonstration of carbohydrates in tissue sections. Such staining procedures include the periodic acid-Schiff reaction, the colloidal iron reaction, and modifications of them. Given the many important biological functions of the sugars in complex glycoconjugates, there is an increasing awareness of histochemical tools for the demonstration of these important compounds. To this end, beside monoclonal antibodies lectins are extremely useful tools for the demonstration of carbohydrates in tissue sections. Despite a rather heterogeneous derivation from viruses, bacteria, protozoa, plants, and animal cells, lectins are best classified on the basis of their carbohydrate specificity. Lectins recognize D-pyranose sugars and can be classified into a limited number of carbohydrate-binding groups depending on their preferential of the following sugars, as follows: mannose/glucose-binding lectins, N-acetylgalactosamine/galactose-binding-lectins, N-acetylglucosamine-binding lectins, fucose-binding lectins, and sialic acid-binding lectins (GOLDSTEIN and PORETZ 1986).

In most instances lectins used as tools in histochemistry are isolated from plants or invertebrates. Lectins which have been used in histochemistry and which may represent future tools in carbohydrate histochemistry have been drawn from the literature (GALLAGHER 1984, GOLDSTEIN and PORETZ 1986, WU et al. 1987, WU and SUGUII 1988, UHLENBRUCK et al. 1988 a), and are listed in Table 12, along with their carbohydrate specificities. Although the combining site of some lectins appears complementary to a single sugar, nominal monosaccharide specificity does not always reflect the exact nature of the carbohydrate-binding site. Many lectins have been found to possess extended binding sites (Table 12), accommodating two to six carbohydrate units. In pioneering studies Kabat and Lemieux and co-workers have defined binding epitopes of plant lectins and antibodies (BOCK et al. 1988, LEMIEUX et al. 1984, 1985). The protein interacts with a tight cluster of two or three hydroxyl groups which may originate in one or two sugar units. This strong polar interaction largely determines the specificity of the binding. In addition to this, there is a more extensive binding region which is nonpolar in nature and extends from the polar grouping. The interaction of this surface with a complementary nonpolar surface of the protein is the main source of the stability of the complex.

Some lectins show marked differences with respect to their anomeric specificity, whereas others appear to be almost identical in this respect.

Examples of the former are the lectins resolving ABH(0) blood expression (Table 13), the *Pisum sativum* agglutinin (PSA), the lectins from *Griffonia simplicifolia*, the *Lotus tetragonolobus* lectin, and concanavalin A (Con A) (GOLDSTEIN and PORETZ 1986). In contrast, soybean agglutinin (SBA) and *Ricinus communis* agglutinin (RCA) will bind to both α and β forms of Gal-NAc or Gal respectively.

8.2 Detection of Lectin-Binding Sites

The presence of lectin receptors on cell surfaces, within the cytoplasm, and in cell organelles can be readily demonstrated by the use of suitable lectin derivatives, in a way generally similar to the techniques employed for the immunohistochemical demonstration of antigenic structures.

Lectins are in general stable enough molecules to withstand conjugation with various labels, including enzymes, fluorochromes, radioisotopes, solid phase matrices, and electron-dense metals, without inactivation of carbohydrate-binding sites. Additionally, antilectin antibodies and biotinylated lectin derivatives can be used in appropriate indirect assays. Figure 21 displays some general approaches for the histochemical detection of lectin-binding sites in histochemistry (VIERBUCHEN et al. 1980, HSU and RAINE 1982, KUHLMANN and PESCHKE 1984, SCHAUMBURG-LEVER et al. 1984a, SKUTELSKY et al. 1987).

Controls for staining specificity include the incubation of the lectin in the presence of an excess (0.2 M) of its specific blocking sugar.

Table 12. Carbohydrate-binding specificities (receptors) of lectins

Lectin	Carbohydrate specificities

I. N-Acetylgalactosamine (GalNAc) binding lectins

a) GalNAcα(1-3)GalNAc (Forssman antigen) specific lectins

Helix pomatia (edible snail) agglutinin (HPA)	GalNAcα(1-3)GalNAcβ(1-3)Galα(1-4)Galβ(1-4)Glc > Forssman specific pentasaccharide GalNAcα(1-3)GalNAc > GalNAcα(1-3)Galβ(1-3)GlcNAc > blood group A specific trisaccharide GalNAc
Dolichos biflorus (horse gram) agglutinin (DBA)	GalNAcα(1-3)GalNAcβ(1-3)Galα(1-4)Galβ(1-4)Glc > Forssman specific pentasaccharide GalNAcα(1-3)GalNAc ≫ GalNAcα(1-3)Galβ(1-3)GlcNAc > blood group A specific trisaccharide GalNAcα(1-3)Gal > GalNAc
Wistaria floribunda agglutinin (WFA)	GalNAcα(1-6) > GalNAcα(1-3)Gal = GalNAcα(1-3)GalNAcβ(1-3)Galα(1-4)Galβ(1-4)Glc > Forssman specific pentasaccharide GalNAcα(1-3)Galβ(1-3)GlcNAc > blood group A specific trisaccharide GalNAcβ(1-3)Galα(1-4)Galβ(1-4)Glc > GalNAcα(1-3)Gal > GalNAc > GalNAcα(1-3)[L-Fucα1-2]Galβ(1-4)GlcNAcβ(1-6)R > Galβ(1-4)GlcNAC > Galβ(1-4)Glc
Amphicarpaea bracteata (hog peanut) agglutinin (ABrA)	GalNAcα(1-3)GalNAc > GalNAcα(1-3)Gal > GalNAc

b) GalNAcα(1-3)Gal (blood group A) specific lectins

Glycine max (soybean) agglutinin (SBA)	GalNAcα or β-linked GalNAcα(1-3)Galβ(1-3)GlcNAc = GalNAcα(1-3)Gal ≫ GalNAcα(1-3)[L-Fucα1-2]Galβ(1-4)GlcNAcβ(1-6)R > Galα(1-6)Glc (Melibiose) > Galβ(1-4)Glc (Lactose)
Phaseolus lunatus (lima bean) agglutinin (LBA)	GalNAcα(1-3)[L-F-Fucα1-2]Galβ(1-4)Glc > GalNAc > Gal
Vicia villosa (hairy vetch) agglutinin (VAA)	GalNAcα(1-3)Gal > GalNAcα(1-6)Gal > GalNAcα(1-3)Galβ(1-3)GlcNAc > GalNAcα(1-3)[L-Fucα1-2]Galβ(1-4)GlcNAcβ(1-6)-R
Griffonia (Bandeiraea) simplicifolia agglutinin A4 (GSA-A4)	GalNAcα(1-3)Galβ(1-3)GlcNAc = GalNAcα(1-3)Gal = Gal (1-6)Gal > GalNAcα1-R > Galα(1-3)Gal, Galα(1-6)Glc > Gal

Table 12 (continued)

Lectin	Carbohydrate specificities

c) GalNAc 1-Ser or Thr (Tn-antigen) specific lectins

Vicia villosa B4 (hairy vetch) agglutinin (VVA-B4)	A fetuin glycopeptide containing two GalNAcα1-0 to Ser or Thre of the peptide chains (two Tn structures) > single Tn structure > Galβ(1-3)GalNAcα1-0-Ser/Thr > NeuACα(2-3)Ga1β(1-3)GalNAcα1-0-Ser/Thre
Salvia sclarea agglutinin (SSA)	Glycopeptides containing two Tn structures > Glycopeptides containing one Tn structure

d) Other GalNAc-binding lectins

Tridacna maxima (röding clam) agglutinin (TMA)	GalNAc > Galβ(1-6)Gal, Galβ(1-4)Glc, Gal > Galα(1-6)Glc
Macrotyloma axillare agglutinin (MAL)	GalNAc
Vicia cracca agglutinin (VCA)	GalNAc
Bryonica diodica agglutinin (BDA)	GalNAc > Gal (1-4)Glc > Gal (1-6)Glc

II. Galactose (Gal) binding lectins

a) Galβ(1-3)GalNAc (Thomson-Friedenreich antigen, T) specific lectins

Bauhinia purpurea alba agglutinin (BPA)	Galβ(1-3)GalNAc > Galα(1-3)GalNAcβ(1-3)Galβ(1-4)Glc > lacto-N-tetraose
BPA	Galβ(1-3)GlcNAc, GalNAcα(1-3)Galβ(1-3)GlcNAc > blood group A specific trisaccharide Galβ(1-4)GlcNAc, Galβ(1-4)Glc > Gal
Arachis hypogaea (peanut) agglutinin (PNA)	Galβ(1-3)GalNAc > Galβ(1-4)GlcNAc > Galβ(1-3)GlcNAc > Galβ(1-4)Glc > Gal ≫ GalNAc
Maclura pomifera (Osage orange tree) agglutinin (MPA)	Galβ(1-3)GalNAc > GalNAc (1-6)Gal > Galα(1-6)Glc > GalNAc > Gal
Vicia graminea agglutinin (VGA)	Galβ(1-3)GalNAcα1-linked (clusters) and desialized blood group N glycopeptide > NANAα(2-3)Galβ(1-3)GalNAcα1-linked > sialized blood group N glycopeptide desialized M blood group glycopeptide ≫ sialized blood group M glycopeptide
Agaricus bisporus (mushroom) agglutinin (ABiA)	Galβ(1-3)GalNAcα-N-O-tosylserine
Artocarpus integrifolia agglutinin, (Jacalin, AIL)	Galβ(1-3)GalNAc > Galβ(1-3)GlcNAc, Galβ(1-4)GlcNAc, Galβ(1-4)Glc

Table 12 (continued)

Lectin	Carbohydrate specificities
Sophora japonica (Japanese pagoda tree) agglutinin (SJA)	Galβ(1-3)GalNAC-N-tosyl-ser > Galβ(1-3)GlcNAc > Galβ(1-3)GlcNAcβ(1-3)Galβ(1-4)Glc > Galβ(1-3)GlcNAcβ(1-6)Gal > Galβ(1-4)GlcNAcβ1-4)Gal > Galβ(1-4)GlcNAc > GalNAc > lactose > Gal
Ricin (RCA2)	[Galβ(1-3)GalNAcα1-Ser/Thr]₄ glycopeptide > Triantennary oligosaccharides containing N-acetyllactosamine at non-reducing end > [Galβ(1-3)GalNAcα1-Ser/Thr] glycopeptide > Galβ(1-3)GalNAcα1-Ser/Thr > GalNAcα1-Ser/Thr

b) Galβ(1-3,4)GlcNAc (human blood group type I and type II carbohydrate precursor chain)
   binding lectins

Lectin	Carbohydrate specificities
Ricinus communis (castor bean) agglutinin (RCA1)	Biantennary oligosaccharides containing N-acetyllactosamine linked units at non reducing end > Galβ(1-4)GlcNAcβ(1-6)Gal > Galβ(1-4)GlcNAc > Galβ(1-3)GlcNAc > Galβ(1-4)Glc
Erythrina crisstagalli agglutinin (ECA)	Tetraantennary and triantennary oligosaccharides containing four or three N-acetyllactosamine branches, respectively > biantennary oligosaccharides containing two N-acetyllactosamine branches > monoantennary oligosaccharide containing one N-acetyllactosamine residue > Galβ(1-4)GlcNAc > Galβ(1-3)GlcNAc > GalNAc > Gal

Datura stramonium (thorn apple) agglutinin DTA

$$
\begin{array}{l}
\text{Galβ(1-4)GlcNAcβ(1-2)} \\
\qquad\qquad | \\
\qquad\qquad \text{Man- > (GlcNAcβ1-4)}_3 > \\
\qquad\qquad | \\
\text{Galβ(1-4)GlcNAcβ(1-6)} \\
\text{Galβ(1-4)GlcNAc > GlcNAcβ(1-4)Glc}
\end{array}
$$

Phaseolus vulgaris (isoagglutinin L4) (red kidney bean) leukoagglutinin (PHA-l)

Tri- and tetraantennary structures containing Galβ(1-4)GlcNAc-(N-acteyllactosamine)- linked units at nonreducing end >

$$
\begin{array}{l}
\text{Galβ(1-4)GlcNAcβ(1-2)Manα(1-3)} \\
\qquad\qquad\qquad\qquad | \\
\qquad\qquad\qquad\qquad \text{Man >} \\
\qquad\qquad\qquad\qquad | \\
\text{Galβ(1-4)GlcNAcβ(1-2)Manα(1-6)} \\
\text{Galβ(1-4)GlcNAcβ(1-2)Man > GlcNAcβ(1-2)Man}
\end{array}
$$

Phaseolus vulgaris (isoagglutinin E4) (red kidney bean) erythroagglutinin (PHA-E-4)

Minimum structural unit:

$$
\begin{array}{l}
\text{Galβ(1-4)GlcNAcβ(1-2)Manα(1-6)} \\
\qquad\qquad\qquad | \\
\qquad\qquad\text{GlcNAcβ(1-4)-Manβ(1-4)GlcNAc-R1} \\
\qquad\qquad\qquad | \\
\text{R3 —— GlcNAcβ(1-2)Manα(1-3)} \\
\qquad\qquad\qquad | \\
\qquad\qquad\qquad \text{R2}
\end{array}
$$

R1 :GlcNAc or Fucα(1-6)GlcNAc
R2, R3: hydrogen atoms or sugars

Table 12 (continued)

Lectin	Carbohydrate specificities
Geodia cydonium agglutinin (GCA)	Galβ(1-4)GlcNAc, Galβ(1-4)Glc, Galβ(1-3)GlcNAc> GalNAc

c) Galα(1-3)Gal (blood group B) specific lectin

Griffonia simplicifolia I-B4 agglutinin (GSA-B4)	Galα(1-3)Gal, melibiose > raffinose > Gal > GalNAc

d) Other galactose-binding lectins

Abrus precatorius (jequirity bean) agglutinin (APA)	Lactose, Gal > melibiose > Fuc
Adenia digitata agglutinin (ADA)	Gal > GalNAc
Crotalaria juncea (sunhemp) agglutinin (CJA)	Lactose > melibiose > raffinose > GalNAc > Gal
Hura crepitans (sand-box tree) agglutinin (HCA)	Lactose > Gal and melibiose
Momordica charantica agglutinin (MCA)	Gal > GalNAc
Psophocarpus tetragonolobus agglutinin	GalNAc/Gal
Viscium album (mistletoe) agglutinin I (VAA I)	Gal > GalNAc

III. N-Acetylglucosamine-specific lectins

a) Chitin oligosaccharide specific lectins

Triticum vulgaris wheat germ agglutinin (WGA)	$(GlcNAc\beta1-4)_5 > (GlcNAc\beta1-4)_4$ and $(GlcNAc1-4)_3 >$ GlcNAcβ(1-4)GlcNAc > Manβ(1-4)GlcNAcβ-)GlcNAc-Asn > N-acetylglucosamine and sialic acid
Phytolacca americana (pokeweed) agglutinin (PAA)	$(GlcNAc\beta1-4GlcNAc)$ n 1-3
Solanum tuberosum (potato) agglutinin (STA)	$(GlcNAc\beta1-4)_4 > (GlcNAc\beta1-4)_3 >$ $(GlcNAc\beta1-4)_2 > GlcNAc$
Griffonia (Bandeiraea) simplicifolia II agglutinin (GSA II)	$(GlcNAc\beta1-4)_4 > (GlcNAc\beta1-4)_3 > GlcNAc\beta(1-4)GlcNAc >$ GlcNAc
Aaptos papillata I agglutinin (APA-I)	$(GlcNAc\beta1-4)_4 > (GlcNAc\beta1-4)_3 >$ Glcβ(1-4)GlcNAc
Cytisus multiflorus agglutinin II (CMA-II)	$(GlcNAc\beta1-4)_3 > GlcNAc\beta(1-4)GlcNAc$
Lycopersicon esculentum (tomato) agglutinin (LEA)	$(GlcNAc\beta1-4)_4 > (GlcNAc\beta1-4)_3 >$ GlcNAcβ(1-4)GlcNAc

Table 12 (continued)

Lectin	Carbohydrate specificities

b) Other GlcNAc-binding lectins

Hordeum vulgare agglutinin

$(-GlcNAc\beta1-4-) > GlcNAc$

Laburnum alpinum agglutinin (LAA)

$GlcNAc\beta(1-4)GlcNAc > cellobiose > lactose > Gal\beta(1-3)GlcNAc$

Rice lectin

Chitin oligosaccharide

IV. D-Mannose-and/or glucose-binding lectins

a) Manα1-linked oligosaccharide-specific lectins

Concanavalia ensiformis (jack bean) agglutinin (Con A)

$(6-1)\alpha Man\beta1-2GlcNAc$
|
R-Man >
|
$(3-1)\alpha Man2-1\beta GlcNAc$

$(6-1)\alpha Man$
|
R-Man >
|
$(3-1)\alpha Man$

$(6-1)\alpha Man(2-1)\beta GlcNAc(4-1)\beta Gal(6-2)\alpha NANA$
|
R-Man >
|
$(3-1)\alpha Man(2-1)\beta GlcNAc(4-1)\beta Gl(6-2)\alpha NANA$
Man

Con A has a great affinity for the trimannosidic core structure

$NeuAc\alpha(2-6)Gal\beta(1-4)GlcNAc\beta(1-2)Man\alpha(1-3)$ ⟍
$\qquad\qquad GlcNAc\beta(1-4)-Man\beta(1-4)GlcNAc\beta(1-4)GlcNAc\beta1-Asn$
$\qquad Gal\beta(1-4)GlcNAc\beta(1-2)Man\alpha(1-6)$ ⟋
$\qquad\qquad Fuc\alpha1,3$ $\qquad\qquad\qquad\qquad\qquad\qquad\qquad Fuc\alpha1,6$

substituted by two GlcNAc residues at the nonreducing end. The affinity is not affected by a bisecting GlcNAc group. Affinity is reduced when the GlcNAc residues are substituted by Gal. Fuc is not important for the binding.

Lens culinaris (lentil) agglutinin (LCA)

In contrast to Con A, L-Fucα1-6 linked to GlcNAc at the reducing end in the complex oligosaccharide structure is an important binding determinant.
Manα1-linked > Glcα1-linked > GlcNAc

Pisum sativum (pea) agglutinin (PSA)

L-Fucα1-6 linked to GlcNAc at the reducing end of the complex oligosaccharide structure is an important binding determinant
Manα1-linked > Glcα1-linked > GlcNAc

Vicia faba (fava bean) agglutinin (VFA)

See LCA and PSA

Table 12 (continued)

Lectin	Carbohydrate specificities

b) Other mannose-and/glucose-binding lectins

Lathyrus sativum
agglutinin (LSA)

Man > Glc

Galanthus nivalis
(snow drop bulb)
agglutinin (GNA)

GNA precipitates highly branched yeast mannans but does not
react with most glycans. Hapten inhibition experiments showed
that D-mannose but not N-acetyl-D-mannosamine is an inhibitor.
Strict requirement for nonreducing end mannose residues

Dioclea grandiflora
agglutinin (DGA)

Man > Fru > Glc

V. L-Fucose-binding lectins

a) Monofucosyl-specific lectins

Ulex europaeus
(furz seed)
agglutinin I (UEA I)

L-Fucα(1-2)Galβ(1-4)GlcNAcβ(1-6)-R >
L-Fucα(1-2)Galβ(1-4)[L-Fucα1-3]GlcNAcβ(1-6)-R
L-Fucα(1-2)Galβ(1-4)GlcNAc >
L-Fucα(1-2)Galβ(1-4)[L-Fucα1-3]Glc

Ulex europaeus
agglutinin II
(UEA II)

L-Fucα(1-2)Galβ(1-4)GlcNAcβ(1-6)R >
L-Fucα(1-2)Galβ(1-4)Glc >
(GlcNAcβ1-4)₄(GlcNAcβ1-3)₃ > (GlcNAcβ1-4)₂

Lotus tetragonolobus
(asparagus pea)
agglutinin (LTA)

L-Fucα(1-6)GlcNAc >
L-Fucα(1-2)Galβ(1-4)[Fucα1-3]GlcNAcβ(1-6)-R >
L-Fucα(1-2)Galβ(1-4)GlcNAcβ(1-6)-R >
L-Fuc

Aleuria aurantia
(orange peel fungus)
agglutinin
(AAurA)

L-Fuc > L-Fucα(1-2)Galβ(1-4)Glc >
Galβ(1-3)[L-Fucα1-4]GlcNAcβ(1-3)Galβ(1-4)Glc >
Galβ(1-4)[L-Fucα1-3]Glc >
L-Fucα(1-2)Galβ(1-3)GlcNAcβ(1-3)Galβ(1-4)Glc

Euonymus europaeus
(spindle tree)
agglutinin

Galα(1-3)[L-Fucα1-2]Galβ(1-4)GlcNAcβ(1-6)-R >
Galα(1-3)[L-Fucα1-2]Galβ(1-3)GlcNAcβ(1-3)Gal >
L-Fucα(1-2)Galβ(1-3)GlcNAcβ(1-3)Galβ(1-4)Glc >
Galα(1-3)[L-Fuc 1-2]Galβ(1-4)[L-Fucα1-3]Glc
Galα(1-3)Gal > Galβ(1-4)GlcNAc

b) Difucosyl specific lectin

Griffonia
simplicifolia IV
agglutinin (GSA-IV)

L-Fucα(1-2)Galβ(1-3)[L-Fuc1-4]GlcNAcβ(1-3)Galβ1-4)Glc >
Galβ(1-3)[L-Fucα1-4]GlcNAcβ(1-3)Galβ(1-4)[L-Fucα1-3]Glc >
L-Fucα(1-2)Galβ(1-4)[L-Fucα1-3]Glc =
L-Fucα(1-2)Galβ(1-4)[L-Fucα1-3]GlcNAc >
Galβ(1-3)[L-Fucα1-4]GlcNAcβ(1-3)Galβ(1-4)Glc

c) Other L-fucosyl-specific lectins

Anquilla anquilla (eel
serum) agglutinin (AAngA)

L-Fucα(1-3)Gal > L-Fuc >
L-Fucα(1-3)Galβ(1-4)GlcNAc

Laccaria amethystina
(mushroom) agglutinin (LAA)

L-Fuc

Table 12 (continued)

Lectin	Carbohydrate specificities
VI. Sialic acid-binding lectins	
Limulus polyphemus (horseshoe crab) agglutinin (LPA)	Neu5Gcα(2-3)GalNAc > Neu5Acα(2-6)GalNAc > Neu5Ac = Neu5Gc
Homarus americanus (lobster) agglutinin (HAA)	NeuAc, NeuGly, and ManNAc
Carcinoscorpius rotunda cauda (Indian horseshoe crab) agglutinin (CRCA)	Neu5Acα(2-6)GalOH > Neu5Acα(2-6)Galβ(1-4)Gal glucuronic acid
Limax flavus (slug) agglutinin (LFA)	NeuAc and NeuGC
Sambucus nigra (elder bark) agglutinin (SNA)	Neu5Acα(2-6)Gal/GalNAc > Neu5Acα(2-3)Gal > Gal

Table 13. Representative lectins with blood group specificities (MYLLYLÄ et al. 1971, BIRD and WINGHAM 1973, BIRD et al. 1971, 1978, JUDD 1980, LIS and SHARON 1986)

Blood group	Lectin	Carbohydrate specificity
A	*Dolichos biflorus*	GalNAc
	Helix pomatia	
	Vicia villosa	
	Vicia cracca	
	Phaseolus lunatus	
	Griffonia simplicifolia I-A$_4$	
B	*Griffonia simplicifolia* I-B$_4$	Gal
H	*Ulex europaeus* I	Fuc
	Lotus tetragonolobus	
	Anquilla anguilla	
TF	*Arachis hypogaea*	Galβ(1-3)GalNAc-O-Ser/Thr
Tn	*Salvia sclarea*	GalNAc-O-Ser/Thr

8.3 Lectin Binding to External, Internal, and Branched Carbohydrate Sequences

Most lectins react with the nonreducing, terminal glycosyl groups of the glycan part of complex carbohydrates; examples include the peanut agglutinin or the *Erythrina cristagalli* agglutinin (ECA) (Fig. 22). On the other hand Con A will bind to mannosyl groups which are internal or external constituents of the glycan chain in complex carbohydrates (Fig. 22). Similar to Con A, three other mannose-specific lectins (*Pisum sativum* agglutinin (PSA), *Lens culinaris* agglutinin (LCA) and the lectin from the fava bean will interact also with the

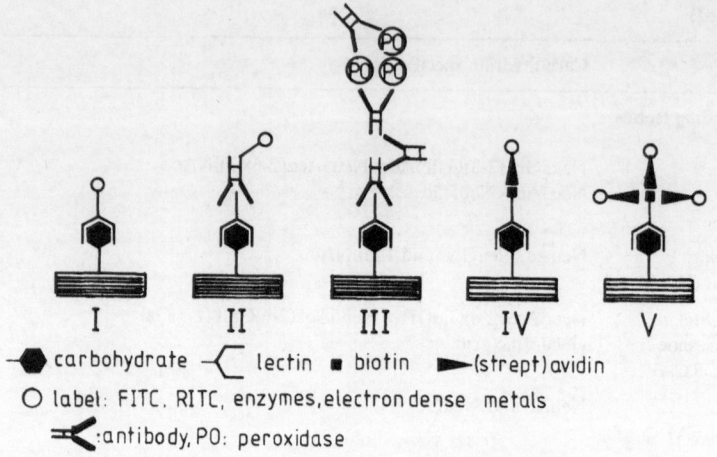

Fig. 21. Schematic representation of some general approaches to the use of lectins in histochemistry: I, direct method; II, and IV, indirect techniques using labeled antilectin antibodies (II) or biotinylated lectin with labeled (strept)avidin (IV); III, four-step unlabeled peroxidase-antiperoxidase technique; V, biotinylated lectin and labeled (strept)avidin complex method

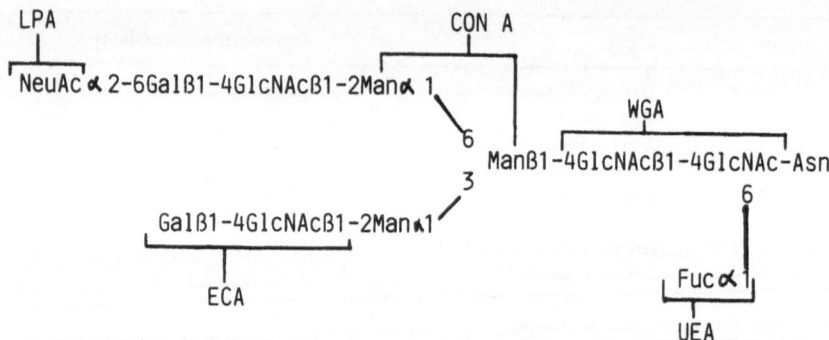

Fig. 22. Hypothetical biantennary oligosaccharide of the complex type, illustrating probable lectin-binding sites. ECA, *Erythrina cristagalli* agglutinin; LPA, *Limulus polyphemus* agglutinin; Con A, concanavalin A; WGA: wheat germ agglutinin; UEA, *Ulex europaeus* agglutinin

nonreducing mannosyl and internal mannosyl groups. N-acetylglucosamine units are typically internal residues of the glycan chains which are recognized by the wheat germ agglutinin (Fig. 22), the potato lectin, and the goarse seed (UEA II) lectin. In accordance with these differences, lectins have been divided into "endo"- and "exo"-lectins with respect to the topography of the lectin binding sites in the glycan chains (Gallagher 1984).

Binding of mannosyl-specific lectins is influenced by the glycosyl residues surrounding the immediate combining site (Fig. 23). These enhancing and inhibiting effects of neighboring carbohydrates are also responsible for the

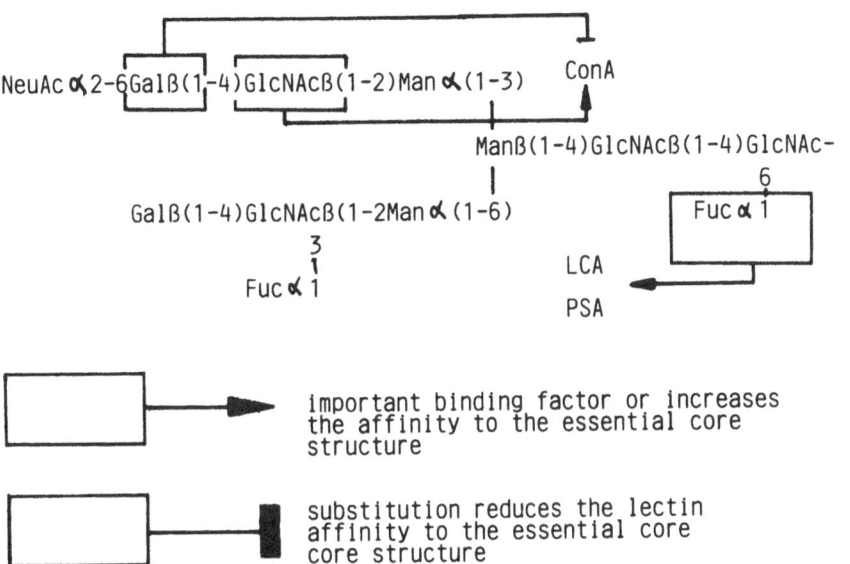

Fig. 23. Effect of carbohydrate residues substituted to the trimannosyl core structure of a hypothetical glycan of the complex type on the binding of concanavalin A (Con A), *Pisum sativum* agglutinin (PSA), and *Lens culinaris* agglutinin (LCA)

differences observed in the histochemical reactivity of the "mannose"-specific lectins.

While Con A has a high affinity for high mannose type glycans which short sequences of α-linked mannose residues, the *Lens culinaris* agglutinin exhibits a preference for complex N-linked glycan parts with long sequences containing fucose (YAMAMOTO et al. 1982).

8.4 Histochemical Reactivity of Lectins

Lectins of the same nominal monosaccharide specificity can bind to different cell structures with different affinities. Figure 24 shows the staining of human liver tissue by four different lectins of the galactose group (Table 14).

The PNA did not bind to native liver tissue. In contrast, the lectin *Artocarpus integrifolia* agglutinin, which has also been reported to possess a high affinity to the PNA receptor, showed distinct labeling of liver cells along the secretory cell pole. The N-acetyllactosamine-specific lectin ECA showed selective labeling of the Kupffer cells. In comparison to these lectins, *Geodia cydonium* agglutinin showed a broad reactivity and labeled all sinusoidal structures. This example demonstrates an important point in lectin histochemistry, namely to employ a group of lectins that possess the same or similar nominal carbohydrate specificity but may exhibit variation in their binding properties,

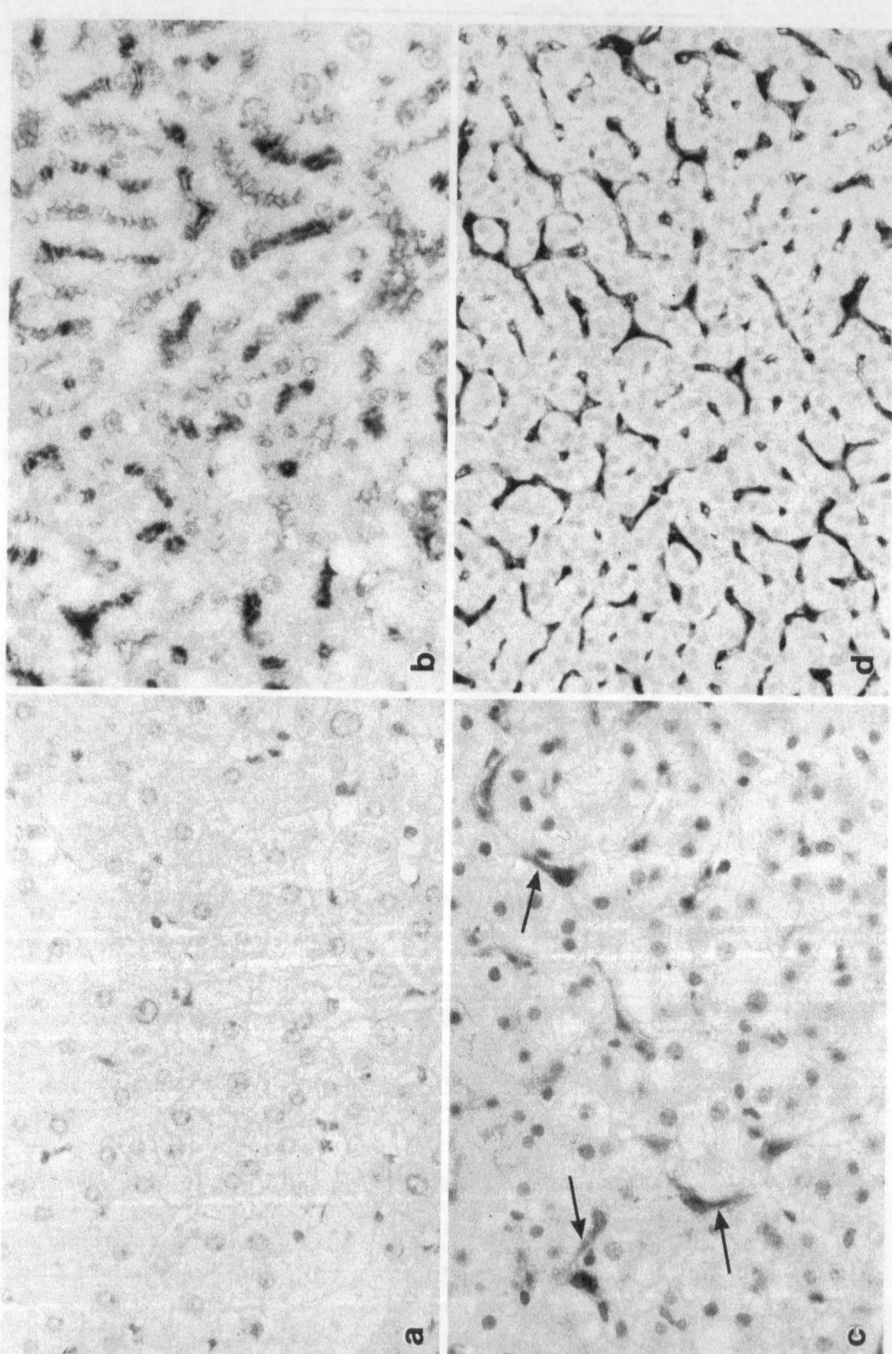

Fig. 24a–d. Reactivity of human liver with different galactose-binding lectin. **a** PNA; **b** *Artocarpus integrifolia* agglutinin; **c** *Erythrina cristagalli* agglutinin (arrows Kupffer cells); **d** *Geodia cydonium* agglutinin

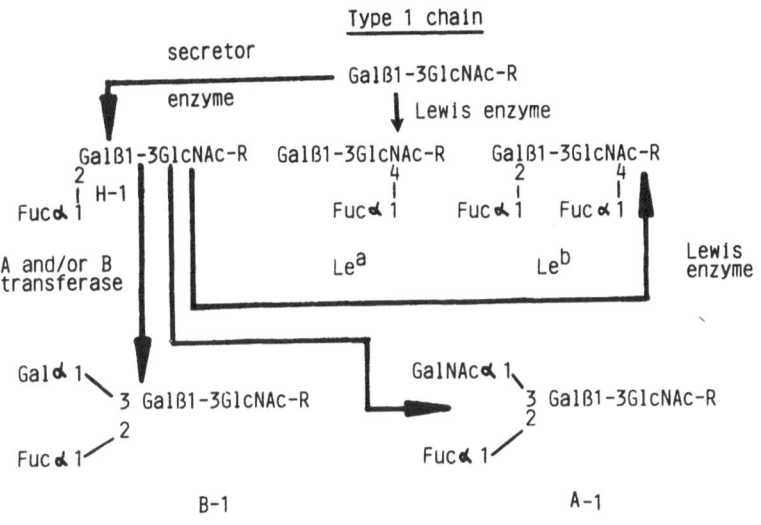

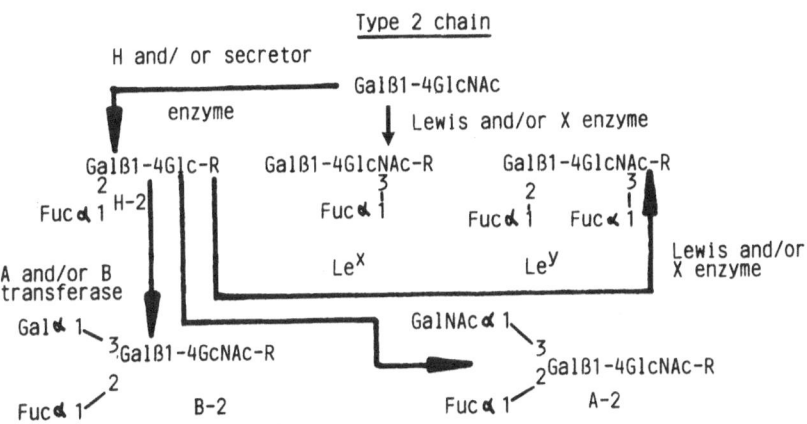

Fig. 25. Schematic representation of ABH and Lewis blood group determinants based on type 1 and type 2 chain precursors

Table 14. Lectin-binding sites of galactose-specific lectins (UHLENBRUCK et al. 1988a)

Structure	Lectin binding sites for
Galβ(1-3)GalNAc	Peanut agglutinin
Galβ(1-3)GalNAc	Jack fruit lectin
Galβ(1-4)GlcNAc	*Erythrina cristagalli* agglutinin
Galβ(1-3)GlcNAc	*Geodia cydonium* agglutinin

Gal, galactose; *GalNAc*, N-acetylgalactosamine; *GlcNAc*, N-acetylglucosamine which can allow fine dissection of glycan structures. One of the earliest applications of lectins was to distinguish between erythrocytes of different blood groups (Table 13).

In this context another example of the use of lectins with the same nominal sugar specificity involves the fucose-specific lectins *Ulex europaeus* agglutinin and *Lotus tetragonolobus* agglutinin. α-L-Fucosyl residues are important constituents of blood group determinants (Fig. 25) and both lectins have been used to agglutinate blood group 0 (H) antigens.

Blood group ABH and the blood groups of the Lewis system are built up by the sequential addition of fucosyl, galactosyl, and N-acetylgalactosamine residues to type 1 (Galβ1-3GlcNAc-R) and type 2 (Galβ1-4GlcNAc-R) blood precursor substances, i.e., backbone carbohydrate of both glycolipids and glycoproteins. Fucosylation of the blood group precursor chain results in the formation of the H antigen, which represents the precursor substance of the A and B blood group determinants (Fig. 25).

In addition to the genes controlling the synthesis of these structures, another gene locus controls the expression of the antigens in secretions and on various cell types of different organs. Thus the classical terms "secretor" and "nonsecretor" indicate the capacity of an individual to secrete such substances in the saliva (RACE and SANGER 1975). Secretor individuals produce ABH-, Lewis[b]-, and Lewis[y] antigens in their saliva, whereas nonsecretors produce Lewis[a] and Lewis[x] antigens (SAKAMOTO et al. 1984).

Beside the H antigen, α-L-fucosyl residues are also found in the Le[x] and the Le[y] antigens (Fig. 25). *Ulex europaeus* I agglutinin binds to terminal L-fucosyl residues of structures derived from type 1 and especially from type 2 chain precursors while *Lotus tetragonolobus* agglutinin binds only to monofucosyl oligosaccharides of type 2. However, neither of the fucose-specific lectins recognizes the fucosyl group in the X hapten or the α1-6 linked fucosyl residues found in the glycan parts of N-linked glycoproteins. In contrast, the lectin purified from the orange peel fungus (*Aleuria aurantia* agglutinin, AAurA) binds to terminal fucosyl residues but does not require a special linkage to the penultimate sugar or a particular chain type (KOCHIBE and FURUKAWA 1980).

The delineation of the presence and structure of blood group substances binding lectins has been in progress for many years. However, there may be differences in the interpretation of serological and histochemical data.

In comparative immunohistochemical studies on the occurrence of blood group ABH isoantigens with blood group specific monoclonal antibodies (MAbs) and blood group-specific lectins in mammary carcinomas there were differences between the immuno- and lectin-histochemical results (VIERBUCHEN et al. 1988d). In contrast to the normal breast epithelium, most mammary carcinomas lacked the ability to form ABH blood antigens (Fig. 26).

However, carcinomas from blood group A patients who did not express the blood group A antigen as demonstrated by the MAb showed expression of an A-like antigen which was detected by the lectin *Dolichos biflorus* agglutinin (DBA) (Fig. 26). Thus the lectin must recognize an epitope different from the A antigen as detected by the MAb.

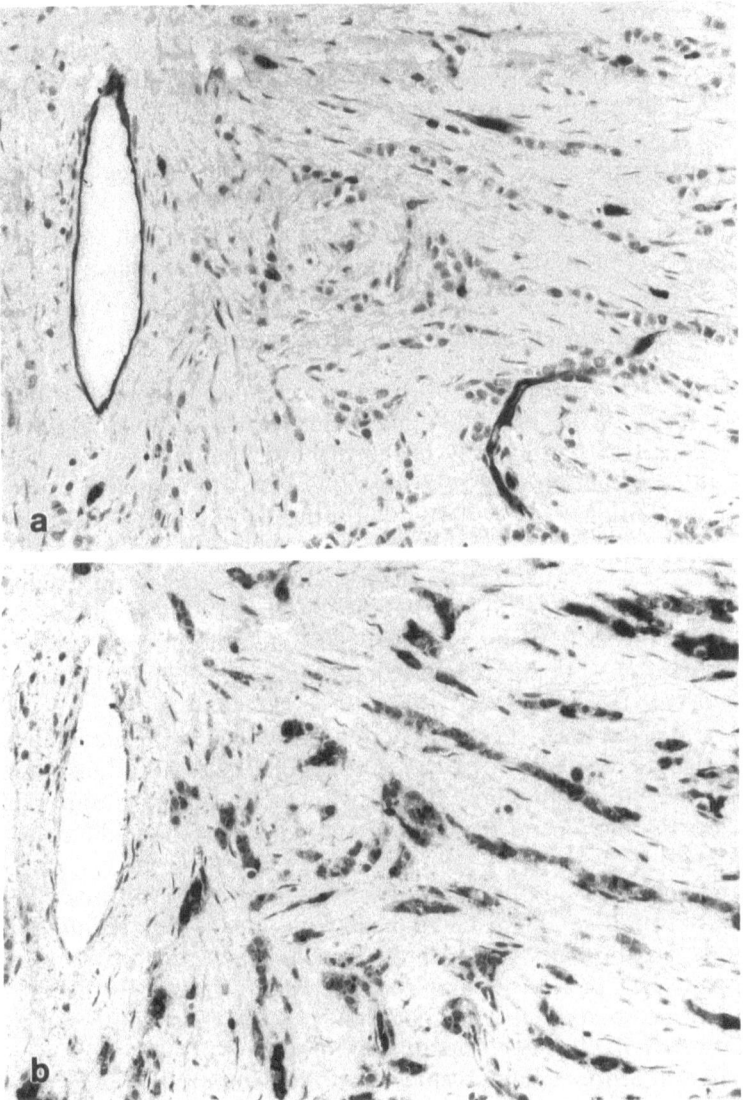

Fig. 26a, b. Different expression of blood group A isoantigen as detected by anti-A MAb (a) and an A-like substance as detected by the lectin DBA (b) in a human mammary carcinoma. **a** Blood group A isoantigen occurs on the endothelial cells while the tumor cells do not bind the MAb. **b** Strong cytoplasmic expression of DBA reactivity in tumor cells

It has been shown that the Tn antigen, which is represented by the structure GalNAc-0-Ser/Thre, can be detected by its interaction with DBA (Bird and Wingham 1973). From the comparative immuno- and lectin-histochemical studies one may conclude that DBA detects the Tn determinant in the absence of blood group A antigen.

Evidence has been presented that the expression of the Tn antigen is due to the lack of the enzyme 3β-galactosyl-transferase (Cartron et al. 1978), suggesting that the expression of the Tn antigen is a somatic mutation in which the addition of the galactose to form the TF antigen does not take place.

8.5 New Lectins and Their Application in Lectin Histochemistry

A lectin from *Erythrina cristagalli* seeds (ECA) has been isolated and purified (Iglesias et al. 1982). It was found that the hemagglutination of blood group 0 erythrocytes was best inhibited by N-acetyllactosamine residues and the β-linkage of an adjacent sugar (Kaladas et al. 1982). This lectin has been introduced as an addition histochemical tool for the detection of N-acetyllactosamine type glycoconjugates (Vierbuchen et al. 1986, 1988c).

After removal of sialic acid residues, the penultimate disaccharide is stainable with the lectin, indicating the presence of the trisaccharide NeuAcα2-3, 6Galβ1-4GlcNAc (Fig. 27). Peanut agglutinin has a high affinity for the disaccharide Galβ1-3GalNAc but also reacts with the Galβ1-3,4GlcNAc disaccharides (Pereira et al. 1976). In some tissues the latter disaccharides are the dominant ones and the PNA reactivity cannot be equated with the Galβ1-3GalNAc sequence (Thomsen-Friedenreich antigen). In these locations comparative and competitive inhibition sutdies with the unlabeled lectins can provide further information on the exact nature of the lectin-binding sites (Vierbuchen et al. 1988c).

Jacalin is a lectin obtained from the *Artocarpus integrifolia* (jackfruit). This lectin has been shown to bind to α-Gal units and possesses a high affinity for the disaccharide Galβ1-3GalNAc (Table 14). Jacalin has been shown to bind to IgA but not to IgG or IgM (Roque-Barreira and Campus Neto 1985). From the diagnostic point of view, the importance of jacalin is that it may be a useful tool for studying IgA nephropathies (Andre et al. 1985) as well as for the characterization of some multiple myelomas and dermatitis herpetiformis (Egelrud 1986).

The three mannose-specific lectins Con A, *Lens culinaris* agglutinin, and *Pisum sativum* agglutinin all bind to internal and external (nonreducing) mannosyl residues. Therefore, these lectins cannot distinguish between internal and external mannosyl residues; i.e., mannose residues which occur in complex type glycoproteins and in high mannose type glycoproteins. Furthermore, for the so-called mannose-specific lectins, mannose is only 2.5–5 times more inhibitory than glucose. In contrast to these lectins, the lectin *Galanthus nivalis* agglutinin (GNA) is not inhibited by glucose. A strict requirement of GNA for terminal nonreducing mannosyl residues is also evident from the fact that the oligosaccharides in which the terminal carbohydrates are blocked by the sub-

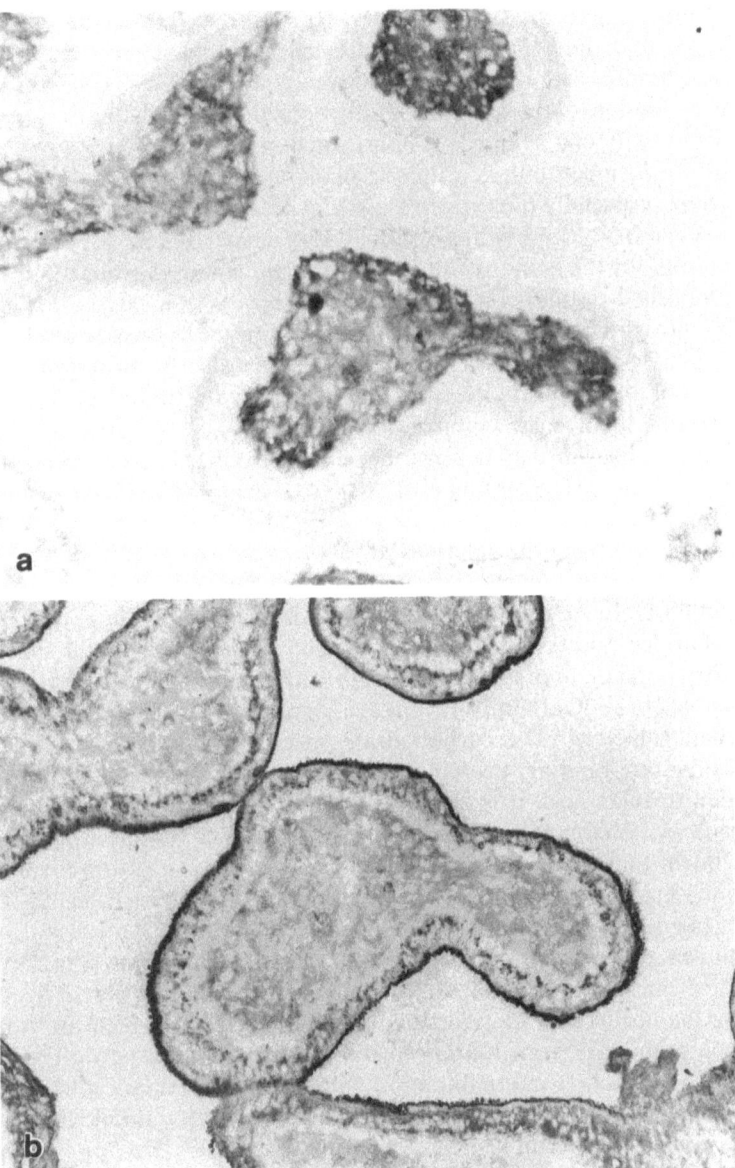

Fig. 27. a Placenta tissue exhibiting native ECA-binding sites in the villous stroma and histiocytes, whereas the cells of the cyto- and syncytiotrophoblast remained unstained. **b** Liberation of masked ECA-binding sites on the trophoblast after enzymatic cleavage of sialic acid

stitution with other sugars are not inhibitory. Results from hapten inhibition studies indicated that the Manα(1-3)Man disaccharide unit is most complementary to the binding site of this lectin. Glycopeptides which possess two Manα(1-3)Man residues show a greater affinity for the lectin than glycopeptides which carry only one Manα(1-3) unit (SHIBUYA et al. 1988).

Galanthus nivalis agglutinin is a unique plant lectin recognizing terminal mannosyl groups, especially those possessing Manα(1-3)Man units. In combination with other mannose-binding lectins, GNA may also be a powerful histochemical tool for the demonstration of terminal mannosyl groups (VIERBUCHEN, unpublished results) (Fig. 28), especially those which carry the high mannose type glycan chains. *E. coli* type 1 lectin appears to have a binding specificity similar to GNA. Therefore, this lectin may also be an important histochemical tool for the in situ investigation of carbohydrates for the microbial lectin which is involved in colonization by urotoxic *E. coli* strains. The lectin isolated from the slug *Limax flavus* has been introduced as a histochemical probe for sialic acid (HEDMAN et al. 1986, ROTH et al. 1984, SCHULTE and SPICER 1984).

A new plant lectin from the elderberry (*Sambucus nigra* agglutinin, SNA) has been shown by immunochemical techniques to bind specifically to terminal NeuACα(2-6)Gal/GalNAc residues of glycoconjugates (SHIBUYA et al. 1987).

Oligosaccharides possessing this sequence showed a 1800–10000 times more inhibitory potency than galactose. On the other hand, oligosaccharides with the NeuNAcα(2-3)Gal/GalNAc linkage were only 30–60 times more inhibitory than galactose. The carbohydrate specificity of SNA is unique among carbohydrate-binding proteins. Although some lectins have been known to bind to sialic acid, it is not clear whether these lectins exhibit any linkage specificity. Preliminary experiments (VIERBUCHEN, unpublished results) have shown that the lectin can be used as a histochemical tool for the demonstration of sialic acid residues (Fig. 29). In the kidney glomerulus, SNA shows an inverse reactivity to that of ECA (Figs. 29, 30).

The occurrence of neuraminidase-resistant SNA-binding sites (Fig. 29), however, indicates either the presence of carbohydrate receptors different from sialic acid or the occurrence of sialic acid residues for SNA which are not cleaved by neuraminidase from *Vibrio cholerae* used in this study. Additional lectin-histochemical studies in combination with various enzymatic pretreatments of tissue sections with glycosidases (galactosidase, hexosaminidase, fucosidase) are needed to determine the exact nature of the SNA-binding sites in tissue sections.

8.6 Pattern of Lectin Binding upon Chemical and Enzymatic Pretreatment of Tissue Sections

Sialic acid residues terminate oligosaccharide side chains in both O- and N-linked glycoproteins. Sialic acid is either linked to penultimate galactose or N-acetylgalactosamine. After removal of terminal sialic acid residues by pre-

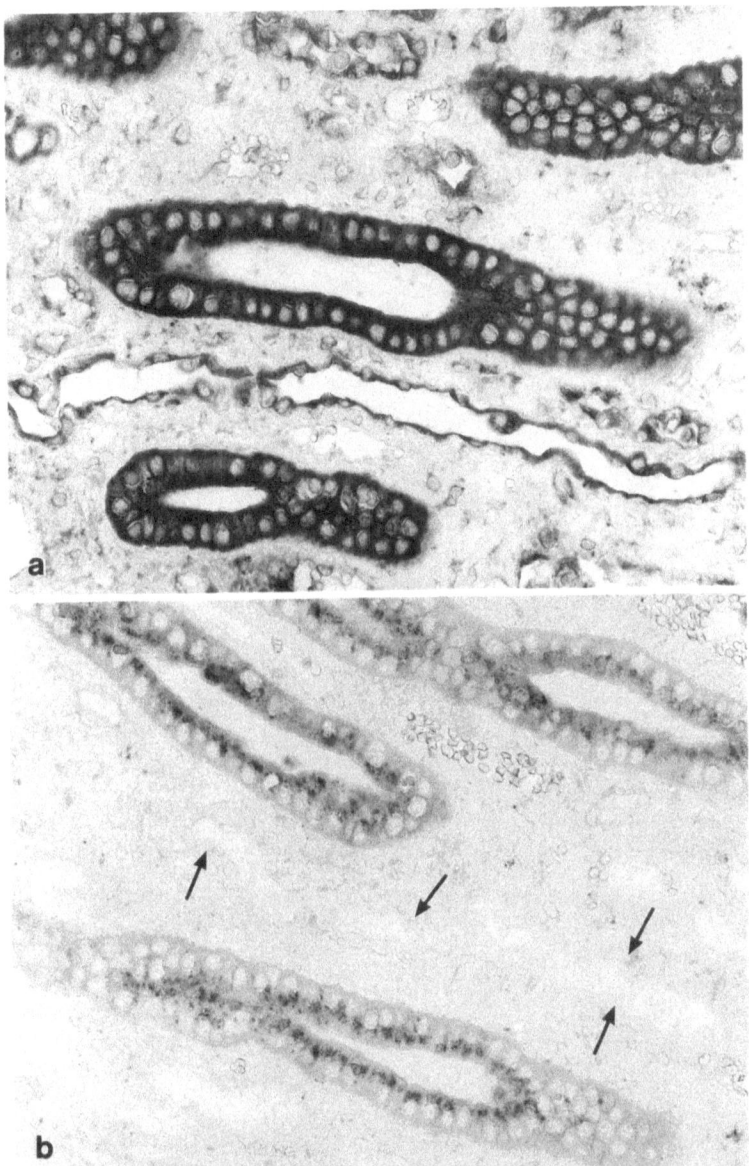

Fig. 28 a, b. Section through the medulla of a human kidney showing a different reactivity of Con A and GNA. While Con A reactivity occurs diffusely in cells of the collecting duct and Henle's loop (**a**), the GNA reactivity (**b**) is confined in a granular manner to the luminal surface of the collecting duct, the cells of the Henle's loop remaining unstained (arrows)

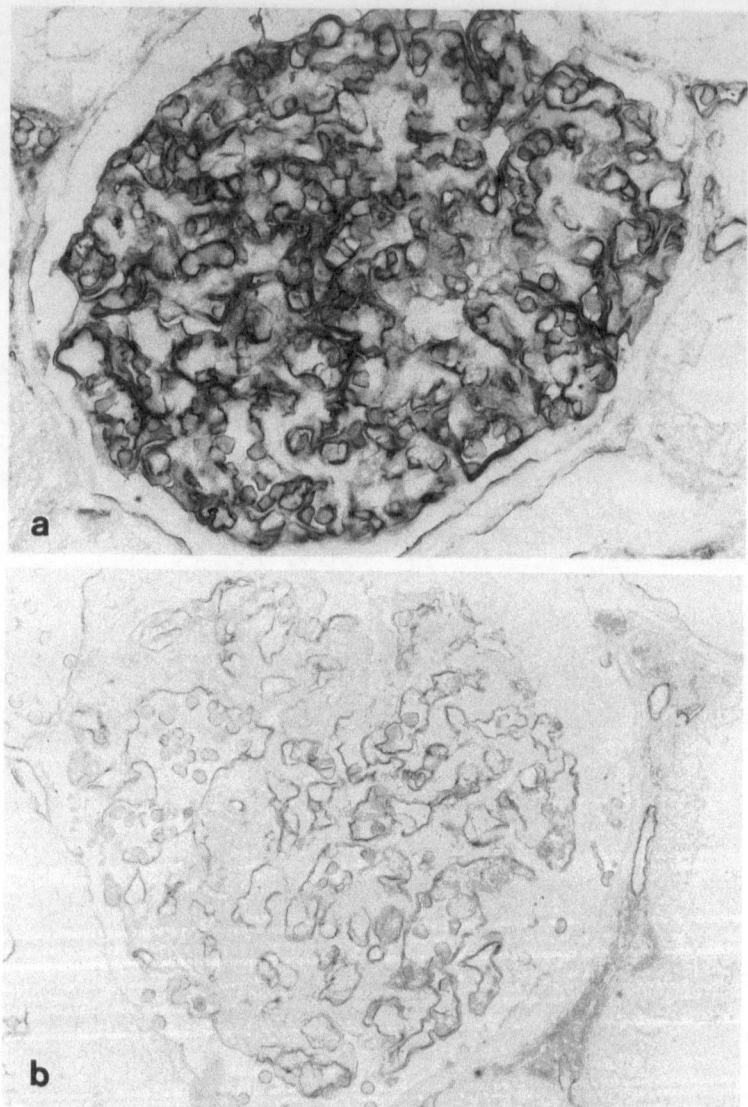

Fig. 29 a, b. Binding sites for the lectin SNA in the kidney glomerulus. **a** Distinct SNA reactivity along the surface of the capillary endothelium, on the erythrocytes, and on the podocytes. **b** Tissue section after pretreatment with neuraminidase, resulting in a loss of SNA reactivity on the podocytes while the capillary endothelium shows a weak residual lectin reactivity

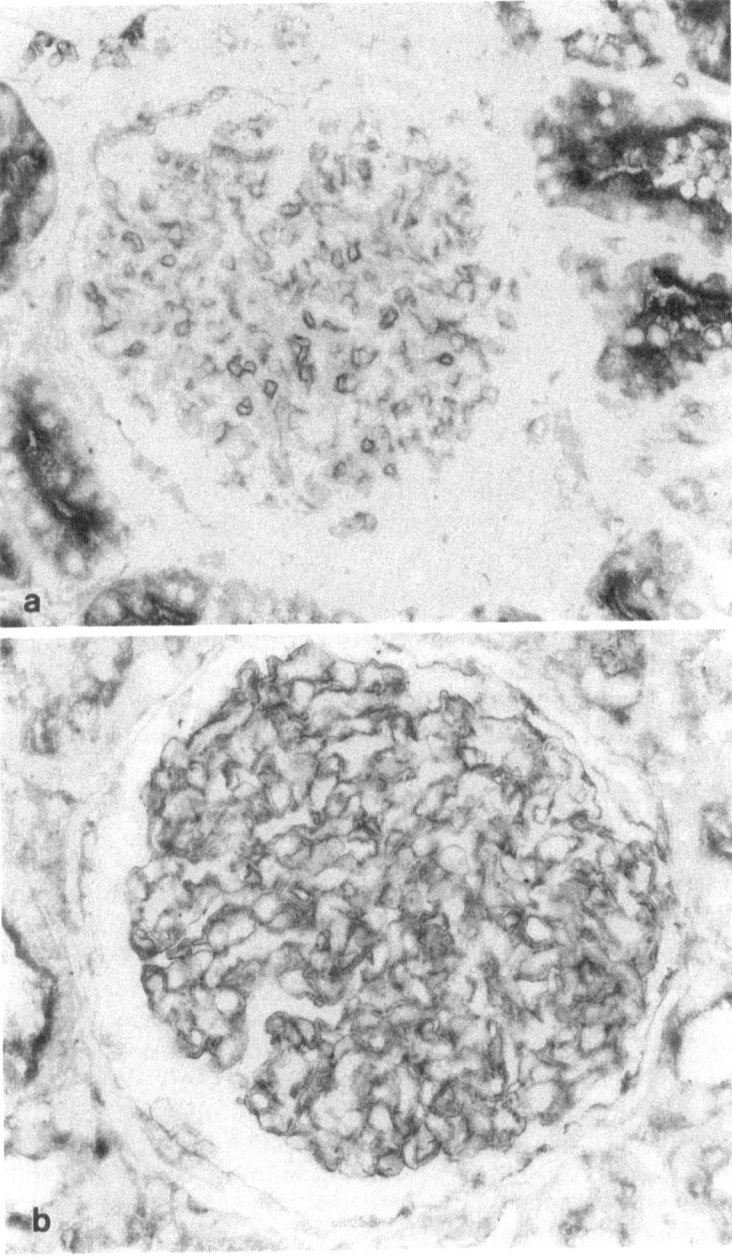

Fig. 30 a, b. ECA reactivity in the glomerulus of the kidney. **a** Native ECA-binding sites occur along the luminal surface of the capillary endothelium and on the erythrocytes. **b** Unmasking of ECA binding sites on the podocytes by enzymatic cleavage of sialic acid residues

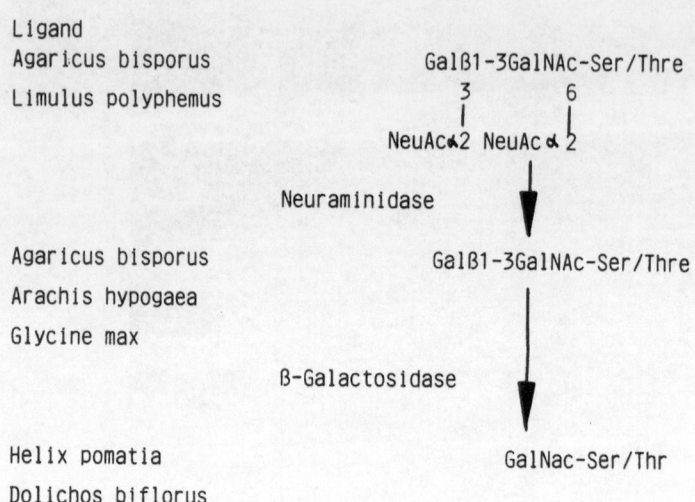

Ligand
Agaricus bisporus
Limulus polyphemus

Neuraminidase

Agaricus bisporus
Arachis hypogaea
Glycine max

ß-Galactosidase

Helix pomatia
Dolichos biflorus

Fig. 31. Illustration of a possible lectin binding sequence to an oligosaccharide chain on subsequent enzymatic degradation

treatment of tissue sections with neuraminidase (VIERBUCHEN et al. 1980) subterminal sugar residues become stainable. Figure 31 shows, for example, the pattern of lectin binding that might be expected upon degradation of an O-linked tetrasaccharide such as is found on the major human erythrocyte membrane glycoprotein (THOMAS and WINZLER 1969).

Histochemical sequences that utilize PNA, ECA, or DBA staining in conjunction with neuraminidase pretreatment of tissue sections make it possible to localize the terminal trisaccharides NeuAcα(2-6,3)Galβ(1-3)GalNAc-R (neuraminidase PNA sequence) and NeuAcα(2-3,6)Galβ(1-4)GlcNAc-R (neuraminidase ECA sequence) as well as the disaccharide NeuAcα(2-3,6)GalNAc-R (neuraminidase DBA sequence). Furthermore, a variety of glycosidases are now available with which sequential dissection of glycan structures can be undertaken. On the basis of combined stepwise digestion of tissue sections with specific glycosidases and subsequent staining with lectins, it has been shown that in UEA I reactive pancreatic acinar cells fucosylated Galβ(1-3,4)GlcNAc and fucosylated Galβ(1-3)GalNAc units exist (ITO et al. 1988).

Ulex europaeus agglutinin I has been widely used as a marker for vascular endothelium. Using a broad library of different lectins in conjunction with sequential enzymatic digestion of tissue sections, it was concluded that the UEA I-binding saccharides were probably part of the outer chains of bisected, biantennary N-linked oligosaccharides (ROBB and STODDART 1987).

Mild periodic oxidation with 1 mM periodic acid oxidizes the polyhydroxyl side chain of terminal N-acetylneuraminic acid but fails to oxidize sialic acids with polyhydroxyl side chains possessing C-9 O-acetyl or C-7,9 di-O-acetyl group analogs. Sites in which 1 mM periodate pretreatment selectively blocks staining by the neuraminidase PNA/DBA sequence are taken to possess subterminal periodate labile Gal or GalNAC residues (SCHULTE and SPICER 1985). In contrast, sites in which 1 mM periodate oxidation does not block subsequent staining with the sialidase PNA/DBA sequence must possess terminal sialic acid residues containing O-acetyl groups on their polyhydroxyl side chains (SCHULTE and SPICER 1985). In these sites inhibitory effects of periodate oxidation can be overcome if O-acetyl substituted side chains are deacetylated by saponification (SCHULTE and SPICER 1985).

Concanavalin A binds to internal and external mannosyl groups with unmodified hydroxyl groups at the C-3, C-4, and C-6 positions. In accordance with the effect of periodate oxidation on Con A reactivity, different classes of Con A-binding sites have been distinguished. Class I Con A reactivity is abolished by periodate oxidation. α-Mannosyl residues that undergo such periodate oxidation would be expected to be in external positions linked at C-2 to their neighboring units (KATSUYAMA and SPICER 1978). Class II Con A-binding sites are characterized by the paradoxical appearance of Con A reactivity after periodate oxidation. This Con A reactivity is thought to be due to the unmasking of stainable mannosyl units by oxidation of vicinal glycols on adjacent sugar residues which results in disruption of bonds between nearby sugar and C-6 mannose hydroxyl (KATSUYAMA and SPICER 1978). Class III Con A binding sites are induced following periodate oxidation but are resistant to subsequent reduction. It is assumed that class III Con A binding sites are due to the elimination of the masking due to a C-3 hydroxyl group of an adjacent fucosyl residue. Reduction does not reverse this C-3 hydroxyl group, which makes the staining irreversible (KAWANO et al. 1984).

Griffonia simplicifolia agglutinin (GSII) is of particular interest and value because it is the only lectin that interacts with terminal nonreducing GlcNAc groups. However, GSII also recognizes glycogen. Therefore, diastase labile GS II reactivity can be attributed to the presence of glycogen (SCHULTE and SPICER 1983, SCHULTE and SPICER 1984, SCHULTE et al. 1985).

8.7 General Applications of Lectins

Lectins represent powerful tools which have been used extensively for preparative and analytical purposes in biochemistry, cell biology (including the study of subcellular domains and cell membranes), immunology, and related areas.

The majority of lectin-histochemical studies performed on pathological specimens have focused on several areas. These include changes in lectin binding during malignant transformation and the course of infectious diseases, and the expression of lectin-binding sites in relation to tissue differentiation and functional states.

8.7.1 Alteration of Lectin Reactivity with Malignancy

Numerous studies have been carried out on the changes in the lectin binding of malignant cells in comparison to normal ones. Furthermore, much work has been undertaken to relate differences in glycosylation, as detected by lectins, to tumor behavior.

Although cases of selective interactions of lectins have been reported (BOLAND et al. 1982, HOWARD et al. 1981, HSU and RAINE 1982, LOUIS et al. 1981, 1983, RAEDLER et al. 1983, REE and HSU 1983, REISSNER et al. 1979), there is as yet no system in which lectin binding can be used reliably for the detection of malignancy.

Another problem in the interpretation of data obtained from lectin-histo-chemical studies on malignant tumors is the heterogeneity of the expression and also in the deletion of lectin-binding sites between the different neoplasms of a given tumor class and even within the same tumor. This explains why it is difficult to relate lectin-histochemical results obtained from tumor tissues to their biological and clinical behavior.

Nevertheless, such investigations do contribute to our knowledge on the occurrence of saccharides in normal and transformed tissues. It has been reported that GS II uniquely interacts with both surface coat and goblet type mucins of colonic carcinomas but not with the corresponding normal cells (NAKAYAMA et al. 1985). Alterations in the expression of fucose-containing glycoconjugates has been reported in gastric carcinomas (McCARTNEY 1987), large intestinal carcinomas (YONEZAWA et al. 1982), endometrial carcinomas (WEST and COPE 1989), and breast carcinomas (WALKER 1984, FENLON et al. 1987). Biochemical studies have demonstrated that UEA I binds to high molecular weight glycoproteins which occur in normal and cancer cells of the ascending and transverse colon, but only to carcinomas of the descending and sigmoid colon and to rectal carcinomas, not to the corresponding normal epithelium (MATSUSHITA et al. 1985). *Dolichos biflorus* agglutinin binds to the cells of the normal colonic mucosa while most colonic carcinomas lack DBA-binding sites (Fig. 32). In many colonic carcinomas the loss of DBA-binding sites is accompanied by the expression of PNA-binding sites in the tumor tissue while normal colonic mucosa lacks PNA-reactivity (Fig. 32) (VIERBUCHEN 1987). Although the expression of PNA-binding sites is not exclusive for malignant neoplasms of the colonic mucosa, carcinomas exhibit a more intense and general expression than adenomas (VIERBUCHEN 1987). In contrast to other findings (KELLOKOMPU et al. 1986), our studies revealed that colonic carcinomas express native as well as sialic acid-masked PNA-binding sites (VIERBUCHEN 1987).

The enhanced expression of PNA-binding sites in colorectal carcinomas indicates an increased expression of glycoconjugates with terminal galactose. This alteration in the expression of glycoconjugates may dictate the metastatic potential of colorectal carcinomas. In this context the hepatic galactose-binding lectin may be involved in the arrest and subsequent growth (mitogenic activity of the liver lectin) of metastatic tumor cells. This conclusion has been supported by recent findings on the metastatic capacity of murine colonic carcinoma cells. After intrasplenic injection of tumor cell subpopulations, liver cell metastasis was found to be proportional to the tumor cell surface galactose expression (YEATMAN et al. 1989).

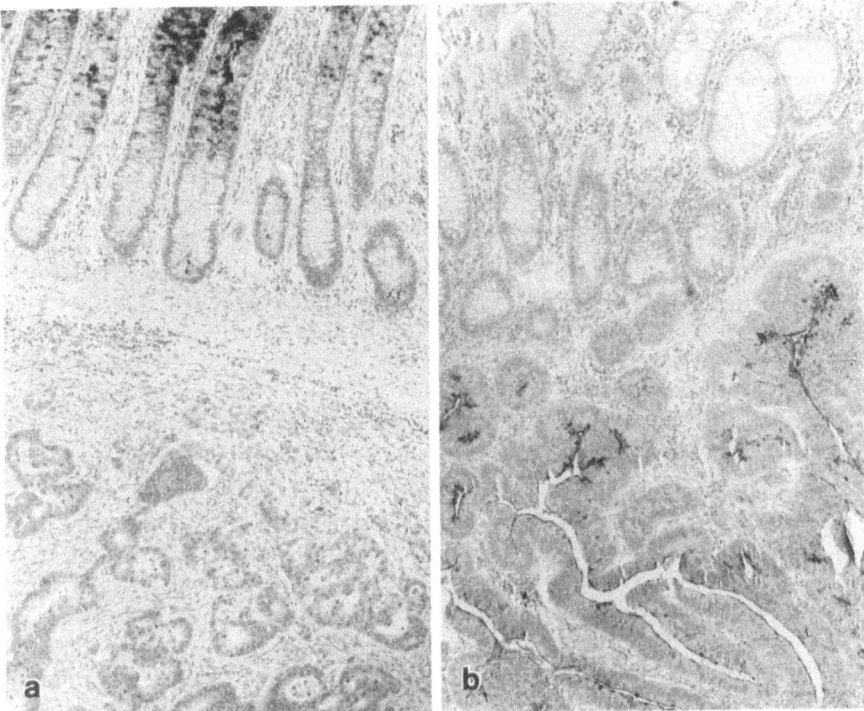

Fig. 32 a, b. Occurrence of binding sites for PNA and DBA in colonic carcinoma. **a** Lack of DBA-binding sites in the colonic carcinoma while the goblet cells of the noncarcinomatous mucosa express DBA reactivity. **b** Expression of PNA binding sites in the same tumor, contrasting with the lack of PNA reactivity of the noncarcinomatous mucosa

The UEA I and Helix pomatia agglutinin (HPA) reactivity of breast carcinomas has been shown to be related to the lymph node status, time to locoregional recurrence, and survival (FENLON et al. 1987). Furthermore, the detection of GalNAc type glycoconjugates in breast cancers by HPA is of value in assessing long term prognosis (LEATHEM and BROOKS 1987). Using monoclonal antibodies and the blood group A specific-lectin DBA, comparative immuno- and lectin-histochemical studies revealed that in many cases breast carcinomas express lectin reactive GalNAc residues but not the blood group A isoantigen. These carcinomas, like the HPA-reactive tumors, showed increased aggressiveness (VIERBUCHEN et al. 1988 d).

It has been presumed that HPA recognizes an undefined biological marker that indicates biological aggressiveness of mammary carcinomas. As described earlier in this chapter, it is reasonable to conclude that in the absence of blood group A antigen the lectins will bind to the *Tn antigen*. The Tn antigen has been reported to be an autoimmunogen and on carcinoma cell membranes a cell adhesion molecule. Tn is occluded and nonreactive in healthy and noncarcinomatous tissues. This carbohydrate antigen has been found to be a carcinoma-associated antigen and an attachment molecule in the binding to healthy cells (SPRINGER 1989). Beside the elucidation of some aspects of breast cancer pathogenesis, these findings have implications for the selection of breast cancer patients for adjuvant therapy.

There has been great interest in the use of lectins, particularly PNA, to predict *steroid hormone receptor status*. It was concluded by KLEIN et al. (1981, 1983) that the expression of distinct patterns of PNA reactivity in breast carcinomas is associated with the occurrence of steroid hormone receptors and response to endocrine therapy (KLEIN et al. 1981, 1983). However, other studies did not reveal a clear association between PNA reactivity and the presence of hormone receptors (BÖCKER et al. 1984, WALKER et al. 1985). On the other hand, studies on experimental breast carcinomas clearly demonstrated an association between the occurrence of PNA-binding sites in the tumor tissue and the hormone dependence of tumor growth (VIERBUCHEN et al. 1983, 1988 b). In rat mammary tumors it has been shown that estrogen induces the formation of PNA-reactive substances while ovariectomy inhibits the expression of PNA-binding sites (VIERBUCHEN et al. 1983 a, 1988 b). Changes in glycoconjugate expression have been lectin-histochemically detected in early embryogenesis (ARITA et al. 1985, KIMBER 1986, SLACK 1985) and during differentiation of embryonal and teratocarcinoma cell lines (ARITA et al. 1985, LEPPANEN et al. 1986). This can be of value in assessing human germ cell tumors (TESHIMA et al. 1984) with regard to the differentiation between embryonal carcinoma cells and somatic or yolk sac components. Comparative lectin-histochemical studies on normal, hyperplastic, adenomatous, and carcinomatous thyroid tissue revealed differences between the carcinomatous and noncarcinomatous tissues. Papillary and follicular carcinomas showed focal reactivity with *Maclura pomifera* agglutinin (MPA), PNA, soy bean agglutin (SBA), UEA I, and the *Solanum tuberosum* agglutinin while noncarcinomatous thyroid tissue was not reactive (SOBRINHO-SIMOES and DAMJANOV 1986). It has been suggested that this lectin reactivity could be due to the binding of the lectins to blood group and blood group-related antigens reexpressed in neoplastic thyroid tissue (VIERBUCHEN et al. 1989 b). It is assumed that the focal and inconstant nature of lectin reactivity excludes the diagnostic use of lectins (SIBRINHO-SIMOES and DAMJANOV 1986).

8.7.2 Use of Lectins in Infectious Disease

There is specific attachment of numerous bacterial strains due to the interaction of the bacterial lectin with the target carbohydrate (see Chapt. 7.2.1).

On the other hand many plant lectins have been used to characterize micro-organisms. A panel of lectins has been used to distinguish between New World *Leishmania* and *Trypanosoma* (SCHOTTELIUS et al. 1988). A correlation is assumed between the PNA-type, dominant north of the Amazon, and cardiomyopathy – the major cause of death there. The WGA type and enteromegalies are very rare or nonexistent there (SCHOTTELIUS et al. 1988).

Using the different affinities of the saccharides in fungi to a library of lectins it is possible to distinguish between *Aspergillus fumigatus, Candida albicans*, and *Rhizopus oryzae*. Furthermore, *Cryptococcus neoformans, Blastomyces dermatitidis*, and *Paracoccidioides brasiliensis* show different lectin binding patterns (STODDART and HERBERTSON 1978 a, 1978 b). The great versatility of lectins as diagnostic tools has also been demonstrated in bacilli. Distinct reaction patterns were detected especially after removal of sialic acid (UHLENBRUCK et al. 1988 b). Group B streptococci are of special interest as they have a sialic acid-galactose-N-acetylglucosamine residue as part of their antigenic carbohydrate structures which also occurs in human glycoconjugates and may be the cause of autoimmune reactions (UHLENBRUCK et al. 1988 b).

In addition to other glycosidases many bacteria produce a neuraminidase that can remove terminal sialic acid residues from the host cells. The in vivo action of pneumococcal neuraminidase has been demonstrated in some cases of hemolytic uremic syndrome (HUS) developing during the course of pneumococcal pneumonia (KLEIN 1981, KLEIN et al. 1980). In these cases of HUS the unmasking of cryptic, i.e., sialic acid-masked Thomsen-Friedenreich antigen, which is represented by the disaccharide Galß(1-3)GalNAc, was demonstrated by the use of PNA in kidney glomeruli and on the erythrocytes. The choroid plexus shows functional and structural similarities to the kidney glomerulus. In cases of pneumococcal meningitis the in vivo action of the neuraminidase has been demonstrated (VIERBUCHEN and KLEIN 1983) by the exposure of PNA-binding sites on the plexus epithelium (Fig. 33).

A pathogenic mechanism has been suggested on the basis of an enzymatically induced auto-immune response to unmasked Thomsen-Friedenreich antigen, because all human sera contain natural antobidies to this antigen. However, neither systemic nor intracisternal application of neuraminidase to mice or rats alone induced the appropriate disease (VIERBUCHEN, unpublished results). Therefore, it has been suggested that the bacterial neuraminidase may uncover carbohydrate structures and thereby make possible the adhesion of the bacteria or bacterial toxins to cryptic receptors (VIERBUCHEN 1987). In a similar way the neuraminidase effects of infective trypomastigotes of *Trypanosoma* cruzi on the glycoconjugates of erythrocytes have been demonstrated (PEREIRA 1983).

8.7.3 Identification of Specific Cell Types

The differences in the occurrence of carbohydrates on the cell surface and within the cytoplasm (cellular organelles) can be used for the isolation and lectin-histochemical characterization of distinct and specialized cell types. UEA I binds to human endothelial cells and represents an endothelial marker more sensitive than the factor VIII-related antigen (ORDONEZ and BATSAKIS 1984). In contrast to the endothelium of blood vessels, human liver and bone marrow sinusoidal lining cells do not stain with UEA I (ROUSSEL 1985). UEA I has been used for the demonstration of vascular invasion of breast carcinomas (LEE et al. 1986). UEA I stains the endothelium in hemangiomas, angiokeratomas, and Kaposi's sarcomas (ORDONEZ and BATSAKIS 1984, HOSAKA et al. 1985). In the pancreas the cells of the exocrine gland express

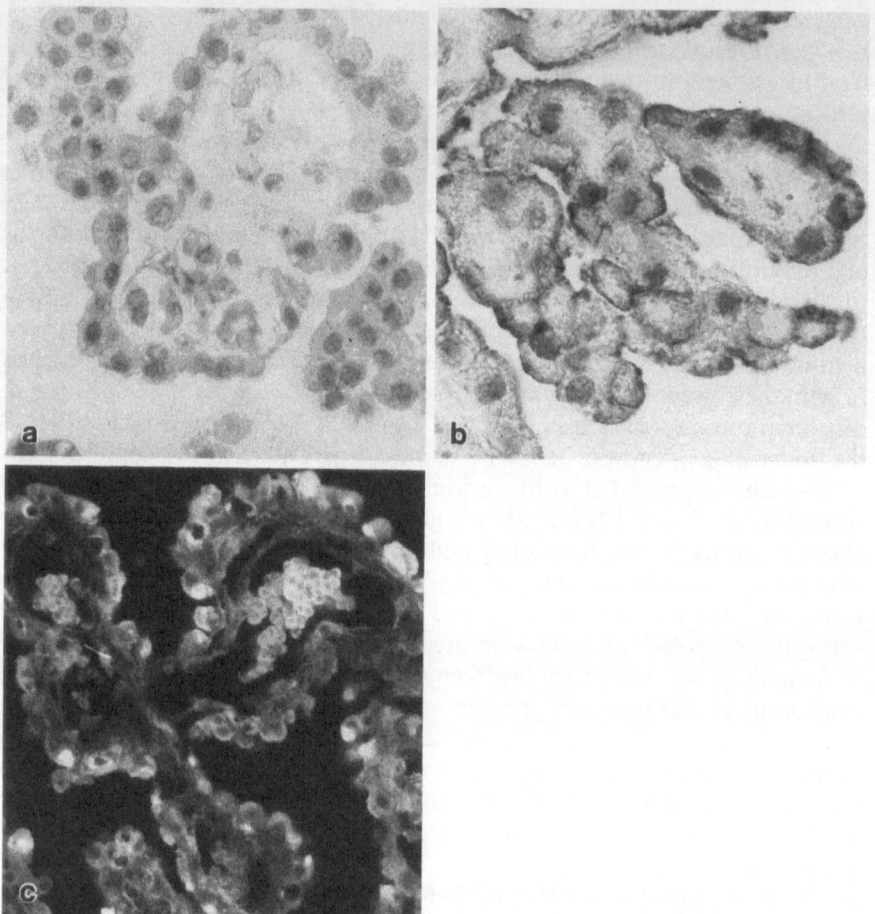

Fig. 33 a–c. PNA reactivity of normal epithelium of the plexus choroideus and in cases of pneumococcal meningitis. **a** Lack of PNA binding sites on the normal plexus epithelium. **b** Unmasked PNA binding sites on the plexus cells in a patient with pneumococcal meningitis. **c** Liberation of PNA-binding sites on the plexus epithelium and on the erythrocytes in another patient with pneumococcal meningitis

native ECA-binding sites while the cells of the endocrine gland possess only sialic acid-masked ECA-binding sites (VIERBUCHEN et al. 1988c).

In the anterior pituitary gland differences between the lectin-binding pattern of the different cell types have been observed. The corticotrophs stain selectively with fucose-binding lectins. The thyrotrophs can be demonstrated selectively by the periodate-oxidation-borohydride reduction Con A sequence (NAKAGAWA et al. 1986c).

Histochemical studies have revealed differences in lectinreactivity between normal, atrophic, and adenomatous tissue of the parathyroid gland (THIELE et al. 1986). Assessment of functional activity of adenomas by clinical parameters displayed a significant correlation with the expression of PNA- and UEA-binding sites.

Lectins have been widely used for the separation especially of lymphatic cells; this may be considered an extension of their use for purification of glycoproteins and glycopeptides. Providing that the cells of the different lymphatic tissues and lymphatic compartments differ in their glycoconjugates, lectins bind differently to individual cells. Cortical thymocytes express native binding sites for ECA, while the medullar thymocytes exhibit only sialic acid-masked ECA-binding sites (VIERBUCHEN et al. 1988c).

A small percentage of peripheral blood lymphocytes do not express ECA-binding sites (HARRIS et al. 1987). Histochemical studies have shown that in lymph nodes the ECA-negative lymphocytes are diffusely scattered in the interfollicular area (VIERBUCHEN et al. 1988c). It was found that the ECA-negative lymphoid cells represent a population enriched with NK cytotoxic/cytolytic activity. Negative selection of mononuclear cells with ECA resulted in a population highly enriched in cells exhibiting NK functions (HARRIS et al. 1987). This technique was found to be advantageous for the study of NK heterogeneity and NK function. Peanut agglutin binds to immature thymocytes of mouse and man and not to the mature cells (DeMAIO et al. 1986). The loss of PNA reactivity in mature (medullar) thymocytes could reflect the loss of the binding glycoprotein (gp170/180), or a modification of its carbohydrate units (DE MAIO et al. 1986). Using lectins the occurrence of hematopoietic stem cells in the peripheral blood and bone marrow has been investigated. Among the different lectins tested, fucose-binding lectins yielded the highest recovery of stem cells from the nonagglutinated fraction (GIRGERT et al. 1987). Lectins have been used to characterize Hodgkin's and Sternberg-Reed cells in Hodgkin's disease (HSU and JAFFE 1984, HSU et al. 1985), to identify histiocytosis X cells (REE and KADIN 1986), and to distinguish between benign and malignant histiocytes (REE and KADIN 1985).

Megakaryocytes are rich in sialic acid residues and sialic acid-binding lectins are good markers (SCHICK and FILMYER 1985). Lectin histochemistry has been used to study the occurrence of carbohydrates in the glomerulus and within the different segments of the nephron (VIERBUCHEN et al. 1980, HENNIGAR et al. 1985a, ORTMANN et al. 1988a, b) as well as in renal tumors (VIERBUCHEN et al. 1980, HENNIGAR et al. 1985b, WICK et al. 1986, ORTMANN et al. 1988a, b). Lectin-histochemical studies have disclosed heterogeneity of intercalated cells along rat kidney collecting ducts (BROWN et al. 1985, SCHUSTER et al. 1986). The significance of this finding remains incompletely understood. However, recent comparative immuno- and lectin-histochemical studies on normal human kidney tissue and renal tumors have revealed that renal oncocytomas most probably originate from intercalated cells of the collecting ducts (ORTMANN et al. 1988a, b).

Lotus tetragonolobus agglutinin (LTA) and PNA have been used to investigate the histogenesis of infantile and adult polycystic kidney disease. Cysts in infantile cystic disease only exhibit PNA reactivity, whereas cysts in adult polycystic disease show LTA as well as PNA reactivity (Faraggina et al. 1985). Expression of lectin-binding sites and blood group isoantigens has been investigated in normal urothelium and during neoplastic transformation of the transitional mucosa (Limas and Lange 1986, Orntoft et al. 1987). Spontaneous binding of PNA was absent in normal transitional mucosa but was observed in 10% of noninvasive and 65% of invasive transitional carcinomas. Comparative studies on the expression of blood group isoantigens showed a gradual deletion of A and H blood group isoantigens and a uniform expressionof Le^b and sialyl Le^a antigen.

In the skin there is a defined zonal lectin-binding pattern. Changes in lectin reactivity occur during differentiation from basal cells to the cornified layer (Elias et al. 1983; Nemamic et al. 1983; Virtanen et al. 1986). Basal cells stains with Griffonia simplicifolia agglutinin I (GSA I), ConA, Lens culinaris agglutinin (LCA), Pisum sativum agglutinin (PSA), Ricinus communis agglutinin (RCA I), and wheat germ agglutinin (WGA). The spinous cell layer shows Con A, Griffonia simplicifolia agglutinin II (GSA II), PNA, RCA I, soy bean agglutin (SBA), Sophora japonica agglutinin (SJA), succhinylWGA (sWGA), and WGA reactivity which is preserved to some extent in the granular cell layer with the addition of binding sites for the UEA (Elias et al. 1983). Both precancerous lesions and the squamous cell carcinomas show an altered lectin reactivity that corresponds to the deletion of ABH blood group isoantigens (Schaumburg-Lever et al. 1984b). Furthermore, altered lectin binding has been observed in lesions of psoriasis, epidermolysis bullusa, pachyonychia, and epidermolytic hyperkeratosis (Kariniemi et al. 1983, Brysk et al. 1984).

In neuropathology lectins have been extensively used to study glycoconjugates in normal and pathologically altered central and peripheral nervous system (Nakagawa et al. 1986a, b, d, Schelper et al. 1985, Streit et al. 1985a, b; Dolapchiewa et al. 1986, Estruch and Damjanov 1986, Gulati et al. 1986). These studies have revealed that the central as well as the peripheral nervous system contains galactose and N-acetylgalactosamine type glycoconjugates (Streit et al. 1985a, b). Some of the neurons that use γ-aminobutyric acid as a transmitter contain terminal GalNAc residues (Nakagawa et al. 1986b). Glycoconjugates of nociceptive primary neurons which contain substance P possess terminal galactose residues (Streit et al. 1986). Studies on the peripheral human nervous system using the lectin ECA have revealed the occurrence of sialic acid-masked N-acetyllactosamine type glycoconjugates in myelinated nerves (Vierbuchen et al. 1988c).

Lectin binding to pathologically altered nervous tissues has been studied in cases of injury, ischemia, infarction, peripheral neuropathy, amyloid plaques, and storage diseases (Schelper et al. 1985, Nishida et al. 1986, Schwechheimer et al. 1984, Kadoto et al. 1986, Szumanska et al. 1986, Faraggina et al. 1981, Alroy et al. 1985, 1986a, b).

Glycoconjugates detected by the Helix pomatia agglutinin and Maclura pomifera agglutinin occurred selectively in altered nerves but were not pathognomonic for any type of injury (Estruch and Damjanov 1986).

Some types of central nervous neoplasms have been analyzed for the occurrence of lectin-binding sites. In tumors with neuronal differentiation a correlation between Con A reactivity and morphological differentiation was observed (SCHWECHHEIMER et al. 1984). Granular cell tumors from different areas of the central nervous system expressed PNA-binding sites (SCHWECHHEIMER et al. 1983).

Studies on the cells of the normal choroid plexus and plexus papillomas revealed a surface coating by sialomucopolysaccharides (MÜLLER et al. 1980). After enzymatic removal of sialic acid residues by neuraminidase, normal plexus epithelium and plexus papillomas disclosed PNA- and RCA I-binding sites (MÜLLER et al. 1980, VIERBUCHEN and KLEIN 1983). The application of lectin-histochemical methods seems suitable to clarify the differential diagnosis between choroid plexus papillomas, metastases of carcinomas, and papillary ependymomas (MÜLLER et al. 1980).

Histochemical assessment of the different fiber types in muscle tissue is essential for the diagnosis of certain muscular and neuromuscular disorders. Fiber types are characterized by the demonstration of oxidative enzymes and the myofibrillar adenosine triphosphatase enzyme activity. Mild digestion of frozen samples of skeletal muscle specimens followed by Con A staining has been proposed for identifying different fiber types (YAMAGAMI et al. 1985). Recently a glycohistochemical method using (neo)glycoproteins for the detection of endogenous carbohydrate-binding proteins (endogenous lectins) has been used for fiber typing on formaldehyde-fixed and paraffin-embedded biopsy specimens (BARDOSI et al. 1989).

Acknowledgement. This work was supported by Deutsche Forschungsgemeinschaft grant VI 106/1-1.
The author wishes to thank Mrs E. Vierkotten for expert technical assistance

References

Achord DT, Brot FE, Gonzalez-Noriega A, Sly WS. Stahl P (1977) Human β-glucuronidase. II. Fate of infused human placentae β-glucuronidase in the rat. Pediatr Res 11:816–822

Achord DT, Brot FE, Bell CE, Sly WS (1978) Human β-glucuronidase: in vivo clearance and in vitro uptake by a glycoprotein recognition system on reticuloendothelial cells. Cell 15:269–278

Alroy J, Orgad U, Ucci AA, et al. (1985) Neurovisceral and skeletal GM1 gangliosidosis in dogs with β-galactosidase deficiency. Science 229:470–472

Alroy J, Ucci AA, Goyal V, Aurilio A (1986a) Histochemical similarities between human and animal globoid cells in Krabbe's disease: lectin study. Acta Neuropathol 71:26–31

Alroy J, Ucci AA, Goyal V, Woods W (1986b) Lectin histochemistry of glycolipid storage diseases on frozen and paraffin embedded tissue sections. J Histochem Cytochem 34:501–505

Amsden A, Ewan V, Yoshida T, Cohen S (1978) Studies on cellular receptor for lymphokines I. Interaction of chemotactic factors with monosaccharides. J Immunol 120:542–549

Andre C, Bertoux F-C, Andre F (1985) Distribution of the two subclasses IgA1 and IgA2 in IgA nephropathy. In: Mucosal Immunity IgA and Polymorphonuclear Neutrophils (Revillard JP, Voisin C, Wierzbiecki N, eds.), pp 173–175, Foundation Franco-Allemande Suresness

Apgar JR, Cresswel P (1982) Expression of cell surface lectins on activated human lymphoid cells. Eur J Immunol 12:570–576

Arita Y, Ogata SI, Ozawa M, Muramatsu T (1985) Distinct properties of receptors for *Dolichos biflorus* agglutinin (DBA) isolated from small intestine and adult mice endodermal cells of early embryo and teratocarcinomas. Differentiation 28:254–259

Ashwell G (1977) The role of cell surface carbohydrates in binding phenomena. Mammalian Cell Membranes 4:57

Ashwell G, Harford J (1982) Carbohydrate-specific receptors of the liver. Ann Rev Biochem 51:531–534

Ashwell G, Morell AG (1974) The role of surface carbohydrates in the hepatic recognition and transport of circulating glycoproteins. Adv Enzymol 41:99–128

Baba T, Yoshida T, Yoshida T, Cohen S (1979) Suppression of cell mediated immune reactions by monosaccharides. J Immunol 122:838–841

Bardosi A, Dimitri T, Wosgim B, Gabius HJ (1989) Expression of endogenous receptors for neoglycoproteins, especially lectins, that allow fiber typing on formaldehyde-fixed, paraffin embedded mucsle biopsy specimens. A glycohistochemical, immunohistochemical, and glycobiochemical study. J Histochem Cytochem 37:989–998

Barondes SH (1978) Developmentally regulated slime mold lectins and specific cell cohesion. In: Lerner RA, Bergsma D (eds.) The molecular basis of cell–cell interaction. Alan R Liss, New York, pp 491–496

Barondes SH (1981) Lectins: their multiple endogenous cellular functions. Ann Rev Biochem 50:207–231

Barondes SH (1984) Soluble lectins: a new class of proteins. Science 223:1259–1264

Barondes SH (1988) Bifunctional properties of lectins: lectins redefined. Trends Biochem Sci 13:480–482

Barzilay M, Monsigny M, Sharon N (1982) Interaction of soy bean agglutinin with human peripheral blood lymphocyte subpopulations: evidence for the existence of a lectin like substance on the lymphocyte surface. In: Madsen TC (ed) Lectins biology, biochemistry, and clinical biochemistry. Walter De Gruyter, Berlin, pp 67–81

Berger EG, Buddecke F, Kamerling JP, Kobata A, Paulson JC, Vliegenhart JFG (1982) Structure, biosynthesis and function of glycoprotein glycans. Experientia 38:1129–1258

Beuth JG, Uhlenbruck G (1985) Leber-Lektine als Mediatoren der Metastasierung maligner Tumoren. Funkt Biol Med 5:269–275

Beuth J, Oette K, Uhlenbruck G (1986) Lektine – Marker Moleküle tumorassoziierter Antigene und Mediatoren der Metastasierung. Med Welt 37:543–545

Beuth J, Ko HL, Oette K, Pulverer G, Uhlenbruck G (1987) Inhibition of liver metastasis in mice by blocking hepatocyte lectins with arabinogalactan infusions and D-galactose. J Cancer Res Clin Oncol 113:51–55

Bird GWG, Wingham J (1973) Seed agglutinin for rapid identification of Tn polyagglutination. Lancet I:677

Bird GWG, Shinton NK, Wingham J (1971) Peristent mixed field polyagglutination. Br J Haematol 21:443–453

Bird GWG, Wingham J, Pierce SR, Oats GD, Pallock A (1978) Tn a "new" form of polyagglutination. Lancet I:1215–1210

Blaese RM, Tosato G, Greene WC, Fleisher TA, Muchmore AV (1983) The role of cell surface lectin–carbohydrate interactions in cellular recognition, cooperation, and regulation. Am J Pediatr Oncol Hematol 5:199–206

Boaz L, Ward H, Pereira MEA (1986) Prolectin activation in *Giardia lamblia*. In: Mirelman D (ed) Microbial lectins and agglutinins. Wiley, New York, pp 301–317

Bock K, Breimer ME, Brignole A et al. (1985) Specificity of binding of a Strain of uropathogenic *Escherichia coli* to gal-alpha-1-4-gal containing glycosphingolipids. J Biol Chem 260:8545–8551

Bock K, Karlsson K-A, Stromberg N, Teneberg S (1988) Interaction of viruses, bacteria and bacterial toxins with host cell surface glycolipids. Aspects on receptor identification and dissection of binding sites. In: Wu A, Adams GA (eds) The molecular immunology of complex carbohydrates, Plenum, New York, pp 153–186

Böcker W, Klaubert A, Bahnsen J et al. (1984) Peanut lectin histochemistry of 120 mammary carcinomas and its relation to tumour type, grading, staging, and receptor status. Virchows Arch A 403:149–161

Boland CR, Montgomery CK, Kim YS (1982) Alterations in human colonic mucin occurring with cellular alteration and neoplastic transformation. Proc Natl Acad Sci (USA) 79:2051–2055

Bolen JB, Consigh RA (1979) Differential absorption of polyoma virions and capsids to mouse kidney cells and guinea pig erythrocytes. J Virol 32:679–683

Boulanger P, Lonberg-Holm (1981) Components of non-enveloped viruses which recognize receptors. In: Lonberg Holm K, Philipson L (eds) Virus Receptors Part 2. Chapman and Hall, London, pp 21–45

Bowles DJ, Hanke DE (1977) Evidence for lectin activity associated with glycophorin, the major glycoprotein of human erythrocyte membranes: implications for the structure of membranes. FEBS Lett 82:34–38

Boyd WC, Shapeleigh E (1954) "Specific precipitating of plant agglutinins (lectin)". Science 119:419

Bracha R, Kobiler D, Mirelman D (1982) Attachment and ingestion of bacteria by trophozoites of Entamoeba histolytica. Infect Immun 36:396–406

Bracha R, Mirelman D (1983) Adherence and ingestion of Escherichia coli serotype O 55 by trophozoites of Entamoeba histolytica. Infect Immun 50:882–887

Bracha R, Mirelman D (1984) Virulence of Entamoeba histolytica trophozoites. Effects of bacteria, microaerobic conditions, and mitronidazole. J Exp Med 160:353–368

Brackenbury R (1985) Molecular mechanisms of cell adhesion in normal and transformed cells. Cancer Metastasis Rev 4:41–58

Brahim F, Osmond DG (1970) Migration of bone marrow lymphocytes demonstrated by selective bone marrow labelling with thymidine-H3. Anat Rec 168:139–160

Brown D, Roth J, Orci L (1985) Lectin gold cytochemistry reveals intercalated cell heterogeneity along rat kidney collecting ducts. Am J Physiol 248:348–356

Bridges K, Harford J, Ashwell G, Klausner RD (1982) Fate of receptor and ligand during endocytosis the asialoglycoproteins by isolated hepatocytes. Proc Natl Acad Sci USA 79:350–354

Brysk MM, Miller J, Hebert AA (1984) Concanavalin A distinguishes among diseases of altered epidermal differentiation. J Invest Dermatol 82:18–20

Bross IDJ, Blumenson LE (1976) Metastatic sites that produce generalized cancer: identification and kinetics of generalisation sites. In: Weiss L (ed) Fundamental aspects of metastasis. North Holland, Amsterdam, pp 359–375

Burness ATH (1981) Glycophorin and sialylated components as receptors for viruses. In: Lonberg Holm K, Philipson L (eds) Virus receptors part 2. Chapman and Hall, London, pp 65–84

Burness ATH, Pardoe LU (1981) The effect of enzymes on the attachment of influenza and encephalomyocarditis virus to erythrocytes. J Gen Virol 55:275–278

Butcher EC, Scollay RG, Weissman IL (1979) Lymphocyte adherence to high endothelial venules. Characterization of a modified in vivo assay, and examination of syngeneic and allogeneic lymphocyte subpopulations. J Immunol 123:1996–2003

Butcher EC, Scollay RG, Weissman IL (1980) Organ specificity of lymphocyte migration: mediation by highly selective lymphocyte interaction with organ-specific determinants on high endothelial venules. Eur J Immunol 10:556–561

Cantor H, Weissman I (1976) Development and function of subpopulations of thymocytes and T lymphocytes. Prog Allergy 20:1–64

Carding SR, Thorpe SJ, Thorpe R, Feizi T (1985) Transformation growth related changes in levels of nuclear and cytoplasmic proteins antigenetically related to mammalian β-galactoside binding lectin. Biochem Biophys Res Commun 127:680–686

Carrer JP, Brisson JR (1984) The three dimensional structure of N-linked oligosaccharides. In: Ginsburg V, Robbins PW (eds) The Biology of Carbohydrates, Vol. 2. Wiley New York, pp 289–331

Carrol SB, Ippolito LM, Dewolf WC (1982) Heparin binding agglutinin on human teratocarcinoma cells. Biochem Biophys Res Commun 109:1353–1359

Cartron JP, Andreu G, Cartron J, Bird GWG, Salmon C, Gerbal A (1978) Demonstration of T-transferase deficiency in Tn polyagglutinable blood samples. Eur J Biochem 92:111–119

Chany-Fournier F, Paulsoin A, Chany C (1978) Isolation, preliminary characterization and interferon antagonistic effects of a mammalian lectin like substance. Proc Natl Acad Sci 75:2333–2337

Cheresh DA, Pierschbacher MD, Herzig MA, Mujoo K (1986) Disialogangliosides are involved in the attachment of human melanoma and neuroblastoma cells to extracellular matrix proteins. J Cell Biol 102:688–696

Critchley DR, Magnani JL, Fishman PH (1981) Interaction of cholera toxin with brush border membranes. Relative roles of gangliosides and galactoproteins as toxin receptors. J Biol Chem 256:8724–8731

Critchley DR, Streuli CH, Kellie S, Ansell S, Patel B (1982) Characterization of the cholera toxin receptor on Balb/c3T3 cells as a ganglioside similar to, or identical to ganglioside GM1. No evidence for galactoproteins with receptor activity. Biochem J 204:209–219

Crittenden SL, Roff CF, Wang JL (1982) Carbohydrate-binding protein 35: identification of the galactose-specific lectin in various tissues of mice. Mol Cell Biol 4:1252–1259

Deas J, Lee LT (1981) Competitive inhibition by soluble erythrocyte glycoproteins of penetration by *Plasmodium falciparum*. Am J Trop Med Hyg 30:1164–1167

DeMaio A, Lis H, Gershoni JM, Sharon N (1986) Identification of Peanut agglutinin-binding glycoproteins on immature human thymocytes. Cell Immunol 99:345–353

Decker JM (1980) Lectin-like molecules on murine thymocytes and splenocytes I. Molecules specific for the oligosaccharides moiety of the fetal alpha globulin fetuin. Mol Immunol 17:803–808

Dennis JC, Waller CA, Schirrmacher V (1984) Identification of asparagine linked oligosaccharides involved in tumor cell adhesion to laminin and type IV collagen. J Cell Biol 99:1416–1423

Dolapchieva S, Ichev K, Ovtscharoff W (1986) Lectin binding sites in axon-myelin-Schwann cell complex. Acta Histochem Cytol 19:253–261

Donita ST, Poindexter NJ, Ginsberg BH (1982) Comparison of the binding of the cholera toxin and *Escherichia coli* toxin to Y1 adrenal cells. Biochemistry 21:660–664

Douy A, Gallot B (1980) Synthesis and ordered structure of amphopathic block copolymers with a saccharide and a peptide block. Biopolymers 19:493–507

Draper RK, Chin D, Simon MI (1978) Diphtheria toxin has the properties of a lectin. Proc Natl Acad Sci (USA) 75:261–265

Duguid JP, Smith IW, Dempster G, Edmunds PN (1955) Nonflagellar filamentous appendages ("fimbriae") and haemagglutinating activity in *Bacterium coli*. J Pathol Bacteriol 70:335–348

Egelrud T (1986) Dermatitis herpetiformis: selective deposition of immunoglobulin A 1 in granular deposits in clinically normal skin. Acta Derm Venerol 66:11–15

Eidels L, Proia RL, Hart DA (1983) Membrane receptors for bacterial toxins. Microbiol Rev 47:596–620

Elias PM, Chaung JC, Orozco-Topete R, Nemanic MK (1983) Membrane glycoconjugate visualization and biosynthesis in normal and retinoid-treated epidermis. J Invest Dermatol 81:81–85

Erb P, Stein AC, Alkan SS, Studer S, Zoumbon E, Gisler RH (1980) Characterization of accessory cells required for helper T cell induction in vitro: evidence for a phagocytic Fc-receptor, and Ia bearing cell type. J Immunol 125:2504–2507

Estruch R, Damjanov I (1986) Lectin histochemistry applied to human nerves. Arch Pathol Lab Med 110:730–735

Eisenberg B, Delosse J, Harford F, Michalek J (1982) Carcinoma of the colon and rectum: the natural history reviewed in 1704 patients. Cancer 49:1131–1134

Ewing J (1928) A treatise on tumors, 3rd ed W.B. Saunders. Philadelphia

Ezekowitz RAB, Austyn J, Stahl PD, Gordon S (1981) Surface properties of bacillus Calmette-Guerin-activated mouse macrophages. Reduced expression of mannose-specific endocytosis, Fc receptors, and antigen F4/80 accompanies induction of Ia. J Exp Med 154:60–76

Facer CA (1983) Erythrocytes sialoglycoproteins and *Plasmodium falciparum* invasion. R Soc Trop Med Hyg 77:524–530

Faraggina T, Churg J, Grishman E et al. (1981) Light and electron microscopic histochemistry of Fabry's disease. Am J Pathol 103:247–262

Faraggina T, Bernstein J, Strauss L, Churg J (1985) Use of lectins in the study of histogenesis of renal cysts. Lab Invest 53:575–579

Fenlon S, Ellis IO, Bell J, Todd JH, Elston LW, Blamey RW (1987) *Helix pomatia* and *Ulex europaeus* lectin binding in human breast carcinoma. J Pathol 152:169–176

Fields BN (1985) Virology, Raven, New York

Fidgen KJ (1975) The action of *Vibrio cholerae* and *Corynebacterium diphtheriae* neuraminidase on the Sendai virus receptor of human erythrocytes. J Gen Microbiol 89:48–56

Fidler IJ (1984) The evolution of biological heterogeneity in metastatic neoplasms. In: Nicolson GL, Milas L (eds) Cancer invasion and metastasis. Biological and therapeutic aspects. Raven, New York, pp 5–26

Firon N, Ofek I, Sharon N (1984) Carbohydrates of the mannose specific fimbrial lectins of enterobacteria. Infect Immun 43:1088–1090

Fishman PH, Moss J, Vaughn M (1976) Uptake and metabolism of gangliosides in transformed mouse fibroblasts. Relationship of ganglioside structure to choleragen response. J Biol Chem 251:4490–4494

Fleisher TA, Greene WC, Blaese RM, Waldmann TA (1981) Soluble suppressor supernatant elaborated by concanavalin A activated human mononuclear cells. II. Characterization of a soluble suppressor of B cell immunoglobulin production. J Immun 26:1192–1197

Forbes JT, Bretthauer RK, Oeltmann TN (1981) Mannose-6-, fructose-1-fructose-6-phosphates inhibit human natural cell cytotoxicity. Proc Natl Acad Sci 78:5797–5801

Ford WL (1975) Lymphocyte migration and immune response Prog Allergy 19:1–59

Ford WL, Simmonds SJ (1972) The tempo of lymphocyte recirculation from blood to lymph in the rat. Cell Tissue Kinet 5:175–189

Fried H, Cahan LD, Paulson JC (1981) Polyoma virus recognizes specific sialyloligosaccharide receptors on host cells. Virology 109:188–192

Gabius HJ, Engelhardt R, Rehms S, Cramer F (1984) Biochemical characterization of endogenous carbohydrate-binding proteins from spontaneous murine rhabdomyosarcoma, mammary adenocarcinoma and ovarian teratoma. J Natl Cancer Inst 73:1349–1357

Gabius HJ, Engelhardt R, Casper J, Scholl HJ, Nagel GA, Cramer F (1985a) Comparison of endogenous lectins in human embryonic carcinoma and yolk sac carcinoma. Tumor Biol 6:471–482

Gabius HJ, Vehmeyer K, Engelhardt R, Nagel GA, Cramer F (1985b) Carbohydrate-binding proteins of tumor lines with different growth properties. I. Differences in their pattern for three clones of rat fibroblasts transformed with a myeloproliferative sarcoma virus. Cell Tissue Res 241:9–15

Gabius HJ, Engelhardt R, Rehm S, Barondes SH, Cramer F (1986a) Presence and relative distribution of the endogenous beta-galactoside-specific lectins in different tumor types of rat. Cancer J 1:19–21

Gabius HJ, Engelhardt R, Rehm S, Deerberg F, Cramer F (1986b) Differences in the pattern of endogenous lectin from spontaneous rat mammary tumors. Tumor Biol 7:71–81

Gabius HJ, Engelhardt R, Sartoris DJ, Cramer F (1986c) Pattern of endogenous lectins of a human sarcoma (Ewing's sarcoma) reveals differences to human normal tissues and tumors of epithelial and germ cell origin. Cancer Lett 31:139–145

Gabius HJ, Debbage PL, Engelhardt R, Osmers R, Lange W (1987) Identification of endogenous sugar binding proteins (lectins) in human placenta by histochemical localization and biochemical characterization. Eur J Cell Biol 44:265–272

Gallagher JT (1984) Carbohydrate-binding properties of lectins: a possible approach to lectin nomenclature and classification. Biosci Rep 4:621–625

Gallatin M, St John TP, Siegelman M, Reichert R, Butcher EC, Weissman IL (1986) Lymphocyte homing receptors. Cell 44:673–680

Gartner TK, Podleski TR (1976) Evidence that the types and specific activity of lectins control fusion on L6 myoblasts. Biochem Biophys Res Commun 70:1142–1149

Gartner TK, Williams DC, Minion FC, Phillips DR (1978) Thrombin-induced platelet aggregation is mediated by a platelet plasma membrane-bound lectin. Science 200:1281–1283

Gervais M, Gallot B (1982) Glycan conformational change in mesomorphic structures of ternary systems. Liposaccharid/phospholipide/waters. Biochem Biophys Acta 688:586–596

Gething MJ, Sambrook J (1982) Construction of influenza hemagglutinating genes that code for intracellular and secreted forms of the protein. Nature 300:598–603

Geuze HJ, Slot JW, Strous GJ, Hasilik A, von Figura K (1985) Possible pathways for lysosomal enzyme delivery. J Cell Biol 101:2253–2262

Geuze HJ, Stoorvogel W, Strous GJ, Slot JW, Bleekenmolen JE, Mellman I (1988) Sorting of mannose-6-phosphate receptor and lysosomal membrane proteins in endocytotic vesicles. J Cell Biol 107:2491–2501

Girgert R, Bruchelt G, Treuner J (1987) Separation of haematopoetic stem cells in peripheral blood by lectins. In: Bog-Hansen TC, Freed DLJ (eds) Lectins, biology, biochemistry, clinical. Biochemistry, vol 6. Sigma Chemical Company, St Louis, pp 447–452

Goldberg DE, Kornfeld S (1981) The phosphorylation of β-glucuronidase oligosaccharides in mouse P 388D cells. J Biol Chem 256:13060–13067

Goldschneider I, McGregor DD (1968) Migration of lymphocytes and thymocytes in the rat. I. The route of migration from blood to spleen and lymph nodes. J Exp Med 127:155–168

Goldstein IJ, Hughes RC, Monsigny M, Osawa T, Sharon N (1980) What should be called a lectin. Nature 285:66

Goldstein IJ, Poretz RD (1986) Isolation, physicochemical characterization, and carbohydrate binding specificity of lectins. In: Liener IE, Sharon N, Goldstein IE (eds) The lectins, properties, functions and applications. Academic, New York, pp 33–247

Gowans JL (1957) The effect of the continuous re-infusion of lymph and lymphocytes on the output of lymphocytes from the thoracic duct of unanaesthetized rats. Br J Exp Pathol 38:67–78

Gowans JL (1959) The recirculation of lymphocytes from blood to lymph in the rat. J Physiol (Lond) 146:54–59

Gowans JL, Knight EJ (1964) The route of re-circulation of lymphocytes in the rat. Proc R Soc Lond [Biol] 159:257–262

Grabel LB (1984) Isolation of a putative cell adhesion mediating lectin from teratocarcinoma stem cells and its possible role in differentiation. Cell Differ 15:121–124

Grabel LB, Rosen SD, Martin GR (1979) Teratocarcinoma stem cells have a cell surface carbohydrate binding component implicated in cell–cell adhesion. Cell 17:477–484

Grabel LB, Glabe CG, Singer MS, Martin GR, Rosen SD (1981) A fucan specific lectin on teratocarcinoma stem cells. Biochem Biophys Res Commun 102:1165–1171

Graupner GA, Vollmers HP, Brichmeyer W, Gabius HJ, Cramer F (1984) Distinct pattern of saccharide-mediated inhibition of homotypic cell aggregation in four human carcinoma cell lines. Hoppe Seyler's Z Physiol Chem 365:992–993

Greene WC, Fleischer TA, Waldmann TA (1981) Soluble suppressor supernatants elaborated by concanavalin A activated human mononuclear cells. I-characterization of a soluble suppressor of T-cell proliferation. J Immunol 126:1185–1191

Griffiths GB, Hoflak B, Simons K, Mellman I, Kornfeld S (1988) The mannose-6-phosphate receptor and the biogenesis of lysosomes. Cell 52:329–341

Guerrant RL, Brush J, Ravdin JL, Sullivan JA, Mandell GL (1981) Interaction between entamoeba histolytica and human polymorphonuclear neutrophils. J Infect Dis 143:83–93

Gulati AK, Zalewski AA, Sharma KB, Ogrowsky D, Sohal GS (1986) A comparison between lectin binding in rat and human peripheral nerve. J Histochem Cytochem 34:1487–1493

Guyot G (1908) Über die bakterielle Hämagglutination (bacterio-hemagglutination). Zentralbl Bakteriol Abt Orig 47:640–653

Hadley TJ, Klotz TF, Miller LH (1986) Invasion of erythrocytes by malaria parasites: a cellular and molecular overview. Ann Rev Microbiol 40:451–457

Hadley TJ, Klotz FW, Pasvol G, Heynes JD, McGinniss MH, Okubo Y, Miller LH (1987) Falciparum malaria parasites invade erythrocytes that lack glycophorin A and B ($M^k M^k$). Strain differences indicate receptor heterogeneity and two pathways of invasion. J Clin Invest 80:1190–1193

Harris DT, Iglesias JL, Argov S, Toomey J, Koren HS (1987) Heterogeneity of human natural killer (NK) cells: enrichment of NK by negative-selection with the lectin from Erythrina cristagalli. J Leukocyte Biology 42:163–170

Hasilik A, Neufeld EF (1980) Biosynthesis of lysosomal enzymes in fibroblasts: synthesis as precursors of higher molecular weight. J Biol Chem 255:4937–4945

Hasilik A, Klein U, Waheed A, Strecker G, von Figura K (1980) Phospharylated oligosaccharides in lysosomal enzymes. Identification of α N-acetylglucosamine-(1)-phospho-(6)-mannose diester groups. Proc Natl Acad Sci USA 77:7074–7078

Haywood AM, Boyer BP (1982) Sendai virus membrane fusion: time course and effect of temperature, pH, calcium, and receptor concentration. Biochemistry 24:6041–6046

Hedman K, Pastan I, Willingham MC (1986) The organelles of the trans domain of the cell. Ultrastructural localization of sialoglycoconjugates using *Limax flavus* agglutinin. J Histochem Cytochem 34:1069–1077

Helenius A, Mellman I, Wall DA, Hubbard A (1983) Endosomes. Trends Biochem Sci 506:245–254

Hennigar RA, Schulte BA, Spicer SS (1985a) Heterogeneous distribution of glycoconjugates in human kidney tubules. Anat Rec 211:376–390

Hennigar RA, Sens DA, Spicer SS et al. (1985b) Lectin histochemistry of nephroblastoma (Wilms Tumor). Histochem J 17:1091–1110

Herberman RB, Nunn ME, Lavrin DH (1975) Natural cytotoxic reactivity of mouse lymphoid cells against syngeneic and allogeneic tumors. Int J Cancer 16:216–229

Herberman RB, Holden HT (1978) Natural cell mediated immunity. Adv Cancer Res 27:305–377

Higgins TK, Sabatino AB, Remold HG, David JR (1978) Possible role of macrophage glycolipids as receptors for migration inhibitory factor (MIF). J Immunol 121:880–886

Hodes RJ, Hatchcock KS, Singer A (1980) Major histocompatibility complex restricted cell recognition. J Exp Med 152:1779–1794

Hoflack B, Kornfeld S (1985) Purification and characterization of a cation-dependent mannose-6-phosphate receptor from murine P3 88D1 macrophages and bovine liver. J Biol Chem 260:12008–12014

Holmgren J, Svennerholm L, Elwing H, Fredman P (1980) Sendai virus receptor: proposed recognition structure based on binding to plastic absorbed gangliosides. Proc Natl Acad Sci USA 77:1947–1950

Hosaka M, Murase N, Orito T, Mori M (1985) Immunohistochemical evaluation of factor VIII related antigens, filament proteins and lectin binding in hemangiomas. Virchows Arch [A] 40:237–247

Howard DR, Ferguson P, Batsakis JG (1981) Carcinoma associated cytostructural antigenic alterations: detection by lectin binding. Cancer 47:2872–2877

Howard RJ, Haynes JD, McGinniss MH, Miller LH (1982) Studies on the role of red blood cell glycoproteins as receptors for invasion by *Plasmodium falciparum* merozoites. Mol Biochem Parasitol 6:303–315

Hsu SM, Jaffe ES (1984) LeuM1 and peanut agglutinin stain the neoplastic cells of Hodgkin's disease. Am J Clin Pathol 82:29–32

Hsu SM, Raine L (1982) Versatility of biotin-labelled lectins and avidin-biotin-peroxidase complex for localization of carbohydrate in tissue sections. J Histochem Cytochem 30:157–161

Hsu MC, Scheid A, Choppin PW (1979) Reconstitution of membranes with individual paramyxovirus glycoproteins and phospholipid in cholate solution. Virology 95:476–491

Hsu SM, Yang K, Jaffe ES (1985) Phenotypic expression of Hodgkin's and Sternberg cells in Hodgkin's disease. Am J Pathol 118:209–215

Huang RTC, Rott R, Klenk HD (1973) On the receptor of influenza viruses. 1. Artificial receptor for influenza virus. Z Naturforsch [C] 28:342–345

Hughes RC (1983) Glycoproteins. Chapman and Hall, London

Iglesias JL, Lis H, Sharon N (1982) Purification and properties of a D-galactose/N-acetylgalactosamine specific lectin from *Erythrina cristagalli*. Eur J Biochem 123:247–252

Ito N, Nishi K, Nakajama M, Yoshiro O, Hirota T (1988) Effects of alpha-L-fucosidase digestion on lectin staining in human pancreas. J Histochem Cytochem 36:503–509

Jaffe EA, Leung LK, Nachmann RL, Levin RI, Mosher DF (1982) Thrombospondin is the endogenous lectin of human platelets. Nature 295:246–248

Jarumilinta R, Maegraith BG (1969) Enzymes of *Entamoeba histolytica*. Bull WHO 41:269–273

Jiang PH, Fournier FC, Galliot BR, Sarragne M, Chany C (1983) Sarcolectin. An interferon antagonist extracted from hamster sarcomas and normal muscles. J Biol Chem 258:12361–12367

Judd WJ (1980) The role of lectins in blood group serology. Crit Rev Clin Lab Sci 1:171–214

Jungery M, Pasvol G, Newbold CI, Weatherhall DJ (1983) A lectin-like receptor is involved in invasion of erythrocytes by *Plasmodium falciparum*. Proc Natl Acad Sci USA 80:1018–1022

Kadoto E, Tanji K, Nishida S et al. (1986) Lectin (UEA-I) reaction of capillary endothelium with reference to permeability in autopsied cases of cerebral infarction. Histol Histopathol 1:219–226

Kahan B (1979) Ovarian localization by embryonal teratocarcinoma cells derived from female germ cells. Somatic Cell Genet 5:763–780

Kaladas PM, Kabat EA, Iglesias JL, Lis H, Sharon N (1982) Immunochemical studies on the combining site of the galactose/N-acetylgalactosamine specific lectin from Erythrina cristagalli seeds. Arch Biochem Biophys 217:624–637

Kamenov B, Kieran MW, Berrington-Nigh J, Longecker BM (1984) Homing receptors as functional markers for classification, prognosis, and therapy of leukemias and lymphomas. Proc Soc Exp Med 177:211–219

Kariniemi AL, Holthöfer H, Miettinen A, Virtanen I (1983) Altered binding of Ulex europaeus I lectin to psoriatic epidermis. Br J Dermatol 109:523–559

Kaszenowski UH, Karme M (1981) Selective inhibition of T-suppressor-cell function by monosaccharides. Nature 289:181–184

Kathan RH, Winzler RJ, Johnson CA (1961) Preparation of an inhibitor of viral hemagglutination from erythrocytes. J Exp Med 113:37–45

Katsuyama T, Spicer SS (1978) Histochemical differentiation of complex carbohydrates with variants of the concanavalin A-horseradish peroxidase method. J Histochem Cytochem 26:233–250

Kawano K, Uehara F, Sameshima M, Ohba N (1984) Application of lectins for detection of goblet cell carbohydrates of the human conjunctiva. Exp Eye Res 38:439–447

Kawasaki T (1982) Studies on mannan-binding proteins. Mammalian lectin specific for mannose and N-acetyl-glucosamine residues. Seikagaku 54:1019–1032

Kawasaki T, Ashwell G (1977) Isolation and characterization of an avian hepatic binding protein specific for N-acetyl-glucosamine terminated glycoproteins. J Biol Chem 252:6536–6543

Kawasaki T, Mizuno Y, Masuda T, Yamashina I (1980) Mannan-binding protein in lymphoid tissues of the rat. J Biochem (Tokyo) 88:1891–1894

Kawasaki N, Kawasaki T, Yamashina I (1983) Isolation and characterization of a mannan-binding protein from human serum. J Biochem 94:937–947

Kellokumpu I, Karhi K, Andersson LC (1986) Lectin-binding sites in normal, hyperplastic adenomatous and carcinomatous human colorectal mucosa. Acta Pathol Immunol Scand 94:271–280

Kennedy DW, Rohrbach DH, Martin GR, Momoi T, Yamada KM (1983) The adhesive glycoprotein laminin is an agglutinin. J Cell Physiol 114:257–259

Keusch GT (1981) Receptor mediated endocytosis of shigella cytotoxin. In: Middlebrook JL, Kohn D (eds) Receptor mediated binding and internalization of toxins and hormones. Academic, New York, pp 95–108

Keusch GT, Jacewicz M (1977) Pathogenesis of Shigella diarrhea VII: Evidence for a cell membrane toxin receptor involving ß1 ends to 4-acetyl-D-glucosamine oligomers. J Exp Med 146:535–546

Kieda C, Monsigny M (1983) Lymphocyte membrane lectin: evidence for the involvement of a beta galactoside receptor in cell–cell interaction, In: Parker JW, O'Brien RL (eds) Intercellular communication in leukocyte function. John Wiley, London, pp 649–652

Kieda C, Bowles DJ, Ravid A, Sharon N (1978) Lectins in lymphocyte membranes. FEBS Lett 94:391–396

Kieda C, Roche AC, Delmotte F, Monsigny M (1979) Lymphocyte membrane lectins: Direct visualization by the use of fluoresceinyl glycosylated cytochemical markers. FEBS Lett 99:329–332

Kieda C, Monsigny M, Waxdal MJ (1982) Endogenous lectins on human peripheral mononuclear leukocytes. In: Bog-Hansen TC, Spengler GA (eds) Lectins, biology, biochemistry, clinical biochemistry, Vol 3. DeGruyter, Berlin, pp 427–433

Kieda C, Monsigny M (1986) Involvement of membrane sugar receptors and membrane glycoconjugate in the adhesion of 3 LL subpopulations to cultured pulmonary cells. Invasion Met 6:347–366

Kieran NW, Longnecker BM (1983) Organ-specific metastasis with special reference to avian systems. Cancer Met Rev 2:165–182

Kiess W, Blickenstaff GD, Sklar MM, Thomas CL, Nissley SP, Sahagian GG (1988) Biochemical evidence that the type II insulin-like growth factor receptor is identical to the cation-independent mannose-6-phosphate receptor. J Biol Chem 263:9339–9344

Kiessling R, Wigzell H (1979) An analysis of murine NK cell as to structure, function, and biological relevance. Immunol Rev 44:165–208

Kimber SJ (1986) Distribution of lectin receptors in postimplantation mouse embryo at 6–8 days gestation. Am J Anat 177:203–220

Kitamura MM, Iwamori M, Nagay Y (1980) Interaction between Clostridium botulinum neurotoxin and gangliosides. BBA 628:328–335

Klein PJ (1981) Die mikroangiopathische Anämie. Gustav Fischer Verlag, Stuttgart

Klein PJ, Vierbuchen M, Bulla M, Wehner B, Schindera F, Lennartz KJ, Fischer J (1980) Hämolytisch urämisches Syndrom (HUS). Histochemischer Nachweis infektiös-toxischer Nierenveränderungen. In: Gessler U, Seybold D (eds) Glomerulonephritis. Thieme, New York, 96–104

Klein PJ, Vierbuchen M, Würz H, Schulz KD, Newman RA (1981) Secretion associated lectin binding sites as a parameter of hormone dependence in mammary carcinoma. Br J Cancer 44:476–478

Klein PJ, Vierbuchen M, Fischer J, Schulz KD, Farrar G, Uhlenbruck G (1983) The significance of lectin receptors for the evaluation of hormone dependence in breast cancer. J Steroid Biochem 19:839–844

Kobayashi Y, Sawada JI, Osawa T (1978) Isolation and characterization of an inhibitory glycopeptide against guinea pig lymphotoxin from the surface of L cells. Immunochemistry 15:61–66

Kobiler D, Mirelman D (1980) Lectin activity in Entamoeba histolytica P throphozoites. Infect Immun 29:221–225

Kobiler D, Mirelman D (1981) Adhesion of Entamoeba histolytica trophozoites to monolayers of human cells. J Infect Dis 144:539–546

Kochibe N, Furukawa K (1980) Purification and properties of a novel fucose-specific hemagglutinin of Aleuria aurantia. Biochemistry 19:2841–2846

Kolb H (1982) A ganglioside receptor on lymphocytes mediates recognition of self. Biochem Biophys Res Commun 105:1488–1495

Kolb H, Shudt C, Kolb-Backofen V, Kolb HA (1978) Cellular recognition by rat liver cells of neuraminidase treated erythrocytes. Exp Cell Res 113:319–325

Korhonen TK, Väisinen-Rhen V, Pere M, Parkinen A, Finne J (1984) Escherichia coli fimbriae recognizing sialyl galactoside. J Bacteriol 159:762–766

Kornfeld S (1986) Trafficking of lysosomal enzymes in normal and disease states. J Clin Invest 77:1–6

Kozutsumi Y, Kawasaki T, Yamashina I (1980) Isolation and characterization of a mannan-binding protein from rabbit serum. Biochem Biophys Res Commun 95:658–664

Kuhlenschmidt T, Lee YC (1984) Specificity of chicken liver carbohydrate binding protein. Biochemistry 23:3569–3575

Kuhlmann WD, Pescke P (1984) Comparative study of procedures for histochemical detection of lectin binding by use of Griffonia simplicifolia agglutinin I and gastrointestinal mucosa of the rat. Histochemistry 81:265–272

Laver WG, Valentine RC (1969) Morphology of the isolated hemagglutinin and neuraminidase subunits of influenza virus. Virology 38:105–119

Leathem AJ, Brooks SA (1987) Predictive value of lectin binding on breast-cancer recurrence and survival. Lancet II:1054–1056

Lee G, Consiglio E, Habig W, Dyer S, Hardegree C, Kohn LD (1978) Structure: function studies of receptors for thyrotropin and tetanus toxin. Lipid modulation of effector binding to the glycoprotein receptor component. Biochem Biophys Res Commun 83:313–320

Lee AKC, DeLellis RA, Wolfe HJ (1986) Intramammary lymphatic invasion in breast carcinoma. Evaluation using ABH isoantigens and endothelial markers. Am J Surg Pathol 10:589–594

Lemieux RU, Wong TC, Liao J, Kabbat EA (1984) The combining site on anti-I-Ma (Group 1). Mol Immunol 21:751–759

Lemieux RU, Venot AP, Spohr U et al. (1985) Molecular recognition. V. The binding of the B human blood group determinant by hybridoma monoclonal antibodies. Can J Chem 63:2664–2668

Lennartz WJ (1980) The biochemistry of glycoproteins and proteoglycans. Plenum, New York

Leppanen A, Kopryuo A, Puro K, Rekonen O (1986) Glycoproteins of human teratocarcinoma cells (PA1) carry both anomers of O-glycosyl-linked D-galactopyranosyl-(1-3)2-acetamido-2-deoxy-D-galactopyranosyl groups. Carbohydr Res 153:87–95

Levine MM, Kapen JP, Black RE, Clements ML (1983) New knowledge on pathogenesis of bacterial enteric infections as applied to vaccine development. Microbiol Rev 47:510–556

Li ZQ, Perkins SJ, Loucheux-Lefebvre (1983) Alpha-1-acid glycoprotein: a small-angle neutron scattering study of human plasma glycoproteins. Eur J Biochem 130:275–279

Limas C, Lange P (1986) T-antigen in normal and neoplastic urothelium. Cancer 58:1236–1245

Lindahl M, Brossmer R, Wadström T (1988) Sialic acid and N-acetyl-galactosamine specific bacterial lectins of enterotoxigenic Escherichia coli (ETEC). In: Wu A, Adams GL (eds.) The molecular immunology of complex carbohydrates. Plenum, York, pp 123–152

Lipsick JL, Beyer EC, Barondes SH, Kaplan ND (1980) Lectins from chicken tissues are mitogenic for Thy-1 negative murine spleen cells. Biochem Biophys Res Commun 97:56–61

Lis H, Sharon N (1986) Lectins as molecules and as tools. Ann Rev Biochem 55:35–67

Liu DY, Petschek KB, Remold HG, David JR (1980) Role of sialic acid in the macrophage's glycolipid receptor for MIF. J Immunol 124:2042–2047

Liu DY, David JR, Remold HG (1982a) Glycolipid affinity purification of migration inhibitory factor. Nature 296:78–80

Liu DY, Petschek KD, Remold HG, David JR (1982b) Isolation of a guinea pig macrophage glycolipid with the properties of the putative migration inhibitory factor receptor. J Biol Chem 257:159–162

Lotan R (1987) Cell adhesion receptors and cancer metastasis. Cancer Bull 39:156–162

Lotan R, Raz A (1983) Low colony formation in vivo and in culture as exhibited by metastatic melanoma cells selected for reduced homotypic aggregation. Cancer Res 43:2088–2093

Louis CJ, Wyllie RG, Chou ST, Sztynda T (1981) Lectin binding affinities of human epidermal tumors and related conditions. Am J Clin Pathol 75:642–647

Louis CJ, Sztynda T, Cheng ZM, Wyllie RG (1983) Lectin binding affinities of human breast tumors. Cancer 52:1244–1250

Marchesi VD, Gowans JL (1964) The migration of lymphocytes through the endothelium of venules in lymph-nodes. Proc R Soc Lond [Biol] 159:282–290

Markwell MAK, Paulson JC (1980) Sendai virus utilizes specific sialyloligosaccharides as host cell receptor determinants. Proc Natl Acad Sci USA 77:5693–5697

Markwell MAK, Svennerholm L, Paulson JC (1981) Specific ganglioside function as host cell receptors for Sendai virus. Proc Natl Acad Sci USA 78:5406–5410

Markwell MAK, Fredman P, Svennerholm L (1984) Receptor ganglioside content of three hosts for Sendai virus MDBK, HeLa, and MDCK cells. Biochim Biophys Acta 775:7–16

Markwell MAK, Portner A, Schwartz AL (1985) An alternative route of infection for virus entry by means of the asialoglycoprotein receptor of a Sendai virus mutant lacking attachment protein. Proc Natl Acad Sci USA 82:978–982

Matsushita Y, Yonezawa S, Nakamura T, Shimizu S, Ozawa M, Maramatsu T, Sato E (1985) Carcinoma-specific specific Ulex europaeus agglutinin I binding glycoproteins of human colorectal carcinoma and its relation to carcinoembryonic antigen. J Natl Cancer Inst 75:219–226

Maynard Y, Baenziger JU (1982) Characterization of mannose and N-acetyl-glucosamine specific lectin present in rat hepatocytes. J Biol Chem 257:3788–3794

McCartney JC (1987) Fructose-containing antigens in normal and neoplastic human gastric mucosa. A comparative study using lectin histochemistry and blood group immunohistochemistry. J Pathol 152:23–30

McGuire EJ, Mascali JJ, Gardy SR, Nicolson GL (1984) Involvement of cell–cell adhesion molecules in liver colonization by metastatic murine lymphoma/lymphosarcoma variants. Clin Exp Metastasis 2:213–222

McKenzie IEC, Clarke A, Parish CR (1977) I a antigenic specificities are oligosaccharides in nature: hapten inhibition studies. J Exp Med 145:1039–1053

Mekada E, Uchida T, Okada Y (1979) Modification of the cell surface with neuraminidase increases the sensitivity of cells to diphtheria toxin and Pseudomonas aeruginosa exotoxin. Exp Cell Res 123:137–146

Mentzer T, Burrakoff S, Faller D (1986) Adhesion of T lymphocytes to human endothelial cells is regulated by the LFA-1 membrane molecule. J Cell Physiol 126:283–290

Meromsky L, Lotan R, Raz A (1986) Implications of endogenous tumor cell and surface lectins as mediators of cellular interactions and lung colonization. Cancer Res 46:5270–5275

Middlebrook JL, Dorland RB (1978) Response of cultured mammalian cells to the exotoxin of *Pseudomonas aeruginosa* and Corynebacterium diphteriae: differential cytotoxins. Can J Microbiol 23:183–189

Middlebrook JL, Dorland RB (1984) Bacterial toxins: cellular mechanisms of action. Microbiol Rev 48:199–221

Miller LH, Haynes JD, McAuliffe FM, Shiroishi T, Durocher J, McGinniss MH (1977) Evidence for differences in erythrocyte surface receptors for the malarial parasites *Plasmodium falciparum* and *Plasmodium knowlesi*. J Exp Med 146:277

Mirelman D, Ofek I (1986) Introduction to microbial lectins and agglutinins. In: Mirelman D (ed.) Microbial lectins and agglutinins. John Wiley, New York, pp 1–19

Mitchell GH, Hadley TJ, McGinnis MH, Klotz FH, Miller LH (1986) Invasion of erythrocytes by *Plasmodium falciparum* malaria parasites. Evidence for receptor heterogeneity. Blood 67:1519–1521

Montreuil J (1980) Primary structure of glycoprotein glycans. Basis for the molecular biology of glycoproteins. Adv Carbohydr Chem Biochem 37:157–223

Montreuil J (1982) Glycoproteins. In: Neuberger A, Van Deenen LLM (eds) Comprehensive biochemistry. Elsevier, Amsterdam, pp 1–188

Morell AG, Gregoriadis G, Scheinberg IH, Hickman J, Ashwell G (1971) The role of sialic acid in determining the survival of glycoproteins in the circulation. J Biol Chem 246:1461–1467

Mori K, Kawasaki T, Yamashina I (1983) Identification of the mannan-binding protein from rat livers as a hepatocyte protein distinct from the mannan receptor on sinusoidal cells. Arch Biochem Biophys 222:542–552

Morita A, Tsao D, Kim YS (1980) Identification of cholera toxin binding glycoproteins in rat intestinal microvillus membranes. J Biol Chem 255:2549–2553

Moss J, Fishman PH, Manganiello C, Vaughn M, Brady RD (1976) Functional incorporation of gangliosides into intact cells: induction of choleragen responsiveness. Proc Natl Acad Sci USA 73:1034–1037

Moss J, Garrison S, Fishman PH, Richardson SH (1979) Gangliosides sensitize unresponsive fibroblasts to *Escherichia coli* heat labile enterotoxin. J Clin Invest 64:381–384

Moss J, Osborne JC, Fishman PH, Nakaya S, Robertson DC (1981) *Escherichia coli* heat labile enterotoxin: ganglioside specificity and ADP ribosyltransferase activity. J Biol Chem 256:12861–12865

Muchmore AV, Blaese RM (1980) Evidence that monocyte mediated cellular recognition phenomena are mediated by receptors with specificity for simple oligosaccharides, In: Unanue ER, Rosenthal AS (eds) Macrophage regulation of immunity. Academic, New York, pp 505–517

Müller W, Klein PJ, Vierbuchen M, Uhlenbruck G (1980) Lectin binding sites in the choroid plexus and choroid plexus papillomas. Neurosurg Rev 3:57–65

Myllylä G, Furuhjelm U, Nordling S, Pirkola A, Tippett P, Gavin J, Sanger J (1971) Persistent mixed field polyagglutinability: electrokinetic and serological aspects. Vox Sang 20:7–23

Nakagawa F, Schulte BA, Spicer SS (1986a) Selective cytochemical demonstration of glycoconjugate containing terminal N-acetylgalactosamine on some brain neurons. J Comp Neurol 243:280–290

Nakagawa F, Schulte BA, Wu JY, Spicer SS (1986b) GABAergic neurons of rodent brain correspond partially with those staining for carbohydrate with terminal N-acetylgalactosamine. J Neurocytol 15:389–396

Nakagawa F, Schulte BA, Sens MA, Kochibe N, Spicer SS (1986c) Lectin cytochemistry of cell types in human and canine pituitary. Histochemistry 85:57–66

Nakagawa F, Schulte BA, Spicer SS (1986d) Lectin cytochemical evaluation of somatosensory neurons and their peripheral and central processes in rat and man. Cell Tissue Res 245:579–589

Nakayama J, Katsuyama T, Ono K, Honda T, Akamatsu T, Hattori H (1985) Large bowel carcinoma-specific antigens detected by the lectin *Griffonia simplicifolia* agglutinin II. Jpn J Cancer Res 76:1078–1084

Nemamic MK, Whitehead JS, Elias PM (1983) Alterations in membrane sugars during epidermal differentiation. Visualization with lectins and role of glycosidases. J. Histochem Cytochem 31:887–897

Neufeld EF, Ashwell G (1980) Carbohydrate recognition systems for receptor-mediated pinocytosis. In: Lennarz WG (ed) The biochemistry of glycoproteins and proteoglycans. Plenum, New York, pp 241–266

Nicolson GL (1982) Cancer metastasis: organ colonization and the cell surface properties of malignant cells. Biochem Biophys Acta 495:113–176

Nicolson GL (1984) Cell surface molecules and tumor metastasis. Exp Cell Res 150:3–22

Nicolson GL, Poste G (1983) Tumor cell diversity and host response in cancer metastasis. Properties of metastatic tumor cells. Curr Probl Cancer 7:1–83

Nishida S, Akai F, Hiruma S, Maeda M, Tanji K, Hashimoto S (1986) Experimental study of WGA binding on the endothelial cell surface in cerebral ischemia. Histol Histopathol 1:69–74

Novogradsky A, Ashwell A (1977) Lymphocyte mitogenesis induced by a mammalian liver protein that specifically bind desialylated glycoproteins. Proc Natl Acad Sci USA 74:676–678

Nowell PC (1960) Phytohemagglutinin: an initiator of mitosis in cultures of normal human leukocytes. Cancer Res 20:462–466

Nozima T, Vasui K, Homma P (1968) Reaction of several saccharides with Japanese encephalitis virus. Acta Virol 12:296–300

Ofek I, Beachey EH (1978) Mannose binding and epithelial adherence of *Escherichia coli*. Infect Immun 22:247–254

Ofek I, Mirelman D, Sharon N (1977) Adherence of *Escherichia coli* to human mucosal cells mediated by mannose receptors. Nature 265:623–625

Oppenheimer SB (1975) Functional involvement of specific carbohydrate in teratoma cells adhesion factor. Exp Cell Res 92:122–126

Ordonez NG, Batsakis JG (1984) Comparison of *Ulex europaeus* I lectin and factor VIII related antigens in vascular lesions. Arch Pathol Lab Med 108:129–132

Orntoft TF, Nielsell MJS, Wolf H, Olsen S, Clausen H, Hakamori S, Dabelstein E (1987) Blood group ABO and Lewis antigen expression during neoplastic progression of human urothelium. Immunohistochemical study of type 1 chain structures. Cancer 60:2641–2648

Ortmann M, Vierbuchen M, Koller K, Fischer R (1988a) Renal oncocytoma. I. Cytochrome C oxidase in normal and neoplastic renal tissue as detected by immunohistochemistry – a valuable aid to distinguish oncocytomas from renal cell carcinomas. Virchows Arch [B] 56:165–173

Ortmann M, Vierbuchen M, Fischer R (1988b) Renal oncocytoma. II. Lectin and immunohistochemical features indicating an origin from the collecting duct. Virchows Arch [B] 56:175–184

Paeietta E, Stockert RJ, Morell AG, Diehl V, Wiernik PH (1986) Lectin activity as a marker for Hodgkin disease cells. Proc Natl Acad Sci USA 83:3451–3455

Paget S (1889) The distribution of secondary growths in cancer of the breast. Lancet I:571–573

Pardoe IU, Burness ATH (1981) The interaction of the encephalomyocarditis virus with the erythrocyte receptor on affinity chromatography columns. J Gen Virol 57:239–243

Parish CR, Jackson DC, McKenzie IFC (1977) Evidence that Ia antigenic specificities are defined by carbohydrates. In McDevitt HO (ed) Ir and Ia antigens. Academic, New York, pp 243–253

Parish CR, Higgins TJ, McKenzie IFC (1978) Comparison of antigens recognized by xenogeneic and allogeneic anti-Ia antibodies. Evidence for two classes of Ia-antigens. Immunogenetics 6:343–354

Pasvol G, Wainscoat JS, Weatherall DJ (1982) Erythrocytes deficiency in glycophorins resist invasion by the malarial parasite *Plasmodium falciparum*. Nature 297:64–66

Pereira MEA (1983) A developmentally regulated neuraminidase activity in *Trypanosoma cruzi*. Science 219:1444–1446

Pereira MEA (1986) Lectins and agglutinins in protozoa. In: Mirelman D (ed) Microbial lectins and agglutinins. John Wiley, New York, pp 297–299

Pereira MEA, Kabat EA, Lotan R, Sharon N (1976) Immunochemical studies on the specificity of the peanut (Arachis hypogaea) agglutinin. Carbohydr Res 37:89–102

Perkins M (1981) Inhibitory effect of erythrocyte components on the in vitro invasion of human malarial parasite (*Plasmodium falciparum*) into host cells. J Cell Biol 90:563–567

Podolsky DK, Carter EA, Isselbacher KJ (1983) Inhibition of primary and metastatic tumor growth in mice by cancer associated glycosyltransferase acceptor. Cancer Res 43:4026–4030

Poretz RD, Riss H, Timberlake JW, Chien SM (1974) Purification and properties of the hemagglutinin from *Sophorica japonica* seeds. Biochemistry 13:2046–2050

Portner A (1981) The HN glycoprotein of Sendai virus: analysis of site(s) involved in hemagglutinating and neuraminidase activities. Virology 115:375–384

Poste G, Allen H, Matta KL (1979a) Cell surface receptors for lymphokines. II. Studies on the carbohydrate composition of the MIF receptor on macrophages using synthetic saccharides and plant lectins. Cell Immunol 44:88–98

Poste G, Kirsh R, Fidler IJ (1979b) Cell surface receptors for lymphokines. I. The possible role of glycolipid as receptors for macrophage migration inhibitory factor (MIF) and macrophage activation factor (MAF). Cell Immunol 44:71–88

Prieels JP, Pizzo SV, Glasgow LR, Paulson JC, Hill RL (1978) Hepatic receptor that specifically binds oligosaccharides containing fucosyl-alpha-1-3-N-acetylglucosamine linkages. Proc Natl Acad Sci USA 75:2215–2219

Race CC, Sanger R (1975) Blood group antigens in man, 6th ed. Blackwell Scientific, Oxford

Raedler A, Schmiegel WH, Raedler E, Arndt R, Thiele WG (1983) Lectin defined cell surface glycoconjugates of pancreatic cancer cells and their nonmalignant counterparts. Exp Cell Biol 51:19–28

Ravdin JL (1986) Pathogenesis of disease caused by *Entamoeba histolytica*: studies of adherence, secreted molecules, secreted toxins, and contact dependent cytolysis. Rev Infect Dis 8:247–260

Ravdin JL, Guerrant RL (1981) Role of adherence in cytopathogenic mechanisms of *Entamoeba histolytica*. Study with mammalian tissue cells culture and human erythrocytes. J Clin Invest 68:1305–1313

Ravdin JL, Murphy CF, Salata RA, Guerrant RL, Hewlett El (1985) N-Acetyl-galactosamine-inhibitable adherence lectin of *Entamoeba histolytica*. Partial purification and relation to amoebic virulence in vitro. J Infect Dis 151:804–815

Raz A, Lotan R (1981) Lectin-like activities associated with human and murine neoplastic cells. Cancer Res 41:3642–3647

Raz A, Meromsky L, Carmi P, Karakash R, Lotan D, Lotan R (1984) Monoclonal antibodies to endogenous galactose-specific tumor cell lectins. EMBO J 3:2979–2983

Raz A, Meromsky L, Lotan R (1986) Differential expression of endogenous lectins on the surface of nontumorigenic, tumorigenic, and metastatic cells. Cancer Res 46:3667–3672

Raz A, Meromsky L, Zvibel I, Lotan R (1987) Transformation related changes in the expression of endogenous cell lectin. Int J Cancer 39:353–360

Ree HJ, Hsu SM (1983) Lectin histochemistry of malignant tumors. I Peanut agglutinin (PNA) receptors in follicular lymphomas and follicular hyperplasia: an immunohistochemical study. Cancer 51:1631–1638

Ree HJ, Kadin ME (1985) Lectin distinction of benign from malignant histiocytes. Cancer 56:2046–2050

Ree HJ, Kadin ME (1986) Peanut agglutinin. A useful marker for histiocytosis X and interdigitating reticulum cells. Cancer 57:282–287

Reissner Y, Bimiaminov M, Rosenthal E, Sharon N, Ramot B (1979) Interaction of peanut agglutinin with normal human lymphocytes and with leukemic cells. Proc Natl Acad Sci USA 76:447–451

Rivier DA, Darekar MR (1975) Inhibition of the adhesiveness of enteropathogenic *E. coli*. Experientia 31:662–664

Robb JL, Stoddart RW (1987) Partial analysis of the UEA I binding oligosaccharide of human endothelial cells. In: Bog-Hansen TC, Freed DLJ (eds) Lectins–Biology, Biochemistry, Clinical Biochemistry. Sigma Chemical Company, ST. Louis, pp 631–633

Roche AC, Barzilay M, Midoux R, Junqua S, Sharon N, Monsigny M (1983) Sugar-specific endocytosis of glycoproteins by Lewis lung carcinoma cells. J Cell Biochem 22:131–140

Rocklin RE (1976) Role of monosaccharides in the interaction of two lymphocyte mediators with their target cells. J Immunol 116:816–820

Roder JC, Karre K, Kiessling R (1981) Natural killer cells. Prog Allergy 28:66–159

Rogers GN, Paulson JC (1983) Receptor determinants of human and animal influenza inoculates: differences in receptor specificity of the H3 hemagglutinin based on species of origin. Virology 127:361–373

Roque-Barreira MC, Campus-Neto A (1985) Jacalin: An IgA binding lectin. J Immunol 134:1740–1743

Rosales-Encina JL, Meza I, Lopez-De-Leon A, Talamas-Rohana P, Rojkind M (1987) Isolation of a 220 kilodalton protein from a virulent strain of *Entamoeba histolytica*. J Infect Dis 156:790–797

Roseman S (1970) The synthesis of complex carbohydrates by multiglycosyltransferase system and their potential function in intracellular adhesion. Chem Phys Lipids 5:270–297

Roth J, Lucocq JM, Charest M (1984) Light and electron microscopic demonstration of sialic acid residues with the lectin from *Limax flavus*. A cytochemical affinity technique with the use of fetuin-gold complex. J Histochem Cytochem 32:1167–1176

Roussel PF (1985) Marguage immunoperoxydase (PAP) des vaisseaux sanguins par trois lectins: *Ulex europaeus* I (UEA I), *Lotus tetragonolobus* (LTA), *Dolichos biflorus* (DBA). CRC Soc Biol 179:59–69

Sahagian GG, Distler J, Jourdian GW (1982) Membrane receptor for phosphomannosyl residues. Methods Enzymol 83:392–396

Sahagian GG, Neufeld EF (1983) Biosynthesis and turnover of the mannose-6-phosphate receptor in cultured chinese hamster ovary cells. J Biol Chem 258:7121–7128

Sakamoto J, Yin BT, Lloyd KO (1984) Analysis of the expression of H, Lewis, X, Y, and precursor blood group determinants in saliva and red blood cells using a panel of mouse monoclonal antibodies. Mol Immunol 21:1093–1098

Salit JE, Gotschlich EC (1977a) Hemagglutination by purified type I *Escherichia coli* pili. J Exp Med 146:1169–1181

Salit JE, Gotschlich EC (1977b) Type I *Escherichia coli* pili. Characterization of binding to monkey kidney cells. Exp Med 146:1182–1194

Schauer R (1988) Sialic acid as antigenic determinant of complex carbohydrates. In: Wu A, Adams GL (eds) The molecular immunology of complex carbohydrates. Plenum, New York, pp 47–72

Schauer R, Fischer C, Lee H, Ruch B, Kelm S (1988) Sialic acids as regulators of molecular and cellular interactions. In: Gabius HJ, Nagel GA (eds) Lectins and glycoconjugates in oncology. Springer, Heidelberg Berlin New York, pp 7–23

Schaumburg-Lever G, Alroy J, Ucci AA, Lever WF (1984a) Distribution of carbohydrate residues in normal skin. Arch Dermatol Res 276:216–223

Schaumburg-Lever G, Alroy J, Gavris V, Ucci AA, Lever WF (1984b) Cell-surface carbohydrates in proliferative epidermal lesions. Distribution of A, B, and H blood group antigens in benign and malignant lesions. Am J Dermatol 6:583–589

Scheid A, Choppin PW (1974) The hemagglutinating and neuraminidase protein of a paramyxovirus: interaction with neuraminic acid in affinity chromatography. Virology 62:125–133

Scheid A (1981) The viral components of myxo- and paramyxo-viruses which recognize receptors. In: Lonberg-Holm K, Philipson L (eds) Virus receptors part 2. Chapman and Hall, London, pp 49–62

Schelper RL, Whitters E, Hart M (1985) True microglia distinguished from macrophages by specific lectin binding. J Neuropathol Exp Neurol 44:332

Schick PK, Filmyer WG (1985) Sialic acid in mature megakaryocytes: detection by wheatgerm agglutinin. Blood 65:1120–1126

Schirrmacher V (1985) Cancer metastasis. Experimental approach, theoretical concepts, and impacts for treatment. Adv Cancer Res 43:1–73

Schirrmacher V, Cheingsong-Popov R, Arnheiter A (1980) Hepatocyte–tumor cell interaction in vitro. I. Conditions for rosette formation and inhibition by anti H2 antibody. J Exp Med 151:984–989

Schlesinger PH, Rodman JS, Doebber TW, Stahl PD, Lee YC, Stowell CP, Kuhlenschmidt TB (1980) The role of extra-hepatic tissues in the receptor-mediated plasma clearance of glycoproteins terminated by mannose and N-acetyl-glucosamine. Biochem J 192:597–606

Schottelius J, Gerken J, Centurion-Lara A (1988) Comparative studies on new world *Leishmania* and *Trypanosoma* by lectin tests and concanavalin A precipitation of ^{125}I labelled cells. J Clin Chem Clin Biochem 26:834–835

Schulman S, Oppenheim JD, Vanderberg JP (1983) Assay of erythrocyte components as inhibitors of *Plasmodium falciparum* merozoite invasion of erythrocytes. Am J Trop Med Hyg 32:666–670

Schulte BA, Spicer SS (1983) Light microscopic detection of sugar residues in glycoconjugates of salivary glands and the pancreas with lectin-horseradish peroxidase conjugates. I. Mouse. Histochem J 15:1217–1238

Schulte BA, Spicer SS (1984) Light microscopic detection of sugar residues in glycoconjugates of salivary glands and pancreas with lectin-horseradish peroxidase conjugates II. Rat Histochem J 16:3–20

Schulte BA, Spicer SS (1985) Histochemical methods for characterizing secretory cell surface sialoglycoconjugates. J Histochem Cytochem 33:427–438

Schulte BA, Spicer SS, Miller RA (1985) Lectin histochemistry of secretory and cell surface glycoconjugates in the ovine submandibular gland. Cell Tissue Res 240:57–66

Schuster VL, Bonsib SM, Jennings ML (1986) Two types of collecting duct mitochondria-rich (intercalated) cells: lectin and band 3 cytochemistry. Am J Physiol 251:347–355

Schwartz RH, Yano A, Paul WE (1978) Interaction between antigen-presenting cells and primed T lymphocytes. Immunol Rev 40:153–180

Schwartz R, Schirrmacher V, Mühlradt PF (1983) Glycoconjugates of murine tumor cell lines with different metastatic capacities. I. Differences in fucose utilization and glycoprotein pattern. Int J Cancer 33:503–509

Schwechheimer K, Möller P, Schnabel P, Waldherr R (1983) Emphasis on peanut lectin as marker for granular cells. Virchows Arch [A] 399:289–297

Schwechheimer K, Weiss G, Möller P (1984) Concanavalin A binding and neuronal differentiation. A light microscopic study on neuronal tumors. Virchows Arch [A] 402:297–306

Sharon N (1984) Surface carbohydrates and surface lectins are recognition determinants in phagocytosis. Immunol Today 5:143–147

Sharon N, Lis H (1972) Lectins. Cell agglutinating and sugar-specific proteins. Science 177:949–959

Sharon N, Lis H (1981) The chemistry and biology of glycoproteins. Chem Eng News 59:21–44

Sharon N, Lis H (1982) Glycoproteins. In: Neurath H, Hill RL (eds), The proteins. Academic, New York, pp 1–144

Shepherd VL, Lee YC, Schlesinger PH, Stahl PD (1981) L-fucose terminated glycoconjugates are recognized by pinocytosis receptors on macrophages. Proc Natl Acad Sci USA 78:1019–1022

Sherblom AP, Sathyamoorthy N, Decker JM, Muchmore AV (1989) IL-2, a lectin with specificity for high mannose glycopeptides. J Immunol 143:939–944

Shibuya N, Goldstein IJ, Brocknert WF, Nsimba-Lubaki M, Peeters B, Peumans WJ (1987) The elderberry (*Sambucus nigra*) bark lectin recognizes the Neu5Ac(alpha2-6)Gal/GalNAc sequence. J Biol Chem 262:1596–1601

Shibuya N, Goldstein IJ, Van Damme EJM, Peumans WJ (1988) Binding properties of a mannose-specific lectin from the snow-drop (*Galanthus nivalis*) bulb. J Biol Chem 163:728–734

Simmons CF, Schwartz AL (1984) Cellular pathways of galactose terminal ligand movement in a cloned human hepatoma cell line. Mol Pharmacol 26:509–519

Simpson DL, Thorne DR, Loh HH (1977) Developmentally regulated lectin in neonatal rat brain. Nature 266:367–369

Simpson DL, Thorne DR, Loh HH (1978) Lectins: endogenous carbohydrate-binding proteins from vertebrate tissue. Functional role in recognition processes. Life Sci 22:727–748

Skutelsky E, Goyal V, Alroy J (1987) The use of avidin gold complex for light microscopic localization of lectin receptors. Histochemistry 86:291–294

Slack JMW (1985) Peanut lectin receptor in the early amphibian embryo: regional markers for the study of embryonic induction. Cell 41:237–247

Sly WS, Fisher HD (1982) The phosphomannosyl recognition system for intracellular and intercellular transport of lysosomal enzymes. J Cell Biochem 18:67–85

Sly WS, Stahl P (1978) Receptor-mediated uptake of lysosomal enzymes. In: Silverstein SC (ed) Transport of macromolecules in cellular systems. Dahlem Konferenzen, Berlin, pp 229–244

Sobrinho-Simoes M, Damjanov I (1986) Lectin histochemistry of papillary and follicular carcinoma of the thyroid gland. Arch Pathol Lab Med 110:722–729

Springer GF (1989) Tn epitope (N-acetyl-D-galactosamine-alpha-O-serine/threonine) density in primary breast carcinoma: a functional predictor of aggressiveness. Mol Immunol 26:1–5

Stahl PD, Gordon S (1982) Expression of a mannosyl-fucosyl receptor for endocytosis on cultured primary macrophages and their hybrids. J Cell Biol 93:49–56

Stahl PD, Schlesinger PH, Rodman SS, Doebber T (1976) Recognition of lysosomal glycosidases in vivo inhibited by modified glycoproteins. Nature 264:86–88

Stahl PD, Rodman JS, Miller MJ, Schlesinger PH (1978) Evidence for receptor mediated binding of glycoproteins, glycoconjugates, and lysosomal glycosidases by alveolar macrophages. Proc Natl Acat Sci USA 73:1399–1403

Stamper HB Jr, Woodruff JJ (1976) Lymphocyte homing into lymph nodes: in vitro demonstration of the selective affinity of recirculating lymphocytes for high endothelial venules. J Exp Med 144:828–833

Steer CJ, Furbish FS, Barranger JA, Brady RO, Jones EA (1978) The uptake of agalactogluco-cerebrosidase by rat hepatocytes and Kupffer cells. FEBS Lett 91:202–205

Stein M, Meyer HE, Hasilik A, von Figura K (1987a) Mr 46000 mannose-6-phosphate specific receptor purification, subunit composition, and chemical modification. Hoppe Seyler's Z Physiol Chem 368:927–963

Stein M, Zijderhand-Bleekemolen JE, Geuze HJ, Hasalik A, von Figura K (1987b) Mr 46000 mannose-6-phosphate specific receptor: its role of lysosomal enzymes. EMBO J 6:2677–2681

Stevens SK, Butcher EC (1982) Differences in the migration of B and T lymphocytes. Organ-selective localization in vivo and the role of lymphocyte-endothelial cell recognition. J Immunol 128:844–851

Stockert RJ, Morell AG, Scheinberg IH (1976) The existence of a second route for the transfer of certain glycoproteins from the circulation into the liver. Biochem Biophys Res Commun 68:988–993

Stoddart RW, Herbertson BM (1978a) The use of lectins in the detection and identification of human fungal pathogens. Biochem Soc Trans 5:233–235

Stoddart RW, Herbertson BM (1978b) The use of fluorescein labelled lectins in the detection and identification of fungi pathogenic for man. Preliminary study. J Med Microbiol 11:315–324

Stoolman LM, Rosen SD (1983) Possible role for cell-surface carbohydrate-binding molecules in lymphocyte recirculation. J Cell Biol 96:722–729

Stoolman LM, Tenfords TS, Rosen SD (1984) Phosphomannosyl receptors may participate in the adhesive interaction between lymphocytes and high endothelial venules. J Cell Biol 99:1535–1540

Streit WJ, Schulte BA, Balentine D, Spicer SS (1985a) Histochemical localization of glactose-containing glycoconjugates in sensory neurones and their processes in the central and peripheral nervous system of the rat. J Histochem Cytochem 33:1042–1052

Streit WJ, Schulte BA, Spicer SS, Balentine D (1985b) Histochemical localization of galactose containing glycoconjugates at peripheral nodes of Ranvier in the rat. J Histochem Cytochem 33:33–39

Streit WJ, Schulte BA, Balentine D, Spicer SS (1986) Evidence for glycoconjugate in nociceptive primary sensory neurons and its origin from the Golgi complex. Brain Res 377:1–17

Stutman O, Dien P, Wisun RE, Lattime C (1980) Natural cytotoxic cells against solid tumors in mice. Blocking of cytotoxicity by D-mannose. Proc Natl Acad Sci USA 77:2895–2898

Suttajit M, Winzler RJ (1971) Effect of modification of N-acetylneuraminic acid on the binding of glycoproteins to influenza virions and on susceptibility for cleavage by neuraminidase. J Biol Chem 246:3398–3404

Szumanska G, Vorbrodt AW, Wisniewski HM (1986) Lectin histochemistry of scrapie amyloid plaques. Acta Neuropathol 69:205–212

Tabas I, Kornfeld S (1980) Biosynthetic intermediates of β-glucuronidase contain high mannose oligosaccharides with blocked phosphate residues. J Biol Chem 255:6633–6639

Tada T, Okumura K (1979) The role of antigen specific T-cell factors in the immune response. Adv Immunol 28:1–87

Takeuchi A, Phillips BP (1975) Electron microscopic studies of experimental *Entamoeba histolytica* infections in the guinea pig. Penetration of the intestinal epithelium by trophozoites. Am J Trop Med Hyg 23:34–48

Teshima S, Hiroshashi S, Shimosato Y, Kishi K, Ino Y, Matsumoto K, Yamada T (1984) Histochemically demonstrable changes in cell surface carbohydrates of human germ cell tumors. Lab Invest 50:271–277

Theze J, Waltenbaugh C, Dorf ME, Benacerraf B (1977) Immunosuppressive factor(s) specific for L-glutaminic acid[50] L-tyrosine (GT) II. Presence of I-J determinants on the GT suppressive factor. J Exp Med 146:287–292

Thiele J, Vierbuchen M, Arnold G, Walgenbach S, Fischer R (1986) Lectin-binding sites in human parathyroid tissue. J Histochem Cytochem 34:1201–1206

Thomas DB, Winzler RJ (1969) Structural studies on human erythrocyte glycoproteins: alkalilabile oligosaccharides. J Biol Chem 244:5943–5946

Thorgeirson UP, Hujanen TT, Liotta LA (1984) Cancer cells, components of basement membranes, and proteolytic enzymes. Int Rev Exp Pathol 27:203–234

Tiffany JM, Blough HA (1971) Attachment of myxoviruses to artificial membranes: electron microscopic studies. Virology 44:18–28

Tomaska LD, Parish CR (1981) Inhibition of secondary IgG responses by N-acetyl-D-galactosamine. Eur J Immunol 11:181–186

Tomaska LD, Parish CR (1982) Inhibition of secondary IgG responses by monosaccharides. Evidence for region control. J Immunogenet 9:63–68

Uhlenbruck G (1987) Bacterial lectins: mediators of adhesion. Zentralbl Bakteriol Mikrobiol Hyg [A] 263:497–508

Uhlenbruck G, Beuth HJ, Oette K et al. (1986) Lektine and Organotropie der Metastasierung. Dtsch Med Wochenschr 111:991–995

Uhlenbruck G, Hanisch FG, Vierbuchen M, Dufhues G (1988a) Love to lectins: personal history and priority hysterics. In: Gabius HJ, Nagel GA (eds) Lectins and glycoconjugates in Oncology. Springer, Heidelberg Berlin New York, pp 49–58

Uhlenbruck G, Fröml A, Böhmer GM, Lüttiken R (1988b) Lectinologic and bacteriophoretic investigations of group B streptococci. J Clin Chem Clin Biochem 26:835

Varghese JN, Laver WG, Colman PM (1983) Structure of the influenza virus glycoprotein antigen neuraminidase at 2.9 A resolution. Nature 303:35–40

Vasil ML, Iglewski BH (1978) Comparative toxicities of diphtherial toxin and Pseudomonas exotoxin A. Evidence for different cell receptors. J Gen Microbiol 108:333–337

Vasta GR, Marchalonis JJ, Kohler H (1984) Invertebrate recognition protein cross-reacts with an immunoglobulin idiotype. J Exp Med 159:1270–1276

Vengris VE, Reynolds FH Jr, Hollenberg MD, Pitha PM (1976) Interferon action: role of membrane gangliosides. Virology 72:486–493

Vierbuchen M (1987) Lektinhistochemie – Ein neues diagnostisches Prinzip in der Biologie und Pathologie. Verh Dtsch Ges Path 71:492–502

Vierbuchen M, Klein PJ (1983) Histochemical demonstration of neuraminidase effects in pneumococcal meningitis. Lab Invest 48:181–186

Vierbuchen M, Klein PJ, Uhlenbruck G, Klein HO, Schaefer HE, Fischer R (1980) The significance of lectin receptors in the kidney and hypernephroma (renal adenocarcinoma). Recent Res Cancer Res 75:68–74

Vierbuchen M, Klein PJ, Rösel S, Fischer J (1983) Peanut agglutinin (PNA) binding sites: a useful marker for hormonal dependence in experimental breast cancer. Cancer Detect Prev 6:207–214

Vierbuchen M, Böhmer G, Uhlenbruck G, Fischer R (1986) Immuno- and histochemical studies on galactan and human blood group related receptors in the bovine lung. Immunobiology. 172:11–20

Vierbuchen M, Dufhues G, Uhlenbruck G, Fischer R (1988a) Endogenous carbohydrate binding proteins in mouse tissues – a topohistochemical study. J Clin Chem Clin Biochem 26:829–830

Vierbuchen M, Klein PJ, Würz H, Kallenberg A, Fischer R (1988b) Endogenous as well as exogenous hormonal modulation of lectin binding sites in normal and neoplastic tissue of rat mammary gland. Acta Histochemica [Suppl] Jena XXXVI:179–190

Vierbuchen M, Uhlenbruck G, Ortmann M, Dufhues G, Fischer R (1988c) Occurrence and distribution of glycoconjugates in human tissues as detected by the Erythrina cristagalli lectin. J Histochem Cytochem 36:367–376

Vierbuchen M, Scharl A, Schnepper U, Kallenberg A, Luyken W (1988d) Blutgruppen- und blutgruppenabhängige Tumormarker im Mammakarzinom – eine vergleichende Untersuchung mit monoklonalen Antikörpern und Lektinen. Verh Dtsch Ges Pathol 72:593

Vierbuchen M, Peters G, Ortmann M, Fischer R (1989a) Topography and mechanism of the adhesion of uropathogenic *Escherichia coli* bacteria in the kidney and renal pelvis. Verh Dtsch Ges Pathol 73:264–267

Vierbuchen M, Schröder S, Uhlenbruck G, Ortmann M, Fischer R (1989b) CA 50 and CA 19-9 antigen expression in normal, hyperplastic and neoplastic thyroid tissue. Lab Invest 60:726–732

Virtanen I, Karininiemi AL, Holthöfer H, Lehto VP (1986) Fluorochrome-coupled lectins reveal distinct cellular domains in human epidermis. J Histochem Cytochem 34:307–315

Von Figura K, Hasilik A (1986) Lysosomal enzymes and their receptors. Annu Rev Biochem 55:167–193

Von Figura K, Weber E (1978) An alternative hypothesis of cellular transport of lysosomal enzymes in fibroblasts: effect of inhibitors of lysosomal enzyme endocytosis on intra- and extracellular lysosomal enzyme activities. Biochem J 176:943

Von Figura K, Gieselmann V, Hasilik A (1984) Antibody to mannose-6-phosphate specific receptor induces receptor deficiency in human fibroblasts. EMBO J 3:1281–1286

Von Figura K, Gieselmann V, Hasilik A (1985) Mannose-6-phosphate specific receptor: a transmembrane protein with a C-terminal extension orientated towards the cytosol. Biochem J 225:543–547

Wadström T (1988) Sialic acid-specific lectins. J Clin Chem Clin Biochem 26:827

Walker RA (1984) The binding of peroxidase labelled lectins to human breast epithelium. III. Altered fucose-binding patterns and their significance. J Pathol 145:109–117

Walker RA, Hawkins RA, Miller WR (1985) Lectin binding and steroid hormone receptors in human breast carcinomas. J Pathol 147:103–106

Wall DA, Hubbard AL (1981) Galactose-specific recognition system of the mammalian liver: receptor distribution on the hepatocyte cell surface. J Cell Biol 90:687–696

Wall DA, Wilson G, Hubbard AL (1980) The galactose-specific recognition system of mammalian liver. The route of ligand internalization in rat hepatocytes. Cell 21:79–93

Wang KY, Kuhlenschmidt TB, Lee YC (1985) Isolation and characterization of the major mannose-binding protein in chicken serum. Biochemistry 24:5932–5938

Warr GW, Vasta GR, Ledford BE, Marchalonis JJ (1983) Evidence for the existence of a diverse family of immunoglobulin-related recognition molecules. J Immunogenet 10:17–23

Weir DM (1980) Surface carbohydrates and lectins in cellular recognition. Immunol Today 1:45–51

Weiss L (1977) A pathobiologic overview of metastasis. Semin Oncol 4:5–19

Weiss L (1983) Random and nonrandom processes in metastasis and metastatic efficiency. Invasion Metastasis 3:193–208

Weiss L, Ward PM (1983) Cell detachment and metastasis. Cancer Metastasis Rev 2:111–127

Weiss MM, Oppenheim JD, Vanderberg JP (1981) *Plasmodium falciparum*: assay for in vitro inhibition of merozoite penetration of erythrocytes. Exp Parasitol 51:400–407

West KP, Cope JL (1989) The binding of peroxidase labelled lectins to human endometrium in normal cycling endometrium and endometrial carcinoma. J Clin Pathol 42:142–147

Wick MR, Manivel C, O'Leary TP, Cherwitz DL (1986) Nephroblastoma. A comparative immunocytochemical and lectin histochemical study. Arch Pathol Lab Med 110:630–635

Woodruff JJ, Katz IM, Lucas LE, Stamper HB Jr (1977) An in vitro model of lymphocyte homing. II. Membrane and cytoplasmic events involved in lymphocyte adherence to specialized high-endothelial venules of lymph nodes. J Immunol 119:1603–1610

Wright SC, Bonavida B (1981) Selective lysis of NK sensitive target cells by a soluble mediator released from murine spleen cells and human peripheral blood lymphocytes. J Immunol 126:1516–1521

Wu A, Sugii S, Herp A (1987) A Table of lectin carbohydrate specificities. In: Bog-Hansen TC, Freed DLJ (eds) Lectins, biology-biochemistry clinical biochemistry. Sigma Chemical Company, St. Louis, pp 723–740

Wu A, Suguii S (1988) Differential binding properties of GalNAc and/or Gal specific lectins. In: Wu A, Adams GL (eds) Molecular Immunology of Complex carbohydrates. Plenum, New York, pp 205–264

Yamagami T, Hosaka M, Mori M (1985) Classification of skeletal muscle fiber by comparison of enzyme histochemistry with lectin binding. Cell Mol Biol 31:241–249

Yamamoto K, Tsuji T, Ossawa T (1982) Requirement of the core structure of complex type glycopeptide for the binding to the immobilised lentil- and pea-lectins. Carbohydr Res 110:282–289

Yasui K, Nozima T, Homma R, Ueda S (1968) Effects on alpha mannosidase on the active site of Japanese encephalitis virus. Acta Virol 13:158

Yeatman TJ, Bland KI, Copeland EM, Kimura AK (1989) Tumor cell-surface galactose correlates with the degree of colorectal liver metastasis. J Surg Res 46:567–571

Yonezawa S, Nakamura T, Tanaka S, Sato E (1982) Glycoconjugates with *Ulex europaeus* agglutinin I binding sites in normal mucosa, adenoma, and carcinoma of the human large bowel. J Natl Cancer Inst 69:777–785

Cytosolic/Nuclear Receptors

Steroid Hormone Receptors

E. V. Jensen

1 Introduction

The steroid hormones comprise a large and important family of cell regulators. These include sex hormones (estrogens, progestins, androgens), adrenal cortical hormones (glucocorticoids, mineralocorticoids), vitamin D, and insect hormones (ecdysteroids). As discussed in Sect. 3.1, these agents combine with intracellular receptor proteins, converting them to functional transcription factors, which then bind in the genome to influence the expression of specific genes. Thus, receptors for steroid hormones, as well as for thyroid hormones and retinoic acid, differ from the receptors described in previous chapters, which are located in the plasma membrane and utilize a signal transduction process (second messenger) to deliver the regulatory signal within the responsive cell.

Although steroid hormones affect the behavior of a large number of cells, certain tissues are particularly sensitive, have a higher content of receptor proteins, and are often called target tissues. For example, the estrogens (estradiol, estrone) and progestins (progesterone) control the growth, differentiation, and function of reproductive and accessory sex tissues in the female (uterus, vagina, oviduct, mammary glands), as do androgens (dihydrotestosterone) in the male (prostate, seminal vesicle, penis, chick comb). But they also influence many other tissues, such as skin, bone, hair, larynx, pituitary, hypothalamus, and behavioral centers in the brain. In reptiles, amphibians, and birds, estrogens induce the synthesis of egg yolk proteins in the liver. Glucocorticoids (cortisol) promote gluconeogenesis and regulate carbohydrate metabolism in many mammalian cells, cause involution of lymphoid tissues such as spleen, thymus, and other components of the immune system, and have profound effects on brain and nerve tissue. Mineralocorticoids (aldosterone) promote sodium transport across membranes in organs such as kidney and toad bladder. The active form of vitamin D (1α,25-dihydroxycholecalciferol or calcitriol) acts in intestinal mucosa, bone, and kidney to enhance calcium transport and utilization. Ecdysteroids (20-hydroxyecdysone) modulate gene expression in essentially all insect cells and, in the absence of juvenile hormone, induce metamorphosis.

Long after the chemical structures and biological functions of steroid hormones had been established, the biochemical mechanism by which they exert their hormonal actions remained a mystery. At one time it was considered that they must somehow influence the behavior of enzymes, either by a direct action on the enzyme itself or through participation in enzymatic processes involved in their own metabolism. Beginning in the late 1950s, studies initiated principally with estrogens in the rat uterus and extended by many investigators established that an important action of all types of steroid hormones is to enhance the synthesis of mRNA (and rRNA), suggesting an effect of these agents on gene expression (for references to early studies see MUELLER et al. 1958; MUELLER 1965; O'MALLEY et al. 1969; JENSEN and DESOMBRE 1972; O'MALLEY and MEANS 1974; LIAO 1975; EDELMAN 1975). Early evidence for an action of steroid hormones at the genomic level was also provided by the ability of ecdysteroids to induce rapid morphological changes in chromosomes of insect larvae, an effect characteristic of the onset of metamorphosis (CLEVER and KARLSON 1960; KARLSON 1963).

During this same period, insight complementing that derived from studying the effects of the hormone on biochemical processes was derived from an opposite approach, namely, by determining what actually happens to the steroid itself as it exerts its hormonal action in target tissues. The synthesis in 1957 of hexestrol and of estradiol labeled with carrier-free tritium made it possible to determine the fate of physiological amounts of administered estrogen in responsive as compared to nonresponsive cells. This led to the discovery of steroid hormone receptors, as described in Sect. 2. When tritium-labeled steroids of other hormonal classes became commercially available in the mid 1960s, many laboratories undertook studies of their interaction with mammalian tissues, both in vivo and in vitro. These established that target cells for all types of steroid hormones contain receptor proteins that effect uptake and retention of the steroid and mediate its hormonal action.

2 Identification of Intracellular Receptors

2.1 Estrogen Receptors

Steroid hormone receptors were first discovered with the estrogens by the striking ability of female reproductive tissues to take up and retain radioactive hormone after the administration of physiological amounts of tritiated hexestrol to young sheep and goats (GLASCOCK and HOEKSTRA 1959) or of estradiol to immature rats (JENSEN 1960; JENSEN and JACOBSON 1960). Estradiol was found to combine reversibly with binding sites and to initiate growth of the immature rat uterus without itself undergoing chemical change (JENSEN and JACOBSON 1962); this indicated that the hormone acts by influencing the properties of the receptor protein rather than by participating in reactions of steroid metabolism, as had once been assumed. Similar results were observed in mouse uterus (STONE 1963; STONE and MARTIN 1964; TERENIUS 1965). 17-Methylestradiol, 17-ethynylestradiol, and hexestrol all are incorporated into rat uterus without chemical change, in contrast to estrone and mestranol (17-ethynylestradiol 3-methyl ether), which are converted to estradiol and 17-ethynylestradiol, respectively, before binding in the target tissues (JENSEN et al. 1966).

Exposure of excised uterine tissue (STONE and BAGGETT 1965; TERENIUS 1966; JENSEN et al. 1967a) or of uterine homogenates (JENSEN et al. 1968) to dilute solutions of tritiated estradiol at physiological temperature gives a hormone-receptor interaction similar to that observed in vivo, including formation of the same steroid-receptor complexes described in the next section. This reaction, as well as the integrity of the complexes, was found to depend on the presence of sulfhydryl groups (JENSEN et al. 1967c). Subsequent studies in many laboratories confirmed and extended observations of the binding of estrogens by a variety of hormone-responsive tissues and tumors (references in RASPÉ 1971; JENSEN and DESOMBRE 1972; LIAO 1975). Measurement of receptor content in human breast cancers as a guide to prognosis and therapy selection is now routine medical practice (W. McGUIRE 1978; DESOMBRE et al. 1979).

The specific uptake and retention of estrogens by reproductive tissues are inhibited, both in vivo (STONE 1964; ROY et al. 1964) and in vitro (JENSEN et al. 1967b), by a class of estrogen antagonists, such as nafoxidine, Parke-Davis CI-628, and tamoxifen, which are themselves weak estrogens but which inhibit

the uterotropic action of estradiol. The correlation observed between the reduction in hormone uptake and the inhibition of uterine growth when different
amounts of nafoxidine are given together with estradiol provided the first
evidence that the estrogen-binding components of reproductive tissues are true
receptors in that their association with steroid is actually involved in hormonal
action (JENSEN et al. 1966, 1972a). In contrast, actinomycin D and puromycin,
substances that also prevent uterine growth response to estrogen (MUELLER
et al. 1961; UI and MUELLER 1963), do not inhibit hormone binding. Thus,
association with the steroid is an early step in the uterotropic process, initiating
a sequence of biochemical events that is blocked at later stages by inhibitors
of RNA and protein synthesis.

After the administration of physiological amounts of tritiated estradiol to
immature rats, two forms of the estrogen receptor were identified in uterus and
other target tissues. As first shown by differential centrifugation of homogenates (NOTEBOOM and GORSKI 1965; R. KING et al. 1965; JENSEN et al.
1967a) and confirmed by autoradiography (STUMPF and ROTH 1966; JENSEN
et al. 1967b), the great majority of the hormone is in the nucleus, bound to
chromatin (R. KING et al. 1966). As shown in Fig. 1, it is extracted with 300
or 400 mM KCl as an estradiol-receptor complex that sediments at about 5 S
in hypertonic sucrose gradients (JENSEN et al. 1967b; PUCA and BRESCIANI

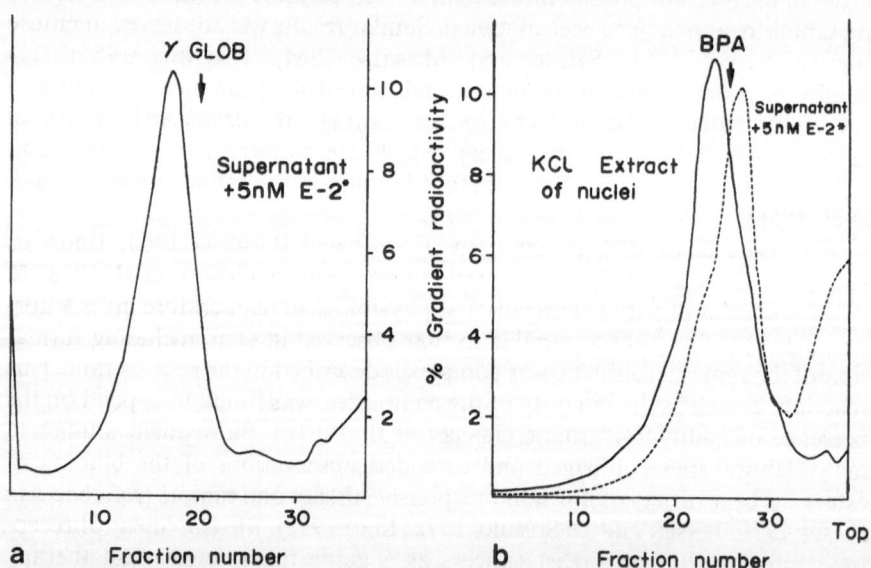

Fig. 1a, b. Sedimentation patterns of radioactive estradiol-receptor complexes from immature rat
uteri. **a** Cytosol fraction of uterine homogenate from untreated animals made 5 nM in tritiated
estradiol and centrifuged in a low-salt sucrose gradient. **b** Comparison of salt-extracted nuclear
complex with cytosol complex obtained from uteri excised 1 h after injection of 100 ng tritiated
estradiol and centrifuged in sucrose gradients containing 400 mM KCl. *Arrows* indicate sedimentation positions of bovine γ-globulin (7 S) and bovine plasma albumin (4.5 S) markers. (JENSEN et al.
1972b)

1968). A smaller amount (20% – 30%) appears in the high speed supernatant or cytosol fraction bound to a protein that sediments at 8–10 S in sucrose gradients of low ionic strength (TOFT and GORSKI 1966) and is dissociated into a 4 S steroid-binding subunit in the presence of salt (ERDOS 1968; KORENMAN and RAO 1968; ROCHEFORT and BAULIEU 1968). In uterine homogenates from untreated rats, essentially all the receptor appears in the cytosol and reacts with estradiol in the cold to form the 8 S/4 S steroid-receptor complex (JENSEN et al. 1967 b; TOFT et al. 1967). Uterine cytosol also contains another binding component of lower affinity and higher capacity (J. CLARK et al. 1978, 1980). The physiological role of this so-called type II receptor is unclear.

2.2 Progestin Receptors

Early attempts to demonstrate specific binding of labeled progesterone by female reproductive tissues proved unsuccessful until it was shown that uptake of progesterone by mouse vagina is dependent on prestimulation by estrogen (PODRATZ and KATZMAN 1968). It has since been established that in many target cells progestin receptor is a gene product of estrogen action. Specific uptake of administered progesterone, markedly increased by estrogen pretreatment, was demonstrated in chick oviduct (O'MALLEY et al. 1969, 1970) and in uteri from ovariectomized guinea pigs (FALK and BARDIN 1970), rabbits (WIEST and RAO 1971), and hamsters (LEAVITT and BLAHA 1972) or from pregnant rats (DAVIES and RYAN 1972). Binding is also seen on exposure of chick oviduct (O'MALLEY et al. 1970) or rat uterus (J. McGUIRE and DEDELLA 1971; SAFFRAN et al. 1973) to labeled progesterone in vitro. In most instances, the substance bound in the target tissue was unchanged progesterone.

Because of the interaction pattern already established for estrogens, studies of progestin receptors moved rapidly to the use of broken cell systems and the search for receptor proteins in the various fractions of tissue homogenates. Specific progesterone-receptor complexes were first observed in the cytosol fraction from chick oviduct (SHERMAN et al. 1970) and from uterus of the guinea pig (MILGROM et al. 1970), rat (MILGROM and BAULIEU 1970; J. McGUIRE and DEDELLA 1971; REEL et al. 1971), and rabbit (WIEST and RAO 1971), either after exposure of tissue to labeled progesterone in vivo or in vitro or on addition of hormone directly to the cytosol. In the next few years many similar findings were reported (references in O'MALLEY and MEANS 1974; JENSEN 1977, 1990). The bound steroid in cytosols obtained from ovariectomized animals usually sediments at about 4 S in hypotonic gradients. After estrogen administration not only is the amount of cytosol receptor increased, but in hypotonic gradients it sediments as a larger entity (6–8 S) that is dissociated by salt to a 3.8–4 S binding unit.

After administration of tritiated progesterone to estrogen-primed ovariectomized rabbits, most of the uterine radioactivity is in the nuclear fraction, from which it is extracted by 400 mM KCl as a complex sedimenting at about 4 S (B. RAO et al. 1973). In similarly treated guinea pigs, predominantly nuclear localization of bound steroid is indicated by autoradiography (SAR and STUMPF 1974). In uterus of the rat (HSUEH et al. 1974) and dog (LESSEY and GORELL 1981), the nuclear complex was demonstrated by exchange of tritiated progesterone for nuclear-bound endogenous hormone. After progesterone injection in vivo or exposure of tissue to labeled hormone in vitro at 37 °C but not at 0 °C, chick oviduct contains hormone in both the cytosol and nuclear fractions of homogenates (O'MALLEY et al. 1969, 1970). Extraction of nuclei with 300 mM KCl gives a progesterone-receptor complex that sediments in hypertonic gradients at about 4 S and thus cannot be distinguished by this criterion from the cytosol complex (O'MALLEY et al. 1971). In both complexes the steroid is unchanged progesterone, and its binding is destroyed by sulfhydryl-blocking reagents (COTY 1980; KALIMI and BANERJI 1981).

Progestin receptors in chick oviduct (SCHRADER and O'MALLEY 1972) and in human breast (LESSEY et al. 1983; GREENE et al. 1988; SHERIDAN et al. 1989) or endometrial (CLARKE et al. 1987) cancer are unique in that they come in two sizes, A and B. These are immunochemically similar (GRONEMEYER et al. 1985, 1987; ESTES et al. 1987). The B protein consists of the A form plus an additional unit at the N terminus, 165 amino acids in human receptor and 128 in chicken (see Sect. 6). Although the A form may be an artifact of proteolytic cleavage, there is evidence that it arises from the alternate initiation of translation of receptor mRNA (GRONEMEYER et al. 1987; CONNEELY et al. 1987b, 1989). The respective roles of these two receptor forms in the interaction with target genes are not yet clear. In contrast to chick oviduct and human cancer, only the full-length receptor is seen in rabbit uterus when precautions are taken to avoid proteolysis (LOOSFELT et al. 1984; LOGEAT et al. 1985a; LAMB et al. 1986).

2.3 Androgen Receptors

Early studies of the uptake of administered androgens by tissues of the rat or chick, using hyperphysiological doses of ^{14}C-testosterone (GREER 1959; BUTENANDT et al. 1960; HARDING and SAMUELS 1962; LAWRENCE and LANDAU 1965) or of tritiated 4-androstene-3,17-dione of moderate specific activity (PEARLMAN and PEARLMAN 1961), showed somewhat higher incorporation of labeled steroid into prostate, seminal vesicle, and chick comb as compared to muscle. However, this was usually less than that in liver, kidney, and adrenal, and the prostate was found to contain a mixture of metabolites, precluding definitive conclusions.

When tritiated testosterone of high specific activity became available, several groups soon demonstrated selective binding of labeled steroid in male reproductive and neural tissues of several species (Table 1; for additional examples see LIANG et al. 1977). Except for dog prostate, the steroid is mainly in the nucleus, as shown by fractionation of homogenates as well as by autoradiography. On incubation with tritiated testosterone at 25°–37 °C in vitro, prostatic tissue shows a similar uptake of steroid, mainly in the nucleus (K.

Table 1. Early demonstration of steroid binding in male reproductive tissues after injection of tritiated androgens

Species	Tissue	Report
Guinea pig [a]	Prostate [b]	Resko et al.1967
Rat [a]	Prostate [c]	Mangan et al. 1968; Tveter and Attramadal 1968, 1969 [d]; K. Anderson and Liao 1968; Bruchovsky and Wilson 1968; Stern and Eisenfeld 1969; Mainwaring 1969a; Tveter 1969; Sar et al. 1970 [d]; Mainwaring and Peterken 1971
	Epididymis	Blaquier et al. 1970; Hansson and Tveter 1971; Ritzén et al. 1971
	Levator ani	Jung and Baulieu 1972
	Pituitary	R. Green et al. 1970; Blaquier et al. 1970; Stern and Eisenfeld 1971; Thieulant et al. 1973; Sar and Stumpf 1973a [d]; Pérez-Palacios et al. 1973
	Hypothalamus	Sar and Stumpf 1973b [d]
Mouse	Prostate	J. Thomas et al. 1970
Dog	Prostate [b]	Kowarski et al. 1969
Dove [a]	Hypothalamus	Zigmond et al. 1972

[a] Castrated male animal
[b] Also seminal vesicle
[c] Some with seminal vesicle also
[d] By autoradiography

Anderson and Liao 1968; Mangan et al. 1968). Target tissues from adult rats castrated for at least 2 days show more hormone binding than those from intact animals, but an increase after castration was not seen in the cock comb.

A major advance in understanding androgen-receptor interaction came with the demonstration that, of the metabolites present in prostate and seminal vesicle after exposure to testosterone in vivo or in vitro, the steroid bound in the nuclei is 5α-androstan-17β-ol-3-one, called dihydrotestosterone or DHT (Bruchovsky and Wilson 1968; K. Anderson and Liao 1968; Mainwaring 1969a). The steroid bound to receptor in the cytosol fraction is also DHT (Mainwaring 1969b; Unhjem et al. 1969). Reduction of testosterone to DHT takes place in many target as well as nontarget tissues. Recognition that DHT binds more tightly to receptor and is the predominant androgen retained in most target cells opened the way for the detailed study of androgen receptors in cell-free systems. However, some androgen-responsive tissues, such as rodent kidney and fetal wolffian duct, show little ability to reduce testosterone to DHT, and in these organs testosterone itself appears to interact with the receptor (Bardin and Catterall 1981). The finding that certain sexual behavioral responses in the rat are stimulated by testosterone but not by DHT (Whalen and Luttge 1971; Beyer et al. 1973) suggested that testosterone acts directly with receptors in the brain, but this situation is complicated by the fact that testosterone, unlike DHT, is converted in brain to estradiol, which may produce the response.

The DHT bound in nuclei of rat prostate exposed to testosterone, either in vivo or in tissue incubations, is solubilized by extraction with salt solutions (BRUCHOVSKY and WILSON 1968; FANG et al. 1969; MAINWARING 1969 a) as a complex sedimenting at 3 S in salt-containing gradients. After similar exposure to hormone, cytosol fractions from prostate (FANG et al. 1969; UNHJEM et al. 1969; MAINWARING 1969 b) or epididymis (HANSSON and TVETER 1971; RITZÉN et al. 1971) contain a macromolecular complex reported for epididymis to sediment at 4 S, and for prostate at 8–9 S dissociated by salt to a 3.5 S unit. Direct addition of DHT, but not testosterone, to cytosol from prostate (FANG et al. 1969; MAINWARING 1969 b; BAULIEU and JUNG 1970; FANG and LIAO 1971; AAKVAAG et al. 1972), epididymis (RITZÉN et al. 1971), or levator ani (JUNG and BAULIEU 1972) from untreated castrated rats, or from cock comb (DUBÉ and TREMBLAY 1974), generates hormone-receptor complexes similar to those obtained in vivo. The prostate and levator ani complexes sediment in hypotonic gradients at 8–9 S (MAINWARING 1969 b; BAULIEU and JUNG 1970; JUNG and BAULIEU 1972; NOZU and TAMAOKI 1975) and are reversibly dissociated by salt to a 4–5 S unit. In contrast, the epididymal complex sediments at 4 S (RITZÉN et al. 1971), and with certain preparations of prostatic cytosol, especially those with long in vivo labeling (MAINWARING and PETERKEN 1971), the complex was found to sediment at 3.5 S in hypotonic medium (FANG et al. 1969).

Certain antiandrogenic compounds, such as cyproterone and its acetate (STERN and EISENFELD 1969, 1971; FANG et al. 1969; WHALEN et al. 1969; GELLER et al. 1969) and flutamide (PEETS et al. 1974; LIAO et al. 1974), block hormone uptake in prostate and seminal vesicle, but not in pituitary and hypothalamus, when given with testosterone or DHT in vivo or in whole tissue incubations. Presumably these agents compete with the hormone for the receptor protein. Similar inhibition is seen when hormone is added directly to cytosol (BAULIEU and JUNG 1970; FANG and LIAO 1971; PEETS et al. 1974; LIAO et al. 1974; MAINWARING et al. 1974). Flutamide is less effective than cyproterone acetate in blocking binding to cytosol receptor in vitro, suggesting that it may need metabolic alteration to become active. Sulfhydryl-blocking reagents prevent hormone binding to cytosol receptor from rat prostate (MAINWARING 1969 b; AAKVAAG et al. 1972) or levator ani (JUNG and BAULIEU 1972), or from cock comb (DUBÉ and TREMBLAY 1974).

2.4 Glucocorticoid Receptors

Initial attempts to demonstrate specific uptake and binding of glucocorticoids in vivo or in vitro, using either ^{14}C-labeled steroids or tritiated hormones of low specific radioactivity, gave results of marginal significance because of the hyperphysiological amounts of hormone needed and the extensive metabolism of the steroid by target tissues such as liver. These early studies did indicate that hormone is taken up by liver (BRADLOW et al. 1954; KENNEY and FLORA

1961; LITWACK et al. 1963) to a greater extent than by kidney, spleen, or skeletal muscle (BELLAMY et al. 1962; DE VENUTO et al. 1962). When tritiated hormones of high specific radioactivity became available, many laboratories showed that glucocorticoid-responsive tissues contain specific hormone-binding components.

The first definitive evidence for glucocorticoid receptor was the binding of cortisol in cytosol from mouse lymphoma, which was increased after adrenalectomy and was higher in steroid-sensitive as compared to steroid-resistant tumors (HOLLANDER and CHIU 1966). In early in vivo studies, selective uptake and retention of administered glucocorticoid was seen in limbic structures (hippocampus, septum) of rat brain (McEWEN et al. 1968, 1969, 1970; McEWEN and PLAPINGER 1970), in rat liver (BEATO et al. 1969, 1970, 1972a) and thymus (BRUNKHORST 1969), and in fetal rabbit lung (GIANNOPOULOS et al. 1973). Uptake by brain is higher in adrenalectomized than in intact rats. Fractionation of brain homogenates showed the binding to be primarily nuclear while in the other tissues, cytosol binding predominates. In several studies, the steroid present in the nuclear fraction was extracted with salt solutions as a macromolecular complex containing unchanged hormone.

In contrast to the sex hormones, where the first evidence for receptor binding came from whole animal studies, most of the early observations of glucocorticoid receptors were made with target cells in culture, soon followed by the use of broken cell systems. On incubation with physiological amounts of various glucocorticoids (cortisol, corticosterone, dexamethasone, triamcinolone acetonide), specific binding in cytosol and nuclear fractions was observed with a variety of animal tissues (Table 2; for additional examples see ROUSSEAU 1975). The ratio of nuclear to cytosolic binding varied, but in all cases studied, the radioactivity in the nuclear fraction was shown to be chemically unchanged hormone, and was much greater after incubation at $25°-37°C$ than at $2°-4°C$.

Complexes similar to those obtained with whole cells are formed by direct addition of hormone in the cold to cytosol fractions from various tissues (Table 2). Radioactive steroid-receptor complexes, obtained either from whole cells or in cell-free systems, were usually detected by gel filtration or by removing unbound steroid with dextran-coated charcoal. In general, gradient centrifugation was not as widely employed with glucocorticoid as with estrogen receptors. In these early experiments, cytosolic complexes were usually reported to sediment at $7-8$ S in hypotonic sucrose gradients and, like the nuclear complex, at $4-5$ S in high-salt gradients (BAXTER and TOMKINS 1971; BEATO and FEIGELSON 1972; CHYTIL and TOFT 1972; BEATO et al. 1973; GARDNER and WITTLIFF 1973; KALIMI et al. 1975). Currently, the cytosol receptor is generally denoted as $8-9$ S (low salt) and the nuclear protein as 4 S (PRATT et al. 1989). Both in whole cells and in cytosol, sulfhydryl-blocking reagents prevent the binding of steroid to receptor (SCHAUMBURG 1970, 1972; BAXTER and TOMKINS 1971; KOBLINSKY et al. 1972; ISHII and ARONOW 1973; GIANNOPOULOS et al. 1973; GARDNER and WITTLIFF 1973), thus distinguishing it from binding proteins of the blood.

Table 2. Early demonstration of glucocorticoid binding by target cells on cytosol fraction

Species	Tissue	Report
Whole cells or tissue slices		
Rat	Thymocyte	MUNCK and BRINCK-JOHNSEN 1968; SCHAUMBURG and BOJESEN 1968; WIRA and MUNCK 1970; SCHAUMBURG 1970; MUNCK et al. 1972
	Hepatoma	BAXTER and TOMKINS 1970; HIGGINS et al. 1973a
	Kidney	FUNDER et al. 1973b
Mouse	Fibroblast	HACKNEY et al. 1970; HACKNEY and PRATT 1971; ISHII et al. 1972; ISHII and ARONOW 1973
	Lymphoma	BAXTER et al. 1971; KIRKPATRICK et al. 1971
Cow	Mammary	TUCKER et al. 1971
Fetal rabbit	Lung, liver	GIANNOPOULOS et al. 1972, 1973
Cytosol		
Rat	Thymocyte	SCHAUMBURG 1972
	Liver	BEATO 1970; BEATO and FEIGELSON 1972; GOPALAKRISHNAN and SADGOPAL 1972; KOBLINSKY et al. 1972; LITWACK et al. 1973
	Hepatoma	BAXTER and TOMKINS 1971; ROUSSEAU et al. 1972a; SINGER et al. 1973
	Kidney	ROUSSEAU et al. 1972b
	Brain	CHYTIL and TOFT 1972
	Mammary	GARDNER and WITTLIFF 1973
Mouse	Fibroblast	PRATT and ISHII 1972; ISHII and ARONOW 1973
	Lymphoma	HOLLANDER and CHIU 1966; ROSENAU et al. 1972; KIRKPATRICK et al. 1972
Fetal rabbit	Lung	BALLARD and BALLARD 1972; GIANNOPOULOS 1973

In contrast to most target tissues, which contain a single glucocorticoid-binding protein, rat kidney shows three distinct kinds of receptors that react both with glucocorticoids and with mineralocorticoids but to different degrees (ROUSSEAU et al. 1972b; FUNDER et al. 1973a, b). The type I binder shows a strong affinity for aldosterone and corticosterone and is called the mineralocorticoid receptor. The type II protein binds dexamethasone (but not cortisol) more avidly than corticosterone or aldosterone and is regarded as the glucocorticoid receptor. The type III protein shows a special affinity for corticosterone as compared to either aldosterone or dexamethasone (FELDMAN et al. 1973); it is also found in certain regions (hippocampus) of the brain (DE KLOET et al. 1975; KROZOWSKI and FUNDER 1983; BEAUMONT and FANESTIL 1983). Because of the similarity in structure of glucocorticoid and mineralocorticoid receptors (see Sect. 6), it is not surprising that there is overlap in their funtional properties. The mineralocorticoid receptor is more sensitive than the glucocorticoid receptor to low levels of cortisol, leading to the suggestion that the dual receptor system may provide a broad glucocorticoid response range from low resting levels to the temporarily elevated concentrations following stress (ARRIZA et al. 1987; EVANS 1989).

The foregoing experiments indicate that glucocorticoids bind without metabolic change to macromolecules in target cells that appear both in the cytosol and nuclear fractions of homogenates. That these are involved in hormonal action was first suggested by the correlation of nuclear binding with enzyme induction in hepatoma cells (BAXTER and TOMKINS 1970; BEATO et al. 1972b; ROUSSEAU et al. 1972a), and by the fact that glucocorticoid-sensitive fibroblasts and lymphomas show greater hormone uptake than the corresponding resistant lines and that adrenalectomy leads to increased binding in target cells.

2.5 Mineralocorticoid Receptors

The first definitive evidence for mineralocorticoid receptors came from autoradiographic studies of toad urinary bladder exposed to tritiated aldosterone (EDELMAN et al. 1963; PORTER et al. 1964) in an in vitro system in which this hormone stimulates sodium transport from the mucosal to the serosal side (CRABBÉ 1961). Radioactivity is incorporated predominantly into the nuclei of mucosal epithelial cells, and the uptake correlates with the effect on transport. The binding of aldosterone in toad bladder was confirmed by measurement of the radioactivity displaceable by excess nonradioactive aldosterone or other active mineralocorticoids (SHARP et al. 1966; AUSIELLO and SHARP 1968). The uptake and retention of aldosterone by rat kidney or duodenum in vivo (SULYA et al. 1963; FANESTIL and EDELMAN 1966; FANESTIL 1968; HERMAN et al. 1968) provided early evidence for the presence of receptors in these tissues. Aldosterone is bound in target cells without chemical alteration (SULYA et al. 1963; FANESTIL and EDELMAN 1966; PASQUALINI et al. 1972).

The labeled steroid incorporated into kidney tissue of the rat (FANESTIL and EDELMAN 1966; SWANECK et al. 1969) and fetal guinea pig (PASQUALINI et al. 1972) was shown by cell fractionation to be predominantly in the nucleus; this has been confirmed by autoradiography (FARMAN and BONVALET 1983). It is bound to chromatin from which it is extracted by salt solutions as a 4 S steroid-protein complex, preferably determined in glycerol gradients (SWANECK et al. 1970; MARVER et al. 1972; PASQUALINI et al. 1972). Also there is a more readily extractable nuclear complex, sedimenting at 3 S, the significance of which has not been established. The corresponding cytosol fractions contain a complex sedimenting at 8–9 S in hypotonic gradients and dissociated into a 4.5 S binding unit in the presence of salt.

Similar cytosol and nuclear complexes are formed after the incubation of kidney or parotid slices with tritiated aldosterone at $25°-37°C$ or by direct addition of the hormone in the cold to kidney cytosol from untreated animals (HERMAN et al. 1968; MARVER et al. 1972, 1974; PASQUALINI et al. 1972; FUNDER et al. 1972). Both the cytosol and nuclear complexes are destroyed by sulfhydryl-blocking reagents (HERMAN et al. 1968). Binding of aldosterone to its receptor, both in vivo and in vitro, is inhibited by a group of related steroids known as spirolactones (FANESTIL 1968; HERMAN et al. 1968; ALBERTI and SHARP 1969; MARVER et al. 1974).

As described in Sect. 2.4, the cytosols of rat kidney and hippocampus contain three types of receptor proteins for adrenal steroids, with differing affinities for aldosterone, corticosterone, and cortisol. A possible physiological significance of this combination of overlapping receptor systems has been proposed (FUNDER and SHEPPARD 1987; ARRIZA et al. 1987; EVANS 1989). In contrast, aldosterone forms a single complex in cytosol from rat hepatoma cells (ROUSSEAU et al. 1972b).

2.6 Vitamin D Receptors

Early studies of the administration of tritiated vitamin D_3 (HAUSSLER et al. 1968; HAUSSLER and NORMAN 1969; LAWSON et al. 1969) or its 25-hydroxy metabolite (HAUSSLER et al. 1971) to hormone-depleted chicks showed that radioactivity becomes bound to nuclear chromatin in cells of the intestinal mucosa as a substance more polar than either of the two steroids injected. This material was soon identified (NORMAN et al. 1971) as $1\alpha,25$-dihydroxyvitamin D_3 ($1\alpha,25$-dihydroxycholecalciferol, now called calcitriol), which is formed in the kidney (FRASER and KODICEK 1970; WONG et al. 1972) by the enzymatic hydroxylation of the 25-hydroxy intermediate produced in the liver (HORSTING and DeLUCA 1969). Unlike vitamin D or its other metabolites, calcitriol is active in nephrectomized animals and is now regarded as the actual hormone.

After administration of tritiated calcitriol in vivo, radioactive steroid is concentrated, without chemical change, in the nuclei of various tissues responsive to vitamin D (Table 3), as shown by either cell fractionation or autoradiographic techniques. The amount of hormone bound in chick intestine correlates with the stimulation of calcium absorption, indicating a physiological role for the binding (BRUMBAUGH and HAUSSLER 1974a).

The chromatin-bound calciferol, produced either in vivo or by incubating calciferol with whole cells or with tissue homogenates from untreated animals, is extracted by salt solutions as a macromolecular complex, sedimenting at 3.1–3.7 S in hypertonic sucrose gradients (LAWSON and WILSON 1974; BRUMBAUGH and HAUSSLER 1975; BRUMBAUGH et al. 1975; WECKSLER et al. 1977; PIKE and HAUSSLER 1983). Hormone-receptor complexes, reported to sediment at 3.0–3.7 S, were demonstrated by direct addition of tritiated calcitriol to cytosol (Table 3; for additional examples see NORMAN et al. 1982; HAUSSLER 1986). In hypotonic gradients, the cytosol complex is reported to sediment at 6.8 or 5.4 S (FELDMAN et al. 1979; COLSTON and FELDMAN 1980). Whereas different types of sulfhydryl-blocking reagents prevent hormone binding to unoccupied cytosolic receptors (WECKSLER et al. 1979b, 1980b; MELLON et al. 1980; COTY et al. 1980), the preformed complex is destroyed by mercurials but not by alkylating agents such as ethylmaleimide or iodacetamide, which suggests that a critical sulfhydryl group is located near the hormone-binding site. In addition to the receptor, which recognizes calcitriol specifically, many cytosols also contain a 6 S protein, probably from the blood, which preferentially binds 25-hydroxycholecalciferol.

Table 3. Early demonstration of calcitriol binding by target tissues or cytosol fraction

Species	Tissue	Report
Vivo or whole tissue		
Chick	Intestine	H. TSAI et al. 1972; WONG et al. 1972; BRUMBAUGH and HAUSSLER 1973, 1974a, 1975; LAWSON and WILSON 1974; ZILE et al. 1978[a]; JONES and HAUSSLER 1979[a]
	Parathyroid	HENRY and NORMAN 1975; WECKSLER et al. 1977; HUGHES and HAUSSLER 1978
	Bone	WONG et al. 1972
Rat	Intestine	T. C. CHEN and DeLUCA 1973; STUMPF et al. 1979[a]
	Kidney	STUMPF et al. 1979[a], 1980[a]
Mouse	Kidney	COLSTON and FELDMAN 1980
	Fibroblasts	PIKE and HAUSSLER 1983
Cytosol		
Chick	Intestine	H. TSAI and NORMAN 1973; BRUMBAUGH and HAUSSLER 1973, 1974b, 1975; HADDAD et al. 1973; LAWSON and WILSON 1974; KREAM et al. 1976; MELLON et al. 1980; WECKSLER et al. 1980a
	Bone	MELLON and DeLUCA 1980
	Parathyroid	HUGHES and HAUSSLER 1978; WECKSLER et al. 1980b
	Kidney	CHRISTAKOS and NORMAN 1979
Rat	Intestine	KREAM et al. 1976, 1977b; WECKSLER et al. 1979b; FELDMAN et al. 1979
	Kidney	CHANDLER et al. 1979
	Bone	KREAM et al. 1977a
Mouse	Kidney	COLSTON and FELDMAN 1979, 1980
	Fibroblast	PIKE and HAUSSLER 1983
	Bone	T. L. CHEN et al. 1979
Human	Intestine	WECKSLER et al. 1979a, 1980a
	Parathyroid tumor	HUGHES and HAUSSLER 1978; WECKSLER et al. 1980b

[a] By autoradiography

2.7 Ecdysteroid Receptors

Because of their induction of chromosomal puffs in salivary glands of insect larvae, ecdysteroids were long considered to act at the DNA level (CLEVER and KARLSON 1960). With the availability of tritiated ecdysteroids, direct evidence was obtained for the specific binding of steroid in nuclei of hormone-responsive cells of insects and crustaceans. On incubation with 20-hydroxyecdysone (formerly β-ecdysone), imaginal disks of *Drosophila melanogaster,* which differentiate in culture in response to ecdysteroids, take up and retain hormone without chemical change by a process sensitive to sulfhydryl-blocking reagents (YUND and FRISTROM 1975). After incubation with tritiated ponasterone A, most of the bound steroid is in the nuclear fraction of homogenates, from which it is extracted by salt solutions as a macromolecular complex (YUND et al. 1978; YUND 1979). Similar results were obtained with ponasterone A or

20-hydroxyecdysone in a cell line (K_c) from *Drosophila* that is responsive to ecdysteroids (MAROY et al. 1978; STEVENS et al. 1980; BECKERS et al. 1980), but not in an insensitive line (IRELAND et al. 1982). Irradiation of *Drosophila* salivary glands previously exposed to ecdysteroid causes the hormone to bind covalently to chromosomes (GRONEMEYER and PONGS 1980); this photochemically induced association appears to be with a single component in the chromosomes considered to be the receptor protein (SCHALTMANN and PONGS 1982).

With imaginal disks and cultured K_c cells, as well as with crayfish tissues (KUPPERT et al. 1978; LONDERSHAUSEN and SPINDLER 1981), direct addition of the hormone to cytosol from untreated tissues gives a steroid-receptor complex, shown with crayfish cytosol to be sensitive to sulfhydryl-blocking reagents. Unless homogenization is carried out in salt-containing buffer, target cell nuclei usually contain a resident population of unfilled receptors that react with hormone in the cold (SAGE et al. 1982; LONDERSHAUSEN et al. 1982). Steroid-receptor complexes, extracted from the nucleus or formed by adding hormone to cytosol or nuclear extract, sediment at 4.3–5 S in hypertonic gradients (LONDERSHAUSEN and SPINDLER 1981; KUPPERT and SPINDLER 1982; OSTERBUR and YUND 1982).

3 Intracellular Interaction Pattern

3.1 Receptor Transformation [1]

3.1.1 Relation of Nuclear to Cytosol Binding

Responsive cells for all classes of steroid hormones are characterized by specific, hormone-binding components that are sulfhydryl-containing proteins identified in both the nuclear and cytosol fractions of tissue homogenates. The relationship between these two forms of the receptor has been the subject of extensive investigation, for it led to the identification of a biochemical role for the steroid in hormonal action. This was first elucidated with the estrogens. A variety of experimental evidence suggested that the estradiol-receptor complex, bound in target cell nuclei of hormone-treated animals, is not produced directly but arises in a two-step process via the initial formation of the cytosol complex (GORSKI et al. 1968; JENSEN et al. 1968). It was then shown that, under the influence of estrogen, the 4 S cytosol receptor actually becomes the 5 S nuclear protein (JENSEN et al. 1971; GSCHWENDT and HAMILTON 1972) and that

---

[1] In the first reports of hormone-induced conversion of native receptor protein to its nuclear-binding form, this phenomenon was known as receptor transformation. With the appreciation of its biological significance, many investigators began to call it activation. More recently there has been a trend back to the use of transformation, reserving the term activation for the acquisition of steroid-binding properties as, for example, by the rephosphorylation of processed receptor.

this change occurs more rapidly in the presence of DNA (YAMAMOTO and ALBERTS 1972). The finding that estradiol given in vivo causes temporary depletion of cytosol receptor is consistent with its utilization to produce nuclear complex (JENSEN et al. 1969; SARFF and GORSKI 1971).

In contrast to the 4 S cytosol receptor, the 5 S "transformed" receptor, produced by the incubation of cytosol with estradiol, resembles that extracted from uterine nuclei of estrogen-treated animals in its ability to bind to isolated nuclei, chromatin, and DNA (JENSEN et al. 1972b; W. McGUIRE et al. 1972; YAMAMOTO and ALBERTS 1972; MILGROM et al. 1973) as well as to phosphocellulose (ATGER and MILGROM 1976). The importance of nuclear binding in biological action was demonstrated in experiments showing that full uterotropic response requires the continued presence of estrogen-receptor complex in the nucleus for a period of several hours (GORSKI and RAKER 1974; J. ANDERSON et al. 1975). It was found that enhancement of RNA polymerase activity in isolated uterine nuclei, previously observed on incubation with estradiol-cytosol mixtures (RAYNAUD-JAMMET and BAULIEU 1969), is elicited by treatment of the nuclei with 5 S estradiol-receptor complex but not with the 4 S form, an effect specific for hormone-dependent tissues and tumors (MOHLA et al. 1972; ARBOGAST and DESOMBRE 1975). Only transformed complex was found to increase the binding of actinomycin D by isolated uterine and breast cancer nuclei (LECLERCQ et al. 1973; VERRIJDT et al. 1985), an effect also produced by estradiol in vivo (LEROY et al. 1972) and which indicates enhanced transcriptional activity (BRACHET and FICQ 1965).

From these and other observations it became apparent that it is the transformed receptor protein that serves as a regulator of gene expression in target cells and that an important role of the hormone is to convert the native form of the receptor to this biochemically functional modification (JENSEN et al. 1971; JENSEN and DESOMBRE 1973). Thus, receptor transformation was recognized as a key step in estrogen action.

It was soon shown that hormone-induced transformation of the native receptor protein to a nuclear binding form is a general phenomenon for all classes of steroid hormones. For progestin receptor, a relationship between cytosol and nuclear hormone-receptor complexes was demonstrated by experiments with chick oviduct tissue in vitro (O'MALLEY et al. 1971; BULLER et al. 1975). On exposure to progesterone in the cold, binding is primarily cytosolic, but on warming the tissue to 37 °C, cytosolic receptor is gradually depleted as nuclear complex increases. Salt-extractable nuclear complex is readily formed by incubating oviduct nuclei at 25 °C with a combination of progesterone and oviduct cytosol but not with either alone. Similarly, incubation of cytosol from guinea pig (MILGROM et al. 1973) or rabbit (FLEISCHMANN and BEATO 1979) uterus or hen oviduct (TOFT et al. 1976) with progestational hormone at 15° – 25 °C transforms the receptor complex, as indicated by its ability to bind to DNA, phosphocellulose, or ATP-Sepharose, respectively.

With androgen receptors, early experiments showed that the uptake of steroid into nuclei of minced prostatic tissue incubated with tritiated testosterone is strongly temperature dependent, with little nuclear concentration taking place at $0°-2°C$ (K. Anderson and Liao 1968; Sar et al. 1970). When treated at 37 °C with either testosterone or DHT, isolated nuclei from prostate (Liao and Fang 1969) or epididymis (Blaquier et al. 1970) do not bind steroid; with prostate it was found that addition of cytosol to the incubation mixture led to the formation of extractable 3 S DHT-receptor complex, similar to that formed in vivo. Subsequent studies showed that cytosols from a variety of androgen-dependent tissues, but not from nontarget tissues, promote the binding of DHT to isolated chromatin or DNA (Mainwaring and Peterken 1971). In rat prostate, androgen-induced depletion of cytosol receptor accompanies nuclear accumulation of steroid-receptor complex in vivo (Steinsapir et al. 1985), with subsequent replenishment of the cytosol protein taking place somewhat more rapidly than in similar experiments with estrogen in uterus.

With glucocorticoids, steroid-receptor complexes, similar to those obtained in whole cells, are formed by treatment of target cell cytosols with hormone in the cold. In contrast, hormone treatment of isolated nuclei does not produce extractable complex unless cytosol is also present and the incubation is carried out at near physiological temperature (Baxter et al. 1972; Ballard and Ballard 1972; Litwack et al. 1973; Kalimi et al. 1973). Cytosol receptor is depleted as nuclear complex is formed (Beato et al. 1973). If the cytosol is first warmed with hormone, extractable complex is produced on subsequent incubation with nuclei in the cold (Munck et al. 1972; Baxter et al. 1972; Higgins et al. 1973 b). Cytosol from resistant lymphomas, which contains little receptor, is relatively ineffective in promoting nuclear binding (Rosenau et al. 1972). By immunocytochemical techniques, nuclear translocation of cytoplasmic receptor was shown to take place in steroid-sensitive but not in steroid-resistant leukemic cells (Antakly et al. 1990). These findings are consistent with the derivation of the nuclear complex from cytosol receptor that has been converted to a DNA-binding form by the action of the hormone. It is further supported by observations that, on incubation of hepatoma cells with hormone, the progressive increase in nuclear complex is accompanied by a decrease in cytosol receptor (Rousseau et al. 1973), and that warming of the cytosol receptor with hormone in vitro effects its conversion to a DNA-binding form (Milgrom et al. 1973). Subsequent studies have established that hormone-induced transformation of glucocorticoid receptors actually takes place in the living cell (Munck and Foley 1979; Marković and Litwack 1980; Miyabe and Harrison 1983).

In studies of the relationship between cytosol and nuclear mineralocorticoid receptors in rat kidney, it was established that, both in tissue slices and in recombined cytosol-nuclear mixtures, the 4 S complex, extractable from the nucleus, is derived from an initial complex of the hormone with the 8–9 S cytosol receptor (Marver et al. 1972, 1974; Funder et al. 1973 b). Unlike cytosol binding, formation of the nuclear complex does not take place readily in the cold, but requires temperatures approaching the physiological. Binding to isolated nuclei occurs only in the presence of cytosol containing receptor, which is depleted as nuclear complex is formed. In these studies, a 3 S nuclear complex, extractable by hypotonic buffer, appeared to be an intermediate between the native complex of the cytosol and the tightly bound nuclear complex. The significance of this "three-step" sequence has not been fully elucidated. Warming rat kidney cytosol with aldosterone converts the receptor to the DNA-binding form (MILGROM et al. 1973).

With vitamin D receptors, studies with homogenates and reconstituted cytosol-nuclear mixtures from chick intestinal mucosa (H. TSAI and NORMAN 1973; BRUMBAUGH and HAUSSLER 1973, 1974b, 1975; PIKE 1982), chick parathyroid (WECKSLER et al. 1977; HUGHES and HAUSSLER 1978), and mouse kidney (COLSTON and FELDMAN 1980) or fibroblasts (PIKE and HAUSSLER 1983) have shown that salt-extractable complex is formed on incubation with tritiated calcitriol only if cytosol from a receptor-containing tissue is also present. Nuclear binding that occurs in the absence of hormone is readily distinguished by its ease of extraction at lower ionic strength (PIKE 1982). In some cases, the formation of nuclear complex was found to be temperature dependent (BRUM-BAUGH and HAUSSLER 1973, 1974b, 1975; WECKSLER et al. 1977; HUGHES and HAULER 1978; COLSTON and FELDMAN 1980), but in other studies (H. TSAI and NORMAN 1973; PIKE 1982) the nuclear complex formed readily in the cold, as also was observed in whole mouse fibroblasts (PIKE and HAUSSLER 1983). It appears that, under some experimental conditions, hormone-induced conversion of vitamin D receptor to its DNA-binding form proceeds at lower temperatures than with other steroid receptors. If enough time is allowed for this reaction to occur, subsequent nuclear binding of the transformed complex will be seen in the cold (HAUSSLER et al. 1981).

The uptake of ecdysteroids into the nuclei of intact imaginal disks (YUND and FRISTROM 1975; YUND 1979) or K_c cells (BECKERS et al. 1980) was found to take place much more rapidly at $20°°-25°C$ than at $0°C$, although eventually the same level of nuclear binding is reached in disks maintained in the cold. That nuclear complex is derived from cytosol receptor is indicated by experiments with crayfish hypodermis in which nuclei, pre-extracted to remove unoccupied receptor, formed nuclear complex when incubated with cytosol containing receptor but not with cytosol heated to destroy the receptor protein (LONDERSHAUSEN et al. 1982).

From the foregoing observations, it is evident that steroid-receptor complexes, bound in target cell nuclei of hormone-treated animals, are derived from the initial formation of a cytosol complex, followed by the hormone-induced conversion of the receptor protein to a form that has acquired affinity for chromatin. With vitamin D and possibly ecdysteroid receptors the temperature dependence of this transformation may be somewhat less than with receptors for other steroid hormones.

Only with estrogen receptor is transformation accompanied by an increase in the sedimentation rate, determined in hypertonic gradients to release the hormone-binding unit from associated macromolecules. Early studies of the transformation process established that the 5 S entity is a dimeric form of a modified 4 S binding unit (LITTLE et al. 1973, 1975; NOTIDES and NIELSEN 1974; NOTIDES et al. 1975). This observation is substantiated by more recent studies (MILLER et al. 1985), and, as discussed below, it is consistent with the interaction of transformed estrogen receptor with a palindromic sequence of amino acids believed to comprise the attachment site in the hormone response elements of target genes (KUMAR and CHAMBON 1988). Even though other transformed receptors do not sediment more rapidly than their native counterparts, transformed glucocorticoid receptor exists as a dimer on gel filtration (WRANGE et al. 1989), and progestin and glucocorticoid receptors appear to interact in a dimeric form with the hormone response elements of target genes (S. TSAI et al. 1988, 1989). The spontaneous dimerization of transformed estrogen receptor is fortunate, for it was the difference in sedimentation rates between the nuclear and cytosol receptors (Fig. 1) that provided the initial clue leading to recognition of the phenomenon of receptor transformation.

3.1.2 Nature of the Transformation Process

Though much has been learned about the transformation reaction with the various classes of steroid hormone receptors (MILGROM 1981; GRODY et al. 1982; SCHMIDT and LITWACK 1982; DAHMER et al. 1984), the exact chemical details of this phenomenon are not completely understood. At physiological pH and ionic strength, transformation requires the presence of hormone and, in most cases, temperatures approaching the physiological. The vitamin D receptor is transformed at lower temperatures than other steroid receptors (WECKSLER and NORMAN 1980; PIKE 1982), suggesting that this conversion may have a lower energy of activation.

In general, receptor transformation is accelerated by dilution or by higher salt concentrations or pH. Conversion to the nuclear-binding form is effected in the absence of steroid by precipitation with ammonium sulfate (DESOMBRE et al. 1972; BULLER et al. 1975; BAILLY et al. 1977), by dialysis or gel filtration (CAKE et al. 1976; GOIDL et al. 1977; BAILLY et al. 1977; SATO et al. 1979), and with progestin receptor, by treatment with salt or heparin (YANG et al. 1982), ribonuclease (T. THOMAS and KIANG 1986), or ATP (SINGH et al. 1986).

Transformation of all classes of steroid hormone receptors is inhibited by molybdate and other transition metal oxyanions (DAHMER et al. 1984; PRATT 1987), which provides a useful means for stabilizing the native form of the receptor. This agent also retards the inactivation of crude receptor preparations in vitro. This action is believed to result from its inhibition of endogenous phosphatases (NIELSEN et al. 1977a), although this interpretation has been questioned for progestin receptor (GRODY et al. 1980). The discovery of molybdate as a stabilizing agent (NIELSEN et al. 1977a; LEACH et al. 1979; TOFT and NISHIGORI 1979) has greatly facilitated receptor purification, as well as studies of structure and function.

As first shown with glucocorticoid receptor (CAKE et al. 1978), pyridoxal 5'-phosphate also blocks nuclear binding of estrogen (MÜLLER et al. 1980) and progestin (NISHIGORI and TOFT 1979) receptors. Although this may be an effect on transformation itself (TRAISH et al. 1980), more likely it involves reaction with a DNA-binding site exposed by transformation (CAKE et al. 1978; TOFT

and NISHIGORI 1979). Transformation of glucocorticoid receptor is prevented
by tolyl-lysyl chloromethane (HUBBARD et al. 1984) and that of estrogen receptor by diisopropyl fluorophosphate (PUCA et al. 1986), both inhibitors of serine proteases. Sulfhydryl groups, required for steroid binding by all classes of
receptors, appear to participate in the transformation of progestin (KALIMI
and BANERJI 1981; PELEG et al. 1988) and glucocorticoid receptors (KALIMI and
LOVE 1980; TIENRUNGROJ et al. 1987a), and it has been suggested that disulfide
bond reduction takes place during transformation of androgen receptors
(WILSON et al. 1986). Phosphorylation or dephosphorylation may be involved
in the transformation process, but as yet no definite role has been established
(PRATT 1987; AURICCHIO 1989). With estrogen receptors, acquisition of nuclear binding ability and dimerization to form 5 S complex appear to be separate
processes that proceed at different rates (BAILLY et al. 1980; MÜLLER et al.
1983).

In early studies of the kinetics of estrogen receptor transformation, the
reaction was found to proceed more rapidly as the concentration is reduced.
This is contrary to expectation for a dimerization process, and it was suggested
that the rate-limiting step for the transformation of estrogen (NOTIDES et al.
1975) and glucocorticoid (VEDECKIS 1983) receptors may be dissociation from
other components present. The phenomenon was elucidated by the finding
that the $8-10$ S forms of untransformed receptors for at least four classes of
steroid hormones consist of a steroid-binding subunit associated with another
protein of molecular weight 90 000 (JOAB et al. 1984; DOUGHERTY et al. 1984;
SCHUH et al. 1985; HOUSLEY et al. 1985) which is present as a dimer (DENIS
et al. 1987; RADANYI et al. 1989) and is lost on transformation (MENDEL et al.
1986a). In the case of glucocorticoids, this component appears to be required
for initial hormone binding by the receptor (BRESNICK et al. 1989). This substance was soon identified as a heat shock protein (CATELLI et al. 1985;
SANCHEZ et al. 1985), which itself does not bind hormone but which appears
to associate with the receptor to obscure the region responsible for binding to
DNA. Interaction with the steroid hormone is considered to displace the heat
shock protein, making the DNA-binding domain of the receptor available for
interaction with target genes (SANCHEZ et al. 1987; BAULIEU et al. 1989; PRATT
et al. 1989; DENIS and GUSTAFSSON 1989). This concept is supported by the
finding that antibodies raised against peptide sequences in the DNA-binding
domain of human glucocorticoid receptor recognize only the transformed
receptor and inhibit its association with DNA (WILSON et al. 1988; URDA et al.
1989). This mechanism is illustrated schematically in Fig. 2 for the transformation of glucocorticoid receptor.

Cytosols containing untransformed receptors for estrogens, androgens and
glucocorticoids contain low molecular weight, heat stable components that
appear to be required for association of the receptor with the heat shock
protein (CAKE et al. 1976; GOIDL et al. 1977; BAILLY et al. 1977; SATO et al.
1979, 1980; SEKULA et al. 1981; LEACH et al. 1982). With glucocorticoid receptors, endogenous factors, which appear to act in a manner similar to added
molybdate, have been identified as a phosphoglyceride (BODINE and LITWACK

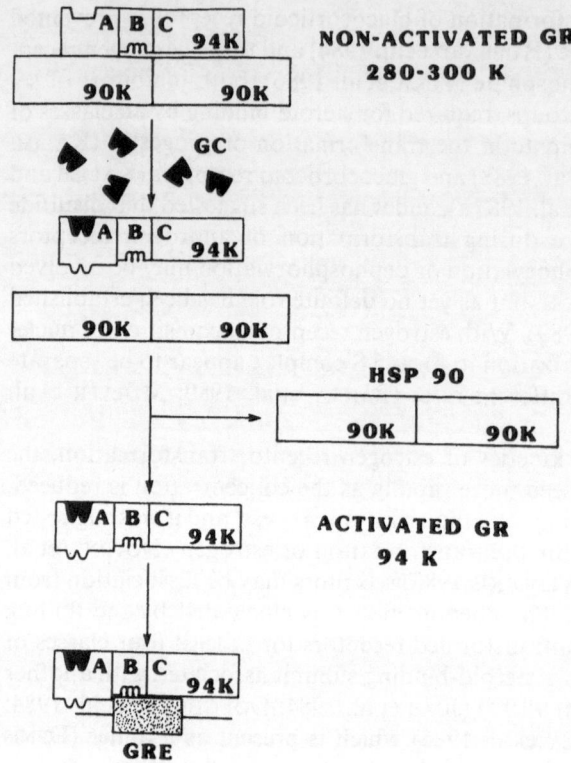

Fig. 2. Proposed model for the transformation (activation) of the glucocorticoid receptor. Non-transformed receptor, present in target cell cytosol, contains a dimer of $M_r \approx 90\,000$ heat shock protein (90 K) that masks the DNA-binding domain (B) of the receptor protein (94 K). Association with the hormone (GC) induces a change in the receptor, resulting in dissociation of the heat shock protein and unmasking the DNA-binding domain, which can then react with glucocorticoid response elements (GRE) of target genes. The steroid-binding and DNA-binding regions, here called A and B, are designated in the conventional description of molecular domains as E and C, respectively (see Fig. 5). (GUSTAFSSON et al. 1989)

1988) and as a metal anion (MESHINCHI et al. 1990). Their removal by dialysis, gel filtration, or chelation eliminates the requirement for hormone in transforming the receptor.

In addition to these small inhibitory factors, glucocorticoid receptor cytosols also contain a macromolecular substance that impedes nuclear binding of the transformed complex (MILGROM and ATGER 1975; SIMONS et al. 1976; LIU and WEBB 1977; ATGER and MILGROM 1978; ISOHASHI et al. 1980). This effect can be counteracted by ATP (HORIUCHI et al. 1981). Hepatomas appear to contain higher levels of this factor than do normal liver cells.

Although the precise details of the transformation phenomenon remain to be elucidated, the concept of dissociation from heat shock protein, coupled with the removal of low-molecular-weight factors that participate in this association, serves to explain many of the heretofore puzzling observations regarding receptor transformation. Still to be elucidated is how reaction with the steroid hormone effects disruption of the macromolecular complex. It has been proposed that the steroid acts directly on the metal anion required for association with heat shock protein (MESHINCHI et al. 1990) or that it stimulates a self-proteolytic activity that is inherent in the receptor protein (PUCA et al. 1986). Perhaps relevant to the latter concept is the "calcium-stabilized" estrogen receptor, formed when uterine cytosol is treated with calcium ions in the presence of EDTA (DESOMBRE et al. 1971). This entity, which sediments at 4.5 S and resembles transformed estrogen receptor in its lack of aggregation and its ability to bind to isolated nuclei, has been shown to result from the action of a proteolytic factor in uterine cytosol that is stabilized by EDTA and activated by calcium ions (PUCA et al. 1972).

3.2 Intracellular Localization

Because untransformed receptors for most steroid hormones usually appear in the cytosol fraction of tissue homogenates, it was originally assumed that the unoccupied receptor is a cytoplasmic protein, and that its hormone-induced conversion to the transformed state was accompanied by migration or "translocation" to the nucleus (JENSEN et al. 1968; GORSKI et al. 1968). Though this scheme was consistent with most earlier experiments, certain observations arose suggesting that much of the native receptor may already be in the nucleus before exposure to hormone.

As described and documented more completely elsewhere (M. WALTERS 1985; JENSEN 1990), unfilled and presumably untransformed receptors for estrogens, progestins, mineralocorticoids, and ecdysteroids are sometimes found in the nuclear fraction of tissue homogenates, whereas unoccupied vitamin D receptor remains in the nuclear pellet in media containing less than physiological salt concentration (M. WALTERS et al. 1982). In contrast to earlier autoradiographic experiments in which rat uterus (JENSEN et al. 1968), exposed to tritiated estradiol at 2 °C, showed relatively little nuclear radioactivity, more recent autoradiographic studies with estradiol in immature rat uteri (P. J. SHERIDAN et al. 1979; MARTIN and SHERIDAN 1982) and with progesterone in rat uteri (P. J. SHERIDAN et al. 1981) or chick oviduct (GASC et al. 1983; ENNIS et al. 1986) found most but not all of the steroid to be bound in the nucleus, even in the cold.

The presence of untransformed steroid receptors in nuclei of target cells was also shown by immunocytochemical experiments with polyclonal or monoclonal antibodies to receptors (see Sect. 5). In contrast to a few reports of substantial immunostaining or immunofluorescence of estrogen receptors in the extranuclear region in rat pituitary cells (MOREL et al. 1981) and in frozen (RAAM et al. 1982; NADJI and MORALES 1986) or embedded (GREENE and JENSEN 1982; MARCHETTI et al. 1987) sections of human breast cancers, most immunocytochemical studies have found only nuclear staining in frozen sections of human breast cancers; human, rabbit, monkey, and rat uterus; rat neurons and anterior pituitary; and human ovarian cancer, as well as in embedded specimens of human breast cancer (W. KING and GREENE 1984; W. KING et al. 1985; for detailed references see JENSEN 1990).

Immunocytochemical studies have detected progestin receptor only in the nuclei of chick oviduct (GASC et al. 1983, 1984; ENNIS et al. 1986); immature rabbit, castrated guinea pig, and human uterus; rabbit pituitary gland; and human breast, ovarian, and endometrial cancer (PERROT-APPLANAT 1985, 1987; CLARKE et al. 1987; PRESS and GREENE 1988; PERTSCHUK et al. 1988). Immunocytochemistry with electron microscopy showed only nuclear localization of estrogen receptor in human uterus (PRESS et al. 1985), although with immature rabbit uterus an immunogold technique detected a small but definite amount of extranuclear progestin receptor in addition to nuclear receptor, apparently associated with ribosomes (PERROT-APPLANAT et al. 1986). Similar nuclear localization of androgen receptor is seen in the prostate of humans (LUBAHN et al. 1988; CHANG 1989a) and intact adult rats (TAN et al. 1988).

In contrast to the findings with gonadal hormones, immunocytochemical studies have shown that untransformed receptors for glucocorticoids and mineralocorticoids are present both in the nucleus and in the cytoplasm of target cells, in many cases with cytoplasmic receptor predominating. With either polyclonal or monoclonal antibodies to glucocorticoid receptor, using either fluorescence or immunostaining techniques, substantial amounts of extranuclear antigen were observed in rat liver and hepatoma (GOVINDAN 1980; ANTAKLY and EISEN 1984; GASC and BAULIEU 1987; WIKSTRÖM et al. 1987); rat and human lymphoid cells (PAPAMICHAIL et al. 1980; ANTAKLY et al. 1989); rat nerve, glial, and pituitary cells (FUXE et al. 1985, 1987; ANTAKLY and EISEN 1984); and human uterine carcinoma (WIKSTRÖM et al. 1987) and leukemia (ANTAKLY et al. 1990). With either frozen or embedded sections of rat tissues nuclear immunostaining was higher in liver, brain, and kidney, while cytoplasmic staining predominated in thymus, spleen, heart, lung, and kidney tubules (TEASDALE et al. 1986). In many of these experiments, adrenalectomy was found to decrease the intensity of nuclear staining, which could be restored by administration of glucocorticoids to the animal or exposure of the cells to hormone in vitro at 37°C but not at 4°C.

Similarly, immunostaining for mineralocorticoid receptor was observed in both the nucleus and the cytoplasm of cells from rat and human (RUNDLE et al. 1989; KROZOWSKI et al. 1989) or rabbit (LOMBÈS et al. 1990) kidney. However, adrenalectomy or hormone administration has no effect on the nuclear-cytoplasmic ratio. The foregoing observations show that in target cells for glucocorticoids and mineralocorticoids a substantial portion of the untransformed receptor resides in the extranuclear region. Recent in vitro mutagenesis experiments have confirmed that the native glucocorticoid receptor is predominantly cytoplasmic (PICARD and YAMAMOTO 1987). Although conventional immunochemical procedures show native vitamin D receptors to be largely in the nuclei in chick intestine (CLEMENS et al. 1988) and human breast cancer

(BERGER et al. 1987), a novel fixation technique indicates that they actually are cytoplasmic in skin fibroblasts (BARSONY et al. 1990).

Nuclear localization of estrogen, progestin, and glucocorticoid receptors in GH_3 rat pituitary tumor cells was also indicated by cell enucleation (WELSHONS et al. 1984, 1985), either with or without cytochalasin B, a fungal metabolite that disrupts microfilaments and permits nuclear removal while maintaining at least partial integrity of the remaining cytoplasts. Nearly all of the estrogen-binding activity was recovered in the nucleoplasts, with only 5% – 10% found in the cytoplast fraction. Similarly, cytochalasin enucleation of human endometrial carcinoma (HEC-50) cells showed 86% of the estrogen receptor to be in the nucleoplast fraction (GRAVANIS and GURPIDE 1986). In contrast, enucleation of bovine and porcine kidney cells revealed only 30% – 40% of the vitamin D receptors to be in the nucleoplasts unless the cells were first exposed to calcitriol, whereupon the nucleoplasts contained 90% of the receptors (S. WALTERS et al. 1986).

The foregoing observations, involving different experimental approaches, leave little doubt that native receptors for gonadal hormones reside predominantly within the nuclear compartment in target cells. This is consistent with the finding that native estrogen receptors have a weak affinity for DNA in vitro (SKAFAR and NOTIDES 1985) and that native progestin receptors contain a nuclear-binding signal sequence, which, when deleted, results in their being cytoplasmic (GUIOCHON-MANTEL et al. 1989). In most cases, the native receptor is so loosely bound that cell disruption in almost any medium results in its extraction into the cytosol. What is not established is whether the native receptor is confined exclusively to the nucleus or whether, as was suggested earlier (WILLIAMS and GORSKI 1972; P. J. SHERIDAN et al. 1979; M. WALTERS et al. 1981; MARTIN and SHERIDAN 1982), a portion exists in the cytoplasm in equilibrium with a nuclear pool.

Because in most experiments essentially no extranuclear immunostaining is observed in target tissues, and only a small amount of binding capacity is found in the cytoplast fraction of enucleated cells, some investigators have concluded that it has been definitely established that untransformed receptors for estrogens, progestins, and androgens are entirely confined to the nucleus, even though this is not true for other steroid hormones. Unfortunately, none of the experimental techniques that demonstrate substantial nuclear localization of native receptors is capable of proving that they are exclusively nuclear.

Immunocytochemistry and autoradiography have the limitation that they often underestimate cytoplasmic components. Both techniques measure concentrations rather than amounts, and in most target cells for gonadal hormones the cytoplasmic volume greatly exceeds the nuclear volume. For example, in a roughly spherical cell with a diameter twice that of its nucleus, if half of the receptor were in the cytoplasm, the immunostaining observed would be 88% nuclear, and a 25% content of cytoplasmic receptor would appear as less than 5% extranuclear and thus be undetectable. Several studies have shown that, as with glucocorticoid receptor, prior exposure to hormone, in vivo or in vitro, markedly increases the intensity of nuclear staining for estrogen receptor

(MCCLELLAN et al. 1984; SAR and PARIKH 1986; PARIKH et al. 1987). This suggests the presence of receptor that is not seen immunochemically until it is converted to the more tightly bound form that either resists loss during experimental manipulation or is concentrated to detectable levels in the nucleus from a dilute state in the cytoplasm. It is also possible that, if they are associated with other macromolecules in a "cytoplasmic anchor" (PICARD and YAMAMOTO 1987; HUNT 1989), native receptors for some steroids may not be immunoreactive until exposure to hormone disrupts this structure.

The results of enucleation experiments are far from compelling, since their indication of a 90% nuclear localization for native glucocorticoid receptor is clearly incompatible with other experimental findings. Moreover, similar experiments found only a small amount of vitamin D receptor in the nucleoplasts unless the cells were first exposed to hormone. These conflicting results may reflect the fact that the cells used in all these experiments were grown in culture with fetal calf serum, known to contain estrogens, progestins, and glucocorticoids but not calcitriol. Because nucleoplasts often contain significant amounts of cytoplasmic components, enucleation techniques can lead to an erroneous assumption of nuclear localization, as with DNA polymerase α in fetal bovine fibroblasts (BROWN et al. 1981).

With such uncertainties in the quantitative assignment of native receptors to the nuclear compartment, and the fact that much of the glucocorticoid and mineralocorticoid receptor is extranuclear, exclusive nuclear localization of untransformed receptors for gonadal hormones is an open question. One cannot ignore the possibility that there is cytoplasmic receptor in equilibrium with a nuclear pool. Considering the ease with which untransformed receptors for gonadal and adrenal hormones move in and out of isolated nuclei, even in the cold, it seems probable that these proteins are diffusible to some extent under physiological conditions in intact cells, as has been demonstrated for progestin receptor by recent deletion mutation experiments (GUIOCHON-MANTEL et al. 1989). In view of the similarities in the biochemical mode of action and the analogies in receptor structure, it seems unlikely that the adrenal steroids would follow an intracellular pathway different from that of other steroid hormones.

In addition to the cytosolic receptor proteins that become tightly bound in the nucleus under the influence of hormone, there are entities in plasma and microsomal membranes of target cells that strongly bind steroid (earlier references in PIETRAS and SZEGO 1979; HAUKKAMAA 1987). More recently, androgen-binding proteins have been identified in prostate microsomes (STEINSAPIR et al. 1985, 1989), and the estrogen-binding entities of uterine microsomes have been studied further (WATSON and MULDOON 1985; LABATE et al. 1986; MULDOON et al. 1988). These substances show some similarity to cytosol receptors but differences in stability and sedimentation properties. Although the microsomal androgen binder is depleted in vivo as nuclear complex is formed (WATSON and MULDOON 1985), neither microsomal substance is transformed to a DNA-binding state, and, unlike the cytosol receptor, the uterine microsomal protein recognizes progesterone as well as estrogen (MULDOON et al. 1988). It has been proposed that the microsomal entities may represent intermediate stages of receptor synthesis (LITTLE et al. 1975), or that they may provide a conduit for hormone traversing the cell on its way to the nucleus (MULDOON et al. 1988).

The estrophilic components observed in plasma membranes of rat hepato-cytes (PIETRAS and SZEGO 1980) and of pig endometrial epithelium (SIERRALTA et al. 1987) show properties similar to those of cytosolic and nuclear receptor proteins. Recent studies with fluorescent estradiol conjugate have demonstrat-ed what appear to be high affinity, low capacity binding sites for estrogen in plasma membranes of hormone-sensitive but not of hormone-insensitive hu-man breast cancer cells (BERTHOIS et al. 1986). Progestin interaction with the plasma membrane of frog oocytes is indicated by photoaffinity labeling of a binding substance in the membrane (SADLER and MALLER 1982). However, in immunocytochemical studies with the electron microscope, monoclonal anti-bodies to classic estrogen or progestin receptors failed to detect antigen in the plasma membranes of human (PRESS et al. 1985) or rabbit (PERROT-APPLANAT et al. 1986) uterine cells. A binding substance for mineralocorticoids has been observed in the plasma membrane of rat kidney cells (FORTE 1972), but its affinity is too weak to be of physiological significance.

Although the principal biological effects of steroid hormones appear to result from the regulation of transcription in target genes, it should be pointed out that there may be other actions that do not involve the genome (for review see LIAO and HIIPAKKA 1984). In particular, certain very rapid effects of estrogens in the uterus, such as hyperemia and water imbibition, are not inhibited by actinomycin D. They may involve the release of histamine (SPA-ZIANI and SZEGO 1959; SZEGO 1965), possibly from uterine eosinophils, which have been shown to concentrate estradiol (TCHERNITCHIN 1973, 1979). Proges-terone has been shown to alter surface membrane properties in frog oocytes (ROBINSON 1979) and to induce oocyte maturation by a process that does not involve its entry into the cell (SCHUETZ 1974; ISHIKAWA et al. 1977; GODEAU et al. 1978).

3.3 Current Status of the "Two-Step" Mechanism

With the recognition that much of the native receptor for steroid hormones is within the nucleus, it was obvious that some revision was needed in the original concept that the untransformed or "cytosol" receptor is an extranuclear entity until interaction with the hormone converts it to the nuclear-bound form. However, the rush to proclaim that the two-step mechanism is no longer valid and that one must search for a new model for steroid hormone action (MARTIN and SHERIDAN 1982; GORSKI et al. 1984, 1986; ENNIS et al. 1986; GORSKI and HANSEN 1987) reflects a misunderstanding of what is meant by "two-step mechanism." This term was originally proposed (JENSEN et al. 1968) to indicate that the 5 S estradiol-receptor complex, found in target cell nuclei after hor-mone administration in vivo, is not generated directly but is derived from an initially formed complex of estradiol with the receptor protein identified in the cytosol (TOFT and GORSKI 1966). When it was found that the 4 S binding unit of the cytosol receptor actually becomes the 5 S nuclear receptor under the influence of estradiol and that only this transformed complex has the ability to bind to chromatin and influence the transcriptional activity of target cell nuclei, the two-step concept took on additional meaning. In this refinement it came to denote the hormone-induced conversion of a native receptor protein to its biochemically functional form that can then interact with target genes (JENSEN and DESOMBRE 1973).

Despite repeated emphasis on receptor transformation followed by genomic binding as the key steps in steroid hormone action, some investigators have regarded translocation from cytoplasm to nucleus as the essence of the two-step mechanism, which it never was. Translocation is a secondary feature in that the transformed receptor complex must assume its genomic binding site from wherever it was originally located. If it is entirely within the nucleus, translocation will be only intranuclear, but if it is partly in the cytoplasm, intracellular translocation will also take place. These are important aspects, but they are details rather than differences in basic mechanism.

Thus, the question of where the untransformed receptor is actually located has no direct bearing on the validity of the two-step mechanism. If there is a cytoplasmic-nuclear equilibrium of receptor distribution, as there appears to be with glucocorticoid receptors, the only refinement needed in the original concept is that the amount of cytoplasmic receptor is less than previously considered but is continually replenished from a nuclear pool (Fig. 3a). If it can be established that native receptors for gonadal hormones are entirely confined to the nucleus, one must explain how the steroid makes its way so rapidly in vivo from the blood transport proteins to the cell nucleus. But once the hormone is in the nucleus, the two-step mechanism of receptor transformation followed by genomic binding would function as previously conceived (Fig. 3b).

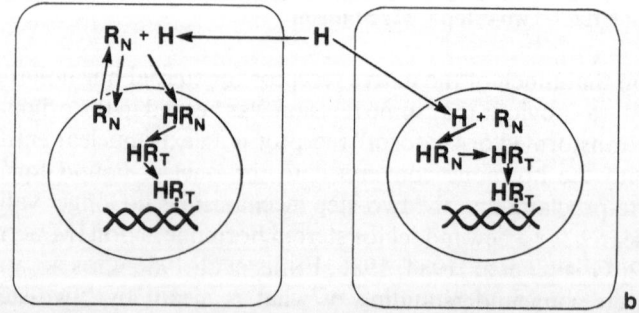

Fig. 3a, b. Schematic representation of hormone-receptor interaction in target cell. H, hormone; R_N, native (untransformed) receptor; R_T, transformed receptor. **a** Extranuclear receptor in equilibrium with nuclear pool of loosely bound native receptor. **b** Native receptor confined to the nucleus

On the basis of considerations here and elsewhere (C. CLARK 1984; S. WALTERS et al. 1986; RINGOLD 1985; GRAVANIS and GURPIDE 1986; PARIKH et al. 1987; GUSTAFSSON et al. 1987), it appears that the iconoclastic fervor concerning the two-step mechanism has been somewhat excessive and presents a distorted impression of the actual situation. Neither of the pathways shown in Fig. 3 represents more than a modification of the original concept. As has been aptly stated, "if steroid receptors are indeed weakly held non-histone nuclear proteins, . . . the change [for the 'two-step' hypothesis] may be more cosmetic than substantive" (SCHRADER 1984).

4 Receptor Phosphorylation

As has been reviewed recently (AURICCHIO 1989), the steroid-binding units of estrogen, progestin, glucocorticoid, and vitamin D receptors, as well as the associated heat shock proteins, are phosphoproteins. The first suggestion that receptors might be phosphorylated came from observations that the ability of cultured thymus cells to bind glucocorticoids varies in proportion to their ATP levels. Cells grown in the absence of oxygen or glucose contain little ATP, and such cells (MUNCK and BRINCK-JOHNSEN 1968), as well as cytosol fractions of their homogenates (BELL and MUNCK 1973), show depressed binding of cortisol, which can be restored by growing the cells aerobically. Inhibition of ATP synthesis by 2,4-dinitrophenol likewise reduces hormone binding (SLOMAN and BELL 1976). These findings, as well as studies with mouse fibroblasts (ISHII et al. 1972), suggested that the hormone-binding substance in the cytosol is continually being generated by an energy-requiring process not involving protein synthesis. This led to the concept of receptor cycling, in which an inert form of the receptor is produced from transformed receptor in the nucleus; it then undergoes energy-dependent reactivation and is recycled back into the receptor system (ISHII et al. 1972; MUNCK et al. 1972; MUNCK and HOLBROOK 1984). Such inactive or "null" receptors were later detected in nuclei of ATP-depleted mouse thymoma cells (MENDEL et al. 1986 b). Similar inhibition of DHT binding in prostatic tissue in vitro by 2,4-dinitrophenol, cyanide, or azide (LIAO and FANG 1969) and enhancement of this binding in prostatic cytosol by ATP or GTP (LIAO et al. 1975) suggested that phosphorylation was involved in androgen receptor function with the possibility of a recycling process (LIAO et al. 1980).

Further evidence for a role of receptor phosphorylation in steroid hormone binding came from several observations. The loss in hormone-binding capacity for glucocorticoids (NIELSEN et al. 1977 a; LEACH et al. 1979), estrogens (AURICCHIO and MIGLIACIO 1980; AURICCHIO et al. 1981 a), and progestins (TOFT and NISHIGORI 1979; GRODY et al. 1980) that occurs on warming homogenates or extracts of target cells is prevented or retarded by phosphatase inhibitors such as molybdate, fluoride, glucose-1-phosphate, or 4-nitrophenyl phosphate. Receptor inactivation occurs on treatment of glucocorticoid receptor with purified alkaline phosphatase (NIELSEN et al. 1977 b). After depletion by warming, hormone-binding ability in mouse fibroblasts (SANDO et al. 1979 a) or rat thymocytes (SANDO et al. 1979 b) is restored by ATP, while the addition of ATP to preincubated cytosol-nuclei mixtures from mouse and calf uterus (AURICCHIO et al. 1981 b) or of ATP, GTP, or cGMP to human endometrial cells, homogenates, or cytosols (FLEMING et al. 1982, 1983) increases the estrogen-binding ability of the receptor.

The foregoing indications for receptor phosphorylation, though suggestive, are still indirect. Early attempts to demonstrate directly that receptors are phosphorylated in target cells were limited by the difficulty in separating receptor from the other phosphorylated proteins present in larger quantities and by endogenous phosphatase and kinase activities. By using purified receptor and an exogenous cAMP-dependent protein kinase, it was first demonstrated that both the A and the B form of the progestin receptor from hen oviduct can undergo phosphorylation with labeled ATP in vitro (WEIGEL et al. 1981). Soon thereafter, purified estrogen receptor from calf uterus, after inactivation with endogenous phosphatase, was shown to be phosphorylated by labeled ATP and an endogenous calcium-stimulated kinase (MIGLIACCIO et al. 1982, 1984). Purified progestin receptor from chick oviduct was likewise phosphorylated by an endogenous magnesium-dependent kinase (GARCIA et al. 1983, 1986 a) or an exogenous cAMP-dependent kinase (SINGH et al. 1986).

Direct evidence for receptor phosphorylation in target cells was first provided by the incorporation of labeled phosphate into progestin receptor on incubation of chick oviduct mince (DOUGHERTY et al. 1982, 1984) or rabbit uterus (LOGEAT et al. 1985 b) with [^{32}P]orthophosphate. Subsequently, many investigators reported phosphorylation of progestin receptor by incubating minces of chick oviduct or rabbit uterus and cultured chick oviduct or T47D human breast cancer cells with labeled orthophosphate (references in AURICCHIO 1989; JENSEN 1990). A and B forms of the receptor are phosphorylated to approximately the same degree (SULLIVAN et al. 1988 a, b), although the B receptors have been reported to have a unique phosphorylation site in addition to those in common with the A form (PURI and TOFT 1986; P. L. SHERIDAN et al. 1989 a). The non-binding 90-kD component of untransformed progestin receptor also undergoes phosphorylation in the same systems (DOUGHERTY et al. 1982, 1984; PURI et al. 1984; PURI and TOFT 1986).

Phosphorylation of estrogen receptor was similarly effected in rat uterine tissue (MIGLIACCIO et al. 1986; AURICCHIO et al. 1987 a) and mouse Leydig tumor cells (SATO et al. 1987). Glucocorticoid receptor, as well as associated heat shock protein, has been phosphorylated in rat liver in vivo (GRANDICS et al. 1984; SINGH and MOUDGIL 1985) and in cultured mouse fibroblasts (HOUSLEY and PRATT 1983; HOUSLEY et al. 1985; SANCHEZ and PRATT 1986; TIENRUNGROJ et al. 1987 b), thymoma (MENDEL et al. 1986 a, 1987; ORTI et al. 1989 a), pituitary tumor (KOVAČIČ-MILIVOJEVIĆ and VEDECKIS 1986), and human breast epithelium (K. RAO and FOX 1987). Similar phosphorylation of vitamin D receptor has been effected in mouse fibroblasts (PIKE and SLEATOR 1985).

In tissues or whole cells, phosphorylation of progestin, glucocorticoid, and vitamin D receptors takes place entirely on serine (PURI et al. 1984; GARCIA et al. 1986a; K. RAO et al. 1987; HAUSSLER et al. 1988; P. L. SHERIDAN et al. 1989a), as is also the case with the associated heat shock protein. With purified oviduct progesterone receptor in vitro, phosphorylation can be effected on either serine (SINGH et al. 1986) or tyrosine (WOO et al. 1986) by using different exogenous kinases. In contrast, estrogen receptor is reported to phosphorylate on tyrosine in rat uterine tissue and in calf uterine extracts using endogenous kinase (MIGLIACCIO et al. 1984, 1986; AURICCHIO et al. 1987a, b), although recent studies have found estrogen receptors in calf uterus and MC7-7 human breast cancer cells to be phosphorylated on serine (DENTON and NOTIDES 1989). Phosphorylation of glucocorticoid receptor and the associated heat shock protein likewise occurs on serine (HOUSLEY and PRATT 1983), although in human breast epithelial cells 11% of the phosphorylation was found to be on tyrosine (K. RAO and FOX 1987).

Phosphorylation of progestin receptors in whole cells or tissues usually is enhanced by the presence of hormone (LOGEAT et al. 1985b; K. RAO et al. 1987; WEI et al. 1987; SULLIVAN et al. 1988a, b; P. L. SHERIDAN et al. 1988, 1989a), although no hormonal stimulation of phosphorylation was seen with cultured chick oviduct cells (GARCIA et al. 1986b, 1987). In mouse fibroblasts, phosphorylation of vitamin D receptor is hormone dependent (PIKE and SLEATOR 1985; HAUSSLER et al. 1988). Phosphorylation in vitro of a partially purified preparation of calf uterine estrogen receptor, using an endogenous kinase, was likewise found to be increased by estrogen (AURICCHIO et al. 1987b). Glucocorticoid is reported to be required for phosphorylation of its receptor in rat liver cytosol (MILLER-DIENER et al. 1985) but not in whole mouse fibroblasts (SANCHEZ and PRATT 1986). With progestin receptor in rabbit uterus (LOGEAT et al. 1985b) and T47D cancer cells (WEI et al. 1987; P. L. SHERIDAN et al. 1988, 1989a), there appear to be two phosphorylation processes: one is hormone independent and occurs with the untransformed receptor, while the second is progestin stimulated and takes place with the transformed receptor, probably after nuclear binding.

The inactivation of receptors by what appear to be endogenous phosphatases and the restoration of hormone-binding ability by ATP suggest that receptor phosphorylation plays a role in their ability to bind hormone. However, expression of cDNA for progestin receptor of chick oviduct in a bacterial system (EUL et al. 1989) or of glucocorticoid receptor in a reticulocyte lysate system (HOLLENBERG et al. 1985) gives recombinant receptor protein that is fully active in hormone binding without need for posttranslational modification. Moreover, both the A and the B form of progestin receptor are synthesized in a nonphosphorylated condition that is fully active, both in binding hormone and in being converted to the transformed state, which then undergoes secondary phosphorylation (P. L. SHERIDAN et al. 1989b). Thus, the general importance of primary phosphorylation in steroid binding by the native receptor is not entirely clear. It is interesting that recombinant human estrogen receptor (S. GREEN et al. 1986) that contains a single amino acid mutation (TORA et al. 1989) shows impaired estrogen binding at 25 °C (but not at 4 °C), which can be corrected by treatment with ATP and an exogenous kinase (MIGLIACCIO et al. 1989).

Likewise, the participation of phosphorylation in receptor transformation and in nuclear binding is not clear. In contrast to transformed progestin receptor, which undergoes additional phosphorylation in rabbit uterus and T47D cells, there appears to be dephosphorylation of transformed estrogen receptor in the nuclei of mouse and calf uterus and of MCF-7 cancer cells (AURICCHIO et al. 1982; DENTON and NOTIDES 1989). It has been proposed that dephosphorylation of a cytosol component may be involved in destabilization of the native glucocorticoid receptor aggregate (REKER et al. 1987). However, there is no net change (a) in phosphorylation of the glucocorticoid receptor itself as it undergoes transformation in target cells (MENDEL et al. 1987) or (b) in the 90-kD heat shock protein as it associates with native receptor or is separated during transformation in vitro (ORTI et al. 1989 a). The native and transformed glucocorticoid receptors appear to be phosphorylated to the same extent (TIENRUNGROJ et al. 1987 b). However, it has recently been found that, in intact mouse thymoma cells, treatment with hormone increases the average number of phosphates on the untransformed and salt-extractable transformed (nuclear) glucocorticoid receptor from three to five, but about 10% of the nuclear-bound receptor is not salt extractable, and this has only three phosphate groups (ORTI et al. 1989 b). Thus, no common pattern has emerged. In spite of the many observations suggesting that phosphorylation and/or dephosphorylation are somehow involved in the nuclear interaction of steroid hormone receptors, and the fact that receptor transformation is blocked by phosphatase inhibitors, the precise biochemical role for phosphorylation in these processes remains obscure.

5 Antibodies to Receptor Proteins

The original recognition of steroid hormone receptors and the earlier knowledge concerning their properties and their function in target cells depended on the use of a radioactive steroid as a marker for the receptor protein to which it binds. Despite the wealth of information obtained using the ligand-binding approach, it has definite limitations, and there was need for an alternative means of receptor detection to complement the use of steroid-binding techniques. For several years, the attempted preparation of specific antibodies to steroid hormone receptors was remarkably unsuccessful for a number of reasons. In target cells the receptors are present in minute amounts, and in crude tissue extracts they are rather labile; before the advent of molybdate stabilization, they readily lost their ability to bind the steroid needed as a marker during purification and concentration to provide an effective immunogen. Moreover, antibodies to steroid hormone receptors generally form nonprecipitating immune complexes, so that most conventional techniques of immuno-

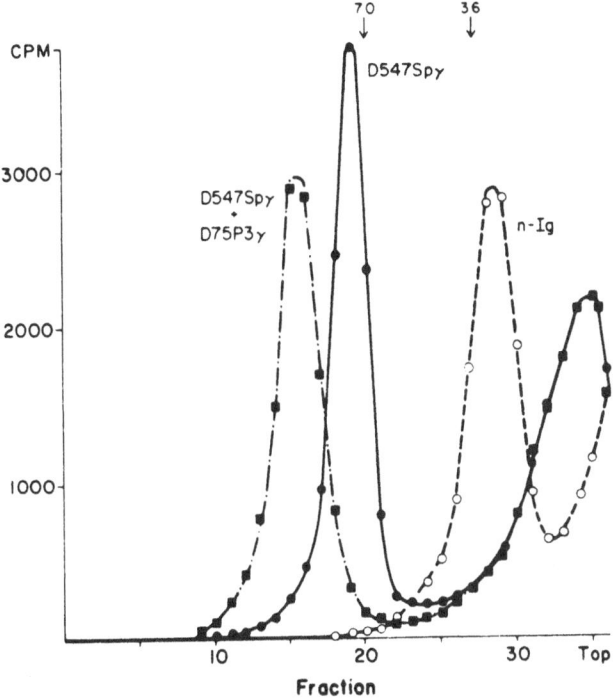

Fig. 4. Use of sedimentation shift to demonstrate reaction of monoclonal antibodies with different epitopes in estrogen receptor. Sedimentation profile of 0.5 pmol tritiated estradiol-receptor complex of MCF-7 breast cancer cell cytosol in sucrose gradients containing 400 mM KCl in the presence of 50 µg control (n-Ig) rat immunoglobulin (o), 40 µg monoclonal antibody D547 (●), or 40 µg D547 followed by 50 µg monoclonal antibody D75 (■). *Arrows* indicate positions of bovine γ-globulin (7.0 S) and ovalbumin (3.6) markers. (GREENE and JENSEN 1982)

chemistry were ineffective in their detection. Fortunately, most antibody preparations that have been described do not interfere with the binding of hormone to its receptor, so the labeled steroid can be used as a marker for the receptor protein in the soluble immune complex. The ability to increase the sedimentation rate of the tritiated hormone-receptor complex (Fig. 4), first used to detect antibodies to estrogen receptor, has provided a valuable method of identifying nonprecipitating immune complexes with other classes of steroid receptors.

5.1 Estrogen Receptors

Although there had been earlier suggestions of antisera to estrogen receptor, it was not until 1977 that definitive antibodies were obtained by immunizing rabbits (GREENE et al. 1977) and later a goat (GREENE et al. 1979) with partially purified transformed receptor from calf uterus. The increase in sedimentation

rate of the tritiated estradiol-receptor complex was used as the principal criterion of its association with one or in some instances two molecules of immunoglobulin (cf. Fig. 4). The polyclonal antibodies against calf uterus were
cross-reactive with estrogen receptor from other mammalian species as well as
from chick oviduct but not with receptors for other steroid hormones. Cross-
reacting polyclonal antibodies were soon reported for estrogen receptor from
calf uterus (RADANYI et al. 1979), human myometrium (COFFER et al. 1980),
and human breast cancer (RAAM et al. 1981). Recently, site-directed polyclonal
antibodies to estrogen receptor were obtained by immunizing rabbits with
synthetic peptides corresponding to sequences in the DNA-binding domain of
the receptor protein (TRAISH et al. 1989); these antibodies recognize native and
transformed estrogen receptor from rat and calf uterus and from human breast
cancer.

The first monoclonal antibodies were obtained with transformed estrogen
receptor of calf uterus and were species specific, recognizing only calf receptor
(GREENE et al. 1980a). One of these antibodies inhibited certain of the criteria
of receptor transformation (FAUQUE et al. 1985, 1989). In contrast, monoclonal antibodies to cytosol receptor from MCF-7 human breast cancer cells were
cross-reacting (GREENE et al. 1980b), as were those obtained with untransformed calf uterine receptor (MONCHARMONT et al. 1982). Of 13 different
monoclonal antibodies, each reacting with a different epitope in the human
estrogen receptor (Fig. 4), eight were found to react with receptor from every
source tested, including avian; four reacted only with mammalian receptor
(rat, calf, monkey, human), and one was specific for receptor from primates
(GREENE et al. 1984). This indicates that some but not all regions of the receptor molecule are conserved over a variety of species. However, these antibodies
do not cross-react with receptors for other steroid hormones. Though they do
not prevent the binding of hormone, six of nine monclonal antibodies tested
recognize epitopes near the hormone-binding site. All polyclonal and monoclonal antibodies recognize both native and transformed receptor, although
the species-specific monoclonal antibodies to calf uterine receptor react much
more strongly with transformed receptor (GREENE et al. 1980a; BORGNA et al.
1984), the form used as immunogen.

Certain of the monoclonal antibodies have been used to: purify estrogen
receptors by immunoaffinity methods; assay receptors in human cancers
(GREENE et al. 1984; W. KING et al. 1985; additional references in JENSEN 1990);
measure occupied receptors in cell nuclei as an alternative to exchange techniques (MONCHARMONT and PARIKH 1983); characterize different domains in
the receptor molecule; localize untransformed receptors in the nucleus as previously described; and clone the cDNA for the human estrogen receptor as
discussed in Sect. 6.

5.2 Progestin Receptors

As methods were developed for receptor stabilization and purification, polyclonal antibodies were prepared to progestin receptor from rabbit (LOGEAT

et al. 1981) and guinea pig (FEIL 1983) uterus and from chick oviduct (RENOIR et al. 1982), as well as to purified A and B forms of the oviduct receptor (TUOHIMAA et al. 1984; GRONEMEYER et al. 1985; BIRNBAUMER et al. 1987). These antibodies react with progestin receptors from other species tested, with the exception of the antibody to rabbit uterine receptor, which recognizes receptor from other mammalian but not from avian species. One preparation, raised against the purified B form of receptor from chick oviduct, reacted only weakly with mammalian receptor; it also gave weak cross-reaction with estrogen receptor from chick oviduct (TUOHIMAA et al. 1984). All the other antibodies proved specific for progestin receptor, and whether raised against the A or the B form of the avian receptor, they recognize both the A and B proteins. Recently site-directed polyclonal antibodies, which react only with transformed receptor, have been raised against a synthetic peptide with the structure of a sequence in the DNA-binding domain (SMITH et al. 1988).

The first monoclonal antibody obtained with avian progestin receptors (RADANYI et al. 1983) turned out to be to the non-steroid-binding component of the untransformed receptor (JOAB et al. 1984). This provided a valuable reagent for demonstrating that this entity is a 90-kD heat shock protein. The antibody does not recognize transformed avian progestin receptors or untransformed human receptors. It cross-reacts with untransformed but not with transformed receptors for estrogens, androgens, and glucocorticoids from avian tissues, indicating the similarity of the heat shock protein in native receptors for different steroid hormones. Subsequently, monoclonal antibodies were prepared directly to the purified 90-kD protein from chick oviduct receptor, two of which recognize the intact native receptor and one of which was found to cross-react with avian glucocorticoid and androgen receptors (SULLIVAN et al. 1985). A monoclonal antibody to partially purified rabbit progestin receptor (NAKAO et al. 1985) was found to recognize a 59-kD protein, common to native progestin, estrogen, androgen, and glucocorticoid receptor from rabbits, but it did not react with progestin receptor from rat, guinea pig, or chick (TAI et al. 1986).

A large number of monoclonal antibodies have been prepared against purified progestin receptor from rabbit uterus (LOGEAT et al. 1983; LORENZO et al. 1988), hen or chick oviduct (DICKER et al. 1984; SULLIVAN et al. 1986), human endometrial cancer (CLARKE et al. 1987), and T47D human breast cancer cells (ESTES et al. 1987; GREENE et al. 1988). In two instances the purified B form was used as the immunogen. As with polyclonal antibodies, monoclonal antibodies raised against rabbit receptor did not react with avian receptor but recognized progestin receptor from several mammalian species (LOGEAT et al. 1983). In contrast to antibodies to rabbit receptor that do not recognize avian receptor, those raised against avian receptor cross-react with that from mammalian tissues, although one monoclonal was obtained that reacted with rabbit but not human receptor (SULLIVAN et al. 1986). Of the many monoclonal antibodies raised against human progestin receptor, some were found to be specific for human receptor, whereas others recognized that from rabbit and, in one instance, from chick as well (ESTES et al. 1987; GREENE et al. 1988). Most monoclonal preparations react with both the A and the B form of the avian or human receptor, but some have been obtained that recognize only the B protein (SULLIVAN et al. 1986; CLARKE et al. 1987; ESTES et al. 1987; GREENE et al. 1988). In no instance did any antibody recognize the A form but not the B form, supporting the view that the two proteins are similar except for an additional moiety in the B form.

In addition to their use in characterizing different components of the receptor protein, antibodies to progestin receptor have been valuable in establishing the nuclear localization of the native receptor, in the immunoassay of progestin receptors in human cancers, and for the cloning of the cDNA for progestin receptor and determining its detailed structure, as described in Sect. 6.

5.3 Androgen Receptors

Because of the special difficulties in purifying androgen receptor in sufficient quantity to serve as an immunogen, the first antibodies to this protein were autoantibodies, obtained from the blood of patients with prostatic disease (LIAO and WITTE 1985). These were detected by their effect on the sedimentation rate of the tritiated androgen-receptor complex from rat prostate; they recognize nuclear and cytosolic androgen receptor but not receptors for estrogen, progestin, or glucocorticoid. By immortalizing lymphocytes from the blood of similar patients, cell lines were developed that produced two monoclonal antibodies that react with androgen receptor from various mammalian species (YOUNG et al. 1988). After the primary structure of the human androgen receptor was determined, it was possible to use peptides corresponding to sequences in the receptor to obtain polyclonal antibodies in the rabbit (LUBAHN et al. 1988; CHANG et al. 1989 b), as well as monoclonal antibodies, using spleen cells from immunized rats (CHANG et al. 1989 b). Recently, purified androgen receptor from human prostate was used to immunize mice and generate a monoclonal antibody, detected by double antibody precipitation, that partially inhibits binding of hormone to the receptor (DEMURA et al. 1989).

The autoantibodies from human sources proved useful in the original cloning of the androgen receptor, and the subsequent preparations have provided means for its immunocytochemical detection and molecular characterization.

5.4 Glucocorticoid Receptors

The first antisera, obtained by immunizing rabbits with partially purified cytosol protein of rat liver (LITWACK et al. 1973; GOVINDAN and SEKERIS 1978; GOVINDAN 1979) or thymocytes (PAPAMICHAIL et al. 1980), were described as precipitating antibodies. In view of the small quantities of steroid receptor present in target tissues, the formation of visible precipitation bands in Ouchterlony gels raises a question about the specificity of these reagents. Well-characterized polyclonal antibodies were later obtained in rabbits with antigen from rat liver (EISEN 1980; OKRET et al. 1981; OKRET 1983; GOVINDAN and GRONEMEYER 1984; BERNARD and JOH 1984) or human lymphoblastoid cells (HARMON et al. 1984), using sedimentation shift, gel filtration, or binding to protein A to identify the soluble immune complexes formed. Although one preparation was specific for rat receptor (EISEN 1980), the others recognized glucocorticoid receptor from at least some other species, but not receptors for other steroid hormones. One preparation reacted only with transformed receptor (BERNARD and JOH 1984), while the others recognized both nuclear and cytosolic forms. More recently, site-directed polyclonal antibodies have been prepared by immunizing rabbits with synthetic peptides corresponding to specific regions of the receptor molecule (HOLLENBERG et al. 1987; WILSON et al. 1988; URDA et al. 1989).

A number of monoclonal antibodies have been prepared to glucocorticoid receptor from rat liver cytosol, either to the native form purified by affinity chromatography (GRANDICS et al. 1982) or to the transformed receptor, purified by elution from DNA-cellulose (WESTPHAL et al. 1982; OKRET et al. 1984; GAMETCHU and HARRISON 1984; TEASDALE et al. 1986). Some monoclonal antibodies (WESTPHAL et al. 1982; OKRET et al. 1984), like many polyclonal antibodies (OKRET et al. 1981; CARLSTEDT-DUKE et al. 1982), recognize only that region of the receptor molecule not involved in steroid or DNA binding (designated C in Fig. 2 and A/B in Fig. 5), but other monoclonal antibodies (TEASDALE et al. 1986) react with the 40-kD proteolytic fragment that lacks this region (WRANGE and GUSTAFSON 1978). Though cross-reactivity varied, most of the monoclonal preparations recognized glucocorticoid receptor from mouse but not from human sources. One monoclonal antibody was unique in cross-reacting with progestin receptor (TEASDALE et al. 1986). A monoclonal antibody recognizing glucocorticoid receptor was also obtained (CAYANIS et al. 1986) by an auto-anti-idiotypic route (JERNE 1974) in which mice are immunized with the steroid coupled to a protein and the hybridomas obtained are screened for the presence of antibodies recognizing the combining site (Fab fragments) of antibodies raised in rabbits against the steroid coupled to a different carrier (cf. GAULTON and GREENE 1986). It recognizes the steroid-binding domain as shown by competition with steroids known to bind to the receptor.

Antibodies to glucocorticoid receptors have proved valuable in detecting receptor in specific cells and tissues, in identifying different domains of the receptor molecule, in cloning the cDNA for the receptor protein and determining its structure, and in studying the intracellular localization of the receptor before and after exposure to hormone.

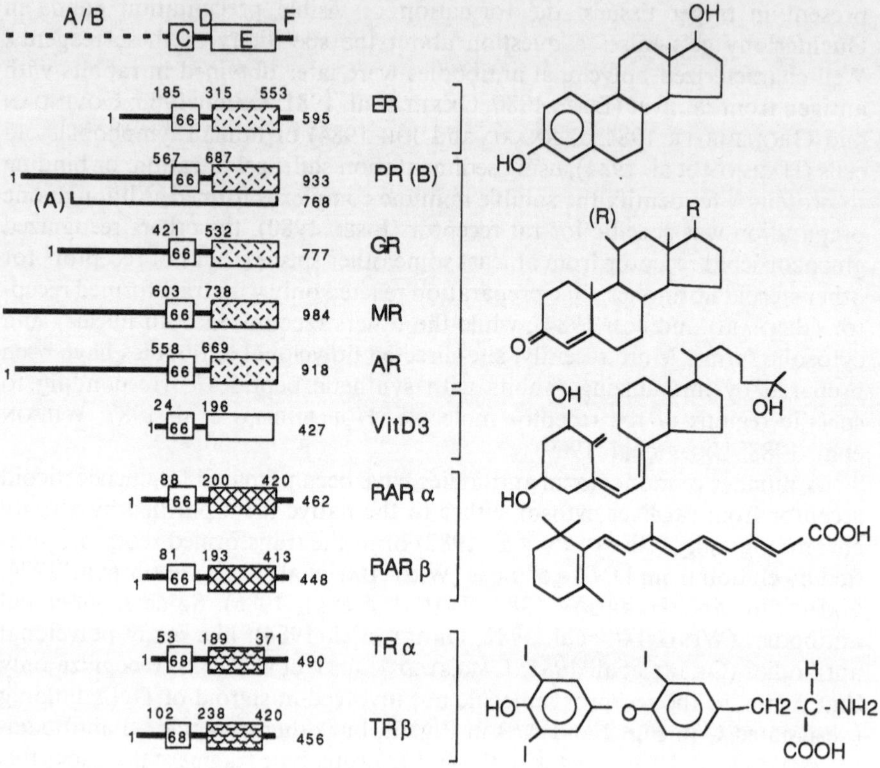

Fig. 5. The human intracellular receptor family and their ligands. The *top diagram* shows the six functional domains: A/B, F, modulating regions; C, DNA-binding region; D, "hinge" region; E, ligand-binding region. *Boxes* indicate highly conserved domains; *thin black lines* are regions of low homology. The position of each domain boundary is given as the number of amino acids from the amino terminus (*1*). Figures in the *C boxes* indicate the number of amino acids in the DNA-binding domain. Receptors are: *ER*, estrogen; *PR*, progestin [(R)=H, R=CH₃CO]; *GR*, glucocorticoid [(R)=OH, R=CH₂OHCO, OH]; *MR*, mineralocorticoid [(R)=OH, R=CH₂OHCO, plus=O at C_{18}]; *AR*, androgen [(R)=H, R=OH]; *VitD₃*, vitamin D; *RAR*, retinoic acid; *TR*, thyroid hormone. It should be noted that androgen receptors in male reproductive tissues react preferentially with DHT, the saturated analog of the AR ligand shown. (Courtesy of Professor PIERRE CHAMBON)

5.5 Mineralocorticoid, Vitamin D, and Ecdysteroid Receptors

Because of the unavailability of purified antigen, polyclonal antibodies to mineralocorticoid receptor have been obtained only recently. After the primary structure was deduced from cloning of the cDNA, polyclonal antibodies were obtained by immunizing rabbits with synthetic peptides corresponding to specific sequences in the hinge region of the molecule (KROZOWSKI et al. 1989; RUNDLE et al. 1989). These antibodies recognize only unoccupied receptor. By the auto-anti-idiotypic approach described in Sect. 5.4, a monoclonal antibody has been obtained that displaces aldosterone and corticosterone from the rabbit receptor (LOMBÈS et al. 1989, 1990). Antibodies recognizing mineralo-corticoid receptor have proved useful in detecting the receptor in kidney cells and in determining its intracellular localization.

Monoclonal antibodies have been prepared to vitamin D receptor from chick (PIKE et al. 1982, 1983) and pig (DAME et al. 1985) intestine. The poly-clonal antibodies in the serum of the immunized rats used for hybridoma production were also characterized (PIKE et al. 1983). Some but not all of these antibodies show cross-reactivity with vitamin D receptor from various tissues of the rat, pig, and human, but do not react with estrogen or glucocorticoid receptors or binding proteins of serum. They recognize nuclear complex, as well as occupied or unoccupied cytosol receptor, and certain monoclonals markedly inhibit the binding of transformed receptor to DNA (PIKE 1984). These antibodies have been used for the immunopurification of receptors, their characterization by immunoblotting, their detection in various tissues, and their intracellular localization by immunocytochemistry.

Although their description had not appeared in the literature by the time this chapter was completed, antibodies to ecdysteroid receptor from *Drosophila melanogaster* have been prepared and used in cloning the cDNA for this protein (D. S. HOGNESS, private communication).

6 Cloning and Structure of Receptors

With the availability of purified receptor proteins for steroid hormones and of specific antibodies that react with them, it became possible to employ tech-niques of molecular biology to clone and express the complementary DNA coding for biosynthesis. From the nucleotide sequences of the cDNAs the amino acid sequences of the receptor proteins were deduced. Expression cloning has been carried out for glucocorticoid receptors from the human (HOLLENBERG et al. 1985; GOVINDAN et al. 1985), rat (MIESFELD et al. 1986), and mouse (DANIELSEN et al. 1986) and for estrogen receptors from human breast cancer (S. GREEN et al. 1986; GREENE et al. 1986), chick oviduct (KRUST et al. 1986; MAXWELL et al. 1987), and frog liver (WEILER et al. 1987). The recombinant protein obtained with the human estrogen receptor was later shown to differ from the natural protein by a single amino acid (TORA et al. 1989).

In the same fashion, the cDNAs for progestin receptor from rabbit (LOOS-FELT et al. 1986), chicken (CONNEELY et al. 1987a; GRONEMEYER et al. 1987), and human (MISRAHI et al. 1987) were cloned and sequenced, as were those for human mineralocorticoid receptor (ARRIZA et al. 1987), human (MCDONNELL et al. 1987) and avian (BAKER et al. 1988) vitamin D receptor, and human and rat androgen receptor (CHANG et al. 1988; LUBAHN et al. 1988; TAN et al. 1988). Although published reports have not yet appeared, cDNA for the ecdysteroid receptor has also been cloned and sequenced (D. S HOGNESS, private commu-nication).

Table 4. Molecular weights of receptor proteins calculated from amino acid composition (kD)

Receptor	Human	Avian
Estrogen	66	66
Progestin (B)	99	86
(A)	81	72
Androgen	99	
Glucocorticoid	94	
Mineralocorticoid	107	
Vitamin D	47.5	55

From the nucleotide sequences of the corresponding cDNAs, the amino acid sequences of the receptor proteins have been deduced. Though they vary in length from 427 to 984 amino acids (Fig. 5), corresponding to the molecular weights listed in Table 4, they show striking homology in certain regions of the molecule and represent a family of gene regulatory agents (S. GREEN and CHAMBON 1986; EVANS 1988). Of considerable interest are the unexpected findings of a relationship of the glucocorticoid (WEINBERGER et al. 1985) and estrogen (S. GREEN et al. 1986) receptors to the viral oncogene erbA and the subsequent demonstration that the c-erbA protein actually is the thyroid hormone receptor (WEINBERGER et al. 1986; SAP et al. 1986).

With the aid of deletion mutants, it was possible to identify individual domains in the receptor molecule and to correlate each with an aspect of receptor function (WEINBERGER et al. 1985; KUMAR et al. 1987; DOBSON et al. 1989). As shown in Fig. 5, each receptor appears to have a steroid-binding region (E) near the C-terminal end, specific for that class of hormone, a cysteine-rich DNA-binding region (C) that is highly conserved, a small "hinge" region (D) joining these two domains, which can be deleted without much effect on function, and a large variable region (A/B) near the N-terminal end that is not essential to function but contributes to maximum activity. Most of the antibodies that have been obtained with progestin (LORENZO et al. 1988) and glucocorticoid (CARLSTEDT-DUKE et al. 1982; WESTPHAL et al. 1982) receptors appear to be directed against the A/B region, in contrast to estrogen receptor, in which most monoclonal antibodies recognize the E region, even though they do not interfere with steroid binding (GREENE et al. 1984).

7 Interaction with Target Genes

When it was recognized that hormone-induced transformation of steroid-receptor complexes causes them to bind to chromatin in the nucleus, attempts were made to identify the so-called acceptor sites to which the transformed receptors become attached (for review see JENSEN 1979). Although some evidence for selective nuclear binding has been reported, it has proved difficult to distinguish a selective interaction with target genes from the large amount of nonspecific binding to DNA in general. With the application of molecular biological approaches to receptor studies and the identification of specific target genes, recombinant systems were constructed in which the effect of steroid-receptor complexes on gene expression could be investigated in greater detail. Although a complete description of the various gene systems that have been studied is beyond the scope of this chapter, some of the general patterns and concepts that have emerged may be mentioned (for reviews see RINGOLD 1985; YAMAMOTO 1985; S. GREEN and CHAMBON 1988; BEATO 1989; BERG 1989; HAM and PARKER 1989; GODOWSKI and PICARD 1989; SAVOURET et al. 1989).

It is now clear that the effect of transformed hormone-receptor complexes on gene expression does not involve interaction at the site of transcription but rather with a "hormone response element" (HRE), an enhancer located in the promoter area in the 5'-flanking region of the target gene. Estrogen receptor was found to bind to the coding strand of DNA in the response element with a 60-fold higher affinity than it does to double-stranded DNA (LANNIGAN and NOTIDES 1989). Palindromic sequences of 13–15 base pairs appear to make up the actual receptor-binding site (STRÄHLE et al. 1987; KLEIN-HITPASS et al. 1988). The hormone response elements for glucocorticoids and progestins appear to be similar in composition and to interact in some degree with both receptors (VON DER AHE et al. 1985), whereas that for estrogens, although related, is somewhat different (KLOCK et al. 1987). Glucocorticoid and mineralocorticoid receptors show close homology in their DNA-binding regions and seem to react with the same gene networks (EVANS 1989). Recognition of the response element appears to reside in the DNA-binding domain of the receptor (region C, Fig. 5), since the chimeric protein formed by replacing this region in the human estrogen receptor with the corresponding region of the human glucocorticoid receptor activates expression of a corticoid-inducible but not estrogen-inducible gene in the presence of estrogen (S. GREEN and CHAMBON 1987).

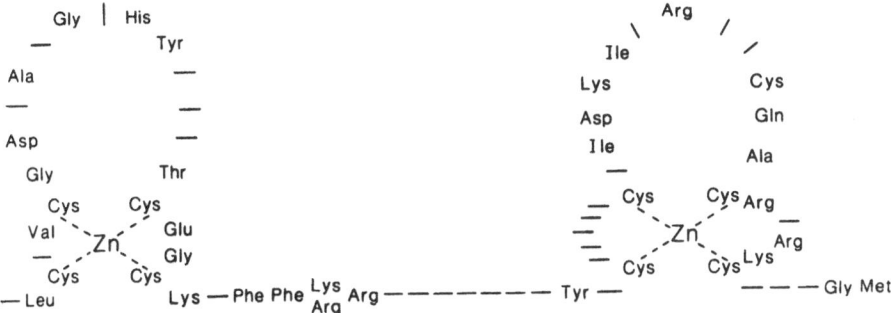

Fig. 6. Hypothetical structure of the DNA-binding domain of steroid hormone receptors with two putative zinc-binding fingers. Each zinc ion forms a tetrahedral coordination complex with cysteine residues. (EVANS 1988)

The precise nature of the interaction of steroid receptors with response elements is currently a subject of extensive study, but many details are still obscure. In intact cells or tissues, the hormone is needed not only to transform the glucocorticoid receptor but also to effect its binding to enhancer elements (BECKER et al. 1986). A similar hormone requirement for DNA binding of the progestin receptor was observed in a cell-free system (BAGCHI et al. 1988). However, after activation by heat, salt treatment or purification, steroid-free receptors for progestins (BAILLY et al. 1986) and glucocorticoids (WILLMANN and BEATO 1986) not only bind to to their specific response elements, but salt-activated progestin receptor stimulates transcription in an in vitro system in the absence of steroid (KLEIN-HITPASS et al. 1990). Moreover, certain mutant receptors for glucocorticoids (GODOWSKI et al. 1987; HOLLENBERG et al. 1987; DANIELSEN et al. 1987) and estrogens (WATERMAN et al. 1988), which lack the steroid-binding domain, have a constitutive ability to interact with DNA in the absence of hormone and stimulate transcription in vitro. It appears that the steroid-binding region somehow prevents the domains for DNA binding and transcriptional activation from functioning and that the presence of hormone relieves this inhibition (EVANS 1988).

Although it has been reported that a single domain in the estrogen receptor is responsible for both DNA binding and activation of the rat prolactin gene (WATERMAN et al. 1988), two gene activation domains, one in the carboxy terminus and the other in the amino terminus, have been identified in the mouse estrogen (LEES et al. 1989) and human glucocorticoid (HOLLENBERG and EVANS 1988) receptors. In the estrogen receptor, the domain in the C-terminus was found to depend on estrogen for its activity, whereas that in the N-terminus is active in the absence of hormone.

The interaction of the receptor with its response element appears to be mediated through a pair of "zinc fingers", located in the highly conserved C region of all classes of receptors (EVANS and HOLLENBERG 1988; FREEDMAN et al. 1988; SEVERNE et al. 1988; S. GREEN et al. 1988; DANIELSEN et al. 1989). Such structures, involving zinc coordinated with appropriately located cysteine or histidine residues (Fig. 6), have provided a general pattern for binding to DNA for a large number of regulatory proteins (KLUG and RHODES 1987). It has been shown that the binding of estrogen receptor to DNA-cellulose depends on the presence of receptor-bound metal that can be chelated by 1,10-phenanthrolene (SABBAH et al. 1987). Even though they do not dimerize very avidly on transformation, progestin and glucocorticoid receptors, like estrogen receptor (KUMAR and CHAMBON 1988), appear to interact with their palindromic response elements in a dimeric form (S. TSAI et al. 1988, 1989).

Just how the binding of transformed steroid-receptor complex to the hormone response element enhances transcription in target genes is still not entirely clear. There is evidence that it may involve the participation of other transcription factors (SCHÜLE et al. 1988; MEYER et al. 1989), including the chicken oviduct upstream promoter (COUP), a protein that resembles the steroid hormone receptor family in its structure (WANG et al. 1987, 1989). It has been observed that association with the steroid increases the rate at which the receptor associates with and dissociates from DNA, thereby enhancing its efficiency in finding the proper location in the target gene (SCHAUER et al. 1989). Stimulation of transcription in a cell-free system by activated progestin receptor appears to result from its enhancement of the formation of a stable preinitiation complex at the target gene promoter (KLEIN-HITPASS et al. 1990).

8 Summary

In the three decades since the original discovery of receptors for steroid hormones, much has been learned about the biochemical processes by which these regulatory agents exert their effects in target tissues. The intracellular receptor proteins are potential transcription factors, needed for optimal gene expression in hormone-dependent cells. They are present in an inactive form until association with the hormone converts them to a functional state that can react with target genes. Transformation of the receptor protein to the nuclear binding form appears to involve the removal of both macromolecular and micromolecular factors that act to keep the receptor from reacting with DNA. Much of the native receptor is present in the nucleus, loosely bound and readily extractable, but for some and possibly all steroid hormones, some receptor is in the cytoplasm, perhaps in equilibrium with a nuclear pool.

Methods have been developed for the stabilization, purification, and characterization of receptor proteins, and through cloning and sequencing of their cDNAs, primary structures for these receptors are now known. This has led to the recognition of structural similarities among the family of receptors for the different steroid hormones and to the identification of regions in the protein molecule responsible for the various aspects of their function. Monoclonal antibodies recognizing specific molecular domains are available for most receptors.

Despite the knowledge that has been acquired, many important questions remain unsolved. How does association with the steroid remove factors keeping the receptor protein in its native state, and how does binding of the transformed receptor to the response element in the promoter region enhance gene transcription? Once it has converted the receptor to the nuclear binding state, is there a further role for the steroid in modulating transcription? Still not entirely clear is the involvement of phosphorylation and/or dephosphorylation in hormone binding, receptor transformation, and transcriptional activation. Less vital to basic understanding but important in the overall picture is whether the native receptors for gonadal hormones are entirely confined to the nucleus or whether there is an intracellular distribution equilibrium.

With the effort now being devoted to this field, and with the application of new experimental techniques, especially those of molecular biology, our understanding of receptor function is progressing rapidly. The precise mechanism of steroid hormone action should soon be completely established.

Acknowledgment. Preparation and publication of this chapter was supported by a grant (RDP-53A) from the American Cancer Society.

References

Aakvaag A, Tveter KJ, Unhjem O, Attramadal A (1972) Receptors and binding of androgens in the prostate. J Steroid Biochem 3:375–384

Alberti KGMM, Sharp GWG (1969) Macromolecular binding of aldosterone in the toad bladder. Biochim Biophys Acta 192:335–346

Anderson JN, Peck EJ Jr, Clark JH (1975) Estrogen-induced uterine responses and growth: relationship to receptor estrogen binding by uterine nuclei. Endocrinology 96:160–167

Anderson KM, Liao S (1968) Selective retention of dihydrotestosterone by prostatic nuclei. Nature 219:277–279

Antakly T, Eisen HJ (1984) Immunocytochemical localization of glucocorticoid receptor in target cells. Endocrinology 115:1984–1989

Antakly T, Thompson EB, O'Donnell D (1989) Demonstration of intracellular localization and up-regulation of glucocorticoid receptor by in situ hybridization and immunocytochemistry. Cancer Res 49:2230s–2234s

Antakly T, O'Donnell D, Thompson EB (1990) Immunocytochemical localization of the glucocorticoid receptor in steroid-sensitive and -resistant human leukemic cells. Cancer Res 50:1337–1345

Arbogast LY, DeSombre ER (1975) Estrogen-dependent in vitro stimulation of RNA synthesis in hormone-dependent mammary tumors of the rat. J Natl Cancer Inst 54:483–485

Arriza JL, Weinberger C, Cerelli G, Glaser TM, Handelin BL, Housman DE, Evans RM (1987) Cloning of human mineralocorticoid receptor complementary DNA: structural and functional kinship with the glucocorticoid receptor. Science 237:268–275

Atger M, Milgrom E (1976) Chromatographic separation on phosphocellulose of activated and nonactivated forms of steroid-receptor complex. Purification of the activated complex. Biochemistry 15:4298–4304

Atger M, Milgrom E (1978) Interaction of glucocorticoid-receptor complexes with rat liver nuclei. Biochim Biophys Acta 539:41–53

Auricchio F (1989) Phosphorylation of steroid receptors. J Steroid Biochem 32:613–622

Auricchio F, Migliaccio A (1980) In vitro inactivation of oestrogen receptor by nuclei. FEBS Lett 117:224–226

Auricchio F, Migliaccio A, Rotondi A (1981 a) Inactivation of oestrogen receptor in vitro by nuclear dephosphorylation. Biochem J 194:569–574

Auricchio F, Migliaccio A, Castoria G, Lastoria S, Schiavone E (1981 b) ATP-dependent enzyme activating hormone binding of estradiol receptor. Biochem Biophys Res Commun 101:1171–1178

Auricchio F, Migliaccio A, Castoria G, Lastoria S, Rotondi A (1982) Evidence that in vivo estradiol receptor translocated into nuclei is dephosphorylated and released into cytoplasm. Biochem Biophys Res Commun 106:149–157

Auricchio F, Migliaccio A, Castoria G, Rotondi A, Di Domenico M, Pagano M, Nola E (1987a) Phosphorylation on tyrosine of oestradiol-17β receptor in uterus and interaction of oestradiol-17β and glucocorticoid receptors with antiphosphotyrosine antibodies. J Steroid Biochem 27:245–253

Auricchio F, Migliaccio A, Di Domenico M, Nola E (1987b) Oestradiol stimulates tyrosine phosphorylation and hormone binding activity of its own receptor in a cell-free system. EMBO J 6:2923–2929

Ausiello DA, Sharp GWG (1968) Localization of physiological receptor sites for aldosterone in the bladder of the toad, *Bufo marinus*. Endocrinology 82:1163–1169

Bagchi MK, Elliston JF, Tsai SY, Edwards DP, Tsai M-J, O'Malley BW (1988) Steroid hormone-dependent interaction of human progesterone receptor with its target enhancer element. Mol Endocrinol 2:1221–1229

Bailly A, Sallas N, Milgrom E (1977) A low molecular weight inhibitor of steroid receptor activation. J Biol Chem 252:858–863

Bailly A, Le Fevre B, Savouret J-F, Milgrom E (1980) Activation and changes in sedimentation properties of steroid receptors. J Biol Chem 255:2729–2734

Bailly A, Le Page C, Rauch M, Milgrom E (1986) Sequence-specific DNA binding of the progesterone receptor to the uteroglobin gene: effects of hormone, antihormone and receptor phosphorylation. EMBO J 5:3235–3241

Baker AR, McDonnell DP, Hughes M et al. (1988) Cloning and expression of full length cDNA encoding human vitamin D receptor. Proc Natl Acad Sci USA 85:3294–3298

Ballard PL, Ballard RA (1972) Glucocorticoid receptors and the role of glucocorticoids in fetal lung development. Proc Natl Acad Sci USA 69:2668–2672

Bardin CW, Catterall JF (1981) Testosterone: a major determinant of extragenital sexual dimorphism. Science 211:1285–1294

Barsony J, Pike JW, DeLuca HF, Marx SJ (1990) Immunocytology with microwave fixed fibroblasts shows 1α,25-dihydroxyvitamin D_3 dependent rapid and estrogen dependent slow reorganization of vitamin D receptors. J Cell Biol (in press)

Baulieu E-E, Jung I (1970) A prostatic cytosol receptor. Biochem Biophys Res Commun 38:599–606

Baulieu E-E, Binart N, Cadepond F et al. (1989) Do receptor-associated nuclear proteins explain earliest steps of steroid hormone function? In: Carlstedt-Duke J, Eriksson H, Gustafsson J-Å (eds) The steroid/thyroid hormone receptor family and gene regulation. Birkhäuser, Basel, pp 301–318

Baxter JD, Tomkins GM (1970) The relationship between glucocorticoid binding and tyrosine aminotransferase induction in hepatoma tissue culture cells. Proc Natl Acad Sci USA 65:709–715

Baxter JD, Tomkins GM (1971) Specific cytoplasmic glucocorticoid hormone receptors in hepatoma tissue culture cells. Proc Natl Acad Sci USA 68:932–937

Baxter JD, Harris AW, Tomkins GM, Cohn M (1971) Glucocorticoid receptors in lymphoma cells in culture: relationship to glucocorticoid killing activity. Science 171:189–191

Baxter JD, Rousseau GG, Benson MC, Garcea RL, Ito J, Tomkins GM (1972) Role of DNA and specific cytoplasmic receptors in glucocorticoid action. Proc Natl Acad Sci USA 69:1892–1896

Beato M (1989) Gene regulation by steroid hormones. Cell 56:335–344

Beato M, Feigelson P (1972) Glucocorticoid-binding proteins of rat liver cytosol I. Separation and identification of the binding proteins. J Biol Chem 247:7890–7896

Beato M, Biesewig D, Braendle W, Sekeris CE (1969) On the mechanism of hormone action XV. Subcellular distribution and binding of [1,2-³H]cortisol in rat liver. Biochim Biophys Acta 192:494–507

Beato M, Braendle W, Biesewig D, Sekeris CE (1970) On the mechanism of hormone action XVI. Transfer of [1,2-³H₂]cortisol from the cytoplasm to the nucleus of rat liver cells. Biochim Biophys Acta 208:125–136

Beato M, Schmid W, Sekeris CE (1972a) Two cortisol-binding proteins from rat liver cytosol. Biochim Biophys Acta 263:764–774

Beato M, Kalimi M, Feigelson P (1972b) Correlation between glucocorticoid binding to specific liver cytosol receptors and enzyme induction in vivo. Biochem Biophys Res Commun 47:1464–1472

Beato M, Kalimi M, Konstam M, Feigelson P (1973) Interaction of glucocorticoids with rat liver nuclei. II. Studies on the nature of the cytosol transfer factor and the nuclear acceptor site. Biochemistry 12:3372–3379

Beaumont K, Fanestil DD (1983) Characterization of rat brain aldosterone receptors reveals high affinity for corticosterone. Endocrinology 113:2043–2051

Becker PB, Gloss B, Schmid W, Strähle U, Schütz G (1986) In vivo protein-DNA interactions in a glucocorticoid response element require the presence of the hormone. Nature 324:686–688

Beckers C, Maróy P, Dennis R, O'Connor JD, Emmerich H (1980) The uptake and release of ponasterone A by the K_c cell line of Drosophila melanogaster. Mol Cell Endocrinol 17:51–59

Bell PA, Munck A (1973) Steroid-binding properties and stabilization of cytoplasmic glucocorticoid receptors from rat thymus cells. Biochem J 136:97–107

Bellamy D, Phillips JG, Jones IC, Leonard RA (1962) The uptake of cortisol by rat tissues. Biochem J 85:537–545

Berg JM (1989) DNA binding specificity of steroid receptors. Cell 57:1065–1068

Berger U, Wilson P, McClelland RA, Colston K, Haussler MR, Pike JW, Coombes RC (1987) Immunocytochemical detection of 1,25-dihydroxyvitamin D_3 receptor in breast cancer. Cancer Res 47:6793–6799

Bernard PA, Joh TH (1984) Characterization and immunochemical demonstration of glucocorticoid receptor using antisera specific to transformed receptor. Arch Biochem Biophys 229:466–476

Berthois Y, Pourreau-Schneider N, Gandilhon P, Mittre H, Tubiana N, Martin PM (1986) Estradiol membrane binding sites on human breast cancer cell lines. Use of a fluorescent estradiol conjugate to demonstrate plasma membrane binding systems. J Steroid Biochem 25:963–972

Beyer C, Larsson K, Pérez-Palacios G, Morali G (1973) Androgen structure and male sexual behavior in the castrated rat. Horm Behav 4:99–108

Birnbaumer M, Hinrichs-Rosello MV, Cook RG, Schrader WT, O'Malley BW (1987) Chemical and antigenic properties of pure 108000 molecular weight chick progesterone receptor. Mol Endocrinol 1:249–259

Blaquier JA, Cameo MS, Charreau EH (1970) Comparative uptake of androstenediol, testosterone and dihydrotestosterone by tissues of the male rat. J Steroid Biochem 1:327–334

Bodine PV, Litwack G (1988) Evidence that the modulator of the glucocorticoid-receptor complex is the endogenous molybdate factor. Proc Natl Acad Sci USA 85:1462–1466

Borgna J-L, Fauque J, Rochefort H (1984) A monoclonal antibody to the estrogen receptor discriminates between the nonactivated and activated estrogen- and anti-estrogen-receptor complexes. Biochemistry 23:2162–2168

Brachet J, Ficq A (1965) Binding sites of ^{14}C-actinomycin in amphibian ovocytes and an autoradiography technique for the detection of cytoplasmic DNA. Exp. Cell Res 38:153–159

Bradlow HL, Dobriner K, Gallagher TF (1954) The fate of cortisone-T in mice. Endocrinology 54:343–352

Bresnick EH, Dalman FC, Sanchez ER, Pratt WB (1989) Evidence that the 90-kDa heat shock protein is necessary for the steroid-binding conformation of the L cell glucocorticoid receptor. J Biol Chem 264:4992–4997

Brown M, Bollum FJ, Chang LMS (1981) Intracellular localization of DNA polymerase α. Proc Natl Acad Sci USA 78:3049–3052

Bruchovsky N, Wilson JD (1968) The intranuclear binding of testosterone and 5α-androstan-17β-ol-3-one by rat prostate. J Biol Chem 243:5953–5960

Brumbaugh PF, Haussler MR (1973) Nuclear and cytoplasmic receptors for 1,25-dihydroxycholecalciferol in intestinal mucosa. Biochem Biophys Res Commun 51:74–80

Brumbaugh PF, Haussler MR (1974a) 1α,25-Dihydroxycholecalciferol receptors in intestine. I. Association of 1α,25-dihydroxycholecalciferol with intestinal mucosa chromatin. J Biol Chem 249:1251–1257

Brumbaugh PF, Haussler MR (1974b) 1α,25-Dihydroxycholecalciferol receptors in intestine. II. Temperature-dependent transfer of the hormone to chromatin via a specific cytosol receptor. J Biol Chem 249:1258–1262

Brumbaugh PF, Haussler MR (1975) Specific binding of 1α,25-dihydroxycholecalciferol to nuclear components of chick intestine. J Biol Chem 250:1588–1594

Brumbaugh PF, Hughes MR, Haussler MR (1975) Cytoplasmic and nuclear binding components for 1α,25-dihydroxyvitamin D_3 in chick parathyroid glands. Proc Natl Acad Sci 72:4871–4875

Brunkhorst WK (1969) Intracellular binding of corticosterone in thymus tissue. Biochem Biophys Res Commun 35:880–886

Buller RE, Toft DO, Schrader WT, O'Malley BW (1975) Progesterone-binding components of chick oviduct. VIII. Receptor activation and hormone dependent-binding to purified nuclei. J Biol Chem 250:801–808

Buller RE, Schwartz RJ, Schrader WT, O'Malley BW (1976) Progesterone-binding components of chick oviduct. In vitro effect of receptor subunits on gene transcription. J Biol Chem 251:5178–5186

Butenandt A, Günther H, Turba F (1960) Zur primären Stoffwechselwirkung des Testosterons. Hoppe-Seylers Z Physiol Chem 322:28–37

Cake MH, Goidl JA, Parchman LG, Litwack G (1976) Involvement of a low molecular weight component(s) in the mechanism of action of the glucocorticoid receptor. Biochem Biophys Res Commun 71:45–52

Cake MH, DiSorbo DM, Litwack G (1978) Effect of pyridoxal phosphate on the DNA binding site of activated hepatic glucocorticoid receptor. J Biol Chem 253:4886–4891

Carlstedt-Duke J, Okret S, Wrange Ö, Gustafsson J-Å (1982) Immunochemical analysis of the glucocorticoid receptor: identification of a third domain separate from the steroid-binding and DNA-binding domains. Proc Natl Acad Sci USA 79:4260–4264

Catelli MG, Binart N, Jung-Testas I, Renoir JM, Baulieu E-E, Feramisco JR, Welch WJ (1985) The common 90-kD protein component of non-transformed '8S' steroid receptors is a heat shock protein. EMBO J 4:3131–3135

Cayanis E, Rajagopalan R, Cleveland WL, Edelman IS, Erlanger BF (1986) Generation of an auto-anti-idiotypic antibody that binds to glucocorticoid receptor. J Biol Chem 261:5094–5103

Chandler JS, Pike JW, Haussler MR (1979) 1,25-dihydroxyvitamin D_3 receptors in rat kidney cytosol. Biochem Biophys Res Commun 90:1057–1063

Chang C, Kokontis J, Liao S (1988) Molecular cloning of human and rat complementary DNA encoding androgen receptors. Science 240:324–326

Chang C, Chodak G, Sarac E, Takeda H, Liao S (1989a) Prostate androgen receptor: Immunohistological localization and mRNA characterization. J Steroid Biochem 34:311–313

Chang C, Whelan CT, Popovich TC, Kokontis J, Liao S (1989b) Fusion proteins containing androgen receptor sequences and their use in the production of poly- and monoclonal anti-androgen receptor antibodies. Endocrinology 123:1097–1099

Chen TC, DeLuca HF (1973) Receptors of 1,25-dihydroxycholecalciferol in rat intestine. J Biol Chem 248:4890–4895

Chen TL, Hirst MA, Feldman D (1979) A receptor-like binding macromolecule for 1α,25-dihydroxycholecalciferol in cultured mouse bone cells. J Biol Chem 254:7491–7494

Christakos S, Norman AW (1979) Studies on the mode of action of calciferol. XVIII. Evidence for a specific high affinity binding protein for 1,25-dihydroxyvitamin D_3 in chick kidney and pancreas. Biochem Biophys Res Commun 89:56–63

Chytil F, Toft D (1972) Corticoid binding component in rat brain. J Neurochem 19:2877–2880

Clark CR (1984) The cellular distribution of steroid hormone receptors: Have we got it right? Trends Biochem Sci 9:207–208

Clark JH, Hardin JW, Upchurch S, Eriksson H (1978) Heterogeneity of estrogen binding sites in the cytosol of the rat uterus. J Biol Chem 253:7630–7634

Clark JH, Markaverich B, Upchurch S, Eriksson H, Hardin JW, Peck EJ Jr (1980) Heterogeneity of estrogen binding sites: relationship to estrogen receptors and estrogen responses. Recent Prog Horm Res 36:89–134

Clarke CL, Zaino RJ, Feil PD, Miller JV, Steck ME, Ohlsson-Wilhelm BM, Satyaswaroop PG (1987) Monoclonal antibodies to human progesterone receptor: characterization by biochemical and immunohistochemical techniques. Endocrinology 121:1123–1132

Clemens TL, Garrett KP, Zhou X-Y, Pike JW, Haussler MR, Dempster DW (1988) Immunocytochemical localization of the 1,25-dihydroxyvitamin D_3 receptor in target cells. Endocrinology 122:1224–1230

Clever U, Karlson P (1960) Induktion von Puff-Veränderungen in den Speicheldrüsenchromosomen von Chironomus tentans durch Ecdyson. Exp Cell Res 20:623–626

Coffer AI, King RJB, Brockas AJ (1980) Antibodies to human myometrial oestrogen receptor. Biochem Int 1:126–132

Colston K, Feldman D (1980) Nuclear translocation of the 1,25-dihydroxycholecalciferol receptor in mouse kidney. J Biol Chem 255:7510–7513

Colston KW, Feldman D (1979) Demonstration of a 1,25-dihydroxycholecalciferol cytoplasmic receptor-like binder in mouse kidney. J Clin Endocrinol Metab 49:798–800

Conneely OM, Sullivan WP, Toft DO et al. (1986) Molecular cloning of the chicken progesterone receptor. Science 233:767–770

Conneely OM, Dobson ADW, Tsai M-J et al. (1987a) Sequence and expression of a functional chicken progesterone receptor. Mol Endocrinol 1:517–525

Conneely OM, Maxwell BL, Toft DO, Schrader WT, O'Malley BW (1987b) The A and B forms of the chicken progesterone receptor arise by alternate initiation of translation of a unique mRNA. Biochem Biophys Res Commun 149:493–501

Conneely OM, Kettleberger DM, Tsai M-J, Schrader WT, O'Malley BW (1989) The chicken progesterone A and B isoforms are products of an alternate translation initiation event. J Biol Chem
264:14062–14064

Coty WA (1980) Reversible dissociation of steroid hormone-receptor complexes by mercurial
reagents. J Biol Chem 255:8035–8037

Coty WA, Schrader WT, O'Malley BW (1979) Purification and characterization of the chick oviduct
progesterone receptor A subunit. J Steroid Biochem 10:1–12

Crabbé J (1961) Stimulation of active sodium transport by the isolated toad bladder with aldosterone in vitro. J Clin Invest 40:2103–2110

Dahmer MK, Housley PR, Pratt WB (1984) Effects of molybdate and endogenous inhibitors on
steroid-receptor inactivation, transformation and translocation. Annu Rev Physiol 46:67–81

Dame MC, Pierce EA, DeLuca HF (1985) Identification of the porcine intestinal 1,25-dihydroxyvi-
tamin D_3 receptor on sodium dodecyl sulfate/polyacrylamide gels by renaturation and immunoblotting. Proc Natl Acad Sci USA 82:7825–7829

Danielsen M, Northrop JP, Ringold GM (1986) The mouse glucocorticoid receptor: mapping of
functional domains by cloning, sequencing and expression of wild-type and mutant receptor
proteins. EMBO J 5:2513–1522

Danielsen M, Northrop JP, Jonklaas J, Ringold GM (1987) Domains of the glucocorticoid receptor
involved in specific and nonspecific deoxyribonucleic acid binding, hormone activation, and
transcriptional enhancement. Mol Endocrinol 1:816–822

Danielsen M, Hinck L, Ringold GM (1989) Two amino acids within the knuckle of the first zinc
finger specify DNA response element activation by the glucocorticoid receptor. Cell 57:1131–
1138

Davies IJ, Ryan KJ (1972) The uptake of progesterone by the uterus of the pregnant rat in vivo and
its relationship to cytoplasmic progesterone-binding protein. Endocrinology 90:507–515

de Kloet ER, Wallach G, McEwen BS (1975) Differences in corticosterone and dexamethasone
binding to rat brain and pituitary. Endocrinology 96:598–609

Demura T, Kuzumaki N, Oda A, Fujita H, Ishibashi T, Koyanagi T (1989) Establishment and
characterization of monoclonal antibody against androgen receptor. J Steroid Biochem
33:845–851

Denis M, Gustafsson J-Å (1989) The $M_r = 90\,000$ heat shock protein: an important modulator of
ligand and DNA-binding properties of the glucocorticoid receptor. Cancer Res 49:2275s–
2281s

Denis M, Wikström A-C, Gustafsson J-Å (1987) The molybdate-stabilized non-activated glucocorticoid receptor contains a dimer of $M_r = 90\,000$ non-hormone-binding protein. J Biol Chem
262:11803–1806

Denton RR, Notides AC (1989) The nuclear form of the estrogen receptor is dephosphorylated.
Abstracts 71st Meeting the Endocrine Society Seattle, p 159

DeSombre ER, Chabaud JP, Puca GA, Jensen EV (1971) Purification and properties of an estrogen-binding protein from calf uterus. J Steroid Biochem 2:95–103

DeSombre ER, Mohla S, Jensen EV (1972) Estrogen-independent activation of the receptor protein
of calf uterine cytosol. Biochem Biophys Res Commun 48:1601–1608

DeSombre ER, Carbone PP, Jensen EV, McGuire WL, Wells SA Jr, Wittliff JL, Lipsett MB (1979)
Steroid receptors in breast cancer. N Engl J Med 301:1011–1012

De Venuto F, Kelleher PC, Westphal U (1962) Interactions between corticosteroids and fractions
of rat liver and muscle cells as determined by "equilibrium fractionation" and equilibrium
dialysis. Biochim Biophys Acta 63:434–452

Dicker PD, Tsai SY, Weigel NL Tsai M-J, Schrader WT, O'Malley BW (1984) Monoclonal antibody
to the hen oviduct progesterone receptor produced following in vitro immunization. J Steroid
Biochem 20:43–50

Dobson ADW, Conneely OM, Beattie W et al. (1989) Mutational analysis of the chicken progesterone receptor. J Biol Chem 264:4207–4211

Dougherty JJ, Puri RK, Toft DO (1982) Phosphorylation in vivo of chicken oviduct progesterone
receptor. J Biol Chem 257:14226–14230

Dougherty JJ, Puri RK, Toft DO (1984) Polypeptide components of two 8 S forms of chicken
oviduct progesterone receptor. J Biol Chem 259:8004–8009

Dubé JY, Tremblay RR (1974) Androgen binding proteins in cock's tissues: properties of ear lobe protein and determination of binding sites in head appendages and other tissues. Endocrinology 95:1105–1112

Edelman IS (1975) Mechanism of action of steroid hormones. J Steroid Biochem 6:147–159

Edelman IS, Bogoroch R, Porter GA (1963) On the mechanism of action of aldosterone on sodium transport: the role of protein synthesis. Proc Natl Acad Sci USA 50:1169–1177

Eisen HJ (1980) An antiserum to the rat liver glucocorticoid receptor. Proc Natl Acad Sci USA 77:3893–3897

Ennis BW, Stumpf WE, Gasc J-M, Baulieu E-E (1986) Nuclear localization of progesterone receptor before and after exposure to progestin at low and high temperatures: autoradiographic and immunohistochemical studies of chick ovidcut. Endocrinology 119:2066–2075

Erdos T (1968) Properties of a uterine oestradiol receptor. Biochem Biophys Res Commun 37:338–343

Estes PA, Suba EJ, Lawler-Heavner J et al. (1987) Immunologic analysis of human breast cancer progesterone receptors. 1. Immunoaffinity purification of transformed receptors and production of monoclonal antibodies. Biochemistry 26:6250–6262

Eul J, Meyer ME, Tora L, Bocquel MT, Quirin-Stricker C, Chambon P, Gronemeyer H (1989) Expression of active hormone and DNA-binding domains of the chicken progesterone receptor in *E. coli*. EMBO J 8:83–90

Evans RM (1988) The steroid and thyroid hormone receptor superfamily. Science 240:889–895

Evans RM (1989) Molecular characterization of the glucocorticoid receptor. Recent Prog Horm Res 45:1–27

Evans RM, Hollenberg SM (1988) Zinc fingers: gilt by association. Cell 52.1–3

Falk RJ, Bardin CW (1970) Uptake of tritiated progesterone by the uterus of the overiectomized guinea pig. Endocrinology 86:1059–1063

Fanestil DD (1968) Mode of spirolactone action: competitive inhibition of aldosterone binding to kidney mineralocorticoid receptors. Biochem Pharmacol 17:2240–2242

Fanestil DD, Edelman IS (1966) Characteristics of the renal nuclear receptors for aldosterone. Proc Natl Acad Sci USA 56:872–879

Fang S, Liao S (1971) Androgen receptors. Steroid and tissue-specific retention of a 17β-hydroxy-5α-androstan-3-one-protein complex by the cell nuclei of ventral prostate. J Biol Chem 246:16–24

Fang S, Anderson KM, Liao S (1969) Receptor proteins for androgens. On the role of specific proteins in selective retention of 17β-hydroxy-5α-androstan-3-one by rat ventral prostate in vivo and in vitro. J Biol Chem 244:6584–6595

Farman N, Bonvalet JP (1983) Aldosterone binding in isolated tubules III. Autoradiography along the rat nephron. Am J Physiol 245F:606–614

Fauque J, Borgna J-L, Rochefort H (1985) A monoclonal antibody to the estrogen receptor inhibits in vitro criteria of receptor activation by an estrogen and an anti-estrogen. J Biol Chem 260:15547–15553

Fauque J, Scali J, Cavaillés V, Borgna J-L (1989) Mapping on the calf estrogen receptor of the binding domain for an antibody interfering with receptor activation. J Steroid Biochem 32:769–780

Feil PD (1983) Characterization of guinea pig anti-progestin receptor antiserum. Endocrinology 112:396–398

Feldman D, Funder JW, Edelman IS (1973) Evidence for a new class of corticosterone receptors in the rat kidney. Endocrinology 92:1429–1441

Feldman D, McCain TA, Hirst MA, Chen TL, Colston KW (1979) Characterization of a cytoplasmic receptor-like binder for 1α,25-dihydroxycholecalciferol in rat intestinal mucosa. J Biol Chem 254:10378–10384

Fleischmann G, Beato M (1979) Activation of the progesterone receptor of rabbit uterus. Mol Cell Endocrinol 16:181–197

Fleming H, Blumenthal R, Gurpide E (1982) Effects of cyclic nucleotides on estradiol binding in human endometrium. Endocrinology 111:1671–1677

Fleming H, Blumenthal R, Gurpide E (1983) Rapid changes in specific estrogen binding elicited by cGMP or cAMP in cytosol from human endometrial cells. Proc Natl Acad Sci USA 80:2486–2490

Forte LR (1972) Effect of mineralocorticoid agonists and antagonists on binding of ^{3}H-aldosterone to adrenalectomized rat kidney plasma membranes. Life Sci 11(I):461–473

Fraser DR, Kodicek E (1970) Unique biosynthesis by kidney of a biologically active vitamin D metabolite. Nature 228:764–766

Freedman LP, Luisi BF, Korzun ZR, Basavappa R, Sigler PB, Yamamoto KR (1988) The function and structure of the metal coordination sites within the glucocorticoid receptor DNA binding domain. Nature 334:543–546

Funder JW, Sheppard K (1987) Adrenocortical steroids and the brain. Annu Rev Physiol 49:397–411

Funder JW, Feldman D, Edelman IS (1972) Specific aldosterone binding in rat kidney and parotid. J Steroid Biochem 3:209–218

Funder JW, Feldman D, Edelman IS (1973a) The roles of plasma binding and receptor specificity in the mineralocorticoid action of aldosterone. Endocrinology 92:994–1004

Funder JW, Feldman D, Edelman IS (1973b) Glucocorticoid receptors in rat kidney: the binding of tritiated-dexamethasone. Endocrinology 92:1005–1013

Fuxe K, Wikström A-C, Okret S et al. (1985) Mapping of glucocorticoid receptor immunoreactive neurons in the rat tel- and diencephalon using a monoclonal antibody against rat liver glucocorticoid receptor. Endocrinology 117:1803–1812

Fuxe K, Cintra A, Agnati LF et al. (1987) Studies on the cellular localization and distribution of glucocorticoid receptor and estrogen receptor immunoreactivity in the central nervous system of the rat and their relationship to the monoaminergic and peptidergic neurons of the brain. J Steroid Biochem 27:159–170

Gametchu B, Harrison RW (1984) Characterization of a monoclonal antibody to the rat liver glucocorticoid receptor. Endocrinology 114:274–279

Garcia T, Tuohimaa P, Mešter J, Buchou T, Renoir J-M, Baulieu E-E (1983) Protein kinase activity of purified components of the chicken oviduct progesterone receptor. Biochem Biophys Res Commun 113:960–966

Garcia T, Buchou T, Renoir J-M, Mešter J, Baulieu E-E (1986a) A protein kinase copurified with chick oviduct progesterone receptor. Biochemistry 25:7937–7942

Garcia T, Jung-Testas I, Baulieu E-E (1986b) Tightly bound nuclear progesterone receptor is not phosphorylated in primary chick oviduct cultures. Proc Natl Acad Sci USA 83:7573–7577

Garcia T, Buchou T, Jung-Testas I, Renoir J-M, Baulieu E-E (1987) Chick oviduct progesterone receptor phosphorylation: characterization of a copurified kinase and phosphorylation in primary cultures. J Steroid Biochem 27:227–234

Gardner DG, Wittliff JL (1973) Characterization of a distinct glucocorticoid-binding protein in the lactating mammary gland of the rat. Biochim Biophys Acta 320:617–627

Gasc J-M, Baulieu E-E (1987) From the structure of steroid receptors to their assessment by immunocytochemistry in target cells. J Steroid Biochem 27:177–184

Gasc J-M, Ennis BW, Baulieu E-E, Stumpf WE (1983) Récepteur de la progestérone dans l'oviducte de poulet: Double révélation par immunohistochimie avec des anticorps antirécepteur et par autoradiographie à l'aide d'un progestagène tritié. C R Acad Sci [D] (Paris) 297:477–482

Gasc J-M, Renoir JM, Radanyi C, Joab I, Tuohimaa P, Baulieu E-E (1984) Progesterone receptor in the chick oviduct: an immunohistochemical study with antibodies to distinct receptor components. J Cell Biol 99:1193–1201

Gaulton GN, Greene MI (1986) Idiotypic mimicry of biological receptors. Annu Rev Immunol 4:253–280

Geller J, van Damme O, Garabieta G, Loh A, Rettura J, Seifter E (1969) Effect of cyproterone acetate on ^{3}H-testosterone uptake and enzyme synthesis by the ventral prostate of the rat. Endocrinology 84:1330–1335

Giannopoulos G (1973) Glucocorticoid receptors in lung. I. Specific binding of glucocorticoids to cytoplasmic components of rabbit fetal lung. J Biol Chem 248:3876–3883

Giannopoulos G, Mulay S, Solomon S (1972) Cortisol receptors in rabbit fetal lung. Biochem Biophys Res Commun 47:411–418

Giannopoulos G, Mulay S, Solomon S (1973) Glucocorticoid receptors in lung. II. Specific binding of glucocorticoids to nuclear components of rabbit fetal lung. J Biol Chem 248:5016–5023

Glascock RF, Hoekstra WG (1959) Selective accumulation of tritium-labelled hexoestrol by the reproductive organs of immature female goats and sheep. Biochem J 72:673–682

Godeau JF, Schorderet-Slatkine S, Hubert P, Baulieu E-E (1978) Induction of maturation in *Xenopus laevis* oocytes by a steroid linked to a polymer. Proc Natl Acad Sci USA 75:2353–2357

Godowski PJ, Picard D (1989) Steroid receptors. How to be both a receptor and a transcription factor. Biochem Pharmacol 38:3135–3143

Godowski PJ, Rusconi S, Miesfeld R, Yamamoto KR (1987) Glucocorticoid receptor mutants that are constitutive activators of transcriptional enhancement. Nature 325:365–368

Goidl JA, Cake MH, Dolan KP, Parchman LG, Litwack G (1977) Activation of the rat liver glucocorticoid-receptor complex. Biochemistry 16:2125–2130

Gopalakrishnan TV, Sadgopal A (1972) Partial purification of cortisol-binding protein from rat liver cytosol and its role in transcription. Biochim Biophys Acta 287:164–186

Gorski J, Raker B (1974) Estrogen action in the uterus: the requisite for sustained estrogen binding in the nucleus. Gynecol Oncol 2:249–258

Gorski J, Hansen JC (1987) The "one and only" step model of estrogen action. Steroids 49:461–475

Gorski J, Toft D, Shyamala G, Smith D, Notides A (1968) Hormone receptors: studies on the interaction of estrogen with the uterus. Recent Prog Horm Res 24:45–80

Gorski J, Welshons W, Sakai D (1984) Remodeling the estrogen receptor model. Mol Cell Endocrinol 36:11–15

Gorski J, Welshons WV, Sakai D et al. (1986) Evolution of a model of estrogen action. Recent Prog Horm Res 42:297–329

Govindan MV (1979) Purification of glucocorticoid receptors from rat liver cytosol. Preparation of antibodies against the major receptor proteins and application of immunological techniques to study activation and translocation. J Steroid Biochem 11:323–332

Govindan MV (1980) Immunofluorescence microscopy of the intracellular translocation of glucocorticoid-receptor complexes in rat hepatoma (HTC) cells. Exp Cell Res 127:293–297

Govindan MV, Sekeris C (1978) Purification of two dexamethasone-binding proteins from rat liver cytosol. Eur J Biochem 89:95–104

Govindan MV, Gronemeyer H (1984) Characterization of the rat liver glucocorticoid receptor purified by DNA-cellulose and ligand affinity chromatography. J Biol Chem 259:12915–12924

Govindan MV, Devic M, Green S, Gronemeyer H, Chambon P (1985) Cloning of the human glucocorticoid receptor cDNA. Nucleic Acids Res 13:8293–8304

Grandics P, Gasser DL, Litwack G (1982) Monoclonal antibodies to the glucocorticoid receptor. Endocrinology 111:1731–1733

Grandics P, Miller A, Schmidt TJ, Litwack G (1984) Phosphorylation in vivo of rat hepatic glucocorticoid receptor. Biochem Biophys Res Commun 120:59–65

Gravanis A, Gurpide E (1986) Enucleation of human endometrial cells: nucleo-cytoplasmic distribution of DNA polymerase α and estrogen receptor. J Steroid Biochem. 24:469–474

Green R, Luttge WG, Whalen RE (1970) Uptake of tritiated testosterone in brain and peripheral tissues of normal and neonatally androgenized female rats. J Comp Physiol Psych 72:337–340

Green S, Chambon P (1986) A superfamily of potentially oncogenic hormone receptors. Nature 324:615–617

Green S, Chambon P (1987) Oestradiol induction of a glucocorticoid-responsive gene by a chimaeric receptor. Nature 325:75–78

Green S, Chambon P (1988) Nuclear receptors enhance our understanding of transcription regulation. Trends Genet 4:309–314

Green S, Walter P, Kumar V, Krust A, Bornert J-M, Argos P, Chambon P (1986) Human oestrogen receptor cDNA: sequence, expression and homology to v-*erb*-A. Nature 320:134–139

Green S, Kumar V, Theulaz I, Wahli W, Chambon P (1988) The N-terminal DNA-binding 'zinc finger' of the oestrogen and glucocorticoid receptors determines target gene specificity. EMBO J 7:3037–3044

Greene GL, Jensen EV (1982) Monoclonal antibodies as probes for estrogen receptor detection and characterization. J Steroid Biochem 16:353–359

Greene GL, Closs LE, Fleming H, DeSombre ER, Jensen EV (1977) Antibodies to estrogen receptor: immunochemical similarity of estrophilin from various mammalian species. Proc Natl Acad Sci USA 74:3681–3685

Greene GL, Closs LE, DeSombre ER, Jensen EV (1979) Antibodies to estrophilin: comparison between rabbit and goat antisera. J Steroid Biochem 11:333–341

Greene GL, Fitch FW, Jensen EV (1980a) Monoclonal antibodies to estrophilin: probes for the study of estrogen receptors. Proc Natl Acad Sci USA 77:157–161

Greene GL, Nolan C, Engler JP, Jensen EV (1980b) Monoclonal antibodies to human estrogen receptor. Proc Natl Acad Sci USA 77:5115–5119

Greene GL, Sobel NB, King WJ, Jensen EV (1984) Immunochemical studies of estrogen receptors. J Steroid Biochem 20:51–56

Greene GL, Gilna P, Waterfield M, Baker A, Hort Y, Shine J (1986) Sequence and expression of human estrogen receptor complementary DNA. Science 231:1150–1154

Greene GL, Harris K, Bova R, Kinders R, Moore B, Nolan C (1988) Purification of T47D human progesterone receptor and immunochemical characterization with monoclonal antibodies. Mol Endocrinol 2:714–726

Greer DS (1959) The distribution of radioactivity in non-excretory organs of the male rat after injection of testosterone-4-C^{14}. Endocrinology 64:898–906

Grody WW, Compton JG, Schrader WT, O'Malley BW (1980) Inactivation of chick oviduct progesterone receptors. J Steroid Biochem 12:115–120

Grody WW, Schrader WT, O'Malley BW (1982) Activation, transformation, and subunit structure of steroid hormone receptors. Endocrine Rev 3:141–163

Gronemeyer H, Pongs O (1980) Localization of ecdysterone on polytene chromosomes of Drosophila melanogaster. Proc Natl Acad Sci USA 77:2108–2112

Gronemeyer H, Govindan MV, Chambon P (1985) Immunological similarity between the chick oviduct progesterone receptor forms A and B. J Biol Chem 260:6916–6925

Gronemeyer H, Turcotte B, Quirin-Stricker C et al. (1987) The chicken progesterone receptor: sequence, expression and functional analysis. EMBO J 6:3985–3994

Gschwendt M, Hamilton TH (1972) The transformation of the cytoplasmic oestradiol-receptor complex into the nuclear complex in a uterine cell-free system. Biochem J 128:611–616

Guiochon-Mantel A, Loosfelt H, Lescop P, Sar S, Atger M, Perrot-Applanat M, Milgrom E (1989) Mechanisms of nuclear localization of the progesterone receptor: evidence for interaction between monomers. Cell 57:1147–1154

Gustafsson J-Å, Carlstedt-Duke J, Poellinger L et al. (1987) Biochemistry, molecular biology, and physiology of the glucocorticoid receptor. Endocrine Rev 8:185–234

Hackney JF, Pratt WB (1971) Characterization and partial purification of the specific glucocorticoid-binding component from mouse fibroblasts. Biochemistry 10:3002–3008

Hackney JF, Gross SR, Aronow L, Pratt WB (1970) Specific glucocorticoid-binding macromolecules from mouse fibroblasts growing in vitro. Mol Pharmacol 6:500–512

Haddad JG, Hahn TJ, Birge SF (1973) Vitamin D metabolites specific binding by rat intestinal cytosol. Biochem Biophys Acta 329:93–97

Ham J, Parker MG (1989) Regulation of gene expression by nuclear hormone receptors. Current Opinion Cell Biol 1:503–511

Hansson V, Tveter KJ (1971) Uptake and binding in vivo of ^{3}H labelled androgen in the rat epididymis and ductus deferens. Acta Endocrinol 66:745–755

Harding BW, Samuels LT (1962) The uptake and subcellular distribution of C^{14}-labeled steroid in rat ventral prostate following in vivo administration of testosterone-4-C^{14}. Endocrinology 70:109–118

Harmon JM, Eisen HJ, Brower ST, Simons SS Jr, Langley CL, Thompson EB (1984) Identification of human leukemic glucocorticoid receptors using affinity labeling and anti-human glucocorticoid receptor antibodies. Cancer Res 44:4540–4547

Haukkamaa M (1987) Membrane-associated steroid hormone receptors. In: Clark CR (ed) Steroid hormone receptors: their intracellular localisation. Ellis Horwood, Chichester, pp 155–169

Haussler MR (1986) Vitamin D receptors: nature and function. Annu Rev Nutr 6:527–562

Haussler MR, Norman AW (1969) Chromosomal receptor for a vitamin D metabolite. Proc Natl Acad Sci USA 62:155–162

Haussler MR, Myrtle JF, Norman AW (1968) The association of a metabolite of vitamin D_3 with intestinal mucosa chromatin in vivo. J Biol Chem 243:4055–4064

Haussler MR, Boyce DW, Littledike ET, Rasmussen H (1971) A rapidly acting metabolite of vitamin D_3. Proc Natl Acad Sci USA 68:177–181

Haussler MR, Pike JW, Chandler JS, Manolagas SC, Deftos LJ (1981) Molecular action of 1,25-dihydroxyvitamin D_3: new cultured cell models. Ann NY Acad Sci 372:502–517

Haussler MR, Mangelsdorf DJ, Komm BS et al. (1988) Molecular biology of the vitamin D hormone. Recent Prog Horm Res 44:263–305

Henry HL, Norman AW (1975) Studies on the mechanism of action of calciferol. VII. Localization of 1,25-dihydroxy-vitamin D_3 in chick parathyroid glands. Biochem Biophys Res Commun 62:781–788

Herman TS, Fimognari GM, Edelman IS (1968) Studies on renal aldosterone-binding proteins. J Biol Chem 243:3849–3856

Higgins SJ, Rousseau GG, Baxter JD, Tomkins GM (1973a) Nuclear binding of steroid receptors: comparison in intact cells and cell-free systems. Proc Natl Acad Sci USA 70:3415–3418

Higgins SJ, Rousseau GG, Baxter JD, Tomkins GM (1973b) Early events in glucocorticoid action. Activation of the steroid receptor and its subsequent specific nuclear binding studied in a cell-free system. J Biol Chem 248:5866–5872

Hollander N, Chiu YW (1966) In vitro binding of cortisol-1,2-^{3}H by a substance in the supernatant fraction of P1798 mouse lymphosarcoma. Biochem Biophys Res Commun 25:291–297

Hollenberg SM, Evans RM (1988) Multiple and cooperative trans-activation domains of the human glucocorticoid receptor. Cell 55:899–906

Hollenberg SM, Weinberger C, Ong ES, Cerelli G, Oro A, Lebo R, Thompson EB, Rosenfeld MG, Evans RM (1985) Primary structure and expression of a functional human glucocorticoid receptor cDNA. Nature 318:635–641

Hollenberg SM, Giguere V, Segul P, Evans RM (1987) Colocalization of DNA-binding and transcriptional activation functions in the human glucocorticoid receptor. Cell 49:39–46

Horiuchi M, Isohashi F, Terada M, Okamoto K, Sakamoto Y (1981) Interaction of ATP with a macromolecular translocation inhibitor of the nuclear binding of "activated" receptor-glucocorticoid complex. Biochem Biophys Res Commun 98:88–94

Horsting M, DeLuca HF (1969) In vitro production of 25-hydroxycholecalciferol. Biochem Biophys Res Commun 36:251–256

Housley PR, Pratt WB (1983) Direct demonstration of glucocorticoid receptor phosphorylation by intact L-cells. J Biol Chem 258:4630–4635

Housley PR, Sanchez ER, Westphal HM, Beato M, Pratt WB (1985) The molybdate-stabilized L-cell glucocorticoid receptor isolated by affinity chromatography or with a monoclonal antibody is associated with a 90–92-kDa nonsteroid-binding phosphoprotein. J Biol Chem 260:13810–13817

Hsueh AJW, Peck EJ Jr, Clark JH (1974) Receptor progesterone complex in the nuclear fraction of the rat uterus: demonstration by ^{3}H-progesterone exchange. Steroids 24:599–611

Hubbard JA, Barrett AJ, Kalimi M (1984) Tosyl-lysyl chloromethane alters glucocorticoid-receptor complex nuclear binding and physical properties. Endocrinology 115:65–72

Hughes MR, Haussler MR (1978) 1,25-Dihydroxyvitamin D_3 receptors in parathyroid glands. Preliminary characterization of cytoplasmic and nuclear binding components. J Biol Chem 253:1065–1073

Hunt T (1989) Cytoplasmic anchoring proteins and the control of nuclear localization. Cell 59:949–951

Ireland RC, Berger E, Sirotkin K, Yund MA, Osterbur D, Fristrom J (1982) Ecdysone induces the transcription of four heat-shock genes in Drosophila S3 cells and imaginal discs. Dev Biol 93:498–507

Ishii DN, Aronow L (1973) In vitro degradation and stabilization of the glucocorticoid binding component from mouse fibroblasts. J Steroid Biochem 4:593–603

Ishii DN, Pratt WB, Aronow L (1972) Steady-state level of the specific glucocorticoid binding component in mouse fibroblasts. Biochemistry 11:3896–3904

Ishikawa K, Hanaoka Y, Kondo Y, Imai K (1977) Primary action of steroid hormone at the surface of amphibian oocyte in the induction of germinal vesicle breakdown. Mol Cell Endocrinol 9:91–100

Isohashi F, Terada M, Tsukanaka K, Nakanishi Y, Sakamoto Y (1980) A low-molecular-weight translocation modulator and its interaction with a macromolecular translocation inhibitor of the activated receptor-glucocorticoid complex. J Biochem 88:775–781

Jensen EV (1960) Studies of growth phenomena using tritium-labeled steroids. Proc 4th Internat Congr Biochem, Vienna 1958, vol 15, p 119

Jensen EV (1977) Receptor proteins: past, present and future. In: Research on steroids, vol 7. Elsevier/North Holland Biomedical, Amsterdam, pp 1–36

Jensen EV (1979) Interaction of steroid hormones with the nucleus. Pharmacol Rev 30:477–491

Jensen EV (1990) Molecular mechanisms of steroid hormone action in the uterus. In: Carsten ME, Miller JD (eds) Uterine function. Plenum, New York, pp 315–359

Jensen EV, DeSombre ER (1972) Mechanism of action of the female sex hormones. Annu Rev Biochem 41:203–230

Jensen EV, DeSombre ER (1973) Estrogen-receptor interaction. Estrogenic hormones effect transformation of specific receptor proteins to a biochemically functional form. Science 182:126–134

Jensen EV, Jacobson HI (1960) Fate of steroid estrogens in target tissues. In: Pincus G, Vollmer EP (eds) Biological activities of steroids in relation to cancer. Academic, New York, pp 161–178

Jensen EV, Jacobson HI (1962) Basic guides to the mechanism of estrogen action. Recent Prog Horm Res 18:387–414

Jensen EV, Jacobson HI, Flesher JW et al. (1966) Estrogen receptors in target tissues. In: Pincus G, Nakao T, Tait JF (eds) Steroid dynamics. Academic, New York, pp 133–157

Jensen EV, DeSombre ER, Jungblut PW (1967a) Interaction of estrogens with receptor sites in vivo and in vitro. Proc 2nd Internat Congr Hormonal Steroids, Milan 1966. Excerpta Medica Foundation, Amsterdam, pp 492–500

Jensen EV, DeSombre ER, Hurst DJ, Kawashima T, Jungblut PW (1967b) Estrogen-receptor interactions in target tissues. Arch Anat Microsc Morphol Exp 56 [Suppl]:547–569

Jensen EV, Hurst DJ, DeSombre ER, Jungblut PW (1967c) Sulfhydryl groups and estradiol-receptor interaction. Science 158:385–387

Jensen EV, Suzuki T, Kawashima T, Stumpf WE, Jungblut PW, DeSombre ER (1968) A two-step mechanism for the interaction of estradiol with rat uterus. Proc Natl Acad Sci USA 59:632–638

Jensen EV, Numata M, Smith S, Suzuki T, Brecher PI, DeSombre ER (1969) Estrogen-receptor interaction in target tissues. Dev Biol 3 [Suppl]:151–171

Jensen EV, Numata M, Brecher PI, DeSombre ER (1971) Hormone-receptor interaction as a guide to biochemical mechanism Biochem Soc Symp 32:133–159

Jensen EV, Jacobson HI, Smith S, Jungblut PW, DeSombre ER (1972a) The use of estrogen antagonists in hormone receptor studies. Gynecol Invest 3:108–122

Jensen EV, Mohla S, Gorell T, Tanaka S, DeSombre ER (1972b) Estrophile to nucleophile in two easy steps. J Steroid Biochem 3:445–458

Jerne NK (1974) Towards a network theory of the immune system. Ann Immunol 125C:373–389

Joab I, Radanyi C, Renoir M et al. (1984) Common non-hormone binding component in non-transformed chick oviduct receptors of four steroid hormones. Nature 308:850–853

Jones PG, Haussler MR (1979) Scintillation autoradiographic localization of 1,25-dihydroxyvitamin D_3 in chick intestine. Endocrinology 104:313–321

Jung I, Baulieu E-E (1972) Testosterone cytosol "receptor" in the rat levator ani muscle. Nature New Biol 237:24–26

Kalimi M, Banerji A (1981) Role of sulfhydryl modifying reagents in the binding and activation of chick oviduct progesterone-receptor complex. J Steroid Biochem 14:593–597

Kalimi M, Love K (1980) Role of chemical reagents in the activation of rat hepatic glucocorticoid-receptor complex. J Biol Chem 255:4687–4690

Kalimi M, Beato M, Feigelson P (1973) Interaction of glucocorticoids with rat liver nuclei. Role of the cytosol proteins. Biochemistry 12:3365–3371

Kalimi M, Colman P, Feigelson P (1975) The "activated" hepatic glucocorticoid-receptor complex. Its generation and properties. J Biol Chem 250:1080–1086

Karlson P (1963) New concepts on the mode of action of hormones. Perspect Biol Med 6:203–214

Kenney FT, Flora RM (1961) Induction of tyrosine-α-ketoglutarate transaminase in rat liver. I. Hormonal nature. J Biol Chem 236:2699–2702

King RJB, Gordon J, Inman DR (1965) The intracellular localization of oestrogen in rat tissues. J Endocrinol 32:9–15

King RJB, Gordon J, Cowan DM, Inman DR (1966) The intranuclear localization of [6,7-³H]-oestradiol-17β in dimethylbenzanthracene-induced rat mammary adenocarcinoma and other tissues. J Endocrinol 36:139–150

King WJ, Greene GL (1984) Monoclonal antibodies localize oestrogen receptor in the nuclei of target cells. Nature 307:745–747

King WJ, DeSombre ER, Jensen EV, Greene GL (1985) Comparison of immunocytochemical and steroid-binding assays for estrogen receptor in human breast tumors. Cancer Res 45:293–304

Kirkpatrick AF, Milholland RJ, Rosen F (1971) Stereospecific glucocorticoid binding to subcellular fractions of the sensitive and resistant lymphosarcoma P1798. Nature New Biol 232:216–218

Kirkpatrick AF, Kaiser N, Milholland RJ, Rosen F (1972) Glucocorticoid-binding macromolecules in normal tissues and tumors. Stabilization of the specific binding component. J Biol Chem 247:70–74

Klein-Hitpass L, Ryffel GU, Heitlinger E, Cato ACB (1988) A 13 bp palindrome is a functional estrogen responsive element and reacts specifically with estrogen receptor. Nucleic Acids Res 16:647–663

Klein-Hitpass L, Tsai SY, Weigel NL et al. (1990) The progesterone receptor stimulates cell-free transcription by enhancing the formation of a stable preinitiation complex. Cell 60:247–257

Klock G, Strähle U, Schütz G (1987) Oestrogen and glucocorticoid responsive elements are closely related but distinct. Nature 329:734–736

Klug A, Rhodes D (1987) 'Zinc fingers': a novel protein motif for nucleic acid recognition. Trends Biochem Sci 12:464–469

Koblinsky M, Beato M, Kalimi M, Feigelson P (1972) Glucocorticoid-binding proteins of rat liver cytosol. II. Physical characterization and properties of the binding proteins. J Biol Chem 247:7897–7904

Korenman SG, Rao BR (1968) Reversible disaggregation of the cytosol estrogen binding protein of uterine cytosol. Proc Natl Acad Sci USA 61:1028–1033

Kovačič- Milivojević B, Vedeckis WV (1986) Absence of detectable ribonucleic acid in the purified, untransformed mouse glucocorticoid receptor. Biochemistry 25:8266–8273

Kowarski A, Shalf J, Migeon CJ (1969) Concentration of testosterone and dihydrotestosterone in subcellular fractions of liver, kidney, prostate, and muscle in the male dog. J Biol Chem 244:5269–5272

Kream BE, Reynolds RD, Knutson JC, Eisman JA, DeLuca HF (1976) Intestinal cytosol binders of 1,25-dihydroxyvitamin D_3 and 25-hydroxyvitamin D_3. Arch Biochem Biophys 176:779–787

Kream BE, Jose M, Yamada S, DeLuca HF (1977a) A specific high-affinity binding macromolecule for 1,25-dihydroxyvitamin D_3 in fetal bone. Science 197:1086–1088

Kream BE, Yamada S, Schnoes HK, DeLuca HF (1977b) Specific cytosol-binding protein for 1,25-dihydroxyvitamin D_3 in rat intestine. J Biol Chem 252:4501–4505

Krozowski ZS, Funder JW (1983) Renal mineralocorticoid receptors and hippocampal corticosterone-binding species have identical intrinsic steroid specificity. Proc Natl Acad Sci USA 80:6056–6060

Krozowski ZS, Rundle SE, Wallace C et al. (1989) Immunolocalization of renal mineralocorticoid receptors with an antiserum against a peptide deduced from the complementary deoxyribonucleic acid sequence. Endocrinology 125:192–198

Krust A, Green S, Argos P, Kumar V, Walter P, Bornert J-M, Chambon P (1986) The chicken oestrogen receptor sequence: homology with v-erbA and the human oestrogen and glucocorticoid receptors. EMBO J 5:891–897

Kumar V, Chambon P (1988) The estrogen receptor binds tightly to its responsive element as a ligand-induced homodimer. Cell 55:145–156

Kumar V, Green S, Stack G, Berry M, Jin J-R, Chambon P (1987) Functional domains of the human estrogen receptor. Cell 51:941–951

Kuppert PG, Spindler K-D (1982) Characterization of nuclear ecdysteroid receptor from crayfish integument. J Steroid Biochem 17:205–210

Kuppert P, Wilhelm S, Spindler K-D (1978) Demonstration of cytoplasmic receptors for the molting hormones in crayfish. J Comp Physiol 128:95–100

Labate ME, Whelly SM, Barker KL (1986) Ribosome-associated estradiol-binding components in the uterus and their relationship to the translational capacity of uterine ribosomes. Endocrinology 119:140–151

Lamb DJ, Kima PE, Bullock DW (1986) Evidence for a single steroid-binding protein in the rabbit progesterone receptor. Biochemistry 25:6319–6324

Lannigan DA, Notides AC (1989) Estrogen receptor selectively binds the "coding strand" of an estrogen responsive element. Proc Natl Acad Sci USA 86:863–867

Lawrence AM, Landau RL (1965) Impaired ventral prostate affinity for testosterone in hypophysectomized rats. Endocrinology 77:1119–1125

Lawson DEM, Wilson PW (1974) Intranuclear localization and receptor proteins for 1,25-dihydroxycholecalciferol in chick intestine. Biochem J 144:573–583

Lawson DEM, Wilson PW, Kodicek E (1969) New vitamin D metabolite localized in intestinal cell nuclei. Nature 222:171–172

Leach KL, Dahmer MK, Hammond ND, Sando JJ, Pratt WB (1979) Molybdate inhibition of glucocorticoid receptor inactivation and transformation. J Biol Chem 254:11884–11890

Leach KL, Grippo JF, Housley PR, Dahmer MK, Salive ME, Pratt WB (1982) Characteristics of an endogenous glucocorticoid receptor stabilizing factor. J Biol Chem 257:381–388

Leavitt WW, Blaha GC (1972) An estrogen-stimulated, progesterone-binding system in the hamster uterus and vagina. Steroids 19:263–274

Leclercq G, Hulin N, Heuson JC (1973) Interaction of activated estradiol-receptor complex and chromatin in isolated uterine nuclei. Eur J Cancer 9:681–685

Lees JA, Fawell SE, Parker MG (1989) Identification of two transactivation domains in the mouse oestrogen receptor. Nucleic Acids Res 17:5477–5488

Leroy F, Preumont AM, Galand P, Brachet J (1972) Increased chromatin acid lability and actinomycin-D binding in endometrial cells under the action of sex steroids. J Endocrinol 52:525–531

Lessey BA, Gorell TA (1981) Nuclear progesterone receptors in the beagle uterus. J Steroid Biochem 14:585–591

Lessey BA, Alexander PS, Horwitz KB (1983) The subunit structure of human breast cancer progesterone receptors: characterization by chromatography and photoaffinity labeling. Endocrinology 112:1267–1274

Liang T, Tymoczko JL, Chan KMB, Hung SC, Liao S (1977) Androgen action: receptors and rapid responses. In: Martini L, Motta M (eds) Androgens and antiandrogens. Raven, New York, pp 77–89

Liao S (1975) Cellular receptors and mechanisms of action of steroid hormones. Int Rev Cytol 41:87–172

Liao S, Fang S (1969) Receptor proteins for androgens and the mode of action of androgens on gene transcription in ventral prostate. Vitam Horm 27:17–90

Liao S, Hiipakka RA (1984) Mechanism of action of steroid hormones at the subcellular level. In: Makin HLJ (ed) Biochemistry of steroid hormones, 2nd edn. Blackwell Scientific, Oxford, pp 633–680

Liao S, Witte D (1985) Autoimmune anti-androgen-receptor antibodies in human serum. Proc Natl Acad Sci USA 82:8345–8348

Liao S, Howell DK, Chang T-M (1974) Action of a nonsteroidal antiandrogen, flutamide, on the receptor binding and nuclear retention of 5α-dihydrotestosterone in rat ventral prostate. Endocrinology 94:1205–1209

Liao S, Tymoczko JL, Castañeda E, Liang T (1975) Androgen receptors and androgen-dependent initiation of protein synthesis in the prostate. Vitam Horm 33:297–317

Liao S, Rossini GP, Hiipakka RA, Chen C (1980) Factors that can control the interaction of the androgen-receptor complex with the genomic structure in the rat prostate. In: Bresciani F (ed) Perspectives in steroid receptor research. Raven, New York, pp 99–112

Little M, Szendro PI, Jungblut PW (1973) Hormone-mediated dimerization of microsomal estradiol receptor. Hoppe-Seylers Z Physiol Chem 354:1599–1610

Little M, Szendro P, Teran C, Hughes A, Jungblut PW (1975) Biosynthesis and transformation of microsomal and cytosol estradiol receptors. J Steroid Biochem 6:493–500

Litwack G, Sears ML, Diamondstone TI (1963) Intracellular distribution of tyrosine-α-ketoglutarate transaminase and 4-C^{14}-hydrocortisone activities during induction. J Biol Chem 238:302–305

Litwack G, Filler R, Rosenfield SA, Lichtash N, Wishman CA, Singer S (1973) Liver cytosol corticosteroid binder II, a hormone receptor. J Biol Chem 248:7481–7486

Liu S-LH, Webb TE (1977) Elevated concentration of a dexamethasone-receptor translocation inhibitor in Novikoff hepatoma cells. Cancer Res 37:1763–1767

Logeat F, Vu Hai MT, Milgrom E (1981) Antibodies to rabbit progesterone receptor: crossreaction with human receptor. Proc Natl Acad Sci USA 78:1426–1430

Logeat F, Vu Hai MT, Fournier A, Legrain P, Buttin G, Milgrom E (1983) Monoclonal antibodies to rabbit progesterone receptor: crossreaction with other mammalian progesterone receptors. Proc Natl Acad Sci USA 80:6456–6459

Logeat F, Pamphile P, Loosfelt H, Jolivet A, Fournier A, Milgrom E (1985a) One-step immunoaffinity purification of active progesterone receptor. Further evidence in favor of the existence of a single steroid-binding unit. Biochemistry 24:1029–1035

Logeat F, Le Cunff M, Pamphile R, Milgrom E (1985b) The nuclear-bound form of the progesterone receptor is generated through a hormone-dependent phosphorylation. Biochem Biophys Res Commun 131:421–427

Lombès M, Edelman IS, Erlanger BF (1989) Internal image properties of a monoclonal auto-anti-idiotypic antibody and its binding to aldosterone receptors. J Biol Chem 264:2528–2536

Lombès M, Farman N, Oblin ME, Baulieu E-E, Bonvalet JP, Erlanger BF, Gasc JM (1990) Immunohistochemical localization of renal mineralocorticoid receptor by using an anti-idiotypic antibody that is an internal image of aldosterone. Proc Natl Acad Sci USA 87:1086–1088

Londerhausen M, Spindler K-D (1981) Characterization of cytoplasmic ecdysteroid receptors in the hypodermis of the crayfish *Orconectes limosus*. Mol Cell Endocrinol 24:253–265

Londerhausen M, Kuppert P, Spindler K-D (1982) Ecdysteroid receptors: a comparison of cytoplasmic and nuclear receptors from crayfish hypodermis. Hoppe-Seylers Z Physiol Chem 363:797–802

Loosfelt H, Logeat F, Vu Hai MT, Milgrom E (1984) The rabbit progesterone receptor. Evidence for a single steroid-binding subunit and characterization of receptor mRNA. J Biol Chem 259:14196–14202

Loosfelt H, Atger M, Misrahi M et al. (1986) Cloning and sequence analysis of rabbit progesterone-receptor complementary DNA. Proc Natl Acad Sci USA 83:9045–9049

Lorenzo F, Jolivet A, Loosfelt H, Vu Hai MT, Brailly S, Perrot-Applanat M, Milgrom E (1988) A rapid method of epitope mapping. Eur J Biochem 176:53–60

Lubahn DB, Joseph DR, Sar M et al. (1988) The human androgen receptor: complementary ribonucleic acid cloning, sequence analysis and gene expression in prostate. Mol Endocrinol 2:1265–1275

Mainwaring WIP (1969a) The binding of [1,2-^{3}H]testosterone within nuclei of the rat prostate. J Endocrinol 44:323–333

Mainwaring WIP (1969b) A soluble androgen receptor in the cytoplasm of rat prostate. J Endocrinol 45:531–541

Mainwaring WIP, Peterken BM (1971) A reconstituted cell-free system for the specific transfer of steroid-receptor complexes into nuclear chromatin isolated from rat ventral prostate gland. Biochem J 125:285–295

Mainwaring WIP, Mangan FR, Feherty PA, Freifeld M (1974) An investigation into the anti-androgenic properties of the non-steroidal compound, SCH 13521 (4′-nitro-3′-trifluoro-methylisobutyrylanilide). Mol Cell Endocrinol 1:113–128

Mangan FR, Neal GE, Williams DC (1968) Subcellular distribution of testosterone in rat prostate and its possible relationship to nuclear ribonucleic acid synthesis. Arch Biochem Biophys 124:27–40

Marchetti E, Querzoli P, Moncharmont B et al. (1987) Immunocytochemical demonstration of estrogen receptors by monoclonal antibodies in human breast cancer: correlation with estrogen receptor assay by dextran-coated charcoal method. Cancer Res 47:2508–2513

Marković RD, Litwack G (1980) Activation of liver and kidney glucocorticoid-receptor complexes occurs in vivo. Arch Biochem Biophys 202:374–379

Maroy P, Dennis R, Beckers C, Sage BA, O'Connor JD (1978) Demonstration of an ecdysteroid receptor in a cultured cell line of *Drosophila melanogaster*. Proc Natl Acad Sci USA 75:6035–6038

Martin PM, Sheridan PJ (1982) Towards a new model for the mechanism of action of steroids. J Steroid Biochem 16:215–229

Marver D, Goodman D, Edelman IS (1972) Relationships between renal cytoplasmic and nuclear aldosterone-receptors. Kidney Int 1:210–223

Marver D, Stewart J, Funder JW, Feldman D, Edelman IS (1974) Renal aldosterone receptors: studies with [^{3}H]aldosterone and the anti-mineralocorticoid [^{3}H]spirolactone (SC-26304). Proc Natl Acad Sci USA 71:1431–1435

Maxwell BL, McDonnell DP, Conneely OM, Schultz TZ, Greene GL, O'Malley BW (1987) Structural organization and regulation of the chicken estrogen receptor. Mol Endocrinol 1:25–35

McClellan MC, West NB, Tacha DE, Greene GL, Brenner RM (1984) Immunocytochemical localization of estrogen receptors in the macaque reproductive tract with monoclonal anti-estrophilins. Endocrinology 114:2002–2014

McDonnell DP, Mangelsdorf DJ, Pike JW, Haussler MR, O'Malley BW (1987) Molecular cloning of complementary DNA encoding the avian receptor for vitamin D. Science 235:1214–1217

McEwen BS, Plapinger L (1970) Association of ^{3}H corticosterone-1,2 with macromolecules extracted from brain cell nuclei. Nature 226:263–265

McEwen BS, Weiss JM, Schwartz LS (1968) Selective retention of corticosterone by limbic structures in rat brain. Nature 220:911–912

McEwen BS, Weiss JM, Schwarz LS (1969) Uptake of corticosterone by rat brain and its concentration by certain limbic structures. Brain Res 16:227–241

McEwen BS, Weiss JM, Schwartz LS (1970) Retention of corticosterone by cell nuclei from brain regions of adrenalectomized rats. Brain Res 17:471–482

McGuire JL, DeDella C (1971) In vitro evidence for a progestogen receptor in the rat and rabbit uterus. Endocrinology 88:1099–1103

McGuire WL (1978) Steroid receptors in human breast cancer. Cancer Res 38:4289–4291

McGuire WL, Huff K, Chamness GC (1972) Temperature-dependent binding of estrogen receptor to chromatin. Biochemistry 11:4562–4565

Mellon WS, DeLuca HF (1980) A specific 1,25-dihydroxyvitamin D$_3$ binding macromolecule in chicken bone. J Biol Chem 255:4081–4086

Mellon WS, Franceschi RT, DeLuca HF (1980) An in vitro study of the stability of the chicken intestinal cytosol 1,25-dihydroxyvitamin D$_3$-specific receptor. Arch Biochem Biophys 202:83–92

Mendel DB, Bodwell JE, Gametchu B, Harrison RW, Munck A (1986a) Molybdate-stabilized nonactivated glucocorticoid-receptor complexes contain a 90-kDa non-steroid-binding phosphoprotein that is lost on activation. J Biol Chem 261:3758–3763

Mendel DB, Bodwell JE, Munck A (1986b) Glucocorticoid receptors lacking hormone-binding activity are bound in nuclei of ATP-depleted cells. Nature 324:478–480

Mendel DB, Bodwell JE, Munck A (1987) Activation of cytosolic glucocorticoid-receptor complexes in intact WEHI-7 cells does not dephosphorylate the steroid-binding protein. J Biol Chem 262:5644–5648

Meschinchi S, Sanchez ER, Martell KJ, Pratt WB (1990) Elimination and reconstitution of the requirement for hormone in promoting temperature-dependent transformation of cytosolic glucocorticoid receptors to the DNA-binding state. J Biol Chem 265:4863–4870

Meyer M-E, Gronemeyer H, Turcotte B, Bocquel M-T, Tasset D, Chambon P (1989) Steroid hormone receptors compete for factors that mediate their enhancer function. Cell 57:433–442

Miesfeld R, Rusconi S, Godowski PJ et al. (1986) Genetic complementation of a glucocorticoid receptor deficiency by expression of cloned receptor cDNA. Cell 46:389–399

Migliaccio A, Lastoria S, Moncharmont B, Rotondi A, Auricchio F (1982) Phosphorylation of calf uterus 17β-estradiol receptor by endogenous Ca^{2+}-stimulated kinase activating the hormone binding of the receptor. Biochem Biophys Res Commun 109:1002–1010

Migliaccio A, Rotondi A, Auricchio F (1984) Calmodulin-stimulated phosphorylation of 17β-estradiol receptor on tyrosine. Proc Natl Acad Sci USA 81:5921–5925

Migliaccio A, Rotondi A, Auricchio F (1986) Estradiol receptor: phosphorylation on tyrosine in uterus and interaction with anti-phosphotyrosine antibody. EMBO J 5:2867–2872

Migliaccio A, Di Domenico M, Green S et al. (1989) Phosphorylation on tyrosine of in vitro synthesized human estrogen receptor activates its hormone binding. Mol Endocrinol 3:1061–1069

Milgrom E (1981) Activation of steroid-receptor complexes. In: Litwack G (ed) Biochemical actions of hormones, vol 8. Academic, New York, pp 465–492

Milgrom E, Atger M (1975) Receptor translocation inhibitor and apparent saturability of the nuclear acceptor. J Steroid Biochem 6:487–492

Milgrom E, Bauilieu E-E (1970) Progesterone in uterus and plasma I. Binding in rat uterus 105000 g supernatant. Endocrinology 87:276–287

Milgrom E, Atger M, Baulieu E-E (1970) Progesterone in uterus and plasma. IV - Progesterone receptor(s) in guinea pig uterus cytosol. Steroids 16:741–754

Milgrom E, Atger M, Baulieu E-E (1973) Acidophilic activation of steroid hormone receptors. Biochemistry 12:5198–5205

Miller MA, Mullick A, Greene GL, Katzenellenbogen BS (1985) Characterization of the subunit nature of nuclear estrogen receptors by chemical cross-linking and dense amino acid labeling. Endocrinology 117:515–522

Miller-Diener A, Schmidt TJ, Litwack G (1985) Protein kinase activity associated with the purified rat hepatic glucocorticoid receptor. Proc Natl Acad Sci USA 82:4003–4007

Misrahi M, Atger M, d'Auriol L et al. (1987) Complete amino acid sequence of the human progesterone receptor deduced from cloned cDNA. Biochem Biophys Res Commun 143:740–748

Miyabe S, Harrison RW (1983) In vivo activation and nuclear binding of the AtT-20 mouse pituitary tumor cell glucocorticoid receptor. Endocrinology 112:2174–2180

Mohla S, DeSombre ER, Jensen EV (1972) Tissue-specific simulation of RNA synthesis by transformed estradiol-receptor complex. Biochem Biophys Res Commun 46:661–667

Moncharmont B, Parikh I (1983) Binding of monoclonal antibodies to the nuclear estrogen receptor in intact nuclei. Biochem Biophys Res Commun 114:107–112

Moncharmont B, Su J-L, Parikh I (1982) Monoclonal antibodies against estrogen receptor: interaction with different molecular forms and functions of the receptor. Biochemistry 21:6916–6921

Morel G, Dubois P, Benassayag C, Nunez E, Radanyi C, Redeuilh G, Richard-Foy H, Baulieu E-E (1981) Ultrastructural evidence of oestradiol receptor by immunochemistry. Exp. Cell Res 132:249–257

Mueller GC (1965) Role of RNA and protein synthesis in estrogen action. In: Karlson P (ed) Mechanisms of hormone action. Academic, New York, pp 228–245

Mueller GC, Herranen AM, Jervell KJ (1958) Studies on the mechanism of action of estrogens. Recent Prog Horm Res 14:95–129

Mueller GC, Gorski J, Aizawa Y (1961) The role of protein synthesis in early estrogen action. Proc Natl Acad Sci USA 47:164–169

Muldoon TG, Watson GH, Evans AC Jr, Steinsapir J (1988) Microsomal receptor for steroid hormones: functional implications for nuclear activity. J Steroid Biochem 30:23–31

Müller RE, Traish A, Wotiz HH (1980) Effects of pyridoxal 5'-phosphate on uterine estrogen receptor. I. Inhibition of nuclear binding in cell-free system and intact uterus. J Biol Chem 255:4062–4067

Müller RE, Traish AM, Wotiz HH (1983) Estrogen receptor activation precedes transformation. Effects of ionic strength, temperature, and molybdate. J Biol Chem 258:9227–9236

Munck A, Brinck-Johnsen T (1968) Specific and nonspecific physiocochemical interactions of glucocorticoids and related steroids with rat thymus cells in vitro. J Biol Chem 243:5556–5565

Munck, A, Foley R (1979) Activation of steroid hormone-receptor complexes in intact target cells in physiological conditions. Nature 278:752–754

Munck A, Holbrook NJ (1984) Glucocorticoid-receptor complexes in rat thymus cells. Rapid kinetic behavior and a cyclic model. J Biol Chem 259:820–831

Munck A, Wira C, Young DA, Mosher KM, Hallahan C, Bell PA (1972) Glucocorticoid-receptor complexes and the earliest steps in the action of glucocorticoids on thymus cells. J Steroid Biochem 3:567–578

Nadji M, Morales AR (1986) In: Immunoperoxidase techniques: a practical approach to tumor diagnosis. American Society of Clinical Pathologists Press, Chicago, pp 76, 172, 173

Nakao K, Myers JE, Faber LE (1985) Development of a monoclonal antibody to the rabbit 8.5 S uterine progestin receptor. Can J Biochem Cell Biol 63:33–40

Nielsen CJ, Sando JJ, Vogel WM, Pratt WB (1977a) Glucocorticoid receptor inactivation under cell-free conditions. J Biol Chem 252:7568–7578

Nielsen CJ, Sando JJ, Pratt WB (1977b) Evidence that dephosphorylation inactivates glucocorticoid receptors. Proc Natl Acad Sci USA 74:1398–1402

Nishigori H, Toft D (1979) Chemical modification of avian progesterone receptor by pyridoxal 5'-phosphate. J Biol Chem 254:9155–9161

Norman AW, Myrtle JF, Midgett RJ, Nowicki HG, Williams V, Popják G (1971) 1,25-Dihydroxy-cholecalciferol: identification of the proposed active form of vitamin D_3 in the intestine. Science 173:51–54

Norman AW, Roth J, Orci L (1982) The vitamin D endocrine system: steroid metabolism, hormone receptors, and biological response (calcium binding proteins) Endocr Rev 3:331–366

Noteboom WD, Gorski J (1965) Stereospecific binding of estrogens in the rat uterus. Arch Biochem Biophys 11:559–568

Notides AC, Nielsen S (1974) The molecular mechanism of the in vitro 4 S to 5 S transformation of the uterine estrogen receptor. J Biol Chem 249:1866–1873

Notides AC, Hamilton DE, Auer HE (1975) A kinetic analysis of the estrogen receptor transformation. J Biol Chem 250:3945–3950

Notides AC, Lerner N, Hamilton DE (1981) Positive cooperativity of the estrogen receptor. Proc Natl Acad Sci USA 78:4926–4930

Nozu K, Tamaoki B (1975) On the role of cytosol receptors in the incorporation of androgens into the prostatic nuclei of rat. J Steroid Biochem 6:57–63

Okret S (1983) Comparison between different rabbit antisera against the glucocorticoid receptor. J Steroid Biochem 19:1241–1248

Okret S, Carlstedt-Duke J, Wrange Ö, Carlström K, Gustafsson J-Å (1981) Characterization of an antiserum against the glucocorticoid receptor. Biochim Biophys Acta 677:205–219

Okret S, Wikström A-C, Wrange Ö, Andersson B, Gustafsson J-Å (1984) Monoclonal antibodies against the rat liver glucocorticoid receptor. Proc Natl Acad Sci USA 81:1609–1613

O'Malley BW, Means AR (1974) Female steroid hormones and target cell nuclei. Science 183:610–620

O'Malley BW, McGuire WL, Kohler PO, Korenman SG (1969) Studies on the mechanism of steroid hormone regulation of synthesis of specific proteins. Recent Prog Horm Res 25:105–160

O'Malley BW, Sherman MR, Toft DO (1970) Progesterone "receptors" in the cytoplasm and nucleus of chick oviduct target tissues. Proc Natl Acad Sci USA 67:501–508

O'Malley, BW, Toft DO, Sherman MR (1971) Progesterone-binding components of chick oviduct II. Nuclear components. J Biol Chem 246:1117–1122

Orti E, Mendel DB, Munck A (1989a) Phosphorylation of glucocorticoid receptor-associated and free forms of the ∼90-kDa heat shock protein before and after receptor activation. J Biol Chem 264:231–237

Orti E, Mendel DB, Smith LI, Munck A (1989b) Agonist-dependent phosphorylation and nuclear dephosphorylation of glucocorticoid receptors in intact cells. J Biol Chem 264:9728–9731

Osterbur DL, Yund MA (1982) Ecdysteroid binding activity in embryos of Drosophila melanogaster. J Cell Biochem 20:277–282

Papamichail M, Tsokos G, Tsawdaroglou N, Sekeris CE (1980) Immunocytochemical demonstration of glucocorticoid receptors in different cell types and their translocation from the cytoplasm to the cell nucleus in the presence of dexamethasone. Exp Cell Res 125:490–493

Parikh I, Rajendran KG, Su J-L, Lopez T, Sar M (1987) Are estrogen receptors cytoplasmic or nuclear? Some immunocytochemical and biochemical studies. J Steroid Biochem 27:185–192

Pasqualini JR, Sumida C, Gelly C (1972) Mineralocorticosteroid receptors in the foetal compartment. J Steroid Biochem 3:543–556

Pearlman WH, Pearlman MRJ (1961) The metabolism of Δ4-androstene-3,17-dione-7-H³; its localization in the ventral prostate and other tissues of the rat. J Biol Chem 236:1321–1327

Peets EA, Henson MF, Neri R (1974) On the mechanism of the anti-androgenic action of flutamide (α-α-α-trifluoro-2-methyl-4'-nitro-m-propionotoluidide) in the rat. Endocrinology 94:532–540

Peleg S, Schrader WT, O'Malley BW (1988) Sulfhydryl group content of chicken progesterone receptor: effect of oxidation on DNA binding activity. Biochemistry 27:358–367

Pérez-Palacios G, Perez AE, Cruz ML, Beyer C (1973) Comparative uptake of [³H] androgens by the brain and the pituitary of castrated male rats. Biol Reprod 8:395–399

Perrot-Applanat M, Logeat F, Groyer-Picard MT, Milgrom E (1985) Imunocytochemical study of mammalian progesterone receptor using monoclonal antibodies. Endocrinology 116:1473–1484

Perrot-Applanat M, Groyer-Picard M-T, Logeat F, Milgrom E (1986) Ultrastructural localization of progesterone receptor by an immunogold method: effect of hormone administration. J Cell Biol 102:1191–1199

Perrot-Applanat M, Groyer-Picard M-T, Lorenzo F et al. (1987) Immunocytochemical study with monoclonal antibodies to progesterone receptor in human breast tumors. Cancer Res 47:2652–2661

Pertschuk LP, Feldman JG, Eisenberg KB et al. (1988) Immunocytochemical detection of progesterone receptor in breast cancer with monoclonal antibody. Cancer 62:342–349

Picard D, Yamamoto KR (1987) Two signals mediate hormone-dependent nuclear localization of the glucocorticoid receptor. EMBO J 6:3333–3340

Pietras RJ, Szego CM (1979) Estrogen receptors in uterine plasma membrane. J Steroid Biochem 11:1471–1483

Pietras RJ, Szego CM (1980) Partial purification and characterization of oestrogen receptors in subfractions of hepatocyte plasma membranes. Biochem J 191:743–760

Pike JW (1982) Interaction between 1,25-dihydroxyvitamin D_3 receptors and intestinal nuclei. Binding to nuclear constituents in vitro. J Biol Chem 257:6766–6775

Pike JW (1984) Monoclonal antibodies to chick intestinal receptors for 1,25-dihydroxyvitamin D_3. Interaction and effects of binding on receptor function. J Biol Chem 259:1167–1173

Pike JW, Haussler MR (1983) Association of 1,25-dihydroxyvitamin D_3 with cultured 3T6 mouse fibroblasts. Cellular uptake and receptor-mediated migration to the nucleus. J Biol Chem 258:8554–8560

Pike JW, Sleator NM (1985) Hormone-dependent phosphorylation of the 1,25-dihydroxyvitamin D_3 receptor in mouse fibroblasts. Biochem Biophys Res Commun 131:378–385

Pike JW, Donaldson CA, Marion SL, Haussler MR (1982) Development of hybridomas secreting monoclonal antibodies to the chicken intestinal 1α,25-dihydroxyvitamin D_3 receptor. Proc Natl Acad Sci USA 79:7719–7723

Pike JW, Marion SL, Donaldson CA, Haussler MR (1983) Serum and monoclonal antibodies against the chick intestinal receptor for 1,25-dihydroxyvitamin D_3. J Biol Chem 258:1289–1296

Podratz KC, Katzman PA (1968) Effect of estradiol on uptake and retention of progesterone by the vagina of the ovariectomized mouse. Fed Proc 27:497

Porter GA, Bogoroch R, Edelman IS (1964) On the mechanism of action of aldosterone on sodium transport: the role of RNA synthesis. Proc Natl Acad Sci USA 52:1326–1333

Pratt WB (1987) Transformation of glucocorticoid and progesterone receptors to the DNA-binding state. J Cell Biochem 35:51–68

Pratt WB, Ishii DN (1972) Specific binding of glucocorticoids in vitro in the soluble fraction of mouse fibroblasts. Biochemistry 11:1401–1410

Pratt WB, Sanchez ER, Bresnick EH, Meshinchi S, Scherrer LC, Dalman FC, Welsh MJ (1989) Interaction of the glucocorticoid receptor with the M_r 90 000 heat shock protein: an evolving model of ligand-mediated receptor transformation and translocation. Cancer Res 49:2222s–2229s

Press MF, Greene GL (1988) Localization of progesterone receptor with monoclonal antibodies to the human progestin receptor. Endocrinology 122:1165–1175

Press MF, Nousek-Goebl NA, Greene GL (1985) Immunoelectron microscopic localization of estrogen receptor with monoclonal estrophilin antibodies. J Histochem Cytochem 33:915–924

Puca GA, Bresciani F (1968) Receptor molecule for oestrogens from rat uterus. Nature 218:967–969

Puca GA, Nola E, Sica V, Bresciani F (1972) Estrogen-binding proteins of calf uterus. Interrelationship between various forms and identification of a receptor-transforming factor. Biochemistry 11:4157–4165

Puca GA, Abbondanza C, Nigro V, Armetta I, Medici N, Molinari AM (1986) Estradiol receptor has proteolytic activity that is responsible for its own transformation. Proc Natl Acad Sci USA 83:5367–5371

Puri RK, Toft DO (1986) Peptide mapping of the avian progesterone receptor. J Biol Chem 261:5651–5657

Puri RK, Dougherty JJ, Toft DO (1984) The avian progesterone receptor: isolation and characterization of phosphorylated forms. J Steroid Biochem 20:23–29

Raam S, Peters L, Rafkind I, Putnum E, Longcope C, Cohen JL (1981) Simple methods for production and characterization of rabbit antibodies to human breast tumor estrogen receptors. Mol Immunol 18:143–156

Raam S, Nemeth E, Tamura H, O'Briain DS, Cohen JL (1982) Immunohistochemical localization of estrogen receptors in human mammary carcinoma using antibodies to the receptor protein. Eur J Cancer Clin Oncol 18:1–12

Radanyi C, Redeuilh G, Eigenmann E et al. (1979) Production et détection d'anticorps antirécepteur de l'oestradiol d'utérus de veau. Interaction avec le récepteur d'oviducte de poule. C R Acad Sci [D] (Paris) 288:255–258

Radanyi C, Joab I, Renoir J-M, Richard-Foy H, Baulieu E-E (1983) Monoclonal antibody to chicken oviduct progesterone receptor. Proc Natl Acad Sci USA 80:2854–2858

Radayni C, Renoir J-M, Sabbah M, Baulieu E-E (1989) Chicken heat-shock protein of $M_r = 90\,000$, free or released from progesterone receptor, is in a dimeric form. J Biol Chem 264:2568–2573

Rao BR, Wiest WG, Allen WM (1973) Progesterone "receptor" in rabbit uterus. I. Characterization and estradiol-17β augmentation. Endocrinology 92:1229–1240

Rao KVS, Fox CF (1987) Epidermal growth factor stimulates tyrosine phosphorylation of human glucocorticoid receptor in cultured cells. Biochem Biophys Res Commun 144:512–519

Rao KVS, Peralta WD, Greene GL, Fox CF (1987) Cellular progesterone receptor phosphorylation in response to ligands activating protein kinases. Biochem Biophys Res Commun 146:1357–1365

Raspé G (ed) (1971) Schering workshop on steroid hormone receptors. Advances in the biosciences, vol 7. Pergamon-Vieweg, Braunschweig

Raynaud-Jammet C, Baulieu E-E (1969) Action de l'oestradiol in vitro: Augmentation de la biosynthèse d'acide ribonucléique dans les noyaux utérine. C R Acad Sci [D] (Paris) 268:3211–3214

Reel JR, Van Dewark SD, Shih Y, Callantine MR (1971) Macromolecular binding and metabolism of progesterone in the decidual and pseudopregnant rat and rabbit uterus. Steroids 18:441–461

Reker CE, LaPointe MC, Kovačič-Millivojević B, Chiou WJH, Vedeckis WV (1987) A possible role for dephosphorylation in glucocorticoid receptor transformation. J Steroid Biochem 26:653–665

Renoir J-M, Radanyi C, Yang C-R, Baulieu E-E (1982) Antibodies against progesterone receptor from chick oviduct. Cross-reactivity with mammalian progesterone receptors. Eur J Biochem 127:81–86

Resko JA, Goy RW, Phoenix CH (1967) Uptake and distribution of exogenous testosterone-1,2-^{3}H in neural and genital tissues of the castrate guinea pig. Endocrinology 80:490–498

Ringold GM (1985) Steroid hormone regulation of gene expression. Annu Rev Pharmacol Toxicol 25:529–566

Ritzén EM, Nayfeh SN, French FS, Dobbins MC (1971) Demonstration of androgen-binding components in rat epididymis cytosol and comparison with binding components in prostate and other tissues. Endocrinology 89:143–151

Robinson KR (1979) Electrical currents through full-grown and maturing Xenopus oocytes. Proc Natl Acad Sci USA 76:837–841

Rochefort H, Baulieu E-E (1968) Récepteurs hormonaux: relations entre les «récepteurs» utérins de l'oestradiol, «8 S» cytoplasmique, et «4 S» cytoplasmique et nucléaire. C R Acad Sci Paris [D] 267:662–665

Rosenau W, Baxter JD, Rousseau GG, Tomkins GG (1972) Mechanism of resistance to steroids: glucocorticoid receptor defect in lymphoma cells. Nature New Biol 237:20–24

Rousseau GG (1975) Interaction of steroids with hepatoma cells: molecular mechanisms of glucocorticoid hormone action. J Steroid Biochem 6:75–89

Rousseau GG, Baxter JD, Tomkins GM (1972a) Glucocorticoid receptors: relations between steroid binding and biological effects. J Mol Biol 67:99–115

Rousseau G, Baxter JD, Funder JW, Edelman IS, Tomkins GM (1972b) Glucocorticoid and mineralocorticoid receptors for aldosterone. J Steroid Biochem 3:219–227

Rousseau GG, Baxter JD, Higgins SJ, Tomkins GM (1973) Steroid-induced nuclear binding of glucocorticoid receptors in intact hepatoma cells. J Mol Biol 79: 539–554

Roy S, Mahesh VB, Greenblatt RB (1964) Inhibition of uptake of radioactive estradiol by the uterus and pituitary gland of immature rats. Acta Endocrinol 47: 669–675

Rundle SE, Smith AI, Stockman D, Funder JW (1989) Immunocytochemical demonstration of mineralocorticoid receptors in rat and human kidney. J Steroid Biochem 33: 1235–1242

Sabbah M, Redeuilh G, Secco C, Baulieu E-E (1987) The binding activity of estrogen receptor to DNA and heat shock protein (M_r 90000) is dependent on receptor-bound metal. J Biol Chem 262: 8631–8635

Sadler SE, Maller JL (1982) Identification of a steroid receptor on the surface of Xenopus oocytes by photoaffinity labeling. J Biol Chem 257: 355–361

Saffran J, Loeser BK, Haas BM, Stavely HE (1973) Binding of progesterone by rat uterus in vitro. Biochem Biophys Res Commun 53: 202–209

Sage BA, Tanis MA, O'Connor JD (1982) Characterization of ecdysteroid receptors in cytosol and naive nuclear preparations of Drosophila K_c cells. J Biol Chem 257: 6373–6379

Sanchez ER, Pratt WB (1986) Phosphorylation of L-cell glucocorticoid receptors in immune complexes: evidence that the receptor is not a protein kinase. Biochemistry 25: 1378–1382

Sanchez ER, Toft DO, Schlesinger MJ, Pratt WB (1985) Evidence that the 90-kDa phosphoprotein associated with the untransformed L-cell glucocorticoid receptor is a murine heat shock protein. J Biol Chem 260: 12398–12401

Sanchez ER, Meshinchi S, Tienrungroj W, Schlesinger MJ, Toft DO, Pratt WB (1987) Relationship of the 90-kDa murine heat shock protein to the untransformed and transformed states of the L cell glucocorticoid receptor. J Biol Chem 262: 6986–6991

Sando JJ, La Forest AC, Pratt WB (1979a) ATP-dependent activation of L cell glucocorticoid receptors to the steroid binding form. J Biol Chem 254: 4772–4778

Sando JJ, Hammond ND, Stratford CA, Pratt WB (1979b) Activation of thymocyte glucocorticoid receptors to the steroid-binding form. The roles of reducing agents, ATP, and heat-stable factors. J Biol Chem 254: 4779–4789

Sap J, Muñoz A, Damm K, Goldberg Y, Ghysdael J, Leutz A, Beug H, Vennström B (1986) The c-erb-A protein is a high affinity receptor for thyroid hormone. Nature 324: 635–640

Sar M, Parikh I (1986) Immunohistochemical localization of estrogen receptor in rat brain, pituitary and uterus with monoclonal antibodies. J Steroid Biochem 24: 497–503

Sar M, Stumpf WE (1973a) Pituitary gonadotrophs: nuclear concentration of radioactivity after injection of [^{3}H]testosterone. Science 179: 389–391

Sar M, Stumpf WE (1973b) Autoradiographic localization of radioactivity in the rat brain after the injection of 1,2-^{3}H-testosterone. Endocrinology 92: 251–256

Sar M, Stumpf WE (1974) Cellular and subcellular localization of ^{3}H-progesterone or its metabolites in the oviduct, uterus, vagina and liver of the guinea pig. Endocrinology 94: 1116–1125

Sar M, Liao S, Stumpf WE (1970) Nuclear concentration of androgens in rat seminal vesicles and prostate demonstrated by dry-mount autoradiography. Endocrinology 86: 1008–1011

Sarff M, Gorski J (1971) Control of estrogen binding protein concentration under basal conditions and after estrogen administration. Biochemistry 10: 2557–2563

Sato B, Nishizawa Y, Noma K, Matsumoto K, Yamamura Y (1979) Estrogen-independent nuclear binding of receptor protein of rat uterine cytosol by removal of low molecular weight inhibitor. Endocrinology 104: 1474–1479

Sato B, Noma K, Nishizawa Y, Nakao K, Matusomoto K, Yamamura Y (1980) Mechanism of activation of steroid receptors: involvement of low molecular weight inhibitor in activation of androgen, glucocorticoid, and estrogen receptor systems. Endocrinology 106: 1142–1148

Sato B, Miyashita Y, Maeda Y, Noma K, Kishimoto S, Matsumoto K (1987) Effects of estrogen and vanadate on the proliferation of newly established transformed mouse Leydig cell line in vitro. Endocrinology 120: 1112–1120

Savouret JF, Misrahi M, Loosfelt H et al. (1989) Molecular and cellular biology of mammalian progesterone receptors. Recent Prog Horm Res 45: 65–120

Schaltmann K, Pongs O (1982) Identification and characterization of the ecdysterone receptor in Drosophila melanogaster by photoaffinity labeling. Proc Natl Acad Sci USA 79: 6–10

Schauer M, Chalepakis G, Willmann T, Beato M (1989) Binding of hormone accelerates the kinetics of glucocorticoid and progesterone receptor binding to DNA. Proc Natl Acad Sci USA 86:1123–1127

Schaumburg BP (1970) Studies of the glucocorticoid-binding protein from thymocytes. I. Localization in the cell and some properties of the protein. Biochim Biophys Acta 214:520–532

Schaumburg BP (1972) Investigations on the glucocorticoid-binding protein from rat thymocytes. II. Stability, kinetics and specificity of binding of steroids. Biochim Biophys Acta 261:219–235

Schaumburg BP, Bojesen E (1968) Specificity and thermodynamic properties of the corticosteroid binding to a receptor of rat thymocytes in vitro. Biochim Biophys Acta 170:172–188

Schmidt TJ, Litwack G (1982) Activation of the glucocorticoid-receptor complex. Physiol Rev 62:1131–1192

Schrader WT (1984) New model for steroid hormone receptors? Nature 308:17–18

Schrader WT, O'Malley BW (1972) Progesterone-binding components of chick oviduct. IV. Characterization of purified subunits. J Biol Chem 247:51–59

Schuetz AW (1974) Role of hormones in oocyte maturation. Biol Reprod 10:150–178

Schuh S, Yonemoto W, Brugge J, Bauer VJ, Riehl RM, Sullivan WP, Toft DO (1985) A 90000-dalton binding protein common to both steroid receptors and the Rous sarcoma virus transforming protein pp60^{v-src}. J Biol Chem 260:14292–14296

Schüle R, Muller M, Kaltschmidt C, Renkawitz R (1988) Many transcription factors interact synergistically with steroid receptors. Science 242:1418–1420

Sekula BC, Schmidt TJ, Litwack G (1981) Redefinition of modulator as an inhibitor of glucocorticoid receptor activation. J Steroid Biochem 14:161–166

Severne Y, Wieland S, Schaffner W, Rusconi S (1988) Metal binding 'finger' structures in the glucocorticoid receptor defined by site-directed mutagenesis. EMBO J 7:2503–2508

Sharp GWG, Komack CL, Leaf A (1966) Studies on the binding of aldosterone in the toad bladder. J Clin Invest 45:450–459

Sheridan PJ, Buchanan JM, Anselmo VC, Martin PM (1979) Equilibrium: the intracellular distribution of steroid receptors. Nature 282:579–582

Sheridan PJ, Buchanan JM, Anselmo VC, Martin PM (1981) Unbound progesterone receptors are in equilibrium between the nucleus and cytoplasm in cells of the rat uterus. Endocrinology 108:1533–1537

Sheridan PL, Krett NL, Gordon JA, Horwitz KB (1988) Human progesterone receptor transformation and nuclear down-regulation are independent of phosphorylation. Mol Endocrinol 2:1329–1342

Sheridan PL, Evans RM, Horwitz KB (1989a) Phosphotryptic peptide analysis of human progesterone receptors. New phosphorylated sites formed in nuclei after hormone treatment. J Biol Chem 264:6520–6528

Sheridan PL, Francis MD, Horwitz KB (1989b) Synthesis of human progesterone receptors in T47D cells. Nascent A- and B-receptors are active without a phosphorylation-dependent posttranslational maturation step. J Biol Chem 264:7054–7058

Sherman MR, Corvol PL, O'Malley BW (1970) Progesterone-binding components of chick oviduct. I. Preliminary characterization of cytoplasmic components. J Biol Chem 245:6085–6096

Sierralta WD, Szendro PI, Kallweit E, Jungblut PW (1987) Biosynthesis and posttranslational finishing of the estradiol receptor. J Steroid Biochem 27:109–113

Simons SS Jr, Martinez HM, Garcea RL, Baxter JD, Tomkins GM (1976) Interactions of glucocorticoid receptor-steroid complexes with acceptor sites. J Biol Chem 251:334–343

Singer S, Becker JE, Litwack G (1973) The principal glucocorticoid binding macromolecule in hepatoma cells in culture is similar to corticosteroid binder II of rat liver cytosol. Biochem Biophys Res Commun 52:943–950

Singh VB, Moudgil VK (1985) Phosphorylation of rat liver glucocorticoid receptor. J Biol Chem 260:3684–3690

Singh VB, Eliezer N, Moudgil VK (1986) Transformation and phosphorylation of purified molybdate-stabilized chicken oviduct progesterone receptor. Biochim Biophys Acta 888:237–248

Skafar DF, Notides AC (1985) Modulation of the estrogen receptor's affinity for estradiol. J Biol Chem 260:12208–12213

Sloman JC, Bell PA (1976) The dependence of specific nuclear binding of glucocorticoids by rat thymus cells on cellular ATP levels. Biochim Biophys Acta 428:403–413

Smith DF, Lubahn DB, McCormick DJ, Wilson EM, Toft DO (1988) The production of antibodies against the conserved cysteine region of steroid receptors and their use in characterizing the avian progesterone receptor. Endocrinology 122:2816–2825

Spaziani E, Szego CM (1959) Further evidence for mediation by histamine of estrogenic stimulation of the rat uterus. Endocrinology 64:713–723

Steinsapir J, Evans AC Jr, Bryhan M, Muldoon TG (1985) Androgen receptor dynamics in the rat ventral prostate. Biochim Biophys Acta 842:1–11

Steinsapir J, Bryhan M, Muldoon TG (1989) Relative binding properties of microsomal and cytosolic androgen receptor species of the ventral prostate. Endocrinology 125:2297–2311

Stern JM, Eisenfeld AJ (1969) Androgen accumulation and binding to macromolecules in seminal vesicles: inhibition by cyproterone. Science 166:233–235

Stern JM, Eisenfeld AJ (1971) Distribution and metabolism of ^{3}H-testosterone in castrated male rats: effects of cyproterone, progesterone and unlabeled testosterone. Endocrinology 88:1117–1125

Stevens B, Alvarez CM, Bohman R, O'Connor JD (1980) An ecdysteroid-induced alteration in the cell cycle of cultured Drosophila cells. Cell 22:675–682

Stone GM (1963) The uptake of tritiated oestrogens by various organs of the ovariectomized mouse following subcutaneous administration. J Endocrinol 27:281–288

Stone GM (1964) The effect of oestrogen antagonists on the uptake of tritiated oestradiol by the uterus and vagina of the ovariectomized mouse. J Endocrinol 29:127–136

Stone GT, Baggett B (1965) The in vitro uptake of tritiated estradiol and estrone by the uterus and vagina of the ovariectomized mouse. Steroids 5:809–826

Stone GM, Martin L (1964) The uptake of tritiated oestradiol and oestrone by the uterus of the ovariectomized mouse following local application. Steroids 3:699–706

Strähle U, Klock G, Schütz G (1987) A DNA sequence of 15 base pairs is sufficient to mediate both glucocorticoid and progesterone induction of gene expression. Proc Natl Acad Sci USA 84:7871–7875

Stumpf WE, Roth LJ (1966) High resolution autoradiography with dry mounted, freeze-dried frozen sections. J Histochem Cytochem 14:274–287

Stumpf WE, Sar M, Reid FA, Tanaka Y, DeLuca HF (1979) Target cells for 1,25-dihydroxyvitamin D_3 in intestinal tract, stomach, kidney, skin, pituitary, and parathyroid. Science 206:1188–1190

Stumpf WE, Sar M, Narbaitz R, Reid FA, DeLuca HF, Tanaka Y (1980) Cellular and subcellular localization of 1,25-(OH)$_2$-vitamin D_3 in rat kidney: comparison with localization of parathyroid hormone and estradiol. Proc Natl Acad Sci USA 77:1149–1153

Sullivan WP, Vroman BT, Bauer VJ, Puri RK, Riehl RM, Pearson GR, Toft DO (1985) Isolation of steroid receptor binding protein from chicken oviduct and production of monoclonal antibodies. Biochemistry 24:4214–4222

Sullivan WP, Beito TG, Proper J, Krco CJ, Toft DO (1986) Preparation of monoclonal antibodies to the avian progesterone receptor. Endocrinology 119:1549–1557

Sullivan WP, Madden BJ, McCormick DJ, Toft DO (1988a) Hormone-dependent phosphorylation of the avian progesterone receptor. J Biol Chem 263:14717–14723

Sullivan WP, Smith DF, Beito TG, Krco CJ, Toft DO (1988b) Hormone-dependent processing of the avian progesterone receptor. J Cell Biochem 36:103–119

Sulya LL, McCaa CS, Read VH, Bomer D (1963) Uptake of tritiated aldosterone by rat tissues. Nature 200:788–789

Swaneck GE, Highland E, Edelman IS (1969) Stereospecific nuclear and cytosol aldosterone-binding proteins of various tissues. Nephron 6:297–316

Swaneck GE, Chu LLH, Edelman IS (1970) Stereospecific binding of aldosterone to renal chromatin. J Biol Chem 245:5382–5389

Szego CM (1965) Role of histamine in mediation of hormone action. Fed Proc 24:1343–1352

Tai P-KK, Maeda Y, Nakao K, Wakim NJ, Duhring JL, Faber LE (1986) A 59-kilodalton protein associated with progestin, estrogen, androgen, and glucocorticoid receptors. Biochemistry 25:5269–5275

Tan J, Joseph DR, Quarmby VE, Lubahn DB, Sar M, French FS, Wilson EM (1988) The rat androgen receptor: primary structure, autoregulation of its messenger ribonucleic acid, and immunocytochemical localization of the receptor protein. Mol Endocrinol 2:1276–1285

Tchernitchin A (1973) Fine structure of rat uterine eosinophils and the possible role of eosinophils in the mechanism of estrogen action. J Steroid Biochem 4:277–282

Tchernitchin A (1979) The role of eosinophil receptors in the non-genomic response to oestrogens in the uterus. J Steroid Biochem 11:417–424

Teasdale J, Lewis FA, Barrett ID, Abbott AC, Wharton J, Bird CC (1986) Immunocytochemical application of monoclonal antibodies to rat liver glucocorticoid receptor. J Pathol 150:227–237

Terenius L (1965) Uptake of radioactive oestradiol in some organs of immature mice. Acta Endocrinol 50:584–596

Terenius L (1966) Specific uptake of oestrogens by the mouse uterus in vitro. Acta Endocrinol 53:611–618

Thieulant ML, Samperez S, Jouan P (1973) Binding and metabolism of [^{3}H]-testosterone in the nuclei of rat pituitary in vivo. J Steroid Biochem 4:677–685

Thomas JA, Smith CG, Mawhinney MG, Knych ET Jr (1970) Subcellular distribution of radioactivity in the prostate gland following the single injection of [1,2-^{3}H]testosterone. Acta Endocrinol 63:505–511

Thomas T, Kiang DT (1986) Ribonuclease-induced transformation of progesterone receptor from rabbit uterus. J Steroid Biochem 24:505–511

Tienrungroj W, Meshinchi S, Sanchez ER, Pratt SE, Grippo JF, Holmgren A, Pratt WB (1987a) The role of sulfhydryl groups in permitting transformation and DNA binding of the glucocorticoid receptor. J Biol Chem 262:6992–7000

Tienrungroj W, Sanchez ER, Housley PR, Harrison RW, Pratt WB (1987b) Glucocorticoid receptor phosphorylation, transformation, and DNA binding. J Biol Chem 262:17342–17349

Toft D, Gorski J (1966) A receptor molecule for estrogens: isolation from the rat uterus and preliminary characterization. Proc Natl Acad Sci USA 55:1574–1581

Toft D, Nishigori H (1979) Stabilization of the avian progesterone receptor by inhibitors. J Steroid Biochem 11:413–416

Toft D, Shyamala G, Gorski J (1967) A receptor molecule for estrogens. Studies using a cell-free system. Proc Natl Acad Sci USA 57:1740–1743

Toft DO, Lohmar P, Miller J, Moudgil V (1976) The properties and functional significance of ATP binding to progesterone receptors. J Steroid Biochem 7:1053–1059

Tora L, Mullick A, Metzger D, Ponglikitmongkol M, Park I, Chambon P (1989) The cloned human oestrogen receptor contains a mutation which alters its hormone binding properties. EMBO J 8:1981–1986

Traish A, Müller RE, Wotiz HH (1980) Effects of pyridoxal 5′-phosphate on uterine estrogen receptor II. Inhibition of estrogen-receptor transformation. J Biol Chem 255:4068–4072

Traish A, Kim N, Wotiz HH (1989) Characterization of polyclonal antibodies to preselected domains of the human estrogen receptor. Endocrinology 125:172–179

Tsai HC, Norman AW (1973) Studies on calciferol metabolism VIII. Evidence for a cytoplasmic receptor for 1,25-dihydroxy-vitamin D_3 in the intestinal mucosa. J Biol Chem 248:5967–5975

Tsai HC, Wong RG, Norman AW (1972) Studies on calciferol metabolism IV. Subcellular localization of 1,25-dihydroxy-vitamin D_3 in intestinal mucosa and correlation with increased calcium transport. J Biol Chem 247:5511–5519

Tsai SY, Carlstedt-Duke J, Weigel NL, Dahlman K, Gustafsson J-Å, Tsai M-J, O'Malley BW (1988) Molecular interactions of steroid hormone receptor with its enhancer element: evidence for receptor dimer formation. Cell 55:361–369

Tsai SY, Tsai M-J, O'Malley BW (1989) Cooperative binding of steroid hormone receptors contributes to transcriptional synergism at target enhancer elements. Cell 57:443–448

Tucker HA, Larson BL, Gorski J (1971) Cortisol binding in cultured bovine mammary cells. Endocrinology 89:152–160

Tuohimaa P, Renoir J-M, Radanyi C, Mešter J, Joab I, Buchou T, Baulieu E-E (1984) Antibodies against highly purified B-subunit of the chick oviduct progesterone receptor. Biochem Biophys Res Commun 119:433–439

Tveter KJ (1969) Subcellular localization of androgen in the rat ventral prostate in vivo. Endocrinology 85:597–600

Tveter KJ, Attramadal A (1968) Selective uptake of radioactivity in rat ventral prostate following administration of testosterone-1,2-^{3}H. Acta Endocrinol 59:218–226

Tveter KJ, Attramadal A (1969) Autoradiographic localization of androgen in the rat ventral prostate. Endocrinology 85:350–354

Ui H, Mueller GC (1963) The role of RNA synthesis in early estrogen action. Proc Natl Acad Sci USA 50:256–260

Unhjem O, Tveter KJ (1969) Localization of an androgen binding substance from the rat ventral prostate. Acta Endocrinol 60:571–578

Unhjem O, Tveter KJ, Aakvaag A (1969) Preliminary characterization of an androgen-macromolecular complex from the rat ventral prostate. Acta Endocrinol 62:153–164

Urda LA, Yen PM, Simons SS Jr, Harmon JM (1989) Region-specific antiglucocorticoid receptor antibodies selectively recognize the activated form of the ligand-occupied receptor and inhibit the binding of activated complexes to deoxyribonucleic acid. Mol Endocrinol 3:251–260

Vedeckis WV (1983) Subunit dissociation as a possible mechanism of glucocorticoid receptor activation. Biochemistry 22:1983–1989

Verrijdt A, Leclercq G, Devleeschouwer N, Danguy A (1985) Tritiated actinomycin-D staining method: a valuable tool to study oestrogen receptor-induced modifications of transcriptional activity in normal and neoplastic cells. Arch Int Physiol Biochim 93:65–73

von der Ahe D, Janich S, Scheiderheit C, Renkawitz R, Schütz G, Beato M (1985) Glucocorticoid and progesterone receptors bind to the same sites in two hormonally regulated promotors. Nature 313:706–709

Walters MR (1985) Steroid hormone receptors and the nucleus. Endocrine Rev 6:512–543

Walters MR, Hunziker W, Norman AW (1981) 1,25-Dihydroxyvitamin D$_3$ receptors: intermediates between triiodothyronine and steroid hormone receptors. Trends Biochem Sci 6:268–271

Walters MR, Hunziker W, Konami D, Norman AW (1982) Factors affecting the distribution and stability of unoccupied 1,25-dihydroxyvitamin D$_3$ receptors. J Recept Res 2:331–346

Walters SN, Reinhardt TA, Dominick MA, Horst RL, Littledike ET (1986) Intracellular location of unoccupied 1,25-dihydroxyvitamin D receptors: a nuclear-cytoplasmic equilibrium. Arch Biochem Biophys 246:366–373

Wang L-H, Tsai SY, Sagami I, Tsai M-J, O'Malley BW (1987) Purification and characterization of chicken ovalbumin upstream promotor transcription factor from HeLa cells. J Biol Chem 262:16080–16086

Wang L-H, Tsai SY, Cook RG, Beattie WG, Tsai M-J, O'Malley BW (1989) COUP transcription factor is a member of the steroid receptor superfamily. Nature 340:163–166

Waterman ML, Adler S, Nelson C, Greene GL, Evans RM, Rosenfeld MG (1988) A single domain of the estrogen receptor confers deoxyribonucleic acid binding and transcriptional activation of the rat prolactin gene. Mol Endocrinol 2:14–21

Watson GH, Muldoon TG (1985) Specific binding of estrogen and estrogen-receptor complex by microsomes from the estrogen-responsive tissues of the rat. Endocrinology 117:1341–1349

Wecksler WR, Norman AW (1980) Biochemical properties of 1α,25-dihydroxyvitamin D receptors. J Steroid Biochem 13:977–989

Wecksler WR, Henry HL, Norman AW (1977) Studies on the mode of action of calciferol. Subcellular localization of 1,25-dihydroxyvitamin D$_3$ in chicken parathyroid glands. Arch Biochem Biophys 183:168–175

Wecksler WR, Mason RS, Norman AW (1979a) Specific cytosol receptors for 1,25-dihydroxyvitamin D$_3$ in human intestine. J Clin Endocrinol Metab 48:715–717

Wecksler WR, Ross FP, Norman AW (1979b) Characterization of the 1α,25-dihydroxyvitamin D$_3$ receptor from rat intestinal cytosol. J Biol Chem 254:9488–9491

Wecksler WR, Ross FP, Mason RS; Norman AW (1980a) Biochemical properties of the 1α,25-dihydroxyvitamin D$_3$ cytosol receptors from human and chicken intestinal mucosa. J Clin Endocrinol Metab 50:152–157

Wecksler WR, Ross FP, Mason RS, Posen S, Norman AW (1980b) Biochemical properties of the 1α,25-dihydroxyvitamin D$_3$ cytoplasmic receptors from human and chick parathyroid glands. Arch Biochem Biophys 201:95–103

Wei LL, Sheridan PL, Krett NL, Francis MD, Toft DO, Edwards DP, Horwitz KB (1987) Immunologic analysis of human breast cancer progesterone receptors. 2. Structure, phosphorylation,
and processing. Biochemistry 26:6262–6272

Weigel NL, Tash JS, Means AR, Schrader WT, O'Malley BW (1981) Phosphorylation of hen
progesterone receptor by cAMP dependent protein kinase. Biochem Biophys Res Commun
102:513–519

Weiler IJ, Lew D, Shapiro DJ (1987) The Xenopus laevis estrogen receptor: sequence homology
with human and avian receptors and identification of multiple estrogen receptor messenger
ribonucleic acids. Mol Endocrinol 1:355–362

Weinberger C, Hollenberg SM, Rosenfeld MG, Evans RM (1985) Domain structure of human
glucocorticoid receptor and its relationship to the v-erb-A oncogene product. Nature 318:670–
672

Weinberger C, Thompson CC, Ong ES, Lebo R, Gruol DJ, Evans RM (1986) The c-erb-A gene
encodes a thyroid hormone receptor. Nature 324:641–646

Welshons WV, Lieberman ME, Gorski J (1984) Nuclear localization of unoccupied oestrogen
receptors. Nature 307:747–749

Welshons WV, Krummel BM, Gorski J (1985) Nuclear localization of unoccupied receptors for
glucocorticoids, estrogens and progesterone in GH_3 cells. Endocrinology 117:2140–2147

Westphal HM, Moldenhauer G, Beato M (1982) Monoclonal antibodies to the rat liver glucocorticoid receptor. EMBO J 1:1467–1471

Whalen RE, Luttge WE (1971) Testosterone, androstenedione and dihydrotestosterone: effects on
mating behavior of male rats. Horm Behav 2:117–125

Whalen RE, Luttge WG, Green R (1969) Effects of the anti-androgen cyproterone acetate on the
uptake of $1,2-^3H$-testosterone in neural and peripheral tissues of the castrate male rat. Endocrinology 84:217–222

Wiest WG, Rao BR (1971) Progesterone binding proteins in rabbit uterus and human endometrium. Adv Biosci 7:251–266

Wikström A-C, Bakke O, Okret S, Brönnegård M, Gustafsson J-Å (1987) Intracellular localization
of the glucocorticoid receptor: evidence for cytoplasmic and nuclear localization. Endocrinology 120:1232–1242

Williams D, Gorski J (1972) Kinetic and equilibrium analysis of estradiol in uterus: a model of
binding-site distribution in uterine cells. Proc Natl Acad Sci USA 69:3464–3468

Willmann T, Beato M (1986) Steroid-free glucocorticoid receptor binds specifically to mouse mammary tumor virus DNA. Nature 324:688–691

Wilson EM, Wright BT, Yarbrough WG (1986) The possible role of disulfide bond reduction in
transformation of the 10 S androgen receptor. J Biol Chem 261:6501–6508

Wilson EM, Lubahn DB, French FS, Jewell CM, Cidlowski JA (1988) Antibodies to steroid
receptor deoxyribonucleic acid binding domains and their reactivity with the human glucocorticoid receptor. Mol Endocrinol 2:1018–1026

Wira C, Munck A (1970) Specific glucocorticoid receptors in thymus cells. J Biol Chem 245:3436–
3438

Wong RG, Myrtle JF, Tsai HC, Norman AW (1972) Studies on calciferol metabolism V. The
occurrence and biological activity of 1,25-dihydroxy-vitamin D_3 in bone. J Biol Chem
247:5728–5735

Woo DDL, Fay SP, Griest R, Coty W, Goldfine I, Fox CF (1986) Differential phosphorylation of
the progesterone receptor by insulin, epidermal growth factor, and platelet-derived growth
factor receptor tyrosine protein kinases. J Biol Chem 261:460–467

Wrange Ö, Gustafsson J-Å (1978) Separation of the hormone- and DNA-binding sites of the
hepatic glucocorticoid receptor by means of proteolysis. J Biol Chem 253:856–865

Wrange Ö, Eriksson P, Perlmann T (1989) The purified activated glucocorticoid receptor is a
homodimer. J Biol Chem 264:5253–5259

Yamamoto KR (1985) Steroid receptor regulated transcription of specific genes and gene networks.
Annu Rev Genet 19:209–252

Yamamoto KR, Alberts BM (1972) In vitro conversion of estradiol-receptor protein to its nuclear
form: dependence on hormone and DNA. Proc Natl Acad Sci USA 69:2105–2109

Yang CR, Mešter J, Wolfson A, Renoir J-M, Baulieu E-E (1982) Activation of chick oviduct progesterone receptor by heparin in the presence or absence of hormone. Biochem J 208:399–406

Young CYF, Murthy LR, Prescott JL et al. (1988) Monoclonal antibodies against the androgen receptor: recognition of human and other mammalian androgen receptors. Endocrinology 123:601–610

Yund MA (1979) Specific binding of 20-hydroxyecdysone to nuclei of imaginal discs of *Drosophila melanogaster*. Mol Cell Endocrinol 14:19–35

Yund MA, Fristrom JW (1975) Uptake and binding of β-ecdysone in imaginal discs of *Drosophila melanogaster*. Dev Biol 43:287–298

Yund MA, King DS, Fristrom JW (1978) Ecdysteroid receptors in imaginal discs of *Drosophila melanogaster*. Proc Natl Acad Sci USA 75:6039–6043

Zigmond RE, Stern JM, McEwen BS (1972) Retention of radioactivity in cell nuclei in hypothalamus of the ring dove after injection of ^{3}H-testosterone. Gen Comp Endocrinol 18:450–453

Zile M, Bunge EC, Barsness L, Yamada S, Schnoes HK, DeLuca HF (1978) Localization of 1,25-dihydroxyvitamin D_3 in intestinal nuclei in vivo. Arch Biochem Biophys 186:15–24

Oncogene and Receptor Expression

Oncogene and Receptor Expression

H. Höfler

1 Introduction

The first evidence that certain genes may be involved in tumorigenesis came from the field of virology based on the original observation of the transplantability of malignant tumors in animals. The breakthrough became possible with the development of the techniques of DNA recombination, sequence analysis, and gene transfer. These methods made possible the isolation of the gene sequences responsible for the transforming potency of the Rous sarcoma virus (v-*src*; STEHELIN et al. 1976). Since that time, more than 25 genes have been found in highly oncogenic (transforming) retroviruses, which are responsible for malignant transformation. In fact, these viral oncogenes (v-*onc*'s) are not very relevant for cancer induction in man, except under experimental conditions and in sporadic cases (TEICH et al. 1984). v-*onc*-related or even homologous genes, however, have been detected in the genome of most eukaryotic species (BISHOP 1983). These highly conserved genes, called proto-oncogenes or cellular oncogenes (c-*onc*'s), are widely distributed throughout the genome of all normal cells. The number of well-characterized c-*onc*'s already exceeds 50 and is still growing. The locations of most c-*onc*'s on human chromosomes have been mapped (STUBBLEFIELD and SANFORD 1987). According to the generally accepted theory, all v-*onc*'s derive from their cellular relatives by recombination of the provirus with coding sequences of the c-*onc*'s. This theory is in line with the fact that not all c-*onc*'s have known viral counterparts (e.g., N-*myc*, L-*myc*, N-*ras*, c-*neu*, *int*-1,2; for details see VARMUS 1984). Generally, c-*onc*'s are expressed in a temporally regulated and tissue-specific fashion. These genes encode a variety of proteins that function as growth factors, protein kinases, proteins homologous to transducing proteins which play a role in membrane signaling, and nuclear proteins which may regulate transcription. Surprisingly, several c-*onc*'s reveal high sequence homologies or are even identical with well-known growth factors (e.g., platelet-derived growth factor) or growth factor receptors (e.g., epidermal growth factor receptor; see below for details). Therefore, the definition of oncogenes and growth factors and their receptors overlaps in many cases, with regard to both nucleotide sequence and function of the encoded proteins. Thus, similarly to growth factors and growth factor receptors, proteins encoded by c-*onc*'s are believed to contribute to the regulation of embryonal development, control of cell regeneration, and differentiation in normal tissue (SLAMON and CLINE 1984; WEINBERG 1985). The demonstration of qualitative and quantitative changes in c-*onc* expression in tumors by the use of recombinant DNA technology has led to the development of a new model of human neoplasia during recent decades. According to this concept, activation of c-*onc* plays an important role in the initiation and progression of human neoplasms, although its precise impact has not yet been unraveled (LAND et al. 1983; BUICK and POLLAK 1984; KNUDSON 1985; WEINBERG 1985; AARONSON and TRONICK 1985; BARBACID 1986a; DUESBERG 1987). In contrast to v-*onc*'s, known as the fastest acting carcinogens and capable of initiation and maintenance of the neoplastic phenotype, activation of one single c-*onc* is insufficient to induce the fully transformed phenotype. Most human neoplasms are thought to be the result of multiple events involving the activation of multiple genes and/or a multistep process (multistage carcinogenesis; WEINSTEIN 1987). Most recently, antagonistic genes ("antioncogenes") which suppress tumor growth have been discovered. Although our knowledge of the genetic background of cell growth regulation has expanded dramatically during recent years, it remains in its infancy.

This review is focused on the possible role of oncogenes in tumorigenesis and tumor progression and methods of detection of activated c-*onc*'s, insofar as they may be of interest for pathologists.

2 Oncogenes and Functions of Their Proteins

The proteins of c-*onc*'s are involved in both currently proposed mechanisms of cellular signal transduction, the activation of tyrosine (serine)-specific protein kinases and coupling of receptors to various effectors by means of G proteins (YARDEN and ULLRICH 1988). A number of c-*onc*'s act as classical (growth factor) receptors, while a smaller group of c-*onc*'s share high degrees of sequence homologies with growth factors (see Sect. 3). Several oncoproteins have been found in the cell nucleus and are thought to regulate gene transcription.

2.1 Oncogenes Encoding Protein Kinases

The vast majority of known c-*onc*'s encode protein kinases (Table 1). Most of the members of this group are receptors acting through tyrosine kinase activation in their cytoplasmic domain. Shared structural features and primary structure revealed the existence of three groups within the tyrosine-specific protein kinase family. Subclass I includes *erb*B-1 (EGF receptor) and the closely related *erb*B-2 (HER2/*neu*) gene. Subclass II is composed of the insulin receptor-related proteins, which share some homology with the c-*onc*'s *src*, *ros*, and *met* (PARK et al. 1986). Subclass III is represented by the proteins of the c-*onc*'s *fms* (CSF-1 receptor) and *kit* (CSF-1 and PDGF receptor-related).

Table 1. Oncogenes encoding tyrosine (serine-)specific protein kinases and related proteins and their cellular localization

Oncogene	Localization of protein	Protein/MW (kilodaltons)
erb B1	Cell membrane	72
erb B2 (*neu*)	Cell membrane	185
src	Cytoplasmic membranes	60
ros	Cytoplasmic membranes	62
fms	Cytoplasm, cell membranes	180
kit	Cytoplasmic membranes	145
yes (*lyn*)	Cytoplasmic membranes	90
fes (*fps*)	Cytoplasm	85/130
abl	Cytoplasm and membranes	150
fgr	Cytoplasmic membranes	72
met	Cytoplasmic membranes	63/110/140/160
onc D (*trk*)	Cytoplasmic membranes	
rel	Cytoplasm	64
mht	?	
sea	?	
mil/*raf*	Cytoplasm	100/78
mos	Cytoplasm	37

Furthermore, the c-*onc*'s *fgr*, *yes*, *abl*, *fes/fps*, *sea*, and *met*, and the fusion genes *trk* and *ret*, have been shown to exhibit tyrosine kinase activities. Proteins encoded by c-*mos* exhibit a cytoplasmic serine kinase activity. The products of c-*raf/mil* act through their serine/threonine kinase activity. For further details and references, see YARDEN and ULLRICH (1988).

2.2 Oncogenes Encoding Guanosine Triphosphate Binding Proteins

Like G proteins, the p21 proteins of the *ras* gene family (Ha-, Ki-, N-*ras*) and the products of the closely related *bas* gene bind guanosine triphosphate (GTP) and hydrolyze GTP to guanosine diphosphate (GDP) (Table 2). The binding of GTP to p21 induces an increased efficiency of protein function. Hydrolysis of GTP to GDP returns the p21 proteins to an inactive state. Single point mutations of the *ras* products (see Sect. 4) are associated with a significant decrease in the intrinsic GTPase function and represent a well-characterized mode of c-*onc* activation. For a recent comprehensive review of the *ras* gene family, see LACAL and TRONICK (1988).

2.3 Oncogenes Encoding Nuclear Proteins

These genes include the *myc* family (c-*myc*, N-*myc*, and L-*myc*) and the related c-*myb*, as well as c-*fos*, the cellular protein p53, c-*ski* and c-*ets* (Table 3). The protein products of *myc* and *myb* have been shown to bind to ds DNA and to nuclear matrix proteins and to regulate transcription. Inhibition of DNA synthesis and DNA polymerase-alpha in human nuclei by anti-c-*myc* antibodies suggests a direct role of c-*myc* proteins in DNA replication.

Furthermore, for c-*myc* and c-*fos* expression products a regulating function of RNA splicing has been proposed. Both c-*fos* and c-*myc* are expressed during the transition from G_0 to the G_1 phase of the cell cycle. In contrast, the expression of the c-*myb* gene occurs later in the cell cycle (S phase). p53 is thought to be involved in SV40 DNA tumor virus-mediated transformation, since the p53 protein forms complexes with the SV40 large T antigen. The proteins of the recently discovered sequences *tat* I of the HTLV I virus and *jun*, which is closely related to activator protein 1 (AP-1), have been shown to act within the nucleus and regulate gene expression. For further reading the excellent reviews by CURRAN (1988) and ERISMAN and ASTRIN (1988) are recommended.

2.4 Oncogenes Encoding Growth Factors

The *sis* oncogene was originally found in a simian sarcoma virus and has a product, $p28^{sis}$, which is found in the cell cytoplasm. The *sis* gene product was characterized as a precursor of PDGF beta (PDGF-2) by HELDIN and WESTER-

Table 2. Oncogenes encoding GTP-binding proteins

Oncogene	Localization of protein	Protein/MW (kilodaltons)
Ha-*ras*	Cytoplasmic membranes	21
Ki-*ras*	Cytoplasmic membranes	21
N-*ras*	Cytoplasmic membranes	21
bas	Cytoplasmic membranes	

Table 3. Oncogenes encoding nuclear (DNA-binding) proteins

Oncogene	Localization of protein	Protein/MW (kilodaltons)
c-*myc*	Nucleus	58–67
N-*myc*	Nucleus	90–200
L-*myc*	Nucleus	
myb	Nucleus	48
fos	Nucleus	39, 55, 135
ets	Nucleus	135
p53	Nucleus	53
ski	Nucleus	125
jun (AP-1)	Nucleus	
tat I	Nucleus	42

Table 4. Oncogenes encoding growth factors

Oncogene	Localization of protein	Protein/MW (kilokaltons)
sis	Cytoplasm	28
hst	Cytoplasm	
int-2	Cytoplasm	

MARK (1986). The secreted p28^{sis} may act either through an autocrine mechanism by stimulation of the specific PDGF receptor of the same cell or by interaction with specific PDGF receptors on neighboring (paracrine function) or distant (exocrine function) cells. Although, c-*sis* is overexpressed in some human osteosarcomas and glial cell tumors, it does not seem to be a general feature of bone, brain, or other tumors so far examined (TEICH 1988). This is as yet the only example of a direct relationship between an oncoprotein and a growth factor. Additionally, polyoma middle T-antigen reveals sequence homologies with the growth factor gastrin and the c-*onc int*-2 exhibits structural relationships to a fibroblast growth factor (DICKSON and PETERS 1987). A further, not yet fully characterized, candidate for an oncogene encoding a growth factor is *hst* ("human stomach cancer gene"; Table 4).

2.5 Others

Although most of the classical oncogenes are well characterized with regard to their gene sequences, expression products, and possible functions, several recently described c-*onc*'s and oncogenic recombinant genes remain to be further elucidated (Table 5). The description of these genes is beyond the scope of this review.

3 Oncogenes Encoding Growth Factor Receptors and Related Proteins

A rapidly growing number of c-*onc*'s are found to encode growth factor receptors and related proteins (Table 6), whereas only a few c-*onc*'s are known to encode growth factors.

3.1 EGF and Related Receptors

The search of computer data bases showed high levels of homology between the putative protein product of v-*erb*B and the amino acid sequence derived

Table 5. Unclassified oncogenes and recombinant genes

dbl
bcl-1,2
tx-1,2,3
tx-4
mcf-2,3,
pvt-1
pim-1
Mlvi-1,2,3
kil
bcr
lca
mam

Table 6. Oncogenes encoding receptors or related proteins

Oncogene	Receptor
erb B1	EGF receptor
erb B2 (*neu*)	EGF receptor related
fms/*fes* (*fps*)	CSF receptor
kit	CSF receptor related
erb A	Thyroid hormone receptor
mas	β-Adrenergic receptor
met	Receptor (insulin related?)

from the receptor for epidermal growth factor (EGF). Based on the predicted structure of the EGF receptor, v-*erb*B would encode a truncated EGF receptor lacking the external EGF-binding domain of the receptor. Generally, truncated receptors lacking the extracellular domain may be responsible for a continuous proliferation signal in transformed cells. A further gene, closely related to c-*erb*B-1 (EGF receptor) was reported by YAMAMOTO et al. (1986): c-*erb*B-2 (HER-2/*neu*) has 50% amino acid homology with the EGF receptor and over 80% of the amino acids in the tyrosine kinase domain are identical. The ligand of c-*neu*, encoding a 185 000-dalton protein-tyrosine kinase with growth factor receptor structure, has not yet been identified. As with the *ras* family, activation by point mutation in the transmembranous domain of c-*neu* was shown to lead to constitutive tyrosine kinase activity. This activation by point mutation, however, was demonstrated only in animal models, not in man. The localization of c-*neu* expression product is of particular interest in connection with the prediction of tumor prognosis (e.g., breast carcinoma; for further details see Sect. 5).

3.2 CSF and Related Receptors

c-*fes* protein is closely related to the oncoprotein of *fms*, which encodes the colony stimulating factor (CSF-1) receptor (SHERR et al. 1985). Within the tyrosine kinase receptor family more recently the c-*kit* gene has been established (YARDEN and ULLRICH 1988). The human proto-oncogene *kit* encodes a transmembranous protein that is also structurally related to the receptor for macrophage colony stimulating factor (CSF-1) and the receptor for PDGF. A ligand for the 145 000-dalton c-*kit* receptor protein has not yet been identified. Similar to the EGF receptor and its related viral counterpart v-*erb*B, the v-*kit* gene shares high sequence homologies with the intracellular (tyrosine kinase) domain of c-*kit*, but lacks the extracellular receptor domain. PDGF receptors can be activated by autocrine pathways involving the oncogene *sis*, which encodes a PDGF precursor. In contrast to receptor activation in normal cells, autocrine activation of PDGF receptors in v-*sis*-transformed cells occurs in intracellular compartments (KEATING and WILLIAMS 1988). The genes of this closely related receptor family (CSF receptor, *fms*, and PDGF receptor) are located on chromosome 5 (5q23–34). A more distant related member of this family is c-*mas*: its product has a tertiary structure which is closely related to β-adrenergic receptor or rhodopsin (LEFKOWITZ et al. 1986; YOUNG et al. 1986).

3.3 Thyroid Hormone Receptor

WEINBERGER et al. (1986) reported that the cDNA sequence of the human c-*erb*A gene, the cellular counterpart of v-*erb*A, indicates that its protein is closely related to a thyroid hormone receptor. This was the first time that a direct relationship between an oncoprotein and a steroid hormone receptor was postulated. In contrast to most other growth factor receptors encoded by c-*onc*'s the *erb*A gene product does not exhibit protein kinase activity.

3.4 Others

Although the c-*onc*'s *abl*, *fgr*, and *yes* do not exhibit receptor activity, they could, according to their structural relationships, phylogenetically originate from receptor genes. We can anticipate that several of the oncogenes which are to be further characterized will probably exhibit growth factor (receptor) functions.

There are several reports linking the expression of c-*onc*'s which do not encode growth factors or growth factor receptors with the mechanism of action of growth factors. Increased production of c-*myc*, for example, is associated with the stimulation of T-lymphocyte and fibroblast proliferation. The *src* family of c-*onc*'s, including *yes*, *fgr*, *ros*, and *abl*, all encode enzymes that might phosphorylate proteins which are normally targets for the growth factor-activated receptor protein-tyrosine kinase and thus provide a continuous intracellular mitogenic stimulus. The biochemical background of the modulation of growth factor receptors by oncogene expression has not yet been completely elucidated (BURGESS 1987).

4 Activation of Proto-oncogenes

The application of the term c-*onc* or proto-oncogene should be restricted to sequences which may be found in all (normal) higher cells. These genes are involved in the control of mitosis and differentiation, and play an important role in embryonic development and tissue repair. Most authors believe that the activation of c-*onc*'s represents a crucial event in cell transformation and for the expression of the malignant phenotype (EGAN et al. 1987). Activation of c-*onc*'s is in most cases caused by mutational events, including transduction, insertional mutagenesis, chromosome translocation or deletion of a suppressor gene, gene amplification, or point mutations. Such altered c-*onc*'s bear tumorigenic potential and are thought to be actively involved in tumor initiation and maintenance. This hypothesis that c-*onc*'s are latent cancer genes, originally based on HUEBNER's oncogene concept (HUEBNER and TODARO 1969), is not generally accepted. The main criticisms of this attractive concept include the facts that (a) not one activated c-*onc* has been isolated that transforms diploid cells, (b) no diploid tumors with activated c-*onc*'s have been found, and (c) the probability of spontaneous transformation in vivo is at least 10^9 times lower than predicted from the mechanisms thought to activate c-*onc*'s (DUESBERG 1987 for review). However, due to the regulatory functions of c-*onc*'s during cell proliferation, c-*onc* (over)expression must be involved in tumor progression, if not as an active then as a passive factor in the complex – and not completely understood – process of tumor cell growth regulation. Consequently, not every overexpression of one or more c-*onc*'s in a given tumor reflects an active (co)factor in tumorigenesis and tumor progression, but rather an unrelated(?) epiphenomenon. Similarly, clear demarcation of a physiological intermittent, regulated c-*onc* overexpression, e.g., during embryonal development and tissue repair, and the constitutive overexpression associated with the activation of a c-*onc*, may be extremely difficult and in some cases impossible (see MARKS 1987 for further details).

4.1 Transduction

Transduction of complete or partial c-*onc* sequences from a cellular into a primarily nononcogenic retroviral genome may result in a highly oncogenic recombinant gene. The activation of the original c-*onc* sequences may be caused by additional mutations or because of the acquisition of strong promoter sequences for DNA transcription. Furthermore, the loss of cellular transcription suppressing sequences during transduction may result in an enhanced expression of the recombinant gene.

4.2 Insertional Mutagenesis

The integration of a sequence derived from a (nononcogenic) retrovirus into a cellular genome in the vicinity of a c-*onc* may be associated with c-*onc* activation and subsequent tumor formation (insertional mutagenesis). Activation of a c-*onc* or a newly formed recombinant gene can occur by insertion of a strong viral promoter (LTRs), transcriptional enhancement over long distances, or the generation of a novel chimeric protein. Most examples of insertional mutagenesis are found in animal tumors: In murine mammary tumors, mouse mammary tumor virus (MMTV) insertions are frequently found near either of the two proto-oncogenes *int*-1 and *int*-2. In man, hepatitis B virus (HBV) is associated with the etiology of hepatocellular carcinoma, and integrated viral DNA can be detected in most tumors. HBV viral DNA sequences have been identified within a recombinant gene homologous to the *erb*A oncogene. Although the latter finding is based on a single case, it suggests the possibility that insertional mutagenesis plays a role in the oncogenicity of HBV virus in man (NUSSE 1987 for review). As in the case of HBV, a close association of the integration of human papilloma virus sequences into the cellular genome and c-*onc* activation is currently under discussion.

4.3 Chromosomal Translocation

Chromosomal translocation of c-*onc*'s leads to gene rearrangement and may be associated with the activation of translocated c-*onc*'s without involvement of viral sequences. Activation is caused by the loss of transcription control, due to either the acquisition of a new highly active promoter or the loss of the influence of a suppressor gene (MATHEW et al. 1987; FRIEND et al. 1988). The description of translocations of the c-*myc* gene in neoplasms of the lymphatic system contributed significantly to the understanding of c-*onc* activation. At the same time DALLA-FAVERA et al. (1982) and TAUB et al. (1982) reported the translocation t(8;14) of the c-*myc* gene into the immunoglobulin heavy chain locus in human Burkitt's lymphoma. According to the original hypothesis, the immunoglobulin-controlling elements cause a constitutive c-*myc* expression. Further studies showed, however, that not all t(8;14) translocations result in c-*myc* overexpression, and that in addition to the translocation mutations in the first (noncoding) c-*myc* exon occur frequently (RABBITTS et al. 1984; RAB-BITTS 1985). These mutations were shown to be associated with a threefold longer half-life of the c-*myc* transcripts, resulting in a cytoplasmic accumulation of c-*myc* mRNA. Other examples for translocation as a mode of *onc* activation include the rearrangement of the c-*abl* gene in chronic myelogenous leukemia (COLLINS et al. 1984): The reciprocal translocation t(9;22) results in a shorter chromosome 22, already known previously as the Philadelphia chromosome. The translocation of the oncogene *bcl*-1 t(11;14) is characteristic of small cell lymphocytic and diffuse large cell lymphomas (TSUJIMOTO et al. 1984a). *bcl*-2 translocation (14;18) is specifically associated with follicular B-cell lymphomas (TSUJIMOTO et al. 1984b).

4.4 Amplification

Another mechanism by which a c-*onc* can be activated is gene amplification. Amplification can range from a fewfold to many hundredfold and is usually associated with an elevated gene expression. Two cytogenetic abnormalities, double minute chromosomes and the occurrence of homogeneously staining regions, may be associated with the increase in the copy number of a particular gene. Even if c-*onc* amplification is not one of the initial steps in tumorigenesis, it most likely contributes to tumor maintenance and progression. Besides the first report of the amplification of the c-*myc* gene in a promyelocytic leukemia cell line HL-60 (COLLINS and GROUDINE 1982), several c-*onc* amplifications have been reported in tumors and some of them have shown significant correlations with tumor prognosis (N-*myc*, c-*myc*, *erb*B-2, inter alia; for details see Sect. 5). Other oncogenes that have been found to be amplified in human tumors include L-*myc*, c-*myb*, Ki-*ras*, Ha-*ras*, c-*abl*, c-*ets*, and *erb*B-1 (ALI-TALO 1987; NISHIMURA and SEKIJA 1987).

4.5 Point Mutation

Point mutation as a mechanism of c-*onc* activation was simultaneously reported for the *ras* gene family by two independent groups (REDDY et al. 1982;

TABIN et al. 1982). Within this gene family (Ha-, Ki-, and N-*ras*) single point mutations preferably occur at two "hot spots" – codon 12 or codon 61 – and result in replacement of one particular amino acid which leads to conformational changes and altered biological activity of the p21 proteins. Further targets of point mutations are codons 13, 59, 63, and 186 (for further details see AARONSON and TRONICK 1985 and LEVINSON 1987). Point mutations of the *ras* family have been found in several human neoplasms: Besides the original description in a human bladder carcinoma cell line T24 (Ha-*ras*), this mode of c-*onc* activation was reported in a high percentage of colorectal carcinomas in man (Ki-*ras*: BOS et al. 1987; FORRESTER et al. 1987). Other examples of c-*onc* activation by point mutation are provided by the c-*onc*'s *erb*B-2/*neu* (BAR-BACID 1986a) and *rsc* (BOLEN et al. 1987). It can be anticipated that with the rapidly growing use of the polymerase chain reaction technique (cf. Sect. 6), point mutations of c-*onc*'s will be detected in many human tumors.

4.6 Other Modes of Activation

Other modes of c-*onc* activation include not yet completely understood post-transcriptional and posttranslational events, resulting either in an increased stability of c-*onc* expression products or in an alteration of *onc* protein conformation associated with increased specific activity (COSMAN 1987; HUNT 1988).

As already mentioned briefly, most neoplasms are thought to be the result of multiple events involving the activation of multiple oncogenes. This concept of multistage carcinogenesis was excellently reviewed by LAND et al. (1983), MARKS (1987), and WEINSTEIN (1987). One has to bear in mind, however, that the environment of the transformed tumor cell and the genetic factors that suppress tumor progression are additional important factors in carcinogenesis. Genes may interact with the activity of growth-regulating (onco)genes, perhaps via *trans*-acting mechanisms, by restraining cell growth and by prevention of clonal cell proliferation. These genes were originally called "anti-oncogenes" because they were originally considered to interact with oncogenes. However, *tumor-suppressing genes* is a more appropriate term, since it includes the possibility of their interaction with genes other than oncogenes. Indeed, only one example of a direct interaction between an oncogene and a tumor-suppressing gene has been reported recently (WHYTE et al. 1988). The best characterized representative of these tumor-suppressing genes is the retinoblastoma (*rb*) gene, which not only seems to be involved in the prevention of retinoblastoma, but has also been found to play a role in the development of osteosarcomas, soft tissue sarcomas, melanomas, breast carcinomas, bladder carcinomas etc. (T'ANG et al. 1988). Currently, the impact of tumor-suppressing genes on tumorigenesis and tumor maintenance must be considered to be even greater than the role of oncogenes (KNUDSON 1985; FRIEND et al. 1988; WEINBERG 1988 for review).

5 Oncogenes and Tumor Prognosis

Reports on the demonstration of activated c-onc's in human tumors are frequent, but some of them are contradictory. YOKOTA et al. (1986) postulated that c-onc alterations are found in more than one-third of advanced tumors. BARBACID (1986 b) found activated c-onc's in 25% of carcinomas. The main reason for these rather low percentages is the fact that the different modes of c-onc activation would require a battery of different detection systems not used in the above-mentioned studies. A possible relationship of c-onc activation and the biological behavior of tumors would be of major interest. Several reports in the recent literature indicate that such a relationship exists.

The classical description of amplification and overexpression of N-myc mRNA in rapidly progressing *neuroblastomas* by SCHWAB et al. (1984) was confirmed by BRODEUR et al. (1986). In an extensive study of 89 cases of neuroblastoma, N-myc amplification was found in 19 of 40 stage IV, 3 of 20 stage III, 2 of 16 stage II, and none of 8 stage I tumors. Furthermore, in stage IVS neuroblastomas N-myc amplification was not found. The rate of gene amplification did not correlate with the rate of mRNA expression in all cases (SCHWAB et al. 1984). The possibility of N-myc activation by prolongation of the N-myc mRNA life span was reported by AMY and BARTOLOMEW (1987). These latter observations have to be taken into consideration when choosing a method for the detection of N-myc activation in neuroblastomas: the demonstration of N-myc DNA amplification may therefore not be as appropriate as the quantitative measurement of N-myc mRNA or its protein (see Sect. 6). Nevertheless, the detection of N-myc DNA amplification or N-myc mRNA overexpression in neuroblastomas is already a widely used tool to gain additional information on the biological behavior of neuroblastomas.

FUNA et al. (1987) reported an overexpression of N-myc mRNA in patients with *small cell lung carcinomas* with extremely poor prognosis, rapid tumor growth, short survival times, and lack of response to chemotherapy. Prognosis-related amplification of the c-myc gene in a subgroup of small cell lung carcinomas has also been reported (WONG et al. 1986).

RIOU et al. (1987) demonstrated in early stages of *squamous cell carcinomas of the uterine cervix* a significant c-myc mRNA overexpression, correlating with an eight times higher incidence of early tumor recurrence.

Rearrangement of oncogenes in *non-Hodgkin lymphomas* is of well-known diagnostic and prognostic significance. The translocation t(8;14) of the c-myc gene in Burkitt's lymphomas and the translocation of bcl-1 t(11;14) and bcl-2 t(14;18) have already been mentioned. Our own studies indicate a close correlation of c-myc mRNA overexpression with the Kiel classification of non-Hodgkin lymphomas. In high grade lymphomas a significant overexpression of c-myc mRNA was found, as compared with low grade lymphomas and cases of lymphadenitis (RADASKIEWICZ et al. 1987).

SLAMON et al.'s (1987) report on the close association of the amplification of c-neu (erbB-2) gene with the rate of recurrence of *breast carcinomas* caused great interest: In 30% of breast carcinomas a 2- to 24-fold amplification of the c-*neu* gene was found. *neu* amplification was significantly associated with early tumor recurrence and poor prognosis. According to several other reports, the incidence of c-*neu* amplification in breast cancers and its usefulness as a predictor of tumor recurrence are still subjects of controversy (MARX 1988). Most recently ZHOU et al. (1989) concluded that no statistically significant correlation exists between the amplification of the c-*neu* gene and the recurrence of tumors. According to the latter authors, analysis of c-*neu* expression alone is not useful because breast cancers utilize multiple genetic mechanisms in progression and metastasis. In line with these findings, the amplification of *bcl*-1, *int*-2, and *hst* loci on chromosome 11q13 was demonstrated in aggressive primary breast tumors (ZHOU et al. 1988; ALI et al. 1989). Several reports have suggested a possible association of Ha-*ras* gene activation with the prognosis of breast tumors. WALKER and WILKINSON (1988) found increased p21^{Ha-ras} expression in 30% of tumors, associated with higher proliferation rates. Point mutations for the *ras* genes were only exceptionally observed in breast tumors, indicating that the activation of the *ras* gene by point mutation is not or is only rarely involved in either the initiation or the metastatic progression of human breast cancer (ROCHLITZ et al. 1989).

FORRESTER (1987) and BOS et al. (1987) reported point mutations of the 12th codon of the Ki-*ras* gene in 39% of primary *colorectal carcinomas*, although without any correlation to sex, age, tumor localization, histological grade, or stage. The relatively high incidence of point mutations in colorectal carcinoma, however, could not be confirmed by other studies (ALEXANDER et al. 1986; MELTZER et al. 1987). Elevated c-*myc* expression was reported in a subgroup of colorectal carcinomas by SIKORA et al. (1987). These authors found relatively low c-*myc* mRNA levels in poorly differentiated tumors and concluded that the quantitative analysis of c-*myc* expression may have prognostic relevance. In our own studies we were able to confirm c-*myc*m RNA overexpression in colorectal carcinomas as compared with normal colorectal mucosa. In contrast to SIKORA's study, however, we found significantly higher c-*myc* mRNA amounts in poorly differentiated tumors (KLIMPFINGER et al. 1989).

A significant overexpression of Ha-*ras* mRNA in *endocrine pancreatic tumors* seems to correlate with their tendency to metastasize, as we demonstrated in a restricted number of cases. Interestingly, overexpression of Ha-*ras* was always accompanied by extremely low levels of c-*fos* mRNA, and vice versa (HÖFLER et al. 1988). In our most recent studies we were able to demonstrate that the overexpression of Ha-*ras* is not associated with point mutations on position 12 of this gene.

Besides the above-mentioned studies in which qualitative and quantitative changes in c-*onc* expression were correlated with the biological behavior of tumors or tumor subgroups, similar attempts were made in many other studies. The latter studies, however, were either based on a too small number of cases or did not reach the appropriate statistical significance due to short observation periods. Furthermore, many studies are not comparable, because different techniques were used for the detection of c-*onc* activation. Finally, the characterization of tumor-specific patterns of c-*onc* expression might be useful to assess the biological behavior of individual tumors and to characterize similar oncological entities which may respond similarly to future therapies.

6 Detection of Oncogene Expression

6.1 Indirect Methods

The original method for the detection of activated c-*onc* was DNA-mediated gene transfer (transfection). This assay is based on the fact that DNA taken from tumor cells can transform selected cells in culture. With this procedure active transforming genes were found in approximaely 15%–20% of all tumors tested, irrespective of their histological type. This method is still widely used to screen the transforming activity of tumor-derived DNA. However, this technique does not allow further characterization of the transforming gene. A further drawback of this assay is the fact that the mouse 3T3 cell line, commonly used in these assays, itself is already abnormal. Therefore, this approach does not sufficiently simulate the situation of tumor initiation in man. A more recent approach utilizes retroviral shuttle vectors for transfection. In contrast to the above-mentioned technique, the DNA used for transfection can be selected and, in some instances, even the site of integration into the target genome can be predicted. Additional techniques include chromosomal transfer resulting in hybridomas and the generation of transgenic animals. For most morphologically orientated pathologists interested in ocogene research, however, transfection assays and the use of transgenic animals play a subordinate role.

6.2 Direct Methods

Methods which allow the direct detection of altered or normal genes and their products have a much greater impact. Gene expression can be studied at the level of nucleic acids (DNA and RNA) and at the protein level, both after extraction (in vitro) and on histological tissue sections or cell or chromosomal preparations (in situ). The advantages of extraction methods include the possibility of quantification of the results, but methods based on morphological detection allow exact cellular and subcellular localization.

Nucleic acid sequence analysis in vitro is approached by a combination of the hybridization technology and enzymatic reactions, including the use of endonucleases, ligases, and polymerases. Generally, DNA and RNA sequences can be analyzed after extraction from fresh, frozen, fixed, and even paraffin-embedded cells or tissue. In the versatile Southern (DNA) blot technique, DNA is cleaved with restriction enzymes, and the fragments are separated according to size in agarose gels and transferred onto nitrocellulose or nylon membranes. The location on the membrane of a restriction fragment containing a particular sequence is determined by hybridization with a specific, isotopically or nonisotopically labeled probe. The related procedure for the detection of specific mRNAs is known as Northern blot analysis. The intensity of the specific bands can be measured by densitometry, thus allowing quantification of a particular DNA or mRNA sequence in a given tissue extract. Southern blot analysis makes possible the detection of gene rearrangements, DNA amplification, and in some instances the characterization of DNA point mutations. The main application of the Northern blot procedure is the assessment of overexpression of a particular mRNA and the characterization of alterations in mRNA size (Fig. 1). Several rapid techniques that do not require nucleic acid purification or immobilization after extraction have been developed recently. Scanning techniques for the detection of single base substitutions (point mutations) at the nucleic acid level include restriction fragment analysis, hybridization under high stringency with allele-specific oligonucleotides (20–30-mers), oligonucleotide ligation assays, RNA A cleavage methods, and denaturing gradient-gel electrophoresis (for review see LAN-DEGREN et al. 1988).

The polymerase chain reaction (PCR; SAIKI et al. 1985) allows the exponential amplification of a DNA (or RNA) segment starting from as little as a single copy. This technique is based on the performance of identical cycles of (a) DNA denaturation, (b) hybridization of oligonucleotide primers flanking the sequence to be amplified, and (c) primer extension by the use of a thermostable (Taq) DNA polymerase. Twenty-five amplification cycles increase the amount of a target sequence by approximately 10^6 fold. The amplified DNA may be analyzed for all given sequence variants by means of the above-mentioned methods. The great impact of PCR in molecular biology is due to the fact that this method can be performed without prior DNA extraction and purification, on cells or tissue samples, including paraffin-embedded tissue sections and with or without microdissection (SHIBATA et al. 1988). Therefore, PCR weds the advantages of morphology and molecular biology in a unique way. It represents the link between classical molecular biology and the morphological detection of genes and their expression products in situ employing in situ hybridization (hybridization histochemistry).

In situ hybridization procedures permit the localization of even single copy DNA sequences exceeding 1.5 kb in length in chromosomal and single cell preparations, and in tissue sections. The visualization of mRNA is more cumbersome: Our experience with the detection of c-onc mRNA expression by in situ hybridization in cells grown in culture and solid tumors, both in man and animal models, indicates that unequivocal detection of oncogene mRNA by in situ hybridization is most successful in overexpressing cells only (Fig. 2). This is probably due to the rapid spontaneous degradation of c-onc mRNA (e.g., 10 min half-life of c-myc mRNA), problems with RNA retention, degradation, and the accessibility of target sequences for probes during the hybridization procedure, and finally, the rather low copy numbers for most c-onc mRNAs in human tissues. Even with the use of isotopically labeled (antisense) cRNA probes, a technique known to be the most sensitive to date, the sensitivity is

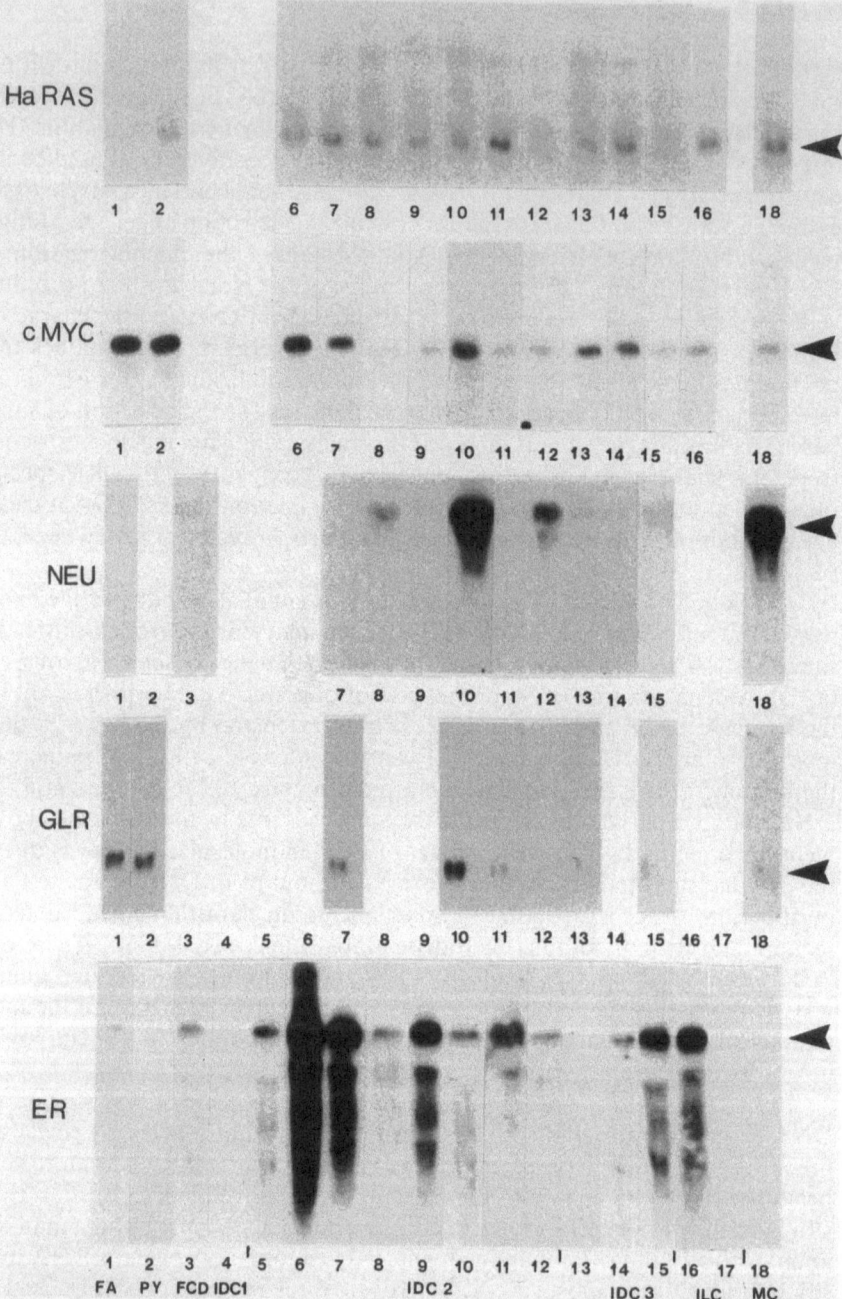

Fig. 1. Analysis of Ha-*ras*, c-*myc*, c-*neu*, glucocorticoid receptor (*GLR*), and estrogen receptor mRNA in different breast tumors by Northern blotting. *FA*, fibroadenoma; *PY*, phyllodes tumor; *FCD*, fibrous cystic disease; *IDC 1–3*, invasive ductal carcinomas with different histological grades; *ILC*, invasive lobular carcinomas; *Mc*, medullary carcinoma. The expression of Ha-*ras* mRNA (1.4 kb class), c-*myc* mRNA (2.2 kb), and *neu* mRNA (4.8 kb) can be compared with the amounts of glucocorticoid receptor mRNA (*GLR*, 2.0 kb) and estrogen receptor mRNA (*ER*, 6.2 kb) in individual cases. Note, for example, the inverse correlation of *neu* and ER mRNA expression in the same tumors!

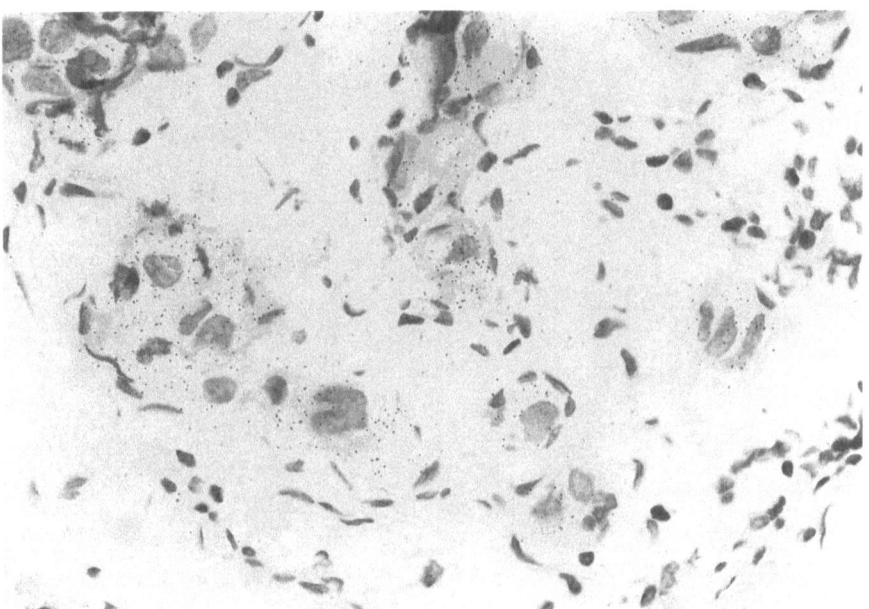

Fig. 2. Demonstration of c-*neu* mRNA expression by in situ hybridization utilizing ^{3}H-labeled cRNA probes in tumor cell complexes of an invasive ductal breast carcinoma (case 12 in Fig. 1). Autoradiography; frozen section, HE counterstained, × 450

limited to the detection of approximately 20 mRNA copies (1 kb in size) per cell (Cox et al. 1984). Particularly for surgical specimens immediate snap-freezing of tissue, optimal tissue pretreatment, and the most sensitive hybridization techniques are required. Because of extensive sequence homologies of c-*onc*'s within the different *onc* families and the limited stringency of in situ hybridization procedures, the results obtained by in situ hybridization have to be controlled very carefully. Currently used in situ hybridization procedures do not allow the detection of point mutations at the DNA and RNA level (for further details see HÖFLER et al. 1988 and HÖFLER 1987).

Because oncogene activation is ultimately always caused by an alteration of the oncoprotein either in structure or in quantity, generally the detection of *onc* expression products should be aspired to at the protein level. As a prerequisite, a large variety of monoclonal and polyclonal antibodies have been raised against synthetic or tumor-derived oncoproteins, and many of them are already available commercially. The application of monoclonal antibodies for Western blot analysis of tissue after protein extraction and electrophoretic separation represents an extremely powerful tool for the elucidation of altered gene expression, including single point mutations. Because of the above-mentioned restrictions on the interpretation of procedures based on the extraction of heterogeneous tissue and the fact that immunohistochemistry represents a most widely used and easy procedure, immunohistological studies are now

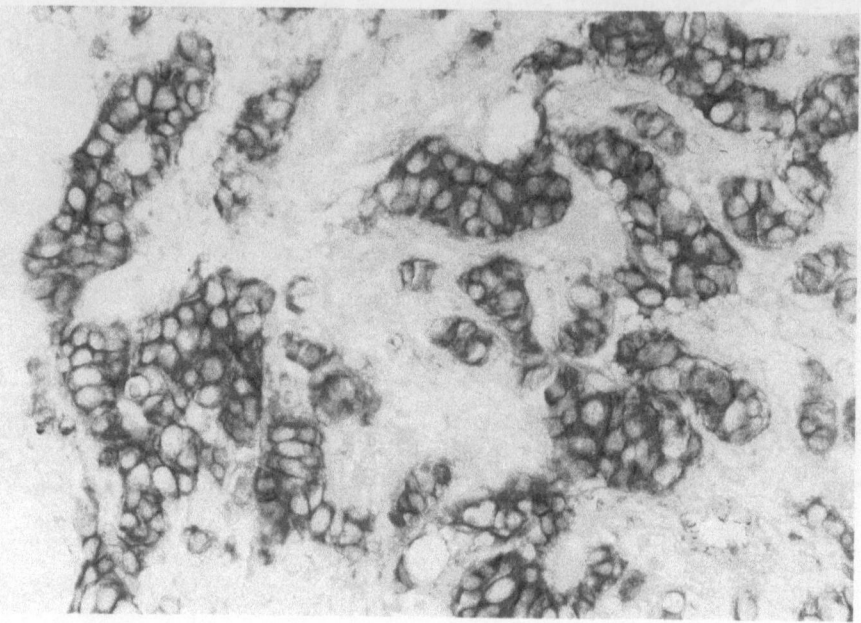

Fig. 3. Demonstration of *neu* protein by immunohistochemistry in an invasive ductal breast carcinoma (same case as in Fig. 2). Monoclonal antibodies to *neu* (HER 2); paraffin section, APAAP, no counterstain, × 500

performed more frequently (Fig. 3). Due to the lack of specificity of some antibodies (ALLEN et al. 1987; VARMA and AUSTIN 1988), immunohistochemical data obtained with antibodies to *onc* proteins require most careful interpretation. Particularly some antibodies to *ras* proteins (e.g., the monoclonal antibody RAP-5) are known to recognize besides $p21^{ras}$ epitopes, sequences which are widely distributed in non-$p21^{ras}$ proteins (for a recent review, see WICK 1989). Additionally, the sensitivity of immunohistochemical detection of *onc* expression is limited, except in some cases with significant overexpression (MORI et al. 1987).

In conclusion, before commencing an experiment for the detection of activated c-*onc*'s in cells or tissue, the appropriate detection systems have to be selected. Because the mode of c-*onc* activation may involve the DNA, RNA, and protein level of gene expression, the characterization of altered oncogene expression at all three levels should be sought.

7 Conclusions and Outlook

Although our knowledge of *onc* expression and interaction with growth factors is still in its infancy, the discovery of oncogenes and tumor-suppressing genes has caused a revolution in cancer research: For the first time we can specify with some precision the genes and their proteins that are involved in the induction of uncontrolled proliferation of a malignant cell. Furthermore, knowledge of c-*onc* actions has contributed to our understanding of cell physiology, in particular cell proliferation and differentiation. The understanding of c-*onc* and growth factor action represents an important aid to tumor diagnosis. It seems likely that a classification of cancers according to the expression pattern of activated c-*onc*'s will become possible, thus leading us beyond the classic morphological approach of tumor characterization. Additionally, the pattern of active c-*onc* expression in a given tumor will probably affect its response to therapeutic agents, both those presently used and the more specific ones we can expect in the future. The development of drugs selectively interacting with c-*onc* expression products can be anticipated. BALTIMORE (1987) predicted, however, that it will be perhaps 20 years or more before we can expect the development of such new drugs which are tailored to the biochemical and structural properties of c-*onc* proteins. Increasing knowledge of c-*onc* activation and the action of tumor-suppressing genes may also have an impact on tumor prevention and the prediction of tumor susceptibility in individual cases. Besides further advances in basic research, standardization of methodology with cross-validation of technology will be necessary to unravel the oncogene-antioncogene puzzle . . . and the pathologist, both as morphologist and as basic scientist, will play a crucial role.

References

Aaronson SA, Tronick SR (1985) The role of oncogenes in human neoplasia. In: DeVita VT et al. (eds) Important advances in oncology 1985. JB Lippincott, Philadelphia, pp 3–15

Alexander RJ, Buxbaum JN, Raicht RF (1986) Oncogene alterations in primary human colon tumors. Gastroenterology 91:1509-1510

Ali IU, Merlo G, Callahan R, Lidereau R (1989) The amplification unit on chromosome 11q13 in aggressive primary human breast tumors entails the *bcl*-1, *int*-2 and *hst* loci. Oncogene 4:89–92

Alitalo K (1987) Amplification of cellular oncogenes in cancer cells. In: Bradshaw RA, Prentis S (eds) Oncogenes and growth factors. Elsevier, Amsterdam, pp 17–23

Allen DC, Foster H, Orchin JC (1987) Immunohistochemical staining of colorectal tissues with monoclonal antibodies to *ras* oncogene p21 product and carbohydrate determinant antigen 19–9. J Clin Pathol 40:157–162

Amy CM, Bartolomew JC (1987) Regulation of N-*myc* transcript stability in human neuroblastoma and retinoblastoma cells. Cancer Res 47:6310–6314

Baltimore D (1987) The impact of the discovery of oncogenes on cancer mortality rates will come slowly. Cancer 59:1985–1986

Barbacid M (1986a) Human oncogenes. In: DeVita VT, Hellman S, Rosenberg S (eds) Important advances in oncology 1986. JB Lippincott, Philadelphia, pp 3–22

Barbacid M (1986b) Oncogenes, mutagenes and cancer. Proc AACR 27:435–452

Bishop JM (1983) Cellular oncogenes and retroviruses. Ann Rev Biochem 52:301–354

Bolen JB, Veillette A, Schwartz AM, DeSeau V, Rosen N (1987) Activation of pp60[c-src] protein kinase activity in human colon carcinoma. Proc Natl Acad Sci USA 84:2251–2255

Bos JL, Fearon ER, Hamilton SR, Verlaan-de Vries M, van Boom JH, van der Erb AJ, Vogelstein B (1987) Prevalence of *ras* gene mutations in human colorectal cancers. Nature 327:293–297

Brodeur GM, Seeger RC, Sather H, Dalton A, Siegel SE, Wong KY, Denman H (1986) Clinical implications of oncogene activation in human neuroblastomas. Cancer 58:541–545

Buick RN, Pollak MN (1984) Perspectives on clonogenic tumor cells, stem cells, and oncogenes. Cancer Res 44:4909–4918

Burgess A (1987) Growth factors and oncogenes In: Bradshaw RA, Prentis S (eds) Oncogenes and growth factors. Elsevier, Amsterdam, pp 123–134

Collins S, Groudine M (1982) Amplification of endogeneous *myc*-related DNA sequences in a human myeloid leukemia cell line. Nature 298:679–681

Collins SJ, Kubonishi I, Miyoshi I, Groudine MT (1984) Altered transcription of the c-*abl* oncogene in K-562 and other chronic myelogenous leukemia cells. Science 225:72–74

Cosman D (1987) Control of mRNA stability. Immunology 8:16–17

Cox KH, DeLeon DV, Angerer LM, Angerer RC (1984) Detection of mRNAs in sea urchin embryos by in situ hybridization using asymmetric RNA probes. Dev Biol 101:485–502

Curran T (1988) The *fos* oncogene. In: Reddy EP, Skalka AM, Curran T (eds) The oncogen handbook. Elsevier, Amsterdam, pp 307–325

Dalla-Favera R, Bregni M, Erikson J, Patterson D, Gallo RC, Croce CM (1982) Human c-*myc* oncogene is located on the region of chromosome 8 that is translocated in Burkitt's lymphoma cells. Proc Natl Acad Sci USA 79:7824–7827

Dickson C, Peters G (1987) Potential oncogene products related to growth factors. Nature 326:833

Duesberg PH (1987) Cancer genes: rare recombinants instead of activated oncogenes (a review). Proc Natl Acad Sci 84:2117–2124

Egan SE, Wright JA, Jarolim L, Anagihara K, Bassin RH, Greenberg AH (1987) Transformation by oncogenes encoding protein kinases induces the metastatic phenotype. Science 238:202–205

Erisman MD, Astrin SM (1988) The *myc* oncogene. In: Reddy EP, Skalka AM, Curran T (eds) The oncogen handbook. Elsevier, Amsterdam, pp 341–379

Forrester K, Almoguera C, Han K, Grizzle WE, Perucho M (1987) Detection of high incidence of K-*ras* oncogenes during human colon tumorigenesis. Nature 327:298–303

Friend SH, Dryja TP, Weinberg RA (1988) Oncogenes and tumor-suppressing genes. N Engl J Med 318:618–622

Funa K, Steinholtz L, Nou E, Bergh J (1987) Increased expression of N-*myc* in human small cell lung cancer biopsies predicts lack of response to chemotherapy and poor prognosis. Am J Clin Pathol 88:216–220

Heldin CH, Westermark B (1986) Platelet-derived growth factor: structure, function and role in autocrine stimulation of growth. In: Tanner W, Gallwitz D (eds) Cell cycle and oncogenes. Springer, Berlin pp 137–144

Höfler H (1987) What's new in "in situ hybridization"? Pathol Res Pract 182:421–430

Höfler H, Childers H, Montminy MR, Lechan RM, Goodman RH, Wolfe HJ (1986) In situ hybridization methods for the detection of somatostatin mRNA in tissue sections using antisense RNA probes. Histochem J 18:597–604

Höfler H, Ruhri C, Pütz B, Wirnsberger G, Hauser H (1988) Oncogene expression in endocrine pancreatic tumors. Virchows Arch [B] 55:355–361

Huebner RJ, Todaro GJ (1969) Oncogenes of RNA tumor viruses as determinants of cancer. Proc Natl Acad Sci USA 64:1087–1094

Hunt T (1988) Controlling mRNA lifespan. Nature 334:567

Keating MT, Williams LT (1988) Autocrine stimulation of intracellular PDGF receptors in v-*sis* transformed cells. Science 239:914–916

Klimpfinger M, Ruhri C, Pütz B, Steindorfer P, Höfler H (Gastroenterology) Oncogene expression in colorectal carcinomas. Gastroenterology (submitted)

Knudson AG (1985) Hereditary cancer, oncogenes, and antioncogenes. Cancer Res 45:1437–1443

Lacal JC, Tronick SR (1988) The *ras* oncogene. In: Reddy EP, Skalka AM, Curran T (eds) The oncogen handbook. Elsevier, Amsterdam pp 257–304

Land H, Parada L, Weinberg RA (1983) Cellular oncogenes and multistep carcinogenesis. Science 222:771–778

Landegren U, Kaiser R, Caskey CT, Hood L (1988) DNA diagnostics – molecular techniques and automation. Science 242:229–237

Lefkowitz RJ, Benovic JL, Kobilka P, Caron MG (1986) β-Adrenergic receptors and rhodopsin: shedding new light on an old subject. Trans Pharmacol Sci 7:444–448

Levinson AD (1987) Normal and activated *ras* oncogenes and their encoded products. In: Bradshaw RA, Prentis S (eds) Oncogenes and growth factors. Elsevier, Amsterdam, pp 74–83

Marks F (1987) What's new in oncogenes and growth factors? Pathol Res Pract 182:831–848

Marx JL (1988) Progress in predicting breast cancer relapses. Science 241:535

Mathew CGP, Smith BA, Thorpe K, Wong Z, Royle NJ, Jeffreys AJ, Ponder BAJ (1987) Deletion of genes on chromosome 1 in endocrine neoplasia. Nature 328:524–526

Meltzer SJ, Ahnen DJ, Battifora H (1987) Protooncogenes abnormalities in colon cancers and adenomatous polyps. Gastroenterology 92:1174–1182

Mori S, Akiyama T, Morishita Y et al. (1987) Light and electron microscopical demonstration of c-*erb*B-2 gene product-like immunoreactivity in human malignant tumors. Virchows Arch [B] 54:8–15

Nishimura S, Sekiya T (1987) Human cancer and cellular oncogenes. Biochem J 243:313–327

Nusse R (1987) The activation of cellular oncogenes by retroviral insertion. In: Bradshaw RA, Prentis S (eds) Oncogenes and growth factors. Elsevier, Amsterdam, pp 59–66

Park M, Gonzatti-Haces M, Dean M et al. (1986) The *met* oncogene: a member of the thyrosine kinase family and a marker for cystic fibrosis. Cold Spring Harbor Symp Quant Biol LI: 967–975

Rabbitts TH (1985) The *myc* protooncogene: involvement in chromosomal abnormalities. Trends Genet 1:327–331

Rabbitts TH, Fortser A, Mamlyn P, Baer R (1984) Effect of somatic mutation within translocated c-*myc* genes in Burkitt's lymphoma. Nature 309:592–597

Radaskiewicz T, Ruhri C, Mosberger I, Pütz B, Höfler H (1987) Overexpression of c-*sis* in non-Hodgkin lymphomas of lymphoblastic and immunoblastic type. Lab Invest 56/1:62 A

Reddy EP, Reynolds RK, Santos E, Barbacid M (1982) A point mutation is responsible for the acquisition of transforming properties by the T24 human bladder carcinoma oncogene. Nature 300:149–152

Riou G, Barrois M, Le MG, George M, Le Doussal V, Haie C (1987) C-*myc* proto-oncogene expression and prognosis in early carcinoma of the uterine cervix. Lancet I:761–764

Rochlitz CF, Scott GK, Dodson JM, Liu E, Dollbaum C, Smith HS, Benz CC (1989) Incidence of activating *ras* oncogene mutations associated with primary and metastatic human breast cancer. Cancer Res 49:357–360

Saiki RK, Scharf S, Faloona F, Mullis KB, Hort GT, Erlich HA, Arnheim N (1985) Enzymatic amplification of beta globin genomic sequences and restriction site analysis for diagnosis of sickle cell anemia. Science 230:1350–1353

Schwab M, Ellison J, Busch M, Rosenau W, Varmus HE, Bishop JM (1984) Enhanced expression of the human gene N-*myc* consequent to amplification of DNA may contribute to malignant progression of neuroblastoma. Proc Natl Acad Sci USA 81:4940–4944

Sherr CJ, Rettnmier CW, Sacca R (1985) The c-*fms* protooncogene product is related to the receptor for the mononuclear phagocyte growth factor, CSF-1. Cell 41:665–676

Shibata DK, Arnheim N, Martin WJ (1988) Detection of human papilloma virus in paraffin-embedded tissue using the polymerase chain reaction. J Exp Med 167:231–236

Sikora K, Chan ST, Evan G, Gabra H, Markham N, Stewart J, Watson J (1987) c-*myc* oncogene expression in colorectal cancer. Cancer 59:1289–1295

Slamon DJ, Cline MJ (1984) Expression of cellular oncogenes during embryonic and fetal development of the mouse. Proc Natl Acad Sci USA 81:7141–7145

Slamon DJ, Clark GM, Wong SG, Levin WJ, Ullrich A, McGuire WI (1987) Human breast cancer: correlation of relapse and survival with amplification of the HER-2/neu oncogene. Science 235:177–182

Stehelin D, Guntaka RV, Varmus HE, Bishop JM (1976) Purification of DNA complementary to nucleotide sequences required for neoplastic transformation of fibroblasts by avian sarcoma viruses. J Mol Biol 101:349–353

Stubblefield E, Sanford J (1987) A general survey of genetics and cancer. Anticanc Res 7:1085–1104

Tabin CJ, Bradley SM, Bergmann CI et al. (1982) Mechanism of activation of a human oncogene. Nature 300:143–148

T'Ang A, Varley JM, Chakraborty S, Murphree AL, Fung YKT (1988) Structural rearrangement of the retinoblastoma gene in human breast carcinoma. Science 242:263–266

Taub R, Kirsch I, Morton C (1982) Translocation of the c-*myc* gene into the immunoglobulin heavy chain locus in human Burkitt's lymphoma and murine plasmacytoma cells. Proc Natl Acad Sci USA 79:7837–7841

Teich MM (1988) Oncogenes and cancer. In: Franks LM, Teich N (eds) Introduction to cellular and molecular biology of cancer. Oxford University Press, New York, pp 200–228

Teich N, Wyke J, Mak T, Bernstein A, Hardy W (1984) Pathogenesis of retroviral induced disease. In: Weiss R, Teich N, Varmus H, Coffin J (eds) RNA tumor viruses, 2nd edn, vol 1. Cold Spring Harbour Laboratory, pp 785–998

Tsujimoto Y, Yunis J, Onorato-Showe L, Erikson J, Nowell PC, Croce CM (1984a) Molecular cloning of the chromosomal breakpoint of B-cell lymphomas and leukemias with the t(11:14) chromosome translocation. Science 224:1403–1406

Tsujimoto Y, Finger LR, Yunis J, Nowell PC, Croce CM (1984b) Cloning of the chromosomal breakpoint of neoplastic B-cells with the t(14;18) chromosome translocation. Science 226:1097–1099

Varma VA, Austin GE (1988) Antibodies to *ras* proteins lack histochemical specificity for neoplastic epithelium in human prostate. Lab Invest 98A p 67

Varmus HE (1984) The molecular genetics of cellular oncogenes. Annu Rev Genet 18:553–561

Walker RA, Wilkinson N (1988) p21*ras* protein expression in benign and malignant human breast. J Pathol 156:147–153

Weichselbaum RR, Beckett M, Diamond A (1988) Some retinoblastomas, osteosarcomas, and soft tissue sarcomas may share common etiology. Proc Natl Acad Sci USA 85:2106–2109

Weinberg RA (1985) The action of oncogenes in the cytoplasm and nucleus. Science 230:770–776

Weinberg RA (1988) Finding the anti-oncogene. Sci Am Sept 1988:34–46

Weinberger C, Thompson CC, Ong ES, Lebo R, Gruol DJ, Evans RM (1986) The c-*erb*-A gene encodes a thyroid hormone receptor. Nature 324:641–646

Weinstein B (1987) Growth factors, oncogenes, and multistage carcinogenesis. J Cell Biochem 33:213–224

Whyte P, Buchkovich KJ, Horowitz JM, Friend SH, Raybuck M, Weinberg RA, Harlow E (1988) Association between an oncogene and an anti-oncogene: the adenovirus E1A proteins bind to the retinoblastoma gene product. Nature 334:124–129

Wick MR (1989) Immunohistochemical detection of *ras* oncogene products. Arch Pathol Lab Med 113:13–15

Wong AJ, Ruppert JM, Eggleston J, Hamilton SR, Baylin SB, Vogelstein B (1986) Gene amplification of c-*myc* and N-*myc* in small cell lung carcinoma of the lung. Science 233:461–464

Yamamoto T, Kamata N, Kawano H et al. (1986) The sequence of the ERB B$_2$ gene. Nature 319:230–234

Yarden Y, Ullrich A (1988) Molecular analysis of signal transduction by growth factors. Biochemistry 27:3113–3119

Yokota J, Tsunetsugu-Yokota Y, Battifora H, Le Fevre C, Cline MJ (1986) Alterations of *myc, myb,* and *ras*^{Ha} proto-oncogenes in cancers are frequent and show clinical correlation. Science 231:261–264

Young T, Waitches G, Birchmeier C (1986) Isolation and characterization of a new cellular oncogene encoding a protein with multiple potential transmembrane domains. Cell 45:711–719

Zhou DJ, Casey G, Cline MJ (1988) Amplification of human int-2 in breast cancers and squamous carcinomas. Oncogene 2:279–282

Zhou DJ, Ahuja H, Cline MJ (1989) Proto-oncogene abnormalities in human breast cancer: c-ERBB-2 amplification does not correlate with recurrence of disease. Oncogene 4:105–108

Special Aspects of Tumor Pathology

Breast Carcinoma

H.-E. STEGNER and W. JONAT

1 Introduction

The determination of estrogen receptor (ER) (JENSEN et al. 1970; MAASS et al. 1972) and progesterone receptor (PR) (McGUIRE and CLARK 1985) is a recognized prerequisite in selecting endocrine therapy in breast cancer. Despite their proven clinical value, the established biochemical steroid receptor assays have a number of inherent disadvantages. The biochemical assays indicate the presence of steroid binding capacity per unit weight of tissue protein, but they cannot identify the site and cellular origin of the receptor. Most breast cancers contain abundant and variable amounts of protein-rich connective tissue which provides a source of error in quantitation of receptor. Interference by endogenous or administered estrogens must be taken into account. In addition, tissue sampling is inaccurate and a critical amount of tumor tissue is needed for exact determination of binding. Moreover, any delay in freezing the tissue reduces the receptor activity, leading to false-negative results. Finally, retrospective analysis of fixed material or paraffin-embedded material is not possible.

Improved mammographic techniques frequently enable small breast tumors and noninfiltrating stages of ductal cancer to be detected when a biochemical assay is not possible because of the lack of appropriate tissue specimens. This is especially true in the case of recurrent breast cancer, when only fine needle aspirates, biopsy specimens, or small amounts of malignant effusions are available.

The production of monoclonal antibodies (mAbs) to steroid hormone receptors has permitted the development of cytochemical assays for direct antigenic recognition of ER and PR in both histological and cytological specimens, expanding the versatility of the surgical pathology laboratory. Since ER and PR are directly visualized within the cells, sampling errors due to such factors as excess stroma or large amounts of nonneoplastic epithelium are avoided. Furthermore, immunohistochemical receptor determination in fine needle aspirates or biopsy specimens spares patients unnecessary surgery for ER determination. On the other hand, the results of immunohistochemical studies of ER, as well as the results of cellular enucleation studies, have led to reevaluation of the distribution of the unoccupied form of the receptor in the intact cell. It now appears that both the steroid-occupied and the steroid-unoccupied form of ER and PR are predominantly nuclear proteins.

This knowledge has changed the conventional concept of steroid action in hormone-dependent neoplastic tissues, especially in human breast cancer as primarily described by JENSEN (1970). The idea of shuttling hormones through the cytoplasm, which was thought to be the major role of their receptors, is no longer valid. Today it is believed that estrogens, in common with other steroid hormones, modulate gene expression in target cells via their interaction with specific nuclear proteins (receptors). The demonstration of receptors only/ mainly within the nucleus fits into this concept. To clarify the role of receptors at the genomic level, the transcriptional level, the translational level, or the functional level, recent advances in molecular biology, and especially in recombinant DNA technology, including restriction enzyme analysis, are now being utilized.

2 Immunocytochemical Demonstration of ER and PR in Histological and Cytological Specimens

Monoclonal antibodies to ER and PR complexes are today available commercially and can be used to localize the receptor protein in both frozen and paraffin-embedded tissue. In a first step towards establishing mAbs to ERs, GREENE et al. (1980) purified ER from calf uterine nuclei. These purified complexes were used to immunize splenic lymphocytes from Lewis rats, which were then fused with three different mouse myeloma lines to yield hybridoma cultures. GREENE et al. described in their first publication three clones secreting rat IgG and seven clones secreting IgM. From this original library of mAbs against ER some are still used in commercial kits for ER immunocytochemical assay (ER-ICA; Abbott Laboratories).

The antibody MI 60–10, developed by MILGROM and co-workers, is murine and is principally directed against the PR of the rabbit, although it also recognizes an epitope of the human receptor (LOOSFELT et al. 1985; NEIS et al. 1989). The visualization of the receptor is achieved through the indirect immunoperoxidase reaction (PAP technique: STERNBERGER et al. 1970). Reduced immunoreaction, especially of ER, is observed in conventional formalin-fixed paraffin sections, the extent of the reduction depending on the mode of tissue preparation and the duration of storage. Compared with the reduction in the ER immunoreaction (ER-ICA) in paraffin-embedded tissue, the decrease in the immunoreaction of PR (PR-ICA) is insignificant. HIORT et al. (1988) have recommended a modification of the standard peroxidase technique to increase the sensitivity of this approach in paraffin sections, the modification entailing the use of cobalt chloride to intensify the color of the diaminobenzidine (DAB) reaction.

Most of the results reported in the following are based on the immunohistochemical determination of ER binding sites. Surprisingly, few reports are available on the determination of the PR.

Using ER-ICA in breast tissues the staining reaction is exclusively confined to the nuclei of epithelial cells. Mesenchymal cells of the interstitial tissue and vascular cells are always negative. Specific immunocytochemical staining of ER can be observed in both benign and malignant cells. A striking feature of the staining pattern is the heterogeneity of staining, a phenomenon characteristic of neoplastic cells but also of normal and hyperplastic cells of the ductolobular unit.

The heterogeneity of ER staining can be interpreted in different ways. Malignant tumors may be composed of a mosaic of cancer cells with different concentrations of ER or of different cell clones that may be either able or unable to produce receptors (ER-positive or -negative cell clones). On the other hand, the expression of ER may be dependent on the phase of the cell cycle and the rate of cell proliferation.

In histological sections of malignant tumors the following different patterns of heterogeneity can be observed: (a) presentation of the minority cell population in clumps or in separate areas of the section; (b) intermingling of positive and negative cells; and (c) a mixture of the above two patterns. The intensity of staining is related to the amount of ER, showing various degrees of density of the brown DAB deposits.

An increasing number of publications have reported that in cytological samples (fine needle aspiration, malignant effusions, bone marrow smears) the immunocytochemical demonstration of ER and PR yields recognition rates comparable to those in histological sections. The results obtained show good correlation with both the conventional biochemical assays and immunohistological findings (McCLELLAND et al. 1987; BURTON et al. 1987; JOHNSON et al. 1987; REINER et al. 1989; HAWKINS et al. 1988; KESHGEGIAN et al. 1988; KUBLER et al. 1988; MASOOD 1989). The qualitative concordance rate is between 85% and 92% with both methods, the sensitivity of the immunocytochemical reaction ranging from 78% to 94%, and the specificity from 71% to 89%.

Detection of ER in cytological samples permits not only definition of the receptor status, but also confirmation of the epithelial nature of suspected tumor cells. Immunocytochemical ER-determination has gained increasing clinical importance and now has a broad spectrum of indications:

- Presurgical determination of receptor status
- Sequential studies following cytotoxic or endocrine treatment
- Receptor determination in small tumors and in aspirates obtained under mammographic direction from suspicious lesions
- Multifocal lesions
- Recurrent disease in which surgical intervention is not necessary or which occurs at locations accessible only to aspiration biopsy
- Malignant effusion (pleura, ascites) from metastatic breast tumors or an occult primary tumor
- Lymph node metastases from metastatic breast tumors or an occult primary tumor
- Bone marrow smears

3 Quantification of the Reaction

Various proposals have been made regarding quantitation of the results of the indirect peroxidase-antiperoxidase (PAP) technique by visual grading of the intensity of the reaction. Subjective estimation and more recently automated image analysis have been tried using a semiquantitative approach based on the intensity of the staining reaction and the proportion of stained cancer cell nuclei. In the score proposed by McCARTY et al. (1985), both the intensity and the distribution of staining are incorporated in a single numerical score. The score, $\Sigma = (i + 1) \times Pi$, consists of the sum of the staining intensities i $(0, 1+, 2+, 3+, 4+)$ multiplied by the percentage of cells in each category of staining.

Table 1. Immunoreactive score (IRS) for semiquantitative estimation of the ER-ICA reaction (REMMELE and STEGNER 1986)

IRS = SI × PP	
SI (staining intensity)	0 = negative
	1 = weak
	2 = moderate
	3 = strong
PP (percentage of pos. cells)	0 = negative
	1 = 10% pos. cells
	2 = 10%–50% pos. cells
	3 = 51%–80% pos. cells
	4 = 80% pos. cells
Maximum IRS: 3 × 4 = 12	

A commission of several working groups of German pathologists and gynecologists evaluated various modifications by interobserver tests and in 1986 recommended the immunoreactive score (IRS) presented in Table 1 (REMMELE and STEGNER 1986). The IRS is defined as the product of estimated prominent staining intensity (SI) and proportion of positive cells (PP). The maximum IRS is 12.

Automated image analysis has been applied to both histological specimens and fine needle aspirates. Nuclear ER is visualized with a commercial immunohistochemical method (ER-ICA: Abbott Labs.) using the indirect PAP technique (brown diaminobenzidine deposits) with a hematoxylin (blue) or fast green counterstain. Various indices of ER positivity can be derived from the integrated density and average density measurements of nuclear DAB. In cytological specimens the image analysis techniques enable editing of overlapping and touching nuclei, stromal nuclei, and debris, thus ensuring that measurements relate to the individual tumor cell nuclei. Each edited field is measured at two wavelengths, 550 nm (green) and 436 nm (blue), to quantitate total and DAB-stained nuclei, respectively.

The distribution histogram of the integrated density of DAB stain per nucleus can then be calculated. The results gained from these methods indicate, as expected from visual interpretation, that a high degree of variability of staining of nuclei exists within and between tumor aspirates, and that the percentage of ER-positive nuclei in the aspirates and the cytosolic ER concentration in the corresponding tumor biopsy as determined by DCC (Dextran coated charcoal method) assay are correlated (HORSFALL et al. 1987, 1989).

Our own results with computer-assisted image analysis were obtained with a microcomputer system (CAS 100, Lombard, Il. USA) using a dual staining two color "nuclear mask" imaging technique. The counterstaining has a major impact on the quality of this system. The method allows discrimination between positive and negative cells within one slide.

Methyl green was selected for the counterstaining because spectral studies (BACUS et al. 1988) have shown that methyl green offers the best spectral separation from DAB, allowing the image to be digitized at two separate wavelengths. Spectral bandpass filters of 650 nm (red) and 500 nm (green) were used. Using control slides with known histoscores, antibody thresholds for both staining intensity and proportion of positive cells can be obtained. Within mathematical models both parameters can be used to calculate the German IRS as described above or the histoscore proposed by McCARTY et al. (1985).

This technique yields easily reproducible results and in the case of ER binding shows excellent specificity and sensitivity compared with biochemical methods. Nonetheless, it requires that the machine operator be familiar with the pathology and the cellular morphological characteristics of the examined tissue samples.

4 Correlation of Biochemical and Immunohistochemical Methods

In comparative biochemical and immunohistochemical studies performed by a great number of investigators the established radioligand binding assay served as a gold standard, although the methods compared have different sources of error and their value in selecting patients responsive to endocrine treatment can only be assessed by prospective clinical trials. In nearly all studies published a significant qualitative relationship has been found between the two types of method, with qualitative concordance rates of between 80% and 90% (DELENA et al. 1988; PERTSCHUK et al. 1985; CUDAHY et al. 1988; PARL and POSEY 1988). Our own results obtained from almost 500 breast cancer tissue samples analyzed by ER-ICA and ER-DCC are in agreement with the above-described results and are shown in Table 2.

Table 2. Biochemical assay (DCC) compared with ER-ICA results in 466 breast cancer tissue samples. The high number of receptor-negative samples is due to a high proportion of advanced and recurrent (aggressive) tumors

	ER-ICA positive	ER-ICA negative
DCC positive	81 (62%)	50
DCC negative	19	316 (94%)

Positive/negative borderline:
DCC method: 20 fmol/mg tissue protein
ER-ICA: histoscore 100 (according to McCARTY et al. 1985).

The concordance is about 60%–70% when a semiquantitative estimation of the ER-ICA assay is used (RASMUSSEN et al. 1988). However, discordant results have been observed in a significant number of cases, ranging between 7% and 35%. Different factors may be responsible for these discrepancies.

A negative hormone binding assay in the presence of stainable receptor can be explained by:

– Competition of endogenous estrogen with [3H] estradiol in premenopausal women
– Paucity of tumor cells in the tissue sample
– Alteration of the receptor molecule at the hormone binding site with preservation of the immunoreactive epitope on the ER molecule.

Biochemically ER-positive cases that fail to stain immunohistochemically can be interpreted as being caused by:

– Modification of the antigenic determinants due to fixation (embedding of tissue in paraffin has been shown to reduce or eliminate specific ER staining)
– Partial proteolysis with formation of receptor fragments that are unrecognizable by the immunohistochemical method but are detectable with the radioligand assay

- Insufficient sensitivity of the mAb for it to detect low concentrations of ER
- Lack of specificity of the mAb (there is molecular heterogeneity of ER, and some forms of ER are not recognized by the antibodies used).

There is some evidence from case-by-case comparison of discordant results that the immunohistochemical assay might be more accurate than the hormone binding assay in identifying the presence of ER in cancer cells (PARL and POSEY 1988). However, its clinical significance in predicting responsiveness to hormonal therapy still has to be fully elucidated.

In the case of ER-negative, PR-positive tissue samples analyzed by the DCC method, the immunohistochemical assay has been shown to improve the accuracy of the receptor result. In a series of 90 patients who were predominantly premenopausal and who had elevated serum estradiol levels that were thought to have led to false-negative results, we were able to demonstrate high amounts of ER by means of ER-ICA. Thus on the basis of our results, the immunohistochemical determination of ER is recommended in premenopausal patients in the early luteal phase with high amounts of endogenous estrogens or in those patients being treated with antiestrogens.

5 Immunohistochemical Steroid Hormone Receptor Detection in Normal Breast Tissues and Benign Proliferative Lesions

The immunohistochemical detection of ERs shows a clear age dependence in the normal parenchyma of the breast. In the ductolobular glandular epithelium of women of reproductive age, ER-positive cells are only rarely to be detected (STEGNER et al. 1986; KÖHLER and BÄSSLER 1987). In older and postmenopausal patients, irregularly distributed receptor-positive cell nuclei of the glandular epithelium are increasingly manifested. The reaction is predominantly inhomogeneous and is not distributed throughout the entire compartment of a ductolobular unit. Stromal cells and myoepithelial cells are receptor-negative. In lobular hyperplasia, a more pronounced ER expression is to be found in the hyperplastic lobuli.

In dysplastic parenchyma (fibrocystic disease), the various proliferative glandular components show a different receptor content. In simple ductal and sclerosing adenosis, receptor-positive cell nuclei can be detected relatively frequently in the luminal cell layer of the ductuli (Fig. 1). The myoepithelial proliferates are regularly negative. The apocrine metaplastic epithelium in adenoses, mastopathic cysts, and papillomas shows either no reaction at all or only a weak one. In ductal epithelial hyperplasia (epitheliosis) and papillomas, the clonal origin of the proliferates is manifested by a frequent relatively homogeneous and segmental reaction (Fig. 2). As in the normal glandular epithelium, an increase in receptor expression with age is shown in dysplastic parenchyma. Owing to the diversity of proliferative forms, there is no clear correlation with the degree of dysplasia (degree of mastopathy, I-III:

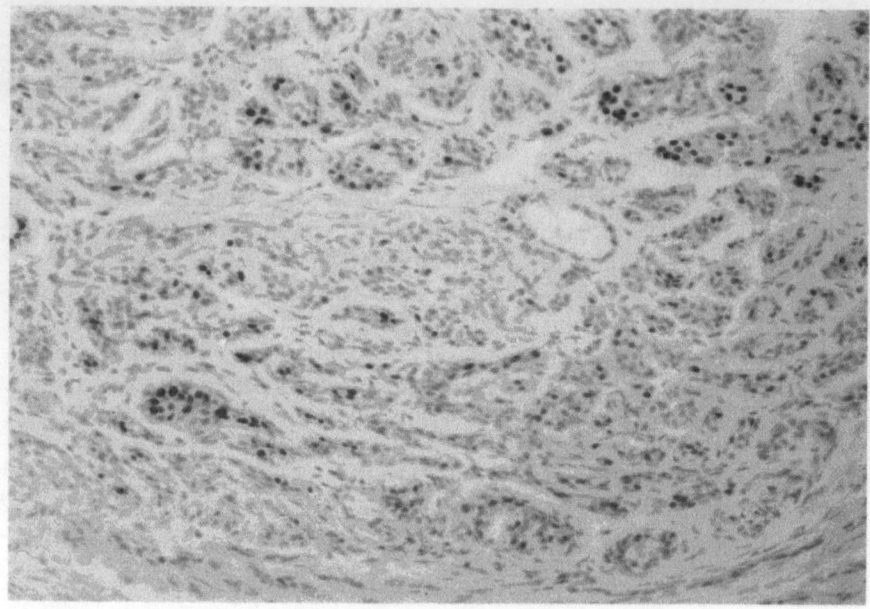

Fig. 1. Immunohistochemical ER demonstration (ER-ICA) in microglandular and sclerosing adenosis

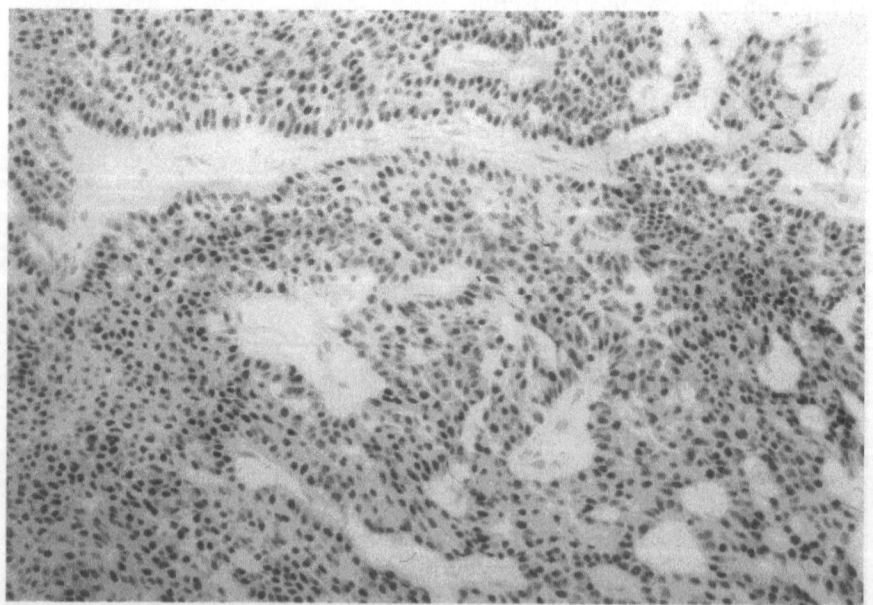

Fig. 2. Immunohistochemical ER demonstration (ER-ICA) in solitary ductal papilloma

PRECHTEL 1972). KÖHLER and BÄSSLER (1987) reported more pronounced ER expression in the regular epithelial hyperplasias of mastopathy grade II than in the type III hyperplasias accompanied by cellular atypia.

6 Immunohistochemical Steroid Hormone Receptor Detection in Malignant Breast Tumors

The immunohistochemical assay is the most appropriate method for ER detection in noninfiltrating tumors of the breast. Comparative studies of preinvasive lesions associated with different types of infiltrating cancer have contributed to a better understanding of the relationships between preinvasive and invasive tumor stages. Like other prognostic morphological criteria of cellular differentiation, the receptor status of intraductal and intralobular neoplasms reflects the inherent malignant potential of the lesion. Within the spectrum of subtypes of both ductal and lobular carcinomas in situ, ER-positive and ER-negative cases can be identified with some correlation to the grade of differentiation in noninfiltrating ductal carcinomas. In contrast to highly differentiated cribriform and papillary ductal carcinomas in situ (DCIS), most intraductal comedocarcinomas have been found to be ER negative by means of ER-ICA (Fig. 3). In highly differentiated DCIS there is in general a more homogeneous reaction of the tumor cell population, whereas in poorly differentiated types of

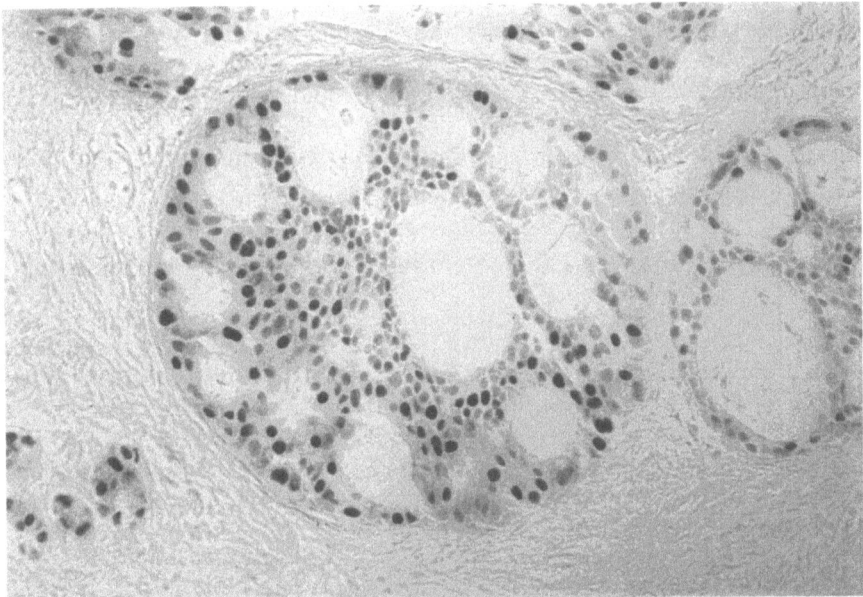

Fig. 3. Immunohistochemical ER demonstration (ER-ICA) in intraductal carcinoma of cribriform type

DCIS a greater tendency for ER heterogeneity can be observed, with various degrees of admixture of negative and positive cells, their peripheral layers frequently showing a greater staining intensity than the more centrally located and degenerating tumor cells. Most of the lobular carcinomas in situ (LCIS) have been found to be ER positive, with a moderate and rather homogeneous reaction of the monomorphic tumor cell population. In infiltrating lobular carcinoma accompanied by the in situ lesion, the reaction has been more pronounced in most cases so far studied. There are only a few references to a correlation between receptor status and the various types of infiltrating breast cancer. As compared with the results of biochemical assays, which failed to demonstrate any constistent relationship in most reports, better correlations could be found in immunohistochemical studies of both cytological and histological specimens. While about 90% of infiltrating lobular carcinomas have been found to be ER positive, only around two-thirds of infiltrating ductal carcinomas (NOS), predominantly of the more highly differentiated glandular types, show an ER-positive reaction in ER-ICA (REINER et al. 1986). Mucoid, papillary, and tubular carcinomas have been found to be preferentially ER positive, whereas virtually all medullary carcinomas and most infiltrating comedocarcinomas (infiltrating ductal carcinomas with an extensive intraductal component) have a negative receptor assay (Figs. 4, 5). The correlation of ER status with various types of malignant breast tumor is shown in Table 3.

Significant positive correlations have been found between receptor status and various pathological features of prognostic importance, including the histological and cytological degree of differentiation (grading) (STEGNER et al. 1986; BERGER et al. 1987; ANDERSEN et al. 1988; KESHGEGIAN et al. 1988; SCHENCK et al. 1988), smaller tumor cell size, and lower levels of either tumor necrosis or lymphocytic infiltration (BERGER et al. 1987; HELLE et al. 1988). ER status has revealed an inverse correlation with epidermal growth factor receptor expression (TOI 1988; WRBA et al. 1988) and expression of p2 *ras* gene products and c-*erb*B2 oncoprotein (QUERZOLI et al. 1988; WRIGHT et al. 1989).

7 Comparison Between ER Content in Primary Tumors and Metastases

From biochemical studies it is known that ER and PR can change if analyzed asynchronously (JONAT et al. 1986). A summary of a German cooperative trial studying ER and PR in asynchronously obtained tissue samples is shown in Table 4. This study, like others reported in the literature, was influenced by the limitations of the biochemical techniques. In the case of immunohistochemistry there are only a few references dealing with the variation of ER in primary tumors and metastases.

Using an immunohistochemical assay on paraffin-embedded tissue in a total of 92 examined regional lymph nodes, ANDERSEN and POULSEN (1988) found an identical status in the primary tumor and metastases. Semiquantified ER content was significantly correlated in the primary tumor and metastases.

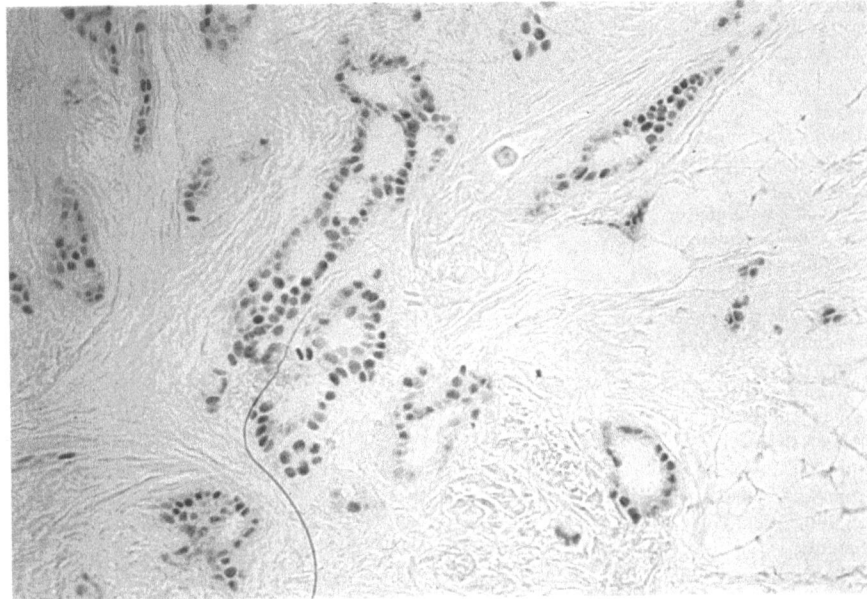

Fig. 4. Immunohistochemical ER demonstration (ER-ICA) in infiltrating tubular carcinoma

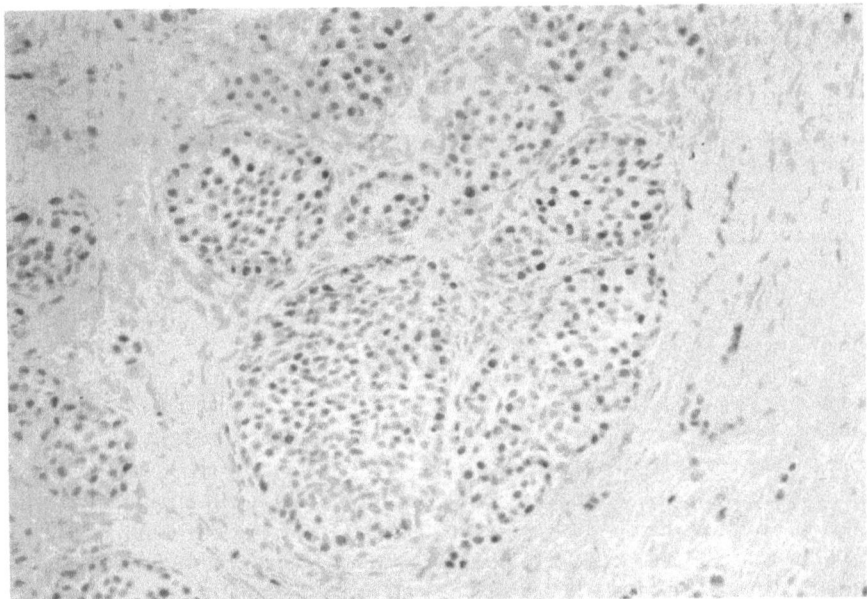

Fig. 5. Immunohistochemical ER demonstration (ER-ICA) in lobular carcinoma in situ (LCIS) associated with infiltrating lobular carcinoma

Table 3. Correlation of ER content (ER-ICA) with primary and metastatic breast carcinomas of different histological type

	ER positive	ER poor[a]	ER negative	Total
Ductal carcinoma (nos)				
– glandular type	68	10	11	89
– solid type	42	8	27	77
Tubular carcinoma	26	4	3	33
Inf. lobular carcinoma	20	6	1	27
Medullary carcinoma	1		9	10
Infiltr. comedocarcinoma	6	2	8	16
Cribriform carcinoma	5			5
Mucinous carcinoma	4	2		6
Papillary carcinoma	2			2
Apocrine carcinoma	1			1
Inflammatory carcinoma			2	2
Paget's disease	1			1
Intraduct., comedo type	3	5	23	31
Intraduct., other types	12	2	6	20
Multiple primary ca.	3		5	8
Sarcoma			6	6
Cutaneous metast.	57	16	32	105
Lymph node metast.	12	3	12	27
Axillary recurrence	4	1	4	9
Lymphangiosis cut.	2	2	4	8
Cystosarcoma phyll.	2		3	5
Fibroadenoma	3	2	2	7
Ductal papilloma	1	4	4	9
	275	67	162	504

[a] Histoscore < 100

Table 4. ER change (DCC) measured asynchronously in primary tumors and recurrent breast cancer: results from a German cooperative study

ER change	No.	No. with change	%	% total
ER+ → ER−	108	44	40.8	35
ER− → ER+	109	31	28.8	

Positive/negative borderline: 20 fmol/mg tissue protein

On the other hand, while Toi et al. (1988), in a comparative study of 25 cases, found the overall status to be consistent in the primary tumor and the nodal metastases, there was a tendency for both the percentage of ER-positive cells and the staining intensity to be lower in nodal metastases as compared with primary lesions. Disparities can be due to tumor heterogeneity or to methodological errors, although ER status as determined by immunohistochemical assay has proven to be very stable in simultaneous biopsies and during the course of the disease (Reiner et al. 1989).

8 Clinical Studies Involving Immunohistochemical Determination of ER and PR

In discussing the clinical importance of receptor determination in breast cancer three aspects currently have to be borne in mind:

1. In patients with metastatic disease, those with receptor-negative tumors rarely respond to endocrine therapy, whereas about two-thirds of those with receptor-positive tumors show objective remissions in response to endocrine manipulations.
2. The consensus conference on adjuvant therapy for breast cancer suggested that receptor status be used in postmenopausal patients in order to select a subgroup to be treated with antiestrogens following operation on the primary tumor with the aim of reducing mortality.
3. Receptor-positive patients survive longer after recurrence compared with receptor-negative ones.

Table 5 summarizes initial reports on ER-ICA results, related to response to endocrine therapy in advanced or recurrent breast cancer (PERTSCHUK et al. 1985; McCARTY et al. 1983; JONAT et al. 1986; McCLELLAND et al. 1987). Compared with conventional biochemical assays (results reported at the consensus development conference on steroid receptors in breast cancer), ER-ICA seems to be more accurate in predicting the clinical outcome. Our first data comparing PR-ICA results and clinical outcome after endocrine therapy in recurrent breast cancer are shown in Table 6. In this retrospective analysis the ability of PR-ICA to select patients for endocrine therapy has thus far not proved as good as that of ER-ICA. Studies analyzing the value of immunohis-

Table 5. Predictive value of ER-ICA: response to endocrine therapy in advanced breast cancer. Endocrine therapy consists of oophorectomy, antiestrogen therapy, or high dosage gestagen therapy

Author	No.	Responding/total number	
		ER-ICA positive	ER-ICA negative
Pertschuk et al. (1985)	43	9/16	2/27
McClelland (1986)[a]	56	21/29	1/27
McCarty (1986)[a]	23	13/14	1/ 9
Jonat et al. (1986)	30	11/18	2/12
Total	152	54/77 70%	6/75 8%
Conventional assays. Bethesda 1979	1336	480/852 56%	27/484 6%

Positive/negative borderline: histoscore 100
[a] personal communication

Table 6. Predictive value of PR-ICA: response to endocrine therapy in advanced breast cancer ($n = 79$ patients) analyzed retrospectively. Endocrine therapy consists of oophorectomy, antiestrogen therapy, or high dosage gestagen therapy

Response	Histoscore positive
Complete remission	7%
Partial remission	45%
No change	8%
Progression	40%

Positive/negative borderline: histoscore 100

tochemical receptor determination of selecting patients for adjuvant treatment procedures or for analyzing the prognostic significance of ER-ICA and PR-ICA are under way.

References

Andersen J, Poulsen HS (1988) Relationship between estrogen receptor status in the primary tumor and its regional and distal metastases. An immunohistochemical study in human breast cancer. Acta Oncol 27: 761–765

Andersen J, Bentzen SM, Poulsen HS (1988) Relationship between radioligand binding assay, immunoenzyme assay and immunohistochemical assay for estrogen receptors in human breast cancer and association with tumor differentiation. Eur J Cancer Clin Oncol 24: 377–384

Bacus S, Julie L, Flowers BS, Michael F, Press MT, Kenneth S, McCarty KS Jr (1988) The evaluation of estrogen receptor in primary breast carcinoma by computer-assisted image analysis. Am J Clin Pathol 90: 233–239

Berger U, Wilson P, McClelland RA, Davidson J, Coombes RC (1987) Correlation of immunocytochemically demonstrated estrogen receptor distribution and histopathologic features in primary breast cancer. Hum Pathol 18: 1263–1267

Burton GV, Flowers JL, Cox EB et al. (1987) Estrogen receptor determination by monoclonal antibody in fine needle aspiration breast cancer cytologies: a marker of hormone response. Breast Cancer Res Treat 10: 287–291

Cudahy TJ, Boeryd BR, Franlund BK, Nordenskjold BA (1988) A comparison of three different methods for the determination of estrogen receptors in breast cancer. Am J Clin Pathol 90: 583–590

De Lena M, Marzullo F, Simone G, Labriola A, Tommasi S, Petroni S, Paradiso A (1988) Correlation between ER-ICA and DCC assay in hormone receptor assessment of human breast cancer. Oncology 45: 308–312

Greene GL, Fitch FW, Jensen E (1980) Monoclonal antibodies to estrophilin: probes for the study of estrogen receptors. Proc Natl Acad Sci USA 77: 157–161

Hawkins RA, Sangster K, Tesdale A, Levack PA, Anderson ED, Chetty U, Forrest AP (1988) The cytochemical detection of oestrogen receptors in fine needle aspirates of breast cancer; correlation with biochemical assay and prediction of response to endocrine therapy. Br J Cancer 58: 77–80

Helle M, Helin H, Antonen J, Krohn K (1988) Human milk fat globule antigen III D5, steroid receptors and histopathologic parameters in breast cancer. Acta Pathol Microbiol Immunol Scand [A] 96: 415–420

Hiort O, Kwan PW, DeLellis RA (1988) Immunohistochemistry of estrogen receptor protein in paraffin sections. Effects of enzymatic pretreatment and cobalt chloride intensification. Am J Clin Pathol 90:559–563

Horsfall D, Jarvis L, Tilley W, Orell S, Cant E (1987) Assessment of estrogen receptor heterogeneity in fine needle aspirates of breast cancer by computerized video image (meeting abstract). Biennale International Breast Cancer Research Conference. March 1–5, Miami, Fl.

Horsfall D, Jarvis LR, Grimbaldestone MA, Tilley WD, Orell SR, (1989) Immunocytochemical assay for oestrogen receptor in fine needle aspirates of breast cancer by video image analysis. Br J Cancer 59:129–134

Jensen EV (1970) The pattern of hormone receptor interaction. In: Griffiths K, Pierrepoint CG (eds) Some aspects of aetiology and biochemistry of prostatic cancer. Alpha, Omega, Alpha, Cardiff, p 151

Johnson H, Mir R, Richer S, Wise L (1987) A comparative study of cytologic smears and frozen-tissue sections in the determination of sex steroid receptor status of breast carcinomas. Surgery 102:628–634

Jonat W, Stegner HE, Maass H (1986) Immunohistochemical measurement of estrogen receptors in breast cancer tissue samples. Cancer Res 46:4296–4298

Jonat W, Kügler G, von Laffert C (1989) Biochemische und immunhistochemische Rezeptorbestimmungen. Act Onkol 46:18–26

Keshgegian AA, Inverso K, Kline TS (1988) Determination of estrogen receptor by monoclonal antireceptor antibody in aspiration biopsy cytology from breast carcinoma. Am J Clin Pathol 89:24–29

Köhler G, Bässler R (1987) Der immunhistochemische Nachweis des Oestrogenrezeptors in benignen Tumoren der Mamma und in Formen der Mastopathie. Pathologe 8:325–333

Kubler HC, Kuhn W, Rummel HH, Krapfl E, Klinga K (1988) Zur praeoperativen immunzytochemischen Oestrogenrezeptorbestimmung an Feinnadelpunktaten beim Mammakarzinom mittels monoklonaler Antikörper. Geburtshilfe Frauenheilkd 48:763–767

Logeat F, Hai MTV, Fournier A, Legrain C, Buttin G, Milgrom E (1983) Monoclonal antibodies to rabbit progesterone receptor: cross reaction with other mammalian progesterone receptors. Proc Natl Acad Sci USA 80:6456

Loosfeld H, Logeat F, Hai MTV, Milgrom E (1985) The rabbit progesterone receptor. J Biol Chem 259:14196

Maass H, Engel B, Hohmeister H, Lehmann F, Trams G (1972) Estrogen receptors in human breast tumors. Am J Obstet Gynecol 133:377

Masood S (1989) Use of monoclonal antibody for assessment of estrogen receptor content in fine-needle aspiration biopsy specimen from patients with breast cancer. Arch Pathol Lab Med 113:26–30

McCarty KS Jr et al. (1983) Clinical and histological correlation of estrogen receptor distribution as determined by immunohistochemistry using monoclonal antibody. Symposium: Estrogen receptor determination with monoclonal antibodies, personal communication

McCarty KS Jr, Miller LS, Cox EB, Konrath J, McCarty KS Sen (1985) Estrogen receptor analysis: correlation of biochemical and immunohistochemical methods using monoclonal antireceptor antibodies. Arch Pathol Lab Med 109:716–721

McClelland RA, Berger U, Wilson P et al. (1987) Presurgical determination of estrogen receptor status using immunocytochemically stained fine needle aspirate smears in patients with breast cancer. Cancer Res 47:6118–6122

McGuire WL, Clark G (1985) Role of progesterone receptors in breast cancer. Semin Oncol XII [Suppl 1]:12–16

Neis KJ, Macher F, Kaul S, Bastert G (1989) Erste Erfahrungen mit einem monoklonalen Antikörper gegen den Progesteronrezeptorkomplex. Geburtshilfe Frauenheilkd 49:109–114

Parl FF, Posey YF (1988) Discrepancies of the biochemical and immunohistochemical estrogen receptor assay in breast cancer. Human Pathol 19:960–966

Pertschuk LP, Eisenberg KB, Carter AC, Feldmann JG (1985) Immunohistologic localization of estrogen receptors in breast cancer with monoclonal antibodies. Cancer 55:1513

Prechtel K (1972) Beziehungen der Mastopathie zum Mammakarzinom. Fortschr Med 90:43–45

Querzoli P, Marchetti E, Bagni A, Marzola A, Fabris G, Nenci I (1988) Expression of p21 *ras* gene products in breast cancer relates to histological types and to receptor and nodal status. Breast Cancer Res Treat 12:23–30

Rasmussen BB, Thorpe SM, Nrgaard T, Rasmussen J, Agdal N, Rose C (1988) Immunohistochemical steroid receptor detection in frozen breast cancer tissue. A multicenter investigation. Acta Oncol 27:757–760

Reiner A, Reiner G, Spona J, Schemper M, Kolb R, Kakesz R, Holzner JH (1986) Vergleich von immunhistochemischem und biochemischem Oestrogenrezeptornachweis beim Mammakarzinom und Beziehungen zur histologischen Tumorklassifikation. Verh Dtsch Ges Pathol 70:243–246

Reiner A, Reiner G, Spona J, Weinlich M, Jakesz R (1989) Estrogen receptor status in simultaneous and sequential biopsy from patients with breast cancer. A comparison between ER-ICA and biochemical analysis (meeting abstract). Proc Annu Meet Am Soc Clin Oncol 8:A80

Remmele W, Stegner HE (1986) Immunhistochemischer Nachweis von Östrogenrezeptoren (ER-ICA) in Mammakarzinomgewebe: Vorschlag zur einheitlichen Formulierung des Untersuchungsbefundes. Dtsch Ärztebl 83:3362–3364

Schenck U, Jutting U, Eiermann W (1988) Hormonrezeptorbestimmung beim Mammakarzinom mit der Dextran-Coated-Charcoal-Methode und mit monoklonalen Antikörpern: Korrelation mit einem zytomorphologischen Grading. Onkologie 11:211–215

Stegner HE, Jonat W, Maass H (1986) Immunhistochemischer Nachweis nukleärer Oestrogenrezeptoren mit monoklonalen Antikörpern in verschiedenen Typen des Mammakarzinoms. Pathologe 7:156–163

Sternberger LA, Hardy PH, Cuculis JJ, Meyer HG (1970) The unlabelled antibody enzyme method of immunohistochemistry. J Histochem Cytochem 18:315–333

Toi M (1988) Comparative study on estrogen receptor, epidermal growth factor and epidermal growth factor receptor using immunocytochemical and biochemical assay in breast cancer. Nippon Geka Gakkai Zasshi 89:725–736

Toi M, Hamada Y, Seto Y et al. (1988) Immunocytochemical study on the variation in estrogen receptors of primary and nodal metastases of breast cancer. Jpn J Surg 18:228–231

Wrba F, Reiner A, Ritzinger E, Holzner JH, Reiner G (1988) Expression of epidermal growth factor receptors (EGFR) on breast carcinoma in relation to growth fractions, estrogen receptor status and morphological criteria. An immunohistochemical study. Pathol Res Pract 183:25–29

Wright C, Angus B, Nicholson S et al. (1989) Expression of c-*erb*-2 oncoprotein: a prognostic indicator in human breast cancer. Cancer Res 49:2087–2090

Prostatic Cancer – Immunohistochemistry of Steroid Hormone Receptors

N. Wernert and G. Seitz

1 Biochemical and Immunohistochemical Findings Concerning Estrogen and Progesterone Receptors in the Prostate and in Common Prostatic Carcinomas

1.1 Findings in the Prostate

Biochemical investigations of estrogen receptor (ER) in the prostate, including cases of benign prostatic hyperplasia (BPH), have yielded both positive (KIR-DANI et al. 1984, 1985; PERTSCHUK et al. 1979; ROBEL et al. 1985; WAGNER et al. 1975) and negative (BASHIRELAHI et al. 1983; PERTSCHUK et al. 1979; ROBEL et al. 1985; WOLF et al. 1985) results. According to ROBEL et al. (1985), ER occurs only in very low concentrations. Progesterone receptor (PR) has also been demonstrated both in central and in peripheral portions in the prostate and in BPH (LÄMMEL et al. 1986; ROBEL et al. 1985; WOLF et al. 1985).

Exact localization of hormone receptors within different cell types is not possible using biochemical methods. To date several authors have failed to demonstrate ER in the human prostate using the ER-ICA test (HARPER et al. 1984, 1986; PERTSCHUK et al. 1985). Using a modified PAP and APAAP method for this test we obtained a clearly positive reaction (WERNERT et al. 1987b, 1988a). We applied an analogous methodology for the immunohistochemical demonstration of PR using a commerical antiserum (WERNERT et al. 1988a).

Both ER and PR are demonstrable immunohistochemically exclusively within the nuclei and in the same cell types. Most of the periglandular stromal cells (fibrocytes and smooth muscle cells) react positively (Fig. 1). However, only some of the interglandular stromal cells are positive for both ER and PR (Fig. 2). Normal basal cells in the prostatic epithelium stain focally positive (Fig. 3). In contrast, basal cell hyperplasia proves extensively positive (Fig. 4). The secretory cylindrical epithelium of the prostate is negative for both receptors. Within the prostatic ducts the urothelium is partially positive (Fig. 5). The urothelium of the prostatic urethra stains only focally. In hyperplastic nodules results are found which are analogous to those in normal parts of the prostate.

No preferential expression was seen on large-field sections through the whole prostate either for ER or for PR in central parts of the prostate or in the outer parts. Throughout the prostate, both receptors were distributed evenly.

Furthermore, ER has been demonstrated immunohistochemically in both the normal canine prostate and the canine prostate affected by BPH (Schulze and Barrack 1987a, b). The results differ slightly from our findings in that ER is predominantly present in the stromal nuclei of the periurethral region. Exclusively in BPH ER can be found within the nuclei of the acinar epithelium.

In essence our findings concerning ER in the human prostate match our results achieved in demonstrating immunohistochemically the ER-associated protein ER-D5 (Seitz and Wernert 1987). This protein is seen within the cytoplasm of the fibromuscular stromal cells (Fig. 6). In normal basal cells it

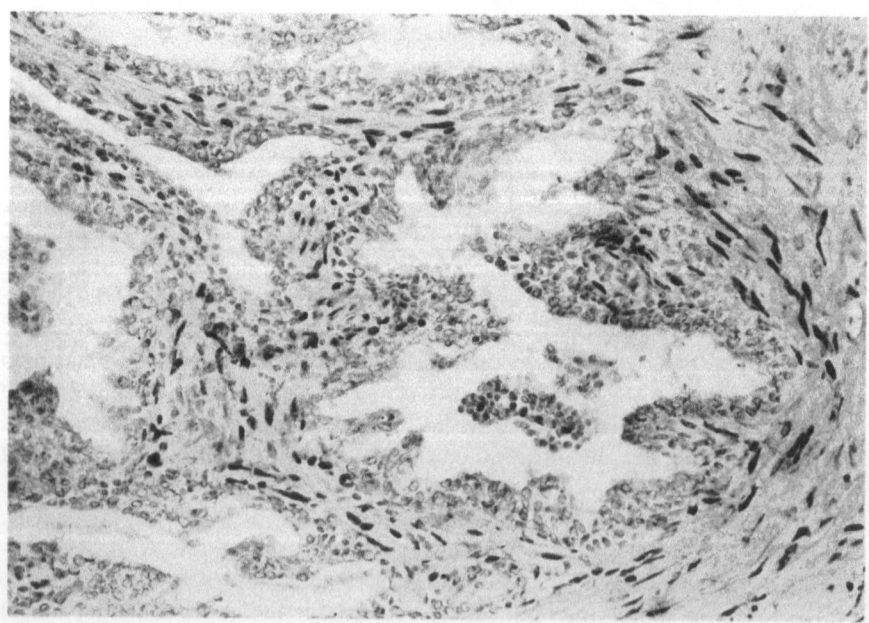

Fig. 1. ER in the nuclei of the periglandular stromal cells. × 250

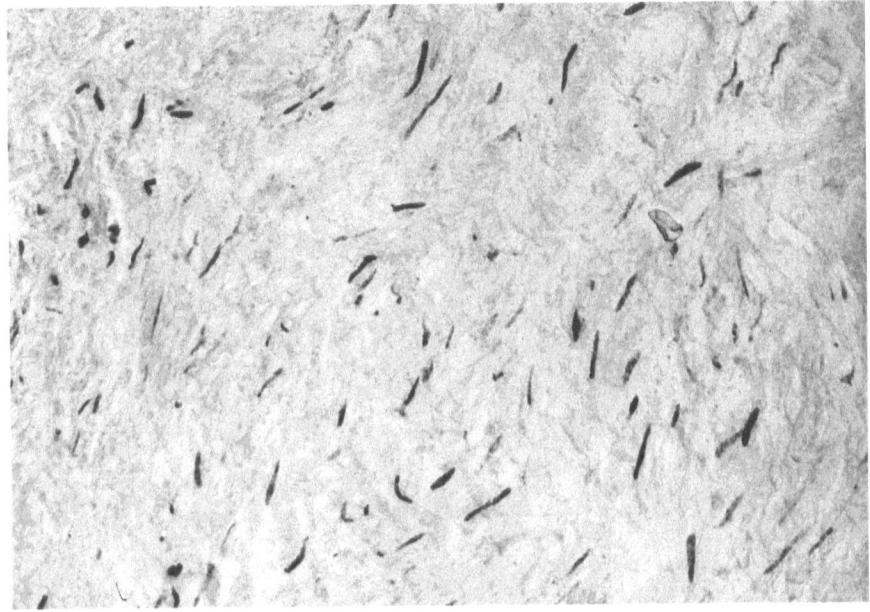

Fig. 2. Some interglandular stromal cells showing intranuclear positivity for PR. × 250

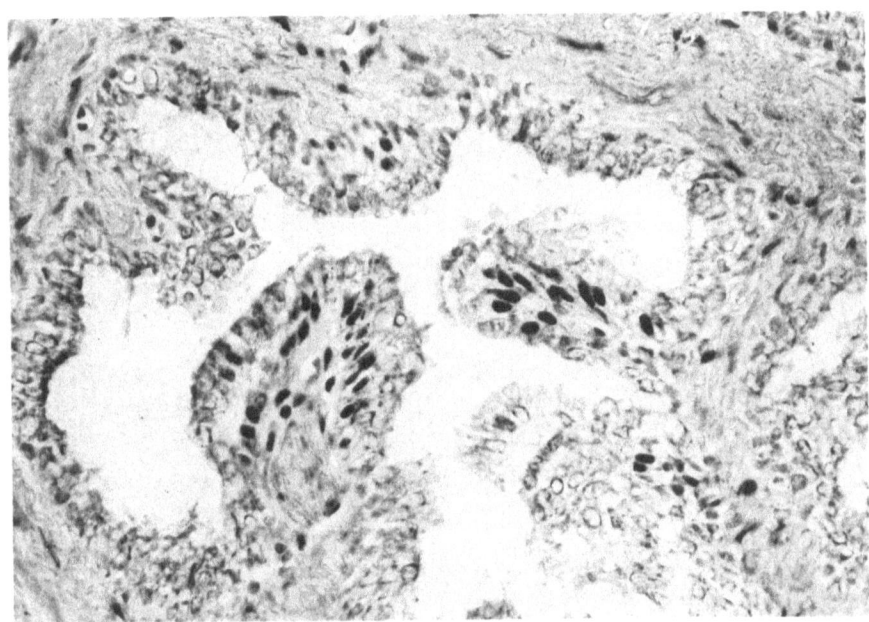

Fig. 3. ER in normal basal cells. × 400

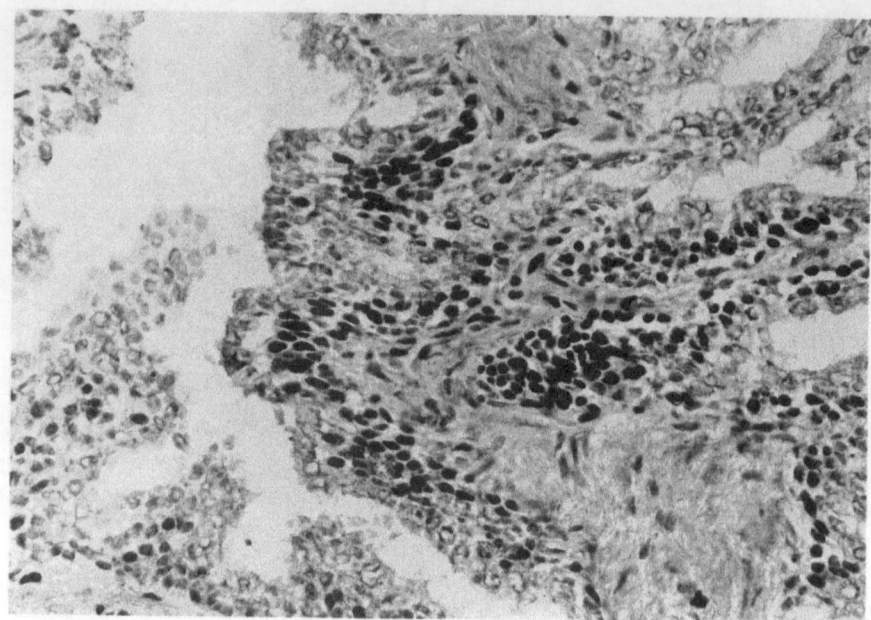

Fig. 4. ER in hyperplastic basal cells. × 400

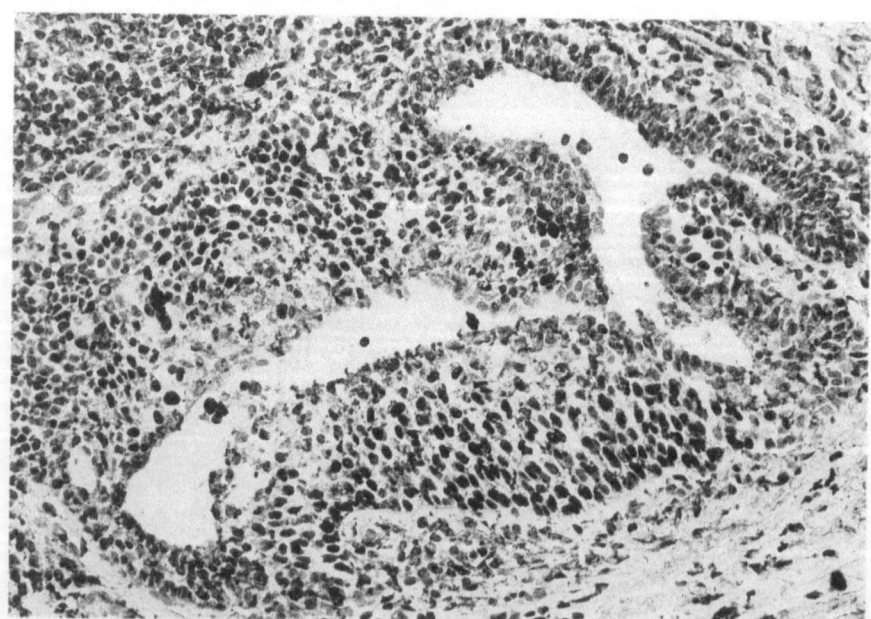

Fig. 5. Urothelium in prostatic ducts showing positivity for ER. × 250

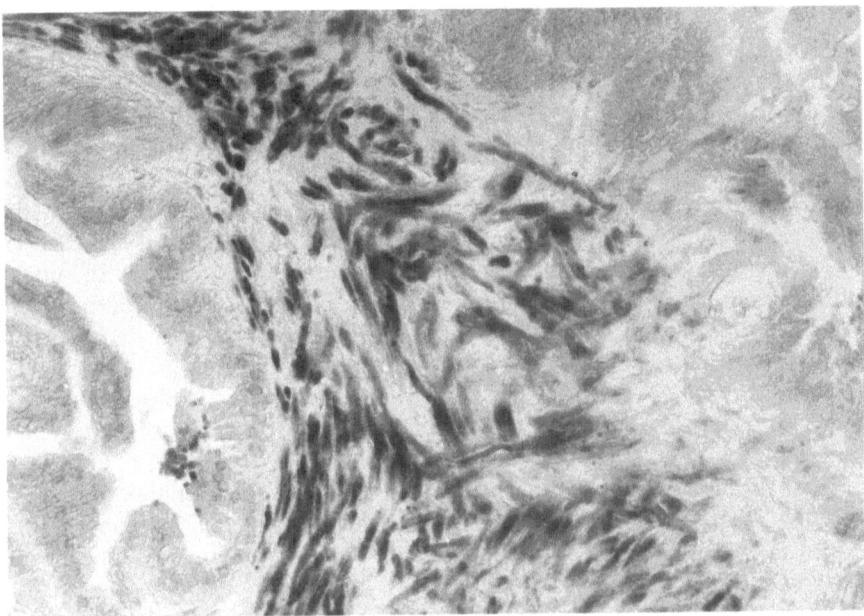

Fig. 6. Stromal cells positive for ER-D5. Epithelium negative (without hematoxylin). × 250

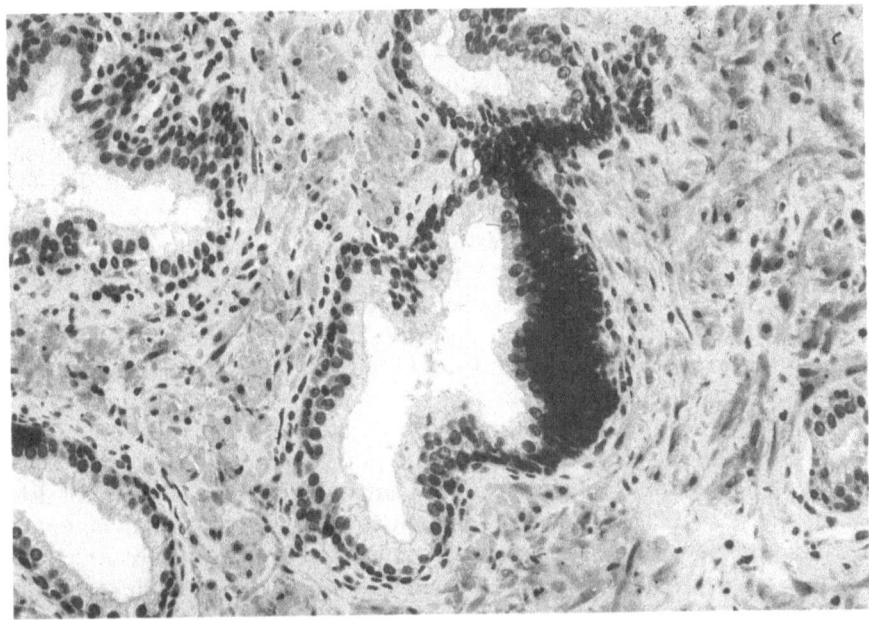

Fig. 7. Basal cell hyperplasia, ER-D5 positive. × 250

is found focally; however, in basal cell hyperplasia (Fig. 7) and in squamous cell metaplasia it is constantly present.

These immunohistochemical findings all accord well with a role for estrogens in the pathogenesis of BPH. For the development of BPH the primary requirements are androgens (TUNN and SCHWEIKERT 1983). BPH is not found after prepubertal castration. The amount of the biologically active testosterone metabolite dihydrotestosterone (DHT) is higher in BPH than in the normal prostate (BARTSCH et al. 1982; TUNN and SCHWEIKERT 1983; TVETER 1974; for review, see SENGE 1983). DHT is predominantly found in the nuclei of the prostatic stroma (BARTSCH et al. 1982). Its increased concentration in BPH is due to a greater activity of the 5α-reductase, which likewise is demonstrable mostly in the prostatic stroma (KRIEG et al. 1981; TUNN and SCHWEIKERT 1983; WILKIN et al. 1980; for review, see SENGE 1983) DHT is thought to play a part in the development of BPH after binding to the androgen receptor in the stroma (SENGE 1983; TVETER 1974).

In addition, estrogens are also believed to play a role in the pathogenesis of BPH. Experimentally a predominantly stromal hyperplasia can be produced in dogs by means of estrogens either in combination with or without androgens (TUNN and SCHWEIKERT 1983, TUNN et al. 1979; for review, see FARNSWORTH 1983). Estradiol and the ER are biochemically demonstrable mainly in the prostatic stroma of human BPH (KIRDANI et al. 1984, KOZAK et al. 1982, SENGE 1983). These findings correspond well with our immunohistochemical results. It has long been known that BPH begins as stromal hyperplasia. The epithelial growth is apparently a secondary phenomenon (FRANKS 1954; MOORE 1943; REISCHAUER 1925). REISCHAUER surmised an inductive influence of the hyperplastic stroma on the epithelium as early as 1925. In the mouse, it has been shown experimentally that the mesenchyma of the urogenital sinus induces the development of the prostate during embryogenesis (SUGIMURA et al. 1985). According to FRANKS et al. (1970), epithelia from human BPH can be maintained in culture only in conjunction with stromal cells. The immunohistochemically obtained results for ER and ER-D5 (SCHULZE and BARRACK 1987 a, b; SEITZ and WERNERT 1987; WERNERT et al. 1987 b, 1988 a) support a role of estrogens in the development of BPH. First a proliferation of the ER-positive stroma may be triggered off by estrogens. The stroma then might induce epithelial growth. Estrogens might cause the primary stromal proliferation directly or indirectly, because they are able to increase conversion of testosterone to DHT in cell cultures of BPH (for review, see LEE and JESIK 1983).

It is well known that estrogens can trigger basal cell hyperplasia, and this fits in well with our demonstration of ER in the basal cells (WERNERT et al. 1988 a). Furthermore, the urothelium in prostatic ducts is positive for ER (WERNERT et al. 1988 a). Therefore patients from our registry with common prostatic carcinomas treated are currently being investigated to ascertain whether these estrogens can cause proliferation of this urothelium and favor the development of transitional cell carcinomas of the prostate.

The biological significance of PR in the human prostate is not known.

1.2 Findings in Prostatic Carcinomas

In the biochemical demonstration of ER in prostatic carcinomas, both positive (HARPER et al. 1986; KIRDANI et al. 1985; LÄMMEL et al. 1986; PERTSCHUK et al. 1979; WAGNER et al. 1975; WOLF et al. 1985) and negative (EKMAN et al. 1979 a, LÄMMEL et al. 1986; PERTSCHUK et al. 1979; WAGNER et al. 1975) results have been obtained. The same is true for PR (EKMAN et al. 1979 a; LÄMMEL et al. 1986). We investigated immunohistochemically the presence of both ER and PR in 12 common prostatic carcinomas using the modified methodology mentioned above (WERNERT et al. 1988 a). All 12 tumors proved completely negative for both receptors. Only the stromal cells between the tumor formations were positive. In only 11 out of 82 prostatic carcinomas (13.4%) could ER-D5 be found immunohistochemically, and as a rule in merely a few cell groups (SEITZ and WERNERT 1987). These results suggest that estrogens do not exert their effect directly on prostatic carcinoma but rather indirectly inhibit its proliferation by reducing the pituitary gonadotropin output and thereby decreasing the serum testosterone level. It is possible that the above-mentioned contradictory positive and negative biochemical findings regarding the demonstration of ER in prostatic carcinomas are due to contamination of tissue homogenates with stromal cells containing ER (KRIEG et al. 1978).

2 Adenoid Cystic-like Prostatic Carcinoma Immunohistochemically Positive for the Estrogen Receptor

There exists a very rare prostatic tumor resembling adenoid cystic carcinoma of the salivary glands (for review, see YOUNG et al. 1988). SESTERHENN et al. (1987) mentioned three such cases from the Armed Forces Institute of Pathology in Washington, D.C. In our prostatic cancer registry we have one such tumor kindly provided by Prof. Kastendieck (Hamburg). It was found within the BPH of a 50-year-old patient. In parts it resembled atypical basal cell hyperplasia (Fig. 8), and in other parts, adenoid cystic carcinoma (Fig. 9). Both morphologically and immunohistochemically (WERNERT 1987 a, 1988 a) the cells showed differentiation towards either basal cells or secretory prostatic epithelium (Fig. 8). ER was demonstrable in the nuclei of the basal cell-type formations (Fig. 10) using a modified PAP and APAAP technique for the ER-ICA test on paraffin sections (SEITZ and WERNERT, unpublished). This agrees well with the occurrence of ER in normal and hyperplastic basal cells of the prostate (WERNERT et al. 1988 a).

As estrogens can induce hyperplasia of normal basal cells, it is very likely that they also stimulate the proliferation of the basal cell-type formations in these rare tumors, which seem to bear a favorable prognosis (YOUNG et al. 1988). We feel that estrogens are strictly contraindicated in their treatment.

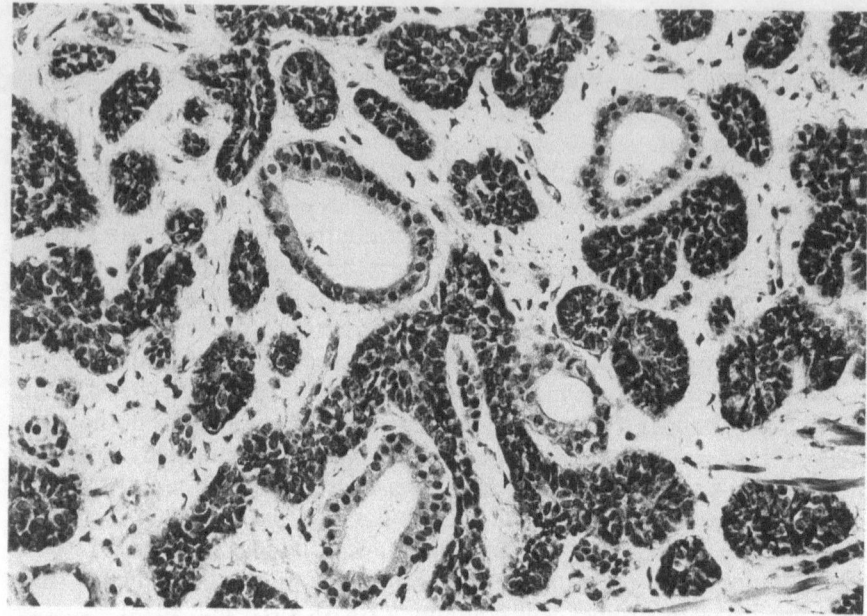

Fig. 8. Adenoid cystic-like carcinoma of the prostate. Parts resembling atypical basal cell hyperplasia. Focal glandular structures with secretory type epithelium. HE, × 250

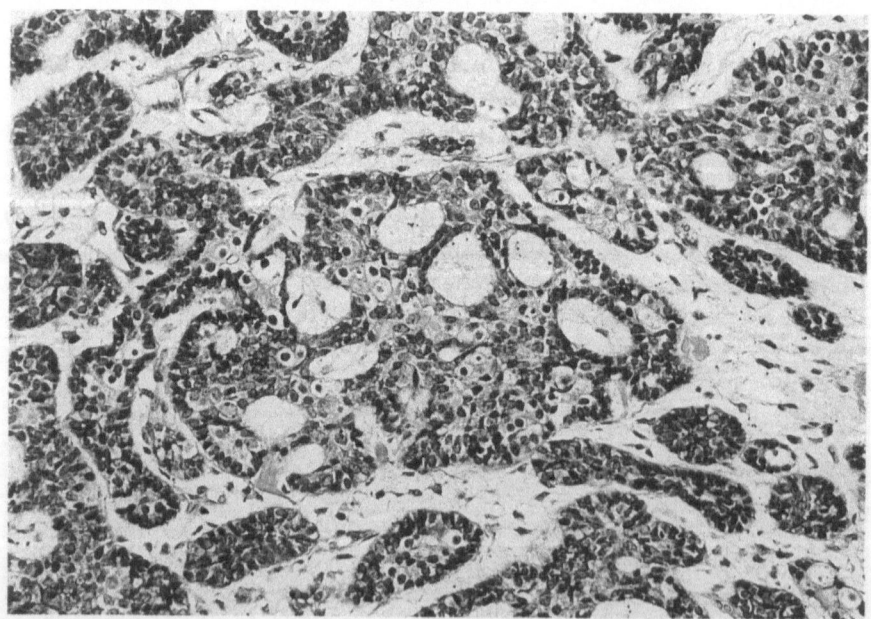

Fig. 9. Adenoid cystic-like carcinoma of the prostate showing cribriform spaces. HE, × 250

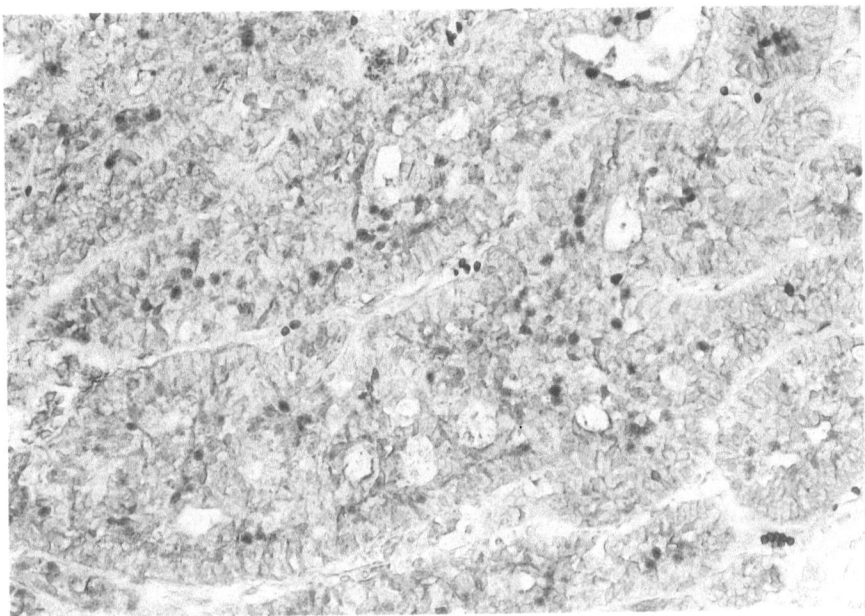

Fig. 10. Adenoid cystic-like carcinoma of the prostate. Basal cell-type formations immunohisto-chemically positive for ER. × 250

3 Squamous Cell Carcinomas of the Prostate Following Estrogen Therapy

Squamous cell carcinomas of the prostate are very rare. In our cancer registry they account for only 0.2% of all prostatic carcinomas (DHOM 1980). Just a small number of cases have been described in the literature (ACCETTA and GARDNER 1982; BENNETT and EDGERTON 1973; VAN BUSKIRK and KIMBROUGH 1954; CORDER and CICMIL 1976; GRAY and MARSHALL 1975; KASTENDIECK and ALTENÄHR 1974; MOYANA 1987; SAITO et al. 1984; SHARMA et al. 1980; SIERACKI 1955; THOMPSON 1942; THOMPSON et al. 1953). The 11 cases in our registry (WERNERT et al. 1988b) and the tumors reported by others can be subdivided into four groups according to their histomorphology and mode of origin:

1. Transitional cell carcinoma of the urinary bladder with malignant squamous areas and involvement of the prostate (WERNERT et al. 1988b)
2. Transitional cell carcinoma of the prostate with malignant squamous parts (GOEBBELS et al. 1985; RHAMY et al. 1973; SEEMAYER et al. 1975; THELMO et al. 1974; WERNERT et al. 1988b)
3. Pure primary squamous cell carcinoma of the prostate (GRAY and MARSHALL 1975; SHARMA et al. 1980; WERNERT et al. 1988b)

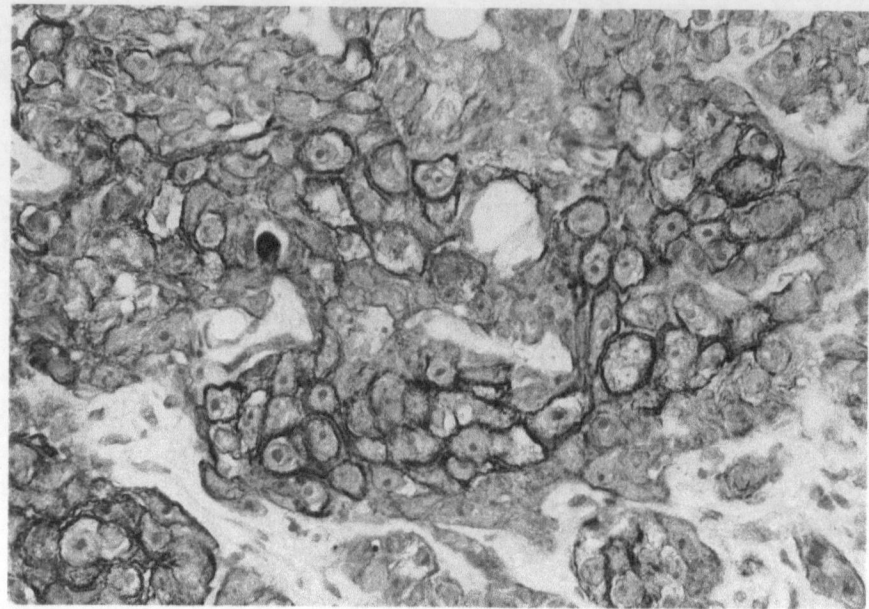

Fig. 11. Squamous cell carcinoma of the prostate showing positivity for keratins from the stratum corneum. × 400

4. Squamous cell carcinoma of the prostate following estrogen treatment of a common prostatic carcinoma (ACCETTA and GARDNER 1982; BENNETT and EDGERTON 1973; KASTENDIECK and ALTENÄHR 1974; MOYANA 1987; SAITO et al. 1984; WERNERT et al. 1988 b).

Two cases from our material belong to the last-mentioned group. In one of these tumors dysplastic squamous cell metaplasia in the preexisting prostate was found, which we regard as precancerous. The stem cell for estrogen-induced squamous cell metaplasia in the prostate is the basal cell which is immunohistochemically positive for ER (WERNERT et al. 1988 a). Therefore in all probability this cell is also the actual cell of origin for squamous cell carcinomas developing after estrogen treatment. This assumption is further supported by the fact that the basal cell is found to share in essence its immunohistochemical features with squamous cell carcinoma when a number of markers are investigated (Fig. 11) (WERNERT 1987; WERNERT et al. 1988 b).

4 The Question of Estrogen-Dependent Growth of Papillary (So-called Endometrioid) Prostatic Carcinoma

Papillary, so-called endometrioid, carcinoma has been regarded by a number of authors as a separate tumor entity arising from müllerian epithelium within

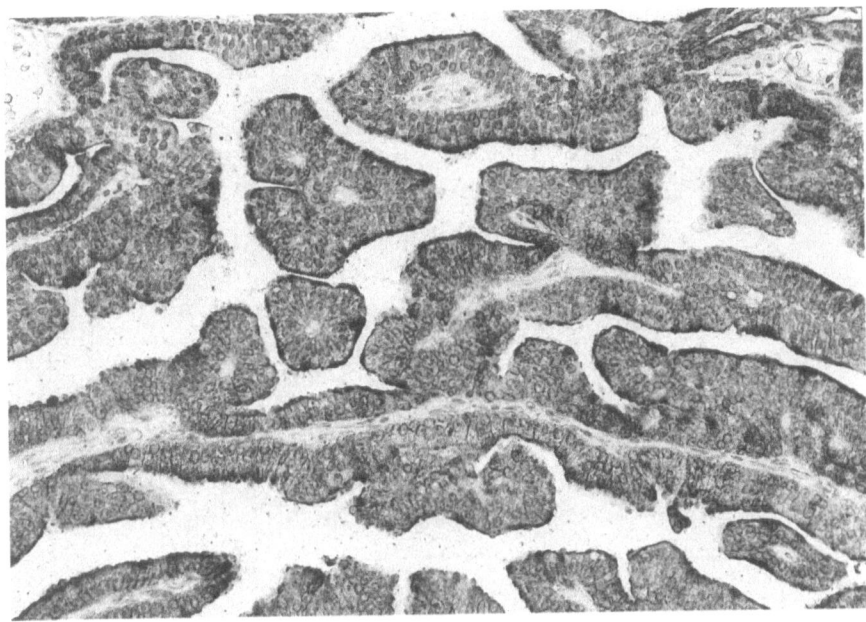

Fig. 12. Papillary prostatic carcinoma positive for PAP. ×250

the prostatic utricle (CARNEY and KELALIS 1973; DUBE et al. 1973; MELICOW and PACHTER 1966; MELICOW and TANNENBAUM 1971; MERCHANT et al. 1976; STEFFENS and LEISTENSCHNEIDER 1983; SUFRIN et al. 1983). Therefore the growth of these tumors has been postulated to be estrogen dependent. As a consequence, estrogen therapy has been regarded as inappropriate (CARNEY and KELALIS 1973; KAUDER et al. 1977; MELICOW and TANNENBAUM 1971; MERCHANT et al. 1976; ROTTERDAM and MELICOW 1975; SUFRIN et al. 1983).

We investigated 51 papillary carcinomas of the prostate (WERNERT 1987; WERNERT et al. 1987a). Forty-two cases were combined with a common prostatic carcinoma, which merged into the papillary tumor parts in 50% of cases. The cells of the papillary tumor parts had either dark or light cytoplasm. The papillary carcinomas did not differ immunohistochemically in any way from common prostatic carcinomas (Fig. 12) (WERNERT 1987; WERNERT et al. 1987a). Findings reported by other authors regarding endometrioid carcinomas are in agreement with our own observations (CARNEY and KELALIS 1973; KUHAJDA et al. 1984; NADJI et al. 1980; PILLARISETTI et al. 1983; ROTTERDAM and MELICOW 1975; STEFFENS and LEISTENSCHNEIDER 1983; SUFRIN et al. 1983; TANNENBAUM 1975; WALKER et al. 1982; YOUNG and LAGIOS 1973; ZALOUDEK et al. 1976). In no case did we find a papillary carcinoma within the utricle. Moreover, according to our results (WERNERT et al. 1986; WERNERT 1987) the epithelial lining of the utricle is identical with that of the prostate both morphologically and immunohistochemically. Of special significance is

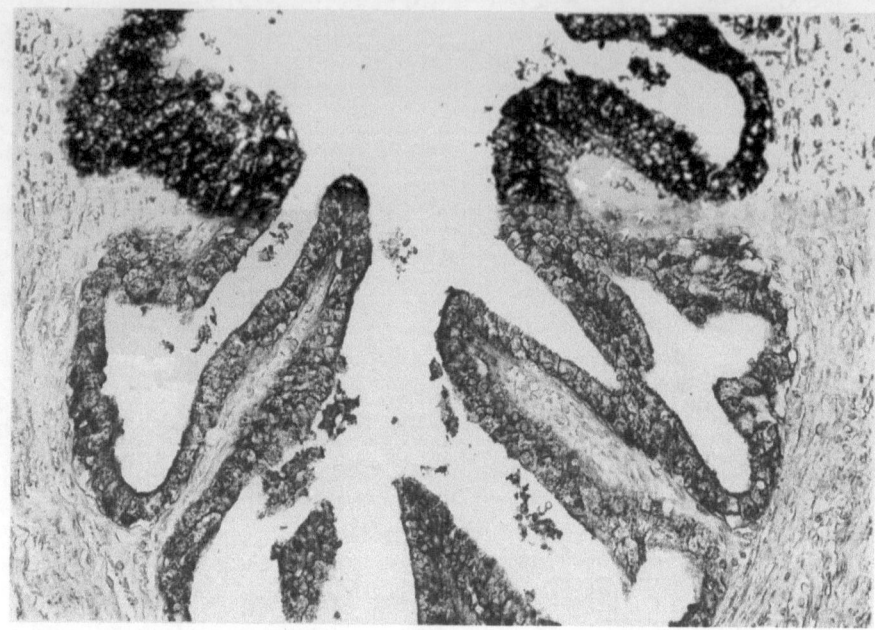

Fig. 13. Prostatic utricle (52-year-old); PSA. Strongly positive secretory epithelium. × 250

the immunohistochemical expression of both prostatic acid phosphatase (PAP) and prostate-specific antigen (PSA) in the cylindrical epithelium (Fig. 13). In contrast to the generally held opinion (Heller and Sprinz 1921; Meyer 1909; Zondek and Zondek 1974), this means that there is no müllerian epithelium within the utricle from which an estrogen-dependent carcinoma analogous to carcinoma of the corpus uteri could arise. In our opinion papillary carcinomas of the prostate on the whole represent mere morphological variants of common prostatic carcinomas and should be treated in the same way. Their growth is certainly not dependent on estrogens.

5 Biochemical and Immunohistochemical Findings Concerning the Androgen Receptor in the Prostate and in Common Prostatic Carcinomas

In 1971 Tveter et al. demonstrated the androgen receptor (AR) in the human prostate. Methodological problems in the biochemical demonstration of AR in prostatic tissue homogenates were overcome by the introduction of new receptor ligands (Bonne and Raynaud 1975; Murthy et al. 1986). In the meantime the biochemical structure of AR has been clarified (Brinkmann et al. 1988). AR has been demonstrated biochemically in the normal prostate

(ROBEL et al. 1985; WAGNER et al. 1975), in BPH (VAN AUBEL et al. 1985; KRIEG et al. 1978; KYPRIANOU and DAVIES 1986; ROBEL et al. 1985), and in prostatic carcinoma (VAN AUBEL et al. 1985; EKMAN 1980; EKMAN et al. 1979 a, b; GHANADIAN et al. 1981; KRIEG et al. 1978; TRACHTENBERG and WALSH 1982). KYPRIANOU and DAVIES (1986) found it in the nuclei of both the epithelium and the stroma in cases of BPH. BOWMAN et al. (1986) demonstrated AR more often in peripheral than in central parts of the prostate. According to VAN AUBEL et al. (1985) there are no quantitative differences in AR content between BPH and carcinoma. These authors, nevertheless, found a considerable variability in the receptor content within different parts of prostatic carcinomas regardless of their degree of differentiation. This suggests that the tumor cells are heterogeneous with regard to their receptor content.

A positive correlation between the AR content in prostatic carcinomas and the response to androgen-depriving therapy has been demonstrated in small groups of patients (EKMAN et al. 1979 b; GHANADIAN et al. 1981; TRACHTENBERG and WALSH 1982, for review; see DE VOOGT and RAO 1983). Nevertheless, biochemical determination of AR values in prostatic carcinomas has not yet become a routine procedure in clinical practice due to the complicated methodology (for reviews, see CONNOLLY and MOBBS 1984 and DE VOOGT and RAO 1983). Moreover there is not yet agreement on the AR concentration at which the growth of a prostatic carcinoma can be regarded as dependent on androgens (for review, see CONNOLLY and MOBBS 1984).

Histochemical demonstration of AR by fluorescein-conjugated testosterone or methyltrienolone (NAITO et al. 1981; PERTSCHUK et al. 1982, 1984) can obviously not be regarded as sufficiently specific (BERNS et al. 1984; DE GOEIJ and BOSMAN 1986; LÄMMEL et al. 1986). Meanwhile direct demonstration of AR on an immunohistochemical basis is still in its infancy (DEMURA et al. 1987). An alternative might be indirect immunohistochemical demonstration by a marker which is produced dependently on androgens. In the breast and breast carcinoma peanut agglutinin (PNA) binding sites are synthesized dependently on estrogens (KLEIN et al. 1983 a, b; VIERBUCHEN et al. 1979). We investigated different immunohistochemical markers (PAP, PSA, different keratins, vimentin, PNA-binding sites, carcinoembryonic antigen, and antichymotrypsin) in common prostatic carcinomas before and after androgen-depriving therapy (WERNERT 1987; WERNERT et al. 1987 b). All antigens were still demonstrable to a variable degree after the treatment in both regressively altered (Fig. 14) and therapeutically uninfluenced tumor portions. Therefore in none of the markers investigated is the expression dependent on androgens. However, regarding the biochemical mechanism of action of the AR within the cell, which influences the transcription on the DNA (for review, see CONNOLLY and MOBBS 1984; HECHTER 1984; MANON et al. 1977; SANDBERG 1980), there are very probably products in prostatic carcinoma whose synthesis is dependent on androgens. An androgen-dependent formation has already been demonstrated for different lectin-binding sites and proteins as well as for the zinc concentration in the nonneoplastic prostate of the dog, the rat, and the rhesus monkey (GAYET et al. 1985; HEYNS et al. 1978; MARTIKAINEN et al.

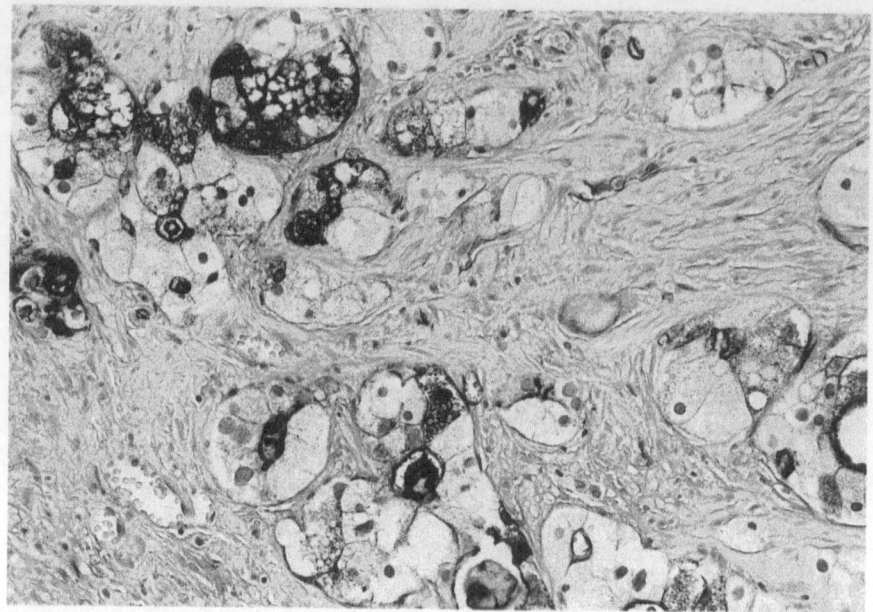

Fig. 14. Prostatic carcinoma treated by androgen deprivation. PNA-binding sites in regressively altered tumor portions. × 250

1986; Orgad et al. 1984; Parker et al. 1978; Srivastava et al. 1984). Therefore further markers, especially lectin-binding sites, must be investigated in prostatic carcinomas before and after androgen-depriving treatment to test them for androgen dependence.

6 The Behavior of the Endocrine Cells in Prostatic Carcinoma Under Hormonal Therapy

Argyrophilic and/or argentaffin endocrine cells can be found in varying numbers in common prostatic carcinomas (Abrahamsson et al. 1987; Azzopardi and Evans 1971; Capella et al. 1981; Fetissof et al. 1983; Feyrter 1951; Kazzaz 974; Wernert 1987). Immunohistochemically these cells are found to contain not only chromogranin (Bier et al. 1988), serotonin, and neuron-specific enolase (Abrahamsson et al. 1987; Wernert 1987) but also a number of hormones such as thyrotropin α-HCG, corticotropin, leu-enkephalin, β-endorphin, somatostatin, glucagon, and calcitonin. The biological significance of these hormones is unclear. None of the 40 patients from the series of Abrahamsson et al. (1987) showed clinical signs of a paraneoplastic endocrine syndrome. These endocrine cells show nuclear atypias and resemble morphologically the entodermal carcinoma cells from which they only differ by virtue of

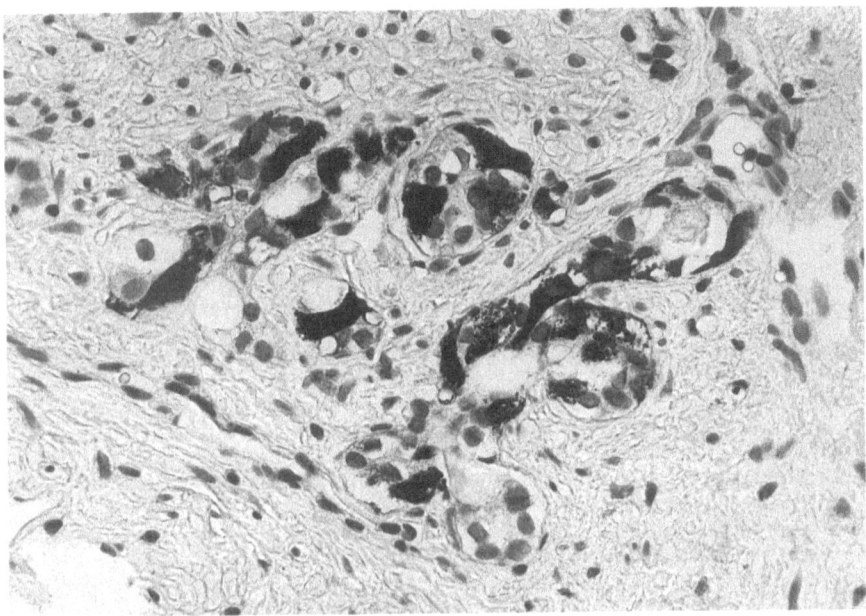

Fig. 15. Prostatic carcinoma treated by estrogens and orchiectomy showing large numbers of chromogranin-positive endocrine tumor cells. × 400

their endocrine immunohistochemical markers. This fits in with the view that these cells are of entodermal and not neural origin, which is further substantiated by the fact that some of them also contain PSA in addition to the endocrine marker serotonin (BONKHOFF and WERNERT, unpublished). Observations in our prostatic cancer registry indicate that these cells can in all probability gain an advantage in proliferation under androgen-depriving therapy (BIER et al. 1988): We found a marked increase in the number of the endocrine cells (Fig. 15) in two pluriform grade III carcinomas after treatment by orchiectomy alone and by orchiectomy combined with estrogens. Individual observations analogous to our findings were made by STRATTON et al. (1986). We suppose that the endocrine cells lack AR and thus grow independently of androgens. This would imply that they continue to proliferate under androgen deprivation, in contrast to AR-positive nonendocrine tumor cells, the growth of which is suppressed. We know of no endocrine symptoms in the aforementioned two patients. We are now beginning to investigate this phenomenon systematically. There is one report of a primary carcinoid of the prostate occurring 10 years after the diagnosis of a prostatic carcinoma which was treated by orchiectomy (WASSERSTEIN and GOLDMAN 1979).

In the most extreme case the proliferation of endocrine tumor cells in prostatic carcinomas under androgen-depriving therapy very likely leads to the formation of a true small cell carcinoma showing the immunohistochemical characteristics of endocrine cells (BIER et al. 1988). There are also reports by Schron et al. (1984) of small cell carcinomas of the prostate occurring after treatment of a common prostatic carcinoma by estrogens with or without orchiectomy.

References

Abrahmsson PA, Wadström LB, Alumets J, Falkmer S, Grimelius L (1987) Peptide hormone- and serotonin-immunoreactive tumor cells in carcinoma of the prostate. Pathol Res Pract 182:298–307

Accetta PA, Gardner Jr WA (1982) Squamous metastases from prostatic adenocarcinoma. Lab Invest 46:2A

Azzopardi JG, Evans DJ (1971) Argentaffin cells in prostatic carcinoma: differentiation from lipofuscin and melanin in prostatic epithelium. J Pathol 104:247–251

Bartsch W, Krieg M, Becker H, Mohrmann J, Voigt KD (1982) Endogenous androgen levels in epithelium and stroma of human benign prostatic hyperplasia and normal prostate. Acta Endocrinol (Copenh) 100:634–640

Bashirelahi N, Young J, Shida K, Yamanaka H, Ito Y, Harada M (1983) Androgen, estrogen, and progesterone receptors in peripheral and central zones of human prostate with adenocarcinoma. Urology 21:530–535

Bennett RS, Edgerton EO (1973) Mixed prostatic carcinoma. J Urol 110:561–563

Berns EMJJ, Mulder E, Rommerts FFG, van der Molden HJ, Blankenstein RA, Bolt-de Vries J, de Goeij TFPM (1984) Fluorescent androgen derivates do not discriminate between androgen receptor-positive and -negative human tumor cell lines. Prostate 5:425–437

Bier B, Seitz G, Wernert N, Dhom G (1988) Differenzierungswandel von Prostatacarcinomen unter Androgenentzug. Verh Dtsch Ges Pathol 72:602

Bonne C, Raynaud JB (1975) Methyltrienolone, a specific ligand for cellular androgen receptors. Steroids 26:227–232

Bowman SP, Barnes DM Blacklock NJ, Sullivan PJ (1986) Regional variation of cytosol androgen receptors throughout the diseased human prostate gland. Prostate 8:167–180

Brinkmann AO, Faber PW, Kuiper GGJM et al. (1988) In: Aumüller G, Krieg M, Senge TH (eds) Klinische und experimentelle Urologie 20. New aspects in the regulation of prostatic function. The androgen receptor: Domain structure and function. W. Zuckschwerdt Verlag München, Bern, Wien, San Francisco:15–27

Capella C, Usellini L, Buffa R, Frigerio E, Solcia E (1981) The endocrine component of prostatic carcinomas, mixed adenocarcinoma-carcinoid tumors and non-tumor prostate. Histochemical and ultrastructural identification of the endocrine cells. Histopathology 5:175–192

Carney JA, Kelalis PK (1973) Endometrial carcinoma of the prostatic utricle. Am J Clin Pathol 60:565–569

Connolly JG, Mobbs BG (1984) Clinical applications and value of receptor levels in treatment of prostate cancer. Prostate 5:477–483

Corder MP, Cicmil GA (1976) Effective treatment of metastatic squamous cell carcinoma of the prostate with adriamycin. J Urol 115:222

De Goeij TFM, Bosman FT (1986) Determination of steroid hormone-dependency of tumors utilizing tissue sections. Survey of histochemical techniques and their application in surgical pathology. J Pathol 149:163–172

Demura T, Kuzumaki N, Oda A, Asano Y, Takayama N, Koganagi T (1987) Establishment of monoclonal antibody to human androgen receptor and its clinical application for prostatic cancers. Eur J Cancer Clin Oncol 23:1237

de Voogt HJ, Rao BR (1983) Present concept of the relevance of steroid receptors for prostatic cancer. J Steroid Biochem 19:845–849

Dhom G (1980) Pathologie des Prostatacarcinoms. Verh Dtsch Ges Urol 32:9–16

Dube VE, Farrow GM, Greene LF (1973) Postatic adenocarcinoma of ductal origin. Cancer 32:402–409

Ekman P (1980) Clinical significance of steroid receptor assay in the human prostate. In: Schröder FH, de Voogt HJ (eds) Steroid receptors, metabolism and prostatic cancer. Excerpta Medica, Amsterdam, pp 208–224

Ekman P, Snochowski M, Dahlberg E, Gustafsson JA (1979a) Steroid receptors in metastatic carcinoma of the human prostate. Eur J Cancer 15:257–262

Ekman P, Snochowski M, Zetterberg A, Högberg B, Gustafsson JA (1979b) Steroid receptor content in human prostatic carcinoma and response to endocrine therapy. Cancer 44:1173–1181

Farnsworth WE (1983) Possible causative factors. In: Hinnmann F Jr (ed) Benign prostatic hypertrophy. Springer, Berlin Heidelberg New York, pp 145–151

Fetissof F, Dubois MP, Arbeille-Bassart B, Lauson Y, Boivin F, Jobard P (1983) Endocrine cells in the prostate gland, urothelium and Brenner tumors. Virchows Arch [B] 42:53–64

Feyrter F (1951) Zur Pathologie des urogenitalen Helle-Zellen-Systems. Virchows Arch 320:564–576

Franks LM (1954) Benign nodular hyperplasia of the prostate: a review. Ann R Coll Surg Engl 14:92–106

Franks LM, Riddle PN, Carbonell AW, Gey GO (1970) A comparative study of the ultrastructure and lack of growth capacity of adult human prostate epithelium mechanically separated from its stroma. J Pathol 100:113–120

Gayet G, Derré P, Samperez S, Jouan P (1985) Induction of thymidine kinase of fetal type by androgens in the rat prostate. Prostate 7:261–270

Ghanadian G, Auf G, Williams G, Davis A, Richards B (1981) Predicting the response of prostatic carcinoma to endocrine therapy. Lancet II:1418

Goebbels R, Amberger L, Wernert N, Dhom G (1985) Urothelial carcinoma of the prostate. Appl Pathol 3:242–254

Gray GF Jr, Marshall VF (1975) Carcinoma of the prostate. J Urol 113:736–738

Harper ME, Sibley PEC, Francis AB, Nicholson RI, Griffiths K (1984) Experience with oestrogen receptor immunocytochemical assay (ER-ICA) for the detection of oestrogen receptors in human prostatic tumours. Symposium on Estrogen Receptor Determination with Monoclonal Antibodies. Monte Carlo, December 14th

Harper ME, Sibley PEC, Barrie Francis A, Nicholson RI, Griffiths K (1986) Immunocytochemical assay for estrogen receptors applied to human prostatic tumors. Cancer Res (Suppl) 46:4288s–4290s

Hechter O (1984) Susceptibility of the prostate cancer cell to different physical, hormonal, and chemical agents. Present status and theoretical prospects for improved prostate cancer therapy. Prostate 5:159–180

Heller J, Sprinz O (1921) Beitrag zur vergleichenden und pathologischen Anatomie des Colliculus seminalis. Z Urol Chir 7:196–258

Heyns W, Van Damme B, De Moor P (1978) Secretion of prostatic binding protein by rat ventral prostate: influence of age and androgen. Endocrinology 103:1090–1095

Kastendieck H, Altenähr E (1974) Das Plattenepithelcarcinom der Prostata als Beispiel einer Tumormetaplasie. Z Krebsforsch 82:335–340

Kauder DH, Lange PH, Gleason DF (1977) Endometrial carcinoma of prostatic utricle. Urology 10:272–275

Kazzaz BA (1974) Argentaffin and argyrophil cells in the prostate. J Pathol 112:189–193

Kirdani R, Pontes E, Murphy G, Sandberg A (1984) Correlation of estrogen and androgen receptor status in prostatic disease measured by high pressure liquid chromatography. J Steroid Biochem 20:401–406

Kirdani R, Emrich L, Pontes E, Priore R, Murphy G (1985) A comparison of estrogen and androgen receptor levels in human prostatic tissue from patients with non-metastatic and metastatic carcinoma and benign prostatic hyperplasia. J Steroid Biochem 22:569–575

Klein PJ, Vierbuchen M, Fischer J, Schulz KD, Farrar G, Uhlenbruck G (1983a) The significance of lectin receptors for the evaluation of hormone dependence in breast cancer. J Steroid Biochem 19:839–844

Klein PJ, Vierbuchen M, Schulz KD, Farrar G, Fischer J, Uhlenbruck G, Fischer R (1983b) Histochemical tumor markers for evaluation of hormone dependence in breast cancer. Cancer Detect Prev 6:199–206

Kozak I, Bartsch W, Krieg M, Voigt KD (1982) Nuclei of stroma: site of highest estrogen concentration in human benign prostatic hyperplasia. Prostate 3:433–438

Krieg K, Grobe I, Voigt KD (1978) Human prostatic carcinoma: significant differences in its androgen binding and metabolism compared to the human beningn prostatic hypertrophy. Acta Endocrinol (Copenh) 88:397–407

Krieg M, Klötzl G, Kaufmann J, Voigt KD (1981) Stroma of human benign prostatic hyperplasia: preferential tissue for androgen metabolism and oestrogen binding. Acta Endocrinol (Copenh) 96:422–432

Kuhajda FP, Gipson T, Mendelson G (1984) Papillary adenocarcinomas of the prostate. Cancer 54:1328–1332

Kyprianou N, Davies P (1986) Association states of androgen receptors in nuclei of human benign hypertrophic prostate. Prostate 8:363–380

Lämmel A, Krieg M, Klosterhalfen H, Bressel M, Voigt KD (1986) Bestimmung von Steroidrezeptoren im Prostatakarzinom: Möglichkeiten und Grenzen. Urologe (A) 25:59–62

Lee C, Jesik C (1983) Effects of castration, estrogen, and androgen administration. In: Hinnmann F Jr (ed) Beningn prostatic hypertrophy. Springer, Berlin Heidelberg New York, pp 229–234

Manon M, Tananis CE, McLoughlin MG, Walsh PC (1977) Androgen receptors in human prostatic tissue: a review. Cancer Treat Rep 61:265–271

Martikainen P, Malmi R, Suominen J (1986) Distribution of glycoconjugates in normal rat ventral prostate and their use as markers of androgen-controlled secretory function in culture. Prostate 8:37–49

Melicow M, Pachter R (1966) Endometrial carcinoma of prostatic utricle (uterus masculinus). Cancer 20:1715–1722

Melicow MM, Tannenbaum M (1971) Endometrial carcinoma of uterus masculinus (prostatic utricle). Report of 6 cases. J Urol 106:892–902

Merchant RF, Grahm AR, Bucher Jr WC, Parker DA (1976) Endometrial carcinoma of prostatic utricle with osseous metastases. Urology VIII:169–173

Meyer R (1909) Zur Entwicklungsgeschichte und Anatomie des Utriculus prostaticus beim Menschen. Arch Mikr Anat 74:844–854

Moore RA (1943) Benign hypertrophy of the prostate. J Urol 50:680–710

Moyana TN (1987) Adenosquamous carcinoma of the prostate. Am J Surg Pathol 11:403–407

Murthy LR, Johnson MP, Rowley DR, Young CYF, Scardino PT, Tindall DJ (1986) Characterization of steroid receptors in human prostate using mibolerone. Prostate 8:241–253

Nadji M, Tabei SZ, Castro A, Ming Chu T, Morales AR (1980) Prostatic origin of tumors. An immunohistochemical study. Am J Clin Pathol 73:735–739

Naito H, Ito H, Wakisaka M, Kambegawa A, Simazaki J (1981) Histochemical observation of R 1881 binding protein in human benign prostatic hypertrophy. Invest Urol 18:337–340

Orgad U, Alroy J, Ucci A, Merk FB (1984) Histochemical studies of epithelial cell glycoconjugates in atrophic, metaplastic, hyperplastic, and neoplastic canine prostate. Lab Invest 50:294–302

Parker MG, Scrace GT, Mainwaring WIP (1978) Testosterone regulates the synthesis of major proteins in rat ventral prostate. Biochem J 170:115–121

Pertschuk LP, Zava DT, Gaetjens E, Macchia J, Wise GJ, Kim DS, Brigati DJ (1979) Histochemistry of steroid receptors in prostatic diseases. Ann Clin Lab Sci 9:225–229

Pertschuk LP, Rosenthal HE, Macchia RJ et al. (1982) Correlation of histochemical and biochemical analyses of androgen binding in prostatic cancer: relation to therapeutic response. Cancer 49:984–993

Pertschuk LP, Macchia RJ, the New York Prostate Cancer Binding Site Group (1984) Histochemical androgen binding assay in prostatic cancer. J Urol 131:1096–1098

Pertschuk LP, Eisenberg KB, Macchia RJ, Feldman JG (1985) Heterogeneity of steroid binding sites in prostatic carcinoma: morphological demonstration and clinical implications. Prostate 6:35–47

Pillarisetti SG, Espinoza CG, Richman AV (1983) Prostatic adenocarcinoma with focal "endometrioid" features: histopathologic and immunochemical findings. Lab Invest 48:68A

Reischauer F (1925) Die Entstehung der sogenannten Prostatahypertrophie. Virchows Arch [A] 256:357–389

Rhamy RK, Buchanan RD, Spalding MJ (1973) Intraductal carcinoma of the prostate gland. J Urol 109:457–460

Robel P, Eychenne B, Blondeau JP, Baulieu EE, Hechter O (1985) Sex steroid receptors in normal and hyperplastic human prostate. Prostate 6:255–267

Rotterdam HZ, Melocow MM (1975) Double primary prostatic adenocarcinoma. Urology 6:245–248

Saito R, Davis BK, Ollapally EP (1984) Adenosquamous carcinoma of the prostate. Hum Pathol 15:87–89

Sandberg AA (1980) Endocrine control and physiology of the prostate. Prostate 1:169–184

Schron DS, Gipson T, Mendelsohn G (1984) The histogenesis of small cell carcinoma of the prostate. An immunohistochemical study. Cancer 53:2478–2480

Schulze H, Barrack ER (1987a) Immunocytochemical localization of estrogen receptors in spontaneous and experimentally induced canine benign prostatic hyperplasia. Prostate 11:145–162

Schulze H, Barrack ER (1987b) Immunocytochemical localization of estrogen receptors in the normal male and female canine urinary tract and prostate. Endocrinology 121:1773–1783

Seemayer TA, Knaack J, Thelmo WL, Wang NS, Nisar Ahmed M (1975) Further observations on carcinoma in situ of the urinary bladder: silent but extensive intraprostatic involvement. Cancer 36:514–520

Seitz G, Wernert N (1987) Immunohistochemical estrogen receptor demonstration in the prostate and prostate cancer. Pathol Res Pract 182:792–796

Senge T (1983) Hormonstoffwechsel und Rezeptoren in der Prostata. In: Helpap B, Senge T, Vahlensieck W (eds) Die Prostata, vol 1. Prostatahyperplasie. PMI, Frankfurt/Main, pp 87–95

Sesterhenn I, Mostofi FK, Davis CJ (1987) Basal cell hyperplasia and basal cell carcinoma. Lab Invest 56:71A

Sharma SK, Malik AK, Bapna BC (1980) Squamous cell carcinoma of prostate. Indian J Cancer 17:134–135

Sieracki JC (1955) Epidermoid carcinoma of the human prostate. Lab Invest 4:232–240

Srivastava A, Chowdhury AR, Setty BS (1984) Zinc content in the epididymis, vas deferens, prostate and seminal vesicles of juvenile rhesus monkeys (Macaca mulatta): effect of androgen and estrogen. Prostate 5:153–158

Steffens J, Leistenschneider W (1983) Utriculuscarcinom der Prostata. Aktuel Urol 14:183–186

Stratton M, Evans DJ, Lampert IA (1986) Effects of oestrogens on endocrine cells in prostatic carcinoma. J Pathol 148:83A

Sufrin G, Ajrawat H, Gaeta J (1983) Endometrial carcinoma of the prostate utricle (Meeting abstract). 78th Annual Meeting, Am Urol Assoc, Las Vegas, Nevada

Sugimura Y, Norman JT, Cunha G, Shannon JM (1985) Regional differences in the inductive activity of the mesenchyme of the embryonic mouse urogenital sinus. Prostate 7:253–260

Tannenbaum M (1975) Endometrial tumors and/or associated carcinomas of prostate. Urology 6:372–375

Thelmo WJ, Seemayer THA, Madarnas P, Mount BMM, Mackinnon KJ (1974) Carcinoma in situ of the bladder with associated prostatic involvement. J Urol 11:491–494

Thompson GJ (1942) Transurethral resection of malignant lesions of the prostate gland. JAMA 120:1105–1109

Thompson GJ, Albers DD, Broders AC (1953) Unusual carcinomas involving the prostate gland. J Urol 69:416–425

Trachtenberg J, Walsh PC (1982) Correlation of prostatic nuclear androgen receptor content with duration of response and survival following hormonal therapy in advanced prostatic cancer. J Urol 127:466–471

Tunn U, Senge T, Schenck B, Neumann F (1979) Biochemical and histological studies on prostates in castrated dogs after treatment with androstanediol, oestradiol and cyproterone acetate. Acta Endocrinol (Copenh) 91:373–384

Tunn UW, Schweikert HU (1983) Endokrinologische Aspekte der Pathogenese der benignen Prostatahyperplasie. In: Helpap B, Senge T, Vahlensieck W (eds) Die Prostata, vol 1. Prostatahyperplasie. PMI, Frankfurt/Main, pp 67–86

Tveter KJ (1974) Some aspects of the pathogenesis of prostatic hyperplasia. Acta Pathol Microbiol Immunol Scand [A] [Suppl] 248:167–174

Tveter KJ, Unjhem O, Attramadal A, Aakvaag A, Hansson A (1971) Androgenic receptors in rat and human prostate. Adv Biosci 7:193–207

Van Aubel OGJM, Bolt-de Vries J, Blankenstein MA, Ten Kate FJW, Schröder FH (1985) Nuclear androgen receptor content in biopsy specimens from histologically normal, hyperplastic, and cancerous human prostatic tissue. Prostate 6:185–194

Van Buskirk KE, Kimbrough JC (1954) Carcinoma of the prostate. J Urol 71:742–747

Vierbuchen MJ, Klein PJ, Uhlenbruck G, Schaefer HE, Fischer R (1979) Hormonabhängige Lectin-Rezeptoren im Brustdrüsengewebe. I. Histochemischer Nachweis von Hormon-induzierbaren Lectin-Rezeptoren in der Brustdrüse der Ratte. Verh Dtsch Ges Pathol 63:682

Wagner RK, Schulze KH, Jungblut PW (1975) Estrogen and androgen receptor in human prostate and prostatic tumor tissue. Acta Endocrinol (Copenh) [Suppl] 193:52

Walker AN, Mills SE, Fechner RE, Perry JM (1982) "Endometrial" adenocarcinoma of the prostatic urethra arising in a villous polyp. Arch Pathol Lab Med 106:624–627

Wasserstein PW, Goldman RL (1979) Primary carcinoid of prostate. Urology 13:318–323

Wernert N (1987) Morphologische und immunhistochemische Untersuchungen zur Orthologie und Pathologie der menschlichen Prostata. Habilitationsschrift Homburg/Saar (FRG)

Wernert N (1988) Bedeutung von Hormonrezeptoren in der Prostata und verschiedenen Prostatacarcinomen. Vortrag für den Sonderforschungsbereich 232 ("Zellrezeptoren") der Universitäten Hamburg und Lübeck, 31. 10. 1988

Wernert N, Kern L, Seitz G, Goebbels R, Dhom G (1986) Morphology and immunohistochemistry of the utriculus prostaticus. Abstracts: Symposium cum Participatione Internationali Progressus in Histochemia Generali, Applicata Atque Diagnostica. Histochemia Myocardii Smolenice (CSSR), 56–57

Wernert N, Lüchtrath H, Seeliger H, Schäfer M, Goebbels R, Dhom G (1987a) Papillary carcinoma of the prostate, location, morphology, and immunohistochemistry: the histogenesis and entity of so-called endometrioid carcinoma. Prostate 10:123–131

Wernert N, Seitz G, Dhom G (1987b) Different markers in conservatively treated prostatic carcinoma and the estrogen receptor in the normal prostate. J Endocrinol Invest 10 (Suppl 2):43

Wernert N, Gerdes J, Loy V, Seitz G, Scherr O, Dhom G (1988a) Investigations of the estrogen (ER-ICA test) and the progesterone receptor in the prostate and prostatic carcinoma on an immunohistochemical basis. Virchows Arch [A] 412:387–391

Wernert N, Goebbels R, Seitz G, Hinkeldey K, Dhom G (1988b) Histogenese von Plattenepithelcarcinomen in der Prostata. Verh Dtsch Ges Pathol 72:451

Wilkin RP, Bruchovsky N, Shnitka TK, Rennie PS, Comeau TL (1980) Stromal 5alpha-reductase activity is elevated in benign prostatic hyperplasia. Acta Endocrinol (Copenh) 94:284–288

Wolf RM, Schneider L, Pontes JEE, Englander L, Karr JP, Murphy GP, Sandberg AA (1985) Estrogen and progestin receptors in human prostatic carcinoma. Cancer 55:2477–2481

Young BW, Lagios MD (1973) Endometrial (papillary) carcinoma of the prostatic utricle – response to orchiectomy. Cancer 32:1293–1300

Young RH, Frierson HF Jr, Mills SE, Kaiser JS, Talbot WH, Bhan AK (1988) Adenoid cystic-like tumor of the prostate gland. A report of two cases and review of the literature on "adenoid-cystic carcinoma" of the prostate. Am J Clin Pathol 89:49–56

Zaloudek C, Williams JW, Kempson RL (1976) "Endometrial" adenocarcinoma of the prostate. Cancer 37:2255–2262

Zondek LH, Zondek T (1974) The prostatic utricle in the fetus and infant. Urol Int 29:458–465

Lineage-Specific Receptors in the Diagnosis of Malignant Lymphomas and Myelomonocytic Neoplasms

M. R. Parwaresch, H. Kreipe, H. J. Radzun, and H. Griesser

1 Introduction

Surface receptors usually represent transmembrane glycoproteins serving the ability of cells to react to environmental conditions. The reaction of ligands with the corresponding receptor is the decisive event promoting transmembrane signals and triggering a complicated chain of intracellular reactions. The overwhelming majority of surface receptors in hematopoietic cells regulate growth, differentiation, defense, adaptation, and other physiological cell functions. Their involvement in neoplastic processes is mostly secondary in nature, although they may be prone to allow external influences upon the dysregulated growth mechanisms.

The ligands involved are mostly derived from sources other than the target cell itself. Typical examples are antigens, immunogenic sequences, and other immunologically relevant ligands such as complement factors (Frank 1987), Fc domains of various immunoglobulins (Igs) (Hogg 1988), IgG multimer, IgA, histocompatibility glycoproteins (Shreffler 1988), and growth factors and hormones, including insulin and corticosteroids. Also of general importance are receptors involved in metabolic processes, such as low-density lipoprotein receptor (Goldstein et al. 1979), which regulates cellular cholesterol synthesis by modulation of the activity rate of 3-hydroxy-3-methylglutaryl-coenzyme A reductase, the main mechanism in the regulation of the blood cholesterol level.

Another biologically highly interesting mode of action is the mechanism designated as autocrine regulation, where the regulatory ligand is produced or at least secreted by the target cell itself. Interleukin-2 (IL-2) is one of the most impressive representatives of growth factors subject to autocrine regulation (Jenkinson et al. 1987; Shimonkovitz et al. 1987). Ample evidence supports the now generally accepted concept that stimulation of antigen-pulsed specifically rearranged T cells by primary signals such as interleukin-1 and -6 triggers the typical cell cycle progression signals, among which IL-2 plays a central role. Such stimulated T cells capable of IL-2 secretion express their own IL-2 receptor.

Table 1. Ligands/receptors on hematopoietic cells arranged according to their biological function

1. *Growth and differentiation*

   T-cell growth factor (IL-2)
   B-cell stimulatory factor I (IL-4)
   B-cell growth factor II (IL-5)
   B-cell differentiation factor (IL-6)
   Lymphopoietin-1 (IL-7)
   Multipotential colony-stimulating factor (IL-3)
   Granulocyte/macrophage colony-stimulating factor (GM-CSF)
   Granulocyte colony-stimulating factor (G-CSF)
   Macrophage colony-stimulating factor (M-CSF)

2. *Specific antigen recognition*

   T-cell antigen receptor
   B-cell surface immunoglobulins

3. *Associated with immune response*

   Complement receptors (C3b, C3d)
   Fc fragment receptors (IgG, IgM, IgA, IgE)
   MHC class I molecules
   MHC class II molecules

4. *Intercellular signal transmitters*

   Cytokines
   Leukocyte function associated antigen (LFA)
   Cellular adhesion molecules

Cautious estimates of the number of surface receptors on various cell types, including those of hematopoetic lineage, have predicted an order of several thousand. The vast majority are hypothetical. However, a small number of these predicted receptors have now been well characterized. In the context of this chapter we can refer to only a small group of these receptors of well-established significance in hematopathological fields.

The diverse ligand–receptor systems have been arbitrarily classified according to their main functional aspects (Table 1). The amount of literature dealing with ligand–receptor systems involved in cell growth and differentiation is steadily growing, partly due to the increasing interest in tumor biology. In the 1980s we have witnessed progress in the recognition of specific antigen receptors of lymphocytes, confirming the prevalence of inborn recognition mechanisms in host defense. Receptor–ligand systems involved in the initiation, acceleration, and enhancement of immunological recognition or in the sequence of events following the antigen–receptor interaction constitute a prominent research field which comprises areas dealing with complement factors, Fc domains of various Igs, and the histocompatibility complex. Furthermore, the group of ligand–receptor systems which serve various exogenous signal transmission pairs, including intercellular communications, is growing. This group comprises cytokines, leukocyte function-associated antigens and/ or cellular adhesion molecules (MAKGOBA et al. 1988; SIMMONS et al. 1988), and the complicated group of receptors preliminarily defined by the various recently introduced monoclonal antibodies, which have been partially classified in clusters of differentiation (CD) (McMICHAEL et al. 1987).

In this article we focus especially on the ligand–receptor systems which have attracted interest in the diagnosis and classification of some of the hematopoietic neoplasms, with special reference to those of lymphoid lineage and myelomonocytic cells.

2 Immunoglobulin and T-Cell Antigen Receptors
– Phenotype and Gene Rearrangement

2.1 Introduction

Over the past few years considerable advances have been made in the diagnosis of lymphoid leukemias and malignant lymphomas. The application of specific monoclonal antibodies helps to define more clearly the lineage and stage of differentiation of malignant lymphoproliferative disorders, while the use of antibodies against immunoglobulin light chains (IgL) $\varkappa$ and λ also helps to define clonality in mature B-cell non-Hodgkin's lymphomas (FOON and TODD 1986). However, if the malignant B-cell population constitutes only a small minority of cells in a lymphoid tumor containing abundant reactive lymphocytes, restriction of IgL surface expression may not be detectable. Clonal surface markers for T-cell disorders still have not been developed.

This makes it difficult to study T-cell-rich lymphoproliferative diseases such as lymphoepiothelioid cell (Lennert's) lymphoma, angioimmunoblastic lymphadenopathy (AILD), and cutaneous T-cell lymphomas. In certain malignant disorders such as non-T-cell acute lymphoblastic leukemia (ALL) and large cell anaplastic CD30 + lymphoma (LCAL), neither clonal- nor, in some instances, lineage-specific cell markers are expressed. Especially in cases of LCAL, aberrant expression of CD2 and/or CD3 has resulted in their classification as histiocytic neoplasms (SUCHI et al. 1987). Immunophenotyping and conventional methods have failed to characterize a clonal subpopulation in Hodgkin's disease (HD) or to define the lineage of the malignant cells.

Molecular probes for immunoglobulin (LEDER 1982) and T-cell receptor (TcR) genes (YANAGI et al. 1984; LEFRANC et al. 1986; TOYONAGA and MAK 1987; TAKIHARA et al. 1988) now provide a necessary tool for overcoming many of these limitations. TcR gene probes in particular represent a diagnostic tool that in combination with morphological and immunocytochemical analyses can aid in the classification of monoclonal T-cell proliferations according to developmental stage. Rearrangement analysis helps in primary diagnosis, defining clonality and lineage, and is useful in monitoring therapy. Early relapse or minimal residueal disease can be detected with high sensitivity since it is possible to detect circulating monoclonal lymphoid populations, which constitute only 1% of the mononuclear blood cells (WALDMANN 1987).

2.2 B- and T-Cell Antigen Receptor Structures and Their Genes

The major role of lymphocytes is the recognition of foreign antigens and distinguishing them from self. The molecules that mediate this distinction are the immunological antigen receptors, which display extensive structural diversity generated by similar genetic mechanisms (TONEGAWA 1985).

The B-lymphocyte receptor is the Ig molecule, which is made up of four polypeptide chains: two heavy chains (μ, δ, γ, ϵ, or α) and two light chains ($\varkappa$ or λ) confined to the arms of the Y-shaped protein. Each polypeptide has both constant (C) and variable (V) regions. The V regions vary from one B-cell clone to another and are folded to create unique antigen-binding sites. When antigen binds to an Ig molecule the carrier B cell may be triggered to undergo cell division, resulting in clonal expansion.

The T-cell receptor can take either of two forms, an α/β or γ/δ heterodimer. Both these forms are associated with the CD3 complex, which participates in signal transduction across the cell membrane. TcR α/β is found on the majority of peripheral T cells and has been shown to be responsible for the recognition of antigen associated with class I and class II MHC surface proteins. TcRγ/δ is found on only 2%–7% of peripheral T cells, which mostly display MHC-nonrestricted cytotoxicity. These T cells are either double-negative (CD4$^-$ CD8$^-$) or sometimes CD8$^+$, and are negative for or express low levels of CD5. TcRγ/δ^+ subsets are more frequently found among intraepidermal and intestinal intraepithelial CD3$^+$ cells (BRENNER et al. 1988).

T lymphocytes

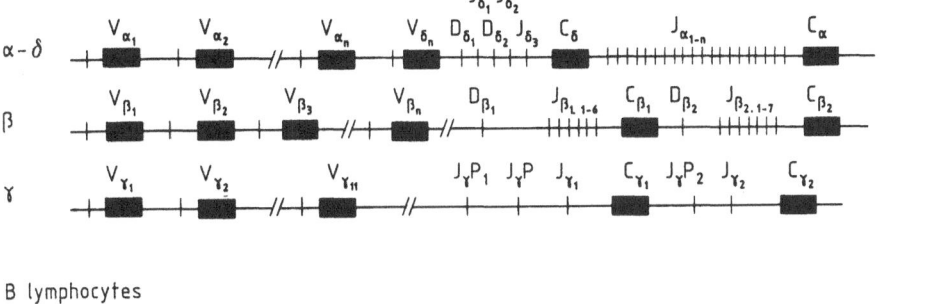

B lymphocytes

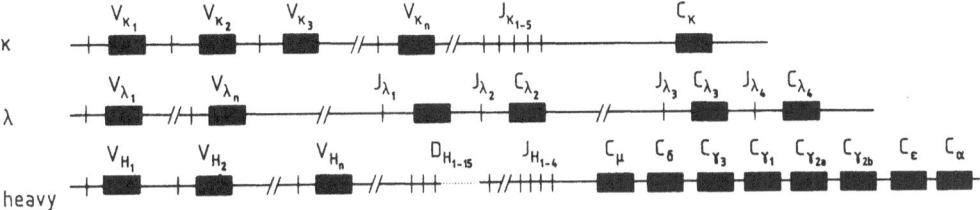

Fig. 1. Germline genomic organization of the human immunoglobulin genes ϰ and λ, heavy chain, and T-cell receptors α, β, γ, and δ

Molecular studies have revealed that TcR genes have Ig-like structural features, V and C regions (Fig. 1). In addition, like Ig genes, noncontiguous V, diversity (D), and joining (J) germline TcR genes are capable of undergoing somatic rearrangements leading to the creation of unique V regions for each maturing T cell. An orderly sequence of recombinational events may ultimately lead to functional rearrangement with transcription of a complete or functional Ig or TcR messenger RNA. Initially, a D segment (in IgH and TcRβ) combines with any one of the J segments (partial rearrangement). In the next step, this DJ segment combines with a V gene sequence, resulting in a variable region gene assembly, which, together with a C region gene, constitutes a completely rearranged Ig or TcR gene (ALT et al. 1987; TOYONAGA and MAK 1987). Ig gene rearrangement and the production of antibody is generally restricted to cells of the B-cell lineage and does not normally occur during differentiation of non-B cells. By analogy, rearrangement of the T-cell receptor genes is largely restricted to cells of T-cell lineage.

2.2.1 Sequential Rearrangement of Ig and TcR Genes During Lymphocyte Ontogeny

During normal B-lymphocyte differentiation and in precursor B-cell lines, the IgH locus is the first to undergo rearrangement (Fig. 2). This process is associated with the surface expression of CD19 and generates the pre-pre-B cells.

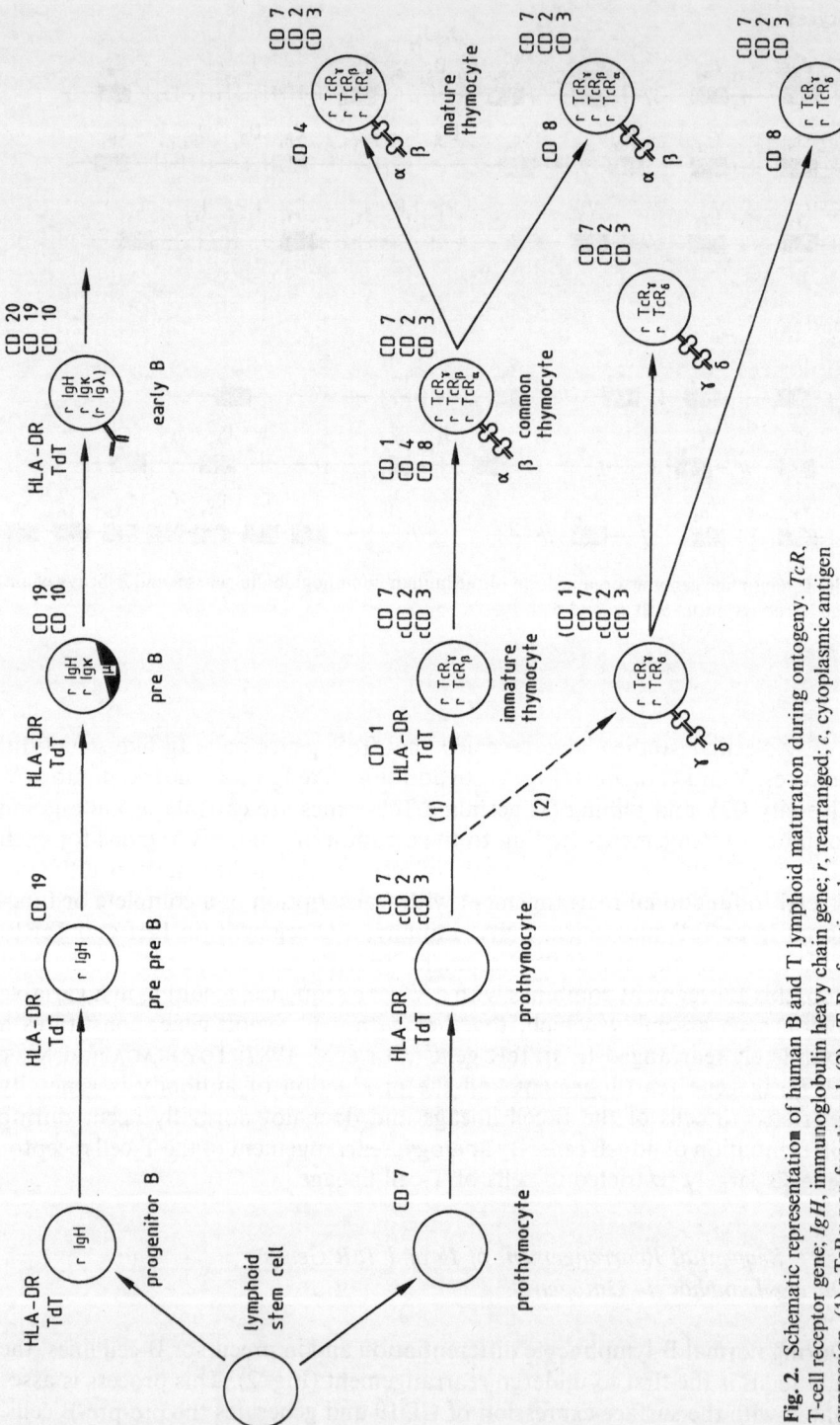

Fig. 2. Schematic representation of human B and T lymphoid maturation during ontogeny. *TcR,* T-cell receptor gene; *IgH,* immunoglobulin heavy chain gene; *r,* rearranged; *c,* cytoplasmic antigen expression. (1) r TcRγ non-functional; (2) r TcRγ functional

If productive, the complete IgH rearrangement results in cytoplasmic Igμ synthesis. Next, an attempt is made to rearrange the Igϰ genes that, if productive, results in the production of a μ-ϰ Ig. In the case of nonproductive ϰ gene rearrangement, Igλ genes are rearranged which may lead to the production of a μ-λ polypeptide. If inefficient or aberrant (nonfunctional open reading frame), no synthesis of light chains occurs and the cell remains at the pre-B stage (DAVEY et al. 1986).

An increasing body of evidence indicates that TcR genes are initially rearranged and expressed in the thymus after the appearance of surface (s) CD7 and cytoplasmic (c) CD2 followed by cCD3 molecule expression (Haynes et al. 1988; Campana et al. 1989). The exact hierarchy of TcR gene activation is not entirely known. Tcβ gene rearrangement precedes sCD3 expression and may occur prior to or together with sCD2 antigen expression (Fig. 2). TcRγ chain gene rearrangement precedes that of TcRβ genes, and the TcRα gene is the last to rearrange. Expression of sCD3 and TcRα/β is observed on large populations of double-positive (CD4$^+$ CD8$^+$) thymocytes. TcRδ genes probably are rearranged shortly after TcRγ genes early in fetal thymic ontogeny, and TcRγ/δ molecules are expressed mainly in double-negative T cells. It is possible that thymocytes may attempt TcRγ rearrangement, and, failing functional rearrangement, may then move on to attempt TcRα rearrangement, thereby deleting the Cδ locus. Cells with a productive TcRγ gene rearrangement would go on to rearrange their δ-chain gene and give rise to the TcRγ/δ+ T cells (WINOTO and BALTIMORE 1989).

2.2.2 Rearrangement Analysis of Malignant Lymphomas and Lymphoid Leukemias

DNA from affected cells in suspension is extracted and then digested with restriction enzymes which cut double-stranded DNA molecules within specific recognition sequences throughout the genome. In germline DNA, cleavage with these enzymes invariably produces fragments of characteristic size. As a result of rearrangement events which mostly are accomplished by a deletion process, the original restriction site is lost and a new restriction site is brought in from upstream sequences. Hence, the germline fragment is replaced by a fragment of different size. Restriction fragments are size-fractionated by gel electrophoresis and then transferred to nitrocellulose or nylon membranes according to the Southern procedure (SOUTHERN 1975). Hybridization to radioactive labeled DNA probes for Ig and TcR genes is done, followed by autoradiography. In polyclonal lymphocyte populations thousands of individually sized fragments different from the germline configuration are produced by the rearrangement processes and show not a single new band but a vertical smudge upon hybridization with an Ig or TcR gene probe (Fig. 3). Only if more than 5% of a given cell population contains the same (clonal) rearrangement can a distinct new band be detected on the autoradiograph.

The fact that TcR and Ig gene rearrangements occur in a random manner during lymphocyte ontogeny, generating a very large number of unique gene

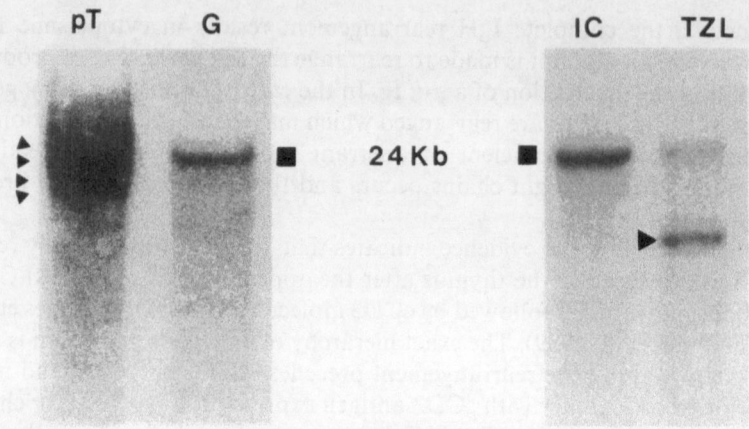

Fig. 3. Southern gel analysis of DNAs from peripheral T lymphocytes (*pT*), granulocytes (*G*), immunocytoma (*IC*), and T-zone lymphoma (*TZL*). The DNAs were restricted with BamHI and hybridized for their T-cell receptor β-chain genes. A smudge containing multiple rearranged bands is seen in pT, while a single (clonally) rearranged band is detected in TZL. ■, germline band; ▶, rearranged band; kb, size in kilobases

combinations, makes these receptor gene probes extremely helpful in the diagnosis of lymphoproliferative lesions. By this means, molecular genetic analyses have confirmed clonality and lineage in lesions such as AILD, lymphoepithelioid cell lymphoma, and LCAL (GRIESSER et al. 1986a). The detection of clonal rearrangements in most cases helps to distinguish nonneoplastic from neoplastic or preneoplastic lymphoproliferations, but clonality does not prove malignancy. Clonal lymphocyte populations have been demonstrated by molecular genetic analysis in benign lymphoproliferative diseases with increased susceptibility to the development of non-Hodgkin's lymphoma. These analyses of lymphomatous lesions in immunodeficient or immunosuppressed patients revealed either oligoclonality in the same tumor or different clonal rearrangement patterns in tissues taken from different sites or at various times. TcRβ gene rearrangements have been detected in skin lesions such as lymphomatoid papulosis, pityriasis lichenoides et varioliformis acuta (WEISS et al. 1987) and other cutaneous T cell-rich pseudolymphomas (GRIESSER et al. 1990) whereas Ig gene rearrangements have been demonstrated in benign lymphoepithelial lesions associated with Sjögren's disease and in systemic Castleman's disease (HANSON et al. 1988). Thus, establishment of a monoclonal population may be one of the many steps leading to neoplastic growth and malignancy. Another issue is the fact that rearrangement of Ig and TcR genes may cross lineage lines. This is especially true for IgH and TcRδ/β genes. In some cases of cALLa-positive pre-B-cell ALL, IgH as well as TcRγ, TcRα, and TcRδ chain rearrangements occur with or without TcRβ chain gene rearrangement, even in the presence of cytoplasmic Igμ chains (HARA et al. 1987, 1988). Furthermore, in cases of acute myeloid leukemia (AML) unexpected IgH and/or TcRβ chain gene rearrangements rarely occur, especially in TdT + leukemic cells (SEREMETIS et al. 1987). So far, functional IgL rearrangement and complete rearrangement of the TcRβ and α loci have been reliable markers for B-cell and T-cell lineage, respectively. However, detection of α-chain gene rearrangements has been limited due to the great distance over which the J segments are dispersed. Since multiple Jα gene probes recently have been isolated, rearrangements in this locus can now be studied (CHAMPAGNE et al. 1988).

The TcRγ gene differs from the β gene in its very limited number of rearranging Vγ genes and in giving rise to a few nongermline bands in polyclonal T cells in DNA hybridization studies. Thus, either an actual T-cell clone with TcRγ rearrangement may be masked by contaminating polyclonal T cells, or different intensities of nongermline bands produced by polyclonal T cells may be mistaken as evidence for a clonal cell population with γ gene rearrangement. To overcome these problems, TcRγ rearrangement results have to be interpreted in comparison with the β gene configuration and in correlation with the T-cell content of the tumor.

In the following, rearrangement studies of lymphoproliferative disorders performed by various groups, including our own, will be discussed (for reviews see: SKLAR et al. 1987; WALDMANN 1987; COSSMAN et al. 1988; GRIESSER et al. 1989).

2.3 Clonal Rearrangements of TcR and Ig Genes in Lymphoproliferative Disease

2.3.1 T-Cell Lymphoma/Leukemia

2.3.1.1 T-Cell Acute Lymphoblastic Leukemias

With the exception of very few cases, one or both alleles of the β-chain genes are rearranged and generally a γ-chain gene rearrangement is detected in the same sample. Exclusive γ-chain gene rearrangement rarely occurs in cases that are CD7$^+$ and CD2$^-$ and CD3$^-$. In a few CD3$^+$ samples, concurrent rearrangement of the α-chain allele and the δ-chain on the other allele has been reported. Rare cases with γ/δ rearrangement but without β-chain gene rearrangement have been recognized. Rearrangement of IgH but not of IgL gene loci occurs in about 10% of cases.

2.3.1.2 Mature T-Cell Malignancies

Virtually all cases of T-cell chronic lymphocytic leukemia (T-CLL), T-prolymphocytic leukemia (T-PLL), adult T-cell leukemia (ATL), mycosis fungoides (MF), and Sézary syndrome have their β-chain genes rearranged together with their γ-genes. Also, a proportion of lymph nodes in patients with MF which histologically show only benign reactive changes (dermatopathic lymphadenopathy) genotypically may show the same rearrangement pattern as the underlying MF (WEISS et al. 1985). It is not yet known whether these findings are relevant for prognosis. Even cases of lymphomatoid papulosis (LP), a chronic disease with a relatively benign course, mostly exhibit clonal β- and γ-chain gene rearrangements. Furthermore, it has been shown that different rearrangement patterns occur in tissues from different sites in the same patient. These observations may favor the notion that LP is a pre-lymphomatous lesion. Likewise, TcRβ genes are rearranged in most non-Hodgkin's T-cell lymphomas, which in nearly all instances coincides with γ-rearrangements. This also holds true for LCAL with a non-B-cell phenotype. Only a few cases have been studied for the δ-chain gene, revealing that about half of them have detectable rearrangements or deletions of this gene locus (TKACHUK et al. 1988). Especially in the AILD type of T-cell lymphomas, IgH gene rearrangement not infrequently occurs in combination with TcR gene rearrangement, which probably correlates with a less favorable clinical course. Recently, a subset of mostly paranasal extranodal, CD3⁻ T-cell lymphomas was described in which TcRγ and β gene rearrangement was undetectable (WEISS et al. 1988).

2.3.1.3 T8/Tγ Lymphocytosis

CD8 lymphocytosis or cytotoxic T-lymphocytosis with neutropenia is a rare disorder with a benign clinical course. The proliferating cells are CD8⁺, mostly CD3⁺, and carry surface receptors for the Fc fragment of IgG. In the vast majority of cases TcRβ-genes have been found to be rearranged. This T-cell disorder, like benign monoclonal gammopathy, represents an example of a clonal disease that clinically may not have a malignant course. A CD4⁺ counterpart of the γ-lymphocytosis with similar characteristics seems to exist (REIS et al. 1988).

2.3.2 B-Cell Lymphoma/Leukemia

2.3.2.1 Non-B, Non-T-Cell Acute Lymphoblastic Leukemia

Non-B, non-T-cell acute lymphoblastic leukemias are perhaps the most heterogeneous group of lymphoproliferative diseases with regard to immunogenotype. Cases which only express HLA-DR and CD19, not CD10 and/or CD20, and which mostly have an incomplete IgH but no or nonfunctional IgL rearrangements, have their TcR genes in a germline configuration. As soon as cALLa expression associated with functional rearrangement of IgH occurs, cross-lineage rearrangements are frequently detected. TcRβ chain gene rearrangements seem to provide higher lineage fidelity than the other TcR genes since β gene rearrangements are detected considerably less often than δ, α, and γ rearrangements in this entity. In contrast to the findings in T-cell neoplasms, the hierarchical order of TcR gene rearrangements is not preserved and simultaneous rearrangement of γ and β-genes with or without α-chain gene rearrangement is rare in B-precursor ALL. Furthermore, unlike T-ALL, where both TcRγ gene loci are rearranged, one allele remains in the germline configuration in cases of non-B and non-T-ALL. For correct lineage information these leukemias especially require a complete genotype using all probes available for Ig and TcR genes.

2.3.2.2 Mature B-Cell Malignancies

Nearly all cases of B-CLL and B-PLL show rearrangement of their IgH and Igϰ, mostly without Igλ chain genes. The most convincing evidence for the B-cell origin of hairy cell leukemia is the detection of IgH and IgL (in 75% Igλ) gene rearrangement in all cases studied so far. TcRβ gene rearrangement occurs in less than 10% of B-CLL cases, but γ-chain gene rearrangement has not been reported.

2.3.2.3 Non-Hodgkin's Lymphoma of B-Cell Type

Among the low-grade B-cell lymphomas, immunocytomas are immunogeno-typically similar to B-CLL cases. Cross-lineage β-chain (but not γ-chain) gene rearrangements are detectable in a considerable proportion of centrocytic lymphomas but are exceedingly rare in high-grade B-cell lymphomas of immunoblastic and centroblastic subtype. On the other hand, lymphoblastic lymphomas have their TcRβ chain genes rearranged in about 40% of the cases in addition to their Ig genes (GRIESSER, unpublished). LCAL cases with a B-cell phenotype very frequently have rearranged their TcRβ or TcRγ genes in addition to IgH and/or IgL genes. Again, as in B-precursor ALL, simultaneous cross-lineage rearrangements of β and γ genes are very rare.

2.3.3 Hodgkin's Disease

Results of molecular genetic studies indicate that in some instances tissues involved by HD show evidence of rearrangements of either Ig or TcR genes. The high number of γ-chain gene rearrangements in early reports most probably is due to the high percentage of polyclonal T cells within these tissues, producing a picture of "pseudoclonality." IgH gene rearrangement has been attributed to the Sternberg-Reed (SR) cells in nodular sclerosing subtypes that are rich in atypical blasts. In a cell separation approach, Ig gene rearrangement also was associated with SR cell-enriched fractions. However, there is no close relationship between the immunophenotype and the genotype of these atypical cells, which makes it difficult to assign T- or B-cell lineage. Between 10% and 20% of HD cases with a low content of SR cells exhibit clonal TcRβ chain gene rearrangements, which is most probably due to the T-cell infiltrate in this tumor. Even though the rearranged bands detected in these cases are mostly less intense than the new bands observed in autoradiographs of T-cell lymphomas, this difference should not be regarded as a differential diagnostic criterion. These findings may lead to the conclusion that minor clonal T-cell populations of unknown significance occasionally predominate among poly-clonal lymphocytes in HD. Thus, the question of whether a clonal population detected in HD represents the neoplastic cells or a clonal response to some antigen is still unsolved (GRIESSER and MAK 1988). Since single IgH or TcRβ gene rearrangements occasionally occur in AML, the detection of rearrange-ments in some SR cell-rich HD cases is not sufficient evidence for a lymphocyt-ic derivation of the tumor cells in every case.

2.4 Concluding Remarks

Rearrangement studies with Ig and TcR gene probes are useful for the detection of clonal cell populations in lymphoproliferative diseases and, in association with the clinical course, help to distinguish reactive conditions from malignant clonal proliferations. Together with immunocytochemical studies, the use of these probes may allow the assignment of nearly all clonal lymphoid disorders to the T- or B-cell lineage and may provide additional information about the possible stage of differentiation from which the tumor cell clones emerge. However, cross-lineage rearrangements and the lack of TcR rearrangements in T-cell neoplasms may complicate genotypic analysis of malignant lymphomas/leukemias.

In most instances, cross-lineage rearrangements differ in their patterns from the appropriate rearrangements. Usually, only one IgH allele (as opposed to two alleles in B-cell neoplasms) rearranges in T-cell neoplasms, with the IgL genes remaining in a germline configuration. In B-cell disorders, chiefly either TcRβ or, more rarely, TcRγ genes are rearranged. Complete (VDJ) rearrangement of TcRβ genes in B-cell neoplasms or of IgH genes in T-cell tumors has not been reported, and no full-length RNA transcripts corresponding to the aberrant gene rearrangements have been detected. Neither TcRγ (GRIESSER et al. 1986b) (and probably TcRδ) nor incomplete DJ joining of TcRβ or IgH gene segments is indicative of lineage commitment. In the experience of our group and others (JAFFE and WARNKE, personal communication), the presence of both B- and T-cell clonal expansions in the same tumor has not yet been observed. Complex Ig and TcR gene rearrangement patterns, which rarely occur in lymphoma cases, may reflect a clonal population originating from an undifferentiated hematopoietic cell that underwent rearrangement of these genes prior to lineage commitment. Another possibility is that the dual genotype may reflect the simultaneous activation in T and B cells of common molecular mechanisms regulating the rearrangement of the Ig and TcR gene loci, which in neoplastic lymphoid cells does not follow the hierarchical order during ontogeny. This may account for the observation that aberrant rearrangements on B-cell neoplasms frequently involve only TcRβ without TcR γ genes. Another event which possibly results in aberrant rearrangement is a chromosomal translocation involving TcR in B-cell and Ig genes in T-cell malignancies (for review, see GRIESSER et al. 1989).

Besides the findings of clonal lymphocyte populations in benign lymphoproliferative disorders, rare cases of malignant T-cell lymphomas have been reported that do not exhibit detectable TcRβ or γ gene rearrangements. This phenomenon has been observed especially in extranodal sites (such as so-called midline lethal granulomas), where the proliferating T cells immunophenotypically are CD2$^+$ but CD3$^-$, and sometimes CD5$^-$ and CD7$^-$ (WEISS et al. 1988). Furthermore, the tumor cells in these cases do not bind βF1, a monoclonal antibody against the cytoplasmic determinant on the TcRβ chain peptide. Thus, the absence of TcR gene rearrangement must be interpreted with appropriate caution as negative evidence for clonality or malignancy in T-cell lesions. In rare instances, correct genotypic analysis of a lymphoproliferative disease requires the application of the whole panel of TcR and Ig gene probes for gene rearrangement as well as transcription analysis. Genotypic data must be interpreted with care and in conjunction with immunophenotypic observations and morphological findings.

3 c-*fms* and M-CSF Expression in Monocytes/Macrophages

3.1 Introduction

As described in the foregoing, B and T lymphocytes can be clearly differentiated by immunophenotyping due to their specific cell surface receptors such as immunoglobulins and T-cell antigen receptors, respectively. Furthermore, because the variability of these receptors is regulated by gene rearrangement and since molecular probes for these genes are available, distinction of clonal (neoplastic) from polyclonal (reactive) proliferations within these lymphocyte subsets is now possible by genotypic analysis. In addition, these techniques have proved helpful in the characterization of anaplastic lymphomas, where a conclusion on T- or B-cell origin cannot be achieved on the basis of conventional methods alone.

Comparable diagnostic difficulties are also met within other non-lymphoid hematopoietic neoplasms. Although a number of monoclonal antibodies are reactive with cells of the myelomonocytic lineage, which enables the immunophenotypic differentiation of granulocytic subsets, monocytes and their derivatives, such as phagocytes, and accessory cells for the humoral and cellular limb of immune response (PARWARESCH et al. 1983; RADZUN et al. 1988 b), none of these reagents could be shown to recognize lineage-specific receptors. Granulocytes and monocytes/macrophages, in contrast to lymphocytes, are engaged in antigen processing only in a nonspecific manner (ROSENTHAL 1980; UNANUE 1976, 1980). Consequently, there is no need to select for genetically diverse individual clones as opposed to the antigen-specific lymphocytes. This is also the reason why methods such as the detection of receptor gene rearrangements, which provide a molecular clonal marker for lymphocytes and their neoplasms, are not available for the identification of transformed cells of the myelomonocytic lineage.

Neoplasms arising from the myelomonocytic cell lineage usually can be easily diagnosed in peripheral blood smears on the basis of morphology and phenotypic markers such as various enzyme activities (LEDER 1967) and expression of differentiation antigens as detected by lineage-specific monoclonal antibodies (BÖDEWADT et al. 1986; KNAPP 1982). But in some forms of chronic myeloproliferative disorders, myelodysplasias, well-differentiated monocytic leukemias, and malignant histiocytosis, diagnostic difficulties soon arise if investigative measures are restricted to morphological, enzyme-cytochemical, and immunophenotypic methods (RADZUN et al. 1989).

In recent years, it has become evident that self-renewal, proliferation, differentiation, and survival of cell populations in different tissue types are under the control of lineage-specific growth factors (DEXTER and ALLEN 1983; METCALF 1986). In this context, the colony-stimulating factors (CSFs) for granulocyte-macrophages (GM-CSF), granulocytes (G-CSF), and macrophages (M-CSF) have attracted considerable attention. The genes for these three factors have been successfully cloned (METCALF 1989). Since CSFs bind in a lineage-specific fashion to their target cells, it was speculated that the corresponding receptors might provide lineage-specific markers. Indeed, recent developments seem to be in keeping with this anticipation. One of these CSF receptors has been identified and molecularly cloned. The cellular counterpart of the transforming retrovirus v-*fms* turned out to be the gene encoding the M-CSF receptor (SHERR et al. 1985).

It could be shown that transforming retroviral oncogenes (v-*onc*) are derived from cellular homologues (c-*onc*) which are involved in normal cell growth and differentiation (BISHOP 1983). Some of these retroviral oncogenes, such as v-*erb*B and v-*fms*, code for transmembrane proteins with tyrosine kinase activity and hence resemble functional receptor molecules occurring physiologically (DOWNWARD et al. 1984; RETTENMIER et al. 1985). With regard to v-*fms*, the coded protein was shown to be homologous with the receptor for M-CSF (SHERR et al. 1985), differing only at the carboxyl termini (COUSSENS et al. 1986). Ample evidence indicates that within the myelomonocytic lineage at least the *fms* protooncogene and M-CSF comprise a receptor-ligand structure that is specifically expressed during monocyte differentiation into macrophages (KREIPE et al. 1986).

In the following we will show to what extent the molecular analysis of c-*fms* and M-CSF expression can contribute to our understanding of normal differentiation and neoplastic transformation within the monocyte/macrophage lineage. In addition, data will be provided suggesting that genotypic information on c-*fms* may be used as a cell-specific marker and exploited to indicate neoplastic transformation within the monocyte/macrophage lineage.

3.2 c-*fms* and M-CSF Expression in Normal Cells of the Monocyte/Macrophage Lineage

Applying Northern blot analysis of separated RNA, c-*fms* transcripts with an approximate length of 4.2 kb can be observed in blood monocytes but not in lymphocytes. Purified granulocytes show low amounts of c-*fms* RNA, most probably due to contaminating monocytes. These results underline that c-*fms* RNA expression is lineage specific in human blood cells. Expression of c-*fms* gene in monocytes/macrophages is modulated during activation and maturation. There is no constant expression of this gene such as occurs with a "housekeeping gene." Modulation of c-*fms* expression in monocytes/macrophages can be observed in vitro as well as in vivo.

In line with these considerations, c-*fms* expression similar to that observed in adherent separated blood monocytes can be detected in the mononuclear

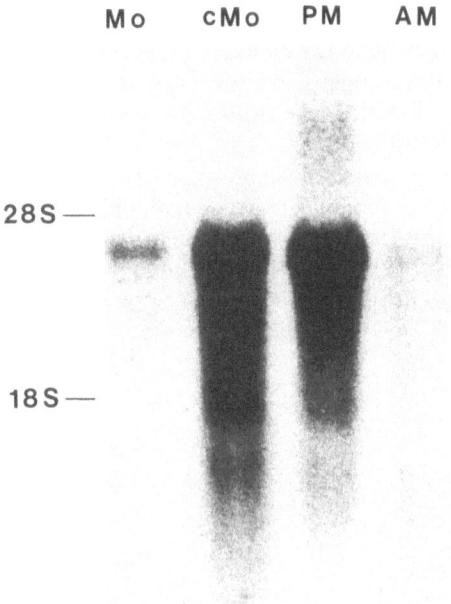

Fig. 4. Different levels of c-*fms* RNA expression can be observed in peritoneal (*PM*) and alveolar macrophages (*AM*) with considerably more *fms* transcripts in PM. Monocytes (*Mo*) show a transient down-regulating of c-*fms* RNA during separation by surface adherence. After 48 h of culture c-*fms* expression increases (*cMo*)

blood cell fraction, which consists of only about 20% monocytes. Obviously, a down-regulation of c-*fms* expression is induced in blood monocytes by surface adherence. The reduction of c-*fms* expression in separated blood monocytes is not irreversible. By further cultivation in vitro over 24 hr, c-*fms* expression can be enhanced. However, prolonged cultivation for 21 days or the addition of various stimuli such as lymphokine-rich supernatants of con-canavalin A-stimulated lymphocytes to the culture medium leads to a down-regulation of c-*fms* expression (RADZUN et al. 1988 a). In addition, multinucleated giant cells generated from cultured blood monocytes reveal a low level of c-*fms* RNA (KREIPE et al. 1988).

Comparing these findings under in vitro conditions with data on resident macrophages of the peritoneal and alveolar cavities occurring in vivo (Fig. 4), peritoneal macrophages show an elevated level of c-*fms* expression as compared with separated blood monocytes, whereas alveolar macrophages are characterized by a low level of c-*fms* RNA expression (KREIPE et al. 1988).

These and similar findings imply that monocyte/macrophage activation is accompanied by modulation of c-*fms* expression. M-CSF receptor expression seems to be the prerequisite for an M-CSF-mediated proliferation of blood monocytes which undergo on average one mitosis during terminal differentiation into peritoneal macrophages (PARWARESCH and WACKER 1984). In addition, M-CSF receptor expression in macrophages regulates their functional activity – as shown for the release of plasminogen activator (LIN et al. 1979) and prostaglandin (KURLAND et al. 1979) –, with prompted down-regulation starting as soon as full maturity is achieved.

Correlating the data on c-*fms* expression with the expression of M-CSF, it can be shown that a reversed pattern of expression exists. The high level of c-*fms* expression in blood monocytes and peritoneal macrophages proceeds with abolition of M-CSF expression. Alveolar macrophages characterized by a reduced level of c-*fms* show a high level of M-CSF expression, as do long-

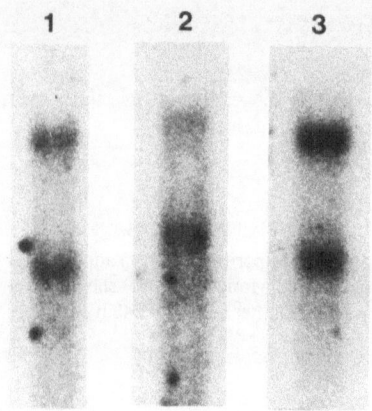

Fig. 5. Low level c-*fms* RNA expression in alveolar macrophages is associated with autostimulatory M-CSF expression, which might induce a down-regulation of M-CSF receptor/c-*fms* gene expression. Two RNA variants of 4.5 kb and 3.3 kb can be detected in each alveolar macrophage sample (*lane 1*, alveolar macrophage; *lanes 2 and 3*, alveolar macrophages from inactive sarcoidosis patients)

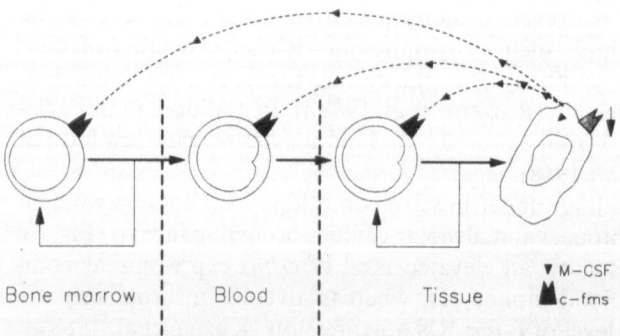

Fig. 6. c-*fms* and M-CSF expression within the monocyte/macrophage lineage. Tissue macrophages produce M-CSF which might provide feedback mechanisms initiating the recruitment of macrophage precursors. During maturation and M-CSF induction a down-regulation of c-*fms* (M-CSF receptor) can be observed

term cultivated blood monocytes, revealing two transcripts of approximately 3.3 kb and 4.5 kb length (Fig. 5) generated by an alternative splicing of M-CSF RNA (LADNER et al. 1987). Since M-CSF has been shown to induce functional activation of monocytes/macrophages, such as the production of interferon and tumor necrosis factor (WARREN and RALPH 1986), alveolar macrophages as a cell population constantly exposed to environmental stimuli are provided via M-CSF with an autostimulatory amplification mechanism of activating signals. In addition, M-CSF expression by alveolar macrophages might provide a feedback mechanism in the generation of monocytes/macrophages from bone marrow precursors, compensating for the turnover and peripheral demise of this macrophage population (Fig. 6).

3.3 c-*fms* and M-CSF Expression in Neoplastic Cells of the Monocyte/Macrophage Lineage

Once it was established that c-*fms* and M-CSF expression regulates self-renewal, proliferation, differentiation, and survival of monocytes/macrophages, it was only logical to examine the extent to which these receptor ligand interactions might be involved in neoplastic growth within this cell lineage. It has been shown that human acute myeloid leukemias (AMLs) are capable of autocrine growth stimulation by simultaneous expression of growth factors and the corresponding receptors (RAMBALDI et al. 1988). Autostimulatory co-expression of M-CSF and c-*fms*, however, seems to be associated with reduced growth capacity of leukemic cells (WANG et al. 1988), while autostimulatory M-CSF production in transfected macrophage cell lines is not sufficient to induce tumorigenicity (ROUSSEL et al. 1988 b). Analyzing human AMLs, most cases reveal a reduced level of c-*fms* and absence of M-CSF expression compared with normal cells of the monocytic lineage. Higher levels of c-*fms* expression can only be observed in AMLs with a more mature monocytic phenotype (Fig. 7) and in myelomonocytic cell lines after induction of monocytic differentiation. As analyzed so far, neither an autostimulatory mechanism nor the activation of c-*fms* as an oncogene appears to provide a mechanism for malignant transformation (PARWARESCH et al. 1990).

3.4 c-*fms* Genotype in Neoplastic Cells of the Monocyte/Macrophage Lineage

In animals, involvement of the M-CSF gene and the c-*fms* gene in leukemogenesis could be demonstrated (GISSELBRECHT et al. 1987; BAUMBACH et al. 1988). Growth factors and growth factor receptor genes have been shown to provide target genes of viral insertional mutagenesis. Applying Southern blotting of DNA, structural changes such as rearrangements or deletions of M-CSF and c-*fms* genes in AML could not be observed (PARWARESCH et al. 1990). Unlike other oncogenes, which are considered dominant cancer genes, c-*fms* might contribute as a recessive gene to the pathogenesis of a preneoplastic

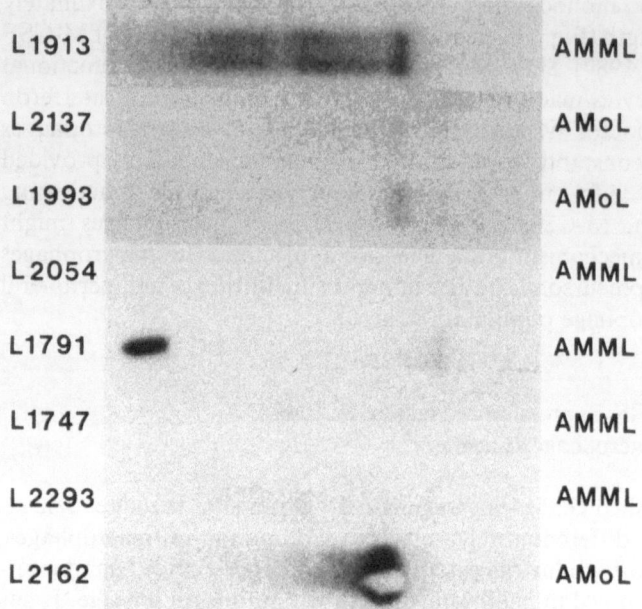

L1913	AMML
L2137	AMoL
L1993	AMoL
L2054	AMML
L1791	AMML
L1747	AMML
L2293	AMML
L2162	AMoL

Fig. 7. c-*fms* gene expression in acute myelomonocytic and acute monocytic leukemias. Only about 50% of acute leukemias reveal c-*fms* transcripts when total cellular RNA is analyzed. Only some cases display c-*fms* RNA to a level comparable to blood monocytes (L2162). Immunohistochemically these leukemias show a more mature monocytic phenotype

hematopoietic disorder. In the $5q^-$ myelodysplastic syndrome, characterized by therapy-resistant anemia and evolution into AML in some patients, one allele of the c-fms gene localized on the deleted long arm of chromosome 5 is lost (NIENHUIS et al. 1985). Most of the leukemia samples we analyzed show a homozygote genotype of the c-*fms* gene as defined by restriction fragment polymorphism (91.3%), compared with the 75% rate of homozygosity observed in normal populations (XU et al. 1985). This indicates that loss of one c-*fms* allele might have occurred. We are currently attempting, by immortalizing T cells from AML patients, to determine whether a loss of heterozygosity within the *fms* gene is operational in the leukemic clones.

3.5 Concluding Remarks

Analysis of c-*fms* and M-CSF gene expression gives an insight into the physiological growth control of monocytes/macrophages and possibly will reveal pathogenetic mechanisms of leukemogenesis. The molecular genetics of c-*fms* and M-CSF gene expression and structure may well contribute to the diagnosis of a group of ill-defined neoplasms ascribed to the monocyte/macrophage lineage. It remains to be shown, however, whether a loss of heterozygosity occurs within the leukemic clones in cases of AML and other neoplasms of the monocyte/macrophage system. Whereas structural changes within the *fms* gene of AML cells have not been demonstrated up to now, point mutations have not yet been investigated. Single base exchanges and small deletions have been shown – though only in nonhematopoietic cells – to convert c-*fms* into a dominant cancer gene (ROUSSEL et al. 1988 a; WOOLFORD et al. 1988). Furthermore, considerable progress in our knowledge of c-*fms* expression or downregulation during normal and transformational events of the monocyte/macrophage lineage is anticipated from systematic analysis of c-*fms* gene regulation such as methylation patterns (CEDAR 1988). This might contribute to our understanding of the molecular regulation of differentiation within hematopoietic cells in general.

Because it is probable that the receptors of the other CSFs for myeloid cells will be cloned in the future, further receptor-ligand structures besides those of T and B lymphocytes and monocytes/macrophages are likely to become available for analysis of hematopoietic neoplasms on the molecular level.

References

Alt F, Blackwell K, Yancopoulos GD (1987) Development of the primary antibody repertoire. Science 238:1079–1083

Baumbach WR, Colston EM, Cole MD (1988) Integration of the BALB/c ecotropic provirus into the colony-stimulating factor-1 growth factor locus in a *myc* retrovirus-induced murine monocyte tumor. J Virol 62:3151–3155

Bishop JM (1983) Cellular oncogenes and retroviruses. Ann Rev Biochem 52:301–354

Bödewadt S, Radzun HJ, Feller AC, Parwaresch MR (1986) Immunophenotyping of acute non-lymphoblastic leukaemias. Virchows Arch [B] 51:79–88

Brenner MB, Strominger JL, Krangel MS (1988) The γ/δ T-cell receptor. In: Dixon FJ (ed) Advances in immunology, vol 43. Academic, Orlando, pp 133–192

Campana D, Janossy G, Coustan-Smith E, Amlot PL, Tian WT, Wong L (1989) The expression of T-cell receptor-associated proteins during T-cell ontogeny in man. J Immunol 142:57–65

Cedar H (1988) DNA methylation and gene activity. Cell 53:3–4

Champagne E, Sagman U, Biondi A, Lewis WH, Mak TW, Minden MD (1988) Structure and rearrangement of the T-cell receptor Jα locus in T-cells and leukemic T-cell lines. Eur J Immunol 18:1033–1038

Cossman J, Uppenkamp M, Sundeen J, Coupland R, Raffeld M (1988) Molecular genetics and the diagnosis of lymphoma. Arch Pathol Lab Med 112:117–127

Coussens L, Van Beveren C, Smith D et al. (1986) Structural alteration of viral homologue of receptor proto-oncogene *fms* at carboxyl terminus. Nature 320:277–280

Davey MP, Bongiovanni KF, Kaulfersch W et al. (1986) Immunoglobulin and T-cell receptor gene rearrangement and expression in human lymphoid leukemia cells at different stages of maturation. Proc Natl Acad Sci USA 83:8759–8763

Dexter TM, Allen TD (1983) The regulation of growth and development of normal and leukaemic cells. J Pathol 141:415–433

Downward J, Yarden Y, Mayes E et al. (1984) Close similarity of epidermal growth factor receptor and v-*erb*B oncogene protein sequences. Nature 307:521–527

Foon KA, Todd III RF (1986) Immunological classification of leukemia and lymphoma. Blood 68:1–31

Frank MM (1987) Complement in the pathophysiology of human disease. N Engl J Med 316:1525–1530

Gisselbrecht S, Fichelson S, Sola B et al. (1987) Frequent c-*fms* activation by proviral insertion in mouse myeloblastic leukaemias. Nature 329:259–261

Goldstein JL, Ho YK, Basu SK, Brown MS (1979) Binding site on macrophages that mediates uptake and degradation of acetylated low density lipoprotein, producing massive cholesterol deposition. Proc Natl Acad Sci (USA) 76:333–337

Griesser H, Mak TW (1988) Immunogenotyping in Hodgkin's disease. Hematol Oncol 6:239–245

Griesser H, Feller AC, Lennert K, Minden M, Mak TW (1986a) Rearrangement of the β chain of the T-cell antigen receptor and immunoglobulin genes in lymphoproliferative disorders. J Clin Invest 78:1179–1184

Griesser H, Feller AC, Lennert K et al. (1986b) The structure of the T-cell gamma chain gene in lymphoproliferative disorders and lymphoma cell lines. Blood 68:592–594

Griesser H, Tkachuk D, Reis MD, Mak TW (1989) Gene rearrangements and translocations in lymphoproliferative diseases. Blood 73:1402–1415

Griesser H, Feller AC, Sterry W (1990) T cell receptor and immunoglobulin gene rearrangements in cutaneous T cell-rich pseudolymphomas. J Invest Dermatol 95 (in press)

Hanson CA, Frizzera G, Patton DF, Peterson BA, McClain KL, Gajl-Peczalska KJ, Kersey JH (1988) Clonal rearrangement for immunoglobulin and T-cell receptor genes in systemic Castleman's disease. Am J Pathol 131:84–91

Hara J, Benedict SH, Mak TW, Gelfand EW (1987) T-cell receptor α-chain gene rearrangements in B-precursor leukemia are in contrast to the findings in T-cell acute lymphoblastic leukemia – comparative study of T-cell receptor gene rearrangement in childhood leukemia. J Clin Invest 80:1770–1777

Hara J, Benedict SH, Champagne E, Takihara Y, Mak TW, Minden M, Gelfand EW (1988) T-cell receptor δ gene rearrangements in acute lymphoblastic leukemia. J Clin Invest 82:1974–1981

Haynes BF, Singer KH, Denning SM, Martin ME (1988) Analysis of expression of CD2, CD3, and T-cell antigen receptor molecules during early human fetal thymic development. J Immunol 141:3776–3784

Hogg N (1988) The structure and function of Fc receptors. Immunol Today 9:185–187

Jenkinson EJ, Kingston R, Owen JJT (1987) Importance of IL-2 receptors in intra-thymic generation of cells expressing T-cell receptors. Nature 329:160–162

Knapp W (1982) Monoclonal antibodies against differentiation antigens of myelopoiesis. Blut 45:301–308

Kreipe H, Radzun HJ, Heidorn K, Parwaresch MR, Verrier B, Müller R (1986) Lineage-specific expression of c-*fos* and c-*fms* in human hematopoietic cells: discrepancies with the in vitro differentiation of leukemia cells. Differentiation 33:56–60

Kreipe H, Radzun HJ, Rudolph P, Barth J, Hansmann ML, Heidorn K, Parwaresch MR (1988) Multinucleated giant cells generated in vitro: terminally differentiated macrophages with down-regulated c-*fms* expression. Am J Pathol 130:232–243

Kurland JS, Pelus LM, Ralph P, Bockman RS, Moore MAS (1979) Induction of prostaglandin E synthesis in normal and neoplastic macrophages: role for colony-stimulating factor(s) distinct from effects on myeloid progenitor cell proliferation. Proc Natl Acad Sci USA 76:2326–2341

Ladner MB, Martin GA, Noble JA, Nikoloff DM, Tal R, Kawasaki ES, White TJ (1987) Human CSF-1: gene structure and alternative splicing of mRNA precursors. EMBO J 6:2693–2698

Leder LD (1967) Der Blutmonozyt. Springer, Berlin Heidelberg New York

Leder P (1982) The genetics of antibody diversity. Sci Am 246:72–83

Lefranc MP, Forster A, Baer T, Stinson MA, Rabbitts TH (1986) Diversity and rearrangement of the human T-cell rearranging γ genes: Nine germline variable genes belonging in two subgroups. Cell 45:237–246

Lin HS, Gordon S (1979) Secretion of plasminogen activator by bone marrow-derived mononuclear phagocytes and its enhancement by colony-stimulating factor. J Exp Med 150:231–245

Makgoba MW, Sanders ME, Luce GEG et al. (1988) ICAM-1, a ligand for LFA-1-dependent adhesion of B, T and myeloid cells. Nature 331:86–88

McMichael A (1987) Leucocyte typing III – white cell differentiation antigens. Oxford University Press, Oxford

Metcalf D (1986) The molecular biology and functions of the granulocyte-macrophage colony-stimulating factors. Blood 67:257–267

Metcalf D (1989) The molecular control of cell division, differentiation commitment and maturation in haematopoietic cells. Nature 339:27–30

Nienhuis AW, Bunn HF, Turner PH, Gopal TV, Nash WG, O'Brien SJ, Sherr CJ (1985) Expression of the human c-*fms* proto-oncogene in hematopoietic cells and its deletion in the 5q-syndrome. Cell 42:421–428

Parwaresch MR, Wacker HH (1984) Origin and kinetics of resident tissue macrophages. Parabiosis studies with radiolabelled leucocytes. Cell Tissue Kin 17:25–39

Parwaresch MR, Radzun HJ, Feller AC, Peters KP, Hansmann ML (1983) Peroxidase-positive mononuclear leukocytes as possible precursors of human dendritic reticulum cells. J Immunol 131:2719–2725

Parwaresch MR, Kreipe H, Felgner J, Heidorn K, Jaquet K, Bödewadt-Radzun S, Radzun HJ (1990) M-CSF and M-CSF receptor gene expression in acute myelomonocytic leukemias. Leukemia Res 14:27–37

Radzun HJ, Kreipe H, Heidorn K, Parwaresch MR (1988a) Modulation of c-*fms* proto-oncogene expression in human blood monocytes and macrophages. J Leukocyte Biol 44:198–204

Radzun HJ, Kreipe H, Zavzava ML, Parwaresch MR (1988b) Diversity of the human monocyte/macrophage system as detected by monoclonal antibodies. J Leukocyte Biol 43:41–50

Radzun HJ, Parwaresch MR, Stingl G, Knapp W (1989) Neoplasms of monocytes, macrophages and dendritic cells. In: Asherson GL, Zembala M (eds) Human monocytes. Academic, London

Rambaldi A, Wakamiya N, Vellenga E, Horiguchi J, Warren MK, Kufe D, Griffin JD (1988) Expression of the macrophage colony-stimulating factor and c-*fms* genes in human acute myeloblastic leukemia cells. J Clin Invest 81:1030–1035

Reis MD, Griesser H, Mak TW (1988) Gene rearrangements in leukemias and lymphomas. In: Hoffbrand AV (ed) Recent advances in haematology, vol 5. Churchill Livingstone, Edinburgh, pp 99–120

Rettenmier CW, Chen JH, Roussel MF, Sherr CJ (1985) The product of the c-*fms* proto-oncogene: a glycoprotein with associated tyrosine kinase activity. Sci Am 228:320–322

Rosenthal AS (1980) Regulation of the immune response – role of the macrophage. N Engl J Med 303:1153–1156

Roussel MF, Downing JR, Rettenmier CW, Sherr CJ (1988a) A point mutation in the extracellular domain of the human CSF-1 receptor (c-*fms* proto-oncogene product) activates its transforming potential. Cell 55:979–988

Roussel MF, Rettenmier CW, Sherr CJ (1988b) Introduction of a human colony stimulating factor-1 gene into a mouse macrophage cell line induces CSF-1 independence but not tumorigenicity. Blood 71:1218–1225

Seremetis SV, Pelicci PG, Tabilio A et al. (1987) High frequency of clonal immunoglobulin and T-cell receptor gene rearrangements in acute myelogenous leukemia expressing terminal deoxynucleotidyl transferase. J Exp Med 165:1703–1712

Sherr CJ, Rettenmier CW, Sacca R, Roussel MF, Look AT, Stanley ER (1985) The c-*fms* proto-oncogene product is related to the receptor for the mononuclear phagocyte growth factor, CSF-1. Cell 41:665–676

Shimonkovitz RP, Husmann LA, Bevan MJ, Crispe IN (1987) Transient expression of IL-2 receptor precedes the differentiation of immature thymocytes. Nature 329:157–159

Shreffler DC (1988) Seventy-five years of immunology: The view from the MHC. J Immunol 141:1791–1798

Simmons D, Makgoba MW, Seed B (1988) ICAM, an adhesion ligand of LFA-1, is homologous to the neural cell adhesion molecule NCAM. Nature 331:624–627

Sklar JL, Weiss LM, Cleary ML (1987) Diagnostic molecular biology of non-Hodgkin's lymphomas. In: Berard CW, Dorman RF, Kaufman N (eds) Malignant lymphoma. Williams & Wilkins, Baltimore, pp 204–224

Southern EM (1975) Detection of specific sequences among DNA fragments separated by gel electrophoresis. J Mol Biol 98:503–517

Suchi T, Lennert K, Tu LY, Kikuchi M, Sato E, Stansfeld AG, Feller AC (1987) Histopathology and immunohistochemistry of peripheral T-cell lymphomas: a proposal for their classification. J Clin Pathol 40:995–1015

Takihara Y, Champagne E, Griesser H et al. (1988) Sequence and organization of the human δ chain gene. Eur J Immunol 18:283–287

Tkachuk D, Griesser H, Feller AC, Lennert K, Mak TW (1988) Rearrangement of the T-cell locus in lymphoproliferative disorders. Blood 72:353–357

Tonegawa S (1985) The molecules of the immune system. Sci Am 253:122–131

Toyonaga B, Mak TW (1987) Genes of the T-cell antigen receptor in normal and malignant T-cells. Ann Rev Immunol 5:585–620

Unanue ER (1976) Secretory function of mononuclear phagocytes. A review. Am J Pathol 83:396–417

Unanue ER (1980) Cooperation between mononuclear phagocytes and lymphocytes in immunity. N Engl J Med 303:977–985

Waldmann TA (1987) The arrangement for immunoglobulin and T-cell receptor genes in human lymphoproliferative disorders. In: Dixon FJ (ed) Advances in immunology, vol 40. Academic, Orlando, pp 247–321

Wang C, Kelleher CA, Cheng GYM et al. (1988) Expression of the CSF-1 gene in the blast cells of acute myeloblastic leukemia: association with reduced growth capacity. J Cell Physiol 135:133–138

Warren MK, Ralph P (1986) Macrophage growth factor CSF-1 stimulates human monocyte production of interferon, tumor necrosis factor, and colony stimulating activity. J Immunol 137:2281–2285

Weiss LM, Hu E, Wood GS, Moulds C, Cleary ML, Warnke R, Sklar J (1985) Clonal rearrangements of T-cell receptor genes in mycosis fungoides and dermatopathic lymphadenopathy. N Engl J Med 313:539–544

Weiss LM, Wood GS, Ellisen LW, Reynolds TC, Sklar J (1987) Clonal T-cell populations in pityriasis lichenoides et varioliformis acuta (Mucha-Habermann disease). Am J Pathol 126:417–421

Weiss LM, Picker LJ, Grogan TM, Warnke RA, Sklar JA (1988) Absence of clonal beta and gamma T-cell receptor gene rearrangements in a subset of peripheral T-cell lymphomas. Am J Pathol 130:436–442

Winoto A, Baltimore D (1989) Separate lineages of T-cells expressing the β and γ/δ receptors. Nature 338:430–432

Woolford J, McAuliffe A, Rohrschneider LR (1988) Activation of the feline c-*fms* proto-oncogene: multiple alterations are required to generate a fully transformed phenotype. Cell 55:965–977

Xu DQ, Guilhot S, Galibert F (1985) Restriction fragment length polymorphism of the human c-*fms* gene. Proc Natl Acad Sci USA 82:2862–2865

Yanagi Y, Yoshikai Y, Leggett K, Clark SP, Alecander J, Mak TW (1984) A human T-cell specific cDNA clone encodes a protein having extensive homology to immunoglobulin chains. Nature 308:145–149

Subject Index

Index of Volumes 79–82 Current Topics in Pathology

J. J. Brooks, J. Q. Trojanowski, V. A. Livolsi, Chondroid Chordoma:
  A Low-Grade Chondrosarcoma and Its Differential Diagnosis
V. Bouropoulou, A. Bosse, A. Roessner, E. Vollmer, G. Edel,
  P. Wuisman, A. Härle, Immunohistochemical Investigation
  of Chordomas: Histogenetic and Differential Diagnostic Aspects
A. Roessner, E. Vollmer, G. Zwaldo, C. Sorg, M. Kolve, D. B.
  v. Bassewitz, P. Wuisman, A. Härle, E. Grundmann, The
  Cytogenesis of Macrophages and Osteoclast-like Giant Cells in Bone
  Tumors with Special Emphasis on the So-Called Fibrohistocytic Tumors

Volume 81: Gastrointestinal Pathology. Edited by G. T. Williams

M. F. Dixon, Progress in the Pathology of Gastritis and Duodenitis
F. Potet, V. Duchatelle, Barrett's Oesophagus
P. Sipponen, Gastric Dysplasia
A. H. Qizilbash, Duodenal and Peri-ampullary Adenomas
C. Fenger, Intra-epithelial Neoplasia in the Anal Canal and Peri-anal Area
M. M. Mathan, V. I. Mathan, Gastrointestinal Biopsy Diagnosis
  in the Tropics
D. A. Levison, P. A. Hall, A. J. Blackshaw, The Gut-Associated
  Lymphoid Tissue and Its Tumours
J. M. Sloan, D. C. Allen, P. W. Hamilton, P. C. H. Watt, The Place
  of Quantitation in Diagnostic Gastrointestinal Pathology
W. V. Bogomoletz, Collagenous Colitis
A. B. Price, Ischaemic Colitis
P. S. Teglbjærg, Intestinal Spirochaetosis
Ph. U. Heitz, P. Komminoth, Biopsy Diagnosis of Hirschsprung's Disease
  and Related Disorders
P. Hermanek, Malignant Polyps – Pathological Factors Governing Clinical
  Management
J. R. Jass, Prognostic Factors in Colorectal Cancer
N. A. Shepherd, H. J. R. Bussey, Polyposis Syndromes – An Update

Volume 82: Pathology of the Nucleus. Edited by J. C. E. Underwood

J. C. E. Underwood, Nuclear Morphology and Grading in Tumours
H. M. H. Kamel, J. Kirk, P. G. Toner, Ultrastructural Pathology
  of the Nucleus
J. Crocker, Nucleolar Organiser Regions
T. J. Stephenson, Quantitation of the Nucleus
P. Quirke, Flow Cytometry in the Quantitation of DNA Aneuploidy
  and Cell Proliferation in Human Disease
P. Dal Cin, A. A. Sandberg, Karyotypic Analysis of Solid Tumors
G. Terenghi, R. A. Fallon, Techniques and Applications of In Situ
  Hybridisation